Treatise on
Invertebrate Paleontology

Prepared under Sponsorship of
The Geological Society of America, Inc.

The Paleontological Society   The Society of Economic Paleontologists and Mineralogists
The Palaeontographical Society   The Palaeontological Association

Raymond C. Moore
Founder

R. A. Robison
Editor

Virginia Ashlock, Jack Keim, Roger B. Williams
Assistant Editors

Part G
Bryozoa
Revised

Volume 1: Introduction,
Order Cystoporata, Order Cryptostomata

By R. S. Boardman, A. H. Cheetham, D. B. Blake, John Utgaard,
O. L. Karklins, P. L. Cook, P. A. Sandberg, Geneviève Lutaud,
and T. S. Wood

The Geological Society of America, Inc.
and
The University of Kansas
Boulder, Colorado, and Lawrence, Kansas
1983

Library of Congress Catalogue Card Number: 53-12913
ISBN 0-8137-3107-0

Published 1983

Distributed by the Geological Society of America, Inc., 3300 Penrose Place,
Boulder, Colo. 80301, to which all communications should be addressed.

The *Treatise on Invertebrate Paleontology* has been made possible by
(1) funding principally from the National Science Foundation of the
United States, from the Geological Society of America through the
bequest of Richard Alexander Fullerton Penrose, Jr., and from the bequest
of Raymond C. and Lilian Moore; (2) contribution of the knowledge
and labor of specialists throughout the world, working in cooperation
under sponsorship of the Geological Society of America, the Paleontological Society, the Society of Economic Paleontologists and Mineralogists, the Palaeontographical Society, and the Palaeontological Association; and (3) acceptance by the University of Kansas of publication
without any financial gain to the University.

TREATISE ON INVERTEBRATE PALEONTOLOGY

Parts of the *Treatise* are distinguished by assigned letters with a view to indicating their systematic sequence while allowing publication of units in whatever order each is made ready for the press. Copies are available on orders sent to the Publication Sales Department, The Geological Society of America, 3300 Penrose Place, Boulder, Colorado 80301.

VOLUMES ALREADY PUBLISHED

Part A. INTRODUCTION, xxiii + 569 p., 371 fig., 1979.

Part C. PROTISTA 2 (Sarcodina, chiefly "Thecamoebians" and Foraminiferida), xxxi + 900 p., 5,311 fig., 1964.

Part D. PROTISTA 3 (chiefly Radiolaria, Tintinnina), xii + 195 p., 1,050 fig., 1954.

Part E. ARCHAEOCYATHA, PORIFERA, xviii + 122 p., 728 fig., 1955.

Part E, Volume 1. ARCHAEOCYATHA, Second Edition (Revised and Enlarged), xxx + 158 p., 871 fig., 1972.

Part F. COELENTERATA, xvii + 498 p., 2,700 fig., 1956.

Part F. COELENTERATA (Supplement 1). Rugosa and Tabulata, xl + 762 p., 3,317 fig., 1981.

Part G. BRYOZOA, xii + 253 p., 2,000 fig., 1953.

Part H. BRACHIOPODA, xxxii + 927 p., 5,198 fig., 1965.

Part I. MOLLUSCA 1 (Mollusca General Features, Scaphopoda, Amphineura, Monoplacophora, Gastropoda General Features, Archaeogastropoda, mainly Paleozoic Caenogastropoda and Opisthobranchia), xxiii + 351 p., 1,732 fig., 1960.

Part K. MOLLUSCA 3 (Cephalopoda General Features, Endoceratoidea, Actinoceratoidea, Nautiloidea, Bactritoidea), xxviii + 519 p., 2,382 fig., 1964.

Part L. MOLLUSCA 4 (Ammonoidea), xxii + 490 p., 3,800 fig., 1957.

Part N. MOLLUSCA 6 (Bivalvia), Volumes 1 and 2 (of 3), xxxviii + 952 p., 6,198 fig., 1969; Volume 3, iv + 272 p., 742 fig., 1971.

Part O. ARTHROPODA 1 (Arthropoda General Features, Protarthropoda, Euarthropoda General Features, Trilobitomorpha), xix + 560 p., 2,880 fig., 1959.

Part P. ARTHROPODA 2 (Chelicerata, Pycnogonida, Palaeoisopus), xvii + 181 p., 565 fig., 1955.

Part Q. ARTHROPODA 3 (Crustacea, Ostracoda), xxiii + 442 p., 3,476 fig., 1961.

Part R. ARTHROPODA 4 (Crustacea exclusive of Ostracoda, Myriapoda, Hexapoda), Volumes 1 and 2 (of 3), xxxvi + 651 p., 1,762 fig., 1969.

Part S. ECHINODERMATA 1 (Echinodermata General Features, Homalozoa, Crinozoa, exclusive of Crinoidea), xxx + 650 p., 2,868 fig., 1967 [1968].

Part T. ECHINODERMATA 2 (Crinoidea), Volumes 1–3, xxxviii + 1027 p., 4,833 fig., 1978.

Part U. ECHINODERMATA 3 (Asterozoans, Echinozoans), xxx + 695 p., 3,485 fig., 1966.

Part V. GRAPTOLITHINA, xvii + 101 p., 358 fig., 1955.

Part V. GRAPTOLITHINA, Second Edition (Revised and Enlarged), xxxii + 163 p., 507 fig., 1970.

Part W. MISCELLANEA (Conodonts, Conoidal Shells of Uncertain Affinities, Worms, Trace Fossils, Problematica), xxv + 259 p., 1,058 fig., 1962.

Part W. MISCELLANEA (Supplement 1). Trace Fossils and Problematica (Revised and Enlarged), xxi + 269 p., 912 fig., 1975.

Part W. MISCELLANEA (Supplement 2). Conodonta, xxviii + 202 p., frontis., 858 fig., 1981.

THIS VOLUME

Part G. Bryozoa (Revised, Volume 1). Introduction, Order Cystoporata, Order Crypto-stomata, xxvi + 626 p., 1,595 fig., 1983.

VOLUMES IN PREPARATION

Part B. Protista 1 (Chrysomonadida, Coccolithophorida, Charophyta, Diatomacea, etc.).
Part J. Mollusca 2 (Gastropoda, Streptoneura exclusive of Archaeogastropoda, Euthyneura).
Part M. Mollusca 5 (Coleoidea).
Part R. Arthropoda 4, Volume 3 (Hexapoda).
Part E Revised. Volume 2. Porifera.
Part G Revised. Bryozoa (additional volumes).
Part L Revised. Mollusca 4 (Ammonoidea).
Part O. Arthropoda 1 (supplement 1, Trilobita).
Part Q. Arthropoda 3 (supplement 1, Ostracoda).
Part W. Miscellanea (supplement 3, worms).

CONTRIBUTING AUTHORS

D. V. Ager, University College of Swansea, Swansea, Wales

T. W. Amsden, Oklahoma Geological Survey, Norman, Oklahoma

W. J. Arkell, deceased

R. L. Austin, University of Southampton, Southampton, England

R. W. Barker, Bellaire, Texas

R. S. Bassler, deceased

R. L. Batten, American Museum of Natural History, New York

F. M. Bayer, Smithsonian Institution, Washington, D.C.

H. H. Beaver, Exxon Production Research Company, Houston, Texas

R. H. Benson, Smithsonian Institution, Washington, D.C.

J. M. Berdan, U.S. Geological Survey, Washington, D.C.

W. A. Berggren, Woods Hole Oceanographic Institute, Woods Hole, Massachusetts

S. M. Bergström, Ohio State University, Columbus, Ohio

Gertruda Biernat, Pánstwowe Wydawnictwo Naukowe, Warsaw, Poland

D. B. Blake, University of Illinois, Urbana, Illinois

R. S. Boardman, Smithsonian Institution, Washington, D.C.

W. A. van den Bold, Louisiana State University, Baton Rouge, Louisiana

Hilbrand Boschma, deceased

A. J. Boucot, Oregon State University, Corvallis, Oregon

D. W. Boyd, University of Wyoming, Laramie, Wyoming

C. C. Branson, deceased

A. J. Breimer, Instituut voor Aardwetenschappen, Vrije Universiteit, Amsterdam, Netherlands

H. K. Brooks, University of Florida, Gainesville, Florida

J. C. Brower, Syracuse University, Syracuse, New York

O. M. B. Bulman, deceased

J. H. Callomon, University College, London, England

A. S. Campbell, St. Mary's College, St. Mary's, California

F. M. Carpenter, Harvard University, Cambridge, Massachusetts

Raymond Casey, Institute of Geological Sciences, London, England

K. E. Caster, University of Cincinnati, Cincinnati, Ohio

v

André Chavan, deceased

A. H. Cheetham, Smithsonian Institution, Washington, D.C.

D. L. Clark, University of Wisconsin, Madison, Wisconsin

A. H. Clarke, Jr., Smithsonian Institution, Washington, D.C.

W. J. Clench, Harvard University, Cambridge, Massachusetts

Eugene Coan, California Academy of Sciences, San Francisco, California

W. S. Cole, Cornell University, Ithaca, New York

S. Conway Morris, Open University, Milton Keynes, England

A. H. Coogan, Kent State University, Kent, Ohio

P. L. Cook, British Museum (Natural History), London, England

L. R. Cox, deceased

Dennis Curry, Middlesex, England

Colette Dechaseaux, Laboratoire de Paléontologie des Vertébrés, Université de Paris, Paris, France

R. K. Dell, Dominion Museum, Wellington, New Zealand

D. T. Donovan, University College, London, England

R. C. Douglass, U.S. Geological Survey, Washington, D.C.

J. W. Durham, University of California, Berkeley, California

F. E. Eames, British Petroleum Company, Middlesex, England

G. F. Elliott, Iraq Petroleum Company, London, England

H. K. Erben, Friedrich Wilhelms Universität, Bonn, West Germany

Harriet Exline, deceased

J. A. Eyer, Professional Geophysics, Inc., Oklahoma City, Oklahoma

R. O. Fay, Oklahoma Geological Survey, Norman, Oklahoma

H. B. Fell, Harvard University, Cambridge, Massachusetts

R. M. Finks, Queens College, Flushing, New York

A. G. Fischer, Princeton University, Princeton, New Jersey

D. W. Fisher, New York State Museum, Albany, New York

C. A. Fleming, New Zealand Geological Survey, Lower Hutt, New Zealand

D. L. Frizzell, deceased

W. M. Furnish, University of Iowa, Iowa City, Iowa

Julia Gardner, deceased

T. G. Gibson, Smithsonian Institution, Washington, D.C.

M. F. Glaessner, University of Adelaide, Adelaide, South Australia

B. F. Glenister, University of Iowa, Iowa City, Iowa

R. E. Grant, Smithsonian Institution, Washington, D.C.

Charles Grégoire, Brussels, Belgium

Fritz Haas, Field Museum of Natural History, Chicago, Illinois

Gerhard Hahn, Philipps Universität, Marburg, West Germany

Anthony Hallam, University of Birmingham, Birmingham, England

Tetsuro Hanai, University of Tokyo, Tokyo, Japan

G. D. Hanna, deceased

Walter Häntzschel, deceased

H. J. Harrington, deceased

A. G. Harris, U.S. Geological Survey, Washington, D.C.

W. H. Hass, deceased

Kotora Hatai, Saito Ho-on Kai Museum of Natural History, Sendai, Japan

W. W. Hay, Joint Oceanographic Institute, Inc., Washington, D.C.

J. E. Hazel, U.S. Geological Survey, Washington, D.C.

J. W. Hedgpeth, Santa Rosa, California

Gunnar Henningsmoen, Institutt for Geologi, Universitet Oslo, Oslo, Norway

L. G. Hertlein, deceased

Ivar Hessland, Universitet Stockholm, Stockholm, Sweden

R. R. Hessler, Scripps Institute of Oceanography, La Jolla, California

Dorothy Hill, University of Queensland, Brisbane, Australia

R. L. Hoffman, Radford College, Radford, Virginia

Helmut Hölder, Universität Münster, Münster, West Germany

L. B. Holthuis, Rijksmuseum van Natuurlijke Historie, Leiden, Netherlands

M. R. House, University of Hull, Kingston upon Hull, England

M. K. Howarth, British Museum (Natural History), London, England

H. V. Howe, deceased

B. F. Howell, deceased

L. H. Hyman, deceased

Valdar Jaanusson, Naturhistoriska Riksmuseet, Stockholm, Sweden

R. M. Jeffords, Exxon Production Research Company, Houston, Texas

J. A. Jeletzky, Geological Survey of Canada, Ottawa, Ontario, Canada

J. G. Johnson, Oregon State University, Corvallis, Oregon

Margaret Jope, Queen's University of Belfast, Belfast, Northern Ireland

O. L. Karklins, U.S. Geological Survey, Washington, D.C.

E. G. Kauffman, University of Colorado, Boulder, Colorado

A. Myra Keen, Stanford University, Palo Alto, California

R. V. Kesling, University of Michigan, Ann Arbor, Michigan

P. M. Kier, Smithsonian Institution, Washington, D.C.

Gilbert Klapper, University of Iowa, Iowa City, Iowa

J. B. Knight, deceased

Jürgen Kullmann, Universität Tübingen, Tübingen, West Germany

Bernhard Kummel, deceased

N. Gary Lane, University of Indiana, Bloomington, Indiana

Aurèle La Rocque, Ohio State University, Columbus, Ohio

M. W. de Laubenfels, deceased

Marius Lecompte, deceased

S. A. Levinson, Exxon Production Research Company, Houston, Texas

Maurits Lindström, Philipps Universität, Marburg, West Germany

Christina Lochman-Balk, New Mexico Institute of Mining and Technology, Socorro, New Mexico

A. R. Loeblich, Jr., University of California, Los Angeles, California

K. E. Lohman, U.S. Geological Survey, Washington, D.C.

N. H. Ludbrook, South Australia Geological Survey, Adelaide, South Australia

Geneviève Lutaud, Laboratoire Cytologie, Université de Paris, Paris, France

A. L. McAlester, Southern Methodist University, Dallas, Texas

Lavon McCormick, Chappell Hill, Texas

D. J. McLaren, Geological Survey of Canada, Ottawa, Ontario, Canada

D. B. Macurda, Exxon Production Research Company, Houston, Texas

Y. T. Mandra, San Francisco State University, San Francisco, California

R. B. Manning, Smithsonian Institution, Washington, D.C.

S. M. Manton, deceased

John Marwick, Havelock North, New Zealand

R. V. Melville, Institute of Geological Sciences, London, England

D. L. Meyer, University of Cincinnati, Cincinnati, Ohio

A. K. Miller, deceased

J. F. Miller, Southwest Missouri State University, Springfield, Missouri

Eugenia Montanaro Gallitelli, Universitá Modena, Modena, Italy

R. C. Moore, deceased

C. P. Morgan, University of Michigan, Ann Arbor, Michigan

H. M. Muir-Wood, deceased

A. H. Müller, Sektion Geowissenschaften der Bergakademie, Freiberg, German Democratic Republic

K. J. Müller, Friedrich Wilhelms Universität, Bonn, West Germany

N. D. Newell, American Museum of Natural History, New York

W. A. Newman, Scripps Institute of Oceanography, La Jolla, California

A. W. Norris, Geological Survey of Canada, Calgary, Alberta, Canada

C. P. Nuttall, British Museum (Natural History), London, England

V. J. Okulitch, University of British Columbia, Vancouver, British Columbia, Canada

A. A. Olsson, Academy of Natural Sciences of Philadelphia, Coral Gables, Florida

A. R. Palmer, Geological Society of America, Boulder, Colorado

K. V. W. Palmer, deceased

Adolf Papp, Paläontologisches Institut, Universität Wien, Wien, Austria

David Pawson, Smithsonian Institution, Washington, D.C.

R. E. Peck, University of Missouri, Columbia, Missouri

B. F. Perkins, University of Texas at Arlington, Arlington, Texas

Alexander Petrunkevitch, deceased

C. W. Pitrat, University of Massachusetts, Amherst, Massachusetts

Christian Poulsen, deceased

A. W. B. Powell, Auckland Institute and Museum, Auckland, New Zealand

H. S. Puri, Tallahassee, Florida

George Radwin, San Diego Natural History Museum, San Diego, California

Franco Rasetti, Universitá di Roma, Rome, Italy

H. Wienberg Rasmussen, deceased

Gerhard Regnéll, Universitet Lund, Lund, Sweden

Manfred Reichel, Universität Basel, Basel, Switzerland

R. E. H. Reid, Queen's University, Belfast, Northern Ireland

R. A. Reyment, Universitet Uppsala, Uppsala, Sweden

F. H. T. Rhodes, Cornell University, Ithaca, New York

Emma Richter, deceased

Rudolf Richter, deceased

Adolf Riedel, Pánstwowe Wydawnictwo Naukowe, Warsaw, Poland

Robert Robertson, Academy of Natural Sciences, Philadelphia, Pennsylvania

R. A. Robison, University of Kansas, Lawrence, Kansas

W. D. I. Rolfe, University of Glasgow, Glasgow, Scotland

C. A. Ross, Gulf Science and Technology Company, Houston, Texas

J. R. P. Ross, Western Washington University, Bellingham, Washington

Michel Roux, Université Paris-Sud, Orsay, France

A. J. Rowell, University of Kansas, Lawrence, Kansas

M. J. S. Rudwick, Vrije Universiteit, Amsterdam, Netherlands

Philip Sandberg, University of Illinois, Urbana, Illinois

W. A. S. Sarjeant, University of Saskatchewan, Saskatoon, Saskatchewan, Canada

O. H. Schindewolf, deceased

Herta Schmidt, Natur-Museum und Forschungs-Institut, Senckenberg, Frankfurt, West Germany

H. W. Scott, University of Illinois, Urbana, Illinois

Klaus Sdzuy, Universität Würzburg, Würzburg, West Germany

Adolf Seilacher, Universität Tübingen, Tübingen, West Germany

R. H. Shaver, Indiana Geological Survey and University of Indiana, Bloomington, Indiana

Hertha Sieverts-Doreck, Stuttgart-Möhringen, West Germany

A. G. Smith, California Academy of Sciences, San Francisco, California

L. A. Smith, Exxon Production Research Company, Houston, Texas

N. F. Sohl, U.S. Geological Survey, Washington, D.C.

I. G. Sohn, U.S. Geological Survey, Washington, D.C.

Tron Soot-Ryen, Hosle, Norway

W. K. Spencer, deceased

James Sprinkle, University of Texas at Austin, Austin, Texas

R. D. Staton, Brown University, Providence, Rhode Island

F. G. Stehli, University of Oklahoma, Norman, Oklahoma

H. B. Stenzel, Louisiana State University, Baton Rouge, Louisiana

L. W. Stephenson, deceased

Leif Størmer, deceased

H. L. Strimple, University of Iowa, Iowa City, Iowa

EDITORIAL PREFACE

THE AIM of the *Treatise on Invertebrate Paleontology,* as originally conceived and consistently pursued, is to present the most comprehensive and authoritative, yet compact statement of knowledge concerning invertebrate fossil groups that can be formulated by collaboration of competent specialists.

The major goal of this revision of Part G is to provide a workable reference to identify Bryozoa above the species level. Introductory papers review some of the current biological concepts of the phylum. The taxonomy is based on the reexamination of critical specimens, a reevaluation and increase in the number of taxonomic characters, and reclas-sification where necessary. Taxonomic characters range from microscopic to colony-wide, requiring new illustration at different magnifications. Systematic descriptions pertaining to two orders, the Cystoporata and Cryptostomata, are included here. The remainder of the bryozoan orders will be covered in subsequent volumes. Most manuscript for this volume was completed in April 1978.

ZOOLOGICAL NAMES

Many questions arise in connection with zoological names, especially including those related to acceptability and to alterations of some that may be allowed or demanded. Procedure in obtaining answers to these questions is guided and to a large extent governed by regulations published (1961) in the *International Code of Zoological Nomenclature*[1] (hereinafter cited simply as the *Code*). The prime object of the *Code* is to promote stability and universality in the use of the scientific names of animals, ensuring also that each name is distinct and unique while avoiding restrictions on freedom of taxonomic thought or action. Priority is a basic principle, but under specified conditions its application can be modified. This is all well and good, yet nomenclatural tasks confronting the zoological taxonomist are formidable. They warrant the complaint of some that zoology, including paleozoology, should be the study of animals rather than of names applied to them.

Several ensuing pages are devoted to aspects of zoological nomenclature that are judged to have chief importance in relation to procedures adopted in the *Treatise*. Terminology is explained, and examples of style employed in the nomenclatural parts of systematic descriptions are given.

A draft of a revised edition of the *Code* was submitted to the meeting of the International Union of Biological Sciences at Helsinki, Finland, in August 1979. This revised edition has not come into force as of this writing, and the existing *Code* of 1961 is, therefore, strictly followed herein.

TAXA GROUPS

Each taxonomic unit (taxon, pl., taxa) belongs to a rank in the adopted hierarchy of classificatory divisions. In part, this hierarchy is defined by the *Code* to include a species-group of taxa, a genus-group, and a family-group. Units of lower rank than subspecies are excluded from zoological nomenclature and those higher than superfamily of the family-group are not regulated by the *Code*.

It is natural and convenient to discuss nomenclatural matters in general terms first and then to consider each of the taxa groups separately. Especially important is the provision that within each taxa group, classificatory units are coordinate (equal in rank), whereas units of different taxa groups are not coordinate.

FORMS OF NAMES

All zoological names are divisible into groups based on their form (spelling). The first-published form (or forms) of a name is defined as original spelling (*Code,* Art. 32) and any later-published form (or forms) of the same name is designated as subsequent spelling (Art. 33). Obviously, original and subsequent spellings of a given name may or may not be identical and this affects consideration of their correctness. Further, examination of original spellings of names shows that by no means all can be distinguished as correct. Some are incorrect, and the same is true of subsequent spellings.

Original Spellings

If the first-published form of a name is consistent and unambiguous, the original spelling is defined as correct unless it contravenes some stipulation of the *Code* (Arts. 26–31), or the original publication contains clear evidence of an inadvertent error, in the sense of the *Code,* or, among names belonging to the family-group, unless correction of the termination or the stem of the type genus is required. An original spelling that fails to meet these requirements is defined as incorrect.

If a name is spelled in more than one way in the original publication, the form adopted by the first reviser is accepted as the correct original spelling, provided that it complies with mandatory stipulations of the *Code* (Arts. 26–31).

Incorrect original spellings are any that fail to satisfy requirements of the *Code*, represent an inadvertent error, or are one of multiple original spellings not adopted by a first reviser.

[1] N. R. Stoll and others (ed. comm.), *International Code of Zoological Nomenclature, adopted by the XV International Congress of Zoology*, xvii + 176 p. (International Trust for Zoological Nomenclature, London, 1961; 2nd edit., xx + 176 p., 1964.)

These have no separate status in zoological nomenclature and therefore cannot enter into homonymy or be used as replacement names, and they call for correction. For example, a name originally published with a diacritic mark, apostrophe, diaeresis, or hyphen requires correction by deleting such features and uniting parts of the name originally separated by them, except that deletion of an umlaut from a vowel in a name derived from a German word or personal name requires the insertion of "e" after the vowel.

Subsequent Spellings

If a name classed as a subsequent spelling is identical with an original spelling, it is distinguishable as correct or incorrect on the same criteria that apply to the original spelling. This means that a subsequent spelling identical with a correct original spelling is also correct, and one identical with an incorrect original spelling is also incorrect. In the latter case, both original and subsequent spellings require correction (authorship and date of the original incorrect spelling being retained).

If a subsequent spelling differs from an original spelling in any way, even by the omission, addition, or alteration of a single letter, the subsequent spelling must be defined as a different name (except that such changes as altered terminations of adjectival specific names to obtain agreement in gender with associated generic names, of family-group names to denote assigned taxonomic rank, and corrections for originally used diacritic marks, hyphens, and the like are excluded from spelling changes conceived to produce a different name). In certain cases species-group names having variable spellings are regarded as homonyms as specified in Art. 58 of the *Code*.

Altered subsequent spellings other than the exceptions noted may be either intentional or unintentional. If demonstrably intentional, the change is designated as an emendation. Emendations may be either justifiable or unjustifiable. Justifiable enemdations are corrections of incorrect original spellings, and these take the authorship and date of the original spellings. Unjustifiable emendations are names having their own status in nomenclature, with author and date of their publication; they are junior objective synonyms of the name in its original form.

Subsequent spellings that differ in any way from the original spellings, other than previously noted exceptions, and that are not classifiable as emendations are defined as incorrect subsequent spellings. They have no status in nomenclature, do not enter into homonymy, and cannot be used as replacement names. It is the purpose of the following chapters to explain in some detail the implications of various kinds of subsequent spellings and how these are dealt with in the *Treatise*.

AVAILABLE AND UNAVAILABLE NAMES

Available Names

An available zoological name is any that conforms to all mandatory provisions of the *Code*. Such names are classifiable in groups which are recognized in the *Treatise*, though not explicitly differentiated in the *Code*. They are as follows:

1. So-called "inviolate names" include all available names that are not subject to alteration from their originally published form. They comprise correct original spellings and commonly include correct subsequent spellings, but include no names classed as emendations. Here belong most genus-group names (including those for collective groups), some of which differ in spelling from others by only a single letter or by the sequential order of their letters.

2. Names may be termed "perfect names" if, as originally published, they meet all mandatory requirements, needing no correction of any kind, but nevertheless are legally alterable in such ways as changing the termination (e.g., many species-group names, family-group names). This group does not include emended incorrect original spellings (e.g., *Boucekites,* replacement of *Boucékites*).

3. "Imperfect names" are available names

that as originally published contain mandatorily emendable defects. Incorrect original spellings are imperfect names. Examples of emended imperfect names are: among species-group names, *guerini* (not *Guérini*), *obrienae* (not *O'Brienae*), *terranovae* (not *terra-novae*), *nunezi* (not *Nuñezi*), *Spironema rectum* (not *Spironema recta,* because generic name is neuter, not feminine); among genus-group names, *Broeggeria* (not *Bröggeria*), *Obrienia* (not *O'Brienia*), *Maccookites* (not *Mc-Cookites*); among family-group names Guembellotriinae (not Gümbellotriinae), Spironematidae (not Spironemidae, incorrect stem), Athyrididae (not Athyridae, incorrect stem). The use of "variety" for named divisions of fossil species, according to common practice of some paleontologists, gives rise to imperfect names, which generally are emendable (*Code,* Art. 45e) by omitting this term so as to indicate the status of this taxon as a subspecies. The name of a variety is always of feminine gender. If the variety is converted into a species or subspecies, the name takes on the gender of the associated genus.

4. "Vain names" are available names consisting of unjustified intentional emendations of previously published names. The emendations are unjustified because they are not demonstrable as corrections of incorrect original spellings as defined by the *Code* (Art. 32c). Vain names have status in nomenclature under their own authorship and date. They constitute junior objective synonyms of names in their original form. Examples are: among species-group names, *geneae* (published as replacement of original unexplained masculine, *geni,* which now is not alterable), *ohioae* (invalid change from original *ohioensis*); among genus-group names, *Graphiodactylus* (invalid change from original *Graphiadactyllis*); among family-group names, Graphiodactylidae (based on junior objective synonym having invalid vain name).

5. An important group of available zoological names can be distinguished as "transferred names." These comprise authorized sorts of altered names in which the change depends on transfer from one taxonomic rank to another, or possibly on transfers in taxonomic assignment of subgenera, species, or subspecies. Most commonly the transfer calls for a change in termination of the name so as to comply with stipulations of the *Code* on endings of family-group taxa and agreement in gender of specific names with associated generic names. Transferred names may be derived from any of the preceding groups except the first. Examples are: among species-group names, *Spirifer ambiguus* (masc.) to *Composita ambigua* (fem.), *Neochonetes transversalis* to *N. granulifer transversalis* or vice versa; among genus-group names, *Schizoculina* to *Oculina* (*Schizoculina*) or vice versa; among family-group names, Orthidae to Orthinae or vice versa, or superfamily Orthacea derived from Orthidae or Orthinae; among suprafamilial taxa (not governed by the *Code*), order Orthida to suborder Orthina or vice versa. The authorship and date of transferred names are not affected by the transfer, but the author responsible for the transfer and the date of his action are recorded in the *Treatise.*

6. Improved or "corrected names" include both mandatory and allowable emendations of imperfect names and of suprafamilial names, which are not subject to regulation as to name form. Examples of corrected imperfect names are given with the discussion of group 3. Change from the originally published ordinal name Endoceroidea (TEICHERT, 1933) to the presently recognized Endocerida illustrates a "corrected" suprafamilial name. Group 6 names differ from those in group 5 in not being dependent on transfers in taxonomic rank or assignment, but some names are classifiable in both groups.

7. "Substitute names" are available names expressly proposed as replacements for invalid zoological names, such as junior homonyms. These may be classifiable also as belonging in groups 1, 2, or 3. The glossary appended to the *Code* refers to these as "new names" (*nomina nova*) but they are better designated as substitute names, since their newness is temporary and relative. The first-published substitute name that complies with the def-

inition here given takes precedence over any other. An example is *Marieita* LOEBLICH & TAPPAN, 1964, as substitute for *Reichelina* MARIE, 1955, *non* ERK, 1942.

8. "Conserved names" include relatively small numbers of species-group, genus-group, and family-group names that have come to be classed as available and valid by action of the International Commission on Zoological Nomenclature exercising its plenary powers to this end or ruling to conserve a junior synonym in place of a rejected "forgotten" name (*nomen oblitum*) (Art. 23b). Currently, such names are entered on appropriate "Official Lists," which are published from time to time.

It is useful for convenience and brevity of distinction in recording these groups of available zoological names to employ Latin designations in the pattern of *nomen nudum* (abbr., *nom. nud.*) and others. Thus we recognize the preceding numbered groups as follows: (1) *nomina inviolata*; sing., *nomen inviolatum*, abbr., *nom. inviol.*; (2) *nomina perfecta*; *nomen perfectum, nom. perf.*; (3) *nomina imperfecta*; *nomen imperfectum, nom. imperf.*; (4) *nomina vana*; *nomen vanum, nom. van.*; (5) *nomina translata*; *nomen translatum, nom. transl.*; (6) *nomina correcta*; *nomen correctum, nom. correct.*; (7) *nomina substituta*; *nomen substitum, nom. subst.*; (8) *nomina conservata*; *nomen conservatum, nom. conserv.* It should be noted that the *Code* does not differentiate between different kinds of subsequent intentional changes of spelling, all of which are grouped as "emendations" (see below).

Additional to the groups differentiated above, the *Code* (Art. 17) specifies that a zoological name is not prevented from availability (a) by becoming a junior synonym, for under various conditions this may be reemployed, (b) for a species-group name by finding that original description of the taxon relates to more than a single taxonomic entity or to parts of animals belonging to two or more such entities, (c) for species-group names by determining that it first was combined with an invalid or unavailable genus-group name, (d) by being based only on part of an animal,

one sex of a species, an ontogenetic stage, or one form of a polymorphic species, (e) by being originally proposed for an organism not considered to be an animal but now so regarded, (f) by incorrect original spelling which is correctable under the *Code,* (g) by anonymous publication before 1951, (h) by conditional proposal before 1961, (i) by designation as a variety or form before 1961, (j) by concluding that a name is inappropriate (Art. 18), or (k) for a specific name by observing that it is tautonymous (Art. 18).

Unavailable Names

All zoological names which fail to comply with mandatory provisions of the *Code* are unavailable names and have no status in zoological nomenclature. None can be used under authorship and date of original publication as a replacement name (*nom. subst.*) and none preoccupies for purposes of the Law of Homonymy. Names identical in spelling with some, but not all, unavailable names can be classed as available if and when they are published in conformance to stipulations of the *Code,* and they are then assigned authorship and take date of the accepted publication. Different groups of unavailable names can be discriminated as follows.

9. "Naked names" include all those that fail to satisfy provisions stipulated in Article 11 of the *Code,* which states general requirements of availability. In addition they include names that, if published before 1931, were unaccompanied by a description, definition, or indication (Arts. 12, 16), as well as names published after 1930 that lacked accompanying statement of characters purporting to serve for differentiation of the taxon, or definite bibliographic reference to such a statement, or that were not proposed expressly as replacement (*nom. subst.*) of a preexisting available name (Art. 13a), or that were unaccompanied by definite fixation of a type species by original designation or indication (Art. 13b). Examples of "naked names" are: among species-group taxa, *Valvulina mixta* PARKER & JONES, 1865 (=*Cribrobulimina*

mixta CUSHMAN, 1927, available and valid); among genus-group taxa, *Orbitolinopsis* SILVESTRI, 1932 (=*Orbitolinopsis* HENSON, 1948, available but classed as invalid junior synonym of *Orbitolina* D'ORBIGNY, 1850); among family-group taxa, Aequilateralidae D'ORBIGNY, 1846 (lacking type-genus), Héilcostègues D'ORBIGNY, 1826 (vernacular not latinized by later authors, Art. 11e(iii)), Poteriocrinidae AUSTIN & AUSTIN, 1843, = family Poteriocrinoidea AUSTIN & AUSTIN, 1842 (neither 1843 nor 1842 names complying with Art. 11e, which states that "a family-group name must, when first published, be based on the name then valid for a contained genus," such valid name in the case of this family being *Poteriocrinites* MILLER, 1821).

10. "Denied names" include all those that are defined by the *Code* (Art. 32c) as incorrect original spellings. Examples are: specific names, *nova-zelandica, mülleri, 10-brachiatus*; generic names, *M'Coyia, Störmerella, Römerina, Westergårdia*; family name, Růžičkinidae. Uncorrected "imperfect names" are "denied names" and unavailable, whereas corrected "imperfect names" are available.

11. "Impermissible names" include all those employed for alleged genus-group taxa other than genus and subgenus (Art. 42a) (e.g., supraspecific divisions of subgenera), and all those published after 1930 that are unaccompanied by definite fixation of a type species (Art. 13b). Examples of impermissible names are: *Martellispirifer* GATINAUD, 1949, and *Mirtellispirifer* GAUTINAUD, 1949, indicated respectively as a section and subsection of the subgenus *Cyrtospirifer, Fusarchaias* REICHEL, 1949, without definitely fixed type species (=*Fusarchaias* REICHEL, 1952, with *F. bermudezi* designated as type species).

12. "Null names" include all those that are defined by the *Code* (Art. 33b) as incorrect subsequent spellings, which are any changes of original spelling not demonstrably intentional. Such names are found in all ranks of taxa. It is not always evident from the original publication whether an incorrect

subsequent spelling is intentional, resulting in a "vain name" which is invalid but available (category 4 above), or unintentional, resulting in a "null name" which is invalid and unavailable. In such cases, the decision of a subsequent author will sometimes have to be arbitrary according to his best judgment.

13. "Forgotten names" are defined (Art. 23b) as senior synonyms that have remained unused in primary zoological literature for more than 50 years. Such names are not to be used unless so directed by ICZN.

Latin designations for the discussed groups of unavailable zoological names are as follows: (9) *nomina nuda*; sing., *nomen nudum,* abbr., *nom. nud.*; (10) *nomina negata; nomen negatum, nom. neg.*; (11) *nomina vetita; nomen vetitum, nom. vet.,* (12) *nomina nulla; nomen nullam, nom. null.*; (13) *nomina oblita; nomen oblitum, nom. oblit.*

VALID AND INVALID NAMES

Important distinctions relate to valid and available names, on one hand, and to invalid and unavailable names, on the other. Whereas determination of availability is based entirely on objective considerations guided by Articles of the *Code,* conclusions as to validity of zoological names may be partly subjective. A valid name is the correct one for a given taxon, which may have two or more available names but only a single correct name, generally the oldest. Obviously, no valid name can also be an unavailable name, but invalid names may include both available and unavailable names. Any name for a given taxon other than the valid name is an invalid name.

A sort of nomenclatorial no-man's-land is encountered in considering the status of some zoological names, such as "doubtful names," "names under inquiry," and "forgotten names." Latin designations of these are *nomina dubia, nomina inquirenda,* and *nomina oblita,* respectively. Each of these groups may include both available and unavailable names, but the latter can well be ignored. Names considered to possess availability conduce to uncertainty and instability, which ordinarily can be removed only by appealed action of ICZN. Because few zoologists care to bother in seeking such remedy, the "wastebasket" names persist.

SUMMARY OF NAME GROUPS

Partly because only in such publications as the *Treatise* is special attention to groups of zoological names called for and partly because new designations are here introduced as means of recording distinctions explicitly as well as compactly, a summary may be useful.

Definitions of Name Groups

nomen conservatum (nom. conserv.). Name unacceptable under regulations of the *Code* which is made valid, either with original or altered spelling, through procedures specified by the *Code* or by action of ICZN exercising its plenary powers.

nomen correctum (nom. correct.). Name with intentionally altered spelling of sort required or allowable by the *Code* but not dependent on transfer from one taxonomic rank to another ("improved name"). (See *Code,* Arts. 26b, 27, 29, 30a(i) (3), 31, 32c(i), 33a; in addition, change of endings for suprafamilial taxa not regulated by the *Code*.)

nomen imperfectum (nom. imperf.). Name that as originally published meets all mandatory requirements of the *Code* but contains defect needing correction ("imperfect name"). (See *Code,* Arts. 26b, 27, 29, 32c, 33a).

nomen inviolatum (nom. inviol.). Name that as originally published meets all mandatory requirements of the *Code* and also is not correctable or alterable in any way ("inviolate name").

nomen negatum (nom. neg.). Name that as originally published constitutes invalid original spelling, and although possibly meeting all other mandatory requirements of the *Code,* cannot be used and has no separate status in nomenclature ("denied name"). It is to be corrected wherever found.

nomen nudum (nom. nud.). Name that as originally published fails to meet mandatory requirements of the *Code* and, having no status in nomenclature, is not correctable to establish original authorship and date ("naked name").

nomen nullum (nom. null.). Name consisting of an unintentional alteration in form (spelling) of a previously published name (either available name, as *nom. inviol., nom. perf., nom. imperf., nom. transl.*; or unavailable name, as *nom. neg., nom. nud., nom. van.,* or another *nom. null.*) ("null

name").

nomen oblitum (nom. oblit.). Name of senior synonym unused in primary zoological literature in more than 50 years, not to be used unless so directed by ICZN ("forgotten name").

nomen perfectum (nom. perf.). Name that as originally published meets all mandatory requirements of the *Code* and needs no correction of any kind but which nevertheless is validly alterable by change of ending ("perfect name").

nomen substitutum (nom. subst.). Replacement name published as substitute for an invalid name, such as junior homonym (equivalent to "new name").

nomen translatum (nom. transl.). Name that is derived by valid emendation of a previously published name as result of transfer from one taxonomic rank to another within the group to which it belongs ("transferred name").

nomen vanum (nom. van.). Name consisting of an invalid intentional change in form (spelling) from a previously published name, such invalid emendation having status in nomenclature as a junior objective synonym ("vain name").

nomen vetitum (nom. vet.). Name of genus-group taxon not authorized by the *Code* or, if first published after 1930, without definitely fixed type species ("impermissible name").

Except as specified otherwise, zoological names accepted in the *Treatise* may be understood to be classifiable either as *nomina inviolata* or *nomina perfecta* (omitting from notice *nomina correcta* among specific names) and these are not discriminated. Names which are not accepted for one reason or another include junior homonyms, senior synonyms classifiable as *nomina negata* or *nomina nuda,* and numerous junior synonyms which include both objective (*nomina vana*) and subjective types; rejected names are classified as completely as possible.

NAME CHANGES IN RELATION TO TAXA GROUPS

Species-group Names

Detailed consideration of valid emendation of specific and subspecific names is unnecessary here because it is well understood and relatively inconsequential. When the form of adjectival specific names is changed to obtain agreement with the gender of a generic name in transferring a species from one genus to another, it is never needful to label the changed name as a *nom. correct.* Likewise, transliteration of a letter accompanied by a diacritical mark in manner now called for by the *Code* (as in changing originally published *bröggeri* to *broeggeri*) or elimination of a hyphen (as in changing originally published *cornu-oryx* to *cornuoryx*) does not require *"nom. correct."* with it.

Genus-group Names

So rare are conditions warranting change of the originally published valid form of generic and subgeneric names that lengthy discussion may be omitted. Only elimination of diacritical marks of some names in this category seems to furnish basis for valid emendation. Is true that many changes of generic and subgeneric names have been published, but virtually all of these are either *nomina vana* or *nomina nulla.* Various names which formerly were classed as homonyms now are not, for two names that differ only by a single letter (or in original publication by presence or absence of a diacritical mark) are construed to be entirely distinct.

Examples in use of classificatory designations for genus-group names as previously given are the following, which also illustrate designation of type species as explained later.

Paleomeandron PERUZZI, 1881, p. 8 [*P. elegans*; SD HÄNTZSCHEL, 1975, p. W91] [=*Palaeomeandron* FUCHS, 1885, p. 395, *nom. van.*].

Vacuocyathus OKULITCH, 1950, p. 392 [*Coelocyathus kidrjassovensis* VOLOGDIN, 1937, p. 478, *nom. nud.*; 1939, p. 237; OD] [=*Coelocyathus* VOLOGDIN, 1934, p. 502, *nom. nud.*; 1937, p. 472, *nom. nud.*].

Cyrtograptus CARRUTHERS, 1867, p. 540, *nom. correct.* LAPWORTH, 1873, *pro Crytograpsus* CARRUTHERS, 1867, ICZN Op. 650, 1963 [*Cyrtograpsus murchisoni*; OD].

As has been pointed out above, it is in many cases difficult to decide whether a change in spelling of a name by a subsequent author was intentional or unintentional, that is, whether it should be classified as *nomen vanum* or *nomen nullum,* and the decision will often have to be arbitrary.

Family-group Names: Use of *"nom. transl."*

The *Code* specifies the endings only for subfamily (-inae) and family (-idae) but all family-group taxa are defined as coordinate, signifying that for purposes of priority a name published for a taxon in any category and based on a particular type genus shall date from its original publication for a taxon in any category, retaining this priority (and authorship) when the taxon is treated as belonging to a lower or higher category. By exclusion of -inae and -idae, respectively reserved for subfamily and family, the endings of names used for tribes and superfamilies must be unspecified different letter combinations. These, if introduced subsequent to designation of a subfamily or family based on the same nominate genus, are *nomina translata,* as is also a subfamily that is elevated to family rank or a family reduced to subfamily rank. In the *Treatise* it is desirable to distinguish the valid alteration comprised in the changed ending of each transferred family-group name by the abbreviation *"nom. trans."* and record of the author and date belonging to this alteration. This is particularly important in the case of superfamilies, for it is the author who introduced this taxon that one wishes to know about rather than the author of the superfamily as defined by the *Code,* for the latter is merely the individual who first defined some lower-rank family-group taxon that contains the nominate genus of the superfamily. The publication containing introduction of the superfamily *nomen translatum* is likely to furnish the information on taxonomic considerations that support definition of the unit.

Examples of the use of *"nom. transl."* are the following.

Subfamily STYLININAE d'Orbigny, 1851
[*nom. transl.* VERRILL, 1864, *ex* Stylinidae D'ORBIGNY, 1851]

Superfamily ANCYLOCERATACEAE Meek, 1876
[*nom. transl.* WRIGHT, 1957, *ex* Ancyloceratidae MEEK, 1876]

Family-group Names: Use of "nom. correct."

Valid name changes classed as *nomina correcta* do not depend on transfer from one category of family-group units to another but most commonly involve correction of the stem of the nominate genus; in addition, they include somewhat arbitrarily chosen modifications of endings for names of tribes or superfamilies. Examples of the use of "*nom. correct.*" are the following.

Family STREPTELASMATIDAE Nicholson, 1889

[*nom. correct.* Wedekind, 1927, p. 7, *pro* Streptelasmidae Nicholson in Nicholson & Lydekker, 1889, p. 297]

Family PALAEOSCORPIIDAE Lehmann, 1944

[*nom. correct.* Petrunkevitch, 1955, p. P73, *pro* Palaeoscorpionidae Lehmann, 1944, p. 177]

Family AGLASPIDIDAE Miller, 1877

[*nom. correct.* Størmer, 1959, p. P12, *pro* Aglaspidae Miller, 1877]

Family-group Names: Replacements

Family-group names are formed by adding letter combinations (prescribed for family and subfamily) to the stem of the name belonging to the genus (nominate genus) first chosen as type of the assemblage. The type genus need not be the oldest in terms of receiving its name and definition, but it must be the first-published as name-giver to a family-group taxon among all those included. Once fixed, the family-group name remains tied to the nominate genus even if its name is changed by reason of status as a junior homonym or junior synonym, either objective or subjective. Seemingly, the *Code* requires replacement of a family-group name only in the event that the nominate genus is found to have been invalid when it was proposed (Arts. 11e, 39), and then a substitute family-group name is accepted if it is formed from the oldest available substitute name for the nominate genus. Authorship and date attributed to the replacement family-group name are determined by first publication of the changed family-group name, but for purposes of the Law of Priority, they take the date of the replaced name. Numerous long-used family-group names are incorrect in being *nomina nuda,* since they fail to satisfy criteria of availability (Art. 11e). These also demand replacement by valid names.

The aim of family-group nomenclature is greatest possible stability and uniformity, just as in other zoological names. Experience indicates the wisdom of sustaining family-group names based on junior subjective synonyms if they have priority of publication, for opinions of different workers as to the synonymy of generic names founded on different type species may not agree and opinions of the same worker may alter from time to time. The retention similarly of first-published family-group names which are found to be based on junior objective synonyms is less clearly desirable, especially if a replacement name derived from the senior objective synonym has been recognized very long and widely. To displace a much-used family-group name based on the senior objective synonym by disinterring a forgotten and virtually unused family-group name based on a junior objective synonym because the latter happens to have priority of publication is unsettling.

Replacement of a family-group name may be needed if the former nominate genus is transferred to another family group. Then the first-published name-giver of the family-group assemblage in the remnant taxon is to be recognized in forming a replacement name.

Family-group Names: Authorship and Date

All family-group taxa having names based on the same type genus are attributed to the author who first published the name for any of these assemblages, whether tribe, subfamily, or family (superfamily being almost inevitably a later-conceived taxon). Accordingly, if a family is divided into subfamilies or a subfamily into tribes, the name of no such subfamily or tribe can antedate the family name. Also, every family containing differentiated subfamilies must have a nominate (*sensu stricto*) subfamily, which is based on the same type genus as that for the family, and the author and date set down for the nominate subfamily invariably are identical with those of the family, without reference to whether the author of the family or some subsequent author introduced subdivisions.

Changes in the form of family-group names of the sort constituting *nomina correcta,* as previously discussed, do not affect authorship and date of the taxon concerned, but in the *Treatise* it is desirable to record the authorship and date of the correction.

Suprafamilial Taxa

International rules of zoological nomenclature as given in the *Code* are limited to stipulations affecting lower-rank categories (subspecies to superfamily). Suprafamilial categories (suborder to phylum) are either unmentioned or explicitly placed outside of the application of zoological rules. The *Copenhagen Decisions on Zoological Nomenclature*[1] (1953, Arts. 59–69) proposed to adopt rules for naming suborders and higher taxonomic divisions up to and including phylum, with provision for designating a type genus for each, hopefully in such manner as not to interfere with the taxonomic freedom of workers. Procedures for applying the Law of Priority and Law of Homonymy to suprafamilial taxa were outlined and for dealing with the names for such units and their authorship, with assigned dates, when they should be transferred on taxonomic grounds from one rank to another. The adoption of terminations of names, different for each category but uniform within each, was recommended.

The Colloquium on Zoological Nomenclature which met in London during the week just before the XVth International Congress of Zoology convened in 1958 thoroughly discussed the proposals for regulating suprafamilial nomenclature, as well as many others advocated for inclusion in the new *Code* or recommended for exclusion from it. A decision which was supported by a wide majority of the participants in the Colloquium was against the establishment of rules for naming taxa above family-group rank, mainly because it was judged that such regulation would unwisely tie the hands of taxonomists. For example, a class or order defined by an author at a given date, using chosen morphologic characters (e.g., gills of bivalves), should not be allowed to freeze nomenclature, taking precedence over another, later-proposed class or order distinguished by different characters (e.g., hinge-teeth of bivalves). Even the fixing of type genera for suprafamiliar taxa might have small value, if any, hindering taxonomic work rather than aiding it. At all events, no legal basis for establishing such types and for naming these taxa has yet been provided.

The considerations just stated do not prevent the editors of the *Treatise* from making "rules" for dealing with suprafamiliar groups of animals described and illustrated in this publication. At least a degree of uniform policy is thought to be needed, especially for the guidance of *Treatise* authors. This policy should accord with recognized general practice among zoologists; but where general practice is indeterminate or nonexistent, our own procedure in suprafamilial nomenclature needs to be specified as clearly as possible. This pertains especially to decisions about names themselves, about citation of authors and dates, and about treatment of suprafamilial taxa which on taxonomic grounds are changed from their originally assigned rank. Accordingly, a few "rules" expressing *Treatise* policy are given here, some with examples of their application.

1. The name of any suprafamilial taxon must be a Latin or latinized uninominal noun of plural form, or treated as such, with a capital initial letter and without diacritical mark, apostrophe, diaeresis, or hyphen. If a com-

[1] Francis Hemming, ed., *Copenhagen Decisions on Zoological Nomenclature. Additions to, and modifications of, the Règles Internationales de la Nomenclature Zoologique.* xxix + 135 p. (International Trust for Zoological Nomenclature, London, 1953).

ponent consists of a numeral, numerical adjective, or adverb, this must be written in full.

2. Names of suprafamilial taxa may be constructed in almost any way. A name may indicate morphological attributes (e.g., Lamellibranchiata, Cyclostomata, Taxoglossa) or be based on the stem of an included genus (e.g., Bellerophontina, Nautilida, Fungiina) or on arbitrary combinations of letters (e.g., Yuania); none of these, however, can be allowed to end in -idae or -inae, reserved for family-group taxa. No suprafamilial name identical in form to that of a genus or to another published suprafamilial name should be employed (e.g., order Decapoda LATREILLE, 1803, crustaceans, and order Decapoda LEACH, 1818, cephalopods; suborder Chonetoidea MUIR-WOOD, 1955, and genus *Chonetoidea* JONES, 1928). Worthy of notice is the classificatory and nomenclatural distinction between suprafamilial and family-group taxa which respectively are named from the same type genus, since one is not considered to be transferable to the other (e.g., suborder Bellerophontina ULRICH & SCOFIELD, 1897; superfamily Bellerophontacea McCOY, 1851; family Bellerophontidae McCOY, 1851). Family-group names and suprafamilial names are not coordinate.

3. The Laws of Priority and Homonymy lack any force of international agreement as applied to suprafamilial names, yet in the interest of nomenclatural stability and the avoidance of confusion these laws are widely applied by zoologists to taxa above the family-group level wherever they do not infringe on taxonomic freedom and long-established usage.

4. Authors who accept priority as a determinant in nomenclature of a suprafamilial taxon may change its assigned rank at will, with or without modifying the terminal letters of the name, but such change(s) cannot rationally be judged to alter the authorship and date of the taxon as published originally. A name revised from its previously published rank is a "transferred name" (*nom. transl.*), as illustrated in the following.

Order **CORYNEXOCHIDA** Kobayashi, 1935

[*nom. transl.* MOORE, 1959, *ex* suborder Corynexochida KOBAYASHI, 1935]

A name revised from its previously published form merely by adoption of a different termination, without changing taxonomic rank is an "altered name" (*nom. correct.*).

Order **DISPARIDA** Moore & Laudon, 1943

[*nom. correct.* MOORE in MOORE, LALICKER, & FISCHER, 1952, p. 613, *pro* order Disparata MOORE & LAUDON, 1943]

A suprafamilial name revised from its previously published rank with accompanying change of termination (which may or may not be intended to signalize the change of rank) is recorded as *nom. transl. et correct.*

Order **HYBOCRINIDA** Jackel, 1918

[*nom. transl. et correct.* MOORE in MOORE, LALICKER, & FISCHER, 1952, p. 613, *ex* suborder Hybocrinites JAEKEL, 1918, p. 90]

5. The authorship and date of nominate subordinate and superordinate taxa among suprafamilial taxa are considered in the *Treatise* to be identical since each actually or potentially has the same type. Examples are given below.

Subclass **ENDOCERATOIDEA** Teichert, 1933

[*nom. transl.* TEICHERT in TEICHERT *et al.*, 1964, p. K128 (*ex* superorder Endoceratoidea SHIMANSKIY & ZHURAVLEVA, 1961, *nom. transl.* TEICHERT in TEICHERT *et al.*, 1964, p. K128, *ex* order Endoceroidea TEICHERT, 1933)]

Order **ENDOCERIDA** Teichert, 1933

[*nom. correct.* TEICHERT in TEICHERT *et al.*, 1964, p. K165, *pro* order Endoceroidea TEICHERT, 1933]

Suborder **ENDOCERINA** Teichert, 1933

[*nom. correct.*, herein, *ex* Endoceratina SWEET, 1958, suborder]

TAXONOMIC EMENDATION

Emendation has two distinct meanings as regards zoological nomenclature. These are: (1) alteration of a name itself in various ways for various reasons, as has been reviewed, and (2) alteration of taxonomic scope or concept in application of a given zoological name.

The *Code* (Art. 33a and Glossary p. 148) concerns itself only with the first type of emendation, applying the term to either justified or unjustified changes, both intentional, of the original spelling of a name. These categories are identified in the *Treatise* as *nomina correcta* and *nomina vana,* respectively. The second type of emendation primarily concerns classification and inherently is not associated with change of name. Little attention generally has been paid to this distinction in spite of its significance.

Most zoologists, including paleozoologists, who have signified emendation of zoological names refer to what they consider a material change in application of the name such as may be expressed by an importantly altered diagnosis of the assemblage covered by the name. The abbreviation *"emend."* then may accompany the name with statement of the author and date of the emendation. On the other hand, many workers concerned with systematic zoology think that publication of *"emend."* with a zoological name is valueless, because more or less alteration of taxonomic sort is introduced whenever a subspecies, species, genus, or other assemblage of animals is incorporated under or removed from the coverage of a given zoological name. Inevitably associated with such classificatory expansions and restrictions is some degree of emendation affecting diagnosis. Granting this, still it is true that now and then somewhat radical revisions are put forward, generally with published statement of reasons for changing the application of a name. To erect a signpost at such points of most significanct change is worthwhile, both as aid to subsequent workers in taking account of the altered nomenclatural usage and as indication that not-to-be-overlooked discussion may be found at a particular place in the literature. Authors of contributions to the *Treatise* are encouraged to include records of all specially noteworthy emendations of this nature, using the abbreviation *"emend."* with the name to which it refers and citing the author and date of the emendation.

Examples from *Treatise* volumes follow.

Order ORTHIDA Schuchert & Cooper, 1932
[*nom. transl. et correct.* MOORE in MOORE, LALICKER, & FISCHER, 1952, p. 220, *ex* suborder Orthoidea SCHUCHERT & COOPER, 1932, p. 43] [*emend.* WILLIAMS & WRIGHT, 1965]

Subfamily ROVEACRININAE Peck, 1943
[Roveacrininae PECK, 1943, p. 465; *emend.* PECK in MOORE & TEICHERT, eds., 1978, p. T921]

STYLE IN GENERIC DESCRIPTIONS

Citation of Type Species

The name of the type species of each genus and subgenus is given next following the generic name with its accompanying author, date, and page reference or after entries needed for definition of the name if it is involved in homonymy. The originally published combination of generic and trivial names for this species is cited, accompanied by an asterisk (*), with notation of the author and date of original publication. An exception in this procedure is made, however, if the species was first published in the same paper and by the same author as that containing definition of the genus that it serves as type; in such case, the initial letter of the generic name followed by the trivial name is given without repeating the name of the author and date. Examples of these two sorts of citations are as follows:

Orionastraea SMITH, 1917, p. 294 [**Sarcinula phillipsi* McCOY, 1849, p. 125; OD].
Schoenophyllum SIMPSON, 1900, p. 214 [**S. aggregatum*; OD].

If the cited type species is a junior synonym of some other species, the name of this latter also is given, as follows:

Actinocyathus D'ORBIGNY, 1849, p. 12 [**Cyathophyllum crenulate* PHILLIPS, 1836, p. 202; M; =*Lonsdaleia floriformis* (MARTIN), 1809, pl. 43, validated by ICZN Op. 419].

In the *Treatise,* the name of the type species is always given in the exact form it had in the original publication; in cases where mandatory changes are required, these are introduced later in the text, mostly in a figure caption.

It is desirable to record the manner of establishing the type species, whether by original designation or by subsequent designation.

Fixation of type species originally. The type species of a genus or subgenus, according to provisions of the *Code,* may be fixed in various ways in the original publication or it may be fixed in specified ways subsequent to the original publication as stipulated by the *Code* (Art. 68) in order of precedence as (1) *original designation* (in the *Treatise* indicated as "OD") when the type species is explicitly stated or (before 1931) indicated by "n. gen., n. sp." (or its equivalent) applied to a single species included in a new genus, (2) defined by use of *typus* or *typicus* for one of the species included in a new genus (adequately indicated in the *Treatise* by the specific name), (3) established by *monotypy* if a new genus or subgenus has only one originally included species (in the *Treatise* indicated as "M"), and (4) fixed by *tautonymy* if the genus-group name is identical to an included species name not indicated as type belonging to one of the three preceding categories.

Fixation of type species subsequently. The type species of many genera are not determinable from the publication in which the generic name was introduced and therefore such genera can acquire a type species only by some manner of subsequent designation. Most commonly this is established by publishing a statement naming as type species one of the species originally included in the genus, and in the *Treatise* fixation of the type species in this manner is indicated by the letters "SD" accompanied by the name of the subsequent author (who may be the same person as the original author) and the date of publishing the subsequent designation. Some genera, as first described and named, included no mentioned species and these necessarily lack a type species until a date subsequent to that of the original publication when one or more species are assigned to such a genus. If only a single species is thus assigned, it automatically becomes the type species and in the *Treatise* this subsequent

monotypy is indicated by the letters "SM." Of course, the first publication containing assignment of species to the genus which originally lacked any included species is the one concerned in fixation of the type species, and if this named two or more species as belonging to the genus but did not designate a type species, then a later "SD" designation is necessary. Examples of the use of "SD" and "SM" as employed in the *Treatise* follow.

Hexagonaria Gürich, 1896, p. 171 [*Cyathophyllum hexagonum Goldfuss, 1826, p. 61; SD Lang, Smith, & Thomas, 1940, p. 69].
Muriceides Studer, 1887, p. 61 [*M. fragilis Wright & Studer, 1889; SM Wright & Studer, 1889].

Another mode of fixing the type species of a genus is action of the International Commission on Zoological Nomenclature using its plenary powers. Definition in this way may set aside application of the *Code* so as to arrive at a decision considered to be in the best interest of continuity and stability of zoological nomenclature. When made, it is binding and commonly is cited in the *Treatise* by the letters "ICZN," accompanied by the date of announced decision and reference to the appropriate numbered Opinion.

It should be noted that *subsequent designation* of a type species is admissible only for genera established prior to 1931. A new genus-group name established after 1930, and not accompanied by fixation of a type species through original designation or original indication, is invalid (*Code,* Art. 13b). Effort of a subsequent author to "validate" such a name by subsequent designation of a type species constitutes an original publication making the name available under authorship and date of the subsequent author.

Homonyms

Most generic names are distinct from all others and are indicated without ambiguity by citing their originally published spelling accompanied by name of the author and date of first publication. If the same generic name has been applied to two or more distinct taxonomic units, however, it is necessary to differentiate such homonyms, and this calls for distinction between junior homonyms and senior homonyms. Because a junior homonym is invalid, it must be replaced by some other name. For example, *Callopora* HALL, 1852, introduced for Paleozoic trepostomate bryozoans, is invalid because GRAY in 1848 published the same name for Cretaceous-to-Holocene cheilostomate bryozoans, and BASSLER in 1911 introduced the new name *Hallopora* to replace Hall's homonym. The *Treatise* style of entry is:

Hallopora BASSLER, 1911, p. 325, *nom. subst. pro Callopora* HALL, 1852, p. 144, *non* GRAY, 1848.

In like manner, a needed replacement generic name may be introduced in the *Treatise* (even though first publication of generic names otherwise in this work is generally avoided). The requirement that an exact bibliographic reference must be given for the replaced name commonly can be met in the *Treatise* by citing a publication recorded in the list of references as shown in the following example.

Mysterium DE LAUBENFELS, herein, *nom. subst. pro Mystrium* SCHRAMMEN, 1936, p. 183, *non* ROGER, 1862 [*Mystrium porosum* SCHRAMMEN, 1936, p. 183].

Otherwise, no mention of the existence of a junior homonym generally is made.

Synonymous homonyms. An author sometimes publishes a generic name in two or more papers of different date, each of which indicates that the name is new. This is a bothersome source of errors for later workers who are unaware that a supposed first publication that they have in hand is not actually the original one. Although the names were separately published, they are identical and therefore definable as homonyms; at the same time they are absolute synonyms. For the guidance of all concerned, it seems desirable to record such names as synonymous homonyms, and in the *Treatise* the junior one of these is indicated by the abbreviation "jr. syn. hom."

Identical family-group names not infre-

quently are published as new names by different authors, the author of the later-introduced name being ignorant of previous publication(s) by one or more other workers. In spite of differences in taxonomic concepts as indicated by diagnoses and grouping of genera and possibly in assigned rank, these family-group taxa are nomenclatural homonyms, based on the same type genus, and they are also synonyms. Wherever encountered, such synonymous homonyms are distinguished in the *Treatise* as in dealing with generic names.

A special, though rare, case of synonymy exists when identical family names are formed from generic names having the same stem but differing in their endings. An example is the family name Scutellidae R. & E. RICHTER, 1925, based on *Scutellum* PUSCH, 1833, a trilobite. This name is a junior synonym of Scutellidae GRAY, 1825, based on *Scutella* LAMARCK, 1816, an echinoid. The name of the trilobite family was later changed to Scutelluidae (ICZN, Op. 1004, 1974).

Synonyms

Citation of synonyms is given next following record of the type species and if two or more synonyms of differing date are recognized, these are arranged in chronological order. Objective synonyms are indicated by accompanying designation "obj.," others being understood to constitute subjective synonyms, of which the types are also indicated. Examples showing *Treatise* style in listing synonyms follow.

Mackenziephyllum PEDDER, 1971, p. 48 [*M. insolitum*; OD] [=*Zonastraea* TSYGANKO in SPASSKIY, KRAVTSOV, & TSYGANKO, 1971, p. 85, *nom. nud.*; *Zonastraea* TSYGANKO, 1972, p. 21 (type, *Z. graciosa*, OD)].

Kodonophyllum WEDEKIND, 1927, p. 34 [*Streptelasma Milne-Edwardsi* DYBOWSKI, 1873, p. 409; OD; =*Madrepora truncata* LINNÉ, 1758, p. 795, see SMITH & TREMBERTH, 1929, p. 368] [=*Patrophontes* LANG & SMITH, 1927, p. 456 (type, *Madrepora truncata* LINNÉ, OD); *Codonophyllum* LANG, SMITH, & THOMAS, 1940, p. 39, *nom. van.*].

Some junior synonyms of either objective or subjective sort may take precedence desirably over senior synonyms wherever uniformity and continuity of nomenclature are served by retaining a widely used but technically rejectable name for a generic assemblage. This requires action of ICZN using its plenary powers to set aside the unwanted name and validate the wanted one, with placement of the concerned names on appropriate official lists.

ABBREVIATIONS

Abbreviations used in this part of the *Treatise* are explained in the following alphabetically arranged list. Standard abbreviations and those found only in the references are not included here.

Afr., Africa
Alb., Albian
Alg., Algeria
AMNH, American Museum of Natural History, New York
Apt., Aptian
Arenig., Arenigian
Ariz., Arizona
Artinsk., Artinskian
Atl., Atlantic

Bathon., Bathonian
Belg., Belgium
Blackriv., Blackriveran
BMNH, British Museum (Natural History), London
Bol., Bolivia
Brit., Britain
Brit. Is., British Isles

C., Central
Cal., California
Can., Canada
Caradoc., Caradocian
Carb., Carboniferous
Carib., Caribbean
Cayug., Cayugan
Cenoman., Cenomanian
Champlain., Champlainian
Chazy., Chazyan
Chester., Chesterian
Cincinnat., Cincinnatian
Co., County
Coll., Collection, -s
Comm., Commission
Coniac., Coniacian
CPC, Commonwealth Palaeontological Collection, Bureau of Mineral Resources, Commonwealth of Australia, Canberra
Cr., Creek
Cret., Cretaceous
Czech., Czechoslovakia

Del., Delaware

Delft, Mineralogisch-Geologisch Museum, Technische Hoogeschool, Delft
Denm., Denmark
Dev., Devonian
Distr., District
Dolbor., Dolborian
Dzhulf., Dzhulfian

E., East
Eden., Edenian
Eifel., Eifelian
Eng., England
Est., Estonia
Eu., Europe

F., Formation
Fla., Florida
FMNH, Field Museum of Natural History, Chicago
Frasn., Frasnian

G. Brit., Great Britain
Ger., Germany
Givet., Givetian
Gotl., Gotland
Gr., Group
Greenl., Greenland
Guadalup., Guadalupian

Helderberg., Helderbergian
HM, Hunterian Museum, University of Glasgow, Glasgow

Ill., Illinois
Ind., Indiana
Indon., Indonesia
Is., Island(s)
Ire., Ireland
ISGS, Illinois State Geological Survey, Urbana
ISGS(ISM), specimen from Illinois State Museum (Springfield), now housed at ISGS, Urbana

Jur., Jurassic

Kans., Kansas
Kazakh., Kazakhstan
Kazan., Kazanian
Kinderhook., Kinderhookian
KUMIP, Kansas University Museum of Invertebrate Paleontology, Lawrence
Ky., Kentucky

L., Lower
lat., latitude, lateral
Leonard., Leonardian
Llandeil., Llandeilian
Llandov., Llandoverian
Llanvirn., Llanvirnian
loc., locality
long., longitude, longitudinal
Ls., Limestone
LSU, Louisiana State University, Baton Rouge
Ludlov., Ludlovian

M, monotypy
M., Middle
Maastricht., Maastrichtian
Mangaze., Mangazeian
Manit., Manitoba
Mass., Massachusetts
Maysvill., Maysvillian.
Mbr., Member
Medit., Mediterranean
Meramec., Meramecian
MGU, Muzej Ministerstva geologii Uzbek. SSU, Tashkent
mi., mile
Mich., Michigan
mid., middle
Mio., Miocene
Minn., Minnesota
Miss., Mississippian, Mississippi
Missour., Missourian
Mo., Missouri

xxiii

Mohawk., Mohawkian
Mt(s)., Mountain(s)
Münster, Geologisch-Paläon-
tologisches Institut der
Westfälischen Wilhelms-
Universität Münster, Müns-
ter-Westfalen

N., North
N.Am., North America
Namur., Namurian
N.Car., North Carolina
Neb., Nebraska
Neth., Netherlands
Niag., Niagaran
nom. conserv., *nomen conservatum* (conserved name)
nom. dub., *nomen dubium*
(doubtful name)
nom. nud., *nomen nudum*
(naked name)
nom. oblit., *nomen oblitum*
(fogotten name)
nom. subst., *nomen substitutum* (substitute name)
nom. transl., *nomen translatum* (transferred name)
Nor., Norway
N.Y., New York
NYSM, New York State
Museum, Albany
N.Z., New Zealand
N.Zemlya, Novaya Zemlya

O., Ocean
obj., objective
OD, original designation
Okla., Oklahoma
Ont., Ontario
Ord., Ordovician
Osag., Osagian
OT, original tautonomy
OUM, Oxford University
Museum, Oxford

Pac., Pacific
Pak., Pakistan
Paleoc., Paleocene
Penin., Peninsula
Penn., Pennsylvanian
Perm., Permian

PGU, Geological Museum of
the Geological Board of the
Maritime Territory, Vladi-
vostok
Philip., Philippines
PIN, Paleontoligicheskij insti-
tut, Akademiya nauk
SSSR., Moscow
Plio., Pliocene
Port., Portugal
prov., province
PSU, Pennsylvania State Uni-
versity (Paleobryozoological
Research Collection), Uni-
versity Park
Pt., Point

Ra., Range
rec., recent
Richmond., Richmondian
RSM, Royal Scottish Museum,
Edinburgh
Russ. platf., Russian platform

S., South
Santon., Santonian
Scot., Scotland
SD, subsequent designation
Sh., Shale
Sib., Siberia
Sib. plat., Siberian platform
Sil., Silurian
SIUC, Southern Illinois Uni-
versity, Carbondale
SM, Sedgwick Museum, Cam-
bridge University, Cam-
bridge; subsequent monotypy
SNIIGGIMS, Muzej Sibirskogo
nauchno-issledovatelskogo
instituta geologii, geofiziki i
mineralnogo syrya, Novosi-
birsk
Spits., Spitsbergen
sp., species
Ss., sandstone
Sta., Station
Stephan., Stephanian
Str., Strait(s)
SU, Department of Geology,
University of Sydney, Syd-
ney

Swed., Sweden
Switz., Switzerland

tang., tangential
Tenn., Tennessee
Terr., Territory
Tournais., Tournaisian
Transcauc., Transcaucasia
transv., transverse
Trenton., Trentonian
TsGM, Central Geological
Museum, Central Geological
and Prospecting Institute,
Leningrad

U., Upper
UI, University of Illinois
Paleontology Museum,
Urbana
Ulster., Ulsterian
UMMP, University of Michi-
gan Museum of Paleontol-
ogy, Ann Arbor
up., upper
USA, United States (America)
USNM, United States
National Museum, Wash-
ington, D.C.
USSR, Union of Soviet Social-
ist Republics

Va., Virginia
Valangin., Valanginian
Vict., Victoria
VNIGRI, Musej, Vsesoyuznyj
neftyanoj nauchno-issledov-
atskij geologorazvedochnoj
institut, Leningrad
Vt., Vermont

W., West
Wash., Washington
Wenlock., Wenlockian
WM, Walker Museum of
Paleontology, University of
Chicago, housed at the
Field Museum, Chicago
Wolfcamp., Wolfcampian

YPM, Peabody Museum, Yale
University, New Haven

REFERENCES TO LITERATURE

The titles of serials cited in the references are abbreviated as recommended in the *Biblio-
graphical Guide for Editors and Authors* (1974, The American Chemical Society, Washington, D.C.); titles of serials not covered in the *Guide* have been abbreviated according to
the standard established in International Standards Organization (ISO) recommendation

833-1974. The names of authors and titles of works in Cyrillic have been transliterated for the most part according to the same standard. A translation of each Cyrillic title is given in brackets at the end of the reference. Full citations of references containing senior homonyms are not included, but may be found in contracted form in S. A. NEAVE, *Nomenclator Zoologicus* (1939–1975, 7 v., Zoological Society, London).

SOURCES OF ILLUSTRATIONS

Most illustrations in this volume are new. Where previously published illustrations are used, the author and date of publication are given in parentheses in the figure caption. Full citation of the publication is provided in the references.

STRATIGRAPHIC DIVISIONS

As commonly cited in the *Treatise,* classification of rocks forming the geologic column is reasonably uniform and firm throughout most of the world as regards major divisions (e.g., series, systems, and rocks representing eras), but it may be variable and unfirm as regards minor divisions (e.g., substages, stages, and subseries), which tend to be provincial in application. A tabulation of commonly cited European and North American divisions is given for systems from the Ordovician to the Permian, which corresponds to the stratigraphic range of bryozoan genera that are diagnosed here.

Generally Recognized Division of Geologic Column

EUROPE	NORTH AMERICA
CENOZOIC ERATHEM	CENOZOIC ERATHEM
QUATERNARY SYSTEM	QUATERNARY SYSTEM
TERTIARY SYSTEM	TERTIARY SYSTEM
MESOZOIC ERATHEM	MESOZOIC ERATHEM
CRETACEOUS SYSTEM	CRETACEOUS SYSTEM
JURASSIC SYSTEM	JURASSIC SYSTEM
TRIASSIC SYSTEM	TRIASSIC SYSTEM
PALEOZOIC ERATHEM	PALEOZOIC ERATHEM
PERMIAN SYSTEM	PERMIAN SYSTEM
Upper Permian Series	Upper Permian Series
Tartarian Stage	Ochoan Stage
Kazanian Stage	Guadalupian Stage
Lower Permian Series	Lower Permian Series
Artinskian Stage	Leonardian Stage
Sakmarian Stage	Wolfcampian Stage
Asselian Stage	
CARBONIFEROUS SYSTEM	
Silesian Subsystem	PENNSYLVANIAN SYSTEM
Stephanian Series	Virgilian Series
	Missourian Series
	Desmoinesian Series
Westphalian Series	Atokan Series
Namurian Series	Morrowan Series
Dinantian Subsystem	MISSISSIPPIAN SYSTEM
	Chesterian Series
Visean Series	Meramecian Series
	Osagian Series
Tournaisian Series	Kinderhookian Series

DEVONIAN SYSTEM
  Upper Devonian Series
    Famennian Stage
    Frasnian Stage
  Middle Devonian Series
    Givetian Stage
    Eifelian Stage
  Lower Devonian Series
    Emsian Stage
    Siegenian Stage
    Gedinnian Stage
SILURIAN SYSTEM
  Pridolian Series
  Ludlovian Series
  Wenlockian Series
  Llandoverian Series
ORDOVICIAN SYSTEM
  Ashgillian Series

  Caradocian Series

  Llandeilian Series
  Llanvirnian Series
  Arenigian Series

  Tremadocian Series[1]

CAMBRIAN SYSTEM

ROCKS OF PRECAMBRIAN ERAS

DEVONIAN SYSTEM
  Upper Devonian Series
    Famennian Stage
    Frasnian Stage
  Middle Devonian Series
    Givetian Stage
    Eifelian Stage
  Lower Devonian Series
    Emsian Stage
    Siegenian Stage
    Gedinnian Stage
SILURIAN SYSTEM
  Pridolian Series
  Ludlovian Series
  Wenlockian Series
  Llandoverian Series
ORDOVICIAN SYSTEM
  Cincinnatian Series (Upper Ordovician)
    Richmondian Stage
    Maysvillian Stage
    Edenian Stage

  Champlainian Series
    (Middle Ordovician)
    Mohawkian Stage
    Chazyan Stage
    Whiterockian Stage
  Canadian Series (Lower Ordovician)

CAMBRIAN SYSTEM

ROCKS OF PRECAMBRIAN ERAS

---

[1] Tremadocian is placed in Cambrian by some authors.

PART G
BRYOZOA
REVISED

Introduction, Order Cystoporata, Order Cryptostomata

By R. S. Boardman, A. H. Cheetham, D. B. Blake, John Utgaard, O. L. Karklins, P. L. Cook, P. A. Sandberg, Geneviève Lutaud, and T. S. Wood

CONTENTS

INTRODUCTION TO THE BRYOZOA

By R. S. Boardman, A. H. Cheetham, and P. L. Cook

[Smithsonian Institution, Washington, D.C.; British Museum (Natural History), London]

Bryozoa constitute a major phylum of invertebrates. Modern Bryozoa are widely distributed in fresh and marine waters, from high altitudes to abyssal depths. They are a dominant component of the sessile fauna of shelf seas and fouling communities. They are also the most abundant fossils in many sedimentary deposits. The fossil record of the phylum extends over the last 500 million years (Ordovician to Holocene) and is characterized by wide distribution, great abundance, and high diversity throughout most of that time.

The Bryozoa are the only phylum in which all known representatives form colonies. A **colony** can consist of a few to tens of millions of minute members called **zooids**. The numbers of zooids in most bryozoan colonies are comparable to the numbers of individuals in a society of ants or a population of ordinary solitary animals. The zooids in a bryozoan colony differ from members of a population or an insect society in being both physically connected and asexually reproduced, but many of their functions are comparable to those of solitary individuals.

Even though Bryozoa are among the most common marine invertebrates in modern seas and in the fossil record, they are not so likely to be recognized as are members of several other major phyla. A bryozoan colony can be so varied in megascopic appearance (see Fig. 7–9, 13–15) as to be practically indistinguishable from some representatives of such other phyla as hydroids, corals, and algae. The distinguishing characters are generally observable only with magnification.

A bryozoan colony is made up of asexually replicated, physically connected zooids. The asexual origin and physical connection of zooids justifies a basic assumption, that genetic makeup is uniform throughout a colony. Nevertheless, morphologic variation is normal among zooids of a colony because of ontogeny, astogeny, polymorphism, and microenvironment. Because of genetic continuity, these sources of variation can be studied within a colony without the complication of differences in genotype, an advantage not available in solitary animals.

Physical continuity allows some zooids, such as nonfeeding polymorphs, to be highly specialized and parts of colonies to develop structures not possible in solitary animals. Feeding zooids in the same colony may differ so in morphology and other functions that, if not physically connected, they could be considered genetically different; many might be placed in distinct taxa.

Colonies can increase in size and their **growth habits** change in response to environmental pressure without any increase in size or change in basic morphology of feeding zooids. This flexibility is an advantage to species in which competition for substrates requires irregular configurations or erect growth. Commonly, an increase in size or change from encrusting to erect growth habit requires structural support accomplished by development of colony-wide skeletal structures, changes in the morphology of some zooids, or both.

Some aspects of bryozoan morphology, especially those related to the colonial state, have not been fully exploited in the study and application of the phylum. The abundance and wide geographic distribution of Bryozoa from the Ordovician to the present, the flexibility of their colony growth habits in response to environmental pressure, and the availability of many morphologic characters for studying their classification and evolutionary trends make Bryozoa potentially highly significant to the study not only of biostratigraphy but of past and present ecology and zoogeography. In general, but with much overlap, the morphology of zooids tends to reveal genetically controlled characters whereas the form of the colony reflects environmental modifications. The lack of significant transportation after death commonly can be detected for many fossil Bryozoa, especially for erect branching colonies preserved nearly intact.

Taken together, these qualities give promise of considerable success not yet realized in the application of Bryozoa study to geologic and zoologic problems.

Acknowledgments.—This section has been improved by technical reviews by W. C. Banta, D. B. Blake, R. J. Cuffey, Eckart Håkansson, O. L. Karklins, G. P. Larwood, Geneviève Lutaud, F. K. McKinney, R. A. Pohowsky, P. A. Sandberg, Lars Silén, R. J. Singh, and T. S. Wood. Useful discussions of some of our ideas were contributed by M. B. Abbott, Krister Brood, Susan Cummings, J. E. Winston, T. G. Gautier, Jean-Georges Harmelin, A. B. Hastings, D. M. Lorenz, C. D. Palmer, J. S. Ryland, and John Utgaard. We by no means intend to imply agreement by all of these colleagues with all of the ideas presented.

L. B. Isham constructed Figures 2 through 4. Other figures were drafted by JoAnn Sanner, who together with D. A. Dean gave other technical assistance.

CHARACTERISTICS OF BRYOZOA

Bryozoa are colonial, aquatic, generally sessile metazoans, regarded as coelomate, with a retractable lophophore and U-shaped digestive tract.

All Bryozoa form colonies (Fig. 1). Each colony consists of one or more kinds of minute zooids and multizooidal parts, and some colonies include extrazooidal parts. Zooids are physically connected, asexually replicated morphologic units that separately perform such major physiologic or structural functions as feeding, reproduction, or support. **Multizooidal parts** include continuous wall layers grown outside existing zooidal boundaries and their enclosed **body cavities**, which become parts of zooids as colonies develop. **Extrazooidal parts** remain outside zooidal boundaries throughout the life of a colony and include walls with or without skeletal layers, skeleton not parts of body walls, and adjacent body cavities.

A colony interacts with the environment as a complete organism comparable to a solitary animal. Internally, however, the zooid corresponds to a solitary animal in that it has systems of organs or other structures that separately perform the major functions of a colony. Zooids differ from solitary animals in being both physically connected and asexually replicated. Therefore, zooids and other parts of a colony are assumed to be genetically uniform.

Colonies characteristically include enormous numbers of replicated zooids, with some notable exceptions in a few taxa, and may be more than one meter in size. The size and growth habit of colonies commonly are highly variable under environmental influence, but in some taxa growth habit and size of colony appear to be narrowly restricted genetically.

Zooids and other parts of colonies are interconnected by cells, tissues, confluent body cavity, or a combination of these, to nourish developing, injured, and nonfeeding zooids, and other parts of colonies incapable of feeding. It is probable that interzooidal connections function in the coordinated nervous behavior observed in some colonies.

Body walls enclose body cavities of zooids, parts of zooids, and all other parts of colonies. Body walls consist of cellular and noncellular layers. Cellular layers can be continuous or can consist of scattered cells. Cellular layers in two of the three major groups of Bryozoa include an inner **peritoneum** lining the body cavity (considered to be a **coelom**) and an outer **epidermis**. A peritoneum also is reported to be present in the third major group but is not part of the body wall. Noncellular layers include outermost cuticular or

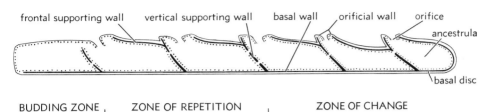

FIG. 1.   Characteristics of the Bryozoa. Diagram of a longitudinal section through an encrusting colony of a fixed-wall stenolaemate bryozoan showing zones of astogenetic change and repetition and basic orientation of zooidal walls. Lophophores, digestive tracts, and some other soft parts have been omitted. The zooid at the proximal end of the colony, extreme right, is the primary zooid (ancestrula). As the colony grows, the expanding exterior wall of the budding zone gains enclosed space that is partitioned into zooids by interior vertical walls. The boundary between zooids runs through the middle of the calcareous layer of interior vertical walls. The cuticle is attached directly to skeletal layers of exterior frontal and basal walls.

gelatinous layers, and in most taxa some calcareous material, the skeleton, between the cuticle and epidermis, or in some taxa between layers of the cuticle. Calcareous layers of zooidal walls and any connected intrazooidal calcareous structures form a zooidal **skeleton**, the **zooecium**. Zooecia of a colony together with any other skeletal parts form a colonial skeleton, the **zoarium**. The entire zoarium is secreted on the external side of the epidermis opposite the body cavity. The skeleton therefore is exoskeletal throughout, even though in some places it is deposited by epidermis that is infolded into existing body cavity.

Body walls are basically of two developmental kinds, exterior and interior (SILÉN, 1944a,b). **Exterior walls** extend the body cavity of zooids and the colony; **interior walls** partition preexisting body cavity into zooids or parts of zooids or extrazooidal structures. Exterior walls include an outermost cuticle or gelatinous layer, which is not necessary and commonly not present as a component of interior walls.

All zooids minimally have body cavities enclosed by body walls (Fig. 1). Body walls can be complete or incomplete so zooidal cavities can be partly open to adjacent zooidal or colony body cavities. **Feeding zooids** must be present at some stage in the lives of all colonies and have in addition to body walls and cavities a protrusible lophophore, an alimentary canal, muscles, a nervous system, and **funicular strands** (Fig. 2–4).

Zooids within a colony can differ distinctly in morphology and function at the same stages of ontogeny and in the same sexual generations. Such zooids are termed **polymorphs**. Polymorphs can be specialized to perform sexual, supportive, connective, cleaning, or defensive functions for example, and can even lack feeding organs entirely.

The body walls of feeding zooids include exterior **orificial walls** and **supporting walls** (Fig. 1). The concepts of orificial and supporting walls are based on comparisons of function and position among taxa and do not necessarily imply homology. Orientation of these zooidal walls relative to zooidal and colony growth directions (**distal**) can differ in major groups.

The orificial wall is exterior and terminal or subterminal. It bears or defines the opening (**orifice**) through which the lophophore is protruded into the environment. It is attached through the orifice to a **vestibular wall** leading to the lophophore and gut (Fig. 2–4), and may or may not be attached to other zooidal walls (Fig. 2). Some kinds of

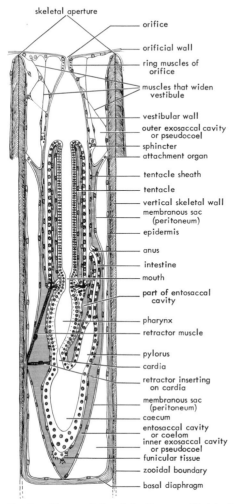

skeletal aperture

orifice

orificial wall

ring muscles of
orifice

muscles that widen
vestibule

vestibular wall

outer exosaccal cavity
or pseudocoel

sphincter

attachment organ

tentacle sheath

tentacle

vertical skeletal wall

membranous sac
(peritoneum)

epidermis

anus

intestine

mouth

part of entosaccal
cavity

pharynx

retractor muscle

pylorus

cardia

retractor inserting
on cardia

membranous sac
(peritoneum)

caecum

entosaccal cavity
or coelom

inner exosaccal cavity
or pseudocoel

funicular tissue

zooidal boundary

basal diaphragm

FIG. 2.   Characteristics of the Bryozoa. Model of
the retracted feeding zooid of a free-walled stenolaemate based on organs of a living tubuliporate
(after Nielsen, 1970, fig. 13) and the zooidal living
chamber of a Paleozoic trepostomate.

polymorphs lacking feeding organs also have orificial walls or their equivalents.

Supporting zooidal walls (Fig. 1) can be either interior or exterior, or a combination, and several kinds may be recognized by their position and orientation relative to the orificial wall. **Basal zooidal walls** are supporting walls that are opposite and generally parallel to orificial walls. All colonies apparently begin with one or more zooids having exterior basal walls. These basal walls form the encrusting base of the colony either alone or by extending distally as multizooidal walls. Zooids budded above the encrusting base of a colony can have exterior or interior basal walls, or can lack basal walls altogether.

Vertical walls are supporting walls that are entirely or in part at high angles to basal and orificial walls, thus giving depth, length, or both to the zooidal body cavity. Vertical zooidal walls can be exterior or interior, or a combination. Exterior vertical walls originate from multizooidal (Gymnolaemata) or extrazooidal (Phylactolaemata) walls. Interior vertical walls originate from interior or exterior zooidal walls, interior extrazooidal walls, or either interior or exterior walls of multizooidal origin (Stenolaemata, Gymnolaemata). Vertical walls may be attached distally to orificial walls, to intervening frontal walls, or a combination, or may terminate beneath orificial walls.

Frontal walls, where present (see Stenolaemata and Gymnolaemata), are exterior supporting walls that originate as zooidal or multizooidal walls. Frontal walls provide a front side to zooids more extensive than the orificial walls alone. Parts of frontal walls can extend beyond the general colony surface to form **peristomes**, which either carry orificial walls at their outer ends or surround orificial walls at their inner ends.

The walls of the vestibule and lophophore are also parts of body walls. The vestibular wall, lophophore, and alimentary canal (Fig. 2–4) apparently originate by infolding of the exterior wall of the colony or internally from the lophophore and gut of existing zooids (see Phylactolaemata). The vestibular wall surrounds a space of variable extent, the **vestibule**, and connects the orificial wall to the tentacle sheath. The vestibule is the passage through which the lophophore is protruded for feeding.

The tentacle sheath and ciliated, coelomate tentacles together constitute the **lophophore**. In position, the lophophore is that part of the body wall of a feeding zooid that begins at the inner end of the vestibule and

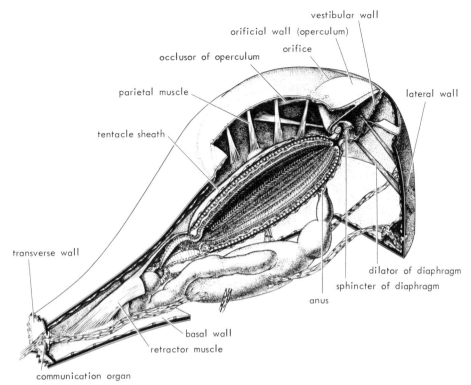

vestibular wall

orificial wall (operculum)

orifice

occlusor of operculum

parietal muscle

lateral wall

tentacle sheath

transverse wall

dilator of diaphragm

sphincter of diaphragm

anus

basal wall

retractor muscle

communication organ

Fig. 3. Characteristics of the Bryozoa. Model of a simple gymnolaemate with the lophophore retracted, based on organs of living cheilostomates (after Nitsche, 1871, fig. 1, 2; Calvet, 1900, pl. 2) and the skeleton of a Mesozoic cheilostomate.

ends at the mouth. The **tentacle sheath** is that part of the body wall that is introverted to enclose the tentacles in the retracted position and is everted to support them in the protruded position (Fig. 3, 4). The boundary between the tentacle sheath and the vestibular wall is generally a sphincter muscle.

A single row of **tentacles** surrounds the mouth in a circular or bilobed pattern. The mouth is opened and closed by muscular action and in a small number of genera is overhung by a fold of body wall (**epistome**). In feeding, the movement of cilia on the tentacles produces currents that concentrate food particles near the mouth.

Protrusion and retraction of the lophophore are accomplished by muscular action. Protrusion involves hydrostatic pressures produced in various ways by muscles modifying the shapes of parts of the body cavity.

Retraction is by direct contraction of retractor muscles.

The digestive tract is complete and recurved, so that the anus opens near the mouth. When the tentacles are protruded, the anus opens on either the distal or proximal side of the tentacle sheath wall below the row of tentacles (Fig. 2–4). The nervous system includes a ganglion near the mouth. Nephridia as well as circulatory and respiratory organs are apparently absent.

In almost all taxa, colonies, but not all zooids, are hermaphroditic; gonads form in zooidal coeloms and are ductless, sex products being released through special openings in the body wall. Embryos are commonly brooded, either within or outside body cavities, to produce ciliated **larvae** or other motile stages. Embryonic fission occurs during brooding in some modern taxa (see Steno-

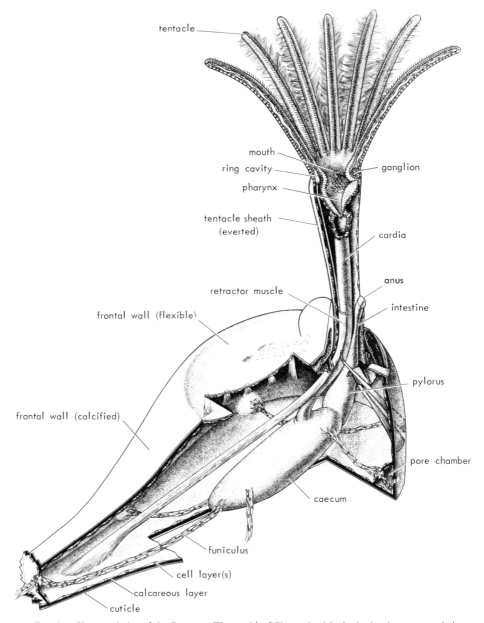

FIG. 4.   Characteristics of the Bryozoa. The zooid of Figure 3 with the lophophore protruded.

laemata). Ciliated larvae include bivalve forms with complete digestive tracts and naked forms lacking digestive tracts.

A larva settles on a substrate and undergoes **metamorphosis** with extensive reorganization of tissues, typically to form a single zooid, the **ancestrula** (Fig. 1). Colonies of some taxa also may be produced asexually by **fragmentation** into groups of functional zooids and in a few taxa by the formation of resistant resting bodies. In most taxa, the ancestrula produced by a larva or the first zooid produced by a resistant resting body differs in size or other morphologic characters from other zooids in a colony. Generally, both types of initial zooids contain feeding organs.

Asexually produced zooids following an ancestrula (Fig. 1) commonly show a morphologic gradient through several generations (**zone of astogenetic change**) leading to one or more kinds of zooids replicated in succeeding generations (**zone of astogenetic repetition**). Newly developing, asexually produced zooids (**buds**) can be initiated either as feeding organs (see Phylactolaemata) or as body walls. Buds initiated as feeding organs can develop from either exterior walls or other developing feeding organs. Buds initiated as body walls develop distally by outward expansion of exterior membranous walls of the colony or of other zooids. Proximally, buds appear as infolds from (1) interior or exterior multizooidal walls, (2) interior walls of other zooids, or (3) interior extrazooidal structures.

Parts of zooids characteristically undergo cyclic phases of degeneration and regeneration in most taxa. Degeneration products commonly form encapsulated masses of degenerating cells, termed **brown bodies**. Parts that degenerate include lophophore, gut, some muscles, and some other nonskeletal parts, varying in different groups.

HISTORICAL REVIEW

The study of Bryozoa has been marked historically by an insufficient number of workers. Approximately 20,000 fossil and living species have been described, but these are undoubtedly a small number in proportion to those that remain to be recognized and investigated. Uneven distribution of studies has left major gaps in our knowledge of Paleozoic and Mesozoic faunas, and a pattern susceptible of interpretation is just emerging from results of work on Cenozoic faunas. Little detailed information is available for the Triassic or Jurassic systems. In North America, there are comparable gaps in the Silurian, Mississippian, Pennsylvanian, Permian, Cretaceous, and upper Tertiary systems. Bryozoa of the Paleozoic Era are relatively well known from the Soviet Union but are generally unknown in Europe. Cretaceous and Cenozoic faunas in Europe have been studied extensively, but revision, synthesis, and comparison with other areas are needed. In southeast Asia, all faunas but those from the Upper Paleozoic are poorly known, and in Australia, faunal studies are scattered throughout most of the Paleozoic and part of the Cenozoic. In South America, Africa, Antarctica, and large parts of Asia, there is little knowledge of any of the fossil record.

Abundant Bryozoa have been found in a few cores of Tertiary sediments in the Atlantic, Pacific, and Indian oceans recovered by the Deep Sea Drilling Project, but the fossil record of Bryozoa in the open oceans is still poorly known.

Living faunas have been investigated extensively throughout the world, but gaps in distributions and the need for revision, synthesis, and comparison of described faunas have delayed an understanding of world biogeographic patterns.

The earliest work in which fossil Bryozoa were described and illustrated is reportedly that of Bassi (1757) (Neviani, 1894; Astrova, 1960a; Annoscia, 1968), but living Bryozoa have been studied for at least 400 years. The early history of the study of Bryozoa (summarized in detail by Harmer, 1930; Hyman, 1959; and Ryland, 1970) was marked by a series of misunderstandings of their nature, resulting in their confusion with plants and coelenterates. The animal nature of Bryozoa seems to have been established with the studies of Ellis (1754, 1755a–c), who considered most of the Bryozoa known to him to be ramified animals, which he called celliferous corallines. Linné (1758), basing his work on Ellis's descriptions and plates, named the Zoophyta as an order of the class

Vermes and considered them to be at least partly of a plant nature. He included bryozoans and coelenterates in this order. The difference between bryozoans and coelenterates was established as the result of observations of a digestive tract with two openings (DE BLAINVILLE, 1820; AUDOUIN & MILNE-EDWARDS, 1828), and of ciliated tentacles (GRANT, 1827). This difference was formalized by establishment of a taxon to which the name Polyzoa of THOMPSON (1830) or Bryozoa of EHRENBERG (1831) was applied. Although the name Polyzoa was published first and a controversy existed for many years (HARMER, 1947; BROWN, 1958), the name has been dropped in most recent literature.

When THOMPSON separated Bryozoa from coelenterates, he placed them with the Mollusca. MILNE-EDWARDS (1843) named the Molluscoidea to include Bryozoa and tunicates, and HUXLEY (1853) added Brachiopoda to this group. The use of Molluscoidea, as either a phylum or a subkingdom, persisted into the twentieth century (CANU & BASSLER, 1920). The most common usage has been as an emended taxon including Bryozoa and Brachiopoda. In some classifications the Phoronida have been included. The names Podaxonia (LANKESTER, 1885), Tentaculata (HATSCHEK, 1888), Vermidea (DELAGE & HEROURARD, 1897), and Lophophorata (HYMAN, 1959) have also been used for varying combinations of these and other phyla.

NITSCHE (1869) distinguished two groups among the Bryozoa as known in the nineteenth century and named them Entoprocta and Ectoprocta. These groups were elevated to phylum rank by HATSCHEK (1888). Some workers (NIELSEN, 1971) still include Entoprocta in the phylum Bryozoa, as was done in the earlier edition of this *Treatise* (BASSLER, 1953). The Entoprocta are excluded from the Bryozoa as recognized here. CUFFEY (1973) included phyla Entoprocta and Ectoprocta in a superphylum Bryozoa.

The shifts in hierarchic level and contents have led some more recent workers to abandon the name Bryozoa for the phylum and to use the name Ectoprocta (HYMAN, 1959;

SCHOPF, 1967, 1968; CUFFEY, 1969). The name Bryozoa is used for the phylum as understood here to exclude the phylum Entoprocta. Reasons for this usage were given by MAYR (1968). Controversy over the name of the phylum contributes little to understanding the bryozoans (SOULE & SOULE, 1968).

The relationships of the Bryozoa to the other phyla have not been established on the basis of a fossil record from which evolutionary trends can be interpreted. Necessary evidence for hypotheses of the origin of Bryozoa, such as that from the Phoronida discussed by FARMER, VALENTINE, and COWEN (1973), would be morphologies intermediate between Bryozoa and other groups that existed at the time of the earliest Bryozoa.

Bryozoa have commonly been divided into two major groups, varying slightly in composition according to the morphologic criteria employed. DE BLAINVILLE (1834) distinguished Bryozoa having bilobed lophophores from those having circular lophophores. GERVAIS (1837) named these groups Polypiaria hippocrepia and Polypiaria infundibulata, respectively. VAN BENEDEN (1848) recognized the hippocrepia division in his study of freshwater Bryozoa, and BUSK (1852) used the name Polyzoa infundibulata in his study of marine Bryozoa. ALLMAN (1856) rejected the classification based on the shape of the lophophore as an artificial grouping and named two new groups, the Phylactolaemata and the Gymnolaemata, based on the possession and lack, respectively, of an epistome overhanging the mouth. ALLMAN's names have generally been accepted in the subsequent literature. Two of the genera placed in the Phylactolaemata and one placed in the Gymnolaemata by ALLMAN, however, were removed by NITSCHE in 1869 to form the phylum Entoprocta.

BORG (1926a) named the Stenolaemata as a third group equal in rank to the Phylactolaemata and Gymnolaemata by dividing the Gymnolaemata into two groups, based on shapes of zooids. He retained the name Gymnolaemata for the major group with a more restricted concept. This three-part division of the phylum has been used by Silén (1944a,b), Ryland (1970), and Boardman and Cheetham (1973).

Marcus (1938a) and Astrova (1960a) retained Allman's two-part division of the Bryozoa into Phylactolaemata and Gymnolaemata as major groups of equal rank. Marcus further proposed a two-part subdivision of the Gymnolaemata at the next lower taxonomic level and named these groups Stenostomata and Eurystomata. The name Stenostomata was proposed as a replacement for Borg's Stenolaemata; the Eurystomata were proposed as a new group.

Cuffey (1973) proposed a formal classification in which Phylactolaemata, Gymnolaemata, and Stenolaemata were retained as groups equal in rank, but arranged in a different two-part division of the phylum (called Ectoprocta by Cuffey). The Phylactolaemata and Gymnolaemata, considered classes by Cuffey, were united in a new superclass Pyxibryozoa. The Stenolaemata formed the only class of a new superclass Tubulobryozoa.

For reasons discussed in the following section, the Phylactolaemata, Gymnolaemata, and Stenolaemata are retained here as taxa of class rank with no further grouping between class and phylum levels.

Busk (1852) subdivided the living marine Polyzoa infundibulata (Gymnolaemata of Allman) into the Cyclostomata (here called Tubuliporata),[1] Cheilostomata, and Ctenostomata. The Tubuliporata and Cheilostomata were soon recognized among fossil Bryozoa (Busk, 1859), and two more divisions were later added, the Trepostomata by Ulrich (1882) and Cryptostomata by Vine (1884). In 1957, Elias and Condra gave the name Fenestrata to a group they removed from the Cryptostomata, and in 1964, Astrova proposed the name Cystoporata for a group she removed from the Paleozoic Tubuliporata, Cryptostomata, and Trepostomata. All seven of these groups are considered here to be orders.

DISTINGUISHING CHARACTERISTICS OF CLASSES

The classification used here at the class level follows the three-part grouping of Borg (1926a), Silén (1944a,b), and some subsequent authors. It may well require modification as additional data become available and therefore is used here as an initial basis for discussion.

Phylum Bryozoa
Class Stenolaemata
    Order Tubuliporata
       (=Cyclostomata of Busk)
    Order Trepostomata
    Order Cryptostomata
    Order Cystoporata
    Order Fenestrata
Class Gymnolaemata
    Order Ctenostomata
    Order Cheilostomata
Class Phylactolaemata

The diagnoses of the classes Stenolaemata and Gymnolaemata are based on our own experience as much as possible. We have relied entirely on the literature and the review by Wood in this volume for the characteristics of the Phylactolaemata. We have, however, attempted to describe phylactolaemate morphology using terminology consistent with that employed for the other two classes. Characterizations of these classes represent our present understanding and include as many characters as this understanding permits. New characters undoubtedly will be added as revision at the generic level proceeds.

---

[1] Because of homonymy with the vertebrate order Cyclostomata Duméril, 1806, *Treatise* policy recommends replacement of the well-known name Cyclostomata Busk, 1852; however, this replacement is not obligatory under the International Code of Zoological Nomenclature. Busk listed Tubuliporina without direct author reference as the only synonym of the Cyclostomata (1852, p. 347). Earlier, in 1847, Johnston had clearly defined the name Tubuliporina as a group name to include the modern species of the Tubuliporidae Johnston, 1838, and the Crisiadae Johnston, 1847 (present-day Crisiidae). Busk renamed the Tubuliporina on the conformation of the aperture rather than any significant change in concept or content. The name Tubuliporina is changed to Tubuliporata Johnston, 1847, to conform to order-level endings and to avoid conflict with the use of Tubuliporina as a suborder. Research on this problem was done by Osborne B. Nye, Jr.

We have attempted to recognize comparable, potentially homologous, phylum-wide structures in these morphologically different classes in order to employ a consistent language to express relative differences and similarities among taxa. Comparability of structures has been evaluated from their modes of growth, functions, and positional similarities. Homologies (comparability due to common ancestry) of most of these structures have not been tested against the fossil record at these higher taxonomic levels. The significance in phylogenetic classification of characters derived from many of these structures, therefore, has yet to be determined. Some characters, such as presence or absence of extrazooidal parts, of polymorphism, and of frontal zooidal walls, might be shown by future study to incorporate iteratively derived states.

We also have attempted to describe and compare the three classes as polythetically as possible. A polythetic classification (SNEATH & SOKAL, 1973, p. 21) results from clustering colonies, populations, or taxa that possess a majority of character states common to a majority of the members of a cluster. A cluster becomes a potential taxon that can be evaluated from data on occurrence in time and space. No one character state or combination of character states must be present for a group to be included in the taxon. The traditional approach to classification in Bryozoa has been monothetic, that is, all members of a taxon have been required to possess a character state or a combination of character states unique to that taxon. Examples of character states used monothetically at high levels of classification include the presence or absence of an episome overhanging the mouth and the presence or absence of intrinsic muscle layers in the body wall. The need for polythetic use of characters at the species level has been recognized for a long time. At higher levels the choice of "defining" monothetic characters has been largely arbitrary and has resulted in at least as much instability as such choice has at lower levels.

The rigorous procedures needed to develop polythetic clusters at the class level require more detailed data than are now available. The polythetic characterizations of the Stenolaemata, the Gymnolaemata, and the Phylactolaemata below include states of 48 morphologic characters (Table 1). Of these characters, 37 are reflected directly or indirectly in skeletons or preserved remnants of soft parts in fossil taxa, and 11 are reflected only in soft parts of living taxa.

In this comparison of the three classes, morphologic similarities were estimated for each pair of classes without making detailed counts of included genera, most of which must be restudied before all character states are available. Two estimates were made, one using as many of the 48 characters as applicable (Fig. 5,A) on the assumption that soft parts of fossil taxa were like those of living representatives of the same class, and the other based only on the 37 characters reflected in the phylactolaemates and in skeletons or remnants of fossil taxa (Fig. 5,B) of the other two classes.

To make the comparison as polythetic as possible, estimates of similarity between each pair of classes were based on the number of character states shared by all taxa in each pair plus the number of states partly shared by overlapping proportions of taxa in each pair. At this stage in our understanding of class characters, the phylogenetic significance of the absence of a character in two of the three classes is not known, so shared absence was given the same weight as shared presence.

Of the 48 characters used to characterize the three classes, 25 provided entirely shared states for a pair of classes and 14 of these characters have unique states in the third class; 21 provided only partly shared states for any pair of classes; and 2 provided only states unique to a class. Fifteen characters provided states partly shared by all three classes.

The proportions of overlap in partly shared states of different characters are estimated to range from few genera to almost all genera within the classes, and these proportions were estimated to the nearest 20 percent (Table 1). The percentages of overlap in four char-

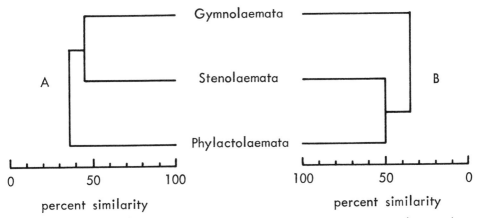

Fig. 5. Distinguishing characteristics of classes. Dendrograms expressing similarities between classes Stenolaemata, Gymnolaemata, and Phylactolaemata based on estimated percentages of morphologically overlapping genera in pairs of classes (Table 1, Fig. 6).——A. Dendrogram based on all 48 characters listed in Table 1.——B. Dendrogram based on 37 characters, omitting those known only in living genera (indicated by asterisk in Table 1).

acters are illustrated diagrammatically in Figure 6. Percentages were employed to remove the effect of the enormous difference in numbers of genera in classes. To make the similarity estimates comparable (scales of similarity in Fig. 5), the sum of shared states plus percentages of overlap for partly shared states was divided by the number of characters applicable to the comparison of each pair of classes.

If all 48 characters are considered polythetically in comparing the three classes (Fig. 5,A), Stenolaemata and Gymnolaemata are more similar to each other than either class is to the Phylactolaemata. This result is apparently in agreement with the two-part arrangement of living taxa employed by MARCUS (1938a) and earlier authors. If the 11 characters that are reflected only in the soft parts of living taxa in all three classes are omitted (Fig. 5,B), the similarity between Stenolaemata and Phylactolaemata becomes greater than that of Gymnolaemata to either. These differences in similarity are small; however, that between any pair of classes falls between 29 and 50 percent with either set of characters employed (Table 1). It seems improbable that these results reflect any clear taxonomic grouping between class and phy-

lum levels. The three classes seem best retained as equally distinct taxa until fossil evidence of their phylogenetic relationships to each other becomes available.

If only character states shared by all taxa in a pair of classes are considered (100 percent in Table 1), the results are similar to those of the polythetic comparison, even though this reduces the number of characters to 25. Stenolaemata and Gymnolaemata share states of 9 characters, 4 of which are reflected in skeletons or remnants of fossil taxa. Stenolaemata and Phylactolaemata share 9, 8 of which are reflected in fossil stenolaemates. Phylactolaemata and Gymnolaemata share 7, 5 of which are reflected in fossil gymnolaemates. This monothetic sharing approach to comparison of classes omits characters derived from interzooidal communication organs, confluent coelom, and budding zones, for example. Present understanding does not justify rejection of such characters at this level, although future discovery of phylogenetic evidence in the fossil record may reveal them to be important only at lower taxonomic levels.

Considering monothetically the characters that are unique to a class, the Phylactolaemata and Gymnolaemata are about equally

TABLE 1. *Morphological Comparison of the Three Major Groups of the Phylum Bryozoa.*
(Percentages to the nearest 20 are estimates of component genera with overlapping character states in each pair of groups; see Fig. 6 and text. Overall morphologic similarity is indicated at the foot of the table; see also Fig. 5. An asterisk marks a character known only in living genera; P, present; A, absent; S, saclike; C, cylindrical.)

Character	Percent Overlap of Character States		
	Stenolaemata–Gymnolaemata	Stenolaemata–Phylactolaemata	Gymnolaemata–Phylactolaemata
1 Outermost layer of exterior walls	100, cuticular	>80, cuticular	>80, cuticular
2 Calcification (P or A)	>80, P	0	<20, A
3* Composition of skeleton	60, calcite	—	—
4 Growth directions of zooids and colony	0	80[a]	0
5 Erect basal zooidal walls (P or A)	40 (20P, 20A)	80, A	20, A
6 Erect basal zooidal walls exterior or interior	20 (<20 int., <20 ext.)	—	—
7 Vertical zooidal wall orientation relative to zooidal growth direction	0	100	0
8 Vertical zooidal walls exterior or interior or a combination	20 (<20 ext., <20 int.)	<20, ext.	<40 (<20 ext., 20 comb.)
9 Completeness of interior vertical zooidal walls	>80, complete	<20, incomplete	0
10 Vertical zooidal walls with endozone and exozone	<20, A	<20, A	100, A
11 Ontogenetic duration of vertical zooidal wall growth	20, early	20, early	100, developed early
12 Shape of zooidal body cavity (S or C)	40 (20S, 20C)	20, C	20, C
13 Frontal zooidal wall (P or A)	20, P	80, A	0
14 Frontal zooidal wall orientation relative to zooidal growth direction	>80, parallel	—	—
15 Flexibility of frontal zooidal wall	0	—	—
16 Orificial wall orientation relative to zooidal growth direction	0	100, transverse	0
17 Orificial wall terminal or subterminal	<20, terminal	100, terminal	<20, terminal
18 Structure of orificial wall	0	100, single membrane	0
19 Orificial wall free or fixed to other zooid walls	20, fixed	20, fixed	100, fixed
20 Ratio of area of orificial wall to cross section of zooidal body cavity	20, smaller	80, same	0
21 Shape of orifice	0	100, simple pore	0
22 Completeness of skeletal margin of aperture	<20	—	—
23 Interzooidal communication organs (P or A)	40, P	60, A	0
24 Interzooidal communication organs in interior or exterior walls	<20, int. only	—	—
25 Extent of confluent body cavity among fully developed zooids	20, A	80[b]	0
26 Retracted position of lophophore and gut during ontogeny	20, constant	20, constant	100, constant

[a] Varies ontogenetically.
[b] Colony-wide.
[c] Multizooidal.

TABLE 1. *(Continued from preceding page.)*

	Character	Percent Overlap of Character States		
		Stenolaemata–Gymnolaemata	Stenolaemata–Phylactolaemata	Gymnolaemata–Phylactolaemata
27	Regeneration and brown bodies (P or A)	100, P	0	0
28	Membranous sac (P or A)	0	0	100, A
29	Parietal muscles (P or A)	0	100, A	0
30	Intrinsic body wall muscle layers (P or A)	100, A	0	0
31*	Diaphragmatic dilator muscles (P or A)	0	0	100, P
32*	Vestibular dilator muscles (P or A)	<20, P	100, P	<20, P
33*	Tentacle number	100, 8–35	<20, <35	<20, ≤35
34*	Tentacle arrangement	100, circular	<20, circular	<20, circular
35*	Epistome (P or A)	100, A	0	0
36	Polymorphism (P or A)	<80 (60P, <20A)	40, A	<20, A
37	Extrazooidal parts (P or A)	80 (20P, 60A)	40, P	20, P
38	Extrazooidal skeleton exterior or interior	60, int.	—	—
39	Brooding of embryos (P or A)	40 (<40P, <20A)	40, P	>80, P
40	Known brooding within or outside of body cavity	<20, within	100, within	<20, within
41*	Single or multiple embryos per zygote	0	0	100, single
42*	Initial zooids produced from larva or directly from embryo	100, larva	0	0
43*	Encapsulated resistant resting bodies produced from funicular strands (statoblasts) or body walls (hibernacula)	—	—	0
44*	Initial zooids produced asexually (P or A)	>80, A	0	<20, P
45	Primary zone of astogenetic change (P or A)	>80, P	100, P	>80, P
46	Extent of budding zones	20ᶜ	20ᵇ	0
47	Initial structures of bud	100, body wall	0	0
48*	Anus on distal or proximal side of tentacle sheath	100, distal	0	0
Percent similarity of pairs of classes:				
All characters		45	42	29
Omitting characters from living genera		39	50	29

distinct. Phylactolaemata all have unique states of 7 characters, 3 of which have contrasting states recognizable in fossil stenolaemates and gymnolaemates (indicated in Table 1 by 0 percent, or 0 percent and "not applicable" under pairs that include the Phylactolaemata). Gymnolaemata have unique states of 6 characters, all of which have contrasting states recognizable in fossil stenolaemates. Stenolaemata have the fewest characters with unique states, 2, one of which is assumed to have a contrasting state in fossil gymnolaemates.

The extreme approach to a monothetic

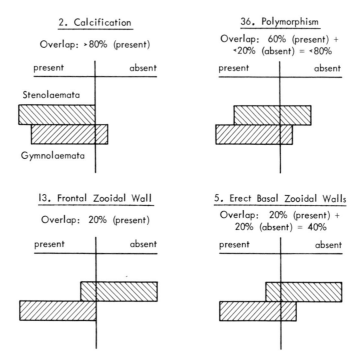

FIG. 6. Distinguishing characteristics of classes. Estimated percentages of stenolaemate and gymnolae-mate genera having overlapping states of four morphologic characters (numbered as in Table 1). Bars of equal length represent 100 percent of the genera in classes, even though numbers of genera in classes are unequal. In two characters shown, calcification and frontal zooidal wall, Stenolaemata and Gymnolaemata overlap only in one state. In the other two characters shown, these two classes overlap in both states. In all four characters shown, all phylactolaemate genera have only one state, absent, and thus overlap one or both other classes to different degrees (see Table 1).

classification is the search for the panacea character or characters having states shared by all taxa of each class and being unique to each class. None of the characters used here separates all three classes.

Even though we obtained some minor differences in the comparisons between any pair of classes, depending on the characters used, with both the polythetic and monothetic approaches, our results are all strikingly different from the two-part arrangement proposed by Cuffey (1973). We found no evidence that Phylactolaemata and Gymnolaemata (forming the superclass Pyxibryozoa of Cuffey) are more similar to each other than either is to the Stenolaemata (the only class in the superclass Tubulobryozoa of Cuffey). This major difference in results may be at least partly explained by the number of different characters used, especially the char-

FIG. 7. Stenolaemate colonies.——*1a,b. Hornera* sp., rec., Westernport, Vict., Australia; fenestrate, free-walled colony with branches connected by crossbars of zooids, zooidal apertures on one side of branches only; *a,* lat. view, *b,* growing surface, USNM 220028, ×2.——*2a,b. Discocytis lucernaria* (Sars), rec., Kvaenang Fjord, Nor., depth of 145–180 m; stalked colony, zooids at ends of rays free-walled, stalk covered by smaller, free-walled polymorphs at high angles to zooids and surface of stalk; *a,* lat. view, *b,* growing surface, USNM 220029, ×4.——*3. Plagioecia* sp., rec., Arctic O.; growing surface of bifoliate colony, zooids free-walled in budding zones near edges of medial multizooidal walls, zooids fixed-walled proximally, developing secondary nanozooids of Silén & Harmelin (1974); USNM 220030, ×4.

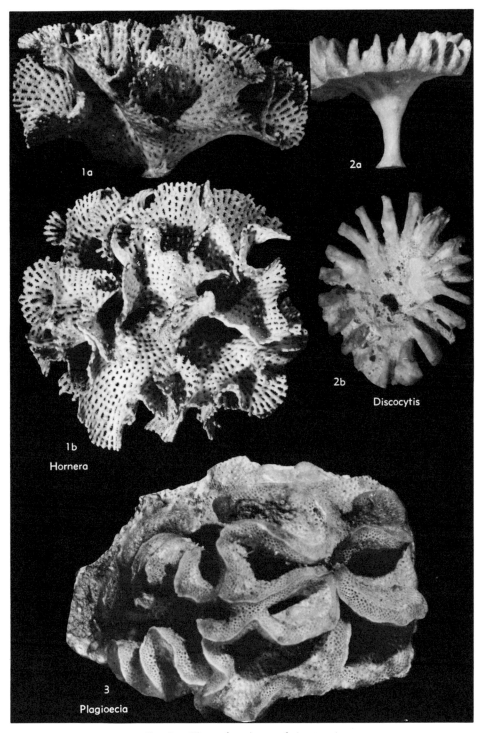

1a

2a

1b
Hornera

2b
Discocytis

3
Plagioecia

Fig. 7. *(For explanation, see facing page.)*

acters that have become available in the past few years (CUFFEY, pers. commun., 1975). Also, some characters and character states were interpreted differently. Even though CUFFEY derived his classification polythetically using the characters available to him, the verbal description in his classification is monothetic, including only the character states shared by all component taxa of each major taxon (CUFFEY, 1973, p. 553). Without the full set of character states for the CUFFEY classification, a detailed comparison with our results is not possible.

Class Stenolaemata.—The Stenolaemata are exclusively marine and have an extensive fossil record (Fig. 7–9). They constitute the overwhelming majority of bryozoans from the Ordovician into the Cretaceous and occur in large numbers in many Tertiary and modern faunas.

An exterior cuticle forms a complete outermost layer around living colonies and presumably occurred around fossil colonies as well. Encrusting basal colony walls are direct lateral extensions of the exterior walls of basal discs of ancestrulae (Fig. 1) and are exterior, multizooidal, and calcified. All basal, vertical, and frontal zooidal walls (Fig. 1) are calcified. Almost all skeletons are calcitic; a few species in the Triassic are aragonitic.

Basal zooidal walls occur in most colonies as parts of encrusting multizooidal colony walls and so are exterior along colony bases. In erect parts of colonies, basal zooidal walls can be parts of multizooidal walls that are either interior (**bifoliate colonies** and probably some **unilaminate colonies**) or exterior (some unilaminate colonies). In some uni-laminate and **dendroid colonies**, basal zooidal walls can be parts of interior walls of other zooids. In erect parts of most dendroid colonies, the inner ends of zooids are pointed and basal walls are absent.

Vertical zooidal walls form elongated conical or tubular shapes. Vertical walls are complete except possibly for those with small skeletal gaps in several Paleozoic species. They are interior walls, except for those in the few uniserial or multiserial species, which are exterior or a combination. Growth directions of vertical walls parallel long axes of zooids (Fig. 10, 11). Zooids can commonly be divided ontogenetically into inner and outer parts. Inner parts (**endozones**) are characterized by one or a combination of growth directions at low angles to colony growth directions or colony surfaces, thin vertical walls, and relative scarcity of intrazooidal skeletal structures. Outer parts (**exozones**) are characterized by growth directions at high angles to colony growth directions or colony surfaces, thicker vertical walls, and concentrations of intrazooidal skeletal structures (Fig. 10, 11).

Frontal zooidal walls (Fig. 1, 11) occur in relatively few stenolaemates, most commonly in species of post-Paleozoic age. They are exterior walls (as in the Gymnolaemata), and so their outermost layer is part of the colony-wide exterior cuticle. Subjacent skeletal layers are structurally continuous with or attached to outermost edges of skeletal layers of interior vertical zooidal walls. Frontal walls range in orientation from nearly parallel with, to perpendicular to, zooidal growth direction in different taxa. Frontal walls in stenolae-

---

FIG. 8. Stenolaemate colonies.——*1a,b. Neofungella* sp., rec., Albatross Sta. 3212, lat. 54°05′30″ N., long. 162°54′ W., S. of Alaska, depth of 90 m; stalked, free-walled colony with stalk covered by exterior terminal diaphragms; *a,* growing surface, *b,* lat. view, USNM 220031, ×4.0.——*2. Corymbopora* sp., Cret. (Cenoman.), Le Mans, Sarthe, France; stalked, branching colony with free-walled autozooids, stalks covered by small, free-walled polymorphs; lat. view, USNM 220032, ×4.0——*3. Frondipora verrucosa* (LAMOUROUX), rec., Medit. Sea, Oran, Alg.; colony of anastomosing branches with clusters of free-walled zooids surrounded by exterior frontal walls of combined free- and fixed-walled zooids; growing surface except for nonzooidal reverse side of branches in lower part of figure, USNM 220033, ×1.5.——*4. Tretocycloecia* sp., up. mid. Yorktown F., Mio., Rice's Pit, Hampton, Va.; free-walled dendroid colony, many branches anastomosing; growing surface, USNM 220034, ×1.5.

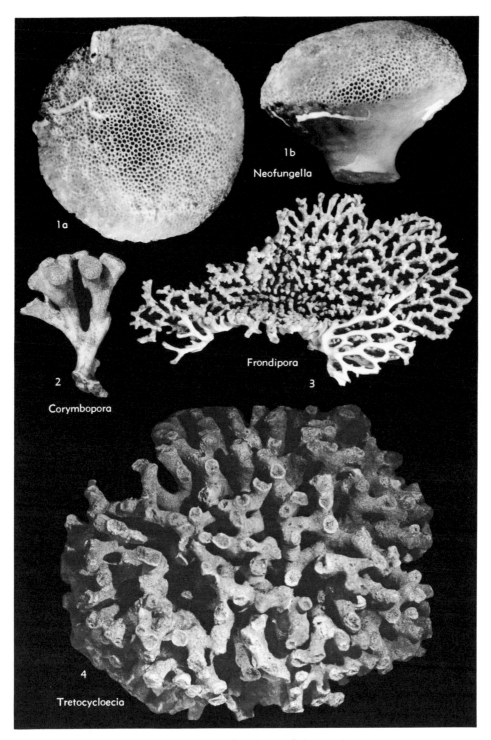

FIG. 8. *(For explanation, see facing page.)*

FIG. 9. *(For explanation, see facing page.)*

mates are entirely calcified and so are inflexible in lophophore protrusion.

Zooecial **apertures** (Fig. 2) are the terminal skeletal openings of zooids. They occur in all stenolaemates, have complete margins, and vary in shape in different taxa. They are the terminations of frontal wall skeleton, vertical wall skeleton where frontal walls are absent, or a combination. Zooids typically elongate through most of their ontogeny by growth of zooidal body walls at apertures. In a few fossil taxa, apertures are covered by exterior, skeletal, hinged structures that apparently performed an operculumlike function.

Orificial walls (Fig. 10, 11) are single membranous exterior body walls that cover skeletal apertures and include the simple circular orifices through which tentacles are protruded. Orificial walls are transverse to zooidal growth direction in most taxa and are terminal, except in the few fossil taxa in which operculumlike structures apparently covered them. Similar orificial walls are assumed for fossil taxa because of the general likeness of supporting zooidal walls and simple skeletal apertures among fossil and recent taxa.

The relationships of orificial walls to zooecial apertures of vertical and frontal walls produce three different kinds of colonies, the third kind being a combination of the first two.

1. In most stenolaemate colonies, frontal walls are absent in feeding zooids and vertical walls support membranous orificial walls (Fig. 10) apparently without direct attachment (**free-walled colonies**). In free-walled colonies, orificial walls are parts of exterior membranous walls that completely cover colonies above their encrusted bases. The membranous covering wall of a colony is held in place by attachment organs within the zooids (Fig. 2). With minor exceptions, all skeletal parts above encrusting colony walls are interior in origin in free-walled colonies and are separated from exterior membranous colony walls by confluent outer body cavities. In some post-Paleozoic taxa, colony-wide exterior cuticle is attached to outer sides of skeletal layers of terminal diaphragms and outer walls of brood chambers.

2. In colonies with frontal walls in feeding zooids (Fig. 1, 11) the colony-wide exterior cuticle is attached to outer surfaces of skeletal layers of the frontal walls, and is also the outer layer of orificial walls, as in all Bryozoa. The cuticular layer of the frontal wall, therefore, fixes individual orificial walls directly to zooecial apertures (**fixed-walled colonies**). Exterior walls on feeding sides of fixed-walled colonies consist primarily of orificial and frontal walls of contiguous zooids. The colony-wide outer body cavity of free-walled colonies is therefore eliminated.

3. In some taxa that have feeding zooids arranged in isolated clusters on colony surfaces, free- and fixed-walled morphologies are combined. The clusters are isolated from each other by exterior frontal walls of their outermost zooids. The apertures of the outermost zooids of each cluster are parts of both frontal and interior vertical walls so that their orificial walls are partly fixed and partly free. The apertures of inner zooids of larger clus-

FIG. 9. Stenolaemate colonies.——*1. Entalophora depressa* (SMITT), rec., Albatross Sta. 2407, Gulf of Mexico; dendroid colony of fixed-walled zooids with long peristomes; USNM 220035, ×4.——*2. Idmidronea atlantica* (FORBES), rec., Thatcher's Is. Light, Mass., depth of 60 m; unilaminate, regularly bifurcating colony of fixed-walled zooids with long peristomes of different lengths within a row, some with flaring ends, reverse side of branches covered by exterior multizooidal wall lacking zooids; USNM 220036, ×4.——*3. Diaperoecia* sp., rec., Australia; unilaminate colony of fixed-walled zooids with some anastomosing branches, reverse side of branches covered by exterior multizooidal wall; USNM 220037, ×4. ——*4.* Tubuliporid, rec., Gulf of Mexico; unilaminate colony of fixed-walled zooids with some anastomosing of branches and crossbars of single or clustered zooids, reverse side of branches covered by exterior multizooidal wall; USNM 220038, ×4.——*5. Lichenopora* sp., rec., Marcial Point, Gulf of Lower Cal., Mexico; complex of free-walled colonies encrusting stick, radial rows formed by zooids with long interior-walled peristomes; USNM 220039, ×5.

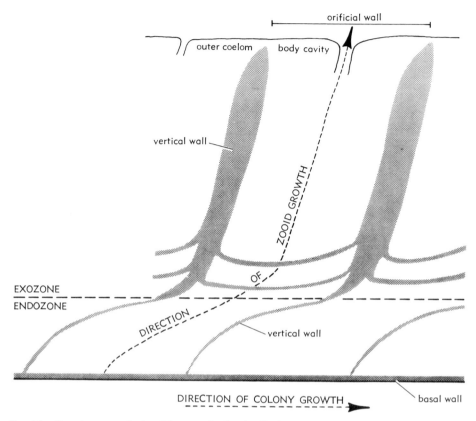

FIG. 10. Stenolaemate colonies. Diagram of a longitudinal section through a zooid of a unilaminate free-walled colony. The frontal wall is absent and the membranous exterior orificial wall is not attached to interior vertical walls (stippled) so that the body cavity is colony-wide around ends of vertical walls. Transverse skeletal diaphragms (stippled) act as floors of the living chamber sequentially with ontogenetic growth. Soft parts are deleted except for the orificial wall and the external cuticle of multizooidal basal wall.

ters are supported entirely by vertical walls so that those zooids are free-walled and an **outer coelomic space** occurs within a cluster. (In Table 1, line 19, the term "fixed" includes both fixed-walled and these combined taxa.)

Physiologic communication among fully developed zooids and between feeding zooids and extrazooidal structures is assumed for all stenolaemates except for a few fixed-walled species of Paleozoic age. Communication must have occurred through confluent outer body cavity around ends of vertical zooidal walls and extrazooidal skeleton in free-walled taxa. In most post-Paleozoic stenolaemates com-

munication is assumed through pores (Fig. 11) in interior vertical zooidal walls. Two means of interzooidal communication, therefore, are assumed for most free-walled taxa of post-Paleozoic age. Additional communication is assumed in a few post-Paleozoic taxa that have communication pores in erect interior median walls, and in a few Paleozoic taxa that have communication pores and gaps in vertical skeletal walls.

In modern stenolaemate species a **membranous sac** (see Fig. 2) surrounds the digestive and reproductive systems in feeding zooids and divides the living chamber into two parts, the **entosaccal cavity** within the

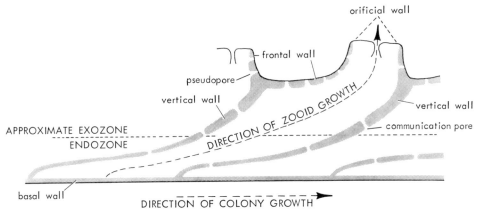

Fig. 11. Stenolaemate colonies. Diagram of a longitudinal section through a zooid of a unilaminate fixed-walled colony. The exterior frontal wall with exterior cuticle attached to skeletal layer (stippled) that contains pseudopores closes interzooidal communication through the outer body cavity of free-walled colonies. Interzooidal communication is assumed through pores in calcified vertical walls. Soft parts are deleted except for the orificial wall and the external cuticle of the multizooidal basal wall.

sac and the **exosaccal cavity** between the membranous sac and the zooidal body wall. A recent study (NIELSEN & PEDERSEN, 1979; first reported by NIELSEN at the 1977 meeting of the International Bryozoology Association) indicates that the membranous sac is peritoneum. Also, body walls of stenolaemates have only one cellular layer, the epidermis (NIELSEN, 1971). Body cavities within sacs, therefore, are surrounded by peritoneum (possibly a mesoderm) and are considered to be coeloms. All body cavities outside of sacs are termed **pseudocoels**, lined either by epidermis or by peritoneum on one side and epidermis on the other.

The membranous sac is attached to the body wall near its inner end by large **retractor muscles** and at its outer end by different kinds of attachment organs or ligaments in different taxa. Membranous sacs contain annular muscles (NIELSEN & PEDERSEN, 1979), which when contracted reduce the volume of the sac, slowly forcing the digestive organs and lophophore outward just far enough to free the tentacles for feeding. The tentacles can be withdrawn quickly by relaxation of the annular muscles and contraction of the powerful retractor muscles.

In feeding zooids of modern species, ten-tacles are arranged in a circle around the mouth, which has no epistome. Tentacle counts have been made for a few taxa and range from 8 to more than 30. The anus reportedly opens on the distal side of the tentacle sheath when tentacles are protruded.

Degeneration-regeneration cycles, which affect most of the functioning organs, occur after the initial growth of feeding zooids. In most taxa, retracted positions of lophophore and gut advance with zooid elongation, apparently by means of degeneration-regeneration saltations. Outward growth of zooids is generally enough for advancing organs to vacate inner parts of zooidal chambers, which can retain the remains of the degeneration process, generally brown bodies. In some fossil taxa vacated chamber space can be partitioned by transverse skeletal **diaphragms**. The last-grown diaphragm apparently formed the base of the living chamber for regenerated organs. In other taxa retracted positions of regenerated organs are fixed in zooecia and any continued elongation occurs in outermost vestibular walls and their enclosing skeleton.

Polymorphs may be larger or smaller than feeding zooids and may have different shapes. One kind, at least, has a reduced lophophore and gut. In some fossil taxa, polymorphs have

Bryozoa

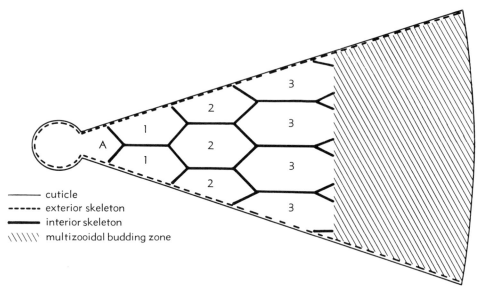

Fɪɢ. 12. Stenolaemate colonies. Idealized diagram of a section parallel to the basal layer of an encrusting colony (after Borg, 1926a, fig. 36) showing position and extent of the confluent, multizooidal budding zone around the basal margin. A is the primary zooid (ancestrula). The fourth generation of zooids is just beginning with the formation of proximal ends of vertical walls.

such small living chambers, however, that functional organs were not possible. Polymorphs may be isolated or contiguous with each other between feeding zooids, may cluster into **maculae** among feeding zooids, may surround clustered feeding zooids, or may form continuous layers on nonfeeding sides or on entire supporting peduncles of colonies.

Extrazooidal skeletal structures are grown by many free-walled colonies and can intervene between zooids in exozones or be colony-wide supporting structures. Most extrazooidal skeleton is interior in origin (see next paragraph for exception), that is, it contributes to the partitioning of preexisting colony body cavity. Growth of extrazooidal skeleton occurs in colony pseudocoels outside of zooidal chambers. The pseudocoel and parts of exterior membranous colony wall opposite extrazooidal skeleton are also considered extrazooidal.

In modern species, brooding of embryos occurs within body cavities of zooidal or extrazooidal **brood chambers** of widely varying shapes and modes of growth. Extrazooidal brood chambers have outer skeletal walls that are exterior in some taxa and interior in others. Skeletal structures in fossils that can be compared directly to known brooding structures in modern species are

Fɪɢ. 13. Cheilostomate colonies.——*1. Cystisella saccata* (Busk), rec., N. Atl., U.S. Fish Comm. sta. 121; heavily calcified, rigidly erect colony with narrow, bilaminate branches and small encrusting base; zooidal orifices open on both sides of branches, covered by thick skeletal deposits of kenozooidal origin proximally; USNM 220040, ×4.0.——*2. Bugula neritina* (Lɪɴɴᴀᴇᴜꜱ), rec., Gulf of Cal., Sonora, Mexico; lightly calcified, flexibly erect colony with narrow, unilaminate branches and basal rootlets; USNM 220041, ×2.0.——*3. Hippoporidra calcarea* (Sᴍɪᴛᴛ), rec., Str. of Fla., Albatross Sta. D2640, depth of 100 m; nodular, multilaminate colony built upon and extending from gastropod shell; outer layers formed by budding in frontal direction; USNM 220042, ×2.0.——*4a,b. Parasmittina nitida* (Vᴇʀʀɪʟʟ), rec., Long Is. Sound; heavily calcified, nodular, multilaminate colony encrusting pebble; outer layers formed by budding in frontal direction; *a,* frontal view, *b,* lat. view, USNM 220043, ×1.5.

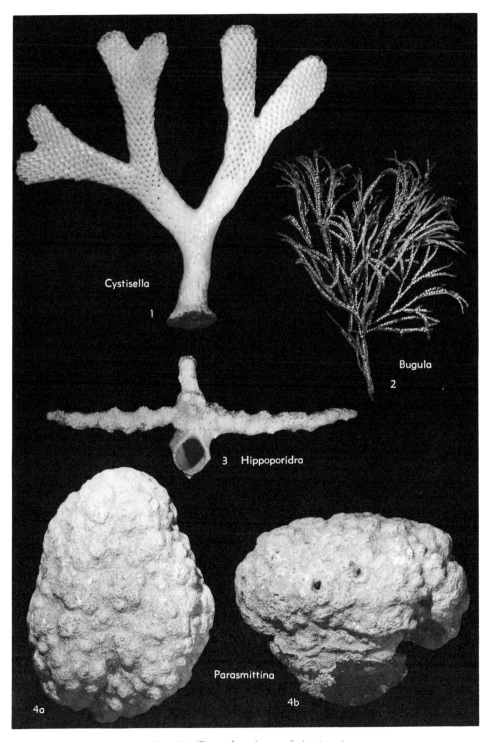

Fig. 13. *(For explanation, see facing page.)*

post-Paleozoic in age. Inferred brood chambers have been reported in a few Paleozoic taxa. Embryonic fission has been reported in modern species in which embryos have been studied. No resistant resting bodies or other asexual generations are known in life cycles.

Metamorphosis of the free-swimming, ciliated, nonfeeding larva in modern species produces the **basal disc**, the encrusting proximal end of the ancestrula (see Fig. 1). The basal disc has an exterior wall calcified from within in most taxa. The ancestrula is completed distally by exterior or interior skeletal walls, or a combination. A single ancestrula occurs in both modern and fossil colonies in species studied. All colonies apparently develop a zone of astogenetic change beginning with the ancestrula and including one to several generations of founding zooids of changing morphology (Fig. 1). The zone of change is followed by a zone of repetition in which similar zooidal morphology is repeated in a potentially endless **zooidal pattern**.

Zooids originate at their inner ends by the appearance of vertical walls growing either from encrusting or erect multizooidal walls, existing walls of zooids, or in a few species from extrazooidal parts. In endozones of both free- and fixed-walled colonies (Fig. 1, 12) vertical walls of most taxa grow into confluent outer body cavities of distal **multizooidal budding zones**. The outermost body walls of these zones are exterior membranous walls grown outside of existing zooids. As colony growth proceeds, multizooidal budding zones advance distally as their proximal regions become parts of zooids, or in many taxa are divided between zooids and extrazooidal parts. In some growth habits of free-walled colonies, budding can occur on all surfaces above encrusting colony walls in both endozones and exozones. In exozones the outer body cavities and outermost exterior membranous walls are parts of established zooids and the confluent cavities available for budding are zooidal. Feeding organs of zooids apparently originate from exterior orificial walls.

Class Gymnolaemata.—The Gymnolaemata include some brackish and freshwater representatives, but the overwhelming majority of members of this class is marine. Gymnolaemates having calcareous body-wall layers produce an abundant fossil record beginning in the Jurassic and extending nearly continuously from the Late Cretaceous onward. Taxa lacking skeletons have been found sporadically distributed as fossils from the Ordovician onward. Late in the Cretaceous, gymnolaemates became the dominant bryozoans in marine communities and remain so in present-day seas.

All exterior body walls have cuticle as the outermost layer in all living gymnolaemate taxa. Cuticles have not been found directly preserved in fossil taxa, but are assumed to have been present. In most taxa calcareous layers occur in some exterior and interior walls of zooids and other parts of zoaria. In a few taxa of major rank, skeletons are lacking, and both exterior and interior walls are stiffened only by cuticular layers, some of which may contain scattered calcareous particles. Where developed, the skeleton may be entirely calcitic or aragonitic or can combine layers of calcite and aragonite within the same zoarium. Zooidal organs are suspended in zooidal body cavities completely enclosed by zooidal body walls (see Fig. 3, 4).

Zooids can be arranged in a great diversity

Fig. 14. Cheilostomate colonies.——*1. Microporina articulata* (Fabricius), rec., Bering Sea, depth of 95 m; well-calcified, flexibly erect colony with jointed subcylindrical branches, base with rootlets, zooidal orifices opening all around branches; USNM 220044, ×2.——*2. Myriapora coarctata* (Sars), rec., N. Pac., Albatross Sta. 2877; well-calcified, rigidly erect colony with subcylindrical branches and small encrusting base, zooidal orifices opening all around branches; USNM 220045, ×2.——*3. Cryptosula pallasiana* (Moll), rec., Long Is. Sound; unilaminate colony encrusting bivalve shell; USNM 220046, ×2. ——*4a,b. Cupuladria biporosa* (Canu & Bassler), rec., Fish Hawk Sta. 7157; cap-shaped, free-living colony, zooidal orifices opening on convex surface, concave basal surface covered by extrazooidal deposits; *a*, frontal view, *b*, lat. view, USNM 220047, ×4.

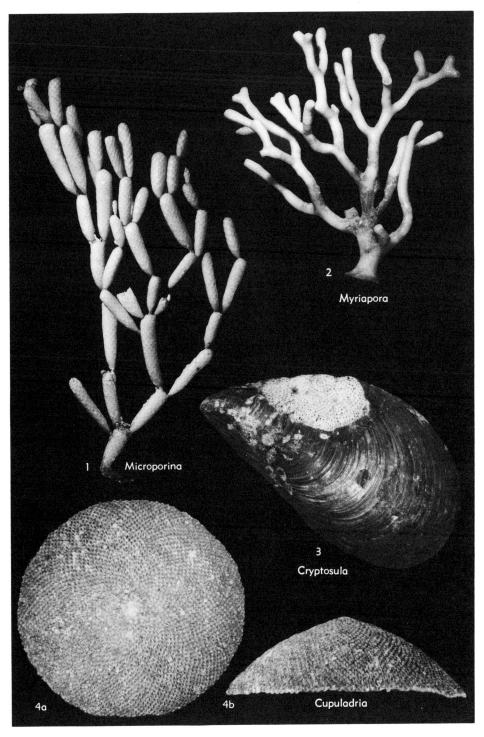

FIG. 14. *(For explanation, see facing page.)*

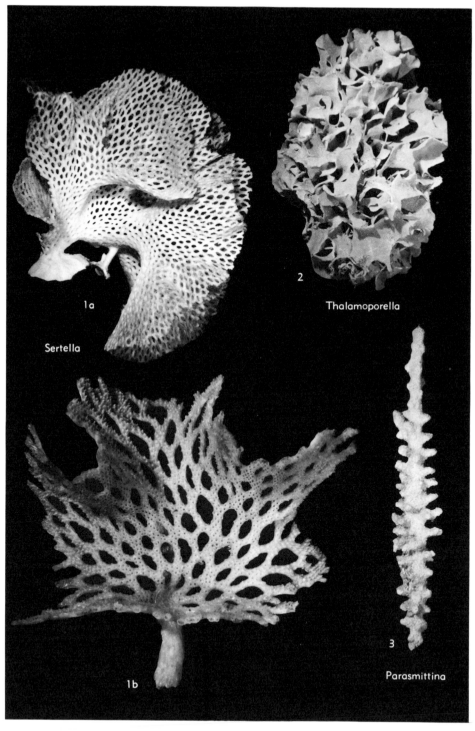

FIG. 15. *(For explanation, see facing page.)*

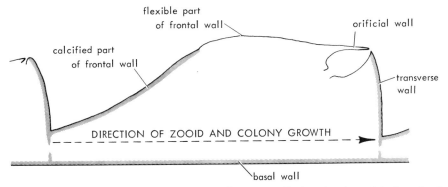

Fig. 16. Cheilostomate colonies. Diagram of a median longitudinal section through body walls of an autozooid of a simple encrusting colony. The exterior frontal wall consists of a calcified proximal part (skeletal layer stippled) protecting zooid organs (not shown, see Fig. 3, 4) and a flexible distal part fixed to the orificial wall and functioning in the hydrostatic mechanism. The exterior basal wall, of multizooidal origin, floors the body cavity. Interior walls are limited to interzooidal communication organs (pore plates) in transverse walls. Soft parts other than cuticle (solid lines) are not shown.

of patterns to form a large variety of colonies. These can include **encrusting** and **free-living colonies** as well as **rigidly erect, flexibly erect**, and **jointed-erect colonies** (Fig. 13–15). Major regions of erect and free-living colonies in some taxa are composed of extrazooidal parts. Principal growth directions of zooids and the colony approximately coincide (Fig. 16, 17).

Basal zooidal walls may be calcified or uncalcified, even within the same colony. In erect unilaminate, bilaminate, or cylindrical branches of colonies, basal zooidal walls most commonly are exterior and include multizooidal layers continuous with those in encrusting bases, but may be interior or absent, with vertical walls meeting at branch axes.

Vertical zooidal walls are calcified in the great majority of taxa and consist of lateral walls elongated subparallel to the direction of zooidal growth and transverse walls oriented subperpendicular to zooidal growth. Zooids budded in specialized directions (for example, frontally budded zooids in subsequent astogenetic zones of some colonies) may have all vertical walls oriented subparallel to the zooidal growth direction. In most taxa lateral walls are exterior and transverse walls include extensive interior components (Fig. 18). In a few taxa vertical walls are all interior, and in a few others having predominantly uniserial growth vertical walls are virtually all exterior. Exterior vertical walls include multizooidal cuticular and, where present, skeletal layers continuous among zooids within a budding series. Interior vertical walls completely separate living chambers of contiguous zooids. Vertical walls are not divided into endozones and exozones and

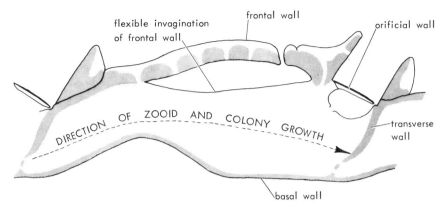

FIG. 17. Cheilostomate colonies. Diagram of a median longitudinal section through body walls of an autozooid of a complex erect subcylindrical colony. The exterior frontal wall consists of three parts. *1.* An outer part is uncalcified except at its proximal end, where it is overlain by the peristome of the proximal zooid. This outer frontal wall is underlain by a separated part of zooidal body cavity and by a protective calcified interior frontal shield (stippled). The frontal shield was formed before invagination of separate cuticle subjacent to it. *2.* A distal extension of frontal wall forms a tubular peristome surrounding and fixed to the orificial wall. The calcified portion of the exterior-walled peristome is structurally continuous with the interior-walled frontal shield. *3.* An invaginated part of the frontal wall has a flexible floor that functions in the hydrostatic system. An exterior basal wall of multizooidal origin floors the principal body cavity. Transverse walls are interior with multiporous pore plates. Soft parts other than cuticle (solid line) are not shown. (For detailed illustrations of this cheilostomate, see Fig. 67; 73; 82,3.)

are completed early in ontogeny to establish a maximum dimension for the zooidal living chamber, which ranges in shape from box- or saclike to cylindrical.

Frontal zooidal walls generally elongate subparallel to the direction of zooidal growth are present in all taxa. Frontal walls are exterior, as in the class Stenolaemata, but in some taxa are associated with subparallel calcified interior walls (Fig. 17). In calcified taxa all or part of the frontal wall (Fig. 16), or an infolded sac derived from it (Fig. 17), remains uncalcified and flexible to function in lophophore protrusion. Frontal walls include cuticular and, where present, some skeletal layers that are continuous among zooids in a budding series.

Orificial walls are subterminal in most taxa, terminal in a few, and consist of one or more movable folds of body wall. The outer side of the orificial wall is fixed to, and includes cuticular layers continuous with, the frontal wall (Fig. 3, 4). The inner side includes cuticular layers continuous with those of the vestibular wall. When closed (Fig. 3, 16, 17),

the orificial wall is subparallel to the direction of zooidal growth and defines a slitlike or puckered orifice. In most taxa, the orificial wall is a single, distally directed flap, stiffened to form an operculum (Fig. 3, 4, 16, 17). In most calcified taxa, marginally incomplete skeletal openings support distal and, in some, lateral margins of the operculum and coincide with these margins of the orifice. In a few taxa margins of skeletal openings may be complete proximally (Fig. 17) and are apparently analogous to skeletal apertures in the class Stenolaemata. Margins of skeletal openings can be formed by transverse or frontal zooidal walls, by structures associated with frontal walls, or by a combination.

Developing zooids at growing tips of budding series or in multizooidal budding zones can have confluent living chambers, but those of fully developed zooids are not confluent. Communication among fully developed zooids and between zooids and extrazooidal parts where present is through **pore plates**, which in modern species are penetrated by cells of special form connected to the body wall and to funicular strands (Fig. 3, 4). Communication organs can occur in interior vertical and basal zooidal walls, in exterior vertical, basal, and frontal walls of zooids that are in contact, and in some intrazooidal walls.

Retracted positions of lophophore and gut are approximately constant at all regenerated phases. Degeneration in modern species results in brown bodies that generally are expelled after regeneration, but are retained in living chambers in some species.

Lophophore protrusion involves contraction of two sets of muscles, parietals and dilators (Fig. 3, 4). **Parietal muscles** traverse the body cavity in bilaterally arranged pairs, from lateral or basal walls to the flexible exposed, overarched, or infolded part of the frontal wall. This flexible part of the frontal wall is depressed by contraction of the parietals, causing the lophophore to protrude. Skeletal evidence of parietal muscles has been found in many fossils. **Dilator muscles** are known only in modern species. **Diaphragmatic dilator muscles** traverse the body cavity in bilaterally or radially arranged groups from lateral or transverse walls or both to the diaphragm. In some taxa vestibular dilators are also present. The diaphragm and vestibule are dilated by contraction of the dilators to allow passage of the lophophore during protrusion.

Feeding zooids of modern species have 8 to 35 tentacles arranged in a circle around the mouth, which has no epistome. The anus opens on the distal side of the tentacle sheath.

Polymorphism is known in the great majority of taxa and is generally reflected in skeletons of calcified taxa. A variety of polymorphs may occur in the same colony to perform sexual reproduction, embryo brooding, and other functions. These polymorphs differ markedly in size, shape, and other morphologic characters from ordinary feeding zooids. Polymorphs may lack lophophore and gut, or have organs different from those of ordinary feeding zooids, with or without feeding ability. Some polymorphs communicate with just one other zooid, in the extreme form being

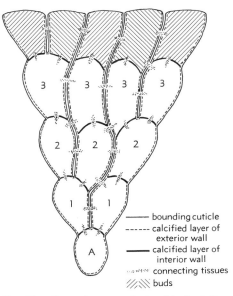

——	bounding cuticle
------	calcified layer of exterior wall
——	calcified layer of interior wall
⸱⸱⸱⸱	connecting tissues
⧄⧅	buds

Fig. 18. Gymnolaemate colonies. Idealized diagram of a section parallel to basal walls of zooids in an encrusting or erect colony (after Silén, 1944b; Banta, 1969) showing positions of buds at distal ends of lineal series. Lateral zooidal walls are exterior walls breached by communication organs (connecting tissues). A is the primary zooid.

almost an appendage of that zooid; some communicate with two or more zooids, either seemingly at random positions in the colony, or in regular positions or clusters.

Extrazooidal parts are known in a few taxa and are apparently limited to calcified groups. Some structures interpreted as polymorphs in uncalcified taxa, however, may prove to be extrazooidal. Proximal parts of rigidly erect colonies in some taxa have outer extrazooidal membranous wall, body cavity, and calcified wall all formed through coalescence of corresponding frontal parts of zooids. Extrazooidal skeletal layers in these taxa succeed, without interruption, zooidal skeletal layers that are parts of either interior or exterior walls. Basal sides of free-living colonies in some taxa and reverse sides of erect colonies in some taxa have outer membranous wall, body cavity, and calcified wall all formed at colony growing edges concurrently with budding of zooids. In some taxa skeletal layers of these extrazooidal parts are parts of compound walls that include interior basal walls of zooids. In other taxa extrazooidal skeletal layers are parts of exterior walls that are in contact with exterior basal walls of zooids.

In the great majority of modern species, embryos are brooded; and in almost all calcified forms that brood, this function is reflected in varying kinds of polymorphic skeletal structures including recognizable brood chambers. Comparable skeletal evidence of brooding has been recognized in the majority of fossil taxa. Fossil species in which evidence of brooding has not been found most commonly are morphologically similar to modern species in which embryos are not brooded. In all but two of the modern genera in which brooding occurs, embryos are held topologically outside the body cavity within chambers of widely differing size, shape, and position, each partly enclosed by the body wall of one or more polymorphic zooids. Embryos in modern species apparently are each produced from a separate egg. Ciliated larvae are naked and lack digestive tracts in most taxa, but are covered with bivalve cuticular shells and have digestive tracts in a few.

Metamorphosis of a larva is followed by development of one (ancestrula) or more primary zooids, which in the great majority of taxa initiate a primary zone of astogenetic change in turn followed by a primary zone of astogenetic repetition. In a few taxa, primary zooids are part of a zone of repetition, with no astogenetic changes in morphology of zooids. Complex astogenetic zonations are known in colonies of some taxa. Initial zooids of some colonies in a few modern freshwater or marine species can be produced asexually from encapsulated resistant resting bodies (**hibernacula**) developed as inswellings or outswellings from body walls of the parent colony.

Zooids are budded most commonly as localized swellings at distal ends of lineally budded series (Fig. 18) bounded by exterior, lateral, frontal, and basal walls of multizooidal origin. Within lineal series zooidal body cavities become separated by ingrowth of interior components of transverse walls and included pore plates, or of pore plates alone, transforming multizooidal structures into zooidal walls. In taxa having all interior vertical walls, budding occurs in laterally confluent multizooidal budding zones similar to those in the class Stenolaemata (Fig. 12). In modern species, budding initiated by outswelling of exterior walls of preexisting zooids or of multizooidal budding zones is followed by infolding of the lophophore and gut from the exterior orificial wall before it has differentiated from the developing frontal wall.

Class Phylactolaemata.—The Phylactolaemata are exclusively a freshwater class. Resistant resting bodies (**statoblasts**) produced by phylactolaemates have been reported as fossils from the Pleistocene and upper Tertiary, but reports of Cretaceous phylactolaemates are problematical and need to be reinvestigated. The few modern species in the class have intercontinental distributions and are the dominant bryozoans in freshwater communities.

Phylactolaemata have all body walls without skeleton, but soft outer noncellular layers of body walls can have adherent foreign particles in most taxa. In some taxa outermost layers of some exterior body walls are gelatinous and thick. In most taxa outermost layers of all exterior body walls are thin cuticles similar in appearance to those in other bryozoan classes.

Zooidal organs are suspended in confluent body cavity (Fig. 19), which can be continuous throughout the colony or divided by widely spaced **septa**. Zooids open in approximately the same direction on the colony surface, which consists of zooidal orificial and exterior vertical walls together with exterior extrazooidal walls. Opposite surfaces in both encrusting and erect colonies consist of exterior extrazooidal walls to which zooidal organs are attached by retractor muscles and the funiculus. Basal and frontal zooidal walls are apparently absent in all taxa. (Walls to which retractor muscles are attached are considered **colony walls** by authors.) The angle between growth directions of zooids and their colony

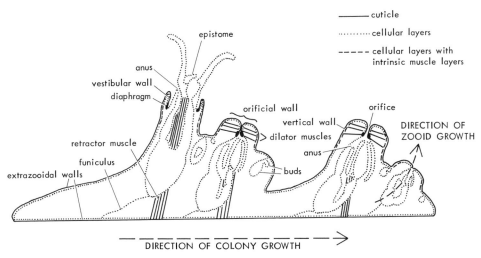

FIG. 19. Phylactolaemate colonies. Diagram of a longitudinal section through generalized encrusting colony with either circular or bilobed lophophore showing interpreted relationships between zooidal and extrazooidal body walls, confluent body cavity, and zooidal organs. The zooid near the proximal end of the colony (left) is one of two or more primary zooids with connected extrazooidal parts, all developed directly from the sexually produced embryo. Primary zooids reportedly do not differ morphologically from more distal, asexually produced zooids toward the right. As the colony grows by distal and outward expansion of the colony body wall, new buds appear as developing zooidal organs both distal to and between preexisting zooids, by infolding from the outer body wall of the colony and from other developing zooidal organs. Orificial and exterior parts of vertical zooidal walls as shown here are subsequently differentiated from the exterior wall of the colony by continued outward expansion to complete the zooid. In some phylactolaemate taxa, incomplete interior vertical zooidal walls (not shown) can grow into the body cavity from the inner ends of exterior vertical walls to separate zooidal body cavities further.

increases ontogenetically to subperpendicular in most taxa. Zooidal and colony body walls include peritoneum as their inner layer. Therefore, body cavities are considered to be coeloms.

Vertical zooidal walls are parallel to the direction of zooidal growth. They are exterior walls in most taxa and a combination of exterior and interior walls in a few taxa (see *Lophopus,* Fig. 141,2). Exterior vertical walls are limited to outer ends of zooids (Fig. 19). Interior vertical walls, where present, are incomplete and extend from inner ends of exterior vertical walls, ending in confluent coelom. Vertical zooidal walls are not divided into endozones and exozones, and are developed early in ontogeny to define the outer part of the zooidal living chamber, which is cylindrical in the part enclosed by exterior vertical walls.

Orificial walls are terminal, perpendicular to the direction of zooidal growth, fixed to exterior vertical zooidal walls, and comparable in area to cross sections of zooidal body cavities. Orificial walls are single membranes containing simple porelike orifices.

Communication among zooids and between zooids and extrazooidal parts of the colony is through confluent coelom. Communication organs have not been reported.

Retracted position of lophophore and gut is approximately constant after exterior vertical zooidal walls are developed. Lophophore and gut degenerate completely, and brown bodies and regeneration apparently do not occur (Wood, pers. commun., 1975).

Lophophores are protruded by contraction of circular and longitudinal **intrinsic body-wall muscles** present in both interior and exterior vertical zooidal walls. A membranous sac and parietal muscles are apparently unnecessary and have not been found. Radially arranged dilator muscles are attached to vestibular walls (**vestibular dilator muscles**) and to the diaphragm (**duplicature muscles**). Tentacles are arranged in a bilobed row, except for one genus in which the pattern is subcircular. Tentacles range in number from 18 to over 100. The mouth has a movable fold of body wall, the epistome, projecting over it from the anal side. The anus opens on the proximal side of the tentacle sheath (Fig. 19).

Polymorphism has not been reported.

Extrazooidal parts are apparently present in all colonies, developed as exterior colony walls concurrently with budding of zooids.

Embryos are brooded in all taxa within body chambers enclosed by infolds from the extrazooidal body wall of the parent colony. Each embryo is produced from a separate egg.

Brooded embryos develop directly into a ciliated motile colony consisting of two or more zooids and associated extrazooidal parts, without metamorphosis. Motile colonies released from parent colonies settle, lose their external cilia, and continue to grow asexually, apparently without a zone of astogenetic change.

New colonies arise most commonly by asexual reproduction through development of encapsulated statoblasts formed internally on funicular strands of zooids. Colonies developed from statoblasts begin with a zooid that can have some morphologic features different from those budded from it, and it thus initiates a zone of astogenetic change.

Budding is initiated by development of lophophore and gut infolded into the confluent coelom from exterior extrazooidal walls or internally from other developing feeding organs (Fig. 19). Orificial, vertical, and vestibular walls of zooids develop subsequently. Buds occur distal to and between preexisting zooids.

NATURE OF BRYOZOAN COLONIES

A colony in Bryozoa consists of physically connected, asexually replicated member zooids with or without connected extrazooidal parts. In this section we are concerned with theoretical aspects of the colony as expressed throughout the phylum. Further descriptions and examples of bryozoan colonies will be found in review and taxonomic sections in this and following volumes.

SOURCES OF MORPHOLOGIC VARIATION WITHIN A COLONY

The basic assumption in this study of Bryozoa is that the colony is genetically uniform. Within sexually produced colonies, only the primary zooid or group of zooids is produced sexually. All other parts of the colony arise from physically continuous mitotic division of cells and secretion of noncellular parts (LUTAUD, 1961). Because of their assumed genotypic uniformity, zooids in a colony might be expected to be morphologically identical. Zooids in a bryozoan colony, however, can differ in some morphologic features. Intracolony morphologic variation follows patterns attributed to four sources (BOARDMAN, CHEETHAM, & COOK, 1970): ontogeny of zooids and, where present, extrazooidal parts; astogeny of the colony; polymorphism of zooids; and microenvironment.

1. **Ontogenetic variation** arises from changes in a zooid (or any extrazooidal part of the colony) during the course of its development, which may or may not continue throughout the life of the zooid. These changes are recognizable within a colony in most Bryozoa as increases in size or complexity among zooids along a gradient extending in a **proximal direction** from growing extremities toward the primary zooids illustrated on the left of Figure 20. Further development of the colony (right side of Fig. 20) transforms younger, less complex

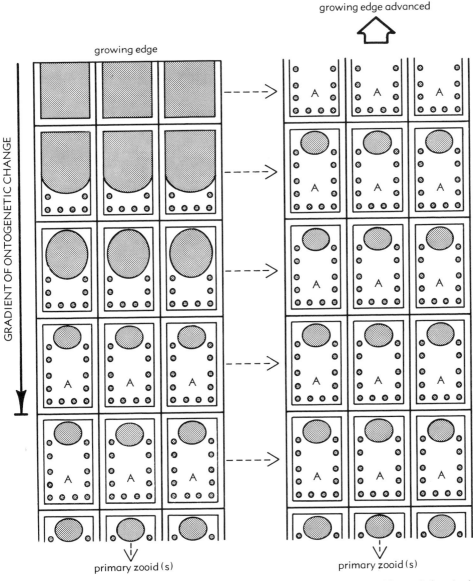

FIG. 20. Colony morphologic variation. Pattern of ontogenetic differences in zooid morphology in the zone of astogenetic repetition of a hypothetical bryozoan colony. In the series shown on the left, zooids have increasing amounts of skeleton from the growing edge to establishment of a fully developed morphology (A) through intermediate morphologies on a gradient directed proximally. Zooids the same distance proximal to the growing edge are identical in morphology, as this diagram assumes no polymorphic or microenvironmental differences. With further growth of the colony, as indicated in the series on the right, zooids of initial and intermediate morphologies have all changed to morphology A, beyond which there is no further ontogenetic change (after Boardman & Cheetham, 1973).

zooids to older, more complex ones (morphology A). Thus, zooids and extrazooidal parts of colonies form a sequential record in proximally directed series of the ontogenetic stages through which the proximal members of a series have progressed.

Bryozoa

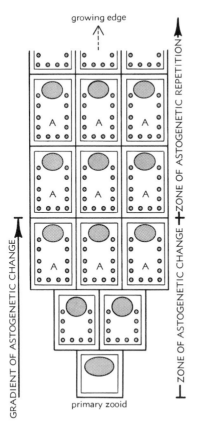

growing edge

GRADIENT OF ASTOGENETIC CHANGE

ZONE OF ASTOGENETIC REPETITION

ZONE OF ASTOGENETIC CHANGE

primary zooid

Fig. 21. Colony morphologic variation. Pattern of astogenetic differences in zooid morphology in the zone of change in a hypothetical bryozoan colony. Zooid morphology changes through one asexual generation of intermediate morphology from the primary zooid to morphology A on a gradient directed distally. Zooids belonging to the same generation are identical, as it is assumed in this diagram that there are no polymorphic or microenvironmental differences. With further growth of the colony, zooids in the zone of change retain the morphology characteristic of their generation (after Boardman & Cheetham, 1973).

2. **Astogeny** (Cumings, 1905, p. 169) is the course of development of the sequence of asexual generations of zooids and any extrazooidal parts that together form a colony. Most bryozoan colonies are developed from a primary zooid or group of primary zooids generally resulting from metamorphosis of a larva. A relatively few colonies arise from either asexually produced resistant resting

bodies or fragmentation. In most Bryozoa, the process of colony founding involves morphologic differences of size and complexity between generations of zooids immediately following the primary zooid or zooids. These differences define a **primary zone of astogenetic change**, which at its distal end develops a pattern capable of endless repetition of zooids (Fig. 21). The primary zone of change comprises the zooids, usually belonging to a few generations, which show morphologic differences from generation to generation in more or less uniform progression distally away from the primary zooids. In a zone of change, therefore, the zooids in each generation in a distally directed series express morphologic characteristics unique to that generation.

The primary zone of astogenetic change is followed distally by a **primary zone of astogenetic repetition** in which large numbers of zooids of repeated morphologies are proliferated, usually through many generations. Morphologic differences attributed to astogeny, therefore, are restricted to zones of change in a colony.

In some Bryozoa, a colony may develop further astogenetic changes in morphology distal to the primary zone of astogenetic repetition. These subsequent zones of change can in turn be followed distally by subsequent zones of repetition in which the morphologic pattern capable of endless repetition is either like or unlike that in the primary zones of repetition. Subsequent zones of change and repetition may be part of the normal budding pattern, as frontal budding in some gymnolaemates (see Cheetham & Cook, this revision), or stimulated by microenvironmental accident, as in patches of intracolony overgrowth common to many stenolaemates (see Boardman, this revision).

3. **Polymorphism** is repeated, discontinuous variation in the morphology of zooids within a colony. Polymorphism may be recognized in the same generation of zooids in a zone of astogenetic change, or in any zooids at the same ontogenetic stage in a zone of astogenetic repetition (Fig. 22).

4. **Microenvironmental variation** is vari-

ation within a colony that cannot be inferred to be an expression of ontogeny, astogeny, or polymorphism. The morphologies of zooids within a colony are the result of continuous reaction by the genotype to the microenvironments of the colony, expressed at any particular time and place in the colony throughout colony growth. Differences in microenvironments during growth of the colony can be expected to produce differences in zooid morphologies (Fig. 23). This morphologic variation may occur in one or more regions of the colony or may affect scattered zooids. An environmental change affecting

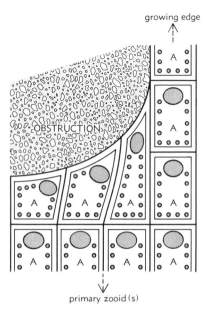

FIG. 23. Colony morphologic variation. Microenvironmental differences in zooid morphology in the zone of astogenetic repetition of a hypothetical bryozoan colony. Zooids belonging to the same generation have slightly differing morphologies, all modifications of the A form. The difference between extremes of morphology is related to growth around the obstruction (after Boardman & Cheetham, 1973).

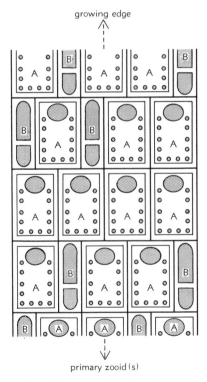

FIG. 22. Colony morphologic variation. Polymorphic difference in zooid morphology in the zone of astogenetic repetition of a hypothetical bryozoan colony. Zooids belonging to the same generation may have either morphology A or morphology B, intermediate morphologies being absent. Polymorphs may also occur in the zone of change in some species (after Boardman & Cheetham, 1973).

the morphology of zooids throughout the colony is a more widespread change than considered here as microenvironmental.

A few of the environmental causes that seem to explain observed morphologic variation within a colony but may also affect the colony as a whole are: crowding by growth of the colony itself or by competitive growth of other organisms, irregularities in the substrate, encrustation by the colony itself or by other organisms, differential turbulence, various forms of breakage, boring, and differential sediment accumulation. Differences in temperature, salinity, light intensity and duration, nutrients, and other environmental factors have been demonstrated to affect morphology, but their effect on intracolony variation is not generally known.

Microenvironmental variation can be recognized as irregular or gradational differences

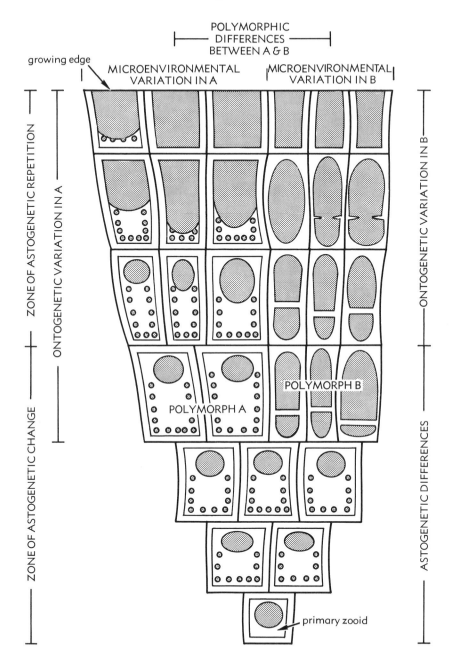

FIG. 24. Colony morphologic variation. Combined patterns of ontogenetic, astogenetic, polymorphic, and microenvironmental differences in zooid morphology in a hypothetical bryozoan colony. Spatial arrangement of polymorphs A and B is for convenience of comparison (although similar to those known in some groups of gymnolaemates); in actual colonies, polymorphs are commonly intermixed in the budding pattern. Note that no two zooids, even those belonging to the same polymorph and the same generation, are identical in the morphologic features shown (after Boardman & Cheetham, 1973).

between zooids; such differences are not necessarily repeated in a colony. Irregular differences can be recognized anywhere in the colony. Gradational differences can be recognized in zooids belonging to the same generation in a zone of astogenetic change, or to the same or different generations (at the same ontogenetic stage) in a zone of astogenetic repetition (Fig. 23).

The artificial restriction of variation within a colony to a single pattern seen in the hypothetical bryozoan colonies of Figures 20–23 is probably never approached very closely in real colonies. All four patterns are commonly combined (Fig. 24), but differences attributable to each source of variation can be separated. As shown in Figure 24, both ontogenetic and **astogenetic differences** in morphology are expressed in a series of zooids parallel to the direction of budding. In most Bryozoa, ontogenetic differences are expressed by generally increasing complexity proximally and astogenetic differences by generally increasing complexity distally. Because of the sequential nature of budding, these differences have relative time significance. Intracolony differences produced by polymorphism and microenvironment are not necessarily sequential.

The hypothetical examples above, and most studies of actual colonies, have emphasized patterns of variation in the morphology of the zooidal body wall, and especially of its skeletal layers. Patterns of variation in the morphology of zooidal organs may not be entirely congruent with those of the body wall. In gymnolaemates and stenolaemates for example, cyclic degeneration and regeneration of the lophophore and associated organs can produce cyclic repetition of ontogenetic gradients, in characters such as tentacle length, proximally from growing extremities. Also, lophophores and other organs may differ with the sex of zooids whose skeletons are not distinguishable (see CHEETHAM & COOK, this revision). Within most colonies, however, ontogenetic, astogenetic, polymorphic, and microenvironmental differences are all reflected in the skeletons of zooids. These differences therefore can be recognized in fossil as well as living taxa and may be taken into consideration in classification.

COLONY CONTROL OF FUNCTION AND MORPHOLOGY

Physical wholeness and assumed genetic uniformity make the colony the unit that survives and contributes to the gene pool in Bryozoa. The colony responds to the environment through its functions, contributed at any level of organization from the entire colony to its member zooids and extrazooidal parts, to their organs, tissues, and cells. Structures at these levels of organization may respond separately to the environment and therefore are subject to natural selection. The colony, however, is comparable to the solitary animal as the viable unit in the environment.

As far as is known, bryozoan colonies perform the following functions at some stage in their development: sexual reproduction; asexual reproduction of zooids; feeding, digestion, and intracolony dispersion of nutrients; formation and evacuation of feces; structural support by the colony itself; growth, and repair of injury; and degeneration and regeneration of zooid organs. These functions include those essential to solitary animals as well as those unique to colony organization. Functionally, therefore, bryozoan colonies and solitary animals are only partly comparable. Other basic functions, such as respiration and excretion of some metabolic wastes, must be assumed but are poorly understood and not considered here.

If only those functions common to both bryozoan colonies and solitary animals are considered, such as feeding and sexual reproduction, solitary animals compare most closely with the member zooids of colonies. Morphologically, solitary animals also compare most closely with zooids. Therefore, zooids are considered to be the basic morphologic and functional units of Bryozoa that correspond most closely to individual solitary animals.

Comparison between bryozoan zooids and

solitary individuals, however, is not exact. Member zooids are not viable by themselves, but grow and function cooperatively with each other and any extrazooidal parts to form viable colonies. It is therefore impossible to separate completely the morphology and functions of zooids from those of the colony.

The degree to which zooids differ from solitary animals morphologically and functionally because of their membership in a colony expresses the degree of control that the colony has over its member zooids (BOARDMAN & CHEETHAM, 1973). Many functions are speculative or unknown, and can become better known through study of living colonies. However, it is generally possible to assess the degree of **colony control** over zooidal functions by inference from the morphology of zooids and extrazooidal parts, that is, by the extent to which zooids in combination with extrazooidal parts differ morphologically from solitary animals (degree of **integration**). Morphologic features under some degree of colony control commonly can be observed to have grown or can be inferred to have functioned cooperatively with adjacent zooids, or as extrazooidal structures separate from zooids.

Structures contained within **zooidal boundaries** that perform functions similar to those of solitary animals reflect a degree of autonomy retained by zooids within their colonies. These structures may be inferred to reflect a degree of **zooidal control**. It is probable, however, that few, if any, zooidal structures in most Bryozoa are grown without some influence from adjacent zooids, extrazooidal parts, or both.

The degrees of morphologic integration expressing colony control may be interpreted on the basis of the following assumptions.

1. The body cavity of the colony (possibly excluding the portion contained by the tentacles and tentacle sheaths of the zooids) is separated from the external environment by body wall having protective cuticle or gelatinous material as its outermost layer.

2. An imperforate cuticular wall is sufficiently impervious to physiologic exchanges to sustain zooid growth and function.

3. An imperforate calcareous wall is sufficiently impervious to physiologic exchanges to permit further zooid growth.

4. Some zooids in a colony are feeding zooids.

5. Growth of cellular tissue and secretion of noncellular layers require a source of nutrients.

6. Nutrients are dispersed through the colony through either cells in mural pores or confluent coelom.

7. Confluent coelom between zooids or between zooids and extrazooidal parts of a colony permits freer physiologic exchange than do cells in mural pores.

8. Extrazooidal parts and zooids lacking feeding organs have direct or indirect access to nutrients from feeding zooids.

9. Morphologic difference between zooids implies physiologic or functional differences, or both.

10. Colony control of zooids and extrazooidal parts may be local or colony-wide in extent.

11. Correlated cyclic growth within a group of zooids is not necessarily a result of colony control, but may be simultaneous separate responses to a cyclic environment.

Integration of vertical zooidal walls.—The walls between zooidal cavities (most commonly vertical walls) may be either interior or exterior. Exterior walls are comparable in mode of growth and morphology to those bounding a solitary individual. These walls have the capability of separating the zooid from the environment, thus expressing **zooidal autonomy**. Interior walls, in their mode of growth and morphology, express colony control in that they partition existing body cavity and have no apparent potential for separating the zooid from the environment. The following combinations of vertical zooidal wall types occur in Bryozoa and are listed in order of increasing integration as the result of colony control.

1. Walls only exterior (a few stenolaemates, most phylactolaemates).

2. Walls partly exterior, partly interior (a

few stenolaemates, most gymnolaemates, few phylactolaemates).

3. Walls wholly interior (most stenolaemates, a few gymnolaemates).

Integration by interzooidal connection.—Soft-tissue connections are generally lacking among solitary animals. Interzooidal connection by zooidal soft tissues and their assumed function in Bryozoa, therefore, express colony control. Connected body cavities of zooids or extrazooidal parts apparently can exchange physiologic substances, and the nature of the connection expresses the degree of colony control. These states are listed in order of increasing integration.

1. No soft-tissue connections between adult zooids (a few stenolaemates).

2. Connections by cells through mural pores (some stenolaemates, all gymnolaemates).

3. Connections by confluent body cavity around ends of complete, interior vertical walls (some stenolaemates).

4. Connections both through mural pores and around ends of interior vertical walls (some stenolaemates).

5. Connections by confluent body cavity through and around incomplete interior vertical walls (a few stenolaemates, some phylactolaemates).

6. Connections by confluent body cavity, no interior vertical walls (most phylactolaemates).

Integration by extrazooidal hard and soft parts.—The presence of extrazooidal hard and soft parts in a colony is an indication of a degree of colony control of growth, because extrazooidal parts are unique to colonial animals, based on the assumption that a solitary individual is internally comparable to a zooid. The development of extrazooidal parts results in a further loss of zooid autonomy from the condition in solitary animals. In Bryozoa, extrazooidal parts are connective or supportive structures outside zooidal boundaries. Extrazooidal parts known in stenolaemates and gymnolaemates form a transitional morphologic series approaching in its variety that of polymorphic zooids. Extrazooidal parts in phylactolaemates form a transitional mor-

phologic series from identifiable colony body wall to identifiable zooidal vertical walls. The following states, listed in order of increasing integration, are known in Bryozoa.

1. Extrazooidal parts absent (many stenolaemates, many gymnolaemates).

2. Extrazooidal parts formed after budding of zooids through coalescence and resorption of zooidal tissue (some gymnolaemates).

3. Extrazooidal parts formed after budding of zooids, extrazooidal in origin (some stenolaemates).

4. Extrazooidal parts formed at the same time as budding of zooids (some stenolaemates, a few gymnolaemates, all phylactolaemates).

Integration through astogeny.—It is assumed that the morphologic differences among zooids imply physiologic and functional differences. Astogenetic differences between zooid generations in a zone of change therefore can be assumed to have been developing toward a repeatable set of physiologies and functions as well as morphologies. The sequence of morphologic as well as inferred physiologic and functional change is an expression of colony control because it is absent in solitary animals. In Bryozoa, the following states, in order of increasing integration, may be present.

1. All zooids of all generations, including the first zooid, of constant morphology (a few sexually produced gymnolaemates, some asexually produced—by fragmentation—stenolaemates and gymnolaemates, all sexually produced phylactolaemates).

2. First zooid or group of zooids different, all others without generational differences (some sexually produced gymnolaemates, asexually produced phylactolaemates).

3. Generational differences between zooids limited to proximal region of colony; that is, the colony has primary zones of astogenetic change and repetition only (many sexually produced stenolaemates, many sexually produced gymnolaemates).

4. Generational differences between zooids present in proximal regions and at least one

distal region of colony; that is, the colony has both primary and subsequent zones of astogenetic change and repetition (many sexually produced stenolaemates, some sexually produced gymnolaemates).

5. Generational differences between zooids on gradient throughout colony; that is, the colony lacks any zone of astogenetic repetition (a few sexually produced gymnolaemates).

Integration through morphologic differences among polymorphs.—Polymorphic differences are an expression of colony control, for polymorphism is one kind of functional reponse to the environment by the colony. In feeding, reproduction, and other basic processes, zooids in a **monomorphic colony** respond virtually as individuals. The response of a polymorph, however, is through its contribution to the colony as a whole, in direct proportion to its functional specialization. The following states, in order of increasing integration, are known in Bryozoa.

1. All zooids of same generation of constant morphology (all phylactolaemates, skeletally many stenolaemates, a few gymnolaemates).

2. Asexually produced zooids polymorphic, all having feeding and sexual reproductive ability (possibly some stenolaemates, some gymnolaemates).

3. Asexually produced zooids polymorphic, some lacking either feeding or sexual reproductive ability (possibly some stenolaemates, some gymnolaemates).

4. Asexually produced zooids polymorphic, some lacking both feeding and sexual reproductive ability (many stenolaemates, most gymnolaemates).

Integration through positional differences of polymorphs.—Another measure of colony control is expressed by polymorph position and structural dependence on other zooids. Polymorphs intercalated randomly in the colony budding pattern probably contribute their specialized functions as separate operating units. Those assembled in repeated groups of one or more kinds of zooids can carry out their specialized functions jointly. These functions include joint production of currents or brooding of larvae in living colonies. **Intrazooidal polymorphs** (zooids changed in morphology and function during life within the same living chambers) and some **adventitious polymorphs** (appendagelike zooids adding functions to those of the supporting zooids) indicate higher degrees of structural dependence on the supporting zooid than polymorphs intercalated in the budding pattern. The following states, listed in order of increasing integration, are known in Bryozoa.

1. All zooids of same generation of constant morphology (all phylactolaemates, skeletally many stenolaemates, a few gymnolaemates).

2. Asexually produced zooids polymorphic, intercalated in the budding pattern randomly (some stenolaemates, some gymnolaemates).

3. Asexually produced zooids polymorphic, intercalated in the budding pattern regularly (some stenolaemates, some gymnolaemates).

4. Asexually produced zooids polymorphic, in repeated groups (many stenolaemates, some gymnolaemates).

5. Asexually produced zooids polymorphic, intrazooidal or adventitious (a few stenolaemates, many gymnolaemates).

USE OF CHARACTERS IN CLASSIFICATION

Classifications consistent with inferred evolutionary history are essential for application to problems in biogeography, biostratigraphy, and other historical aspects of biology. Therefore, evolutionary classifications of Bryozoa must be attempted, even though no definitive classification is likely to be established. Even if it were possible to know the evolutionary history of Bryozoa, more than one classification could be consistent with that history. Taxa are segments of a lineage or grouping of lineages, and the boundaries between taxa can only be placed arbitrarily through the continuum, even if some lineages evolved so rapidly that few generations of intermediates existed. The only nonarbitrary rule for the placement of taxonomic boundaries is that a taxon must not combine lineages having separate evolutionary histories as inferred from their distribution in time and space. Given these restrictions, evolutionary classifications can only be approximations that are subject to improvement.

The evolutionary significance of a classification increases with increased use of genetically controlled characters. This does not mean that characters of unknown genetic significance, such as the presence or absence of polymorphism, cannot be used in a classification, but only that those inferred to lack genetic control, such as the irregular two-dimensional shapes of individual encrusting colonies on rough substrates, should not be used.

In bryozoans, taxonomic characters are derived from morphologic features that must reflect varying proportions of genetic and environmental control. Estimates of the proportions of genetic and environmental control are among the most difficult interpretations to make in evolutionary taxonomy. The only direct and convincing approach to the problem seems to be through experimentation with the breeding and growing of colonies, ideally in their natural habitat. Until studies of living colonies are accomplished for many taxa in many environments, the taxonomist must continue to approach the matter indirectly. For many fossil taxa, of course, such approaches will always be indirect.

All modern classifications or proposed evolutionary arrangements of Bryozoa have been based on morphologic differences expressed as states of taxonomic characters. Some have used only morphologic characters of living forms (e.g., Borg, 1926a; Marcus, 1938a;

Silén, 1942), or of living and fossil forms without reference to the independent evidence of position in time (Cuffey, 1973). One classification is based on inferred position in time of soft-part morphology not available in the fossil record (Jebram, 1973b). Some proposed evolutionary classifications have considered morphologic differences in a time-space framework (e.g., Bassler, 1953; Astrova, 1960a; Ryland, 1970).

Classifications that attempt to express evolutionary relationships depend on the nature and number of characters used as well as the independent evidence of position in time. Improved evolutionary arrangements and new classifications can be achieved by the addition of taxa and taxonomic characters, by improved understanding of stratigraphic relationships, and by new approaches to character analysis and taxonomic philosophy. Material that has not been employed in classifications of the Bryozoa is now available in each of these areas. The procedure for taxonomic character analysis suggested below is based on a new synthesis of the nature of the bryozoan colony and an evolutionary taxonomic philosophy that has not been tried in classifications of Bryozoa.

TAXONOMIC CHARACTER ANALYSIS

In all groups of Bryozoa, the high level of organization of both zooids and colony makes available many morphologic characters for taxonomic study. A character having potential taxonomic importance has states, which are morphologic properties by which organisms differ. Characters may show many states, a wide variety of differences, or few, the simplest being the two-state character of "present" or "absent."

The taxonomic process begins with observations of the more obvious intracolony and intercolony morphologic differences in structures that are initially assumed to be comparable. Initial observations are followed by a three-part character analysis, which tests the evolutionary potential of all available morphologic differences, using biologic processes, assumptions, and principles. In addition to expressing morphologic differences, evolutionary characters and their states should satisfy three major requirements. First, a character should be morphologically independent to the extent that its observable states are not partly determined by states of other characters within the taxon being considered. Second, a character influenced by ontogeny, astogeny, or polymorphism should have separable states that are comparable from colony to colony. Third, a character should be genetically controlled to the extent that its observable states correlate with genetic differences among colonies.

Biological analysis.—The first step in obtaining characters of evolutionary significance is to recognize as many characters as possible that are morphologically, but not necessarily genetically, independent of each other. Morphologic independence of many characters generally adds detail and sensitivity to the resulting classification while guarding against redundancy of characters and morphologic ambiguity of character states. Independent characters are most likely to be recognized by detailed study of the morphology of the whole colony and its parts, and interpretations of mode of growth and function of that morphology. It is generally assumed that morphologic features have biological significance in growth and functions of the colony. Some structures possibly have changed or lost their original function during evolution, but direct evidence of vestigial structures has not been recognized in Bryozoa.

Characters appropriate to any level of the taxonomic hierarchy may be derived from morphologic features at organizational levels of cell, tissue, organ, zooid, unified grouping of zooids, extrazooidal part, or entire colony. Whether characters are independent can be determined only by comparison among colonies and taxa of the character states of comparable, potentially homologous morphologic structures. Improvement in our understanding of the comparability of structures will result only from application of the most revealing study techniques available to comparative morphology, and from more detailed interpretations of mode of growth and function. At this stage in the study of Bryozoa, advancements in biological analysis generally will result in an overall increase in the number of morphologically independent characters to be considered in classifications.

Morphologic features from which independent characters can be derived include orificial and frontal walls in most gymnolaemate bryozoans. These features are morphologically continuous (Fig. 3), but perform different functions and therefore form the basis for two separate sets of characters. In stenolaemate bryozoans, vertical walls may be distinguished from frontal walls in zooids by their microstructure and mode of growth, and by partial functional differences (Fig. 11), allowing them to be recognized as separate features providing separate sets of independent characters.

Intracolony analysis.—The second step in taxonomic character analysis in Bryozoa is to recognize, for each independent character, states that have been separated from, or have taken into account, intracolony variation. A set of separated states of characters must be recognized and expressed for each colony. These may come from the generally recognizable morphologic patterns of ontogeny, astogeny, polymorphism, and narrowly determined microenvironmental modifications within each colony (BOARDMAN, CHEETHAM, & COOK, 1970). Character states separated into major stages of these patterns can be compared directly from colony to colony (see Sources of Morphologic Variation).

It is obvious that all morphologic variations within a colony must fall within the potential range of expression of the presumed uniform colony genotype, and in this sense all morphologic variation is genetically based. Genetically controlled variation as used here, however, applies only to morphologic differences that reflect differences in genotype. Because of the asexual mode of growth of the colony, variation due to differences in genotype is assumed not to occur within a colony but only among colonies, except for somatic mutations, which have not been recognized in Bryozoa. ˙

Intracolony analysis begins with recognition of whether zooids are monomorphic or polymorphic. Each set of polymorphs has a set of character states at least partly different from the sets of other polymorphs. Each character is studied for astogenetic and ontogenetic changes. Some characters change from generation to generation in zones of astogenetic change, but others may be constant from generation to generation whether in a zone of change or a zone of repetition. Similarly, some characters change continuously throughout the life of a zooid, but others are either constant throughout life or may become constant at different ontogenetic stages. The characters that are constant ontogenetically and astogenetically (and as nearly as determinable, microenvironmentally) may be expressed as one state for each polymorph in each colony. Some but not all of these characters may also be constant for all polymorphs in the colony. For example, constant microstructure of vertical walls throughout a colony is a single state representing the entire colony. Likewise, constant calcitic or constant aragonitic composition of all calcified walls in a colony are single states representing entire colonies.

Characters that change in generational patterns indicating either ontogeny or astogeny can be expressed as series of states for each colony. Intervals of these series then serve as the separated states for the colony. For example, maximum extent of vertical walls can be reached early in zooid ontogeny, or these walls can increase throughout the life of the zooid. Aragonite layers may be added to initial calcite wall layers during zooid ontogeny. The size of zooids can increase from generation to generation in a zone of astogenetic change. Thus, the ontogenetic extension of vertical walls, the mixed composition of calcareous walls, and the astogenetic increase in zooid size can be divided into separated character states.

Young living colonies, or modern or fossil colonies that died in early stages of life, commonly show only parts of the series of separated character states present in fully developed colonies. States characteristic of zones of astogenetic repetition or of later stages of zooid ontogeny may be missing. If certain polymorphs or extrazooidal parts are present only in zones of astogenetic repetition or after zooids have reached a certain ontogenetic stage, these too may be missing in young colonies.

Fragments of colonies also commonly lack parts of series of separated character states. In different fragments of a colony, one may find states of different ontogenetic or astogenetic stages, of different sets of polymorphs, or of extrazooidal parts characteristic of the whole colony. If a sufficient number of fragments presumably from the same population is available, their overlapping patterns of variation permit at least tentative reconstruction of the separated character states of whole colonies. In many fossil bryozoan taxa, reconstruction from colony fragments has provided the only basis for recognition of separated character states.

Intercolony analysis.—The third step in taxonomic character analysis in Bryozoa is to attempt to recognize, for each independent character, states that more nearly express genetic rather than environmental difference between colonies. The first two analyses reduce the sources of variation to genetic and environmental differences between colonies. Unfortunately, environmental variation cannot be accounted for in the comparison of colonies to the same degree as ontogenetic, astogenetic, and polymorphic differences. Different colonies, even within the same community, may have been subject to differing environments and therefore record different morphologic reactions. Some species exhibit encrusting growth on an "unlimited" substrate and erect growth where the extent of substrate is or becomes severely limited; an example is illustrated and described by Cook (1968a, p. 124, pl. 1, fig. c,d). A colony growing in a changing environment may combine both modes of growth (Cook,

1968a, p. 124), which demonstrates that the variation between colonies can be of the same kind as that within a colony. Environmental differences thus produce two kinds of morphologic variation in Bryozoa: that expressed by the colony as a whole (colony-wide environmental variation) and that observable within a colony (microenvironmental variation).

Recognition of direct environmental modification of the states of a character does not necessarily rule out the use of that character in deriving a classification. The limits within which the states of a character can express direct environmental modification are assumed to be genetically controlled. Differences of limits within the same range of environments can be inferred to reflect genetic differences of potential taxonomic value. For example, two species might exhibit different but overlapping series of growth habits developed within the same range of environments. The growth habit most commonly developed within each species in the same environmental range, moreover, could fall within the overlap between species and that modal growth habit could be under genetic control. The observed differences between the growth habits of individual colonies themselves, however, would not be directly correlated with genetic differences and thus would have no evolutionary significance as a basis for further taxonomic subdivision.

Proportions of genetic and colony-wide environmental control of many single morphologic characters may be estimated indirectly based on the following assumptions.

1. Characters are assumed to be closely controlled genetically if they remain relatively constant through significant intervals of geologic time, or if their patterns of transitional change are not significantly modified by inferred environmental changes through intervals of geologic time. In either case, inference of genetic control is strengthened by increased independent evidence that the bryozoan successions were subjected to changing environments. Of course, a character that changed through a significant interval of time in correlation with environmental changes can also be closely controlled genetically, but this genetic control would be difficult to distinguish from environmental modification.

2. Some characters derived from structures grown within exterior walls are assumed to reflect increased degrees of genetic control because they are sheltered from some kinds of environmental interference by the comparative stability of the internal environment of the body cavity.

A corollary is that because colony-controlled (integrated) structures are commonly grown within exterior walls, many characters derived from these structures also show greater degrees of genetic control.

3. Microenvironmental modifications are generally recognizable and serve as a basis for estimating the kinds of morphologic differences that might be caused by colony-wide environmental differences.

4. The colony growth habit of many species varies and is assumed to be closely controlled by environment within genetically set limits. Parts of zooids and extrazooidal features that are affected by changes of growth habit can also be assumed to be directly modified by the environment. Environmental modifications are assumed to be especially pronounced in features that are structural in function, relating to the strength of colonies in the different growth habits.

5. Environmental control is assumed for certain modifications of characters not necessarily associated with differences in growth habits of colonies. Such modifications are observable in colonies subjected to environmental changes of short duration relative to colony life, or in colonies that lived in more than one environment. It is assumed that structures in which modifications independent of colony growth habit are observable can be interior or exterior, or colony controlled or zooid controlled. If these modifications are developed throughout whole colonies and are comparable to those developed microenvironmentally in other closely associated colonies, the inference of environmen-

tal influence is more convincing (assumption 3 above).

6. It is assumed that the proportions of genetic and environmental control of any potential taxonomic character can be different in different taxonomic groups under similar environmental circumstances.

Development of increasing colony control appears to have conferred a selective advantage, as suggested by the trend in some stocks toward higher degrees of integration. Some early forms in evolving stocks of major taxonomic rank exhibited so low a degree of integration that member zooids may have functioned nearly as solitary animals (e.g., corynotrypids in stenolaemates, BOARDMAN, this revision; and *Pyriporopsis, Arachnidium,* and similar forms in gymnolaemates, CHEETHAM & COOK, this revision). Evolution of some of these forms can be inferred to have proceeded toward higher degrees of integration. Study of the major branch of gymnolaemates, the cheilostomates, has suggested an increase in colony integration from the Jurassic to the present (BOARDMAN & CHEETHAM, 1973, p. 178–191). Body walls of the earliest cheilostomates were almost entirely exterior, immediately adjacent to the environment. Through time, other cheilostomates appeared with greater proportions of interior vertical walls. The concept of interior zooidal walls in Bryozoa requires that these walls grow under the protection of the colony and not the immediate influence of the environment. Features of their internal construction, therefore, such as lack of cuticle in stenolaemates and microstructure of zooecial boundaries and skeleton, may be less dependent on the environment and more reflective of the genotype.

Physiologic communication among zooids and between zooids and extrazooidal structures of a stenolaemate colony apparently can be through pores or around ends of interior vertical walls, under the protection of the exterior wall and within the body cavity of the colony. The two kinds of connections are used separately or together, in different combinations with other structures for apparent selective advantage, expressed by the functional performance of the colony as a whole. Characters of communicational function in a colony, therefore, might well show more genetic than environmental control and be subject to natural selection in the evolutionary process.

In all but the simplest gymnolaemates, contiguous exterior walls are breached by interzooidal communication organs. Some colonies with significant proportions of exterior vertical walls can therefore have a higher degree of integration than the simplest uniserial forms. Communication organs in both exterior and interior walls are within the body of the colony and thus should show more genetic than environmental control.

Astogenetic differences leading to an ever-repeatable budding plan and functions are expressions of colony control. Astogenetic development of the zone of repetition has selective advantage in that it allows colonies to become larger without increasing the size of member units. The general sequential patterns in zones of change to zones of repetition for a species are constant enough to suggest genetic control. In the gymnolaemate order Cheilostomata, evolutionary trends toward development of increasingly complex astogenetic change further suggest genetic control.

The diversification of functions made possible by the development of polymorphic zooids provides a selective advantage, especially in uniform environments (SCHOPF, 1973). In the gymnolaemate order Cheilostomata, the development and refinement of polymorphs in many evolving stocks appears to be an expression of this selective advantage.

Extrazooidal parts are generally structural in function and add to the strength or flexibility of a colony. They are therefore probably subject to considerable modification from the immediate environment, even though they form under direct colony control. Extrazooidal parts may well provide colony protection for zooids, however, so that zooidal characters can be relatively independent of environmental influence and more nearly reflect the genotype. The regularity of arrangement of zooids in cupuladriid cheilostomates may reflect this kind of control.

Examples of characters in stenolaemates assumed to be environmentally modified without change in colony growth habit include such small variations in growth characteristics of zooidal and extrazooidal structure during their ontogeny as in thickness of different skeletal layers of walls, in overall thickness of wall segments, or in spacing and thickness of basal diaphragms. These modifications appear to be based on growth rates controlled by short-term environmental changes during the life of the colony. These changes are microenvironmental if they are restricted to parts of colonies. If similar-appearing growth changes are colony-wide, they are generally considered to be environmentally controlled, but with less certainty.

Summary.—The aim of the three-part character analysis is to obtain genetically controlled states for as many independent characters as possible for each colony. At the end of this procedure, the list of states for each colony will include some for which the degree of genetic control has been inferred with a high degree of confidence. Others, for which the degree of genetic control has been inferred with less confidence, may or may not be used in classification on the judgment of the investigator.

The effort to distinguish between genetic and environmental control of character states is greatly facilitated by application of the concept of colonies in bryozoans as opposed to the concept of solitary animals. The ability to recognize separately those morphologic differences in bryozoan colonies that result from ontogeny, polymorphism, and microenvironment means that character states controlled by these variables can be "cleanly" removed from consideration without overlapping morphologic confusion with the genetic and colony-wide environmental effects being studied.

GENERAL FEATURES OF THE CLASS STENOLAEMATA

By R. S. Boardman

[Smithsonian Institution, Washington D.C.]

The Stenolaemata are here considered to make up one of three classes of the phylum Bryozoa. Members of the class are characterized by feeding zooids with complete interior vertical walls (Fig. 25, 26) that are commonly elongated to enclose tubular, conical, or sac-shaped body cavities. Vertical walls are elongated parallel to the direction of zooidal growth. Vertical walls of all zooids have skeletal layers, as do basal and frontal walls (Fig. 26) where they occur. In most taxa, zooids open at high angles to colony surfaces, and zooecial apertures are comparable in area to cross sections of living chambers. Zooecial apertures and the terminal membranous orificial walls that cover them in living colonies are transverse to zooidal length. Tentacles are protruded through circular porelike orifices by the action of a membranous sac that surrounds the lophophore and gut in recent stenolaemates.

The class Stenolaemata produced virtually all of the vast accumulation of fossil bryozoans from the Early Ordovician into the Early Cretaceous, a time interval lasting nearly 400 million years. During that interval the class Gymnolaemata is represented by a few scattered species of ctenostomates beginning in the Ordovician and of cheilostomates beginning in the Jurassic (see CHEETHAM & COOK, this revision, General Features of the Gymnolaemata). Stenolaemates are the most abundant fossil group in many rock units throughout the stratigraphic column and the continuity of their stratigraphic occurrences is comparable to that of other major groups of fossils. During the Late Cretaceous, the stenolaemates began to lose their predominance within the phylum to the class Gymnolaemata. Stenolaemate numbers and diversity have apparently been on a slow decline since the Cretaceous. Stenolaemates can be found living in large numbers, however, in many marine communities (e.g., the Medi-

terranean Sea; HARMELIN, 1974, 1976).

The class includes four (Blake, this revision) to six (SHISHOVA, 1968) orders, depending on the classification used. Five orders are recognized here. The Trepostomata, Cystoporata, Cryptostomata, and Fenestrata all appeared during the Ordovician, all were prolific at times during the Paleozoic Era, and all are generally considered to have become extinct during or just after the Permian. The Tubuliporata (formerly Cyclostomata) also appeared in the Ordovician, but remained unimportant in numbers and diversity until the Mesozoic and Cenozoic eras, when they occurred in large numbers.

Unfortunately, the Paleozoic and post-Paleozoic taxa have been studied using different preparation techniques and taxonomic characters. The present literature tests neither the assumed Permian and Triassic extinctions of Paleozoic stocks, nor the generally accepted monophyletic origin of the post-Paleozoic Tubuliporata. One of the questions of highest priority to improved understanding of the class Stenolaemata is the piecing together of its evolutionary history across the Paleozoic-Mesozoic boundary, using modern taxonomic procedures and as many taxonomic characters as are available.

Stenolaemate bryozoans apparently have been entirely marine throughout their history. In Paleozoic rocks their numbers are largest in calcareous shales, mudstones, and some limestones. Colonies that grew erect are commonly preserved broken but unscattered in shales and mudstones, indicating little or no transportation after death (e.g., BOARDMAN, 1960).

Growth habits of colonies of many species of bryozoans have long been assumed to be modified significantly by different environments (e.g., ULRICH, 1890; STACH, 1936). A thorough review of the literature of steno-

Bryozoa

laemate ecology and paleoecology was published by Duncan (1957). Experimental studies are just beginning to emphasize the effects of different environments on colony growth habits and correlated changes of internal morphology within the same species of living stenolaemates (for examples, see Harmelin, 1973, 1974, 1975, 1976).

Details of skeletal structures are commonly well preserved in fossil stenolaemates of all ages and provide many taxonomic characters that can be inferred to be genetically controlled. Skeletal structures furnish evidence of modes of growth, functional morphology, and intra- and intercolony morphologic variation, especially where their relationships with soft parts can be inferred with confidence.

A surprising number of indications or actual fragmentary remains of soft parts occur throughout the fossil record of the stenolaemates and some very general comparisons can be made with the complete soft parts of modern species. Unfortunately, the soft parts of most modern species and their growth and functional relationships with skeletal counterparts are poorly known. For example, recent sectioning of a few randomly selected taxa has revealed four different morphologies affecting the protrusion of tentacles (Boardman, 1973, 1975). Only one of these had previously been reported. Most of the character states derived from soft parts that are assumed to be characteristic of the order Tubuliporata are known from relatively few species and therefore should be investigated further.

Independent, apparently genetically controlled taxonomic characters within colonies that are carefully collected from vertical sequences commonly show transitional changes. Not enough of these detailed studies have been published, however, to demonstrate many evolutionary patterns and detailed morphologic trends. Unfortunately, the study of stenolaemate bryozoans has not been advanced enough for a general realization of their potential value in applied problems of ecology, zoogeography, and biostratigraphy.

Acknowledgments.—Special acknowledgment is due D. A. Dean for his outstanding preparations of specimens and the photographs. Figures were prepared by L. B. Isham and JoAnn Sanner. Preserved specimens were generously volunteered by M. B. Abbott and P. L. Cook (Cape Cod), D. P. Gordon and P. K. Probert (New Zealand), J. G. Harmelin (Mediterranean), J. B. C. Jackson (Jamaica), Geneviève Lutaud (English Channel), R. E. Wass (Australia), and some of the specimens of Cretaceous age from northern Europe by Ehrhard Voigt. Specimens were loaned by P. L. Cook and J. G. Harmelin. The manuscript was reviewed and much improved by D. B. Blake, A. H. Cheetham, O. L. Karklins, F. K. McKinney, Claus Nielsen, C. D. Palmer, and J. E. Winston.

HISTORICAL REVIEW

CLASSIFICATION

The concept of tubular Bryozoa, now the class Stenolaemata, began formally with the establishment of the Tubuliporina Johnston (1847, p. 265), placed under Polyzoa infundibulata, and based on studies of recent Bryozoa only. The group was characterized by Johnston as "Polypidoms calcareous, massive, orbiculated or lobed or divided dichotomously; the cells long and tubular, with a round prominent unconstricted aperture." The characterization was accompanied by a drawing of an unmistakable tubuliporid and descriptions of a number of appropriate taxa.

Later, Busk (1852, p. 346) established the Cyclostomata as a suborder, basing the name on recent Bryozoa ". . .having a round, simple opening to the cell. . . ." Busk recognized

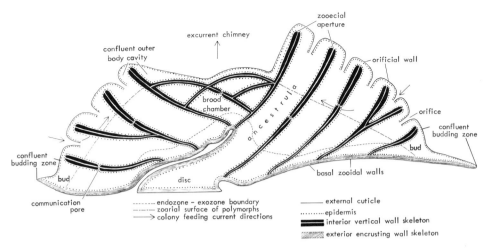

FIG. 25. General features of the Stenolaemata. Diagram of a longitudinal section through the center of a free-walled lichenoporid colony. The plane of section lies within feeding zooids radially arranged in both directions from the colony center. Polymorphs (not shown) are in radial rows between rows of feeding zooids and form a part of the zoarial surface at the lower level indicated by the unevenly dashed line. Arrows parallel the flow of feeding currents past orifices, up to the center of the colony on the polymorph surface, and out through the chimney in the center of the colony. The basal encrusting colony wall is multizooidal, originating in the multizooidal budding zone, which is confluent around the outer margin of the colony. In lichenoporids, budding of most zooids occurs from basal colony walls in endozones in the confluent budding zones. A few zooids are budded from zooidal walls of exozones into confluent outer body cavities, as indicated by bifurcations of vertical walls. (In exozones outer body cavities are divided into zooids so that budding space is zooidal, not multizooidal.) Zooids growing to the right of the colony center form a wedge in the primary direction of encrusting growth. The fold on the left side of the disc provides encrusting colony wall for a wedge of zooids growing in the secondary encrusting growth direction. White lines in the center of interior vertical walls depict zooidal boundaries, indicating that vertical walls are compound.

his suborder as "... coinciding very nearly with the Tubuliporina..." of JOHNSTON. Unfortunately, BUSK's name Cyclostomata had earlier been used in the classification of fishes (DUMÉRIL, 1806). Nevertheless, the name Cyclostomata has been adopted in the classification of both living and comparable fossil tubular bryozoans and Tubuliporina has been ignored. *Treatise* policy recommends that the name Cyclostomata BUSK, 1852, be considered a junior homonym. To replace it, the name Tubuliporina is changed here to Tubuliporata JOHNSTON, 1847, to conform to order-level endings and to avoid conflict with the use of Tubuliporina as a suborder.

Many of the first tubular bryozoans of Paleozoic age to be described were thought to be corals by some paleontologists (e.g., NICHOLSON, 1879, 1881; VINE, 1884, p. 182;

WAAGEN & WENTZEL, 1886, p. 885). Others considered some of the same genera to be bryozoans (e.g., ROMINGER, 1866; LINDSTRÖM, 1876; DOLLFUS, 1875; and ZITTEL, 1880). The controversy was so confused by inadequate understanding of the taxa considered to be critical to the problem that the arguments are nearly impossible to follow in detail. Most of the genera of Paleozoic age involved in the controversy were considered to be bryozoans and placed in the new suborder Trepostomata by ULRICH (1882, p. 151). Trepostomates were finally accepted as bryozoans based largely on the work of ULRICH (from 1882 through 1893), CUMINGS (1912), and CUMINGS and GALLOWAY (1915).

CUMINGS' work was especially convincing. He based his interpretation on the shape of the zooecium of the ancestrula and the arrangement of the first few zooids. Similarities of ancestrulae in the trepostomate colonies of Paleozoic age and in species of undoubted tubuliporate bryozoans placed in the genus *Heteropora* (CUMINGS, 1912, p. 366) suggested that tubuliporates were the ". . .recent Bryozoa most closely related to the Paleozoic Trepostomata. . ." (CUMINGS & GALLOWAY, 1915, p. 350).

GREGORY (1909, p. 122–126) recognized some of the same morphologic features in both the Paleozoic trepostomates and post-Paleozoic tubuliporates and therefore placed some Mesozoic and Cenozoic tubuliporates in the Trepostomata. These similarities, which are now considered to characterize the class Stenolaemata, include long, tubular, parallel zooecia; size of zooecial cross section; presence of zooecial bends; and thicker walled outer segments of zooecia.

BORG (1926a, p. 489) argued that the Cyclostomata (including the Trepostomata and what is here called Tubuliporata), Phylactolaemata, and Gymnolaemata (including the Paleozoic Cryptostomata of this revision) probably had common ancestors but no ". . . lineal relation to one another." For that reason, he raised the Cyclostomata to the same taxonomic level as his Phylactolaemata and Gymnolaemata rather than leaving them in the next lower hierarchical level with the Cheilostomata, Ctenostomata, Cryptostomata, and Trepostomata as interpreted by earlier authors. BORG (1926a, p. 490) concluded that ". . . it seems to me necessary to form a new order for the Cyclostomata, coordinate with the two older orders [now considered classes] Gymnolaemata and Phylactolaemata. I propose that this new order should be termed Stenolaemata." He diagnosed the order as follows:

"Zooids narrow, cylindrical, tapering proximally, with terminal opening; cystids with calcified walls; polypide enclosed in a membranous sac acting as a hydrostatic apparatus, embryonic development within the membranous sac of a fertile polypide which itself degenerates, either in gonozoids, or in a coelomic space between the zooids; polyembryony."

BORG's (1926a, p. 490) classification included three orders, Phylactolaemata, Stenolaemata, and Gymnolaemata. The order Gymnolaemata contained three suborders: Cryptostomata, Cheilostomata, and Ctenostomata.

BORG suggested that the Stenolaemata should be divided into two suborders, the newly restricted Cyclostomata and the Trepostomata; however, he did not actually divide them until 1944 (p. 18, 19), when he reclassified genera and families so that his restricted Cyclostomata included only fixed-walled species (simple-walled species of BORG, single-walled species of subsequent authors, and fused-walled species of BOARDMAN, 1975; see Fig. 26). His Trepostomata apparently included all free-walled stenolaemates of all ages (the double-walled forms of BORG and subsequent authors; see Fig. 25) minus the cryptostomates. BORG's classification within the Stenolaemata has not been followed by subsequent authors.

The Paleozoic order Cryptostomata was defined by VINE (1884, p. 196) to include small ribbon-shaped bifoliate genera and small dendroid (branches circular in cross section) genera that were thought to have "orifice of cell surrounded by vestibule, concealed." This inferred inner position of the orifice was thought to be near the **hemisepta** (shelflike skeletal structures within the zooecia) that occur in some of the included genera. The presumed inner orifice caused the cryptostomates to be compared with the cheilostomates (ULRICH, 1890, p. 333; CUMINGS, 1904, p. 76; BORG, 1926a, p. 481; BASSLER, 1953, p. G119) and to be placed in the same grouping with the cheilostomates and ctenostomates (BORG, 1926a, p. 490). Evidence from modern tubuliporates with similar appearing hemiseptumlike structures (see Fig. 39,4) suggests that the orifice was not at the inner position of the hemiseptum but at the outermost zooecial aperture (Fig. 25). The remainder of the skeleton and the inferred mode of growth are comparable with those of free-walled stenolaemates, and transi-

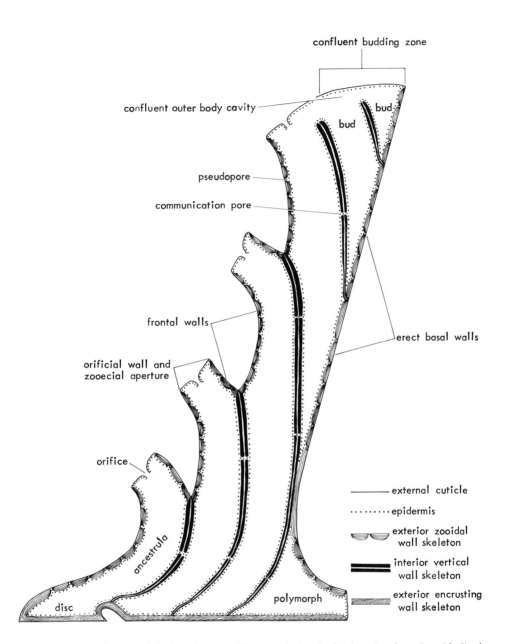

Fɪɢ. 26. General features of the Stenolaemata. Diagram of a longitudinal section through an idealized erect fixed-walled tubuliporate colony. The basal encrusting colony wall is multizooidal at least until it reaches basal polymorphs. The confluent budding zone in the distal end of the erect part of the colony includes outer confluent body cavity, covering exterior membranous wall, and the buds themselves. Calcification of exterior frontal walls at outer ends of interior vertical walls eliminates the outer body cavity of budding zone. Pseudopores in exterior walls do not penetrate exterior cuticle; most communication pores in interior vertical walls are open. The white line in the center of interior vertical walls depicts the zooidal boundary, indicating that vertical walls are compound. Peristomes are the outermost extensions of exterior frontal zooidal walls beyond more general colony surface.

tional morphology is a source of taxonomic confusion in distinguishing some cryptostomates from some trepostomates. The cryptostomates, therefore, have now been placed in the Stenolaemata (BOARDMAN & CHEETHAM, 1969; RYLAND, 1970).

In 1890 ULRICH (p. 349–362) removed the fenestrate (reticulate growth habit) genera of Paleozoic age from the tubuliporates and placed them in the Cryptostomata with the small bifoliate and dendroid forms. ELIAS and CONDRA (1957, p. 35) suggested a return to VINE's original two-part concept of the cryptostomates and elevated the fenestellids to the order Fenestrata. SHISHOVA (1968) removed the dendroid forms from the cryptostomates and made them an order, the Rhabdomesonata. In 1964 ASTROVA removed most of the Paleozoic genera from the Tubuliporata, added some genera that had been in the Trepostomata and Cryptostomata, and combined those genera into a new order, the Cystoporata (see UTGAARD, this volume).

As considered here, the class Stenolaemata [=Stenostomata MARCUS, 1938a] includes the following orders: order Tubuliporata JOHNSTON, 1847; order Trepostomata ULRICH, 1882; order Cryptostomata VINE, 1884 (see BLAKE and KARKLINS, this revision); order Fenestrata ELIAS and CONDRA, 1957; and order Cystoporata ASTROVA, 1964 (see UTGAARD, this revision).

METHODS OF STUDY

Students of the Stenolaemata may be divided into two schools based on preparation techniques and the resulting taxonomic characters employed. The earlier school relied primarily upon those characters that can be observed from outer surfaces or broken sections of zoaria. A second and later school uses characters occurring throughout zoaria. The second school began with the preparation of thin sections cut through zoaria of Paleozoic age in orientations standardized relative to the zooecia. Thin sections reveal morphologic details of colony interiors, adding greatly to the number of potential taxonomic charac-

ters.

Reliance on external characters.—The almost exclusive use of external characters has persisted in western studies of fenestellids (Paleozoic in age), and remains dominant in the taxonomy of post-Paleozoic and recent stenolaemates, the Tubuliporata. CANU and BASSLER, as early as 1920, published drawings and photographs of sections of zoarial interiors of many species of the Tubuliporata at low magnifications. Because they used only generalized zooecial shapes and arrangements from sections and relied mostly on external morphology, little taxonomic advantage was achieved. Only some of the most recent taxonomic papers on the Tubuliporata have employed thin sections more fully (e.g., VISKOVA, 1972, 1973; HARMELIN, 1974; HINDS, 1975; TILLIER, 1975; and NYE, 1976).

Use of external and internal characters.—Early sectioning techniques of the second school have provided the basis for modern sectioning refinements that, in combination with greatly improved light and electron microscopes, produce detailed information on the entire colony. Advantages of the use of sectioning include: first, the potential availability of all taxonomic characters of complete stenolaemate zoaria; second, the relative increase in numbers of characters from internal morphology of zooids and extrazooidal structures over characters concerning colony growth habit; and third, the availability of biological evidence concerning such subjects as mode of growth, functional morphology, reproduction, and feeding.

Oriented sections were first used by NICHOLSON (1876, 1879, 1881) in Scotland and DYBOWSKI (1877) in Russia. It was immediately recognized that new taxonomic characters derived from zooarial interiors differentiated many new taxa from specimens that were either externally poorly preserved, embedded in a hard rock matrix, or had similar colony surfaces.

Sectioning was adopted immediately by ULRICH, whose major monographs of American Paleozoic Bryozoa (1882, 1890, 1893), together with NICHOLSON's monographs, established the necessity for deriving taxonomic characters from both external and internal morphology in Paleozoic Bryozoa. Likewise, NEKHOROSHEV and NIKIFOROVA began work in the early 1900's (ASTROVA in SARYCHEVA, 1960) and established, with the help of other workers, the oriented-section approach to Paleozoic Bryozoa in Russia.

Details of colony interiors, as seen in sections of Bryozoa of Paleozoic age, were especially effective in providing information on zooecia. The importance of zooecial characters was recognized immediately by the first taxonomists to make sections. As a result, taxonomic emphasis shifted from zoarial characters to zooecial characters. "Paleontologists, indeed, have now universally recognized that, in such difficult forms as the Monticuliporoids, the microscopic structure is the chief element in the determination of species; since surface characters may not be recognizable, or may vary greatly according to the state of preservation of the specimens, or other similar circumstances, while mere external form is a more treacherous and delusive guide" (NICHOLSON, 1881, p. v). And ". . . it cannot be questioned that differentiations in the cell or actual home of the polypide are more trustworthy structural variations than the form of the zoarium" (ULRICH, 1890, p. 326).

In an exchange of letters in *Science* in 1887 and 1888 between JAMES and FOERSTE, FOERSTE (1887, p. 225) presented philosophical arguments for the "new" study of internal characters of Paleozoic Bryozoa, which are as challenging today as they were then. "Theoretically development has proceeded in two lines,—one internal, to accommodate itself to the needs of internal function; and one external, to accommodate itself to environment, to the world with which the being comes in contact. Variations of function are far less frequent than those of environment: hence internal structure may still be very similar when external features have already extensively varied. Hence internal structure usually furnishes the reliable characters, which distinguish genera and higher

groups; external features are used for specific determination. . . . It remains to be seen what characters of specific importance cannot be shown in microscopic slides."

The earliest study of thin sections of skeletons was done at relatively low magnifications. ULRICH and BASSLER routinely used hand lenses instead of the microscopes that were available to them. Their observations were necessarily deficient in the description of small-scale characters and their biological interpretations were restricted. Nevertheless, their work and that of their contemporaries on Bryozoa of Paleozoic age was a major improvement because of the addition to the classification of many internal characters.

Another practice commonly employed in this early use of thin sections was based on the assumed correlation between external and internal characters. Often, free specimens from a stratigraphically and geographically restricted fauna were sorted into "species" groupings on external appearance. Only one to several fragments of colonies were actually sectioned from each of those groupings. Early descriptions emphasized internal characters observed from those few sections and were thought to be adequate to distinguish species. Subsequent sectioning of the unsectioned paratype suites of trepostomate species commonly reveals several taxa at the genus and species level because of the prevalence of external homeomorphy. Also, ranges of transitional morphologic variation within species commonly appear greater than first supposed.

An unfortunate result of the thin-sectioning technique itself is the still common custom of describing character states as seen in the two dimensions of thin sections without conversion to their actual three-dimensional condition. Much confusion and misinformation have resulted, adding to the difficulty of biologic understanding and taxonomic application.

At the turn of the century, lack of knowledge of living species was a formidable handicap to biological interpretation of fossil stenolaemates of all ages. CUMINGS (1904,

1905, 1912) and Cumings and Galloway (1915), using standard microscopes of that time, worked out some ingenious approaches to biologic interpretation for Paleozoic stenolaemates, which can be applied inferentially. They unfortunately were not followed until the 1960's when their approaches furnished the foundation for many of the present-day refinements of biologic interpretation of stenolaemates of Paleozoic age.

Beginnings were made on the study of soft parts of stenolaemates in early papers, especially by Harmer (1896, 1898). Later, papers by Borg (e.g., 1926a, 1933, 1944) on recent tubuliporates developed much new information with evidence from enough taxa to indicate the general applicability of some basic features for the entire class. Borg, however, did little work on the skeletons overall (see Borg, 1933, for an exception) and their more detailed relationships to corresponding soft parts. Unfortunately, these excellent beginnings to the study of soft parts of modern stenolaemates have not been continued by zoologists.

A large gap exists between the philosophies and procedures employed in most existing taxonomy of stenolaemates and those philosophies and procedures that have been available beginning in the early 1940's. The selection and treatment of stenolaemate characters at higher taxonomic levels have been based on a minimum of biologic interpretation and are largely arbitrary. Many structures and their characters, both external and internal in colonies, are those most readily observed and described. The taxonomic value of a newly recognized structure or character is commonly judged in proportion to its visual prominence, without inquiring into its possible mode of growth, functional significance, degree of inferred genetic control, or possible occurrence in known taxa that might be related.

Even with use of too few fragments of colonies and too few characters at the species level, it is possible to differentiate some species within local faunas of living stenolaemates or local fossil faunas through restricted time intervals. Many taxonomists have necessarily concentrated on relatively local faunas, and relatively few characters and character states have seemed adequate. Each species recognized, however, should be distinct from all others of the world through time. This seems an overwhelming and perhaps impossible goal that can only be approximated, with each generation hopefully adding improvements.

APPROACH TO TAXONOMIC CHARACTERS

All modern methods for constructing phylogenetically based classifications begin with as many independent taxonomic characters as possible. Although these taxonomic characters should be largely genetically controlled, in practice, they are derived from morphologic structures that initially must be assumed to reflect varying proportions of genetic and environmental control. Unfortunately, estimates of degrees of genetic and environmental control expressed by taxonomic character states are among the most difficult interpretations to make in taxonomy.

Such estimates in modern stenolaemates rely upon some understanding of the biology of the entire colony, including its mode of growth, detailed morphology, astogeny, ontogeny, polymorphism, functional morphology, and environmental modifications. The most convincing estimates are arrived at through study and experimentation with living colonies in their natural habitats. Relatively little is known about the basic biology of living stenolaemates, and that little has yet to be applied to classifications to improve their phylogenetic content. Extrapolations of comparable biologic and taxonomic approximations backward into geologic time require study of as much of the fossilized skeleton of the colony as possible.

COMPARATIVE STUDY OF RECENT AND FOSSIL STENOLAEMATES

The major approach to biologic interpretation of extinct stenolaemates is basically uniformitarian morphologic comparison with living species. The assumptions of the uniformitarian approach used here are listed below.

1. Comparable morphology in fossil and living taxa is assumed to indicate similarity in function and mode of growth. Conversely, different morphologies are generally assumed to indicate modified or different functions. In general, the older the fossil taxa being compared with living species and the greater the morphologic differences, the less assured is the correctness of the biologic interpretation.

2. A few similar functions can be carried on by different morphologies in living colonies, and restricted numbers of these functions can be inferred for fossil taxa. For example, **excurrent chimneys** are localized currents, created by colonies, which carry water and rejected particles away during the feeding process (Fig. 25). They can be set in motion by a number of different morphologies on colony surfaces (see below).

3. Differences in morphology of hypothetical soft parts of fossil taxa should be expected at least to the degree that they occur in comparable living taxa. For example, the general morphology of feeding organs of an exceptionally preserved fossil specimen should not be assumed for its entire family or order if corresponding organs are of several kinds in living species within families or orders (see below).

4. Modes of growth and functions unknown in living species can be expected to have occurred in extinct taxa and can be usefully suggested if the fossil evidence is convincing. Many biologic interpretations unknown in living forms, however, will be necessarily speculative in fossil taxa in proportion to degree of departure from living analogues.

The correctness of many biological interpretations of fossil taxa based on morphologic comparison with recent taxa seems unknowable. These interpretations, therefore, must remain open to question and can change as additional evidence is obtained.

PREPARATION TECHNIQUES

Published preparation techniques make it possible to describe interiors of bryozoan colonies with as much accuracy and detail as exteriors. Three-dimensional relationships and microstructural details of both skeletons and preserved tissues and organs in living position can be determined with certainty.

The time and effort to prepare standard thin sections of skeletons has been cut in half by the use of slides of standard glass-slide thickness made entirely of cellulose acetate (Boardman & Utgaard, 1964). Ground surfaces of specimens are oriented, given a high polish, etched lightly with formic acid, dried thoroughly, flooded with acetone, and placed gently on a blank slide. The impression that is left is a replica that is suitable for qualitative and quantitative studies, records of serial sections, light photography if thin sections cannot be made, identification of small fragments as in well cuttings (Merida & Boardman, 1967), and scanning-electron microscopy (F. M. Bayer, pers. commun.).

Epoxy resins have greatly improved the quality of thin sections (Nye, Dean, & Hinds, 1972). The resins permit a tighter bonding between highly polished specimens and glass slides. More importantly, thin solutions of the resins can impregnate preserved specimens in a vacuum so that hard and soft parts can be sectioned together in living positions in stenolaemates (see Fig. 39, 40, 43–45; Boardman, 1971, 1973, 1975; Boardman & Cheetham, 1973; Boardman & McKinney, 1976; Harmelin, 1976).

The quickest method to determine most three-dimensional relationships within a colony is to use thicker sections with a stereoscopic microscope and transmitted light (e.g., Boardman & Cheetham, 1969, pl. 29, fig. 1). This is especially useful for seeing zooidal patterns or studying structures parallel to zooidal length in **longitudinal sections** where it is difficult to determine if a structure actually ends or merely passes out of the plane of the section. Another useful method for three-dimensional observation retains the chambers and removes the skeletons so that the general arrangements of colony interiors can be observed through the voids that were formerly walls (HILLMER, 1968).

Electron microscopy provides more sensitive and detailed information than can be obtained from light microscopes, especially for investigating modes of skeletal and soft part growth (SANDBERG, this revision; BROOD, 1972; TAVENER-SMITH & WILLIAMS, 1972).

MAJOR PARTS OF COLONIES

ASTOGENETIC ZONES

Stenolaemate colonies can be divided into at least two parts (Fig. 27) based on overall colony development (astogeny). The first or founding part of a colony includes the ancestrula (Fig. 25, 26) and one or more generations of asexually produced founding zooids. The morphology of each generation of founding zooids differs to some extent from the last, and so the first part of a colony is the primary zone of astogenetic change (BOARDMAN, 1968; BOARDMAN, CHEETHAM, & COOK, 1970).

The second part of a colony is attained by the generation that first repeats the morphology of the zooids of the preceding generation. Generations in the second part display morphologically comparable zooids of one or more kinds, which appear in one or more patterns capable of endless repetition. This second part is the primary zone of astogenetic repetition and constitutes the larger part of most stenolaemate colonies.

In most stenolaemates the founding zooids of the zone of change are covered by subsequent generations of zooids (see Fig. 53). The morphology and patterns of founding zooids, therefore, are relatively difficult to determine. Detailed studies of stenolaemates with covered zones of change (e.g., CUMINGS, 1904, 1905, 1912; BORG, 1933, text-fig. 28; 1941; BOARDMAN, & MCKINNEY, 1976; MCKINNEY, 1977c) are few, and taxonomic characters from zones of change generally are not included in classifications. Most of the morphology discussed here is from zones of astogenetic repetition.

ZOOIDS AND MULTIZOOIDAL AND EXTRAZOOIDAL PARTS

Stenolaemate colonies are made up of zooids and multizooidal parts, and many have extrazooidal parts. Zooids within a colony are of two or more kinds, the sexually produced ancestrula, asexually produced feeding zooids, and in many taxa, asexually produced polymorphs.

Minimally, zooids include body walls that enclose body cavities (BOARDMAN & CHEETHAM, 1973, p. 124). In recent colonies, feeding zooids have, in addition to body walls and body cavities, a protrusible lophophore, an alimentary canal, a membranous sac surrounding the alimentary canal and lophophore in retracted position, muscles to move the lophophore in and out, a nervous system, and, apparently, funicular strands (Fig. 2). In the zone of change, the founding zooids include feeding zooids that show some morphologic change from generation to generation. In a zone of repetition, feeding zooids generally have the same morphology at comparable ontogenetic stages, unless disturbed by microenvironmental differences.

Polymorphs are zooids that differ distinctly in morphology and function from ordinary feeding zooids at the same stage of ontogeny and in the same generation within a colony. Polymorphs may or may not be feeding zooids and can occur both in zones of change and zones of repetition.

In fossil stenolaemates, skeletons (zooecia) of feeding zooids can be identified with reasonable accuracy. Within the order Tubuliporata, zooecia of most living species are

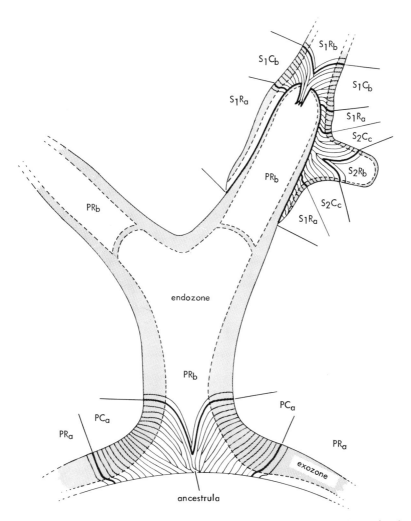

FIG. 27.   Major parts of colonies. Idealized diagram of a hypothetical stenolaemate colony in longitudinal section illustrating the concepts of ontogeny and astogeny. Zooids are drawn in critical regions only. ASTOGENY. The primary zone of astogenetic change, PC_a, includes the ancestrula and succeeding generations of zooids of progressively changing form, which give rise concurrently to two primary zones of repetition: the encrusting growth habit at the base of the colony, PR_a, and the erect growth habit, PR_b. Survival of few zooids in a localized region suffering microenvironmental interruption can give rise to a subsequent zone of change, S_1C_b, which is produced asexually and lacks an ancestrula. The subsequent zone of change in erect colonies commonly produces two subsequent zones of repetition, one encrusting to form an intracolony overgrowth, S_1R_a, and one erect to continue extension of the branch, S_1R_b. A second type of subsequent zone of change, S_2C_c, can develop asexually within an encrusting overgrowth to provide transition to zone of repetition of another branch, S_2R_b. ONTOGENY. Progressively older ontogenetic stages are generally expressed by increasing lengths of zooids. Operationally, ontogenetic stages within a colony are proportional to widths of exozones, the outer regions shown in gray. Widths of exozones generally decrease progressively from the oldest zooids of the colony in the primary zone of change, PC_a, and are in approximate proportion to growth time. The exozone under intracolony overgrowth is narrower than the uninterrupted exozone of the left branch. The narrow exozones crossing endozones of the two branches depict abandoned growing tips, typical of most trepostomates.

Bryozoa

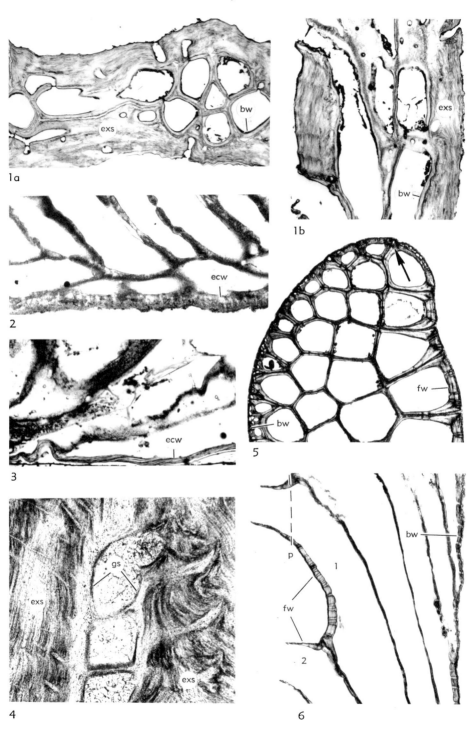

Fig. 28. (For explanation, see facing page.)

comparable to those of fossil species, and there is little doubt as to which kind of zooecium in fossil species contained the feeding organs.

In Paleozoic species, zooecia of feeding zooids of many taxa are not directly comparable morphologically to those of living species. In Paleozoic species with monomorphic zooecia in zones of repetition, some of those zooids must have contained feeding organs for at least a part of their ontogeny. Operationally, all of the zooecia in monomorphic colonies are considered to have been skeletons of feeding zooids. In Paleozoic colonies containing two or more kinds of zooecia, the commonly occurring kind that compares most closely with the zooecia of related monomorphic forms is considered to have contained feeding organs. Further, living chambers of assumed feeding zooids of Paleozoic species are comparable in number, diameter, and in length to living chambers of feeding zooids of living species. In living species, the most common kind of larger zooid contains the feeding organs. Generally, non-feeding polymorphs are smaller than feeding zooids. The assumption is made that the same was generally true for Paleozoic species, although many taxa have polymorphs in maculae (regularly spaced clusters of polymorphs, see Fig. 59) which are larger than zooecia of assumed feeding zooids between maculae.

The second kind of structural part of stenolaemate colonies is the multizooidal structure, which is grown outside of zooidal boundaries, can be colony-wide in extent, and eventually becomes part of a zooid or zooids. The most common multizooidal structures in stenolaemates are **confluent budding zones** of clustered buds and the **encrusting colony walls** from which zooids bud (Fig. 25, 26).

The third kind of structure is the extrazooidal part, which is also grown outside of zooidal boundaries but remains outside of zooidal boundaries throughout the life of a colony. Extrazooidal parts are generally larger than single zooids, and occur in many stenolaemate taxa, commonly providing at least structural support.

MORPHOLOGY AND FUNCTION OF ZOOIDS

Zooids contain complexes of both skeletal and soft parts. Differentiation of parts of skeletons is attempted here so that a set of independent taxonomic characters can be obtained from each part.

BASAL ZOOIDAL WALLS

Basal zooidal walls are body walls at inner ends of zooids opposite orificial walls (Fig. 25). They occur in most colonies as parts of

---

FIG. 28. Stenolaemate morphology.——*1a,b. Hornera* sp., rec., Flinders Is., Vict., Australia; erect, unilaminate, fenestrate colony with basal zooidal walls (bw) covered by laminated extrazooidal skeleton (exs) on reverse sides of branches; *a,b,* transv., long. secs. of same specimen, USNM 250057, ×100. ——*2. Lichenopora* sp., rec., Medit. Sea, Oran, Alg.; granular microstructure in both encrusting colony wall (ecw) and vertical zooecial walls; long. sec., USNM 250058, ×100.——*3. Lichenopora* sp., rec., Galapagos Is.; laminae in encrusting colony wall (ecw) dip proximally toward ancestrula to left, requiring simultaneous edgewise growth; long. sec., BMNH specimen, ×150.——*4. Archimedes* sp., Miss. (Chester.), near W. Lighton, Ala.; laminated extrazooecial skeleton (exs) surrounding granular zooecial skeleton (gs); long.-transv. sec., USNM 182789, ×100.——*5. Idmonea californica* d'Orbigny, rec., Pac. O. at La Jolla, Cal.; erect unilaminate zoarium with exterior basal zooidal walls (bw) and exterior frontal zooidal walls (fw); arrow at junction of basal zooidal walls to left and frontal walls to right, transv. sec., USNM 186545, ×50.——*6. I. californica,* same data as 5; indicated frontal zooidal walls (fw) belong to zooecia 1 and 2, peristome (p) to left of dashed vertical line, long. sec., USNM 186546, ×50.

encrusting multizooidal colony walls. In erect parts of colonies, basal zooidal walls (erect basal walls of Fig. 26) may originate from multizooidal colony walls or walls of older zooids. The part of a multizooidal or zooidal wall subsequently enclosed by a developing zooid forms the basal wall of that zooid.

Encrusting colony walls originate as single structures grown by the colony generally distal to developing buds at growing colony margins (see Fig. 39,5; 60,1–4). Encrusting colony walls become multizooidal as they are divided into basal zooidal walls by zooids spreading outward as colonies develop encrusting growth habits or basal attachments. Encrusting walls are simple exterior walls that occur in most taxa. These walls extend body cavities of colonies (exterior walls) and are consequently calcified on edges and inner surfaces only (**simple skeletal walls**). They consist of an outermost cuticle, skeletal layers, and epidermis.

Most encrusting colony walls have a laminated structure in skeletal layers in which the laminae dip proximally back toward the ancestrula (Fig. 25; 26; 28,3). This direction of dip requires that all of the laminae at the growing edges are calcified simultaneously by **edgewise growth** (addition of calcite on edges of crystals and individual laminae; Fig. 29,2,3; Boardman & Towe, 1966; Boardman & Cheetham, 1969, pl. 28). A few taxa show different microstructures in encrusting colony walls (Fig. 28,2) and such differences have taxonomic value. **Pseudopores** (Fig. 33) typical of calcified layers of other exterior walls have not been found in encrusting colony walls.

Electron microscopy has revealed that in some species, at least, the calcified part of the encrusting wall has two microstructural layers (Tavener-Smith, 1969b; Tavener-Smith & Williams, 1972).

Erect basal zooidal walls can be simple and exterior in some unilaminate colonies (Fig. 26) and compound and interior in others. **Compound skeletal walls** are calcified on edges and both sides simultaneously (Fig. 29,1–3), and so they are necessarily interior

body walls that partition preexisting body cavity.

Unilaminate colonies with erect exterior basal zooidal walls (Fig. 26) are apparently restricted to post-Paleozoic taxa. These erect walls (Fig. 28,5,6) apparently are multizooidal in origin. Many of these taxa form unilaminate colonies in early stages of ontogeny near growing tips and subsequently develop overgrowing layers of polymorphs on reverse sides proximally towards colony bases (Hinds, 1975).

Unilaminate colonies with compound interior basal zooidal walls in erect parts occur in both Paleozoic and post-Paleozoic taxa. Most of these basal walls appear to be zooidal rather than multizooidal in origin. The Paleozoic order Fenestrata is partly characterized by zooidal walls of nonlaminated skeleton (Fig. 28,4). Laminated skeleton covers nonlaminated zooecia, is generally continuous over at least the reverse sides of fenestrate fronds, and is extrazooidal. Recent hornerids have the same relationship of erect interior basal zooidal walls (Fig. 30,2), which on reverse sides of colonies are covered by an outer layer of extrazooidal skeleton proximally in later growth stages (Fig. 28,1a,b).

In bifoliate colonies erect basal zooidal walls originate as multizooidal compound interior walls that extend bladelike through the centers of colonies beyond zooidal boundaries distally (Fig. 30,3,4). These **median walls** provide budding surfaces for vertical zooidal walls on both sides so that feeding zooids are back to back to form colonies of generally flattened branches or expansions of different shapes (see Karklins and Utgaard, this revision).

Evidence indicating that median walls of bifoliate colonies are interior walls includes intermittent development of median walls with interior vertical zooidal walls within colonies (Fig. 30,1) and apparent lack of connections between exterior cuticle and median walls. Connections between exterior cuticle and median walls seem unlikely because of gaps between exterior encrusting walls and proximal ends of median walls in some gen-

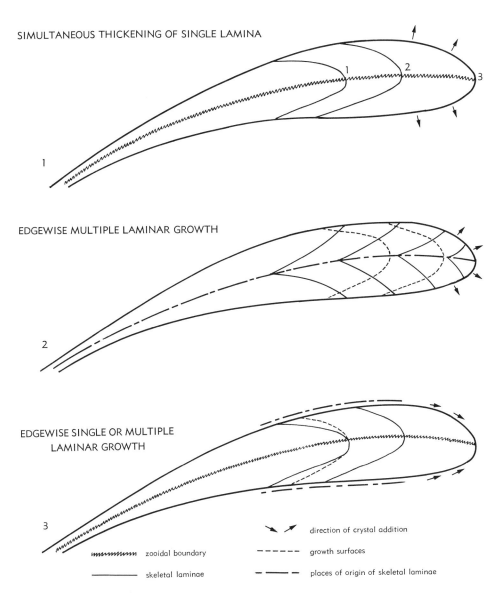

SIMULTANEOUS THICKENING OF SINGLE LAMINA

EDGEWISE MULTIPLE LAMINAR GROWTH

EDGEWISE SINGLE OR MULTIPLE LAMINAR GROWTH

direction of crystal addition

zooidal boundary

growth surfaces

skeletal laminae

places of origin of skeletal laminae

FIG. 29. Stenolaemate morphology. Diagrams of single vertical skeletal walls of adjacent zooids in longitudinal section illustrating hypothetical patterns of calcification.——*1.* Compound wall with laminae arched convexly in direction of growth to right in figure. Laminae grow singly on outermost skeletal surface adjacent to depositing epidermis so that laminae 1 and 2 are parts of a continuous series of laminae that reflect growth surfaces of earlier ontogenetic stages. Places of origin of skeletal laminae and growth surfaces are identical.——*2.* Compound wall with laminae pointing opposite to direction of growth. Walls extend in length by growth of laminae simultaneously on outer edges. Laminae are at high angles to depositing epidermis on skeletal surface and so are not growth surfaces. Places of origin of skeletal laminae are at inner ends of laminae near zooidal boundaries.——*3.* Compound wall with laminae arched convexly in direction of growth. Laminae theoretically can grow singly or several simultaneously by growth of laminae on outer edges. Growth surface (dashed line) is not quite parallel to laminae if laminar growth is multiple. Places of origin of skeletal laminae are at inner ends of laminae at skeletal surfaces.

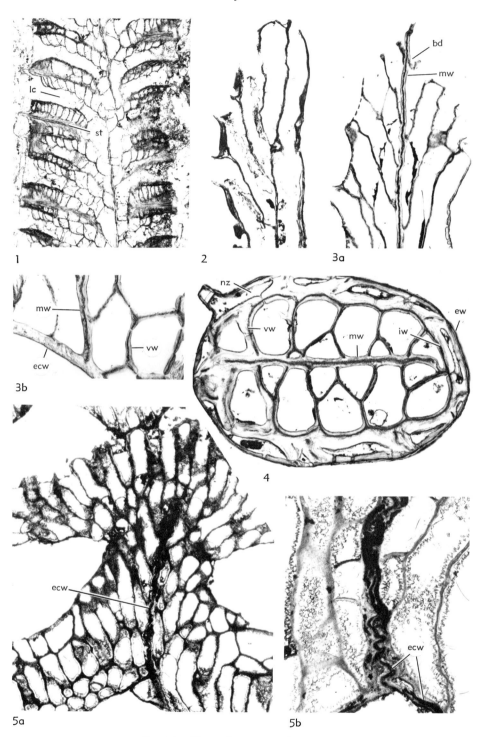

Fig. 30. *(For explanation, see facing page.)*

era (KARKLINS, this revision), skeletal layers of encrusting walls between basal cuticle and median wall at colony bases in other genera (Fig. 30,*3b*), and intervening skeletal layers of exterior zooidal walls at colony margins (Fig. 30,*4*).

The Oligocene-Pliocene genus *Alveolaria* forms subspherical colonies consisting of an open network of thin encrusting layers and cone-shaped expansions (Fig. 30,*5a*). Abutting encrusting layers bend and project vertically for short invervals in a sinuous, back-to-back contact simulating median walls (Fig. 30,*5b*). The colonies, therefore, appear externally to be bifoliate, but their irregular internal structure with the ever-present exterior basal colony wall suggests that they would be more accurately described as a complex of encrusting and cone-shaped growth habits.

VERTICAL ZOOIDAL WALLS

Vertical zooidal walls are body walls that grow parallel to the long axes of zooids to form either elongate conical or tubular body cavities or shorter sac-shaped body cavities. Thus, they provide the depth and length to zooid living chambers (Fig. 25, 26). Vertical walls have an epidermis but apparently no peritoneum (NIELSEN, 1971). The skeletal layers of vertical walls are continuous except for small skeletal gaps and **communication pores** in one Paleozoic suborder (see UTGAARD, this revision), and communication pores in most post-Paleozoic species (Fig. 25, 26).

Most vertical walls are interior, that is, they partition existing body cavity of colonies. Vertical walls that are exterior, or a combination of exterior and interior, occur only in the few uniserial and multiserial encrusting species (Fig. 31,*1–4*). Vertical walls generally bud from encrusting or erect multizooidal colony walls or from vertical walls of existing zooids. Vertical walls bud from extrazooidal structures in a few cystoporates (see UTGAARD, this revision) and from exterior walls of peristomes (Fig. 26) in a few tubuliporates (HARMELIN, 1976, fig. 7).

Zooids can commonly be divided into: (1) inner parts (endozones) characterized by growth directions at low angles to that of the colony or to the colony surface, thin vertical walls, and relative scarcity of intrazooidal skeletal structures; and (2) outer parts (exozones) characterized by growth directions at high angles to that of the colony or to the colony surface, thicker vertical walls, and concentrations of intrazooidal skeletal structures (Fig. 25).

Vertical zooidal walls are contiguous in most taxa and microstructure of the combined skeletal layers of two contiguous zooids in sections displays bilateral symmetry (Fig. 31,*5–7a*; 32,*1–4*). Exceptions include acanthocladiid fenestratids (GAUTIER, 1972, 1973) and the development of lunaria in cystoporates (UTGAARD, this revision). Bilateral symmetry is interpreted to mean that the walls are compound, that is, that they were grown cooperatively on edges and both sides from chambers of adjacent zooids. Zooidal bound-

---

FIG. 30. Stenolaemate morphology.——*1. Peronopora decipiens* (ROMINGER), lectotype, Corryville Mbr., McMillan F., Ord. (Cincinnat.), Cincinnati, Ohio; cystiphragms are the overlapping curved partitions in series above each living chamber (lc) and styles (st) project beyond zoarial surface into overgrowth to left; long. sec., UMMP 6676-3, ×30.——*2. Hornera* sp., rec., Arctic O.; growing tip of unilaminate, free-walled colony; long. sec., BMNH, Blacken Coll. 2.6, ×50.——*3a,b. Plagioecia* sp., rec., Pac. O. at La Jolla, Cal.; *a*, growing tip of bifoliate colony with median wall (mw) and developing buds (bd), long. sec., USNM 250059, ×50; *b*, junctions between encrusting colony wall (ecw), erect median wall (mw), and vertical walls (vw) showing microstructure, long. sec. same colony, ×100.——*4. Diplosolen intricaria* (SMITT), rec., depth of 200–240 m, Barent Sea, 60 mi. N. of North Cape; bifoliate colony showing interior walls (iw) between end of median wall (mw) and outermost exterior wall (ew), microstructure of vertical walls (vw), and nanozooids (nz) around margin of colony; transv. sec., BMNH specimen, ×100. —— *5a, b. Alveolaria semiovata* BUSK, Plio., Broom Hill, Suffolk, Eng.; exterior encrusting colony walls (ecw) in sinuous, back to back contact; *a,b,* long. secs., USNM 250060, ×30, ×100.

aries, therefore, are necessarily within compound vertical walls between adjacent zooidal body cavities and extend generally along centers of bilateral symmetry.

Zooidal boundaries or boundary zones of vertical walls (Fig. 25; 26; 29,*1,3*) are indicated microstructurally by abutting laminae from contiguous vertical walls (Fig. 31,*5*); thin, organic-rich partitions (Fig. 32,*1–4*); or thicker zones of granular appearing admixtures of organic material and small calcite crystals (Fig. 30,*3b,4*; 33,*1*). Some contiguous vertical walls, however, may have undifferentiated laminate or granular microstructure extending across centers of bilateral symmetry so that zooidal boundaries are not indicated microstructurally (Fig. 31,*6,7*) and must be located arbitrarily. Boundary zones of vertical walls lack the longitudinal canals or spaces that Ross (1976, p. 353) suggested ". . .provide the framework for growth and resorption of the body wall. . . ."

Organic-rich partitions occur at zooidal boundaries of interior vertical walls in a number of modern and fossil species of stenolaemates (e.g., Fig. 32,*1–4*; 33,*3*; 34; 42,*5,6*; 44,*4a,b*). In thin sections under a light microscope the partitions appear to be noncellular organic membranes or cuticles. The partitions have been recognized by Harmelin (1974; 1976, pl. 16, fig. 4, 8) as surfaces of discontinuity. Otherwise, organic partitions,

membranes, or cuticles have not been reported in interior vertical walls of stenolaemates (e.g., Borg, 1926a, p. 192; Brood, 1972, p. 28; Boardman & Cheetham, 1973, p. 138).

Organic-rich partitions in vertical walls of stenolaemates are parts of interior walls formed within body cavities of colonies and are considered to be interior in origin, as contrasted with exterior cuticles, which are the outermost layers of exterior walls and are adjacent to the environment. Extensive investigation using electron microscopy is necessary to determine the exact nature of the interior partitions.

FRONTAL ZOOIDAL WALLS

One of the major evolutionary advances of many post-Paleozoic tubuliporates is the more extensive skeletal reinforcement of exterior walls. Such reinforcement provides structural advantages and makes possible a greater variety of colony growth habits in post-Paleozoic species. Calcified layers of exterior walls are attached to inner surfaces of parts of colony-wide exterior cuticles. These structurally reinforced exterior walls form basal zooidal walls at inner ends of zooids on reverse sides of some erect unilaminate species, and frontal zooidal walls at outer ends of zooids (Fig. 26) of many species of several different growth habits.

The calcified layers of frontal walls are

Fig. 31.   Stenolaemate morphology.——*1,2. Corynotrypa inflata* (Hall), Bellevue Mbr., McMillan F., Ord. (Cincinnat.), Cincinnati, Ohio; *1,* uniserial zoarium showing connecting pore between zooecia, long. sec., USNM 186554, ×100; *2,* exterior vertical wall (evw), arrows indicate connecting pores to younger zooecia at bifurcation of uniserial zoarium, sec. parallels base of zoarium, USNM 186553, ×100.—— *3,4.* Stomatoporid tubuliporates; *3,* Bellevue Mbr., McMillan F., Ord. (Cincinnat.), Cincinnati, Ohio; encrusting, single-layered zoarium with exterior (evw) and interior (ivw) vertical walls, small circles are pseudopores in frontal walls, external apertural view, USNM 186556, ×100; *4,* Waynesville F., Ord. (Cincinnat.), Oregonia, Ohio; arrow points to zooecial boundary between interior vertical walls; sec. parallel to zoarial base, USNM 186558, ×100.——*5. Amplexopora septosa* (Ulrich), Mount Hope Sh. Mbr., Ord. (Cincinnat.), Covington, Ky.; compound vertical walls showing bilateral symmetry about zooecial boundary (zb) of abutting laminae, living chamber (lc) intact, protected by overgrowth (ov) and floored by basal diaphragm (bd); long. sec., USNM 138287, ×100.——*6. Rhombotrypella* sp., 9 m above Torpedo Ss. Mbr. of Ochelata F., Penn., Washington Co., Okla.; zooecial boundaries not indicated microstructurally in vertical walls; long. sec., USNM 204859, ×50.——*7a,b. Siphodictyum irregularis* Canu & Bassler, Cret. (Apt.), Faringdon, Eng.; zooecial boundaries not indicated microstructurally in vertical walls, smaller zooecia polymorphs; *a,b,* long., tang. secs., USNM 248243, ×100.

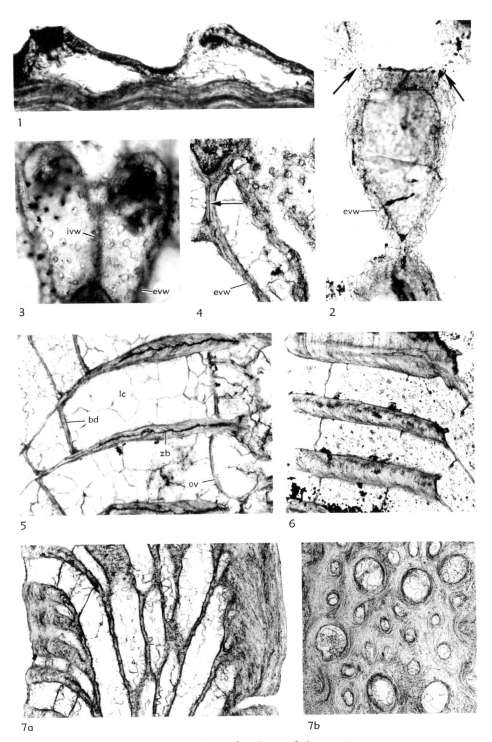

FIG. 31. *(For explanation, see facing page.)*

structurally continuous with or attached to outer ends of one or more calcified layers of the supporting vertical walls of single zooids. Like vertical walls, frontal walls reportedly have an epidermis and no peritoneum. Outer ends of frontal walls are zooecial apertures. Most commonly, calcified layers of frontal zooidal walls are restricted as structural units to single zooids (Fig. 28,5,6), and their calcification is considered to be largely zooidally controlled. Frontal walls occur in the few Paleozoic tubuliporates (BOARDMAN & CHEETHAM, 1973; BROOD, 1973, 1975b) and in many tubuliporates of post-Paleozoic age.

In zooids of recent colonies, frontal walls grow and calcify after the termination of growth of supporting vertical walls. Reportedly, the epidermis of the vertical walls joins with that of the exterior membranous walls (BORG, 1926a, p. 322) at proximal ends of budding zones. After that contact, the epidermis produces zooidal skeletal layers on inner sides of the exterior cuticle to form frontal zooidal walls. Calcification of frontal walls takes place on edges and inner sides only (edgewise growth of simple walls; Fig. 34,1d), and so their skeletal microstructure lacks bilateral symmetry (Fig. 32,1; 35,4).

In a few taxa, calcified layers of frontal zooidal walls form continuous units extending across all of the feeding zooids of colonies proximal to distal budding zones (Fig. 33,4,5). Terminated vertical walls abut inner calcified surfaces of the frontal walls. Calcification of these continuous frontal walls apparently takes place from individual living chambers after zooids have established their vertical walls. The colony-wide frontal walls,

therefore, are considered to have been zooidally controlled. Each zooid has an aperture in these frontal walls.

In a few forms, the juncture between vertical and frontal walls is of a type apparently transitional between connections indicating zooidal frontal walls and connections indicating colony-wide frontal walls. The vertical walls nearest zooidal boundaries apparently reached exterior cuticles, and some inner skeletal layers of the vertical walls abut layers of the frontal walls (Fig. 36,2,4a).

The microstructure of skeletal layers of frontal zooidal walls is commonly correlated with that of supporting vertical walls. If the vertical wall of a zooid has an outer calcified granular layer in the zooidal boundary zone and a laminated layer lining its zooecial chamber (the basic tubuliporidean wall of BROOD, 1972, p. 33; HINDS, 1975, p. 877), the vertical bilaterally symmetrical compound walls of adjacent zooids divide in half at the frontal zooidal walls (Fig. 32,2; 33,1). Each half extends into the calcified parts of the frontal walls of adjacent zooids so that the skeletal portion of the frontal zooidal wall includes an inner laminated skeletal layer and an outer granular skeletal layer. In some taxa with frontal walls the granular layer is replaced by a laminated skeletal layer with the laminae oriented at high angles to growth surfaces (Fig. 33,2,3).

A very different kind of frontal wall is formed in at least one species in which only the calcified zooecial linings of the vertical walls extend outward to form the peristomes that make up most of the frontal walls (Fig. 34,1a,c).

FIG. 32. Stenolaemate morphology.——*1. Diaperoecia indistincta* CANU & BASSLER, rec., 28–30 m, Medit. Sea off Riou Is., Marseille, France; short living chamber with constant retracted position during ontogeny, basal zooidal wall (bw), vertical wall (vw), frontal wall (fw), organic-rich partitions in both vertical walls (op) and hemisepta (oph); long. sec., USNM 250062, ×150.——*2. Idmonea californica* D'ORBIGNY, Pleist., Dead Man Is., San Pedro, Cal.; organic partition (op) in vertical wall (vw), frontal wall (fw); long. sec., USNM 250063, ×150.——*3,4. Cinctipora elegans* HUTTON, rec., 110 m, off Otago Heads, South Is. N.Z.; organic-rich partition (op) in vertical walls; *3,* transv. sec., USNM 250064, ×100, *4,* long. sec., USNM 250065, ×50.——*5a,b. Hornera* sp., rec., a fjord in East Finmark, Nor., 215 m; vertical walls (vw) with laminae convex outward to right (*5a*), extrazooidal skeleton (exs) between zooids; *a,b,* long., tang. secs. same colony, BMNH, Norman Coll., ×100.

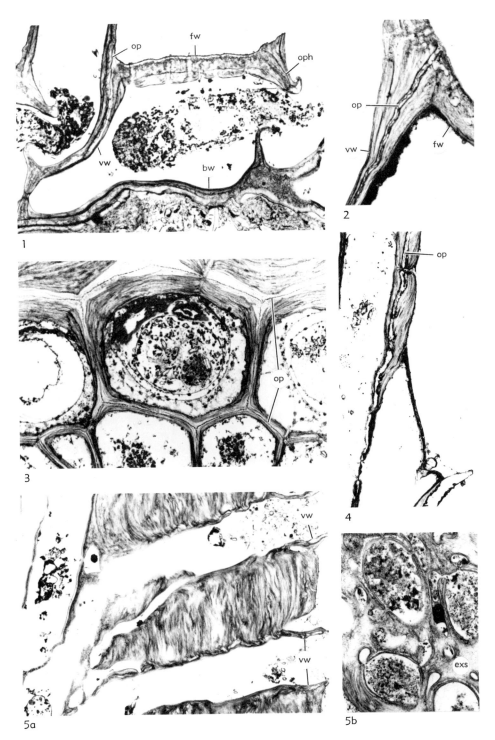

FIG. 32. *(For explanation, see facing page.)*

Organic-rich partitions at zooidal boundaries of vertical walls occur with most of the variations of frontal wall microstructure illustrated here (e.g., Fig. 32,*1,2*; 33,*3*; 34; 55,*4,5*). The interior partitions apparently attach to exterior cuticles at junctions of vertical and frontal walls.

Taxa may be arranged in a morphological series showing transitional differences in length of frontal walls restricted to single zooids. The shortest frontal walls are little more than terminal diaphragms containing apertures (Fig. 33,*4,5*). Longer frontal walls commonly occur with peristome extensions (Fig. 28,*6*; 33,*1*). The longest frontal walls of single zooids may extend virtually along the entire length of an erect colony (Fig. 35,*1,3a,b*). (Exterior walls on the right sides of Fig. 35,*3a,b* are frontal walls or terminal diaphragms because zooids budded from the center of the branch and grew in all directions so that the exterior walls on the right sides of the figures are attached to outer ends of vertical walls of zooids. In contrast, the exterior wall at the bottom of Fig. 35,*2* is an erect basal wall from which zooids budded).

The lengths of single frontal walls from their proximal margins to the bases of possible peristomes are largely determined by the angles that vertical walls of zooids make with surfaces of colonies. The shortest diaphragmlike frontal walls are formed by zooids whose vertical walls intersect the surface of a colony nearly at right angles. As the surface angles of vertical walls decrease, lengths of frontal walls increase because frontal walls are needed to complete outer sides of the calcified walls of living chambers.

Most frontal walls have pseudopores (Fig. 26) that penetrate all or parts of skeletal layers but not exterior cuticles (TAVENER-SMITH & WILLIAMS, 1972). Pseudopores can be few and scattered (Fig. 33,*1*) or more closely spaced than communication pores of supporting vertical walls (Fig. 35,*4*).

ZOOECIAL APERTURES AND ORIFICIAL WALLS

Zooecial apertures are generally simple terminal skeletal openings of zooecia that are oriented transverse to zooidal growth directions. They terminate frontal wall skeleton (Fig. 26), vertical wall skeleton in taxa in which frontal walls are absent (Fig. 25), or a combination of both. Zooids elongate through most of their ontogeny by growth at apertures.

Orificial walls (Fig. 25, 26) are membranous body walls that cover zooecial apertures, and they are therefore also generally transverse to zooidal growth direction. They are the outer terminal part of the complete zooid (Fig. 37), except in the few fossil taxa in which orificial walls were apparently covered by operculumlike structures. In living species orificial walls are single membranous walls that contain simple circular pores (the orifices) through which tentacles are protruded. Orificial walls are part of the exterior walls

FIG. 33. Stenolaemate morphology.——*1*. Idmoneid tubuliporate, rec., Kara Sea, USSR; vertical walls (vw) with thicker boundary zones of organic-rich granular calcite (arrows) continuing as outer skeletal layers of frontal walls (fw); long. sec., USNM 186552, ×100.——*2*. *Spiropora verticellata* (GOLDFUSS), Cret. (up. Maastricht.), Stevns Klint, Seeland, Denm.; vertical (vw) and frontal (fw) walls with two skeletal layers of laminae oriented at high angles to growth surfaces, small tubes cut transversely at center of branch are inner ends of zooecia at bud stage, indicating central grouping of buds at growing tips; transv. sec., USNM 250066, ×100.——*3*. Fixed-walled tubuliporate, rec., 280 m, 51°22.52′ S., 73°8.64′ E., Kerguelen Ridge, S. Indian O.; vertical wall showing organic-rich partition (op) at zooidal boundary and skeletal laminae oriented at high angles to growth surfaces, frontal wall (fw) with pseudopores (ps); long. sec., USNM 186551, ×150.——*4,5*. *Diplocava incondita* CANU & BASSLER, Cret. (Valangin.), Ste Croix, Switz.; *4*, showing frontal walls (fw) containing pseudopores apparently extending across ends of granular vertical walls (vw), restricted aperture (ap); long. sec., USNM 216475, ×100; *5*, showing restricted aperture (ap) formed by frontal wall (fw) with pseudopores; tang. sec., USNM 216476, ×100.

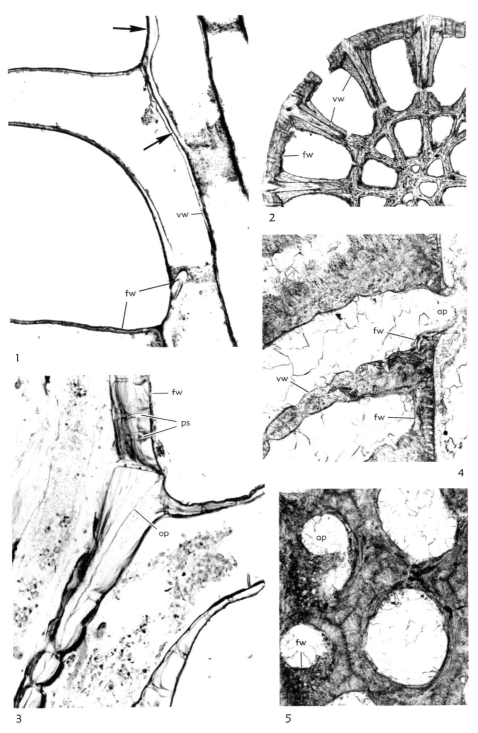

FIG. 33. *(For explanation, see facing page.)*

of feeding zooids and the cuticle of the orificial walls is part of the colony-wide exterior cuticle. Similar orificial walls are assumed for most fossil forms because of the uniformity of simple zooecial apertures, which had to be covered during life.

Borg (1926a, p. 483) considered orificial walls of tubuliporates (terminal walls of stenolaemates in Borg's terminology) to be homologous to the ". . .frontal side of the zooid in the Cheilostomata and Ctenostomata. . . ." Similarities in general position, extent, and mode of growth, however, suggest that until more is known about the evolution between orders and classes, frontal walls as defined above for the stenolaemates are more nearly analogous to frontal walls of cheilostomates and ctenostomates.

In a few Jurassic and Cretaceous tubuliporates (the melicerititids) apertures are closed by calcareous plates (Fig. 36,*2–4*) interpreted to have been **opercula** (Levinsen, 1912). Apertures of all feeding zooids except those in growing tips of branches are covered by the plates. The plates, therefore, must have been hinged to open when zooids were feeding (Fig. 38). The opercula were most likely hinged on their straight proximal margins to the stationary parts of frontal walls. In some colonies opercula have longitudinal ridges on inner sides (Fig. 36,*4b*), possibly for some kind of muscle attachment.

The opercula apparently were developed as parts of frontal walls because the outer sides of the opercula and stationary parts of the frontal walls are aligned. The opercula are exterior structures and some have what appear to be pseudopores, a common feature of frontal walls.

ORIFICIAL WALLS AND COLONY ORGANIZATION

In stenolaemates, the relationships of orificial walls to zooecial apertures of vertical and frontal walls produce two types of colony organization and an intermediate organization that combines the two.

Free-walled colonies.—Free-walled colonies (Fig. 25) are loosely covered by membranous exterior walls not attached at apertures of feeding zooids so that confluent outer body cavities (Borg, 1926a, p. 196) are produced. With minor exceptions, membranous exterior walls of a free-walled colony are attached to skeleton only at encrusting bases of colonies and within living chambers of zooids. The living chamber attachments are at attachment organs (Fig. 2). Orificial and vestibular walls are parts of the membranous exterior walls that extend into zooids to the attachment organs (Fig. 39,*1–3,5*).

In free-walled colonies skeletal walls are largely interior above encrusting colony walls; as they grow they partition colony-wide body cavities established by the advancing membranous exterior walls. Exceptions include calcified exterior walls of terminal diaphragms and brood chambers that interrupt confluent outer body cavities in some postPaleozoic tubuliporates.

Apparent advantages of the free-walled arrangement include colony-wide distribution of nutrients through confluent outer body cavities (Borg, 1926a, p. 204; Boardman & Cheetham, 1969, text-fig. 1; 1973, p. 132) and the possibility of growth of all outer skeletal surfaces throughout colony life. Parts of colonies suffering accident are commonly

Fig. 34. Stenolaemate morphology.——*1a–d. Heteropora? pacifica* Borg, rec., 21–25 m, vicinity of Middleton Is., S. Alaska; *a,* zooecial lining of vertical wall (vw) extended outward to form frontal wall (fw) and peristome (p), terminal diaphragm (td) with closely spaced pseudopores; *b,* frontal wall (fw) similar in microstructure to terminal diaphragm (td); *c,* frontal wall formed distally by extension of zooecial lining (zl) and proximally by combination of thicker wall (arrow) microstructurally comparable to terminal diaphram (td) and zooecial lining; *d,* external cuticle (c) and partly grown skeletal layers of terminal diaphragms (td); all long. secs. from same colony, USNM 186549, all ×100 except *a* ×150.

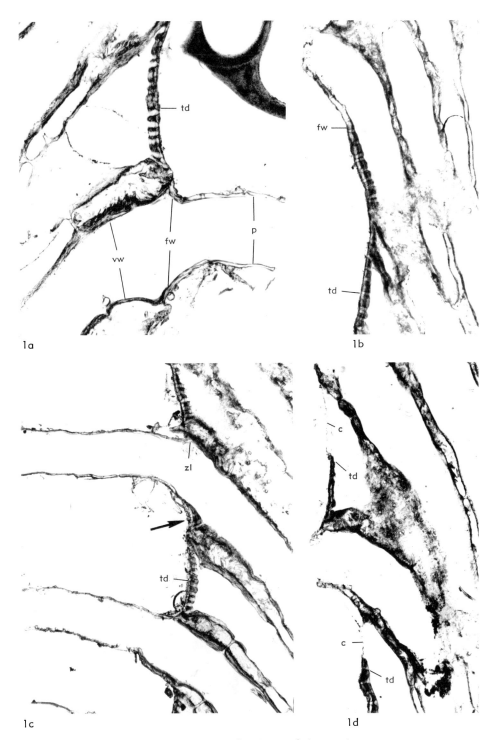

FIG. 34. *(For explanation, see facing page.)*

regenerated under the membranous covering by overgrowth originating from adjacent undamaged zooids.

This is the group of stenolaemates that BORG (1926a, p. 473, fig. 55; 1933, fig. 26) and subsequent authors have called double-walled. The overwhelming majority of Paleozoic Bryozoa and many post-Paleozoic taxa are free-walled stenolaemates.

Fixed-walled colonies.—Stenolaemate colonies are termed fixed-walled if orificial walls of feeding zooids are attached at apertures so that confluent outer body cavities between zooids are eliminated (Fig. 26). The great majority of fixed-walled stenolaemates has frontal walls. Skeletal layers of frontal walls are attached at outer ends of vertical zooidal walls and terminate at apertures. The outermost cuticles of fixed-walled colonies are attached to outer surfaces of the calcareous layers of frontal walls up to apertures, which eliminates outer confluent body cavities.

Communication among feeding zooids of fixed-walled colonies apparently can occur only through pores in vertical walls. Feeding zooids in species without communication pores therefore are presumably without physiologic connection after their zooecia are completed. A probable advantage is gained, however, by the skeletal reinforcement of exterior walls (BOARDMAN & CHEETHAM, 1973, p. 158, 159). The exterior walls on feeding sides of most fixed-walled colonies consist of membranous orificial walls, calcified frontal walls of contiguous zooids (some can be polymorphs) and in many colonies, outer brood-chamber walls.

Fixed-walled stenolaemates are a part of the group of Tubuliporata described by BORG as having simple walls, without clearly designating which walls were simple (frontal walls) (BORG, 1926a, fig. 1, p. 473). They are also part of the group called single-walled by several subsequent authors, or fused-walled (BOARDMAN, 1975, p. 598). The terms simple- or single-walled, and double-walled of BORG are not used here because one kind of colony does not have double the number of walls of the other, as implied by that terminology. The most significant biological differences between the two kinds of colonies seem to be the fusion or lack of fusion of interior and exterior walls (HINDS, 1975, p. 876) and the resulting effects on physiological communication by the free or fixed condition of orificial walls.

Fixed-walled tubuliporates are extremely rare in the Paleozoic (BOARDMAN & CHEETHAM, 1973, p. 159; BROOD, 1975b, p. 69) and common in post-Paleozoic bryozoan faunas.

Combined free- and fixed-walled colonies.—Some feeding zooids in colonies of a few post-Paleozoic tubuliporates have orificial walls that are partly free and partly fixed (BOARDMAN, 1975, p. 601). The combined free- and fixed-walled morphology occurs in colonies of taxa in which apertures of feeding zooids are clustered and the clusters are surrounded by the combined frontal walls of the outermost zooids in the clusters (Fig. 35,3a,b). In these colonies both frontal and vertical walls are generally long and are nearly parallel to colony surfaces. Clusters of apertures vary in cross-sectional shape from circular to irregular in different taxa and may

FIG. 35.   Frontal walls.——*1. Fasciculipora* sp., rec., McMurdo Sound, Antarctica; zooid with aperture (ap) consisting partly of exterior frontal wall (fw) and partly of interior vertical wall (vw); long. sec., USNM 179007, ×50.——*2. Idmonea californica* D'ORBIGNY, rec., Pac. O. off La Jolla, Cal.; erect zoarium with buds starting from exterior basal zooidal walls (bw); transv. sec., USNM 250067, ×50. ——*3a,b. Frondipora verrucosa* LAMOUROUX, rec., Naples Bay, Italy; budding of vertical walls (vw) from central region of branch outward in all directions so that all exterior walls are frontal (fw), clusters of combined free- and fixed-walled zooecia open to left; *a,b,* transv., long. sec. from same zoarium, USNM 250068, ×30.——*4.* Fixed-walled tubuliporate, rec., 285 m, 51°22.52′ S., 73°8.64′ E., Kerguelen Ridge, S. Indian O.; communication pores (cp) in interior vertical walls and pseudopores (ps) in exterior frontal walls; long. sec., USNM 186551, ×150.

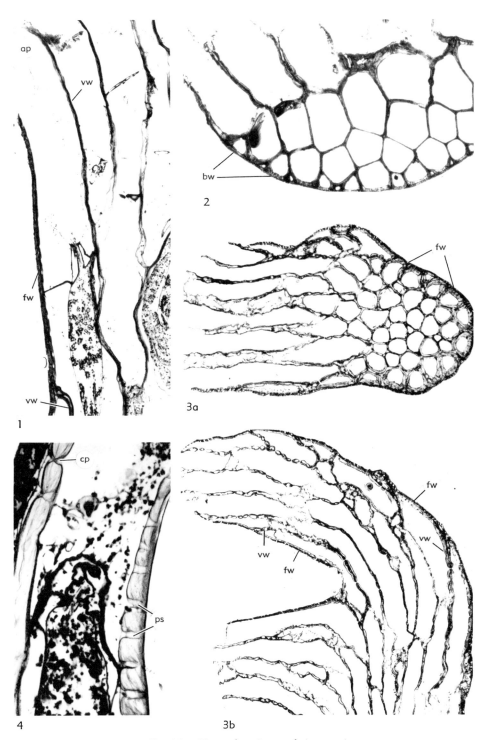

FIG. 35. *(For explanation, see facing page.)*

be formed by few to many contiguous feeding zooids.

The outermost feeding zooids of these clusters have apertures consisting of a combination of exterior frontal walls and interior vertical walls (Fig. 35,*1*). Their orificial walls are attached to the exterior frontal parts of apertures and apparently are free over the interior vertical-walled parts of apertures.

Only the outermost zooids of these clusters have both frontal and vertical walls. Zooids farther inside clusters have apertures consisting entirely of interior vertical walls and are free-walled. Zooids within clusters can presumably communicate with each other through the outer body cavity around the ends of their vertical walls, but clusters are prevented from communicating with other clusters in this manner by intervening exterior frontal walls. Communication among all zooids in these colonies apparently can occur, however, through pores in vertical walls.

A number of genera of fixed-walled tubuliporates develop colonies with apertures of contiguous zooids arranged singly in rows (see Fig. 61,*4a*). Frontal walls of the zooids within a row are extended into peristomes in unworn colonies of modern species. Zooecial apertures occur at outer ends of the exterior-walled and calcified peristomes so that outer body cavities do not occur among zooids in these linear clusters. Zooecia in some fossil zoaria with similar zooecial patterns lack peristomes (HINDS, 1975, p. 881). The peristomes may have been removed by wear or may not have developed. If peristomes were not developed, these fossil colonies could have had combined free and fixed walls with contiguous interior vertical walls forming the parts of apertures within the linear clusters.

SKELETAL STRUCTURES OF LIVING CHAMBERS

The enclosing skeletons of living chambers (Fig. 37) of feeding zooids and polymorphs in both fossil (BOARDMAN, 1971, p. 5) and modern stenolaemates can be the parts of colony skeletons that reveal most about the biology of colonies. **Living chambers** are the outermost parts of zooidal body cavities into which zooidal organs retract. Certainly, living chambers and their skeletons deserve description as an entity in standard taxonomic works. Unfortunately, no part of stenolaemate colonies of all ages has been more ignored historically.

In free-walled fossil taxa of Paleozoic age, many living chambers are floored by **basal diaphragms** and are most likely to be found intact behind overgrowths that protect outer ends of vertical walls from abrasion (Fig. 30,*1*; 31,*5*; 36,*1*; 37). Living chambers also can be recognized behind interior **terminal diaphragms** (see Fig. 43,*2,3*; also discussion of terminal structures).

In post-Paleozoic taxa, living chambers are generally recognizable because of the prevalence of exterior terminal diaphragms. Skeletal terminal diaphragms in post-Paleozoic species have a different microstructure than basal diaphragms and apparently terminate further zooidal growth (Fig. 34,*1a,c*).

Skeletal structures of living chambers can be divided into: basal structures, including zooidal walls or diaphragms that act as floors of living chambers and any structures that project from them; lateral structures, which

---

FIG. 36.   Stenolaemate morphology.——*1. Amplexopora pustulosa* ULRICH, Waynesville Sh., Ord. (Richmond.), Hanover, Ohio; complete living chambers (lc) protected by encrusting overgrowth, abandoned living chambers within zooecia (alc) capped by terminal diaphragms; long. sec., USNM 250069, ×50. ——*2,3. Meliceritites* sp., Cret. (Cenoman.), Le Mans, Sarthe, France; *2,* part of vertical wall (vw) abutting frontal wall (fw), funnel-shaped structure partly attached to operculum (o), long. sec., USNM 250070, ×100; *3,* frontal walls (fw) and opercula (o), both with pseudopores, tang. sec., USNM 216480, ×100. ——*4a,b. Meliceritites* sp., same data as *2; a,* parts of vertical walls (vw) abutting frontal walls (fw), opercula (o), long. sec.; *b,* transv. sec. showing spiral arrangement of zooecia around axial cylinder in center of branch, opercula (o) with ridges on inner sides; USNM 216481, ×100.

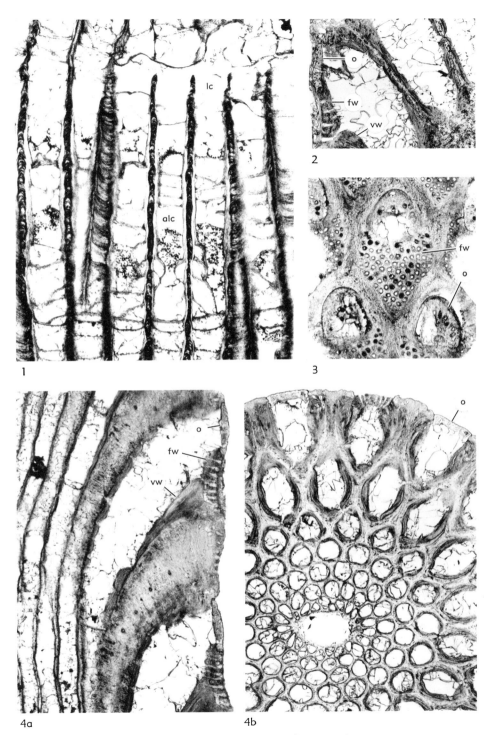

FIG. 36. *(For explanation, see facing page.)*

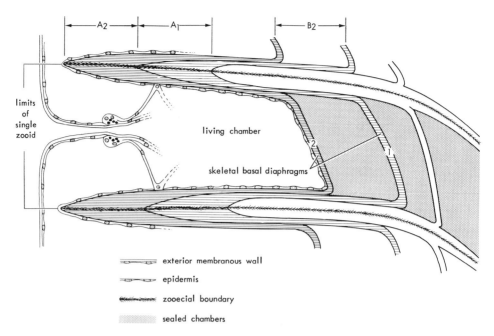

Fig. 37. Stenolaemate morphology. Diagram of a longitudinal section through the exozone of the hypothetical zooid of a free-walled Paleozoic trepostomate; growth direction is to the left, and laminae (not shown) from adjacent zooids point in that direction. The ontogenetically oldest part of skeleton is to the right; the youngest part, which both lines and extends living chamber, is to the left. The youngest basal diaphragm (2) forms the floor of the living chamber and it and older diaphragms seal off abandoned chambers that presumably lacked living tissue. A_1 and A_2 are the hypothetical extent of vertical wall growth during the degeneration part of the last two cycles. B_2 is the resulting displacement of the base of the latest living chamber (see text for further explanation). Zooid includes the terminal exterior membranous wall (orificial wall), body cavity, and skeleton (see brackets defining single zooid).

occupy positions opposite feeding organs as they move in and out, including structures that project from vertical or frontal zooidal walls; and terminal or subterminal diaphragms, which seal living chambers from the environment.

Basal structures and ontogeny.—In those taxa having relatively short zooidal chambers, retracted positions of feeding organs are constant throughout colony life. Inner ends of these shorter chambers can be made up of vertical walls, or combinations of vertical and basal walls (Fig. 32,*1*). Any elongation of short zooids occurs in outermost membranous vestibular walls and at outer ends of enclosing vertical or frontal walls. Brown bodies, which are encapsulated degenerated cells resulting from the cyclic degeneration of most of the organs of zooids (Fig. 40,*3b*),

presumably would be disposed of regularly for lack of storage space.

In stenolaemates with longer zooidal chambers, in contrast, the living chambers and retracted positions of organs advance with skeletal elongation (Fig. 40,*3a,b*), presumably by means of degeneration-regeneration saltations. Outward ontogenetic growth of zooids is enough for advancing organs to vacate inner parts of zooidal chambers.

In many free-walled fossil taxa, vacated regions of zooidal chambers are partioned by transverse basal diaphragms in ontogenetic series (Fig. 31,*5*; 36,*1*; 37). Diaphragms are membranous or skeletal partitions that extend across entire zooidal chambers. The outermost basal diaphragm of a zooid must have acted as the floor of the living chamber for the functional organs of the last regenerated

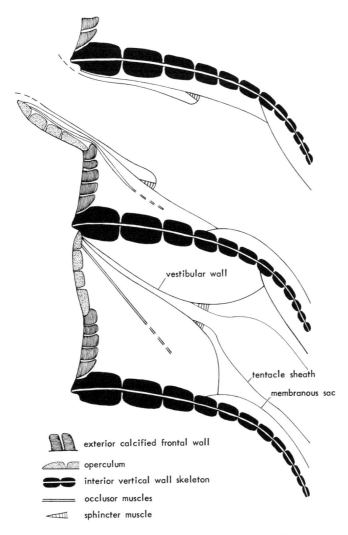

labels within figure:
vestibular wall

tentacle sheath

membranous sac

exterior calcified frontal wall

operculum

interior vertical wall skeleton

occlusor muscles

sphincter muscle

Fig. 38. Zooecial apertures and orificial walls. Reconstruction of a longitudinal section through the outer parts of two zooids of a melicerititid tubuliporate showing hypothetical relationship between opercula of melicerititid and generalized vestibular walls and tentacle sheaths, based on recent tubuliporates. The lower zooid shows the operculum in the closed position seen in fossil specimens (Fig. 36,*4a*). The upper zooid shows a hypothetical open position when tentacles (not shown) were protruded. There is no direct evidence of occlusor muscles to close the operculum. Exterior cuticle presumably covers outer calcified surfaces of frontal walls, opercula, and outer exposed ends of vertical walls.

part of the cycle. Basal diaphragms bend outward where they join enclosing vertical zooecial walls to be continued as skeletal linings, of varying thickness and extent, of living chambers. This outward bend indicates that the diaphragms were deposited by an epidermis on their outer sides at living chamber bases. Paleozoic taxa lacking basal skeletal diaphragms could have had basal membranous diaphragms of similar function, which were not preserved (BOARDMAN, 1971, p. 11).

Both the outward shift of zooidal organs and the spacing of basal diaphragms ontogenetically are apparently results of the

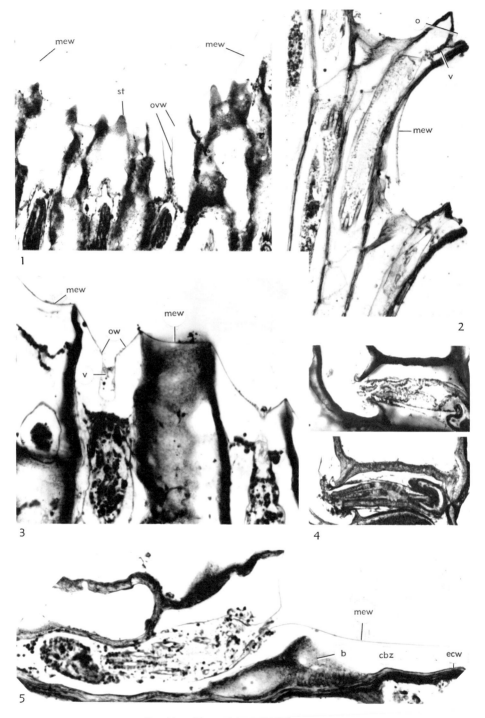

FIG. 39. *(For explanation, see facing page.)*

degeneration-regeneration cycle of feeding zooids (BOARDMAN, 1971, p. 18). More or less continuous growth of vertical zooidal walls during periods of degeneration is assumed. As illustrated in Figure 37, the laminated skeletal microstructure indicates that the newest growth of a vertical wall (A_2) and the outermost diaphragm (2) grew simultaneously as a single skeletal unit. The distance (B_2) between the last two diaphragms (1 and 2) is equal to the distance (A_1) that the vertical wall grew in the previous cycle. When the new organs regenerated they were displaced outward by the distance (A_2) that the vertical wall grew during the newest cycle.

In free-walled fossil taxa lacking communication pores in vertical walls, segments of zooecial chambers enclosed by skeletal diaphragms are assumed to have been sealed physiologically and to have lacked living tissue (Fig. 37). Nutrients for continued growth of vertical walls at outer ends of degenerated zooids presumably would come through the outer body cavity from other feeding zooids of the colony, or were stored within the degenerated zooids themselves. In post-Paleozoic free-walled forms, outer body cavities, and in most taxa, communication pores in vertical walls are both apparently available for transfer of nutrients to regenerating zooids.

In support of the interpretation of the relationship between the spacing of basal diaphragms and the degeneration-regeneration cycle, a few exceptionally preserved specimens of Paleozoic age have a one-to-one relationship between basal diaphragms and presumed fossilized brown bodies (Fig. 40,*1*). Recent tubuliporates have not been found as yet with calcified basal diaphragms in regularly spaced ontogenetic series, but their accumulated brown bodies in inner ends of living chambers can be as many as twenty or more (Fig. 40,*3b*), which compares in number with basal diaphragms in ontogenetically older zooecia of many Paleozoic trepostomates (see Fig. 27 for distribution of ontogenetic stages in a colony).

Lateral skeletal projections.—**Lateral skeletal projections** in living chambers occupy positions opposite feeding organs and are structures of chamber walls that generally reduce or contort living space available to feeding organs. Lateral projections can be shelflike, spinose, or cystose.

Shelflike skeletal projections have been designated by different terms depending on their number and relative position in zooecia. Shelves may be calcified from one or both sides. **Hemisepta** are shelves that generally occur singly on the proximal sides of zooecia or in one or two pairs in alternate positions on proximal and distal sides of zooecia. Proximal and distal hemisepta commonly have different dimensions. Hemisepta have been one of the main polythetic characters (occurring in some taxa but not in others) in the cryptostomates (see BLAKE, this revision) and have only recently been discovered in living (HARMELIN, 1976) and fossil (HINDS, 1973,

---

FIG. 39. Membranous walls of tubuliporates.——*1. Densipora corrugata* MACGILLIVRAY, rec., 5-m wavecut platform, Western Port Bay, W. end Phillip Is., Australia; membranous exterior wall (mew) of freewalled colony supported by skeletal styles (st) leading to orificial-vestibular wall (ovw) of feeding zooid; long. sec., USNM 250071, ×100.——*2. Mesonea radians* LAMARCK, rec., Great Barrier Reef, Low Is., Australia; free-walled colony showing membranous exterior wall (mew), vestibule (v), and orifice (o); long. sec., BMNH specimen, ×150.——*3. Plagioecia dorsalis* (WATERS), rec., 70 m, off Riou Is., Marseille, France; membranous exterior wall (mew) of free-walled colony, orificial wall (ow), vestibule (v); long. sec., Harmelin Coll., ×200.——*4. Diaperoecia indistincta* CANU & BASSLER, rec., 25–35 m, Port Cros, La Palud, France; retracted position of feeding organs behind hemisepta, gut bends in opposite directions in two zooids from same colony; long. sec., Harmelin Coll., ×100.——*5. Plagioecia* sp., rec., 22 m, off Riou Is., Marseille, France; confluent budding zone distal (to the right, and extending well beyond right margin of figure) of feeding zooid, including multizooidal encrusting colony wall (ecw), confluent budding zone (cbz), membranous exterior wall (mew), and developing bud (b); long. sec., Harmelin Coll., ×200.

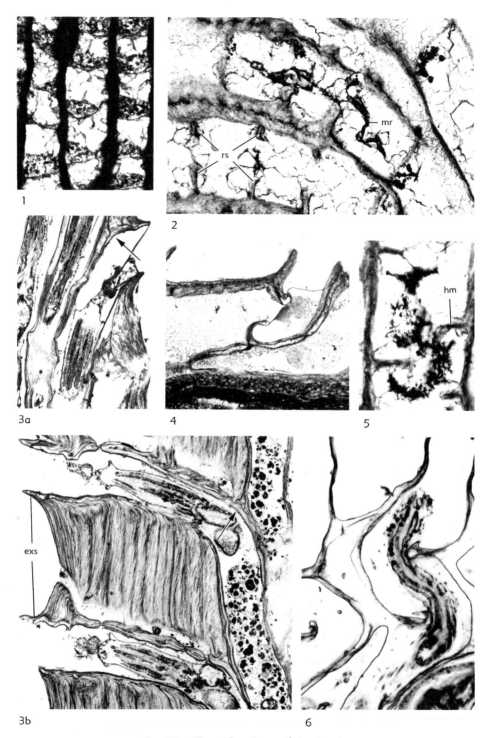

FIG. 40. *(For explanation, see facing page.)*

p. 302) tubuliporates. Retraction of feeding organs behind hemisepta (Fig. 39,4) makes it evident that retracted positions are constant in these short zooecia during their ontogeny. Also, openings between hemisepta are covered by a thickened, apparently protective organic diaphragm during degenerated stages (Fig. 40,4).

Hemiphragms are skeletal shelves of comparable dimensions within a zooid, which alternate in ontogenetic series from opposite sides of chamber walls. In modern species, comparable structures demonstrate how active feeding organs can bend around the projections as they move in and out of living chambers (Fig. 40,6). Fossilized indications of inferred feeding organs have the same relationships to comparable skeletal projections (Fig. 40,5).

Ring septa are centrally perforated diaphragms (Fig. 31,6; 40,2) that have been found in only a few Paleozoic taxa. They originate as lateral structures, outward from basal diaphragms. As ontogeny continues, however, the openings in ring septa may be closed skeletally, suggesting that they eventually acted as living chamber floors (GAUTIER, 1970, p. 9).

Protection from predation might be a function of the chamber constrictions caused by hemisepta, hemiphragms, and ring septa during either the feeding or degenerated state. A lateral shelflike structure projects inward from the frontal and vertical walls of feeding

zooids of a tubuliporate of Cretaceous age, which could serve this same function (Fig. 41,5a–c).

Inward-projecting **mural spines** from zooidal chamber walls are common in stenolaemates of all ages. They may be scattered without noticeable pattern or aligned in rows parallel to zooidal growth (Fig. 41,2). Mural spines may be of different shapes, and more than one shape may occur in the same zooid (Fig. 41,4). The use of spines is not clear (HARMELIN, 1976). In one tubuliporate colony they serve as skeletal supports for presumed attachment ligaments (Fig. 41,1), but this does not seem to be generally true, especially for randomly scattered spines. Spines also occur in brood chambers (Fig. 41,3; BROOD, 1972, p. 70). Certainly more than one function is possible.

Skeletal **cystiphragms** form inwardly curved cysts or collars that extend partly or entirely around living chambers in some Paleozoic taxa. Cystiphragms are calcified on their outer surfaces only, and so are simple partitions. Cystiphragms generally are overlapping in repeated ontogenetic series in zooidal body cavities, causing living chambers to be roughly cylindrical or funnel-shaped and greatly reduced in diameter (Fig. 30,1). Overlapping cystiphragms are closed and show no indication of enclosed soft parts (CUMINGS & GALLOWAY, 1915, p. 354). Cystiphragms have not yet been found in modern species, so no function other than reduction

FIG. 40. Stenolaemate morphology.——*1. Trachytoechus* sp., Dev. (Erian, Petoskey Ls.), Petoskey, Mich.; one-to-one ratio between basal diaphragms and presumed fossilized brown bodies; long. sec., USNM 37518, ×30.——*2. Tabulipora ramosa* (ULRICH), Glen Dean F., Miss. (Chester.), Falls of the Rough, Grayson Co., Ky.; ring septa (rs) and remains of membrane (mr), possibly of membranous sac; long. sec., USNM 167706, ×100.——*3a,b.* Hornerid tubuliporate, rec., Arctic O.; ontogenetic variation showing increase in width of exozone and outward shift of retracted position of feeding organs with increased age within one colony; arrow in each figure points to zooecial bend position, which remains fixed during ontogeny; for intermediate growth stage from same colony, see Figure 44,5; laminated skeleton between zooids is extrazooidal (exs); *3a,* long. sec., *b,* accumulation of brown bodies in base of living chamber, another indicating of advanced growth stage, long. secs., both BMNH specimen, ×100.——*4. Diaperoecia indistincta* CANU & BASSLER, rec., 110 m, Medit. Sea, Levant, Magaud, France; opening between hemisepta covered by organic diaphragm during degenerated stage; long. sec., Harmelin Coll., ×100.——*5. Hemiphragma* sp., Maquoketa Gr., Ord. (Richmond.), Wilmington, Ill.; brown granular deposit with flask shape typical of feeding organs bending around hemiphragms (hm); long. sec., ×100.——*6. Tubulipora ziczac* HARMELIN, rec., 30 m, Port Cros, Gabinière, France; tentacles of feeding zooid bending around hemiphragms; long. sec., Harmelin Coll., ×100.

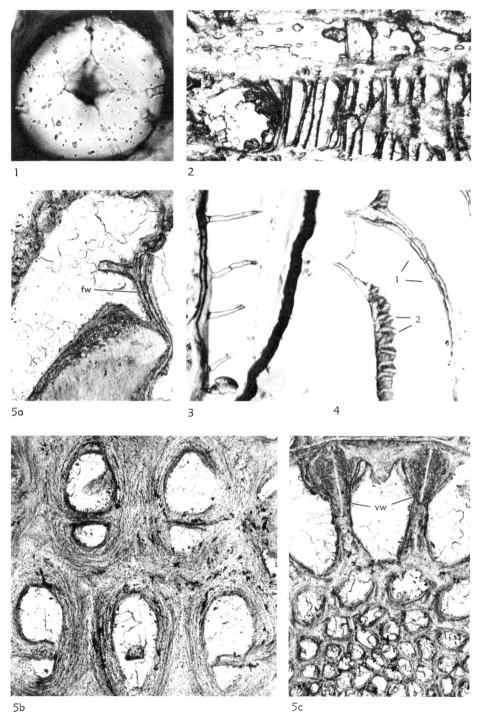

F IG. 41.   *(For explanation, see facing page.)*

of living chamber volume is suggested.

Terminal structures.—Membranous or skeletal terminal and subterminal diaphragms seal living chambers from the surrounding environment because of their position at or near skeletal apertures. Terminal diaphragms are calcified from one side only. It is assumed from examples in a few modern specimens (Fig. 34,1*d*) that skeletal structures calcified on one side are positioned by earlier formed membranes of similar configuration upon which subsequent calcification takes place. Edgewise calcification seems to be the method of skeletal growth.

In post-Paleozoic stenolaemates it is assumed that zooids sealed by terminal diaphragms are in a degenerated state. Growing zooids in the degenerated state of the normal degeneration-regeneration cycle are routinely closed at apertures by membranous terminal diaphragms, which presumably are readily removed by the succeeding regeneration part of the cycle. It also seems probable that the growth of calcified terminal diaphragms terminates the feeding and outward growth of zooids. No indications have been seen of resorption of terminal calcified diaphragms and continued outward growth of vertical walls.

In post-Paleozoic stenolaemates, calcified terminal and subterminal diaphragms bend inward at junctions with vertical walls of chambers (Fig. 42,5,6), indicating that skeletal growth occurs on inner sides of membranous diaphragms within closed living chambers. Calcified terminal diaphragms are exterior walls because they wall off body cavity from the external environment. As in other exterior walls, the skeletal layers are fastened directly to inner sides of outermost cuticles. Apparently, communication pores in vertical walls allow transfer of nutrients among zooids so that degenerated zooids can grow calcified diaphragms after their apertures are closed by membranous diaphragms. Pseudopores are generally abundant in calcified layers of terminal diaphragms but may be few or lacking. Membranous diaphragms can be many near skeletal apertures (Fig. 42,5), and more than one can be calcified in a zooid in reverse order, the inner one later (Fig. 42,6). Apparently, multiple terminal diaphragms indicate that the active outer boundary of a zooid is retreating inward.

In some post-Paleozoic taxa calcified layers of exterior terminal diaphragms are continuations of calcified layers of interior vertical walls just as they are in some frontal zooidal walls (Fig. 42,2). Also, some terminal diaphragms form continuous structural units extending across apertures of a number of zooids (Brood, 1972, fig. 30B). Vertical zooidal walls abut the subsequently formed calcified exterior walls of the terminal diaphragms (Fig. 42,1). The essential morphologic difference between these exterior terminal diaphragms and frontal walls, which can also form structural units across a number of zooids (see above), is the lack of apertures in the terminal diaphragms.

In Paleozoic taxa, diaphragms that are terminal or subterminal to zooidal living chambers (Boardman, 1971, p. 18) bend outward at junctions with vertical walls of chambers

Fig. 41. Lateral skeletal projections.——*1. Lichenopora* sp., rec., Pac. O. off La Jolla, Cal.; dried specimen showing feeding zooid in which membranes attach membranous sac to mural spines; thick tang. sec., USNM 250072, ×400.——*2. Hallopora* sp., Waldron F., Sil. (Niag.), Nashville, Tenn.; alignment of transversely cut mural spines parallel to direction of zooidal growth in zooidal chamber; long. sec., USNM 167698, ×200.——*3. Mecynoecia delicatula* (Busk), rec., 28–30 m, Medit. Sea off Riou Is., Marseille, France; interior of brood chamber of gonozooid with large spines; long. sec., USNM 250073, ×100.——*4. Pustulopora* cf. *P. purpurascens* Hutton, rec., 36 m, off Poor Knights Is., N.Z.; two kinds of mural spines (1 and 2) in living chamber of feeding zooid; long. sec., USNM 216483, ×100.——*5a–c.* Salpinginid tubuliporate, Cret. (Cenoman.), Essen, W. Ger.; shelflike projections connected to both frontal (fw) and vertical walls (vw) in living chambers, cluster of inner ends of zooecia in centers of branches indicate buds clustered centrally at growing tips of branches; *a–c,* long., tang., transv. secs., USNM 213325, ×100.

(Fig. 43,*3*) and communication pores of vertical walls are lacking. As a result, skeletal growth is assumed to have occurred only on outer sides of membranous diaphragms where nutrients are available from adjacent zooids. Because these terminal diaphragms are calcified on their outer surfaces, they are interior structures that formed within outer body cavities protected by outermost membranous exterior walls of free-walled colonies.

Terminal and basal diaphragms in Paleozoic taxa have comparable microstructure and generally differ only in function and position relative to the skeletal aperture when developed. Continued zooidal growth is common beyond terminal diaphragms, so that irregular alternations of basal and terminal diaphragms are found in older zooecia, and abandoned living chambers can be difficult to distinguish (Fig. 36,*1*; 43,*2*; BOARDMAN & MCKINNEY, 1976, p. 66). In most taxa having numerous diaphragms in zooecia, living chambers are generally longer than the spacing between successive basal diaphragms (Fig. 37). Spacing comparable to living chamber length suggests that outer diaphragms of those intervals may have originally been terminal. Some terminal diaphragms appear to serve as basal diaphragms after vertical chamber walls have grown outward enough to house new feeding organs

(Fig. 43,*2*).

Communication pores or larger gaps occur in vertical walls in one suborder of early Paleozoic age. Correlated with such communication potential are diaphragms in two genera that bend inward at vertical wall junctions, indicating growth on inner diaphragm surfaces (UTGAARD, this revision) and implying transfer of nutrients through communication pores. Some of these diaphragms could have served as terminal diaphragms (UTGAARD, 1968b, p. 1446) although they are subterminal or intermediate in position along zooecial length. Apparently they are interior in origin.

Sequential skeletal growth.—The relative time of formation of laminated basal, lateral, or terminal skeletal structures during the ontogeny of the same or adjacent zooecia can be determined by structural continuity of skeletal structures with each other or with the enclosing laminated zooecial walls (BOARDMAN, 1971, p. 14, 15).

Relative time of formation of skeletal structures that abut others can generally be concluded by determining which of the two, the abutting or the abutted, is the supporting structure. In most tubuliporates terminal diaphragms of separate zooids abut supporting vertical zooidal walls and are formed after the vertical walls (Fig. 42,*5,6*). In other tu-

---

FIG. 42. Stenolaemate diaphragms.——*1. Diplocava incondita* CANU & BASSLER, Cret. (Valangin.), Ste Croix, Switz.; vertical walls (vw) abut subsequently formed terminal diaphragms (td) and frontal walls (fw); long. sec., USNM 250074, ×50.——*2. Diplosolen intricaria* (SMITT), rec., 200–235 m, 100 km N. of North Cape, Barent Sea; extension of vertical wall (vw) forms calcified terminal diaphragm (ctd), membranous terminal diaphragm (mtd) at zooidal aperture; long. sec., BMNH specimen, ×100. —— *3. Hemiphragma* sp., Bromide F., Ord. (Champlain.), Spring Cr., Arbuckle Mts., Okla.; laminae of hemiphragms (hm) extend outward and become part of vertical wall (vw) for interval of two to three diaphragms (md) in adjacent mesozooecium; all mesozooecial diaphragms having at zooecial boundary laminae abutting laminae connected to a hemiphragm are outward in position (to right) from that hemiphragm; long. sec., USNM 167709, ×100.——*4. Leptotrypella (Pycnobasis) pachyphragma* BOARDMAN, Wanakah Sh. Mbr., Ludlowville Sh., Dev. (Erian), Deep Run, Canandaigua Lake, N.Y., paratype; superposition of laminae in zooecial lining attached to sequence of progressively younger diaphragms to right in figure; long. sec., USNM 133919, ×50.——*5.* Heteroporid tubuliporate, rec., Pac. O.; laminae from calcified diaphragm (cd) turn inward at junction with vertical wall, membranous diaphragms (md) in closely spaced cluster outward from calcified diaphragm; long. sec., BMNH specimen, ×150.——*6. Heteropora? pelliculata* WATERS, rec., Neah Bay, Wash.; laminae from calcified diaphragms (cd) turn inward at junctions with vertical walls, superposition of laminae at junctions with vertical walls indicates that inner diaphragm developed after outer diaphragm in same zooid; long. sec., USNM 186550, ×100.

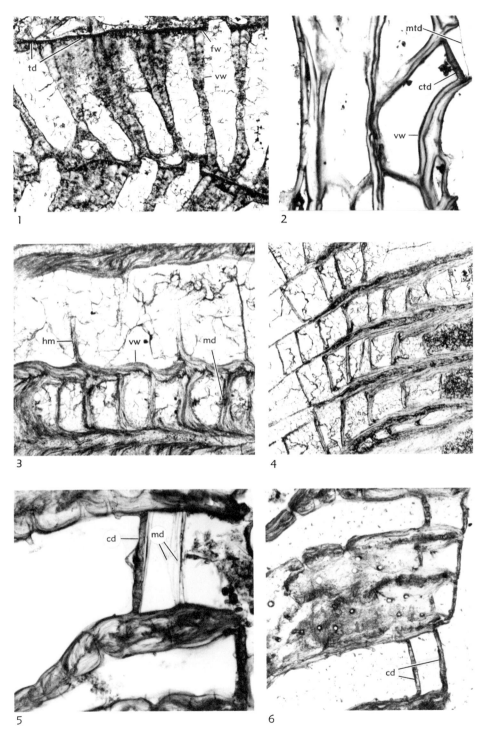

Fig. 42. *(For explanation, see facing page.)*

buliporates terminal diaphragms extend across apertures of a number of adjacent zooids and the previously formed supporting vertical walls abut the younger diaphragms (Fig. 42,*1*). The mode and sequence of growth of this combination is more difficult to understand, and growing tips that actually show the sequence of formation would be helpful.

Superposition in zooecial linings of layers of skeletal laminae attached to skeletal structures within zooecial chambers necessarily indicates relative time of formation of the structures. Basal diaphragms calcified on outer surfaces in zooecia of Paleozoic age (Fig. 42,*4*) are progressively younger outwardly as indicated by superposition of laminae of attached linings. Superposition of layers connected to terminal diaphragms calcified on inner surfaces, however, indicates that terminal diaphragms can be relatively younger inwardly (Fig. 42,*6*) in zooecia of post-Paleozoic age.

Relative time of formation can be determined for skeletal structures within zooidal chambers that are connected by zooecial wall laminae to zooecial boundaries in the same or adjacent zooecia. The outward growth of zooecial walls results in the outward migration of skeletal apertures at zooecial boundaries. If one structure connected to wall laminae intersects the zooecial boundary farther out than another structure in the same or an adjacent zooecium, the one farthest out was developed later. In application, if basal dia-

phragms of feeding zooids are coordinated with degeneration-regeneration cycles, nonalignment of diaphragm laminae of adjacent zooecia along their common boundaries indicates that the cycles were not in unison in those zooids and a degree of zooidal control is expressed. Other structures that can be aligned at zooecial boundaries with basal diaphragms, such as cystiphragms, ring septa (GAUTIER, 1970, pl. 4, fig. 2), or hemiphragms in most Paleozoic species, are also interpreted to be expressions of degeneration-regeneration cycles. Regularly spaced diaphragms in adjacent polymorphs, which are differently spaced than basal diaphragms or hemiphragms, would be controlled by some other cycle, or perhaps be more strongly controlled by environment (Fig. 42,*3*).

BODY CAVITIES AND FEEDING ORGANS OF ZOOIDS

Organs of feeding zooids and polymorphs are the least-known parts of recent stenolaemates. The most detailed coverage of functional organs of a relatively few species are available in papers by BORG and by NIELSEN. Additional information included here is made possible by a sectioning technique (NYE, DEAN, & HINDS, 1972) that produces thin sections containing both skeletal and soft parts in place. Approximately 45 different kinds of tubuliporates have been sectioned. The morphologic differences in soft parts among

FIG. 43. Zooidal soft parts.——*1. Fasciculipora* sp., rec., McMurdo Sound, Antarctica; membranous sac (ms), tentacle sheath (ts), perimetrical attachment organ (pao), and vestibule (v) within vertical (vw) and frontal (fw) walls of zooid; long. sec., USNM 179007, ×150.——*2. Prasopora simulatrix* ULRICH, Ord. (Trenton.), Can.; flask-shaped chambers with calcified walls below and above diaphragm (td), which could have served as both terminal and basal diaphragm, skeletal cystiphragms (sc) reduce volume of living chambers; long. sec., USNM 167688, ×100.——*3. Tetratoechus crassimuralis* (ULRICH), Maquoketa Gr., Ord. (Richmond.), Wilmington, Ill., paralectotype; brown granular deposit in the general shape of feeding organs in living chamber floored by basal diaphragm (bd) and protected by terminal diaphragm (td); long. sec., USNM 204875, ×100.——*4.* Tubuliporid tubuliporate, rec., washed in at Manomet Bay, Cape Cod, Mass.; membranous sac (ms) and tentacle sheaths (ts) in two feeding zooids of a narcotized colony; the tentacle sheath is attached to the membranous sac at points 1 and to the base of the tentacles at points 2; in the zooid to the right, tentacles are protruded far enough that point 2 has moved outward past point 1, causing tentacle sheath to turn inside out; points 1 and membranous sac appear to remain in place; long. sec., USNM 216485, ×150.——*5. Diaperoecia indistincta* CANU & BASSLER, rec., 30 m, Medit. Sea, Port Cros, Gabinière, France; tentacles partly protruded past hemisepta, minute strands connecting membranous sac (ms) to zooecium; long. sec., Harmelin Coll., ×150.

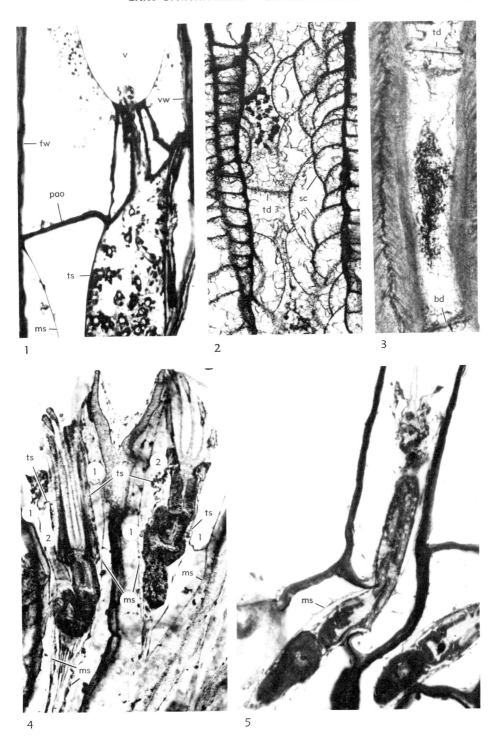

Fig. 43. *(For explanation, see facing page.)*

taxa are striking (Fig. 39, 40, 43–45). The degree of correlation between skeletal and soft-part morphology is an especially important question for classification and that question has yet to be investigated.

All tubuliporates sectioned to date have a membranous sac (BORG, 1926a, p. 207). The sac divides the body cavity into two parts, the entosaccal cavity surrounding the digestive and reproductive systems, and the exosaccal cavity between the membranous sac and the zooidal wall (see Fig. 2). Membranous sacs are attached to the skeletal walls of living chambers near their inner ends by the inner ends of large retractor muscles. The sacs also are attached to skeletal walls of chambers at or near their outer ends and to membranous vestibular walls, which continue out to orificial walls.

A recent study (NIELSEN & PEDERSEN, 1979; first reported by NIELSEN at 1977 meeting of International Bryozoology Association) interpreted the membranous sac to be peritoneum. Also, body walls of stenolaemates have only one cellular layer, the epidermis (NIELSEN, 1970). The entosaccal cavities within sacs, therefore, are surrounded by peritoneum (possibly a mesoderm) and are considered to be coeloms. The exosaccal cavities and all body cavities outside of zooids are pseudocoels, lined either by epidermis, or by peritoneum on one side and epidermis on the other.

A mechanism for tentacle protrusion in the genus *Crisia* was also reported by NIELSEN and PEDERSEN (1979). Their histological work has revealed a series of fine annular muscle cells in the membranous sac. NIELSEN and PEDERSEN suggested that tentacle protrusion in *Crisia* is caused by the contraction of three sets of muscles: (1) longitudinal muscles from the orificial wall to the sphincter muscle at the base of the vestibule, which pull the orificial wall inward; (2) longitudinal muscles in the tentacle sheath, which pull the mouth end of the gut outward; and (3) annular muscles of the membranous sac, which squeeze the feeding organs outward.

The presence of membranous sacs surrounding feeding organs in all preserved tubuliporates studied to date suggests that the sac with its annular muscles is a basic part of tentacle protrusion throughout the class. The system of tentacle protrusion must be confined to living chambers of feeding zooids because all vertical and frontal walls are skeletal and inflexible in the class and cannot enter into zooidal volume changes as frontal walls do in gymnolaemates. Further, living chambers in most stenolaemates, because of outer body cavities and/or open communication pores, are not sealed off from each other. The living chamber, itself, therefore, cannot confine body fluids to single zooids so that differential pressures can be produced to push tentacles out. The membranous sac is the obvious confining organ.

The presence of feeding organs and membranous sac in a large brood chamber (Fig. 45,1) also provides a bit of presumptive evidence. Presumably the tentacles were able to protrude, regardless of what the soft parts were doing there. Certainly, volume restriction and control by the membranous sac in an otherwise oversized chamber must have been a necessary factor in the process.

FIG. 44. Feeding organs.——*1,2. Lichenopora* sp., rec., Galapagos Is.; *1,* radially arranged ligaments attaching top of membranous sac to zooecium, tang. sec., BMNH specimen, ×200; *2,* orificial-vestibular wall (ov), ligament (lg), and mass of eggs (e), long. sec., BMNH specimen, ×150.——*3. Idmidronea atlantica* (FORBES), rec., 24 m, Medit. Sea off Riou Is., Marseille, France; vertical wall (vw) and frontal walls (fw) enclosing outer ends of tentacles and horney cap (hc); long. sec., Harmelin Coll., ×200.—— *4a,b.* Crisinid tubuliporate, rec., 320 m, Nausen Is., W. Palmer Penin., Antarctica; *a,* horny cap (hc) rotated a few degrees, presumably to provide an exit (arrow) for tentacles; *b,* horny caps (hc) in presumed fully retracted position, organic-rich partitions (op) in interior vertical walls; long. secs. from same colony, USNM 216489, ×150.——*5.* Hornerid tubuliporate, rec., Arctic O.; membranous sac (ms) and enclosed retracted feeding organs form flask shapes comparable to flask-shaped chambers of species of Paleozoic age (see Fig. 46); long. sec., BMNH specimen, ×150.

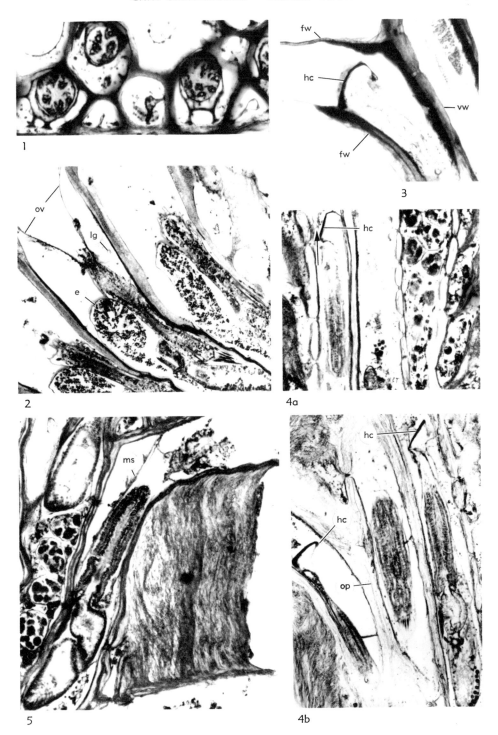

FIG. 44. *(For explanation, see facing page.)*

The region near the outer end of the membranous sac and the fixed end of the tentacle sheath has at least four variations in morphology and attachment of soft parts to skeleton in different taxa. BORG (1926a, p. 209) reported that this attachment was accomplished by eight radially arranged ligaments placed just inward from the outer ends of membranous sacs (Fig. 44,*1*). He apparently assumed that the eight ligaments were present throughout the order.

Most of the colonies sectioned here, however, including both free- and fixed-walled species, have membranous sacs and tentacle sheaths attached by single collarlike membranes (Fig. 43,*1*). These membranes, termed **perimetrical attachment organs** (BOARDMAN, 1973, p. 235), are attached at their inner perimeters to tentacle sheaths and at outer perimeters both to outer ends of membranous sacs and to skeletal body walls. In some species, at least, the attachment organ is attached to walls by many very short ligaments (Fig. 45,*2–4*), most easily detected by the narrow gap between the attachment organ and skeletal wall in longitudinal sections (Fig. 45,*3*).

The perimetrical attachment organ divides the exosaccal body cavity of a zooid transversely into inner and outer portions (Fig. 2). BORG's inferences on tentacle protrusion (1926a, p. 241) were based on exchange of body fluid from outer to inner parts of the exosaccal cavity through spaces between radial ligaments.

In a few fixed-walled species, membranous sacs are attached to chamber walls by a number of minute strands at many different levels (Fig. 43,*4,5*). As tentacles protrude, membranous sacs and orificial-vestibular membranes stay in place. The tentacle sheath surrounds the tentacles in the retracted position and is attached at the base of the tentacles and outer end of the membranous sac. The sheath turns inside out (Fig. 43,*4*) as the tentacles protrude to provide the necessary outward extension, as apparently in all Bryozoa.

The fourth variation in morphology affecting tentacle protrusion is a stiffened horny, uncalcified valve or cap in each feeding zooid of a single free-walled species (Fig. 44,*4a,b*). The cap is attached to the tentacle sheath on one side and apparently the outer end of the membranous sac on the other. No prominent attachment organ has been seen so presumably the membranous sac is attached to chamber walls by minute strands. The cap must act as a flutter valve by rotating about a central axis to allow space for the tentacles to protrude (Fig. 44,*4a*). Normal membranous vestibular walls pass under an indentation in the cap margin when the valve is closed. An apparent cap of similar appearance (Fig. 44,*3*) has been reported (HARMELIN, 1976, pl. 32, fig. 4–7) from a single fixed-walled species; however, subsequent sectioning of other specimens of the species from the same locality has failed to reveal others.

In all Bryozoa, the anus opens through the tentacle sheath below the ring of tentacles. In the classes Stenolaemata and Gymnolaemata the anus reportedly opens on the distal side (toward the colony growing direction) when the tentacles are protruded (e.g., BORG, 1926a, p. 219; JEBRAM, 1973b) and on the

FIG. 45.   Zooidal soft parts.——*1. Lichenopora* sp., rec., "Crab Ledge" E. of Chatham, Mass.; feeding organs surrounded by membranous sac (ms) in large brood chamber; long. sec., USNM 250075, ×100. ——*2–4. Cinctipora elegans* HUTTON, rec., 110 m, off Otago Heads, South Is., N.Z.; *2,* membranous sac (ms), tentacle sheath (ts), sphincter muscle (sm), section cuts perimetrical attachment organ (pao) through short ligaments shown in *4,* long. sec., USNM 250064, ×150; *3,* section cuts perimetrical attachment organ between ligaments, indicated by narrow gap between organ and vertical wall (vw) on both sides, long. sec., USNM 250076, ×100; *4,* perimetrical attachment organ removed from zooid showing approximately 24 short ligaments, USNM 250077, ×150.——*5a–c. Discocytis lucernaria* (SARS), rec., Kara Sea; *a,* sphincter muscle (smm) of mouth at base of tentacles (t); *b,* membranous sac (ms), tentacle sheath (ts), perimetrical attachment organ (pao), sphincter muscle (sm) at top of tentacle sheath; *c,* extreme length of feeding organs of species; all long. secs., USNM 250078, *a,b,* ×150, *c,* ×50.

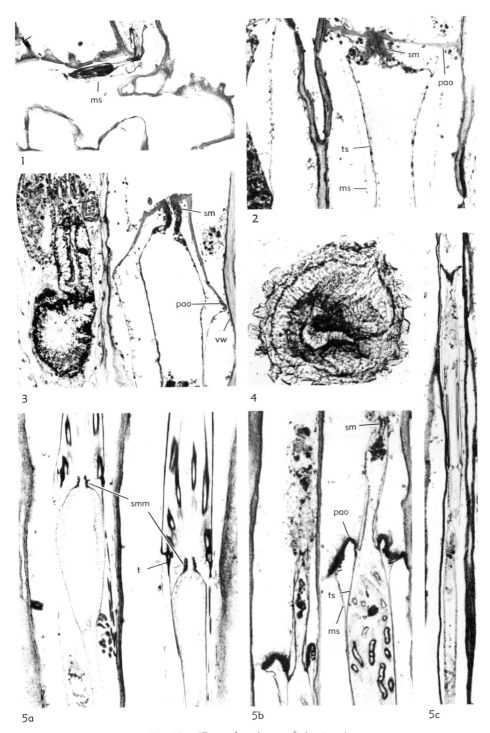

FIG. 45. *(For explanation, see facing page.)*

proximal side in the freshwater Phylactolaemata (WOOD, this revision). In two zooids from the same stenolaemate colony (Fig. 39,4), however, the gut appears to bend in opposite directions in the retracted position. If so, and if one of the lophophores does not twist during protrusion, the anal openings will be on opposite sides when the two lophophores are protruded. This apparent discrepancy suggests that observations need to be made on additional modern stenolaemates before the character state can be used with confidence in classification.

FOSSIL INDICATIONS OF SOFT PARTS

Granular **brown deposits** of iron oxide and some deposits of pyrite presumably represent remains of organic material and have been reported in a number of Paleozoic Bryozoa (e.g., DYBOWSKI, 1877, p. 76, pl. 2, fig. 46; CUMINGS & GALLOWAY, 1915; BOARDMAN, 1971; CORNELIUSSEN & PERRY, 1973; UTGAARD, 1973; BOARDMAN & MCKINNEY, 1976). Most deposits are shapeless or too scattered to be interpreted usefully. Some can be interpreted as having been functional organs (Fig. 40,5; 43,3; 46,1,4a) or brown bodies (Fig. 40,1) of feeding zooids depending upon shape and position in skeletal chambers. Most deposits occur under protective skeletal overgrowths or in skeletally isolated, abandoned chambers between dia-

phragms.

Remains of actual membranes occur in colonies of stenolaemates throughout most of the Paleozoic and are noticeably more common in later Paleozoic species. Again, the majority of membranous remains in zooids are fragmentary and provide little evidence of their biological significance; however, a few could represent the walls of membranous sacs (Fig. 40,2; MCKINNEY, 1969) or orificial-vestibular membranes (Fig. 46,5; BOARDMAN, 1971, fig. 6). A single zooecium of a Late Ordovician specimen shows what appears to be a transverse section across a retracted tentacle crown bearing 10 tentacles (Fig. 46,2; compare with tentacle crown of modern species, Fig. 46,3; BOARDMAN & MCKINNEY, 1976, p. 65).

The most biologically significant finds of membranous remains in colonies of Paleozoic age are of exterior membranous walls of free-walled colonies (Fig. 46,4a–c; BOARDMAN, 1973; BLAKE, this revision). The presence of these delicate walls, added to the skeletal evidence, supports BORG's theory that the free-walled (double-walled) mode of growth found in many modern tubuliporates occurred also in the earliest known Bryozoa of Ordovician age and was the mode of growth for the great majority of Paleozoic taxa.

A third type of preserved indication of soft parts is skeletal and therefore has the potential for retaining living shapes of soft parts. These skeletal structures occur within zooecia

---

FIG. 46. Fossilized soft parts.——*1. Dittopora colliculata* (EICHWALD), Ord. (Wassalem Beds, D3), Uxnorm, Est.; granular brown deposit presumably reflecting generalized shape of feeding organs; long. sec., USNM 250079, ×50.——*2. Tetratoechus crassimuralis* (ULRICH), Maquoketa Gr., Ord. (Richmond.), Wilmington, Ill.; ring of 10 inwardly tapered wedges of brown granules interpreted as tentacles cut transversely by section; tang. sec., USNM 204872, ×150.——*3.* Heteroporid tubuliporate, rec., Pac. O.; tentacles cut transversely by section; tang. sec., BMNH specimen, ×150.——*4a–c.* Dendroid trepostomate, Waynesville F., Ord. (Richmond.), Hanover, Ohio; *a–c,* remnants of exterior membranous walls (arrows), brown granular deposit in generalized shape of feeding organs in *4a;* long. secs., USNM 179006, ×100.——*5. Leptotrypella? praecox* BOARDMAN, Horlick F., L. Dev., Ohio Ra., Antarctica, holotype; remnants of soft parts, probably an orificial-vestibular wall; long. sec., USNM 144807, ×200.——*6. Leptotrypella furcata* (HALL), Windom Mbr., Moscow F., Dev. (Erian), Menteth Cr., Canandaigua Lake, N.Y.; flask-shaped chamber containing granular brown deposits; long. sec., USNM 133901, ×100.—— *7. Prasopora grayae* NICHOLSON & ETHERIDGE, Craighead Ls., Ord., Craighead Quarry near Girvan, Ayrshire, Scot.; flask-shaped chamber containing granular brown deposits; long. sec., RSM 1967-66-406, ×100.

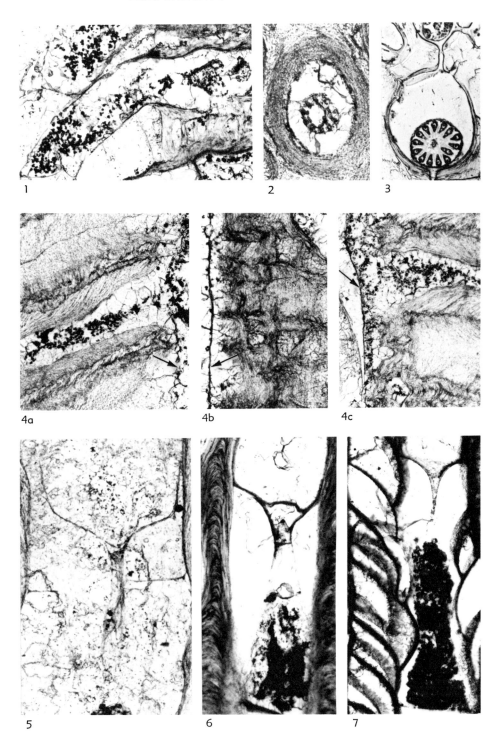

FIG. 46. *(For explanation, see facing page.)*

Bryozoa

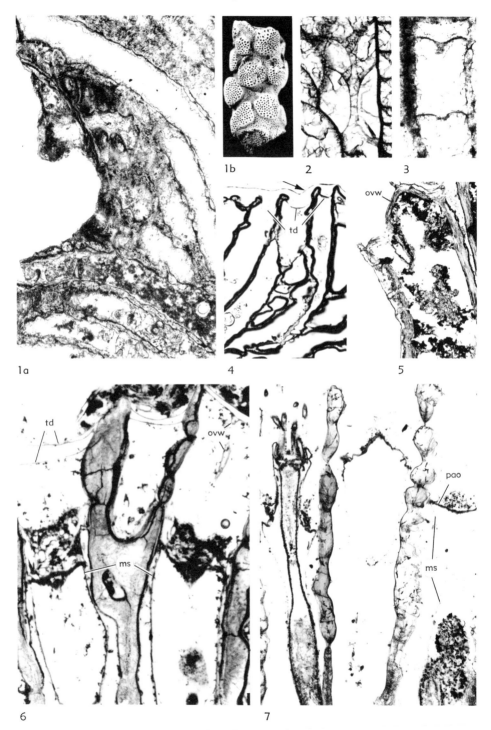

F1G. 47. *(For explanation, see facing page.)*

and generally form inner **flask-** or **funnel-shaped chambers** containing granular brown deposits (Fig. 43,*2*; 46,*6*,*7*). The calcareous laminae of the walls of the flask-shaped chambers continue into surrounding zooecial walls as do the laminae of other skeletal structures within zooidal chambers. The chambers are commonly floored by basal diaphragms and covered by terminal diaphragms.

Cumings and Galloway (1915, p. 354) interpreted flask-shaped chambers to be products of degeneration. Walls of the chambers were thought to be new skeletal body walls housing the shrunken nonfunctional remains of the degeneration process. Boardman (1971, p. 26), Corneliussen and Perry (1973, p. 159), and Utgaard (1973, p. 339) interpreted the walls of the chambers to be skeletal body walls of smaller regenerated intrazooidal polymorphs formed after the organs of the original feeding zooids had degenerated.

A third interpretation (Boardman & McKinney, 1976, p. 66) suggests that the flask shapes were not chambers, but were what they resemble in shape in living tubuliporates, that is, remnants of orificial-vestibular walls (Fig. 39) or membranous sacs (Fig. 44,*5*; 53,*3*) of normal feeding zooids in more or less retracted positions.

Evidence for the third interpretation begins with the discovery that in modern tubuliporates orificial-vestibular walls and membranous sacs retain their functional shapes during at least part of the degeneration process (Fig. 47,*6*,*7*). It is possible, therefore, that in taxa of Paleozoic age, orificial-vestibular walls, membranous sacs, and attachment organs could also have remained in place during part of the degeneration process. These organs then would have been nonfunctional and the zooids dormant so that loss of flexibility due to calcification would not be a problem. Calcification on these static membranes presumably occurred similarly to calcification of membranes of diaphragms and cystiphragms and would have been attached to calcified layers of enclosing zooidal walls in the same manner.

Further evidence for the degeneration hypothesis was found by Walter and Powell (1973) in a fixed-walled tubuliporate species of Jurassic age. Their specimens were interpreted to contain calcified orificial-vestibular walls (compare Fig. 39 and 47,*5*). Comparison with modern tubuliporates leaves no reasonable doubt that the calcified funnels in the Jurassic specimens are calcified orificial-vestibular walls, and that the walls had ceased to function in the feeding process. They were probably acting as terminal diaphragms to protect the living chambers after zooidal growth was completed.

Two skeletal structures similar in shape to orificial walls have since been found in the same zooecium of a tubuliporate species of Cretaceous age (Fig. 47,*3*). Considering the

---

Fig. 47. Soft-part morphology.——*1a,b. Plethopora verrucosa* (Hagenow), Cret. (Maastricht.), St. Pietersberg, Neth.; smaller polymorphs surrounding circular clusters of feeding zooids; *a,* long. sec., *b,* external view, USNM 250080, ×50.0, ×3.5.——*2. Prasopora simulatrix* Ulrich, Ord. (Trenton.), Trenton Falls, N.Y.; double-funneled flask-shaped chamber; long. sec., USNM 167685, ×100.0.——*3. Defranciopora neocomiensis* Canu & Bassler, Cret. (Valangin.), Ste Croix, Switz., syntype; tops of two calcified funnels shaped like orificial walls; long. sec., USNM 250081, ×100.0.——*4. Disporella neopolitana* (Waters), rec., 21 m, Medit. Sea, Plane, near Marseille, France; colony-wide membranous exterior wall (arrow) above degenerated zooids covered by membranous terminal diaphragms (td); long. sec., peel USNM 204876, ×50.0.——*5. Mesenteripora wrighti* Haime, M. Jur., King's Sutton, Northamptonshire, Eng.; calcified orificial-vestibular wall (ovw) closing aperture of zooid; long. sec., OUM, Walford Coll., ×150.0.——*6. Disporella separata* Osburn, rec., South Coronados Is., Baja Cal., Mexico; partly degenerated zooids with membranous terminal diaphragms (td), orificial-vestibular walls (ovw), and membranous sacs (ms); long. sec., USNM 167679, ×200.0.——*7. Neofungella* sp., rec., 133 m, off Victor Hugo Is., W. coast Palmer Penin., Antarctica; intact feeding zooid with tentacles partly protruded on left, partly degenerated zooid to right showing membranous sac (ms) and perimetrical attachment organ (pao) forming a flask-shaped chamber; long. sec., USNM 250082, ×100.0.

Bryozoa

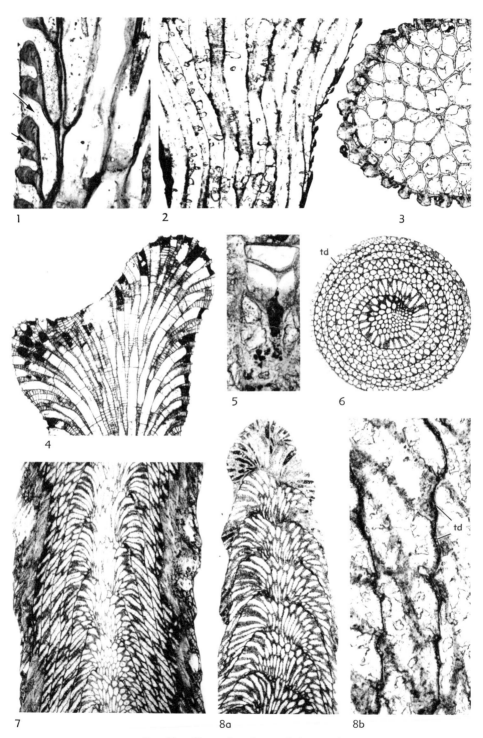

FIG. 48. *(For explanation, see facing page.)*

flexibility displayed in stenolaemates, the inner structure could well have functioned as a basal diaphragm for the organs of the next cycle represented by the outer structure.

The major difference between flask-shaped structures of Paleozoic and those of Jurassic age is that calcification took place on the outer sides of membranes that were originally either exterior or interior in Paleozoic colonies and on inner sides of exterior cuticle in Jurassic species. The Paleozoic flask-shaped walls were interior walls at the time of calcification and the Jurassic walls were exterior. In Jurassic fixed-walled species (Fig. 47,5), communication pores in the vertical walls provided the possibility, at least, for a continuing supply of nutrients within otherwise dormant zooids so that calcification within living chambers could occur. In most Paleozoic free-walled species communication pores are lacking. Body cavities and exterior membranous walls necessarily occurred outward from the membranes being calcified and nutrients presumed necessary for continued growth apparently came from other regions of the colonies through outer body cavities (Fig. 1; 46,4a–c). The exterior membranous walls were probably colony-wide (BOARDMAN & MC-KINNEY, 1976, fig. 13), similar to the exterior wall in the dormant free-walled recent colony (Fig. 47,4).

It is not clear for most flask-shaped structures in Paleozoic colonies whether the walls of the flasks represent orificial-vestibular walls or the membranous sacs attached at their outer ends to form the flask shapes. Either possibility produces comparable shapes in some recent tubuliporates (compare Fig. 39,3,5 with 53,3). Also, the best explanation for the occasional multiple funnels in Paleozoic colonies with mostly single-funnel flasks (Fig. 47,2) has been found in only one zooid of a recent tubuliporate (BOARDMAN, 1971, pl. 1, fig. 5a). In that single zooid a double orificial-vestibular wall developed a similar configuration to double funnels interpreted to be membranous in a Devonian specimen (BOARDMAN, 1971, pl. 1, fig. 2) and to the numerous multiple calcified funnels of Paleozoic age.

POLYMORPHISM

In stenolaemates many taxa have polymorphs and other taxa are entirely monomorphic, at least skeletally. Polymorphs may be isolated or contiguous with each other between feeding zooids and may be numerous enough to isolate feeding zooids from each other (Fig. 31,7a,b). Polymorphs can be arranged regularly (see Fig. 55,4a,b) or irregularly relative to feeding zooids. Polymorphs of one or more kinds may be clustered into maculae (see Fig. 59 and related text) surrounded by feeding zooids, or polymorphs may surround clusters of feeding zooids (Fig. 47,1a,b). Polymorphs cover reverse sides of colony branches or entire supporting stalks (Fig. 48,1–3). Intrazooidal polymorphism occurs where polymorphs

FIG. 48. Polymorphism.——*1.* Crisinid tubuliporate, rec., Philippines expedition of the Albatross, loc. D5559, coll. 1909; small pores (shorter arrow) and polymorphs (longer arrow) on left, feeding zooids open to right; long. sec., USNM 186566, ×150.0.——*2,3. Corymbopora menardi* (MICHELIN), Cret. (Cenoman.), Le Mans, Sarthe, France; small polymorphs on outside of main supporting stalks of zoarium, larger feeding zooids within stalk; *2,* long. sec., USNM 213332, ×30.0; *3,* transv. sec., USNM 213331, ×50.0.——*4,5. Hallopora elegantula* (HALL), Rochester Sh., Sil. (Niagar.), Rochester, N.Y.; *4,* mesozooecia indicated by closely spaced diaphragms and small cross-sectional areas, followed intrazooidally by larger feeding zooecia and widely separated diaphragms, long. sec., USNM 250083, ×7.5; *5,* flask-shaped skeletal structure in small exilazooecium, long. sec., USNM 250084, ×100.0.——*6–8. Terebellaria ramosissima* LAMOUROUX, Jur. (Bathon.), Ranville, France; *6,* spiral budding pattern and connected terminal diaphragms (td), transv. sec., USNM 250085, ×7.5; *7,* terminal diaphragms covering outer ends of zooecia in older part of colony, long. sec., USNM 250086, ×7.5; *8a,b,* tip showing zooecia growing proximally over older zooids in progressively younger cycles and zooecia from same colony with terminal diaphragms (td), long. secs., USNM 250087, ×7.5, ×50.0.

develop within zooecia of regular feeding zooids, either before or after zooids were capable of feeding.

Polymorphs vary widely in morphology and function in stenolaemates. Terms applied to differentiate kinds of polymorphs have been based primarily on soft-part morphology and assumed function in some modern tubuliporates (e.g., nanozooid, kenozooid, gonozooid), or skeletal morphology and position within the colony in both modern and fossil stenolaemates (e.g., dactylethra, firmatopore, nematopore, tergopore, mesozooecium, exilazooecium). Unfortunately, morphology and function together are not well enough known or defined for some of these terms to be used to advantage.

The term **kenozooid**, for example, was defined as a polymorph lacking lophophore and gut, muscles, and orifice (LEVINSEN, 1902, p. 3; 1909, p. v). BORG used the term for any polymorph that functioned as a rhizoid or spine (1926a, p. 239), or later (1933) for any smaller polymorph with aperture that was open or covered by a calcified terminal diaphragm regardless of its soft parts (Fig. 49,8) or possible function.

Dactylethrae (GREGORY, 1896, p. 12) are defined as aborted, shorted zooecia closed externally, as in the Jurassic genus *Terebellaria*. They have been interpreted as a type of kenozooid (e.g., BASSLER, 1953, p. G9;

BROOD, 1972, p. 49). Sections of topotypes of the type species, *T. ramosissima*, suggest that they are zooecia of feeding zooids covered by terminal diaphragms forming continuous skeletal walls across apertures (Fig. 48,6–8).

Nanozooids (Fig. 49, 5–7,9) are exceptionally well known both morphologically and functionally. Nanozooids were named and their soft parts described by BORG (1926a, p. 188, 232–239) from the recent genus *Diplosolen*. BORG reported a lophophore with a single tentacle, muscular system, reduced alimentary canal, membranous sac, and no reproductive structures. The single tentacles are relatively long and have been observed cleaning colony surfaces (SILÉN & HARMELIN, 1974).

In *Diplosolen*, nanozooids are restricted to an outer position in the colony and occur singly between feeding zooids (Fig. 30,4; 49,5). Nanozooids bud in distal confluent budding zones where the compound interior vertical walls of contiguous feeding zooids divide into two compound walls (Fig. 49,5). The outer walls of both feeding zooids and nanozooids are simple exterior frontal walls. The frontal wall of a nanozooid grows distally from its vertical wall, which is contiguous with the vertical wall of the proximal feeding zooid. The nanozooid tentacle protrudes through a small aperture in the frontal wall.

FIG. 49. Polymorphism.——*1. Meliceritites* sp., Cret. (Santon.), Coulommiers, France; two polymorphs in profile, which together form aviculariumlike structure in *3,4;* upper polymorph closed off by opercular shelf (os), lower polymorph apparently produced opercular shelf and large operculum (missing) hinged on frontal wall (fw); long. sec., USNM 216482, ×50.——*2–4. Meliceritites* sp., Cret. (Coniac.), Villedieu, France; *2,* opercular shelf (os) and living chamber (lc) of lower polymorph, frontal wall, and operculum removed by sectioning, tang. sec., USNM 216479, ×50; *3,* polymorph at zoarial surface minus operculum, external view, USNM 216477, ×30; *4,* polymorph with operculum in place, external view, USNM 216478, ×30.——*5. Diplosolen* sp., rec., Popoff Str., Alaska; budding position of nanozooid (nz) at division of interior vertical walls (vw) of two supporting feeding zooids; corresponding walls of distal supporting zooids where nanozooids not formed are parts of exterior frontal walls (fw); long. sec., USNM 250088, ×100.——*6. Plagioecia dorsalis* (WATERS), rec., 70 m, off Riou Is., Marseille, France; intrazooidal nanozooid formed subsequently in outer end of feeding zooid; long. sec., Harmelin Coll., ×200.——*7. Plagioecia* sp., rec., Pac. O. at La Jolla, Cal.; intrazooidal nanozooids with small apertures at outer ends; long. sec., USNM 250059, ×100.——*8.* Heteroporid tubuliporate, rec., Pac. O.; smaller polymorphs (pm) on either side of feeding zooid; long. sec., BMNH specimen, ×100.——*9. Diplosolen intricaria* (SMITT), rec., 200–235 m, 100 km N. of North Cape, in Barents Sea; sequence of growth of walls of feeding zooids (fz) and nanozooids (nz) at growing tip; buds (bd); long. sec., BMNH specimen, ×50.

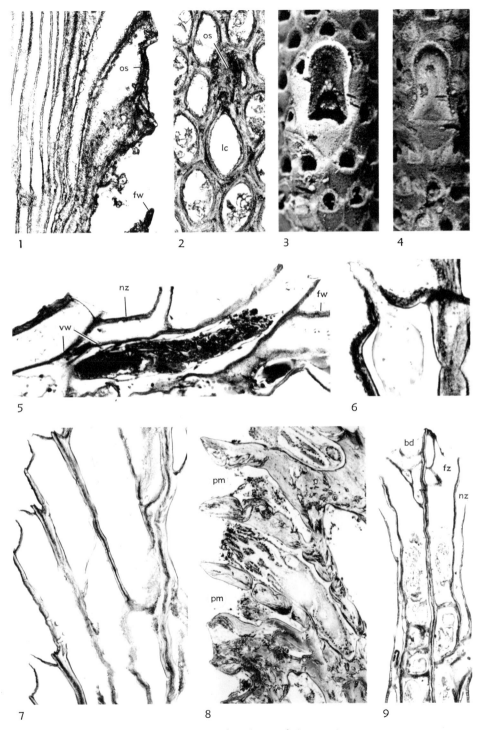

Fig. 49. *(For explanation, see facing page.)*

Bryozoa

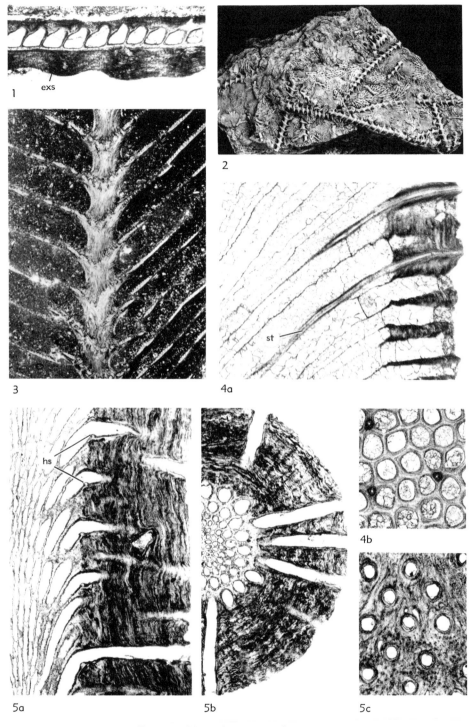

FIG. 50. *(For explanation, see facing page.)*

Two other types of nanozooids have been discovered (SILÉN & HARMELIN, 1974) in another genus, *Plagioecia*. The one of more general interest develops within the zooecium of a degenerated feeding zooid (Fig. 49,6,7) and is an example of intrazooidal polymorphism. An exterior frontal wall develops in the skeletal aperture of the feeding zooid much like a terminal diaphragm except that it contains the smaller aperture of the nanozooid.

Mesozooecia provide another example of intrazooidal polymorphism in a few of the many Paleozoic trepostomates in which they occur. Mesozooecia are skeletons of **mesozooids**, generally small, space-filling polymorphs between zooecia of feeding zooids in exozones. They are closely tabulated out to their distal ends (Fig. 42,3) so that no room is available for functional organs. In one group, the halloporids, zooids bud as mesozooids in endozones at growing tips, are transformed to feeding zooids intrazooidally (Fig. 48,4), and in later growth stages can revert to mesozooids.

Exilazooecium is a term used for skeletons of polymorphs in colonies of Paleozoic age that have few or no basal diaphragms in their chambers. The available chamber space allows for possible organs. A flask-shaped skeletal structure occurring in one of the few exilazooecia seen in the genus *Hallopora* (Fig. 48,5) suggests that at least some **exilazooids** did have functional organs.

In some species of the meliceritids occur operculate polymorphs that superficially resemble the avicularia of cheilostomate Bryozoa (Fig. 49,3,4). Each polymorph occupies two enlarged zooidal spaces on colony surfaces in these species, and internally, at least two polymorphs were involved, one above the other. The more proximal polymorph grew a thickened interior vertical wall that covered the upper polymorph and functioned as an opercular shelf, apparently for the operculum to close against (os, Fig. 49,1,2). The operculum was hinged on the frontal wall (fs, Fig. 49,1,3,4) of the more proximal polymorph, so apparently was produced by the polymorph. These operculate tubuliporates are extinct and the function of the polymorphs is unknown.

EXTRAZOOIDAL PARTS

Parts of colonies formed outside of zooidal boundaries are considered either multizooidal or extrazooidal. Body cavities or walls that are formed outside of zooidal boundaries and subsequently become parts of zooids are termed multizooidal. Body cavities or structures that develop outside of zooidal boundaries and remain outside of those boundaries throughout the life of a colony are termed extrazooidal.

Extrazooidal parts occur in many stenolaemates and range from small spinelike skeletal growths between zooidal walls to structures that are virtually colony-wide. Because

---

FIG. 50. Extrazooidal parts.——*1. Archimedes wortheni* (HALL), Warsaw Ls., Miss. Warsaw, Ill., lectotype; extrazooidal skeleton (exs) on reverse side of fenestrate branch; long. sec., AMNH 7525, ×30.0.——*2. Archimedes proutanus* ULRICH, Miss. (Chester.), Sloans Valley, Ky., syntypes; spiral axial supports of colonies surrounded by broken fronds of this and other species of fenestrates; external view, USNM 43737, ×0.5.——*3. Archimedes* sp., Miss. (Chester.), 19 km S. of West Lighton, Ala., near Fox Trap Cr.; section through a spiral extrazooidal support showing relationship with fenestrate fronds extending distally outward; long. sec., USNM 182789, ×4.0.——*4a,b. Dekayia aspera* MILNE-EDWARDS & HAIME, Fairmount Ls. Mbr., Fairview F., Ord. (Maysvill.), Covington, Ky.; *a,* beginning of style (st) in endozone, long. sec.; *b,* three styles cut transversely from same zoarium, tang. sec.; both USNM 250089, ×30.0.——*5a–c. Pustulopora verrucosa* ROEMER, Cret. (Santon.), Crosz Bülten, Ger.; laminae of extrazooidal skeleton concave outward between zooecial walls and minute styles, hemisepta (hs); *a–c,* long., transv., tang. secs., all USNM 250090, ×30.0.

Bryozoa

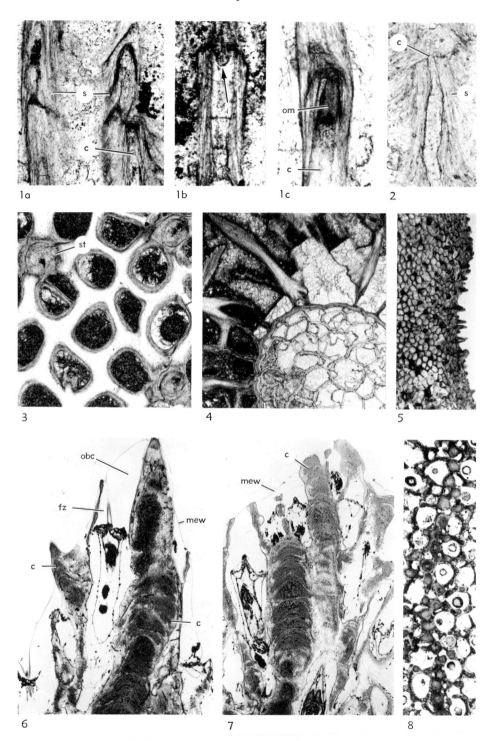

F<small>IG</small>. 51. *(For explanation, see facing page.)*

they are not parts of feeding zooids, their growth depends upon transfer of nutrients. Apparently most extrazooidal skeletal structures in stenolaemates are interior in origin; however, outer skeletal walls of extrazooidal brood chambers in many post-Paleozoic taxa are exterior walls. Interior extrazooidal parts are connected to zooids by outer body cavities protected by exterior membranous walls. The outer body cavities and exterior membranous walls opposite extrazooidal skeleton are also considered to be extrazooidal.

In some taxa, extrazooidal skeleton provides supports for erect colonies, for example on the reverse sides of free-walled unilaminate colonies such as recent hornerids (Fig. 28,1a,b) and Paleozoic fenestellids (Fig. 50,1), as cross supports to form fenestrules in some fenestrate growth, or as massive marginal (McKinney, 1977a) or axial (Fig. 28,4; 50,2,3) colony-wide supports in other fenestellids (see Blake, Karklins, Utgaard, this revision, for many examples of extrazooidal skeleton).

In some taxa, extrazooidal skeleton intervenes between zooids in exozones, either in irregular patches, in spaces longitudinally along colonies distally, or completely surrounding the zooids. This intervening extrazooidal skeleton can be either vesicular or solid (Fig. 32,5a,b; 40,3a,b; 50,5a,b). The distinction between zooidal and extrazooidal skeleton is evident where microstructural boundaries of zooids are apparent (Fig.

32,5a,b). Where zooidal boundaries are not clearly indicated microstructurally, extrazooidal skeleton between zooids can be distinguished by reversals in orientation of laminae from convex outward in zooidal walls to concave outward in extrazooidal skeleton (Fig. 40,3a,b; 50,5a–c). The controlling criterion for distinguishing extrazooidal skeleton is that it was calcified by epidermis that cannot be associated with a particular zooid.

Styles (acanthopores of authors; acanthostyles of Boardman & McKinney, 1976, p. 28; stylets of Blake, this revision) are elongate rodlike structures that form spinose projections on zoarial surfaces of many Paleozoic stenolaemates and at least one modern free-walled genus. Styles extend approximately parallel to zooecial walls and have their origin in exozones (Fig. 30,1), or less commonly in endozones (Fig. 50,4a,b). Styles form spinose projections throughout ontogenetic development and extend in length at growth rates comparable to or exceeding those of surrounding vertical zooidal walls.

Styles are interpreted to have had central skeletal cores in living colonies. The cores are nonlaminated in most taxa but may be laminated or a combination in some (Fig. 51). Cores are centered on zooecial boundaries or are surrounded by extrazooidal skeleton, and are considered to be extrazooidal. Nonlaminated cores are commonly continuous rods (Blake & Towe, 1971) but in some taxa may be divided into segments by laminae from

---

Fig. 51. Stenolaemate styles.——*1a–c. Leptotrypella? praecox* Boardman, Horlick F., L. Dev., Ohio Ra., Antarctica, holotype; styles showing nonlaminated granular cores (c), laminated sheaths (s); *a*, ×100; *b*, laminated zooecial wall grown on broken style (arrow), ×200; *c*, organic matter (om) at end of core, followed by subsequent outward growth of style, ×200; all long. secs., USNM 144807.——*2. L.? praecox*, same data as *1* but paratype; broken style with core (c) extending beyond sheath (s); USNM 250091, ×200. ——*3. Polycylindricus asphinctus* Boardman, Wanakah Sh. Mbr., Ludlowville F., Dev. (Erian), Elma, N.Y., holotype; large styles (st) with laminated cores; tang. sec., USNM 133916, ×50.——*4. Polycylindricus clausus* Boardman, Centerfield Ls. Mbr., Ludlowville F., Dev. (Erian), Paines Cr., Cayuga Lake, N.Y., paratype; long styles with laminated cores; transv. sec., USNM 133922, ×30.——*5. P. asphinctus*, same data as *3* but Big Tree Shale Pit, Erie Co., N.Y.; surface expression of styles; USNM 158321, ×5.——*6–8. Densipora corrugata* MacGillivray, rec., 5-m wave-cut platform, Western Port Bay, W. end Phillip Is., Australia; *6*, styles with sparsely laminated cores (c) covered by membranous exterior walls (mew), outer body cavity (obc), feeding zooid (fz), long. sec., USNM 250092, ×100; *7*, styles with cores (c), membranous exterior wall (mew), long. sec., USNM 250071, ×100; *8*, ridge of large styles, tang. sec., USNM 250093, ×50.

surrounding sheaths or zooidal skeleton (BLAKE, 1973b).

Laminated sheaths surround the skeletal cores and in most taxa the sheaths are microstructurally continuous with adjacent zooecial walls or extrazooidal skeleton. Laminae of the sheaths bend outward against the cores to form cone-in-cone patterns (Fig. 51,*1a,2*), so that both cores and enclosing sheaths extend beyond zooecial walls to form spines on colony surfaces. Style sheaths in some taxa can be considered extrazooidal, but because of microstructural continuity with zooidal walls, sheaths in many taxa are not clearly either zooidal or extrazooidal.

Styles were first interpreted to have been hollow tubes during colony life, hence the term "acanthopore." More recently, authors have interpreted the cores as filled with skeletal material during life (e.g., TAVENER-SMITH, 1969b; ARMSTRONG, 1970; BROOD, 1970; BLAKE, 1973b). For recent interpretations of styles as hollow tubes and kenozooids during colony life, see ASTROVA (1971, 1973).

There is much evidence for the interpretation of style cores as skeletal in living colonies.

1. Styles are present as long spines beyond the ends of zooecial walls (Fig. 30,*1*; 51,*3–8*).

2. Laminae of sheaths extend outward against the cores of styles. Laminae are added to outer surfaces of sheaths as indicated by progressive thickening of sheaths toward the bases of styles and by structural continuity with surrounding zooidal walls or extrazooidal skeleton. A solid projecting core of some kind would appear necessary to deflect the depositing epidermis outward beyond zooidal walls to form the sheaths. In recent taxa, laminae surrounding demonstrable pores turn not outward but inward, into the pores relative to direction of thickening of surrounding wall (Fig. 35,*4*).

3. The microstructure of nonlaminated style cores is relatively constant, and it compares with the microstructure of such skeletal parts as most lunaria in cystoporates and

zooecial walls of fenestellids. Microstructures of styles differ from those of fillings of adjacent abandoned chambers (Fig. 50,*4a,b*). In addition to being commonly nonlaminated, many cores contain minute pyrite crystals in varying proportions thought to be indications of organic-rich skeletal material. The chemical composition of the cores in two species of *Stenopora,* a trepostomate, is more complex than that of the secondary calcite of living chambers (ARMSTRONG, 1970, p. 584), presumably reflecting the differences in microstructure.

4. In well-preserved specimens of Devonian age containing membranous structures (BOARDMAN, 1971, p. 9), many styles were either broken or stopped growing for some less obvious reason. The outer ends of many cores of terminated styles contain brown material, suggesting a concentration of organic matter at core ends (Fig. 51,*1a,c*) and progressive calcification inwardly. Laminated skeleton of a vertical zooidal wall rests on the nonlaminated core of one broken style, which apparently was present when wall growth was renewed (Fig. 51,*1b*). Another style broke, leaving the core extending beyond the laminated sheath (Fig. 51,*2*).

5. Zoaria of Paleozoic age are commonly preserved in terrigenous mudstones or shales. Many styles are broken or worn off at zoarial surfaces. If they had been hollow tubes they would certainly have been routinely filled with terrigenous material after death, just as are open chambers of zooecia. Terrigenous material has not been observed in cores of styles by the author. (For contrasting observations, see ASTROVA, 1973, p. 7.)

Colonies of the recent free-walled tubuliporate genus *Densipora* develop sinuous ridges supported by single rows of large styles (Fig. 51,*6–8*). The styles consist mostly of laminated cores. Several smaller styles with granular cores occur at each zooecial aperture. Unfortunately, the zooecial walls appear granular rather than laminated and intergrowth relationships are not clear.

Sections of *Densipora* with soft parts intact provide some insight into the mode of growth and function of styles in colonies of Paleozoic age. The styles of *Densipora* are part of the interior skeleton and are within the exterior membranous walls of the colonies (Fig. 51,6,7). In order to grow in length, styles throughout the history of the phylum necessarily have had epidermis and outer body cavity between their outer ends and exterior colony walls. With intervening body cavities there is no evidence that exterior colony walls could have been fastened to or held in place by styles, as has often been suggested.

The only suggested function of styles is to raise exterior membranous walls above zooecial apertures and skeletal surfaces. The raising of membranous walls (Fig. 39,1; 51,6,7) increases the volumes of outer body cavities and, presumably, colony-wide communication through those cavities. Outer body cavities obviously provided adequate communication in colonies of Paleozoic age that lacked both styles and communication pores in vertical walls. The apparent disappearance of styles when communication pores developed in post-Paleozoic stenolaemates, however, suggests that there might have been some communication advantage associated with styles.

MODE OF GROWTH, MORPHOLOGY, AND FUNCTION OF COLONIES

SEXUAL REPRODUCTION

Colonies are reportedly bisexual, in modern tubuliporates as in other bryozoan classes. Zooids within many colonies are also bisexual but apparently in some taxa are unisexual. Both male and female reproductive cells originate in the peritoneum of confluent budding zones. Both kinds of cells become attached to zooids and develop within body cavities (BORG, 1926a, p. 336–343).

Sperm cells begin multiplication inside membranous sacs of feeding zooids within a thin peritoneum attached to the funiculus near the inner end of the gut. In some species, concentrations of spermatozoa are large and expand outward along the gut (Fig. 44,2) or inward to the funiculus. Release of spermatozoa occurs through the ends of tentacles in at least two species of tubuliporates, a method of sperm release more generally observed in gymnolaemates (SILÉN, 1972). After the spermatozoa escape, zooidal feeding organs degenerate (BORG, 1926a, p. 336–341).

Eggs that do not become associated with zooids degenerate in confluent budding zones. Only one or two eggs attach to a single zooid and, within a colony, most of those also degenerate. In fertile zooids, eggs are surrounded by a thin peritoneum and begin development inside membranous sacs. In some species feeding organs of fertile zooids never fully develop and never become functional (BORG, 1926a, p. 410–416).

Eggs are fertilized internally, within the membranous sacs. How released sperm enter body cavities of maternal zooids is not known. Cross breeding is generally assumed. As soon as eggs are fertilized, the feeding organs of maternal zooids degenerate (BORG, 1926a, p. 412–419).

The embryology of modern stenolaemates is characterized by **embryonic fission (polyembryony)** in the species studied. One or rarely two primary embryos develop in a fertile zooid. Primary embryos divide to form secondary embryos, and in some species tertiary embryos are developed, presumably all with the same genetic makeup (HARMER, 1893). Embryonic fission counteracts the reproductive disadvantage of a small number of primary embryos and necessitates large brood chambers, some of which reportedly can hold as many as 100 embryos at one time.

Brood chambers are all coelomic cavities and have many forms and modes of development in tubuliporates. In many taxa they are single inflated polymorphs (gonozooids) large enough to accommodate the developing embryos (Fig. 52,8; BORG, 1926a, p. 345–

357; 1933, fig. 27). In many tubuliporates one or more fertile zooids give rise at their distal ends to large, highly inflated extrazooidal brood chambers on colony surfaces (Fig. 52,*5,7*; BORG, 1926a, p. 357–396). In some taxa middle segments of body walls of several adjacent fertile zooids are resorbed allowing eggs to escape into the space produced by the resorption (Fig. 52,*1*; BORG, 1933, fig. 28). In a few taxa these extrazooidal chambers formed by resorption can be floored by zooidal diaphragms and roofed by undisturbed outer ends of zooidal walls so that the chambers are not visible on colony surfaces (BORG, 1933, fig. 29).

The outer walls of extrazooidal brood chambers are simple calcified exterior walls in fixed-walled and many free-walled tubuliporates (Fig. 52,*3,4,6*). In some taxa of free-walled colonies the skeletal walls of brood chambers are interior walls with an outer body cavity and membranous exterior walls outward from the interior skeletal wall (Fig. 52,*1,5,7*; BORG, 1926a, fig. 92). Brood chamber apertures are developed for release of larvae (Fig. 52,*1,6–8*).

In Paleozoic stenolaemates skeletal indications of inferred brood chambers have been reported in a few taxa of two orders, the Cystoporata (see UTGAARD, this revision) and the Fenestrata (e.g., TAVENER-SMITH, 1966; STRATTON, 1975). In both orders the inflated chambers are skeletal blisters attached to outer ends of zooecia, similar in position to generally larger brood chambers of most post-Paleozoic tubuliporates.

Yet to be investigated is whether or not all modern tubuliporates have large gonozooids or brood chambers, and if not, whether they undergo polyembryony. If it were found that large brood chambers are necessary to accommodate the multiple embryos resulting from polyembryony, as it would seem, fossil taxa such as most Paleozoic species that lack skeletal indications of comparably large chambers could be assumed to have not undergone polyembryony in their reproductive cycles.

According to NIELSEN (1970), the released larvae are rounded, radially symmetrical, lack a gut, and are ciliated. They swim for a short period, apparently measured in minutes to a few hours. At metamorphosis, a posterior evagination produces an adhesive organ in contact with the substrate and an anterior evagination brings the exterior cuticle to the surface. The ciliated outer layer is turned inward by the evaginations and the ciliated cells disintegrate. The exterior cuticle covers the body, calcification begins on the inner sides of the body, and the basal disc of the first adult member of the colony, the ancestrula, is formed.

The ancestrula, the primary zooid of stenolaemate colonies (Fig. 25, 26), generally begins with an encrusting hemispherical or disc-shaped body (Fig. 52,*2*). Basal discs have exterior walls consisting minimally of an outermost cuticle, epidermis, and peritoneum.

FIG. 52. Brood chambers.——*1. Lichenopora* sp., rec., "Crab Ledge," E. of Chatham, Mass.; feeding zooid (fz), extrazooidal brood chamber (bc) with embryos in sac (es), interior skeletal wall (isw), and brood chamber aperture (bca); long. sec., USNM 250094, ×100.——*2.* Tubuliporid tubuliporate, rec., Manomet Pt., Cape Cod Bay, Mass.; young colony showing basal disc (bd) of ancestrula; sec. parallel to encrusting colony base, USNM 250095, ×30.——*3. Densipora corrugata* MacGILLIVRAY, rec., 5-m wavecut platform, Western Port Bay, W. end Phillip Is., Australia; feeding zooid in middle of brood chamber, which has exterior outer walls (ew); long. sec., USNM 250093, ×100.——*4. Plagioecia sarniensis* (NORMAN), rec., 28–30 m, Medit. Sea off Riou Is., Marseille, France; brood chamber, its exterior outer wall (ew), and aperture (bca); long. sec., USNM 250096, ×150.——*5. Hornera* sp., rec., Poor Knights Is., N.Z.; extrazooidal brood chamber with interior skeletal wall (isw) on reverse side of colony, feeding zooids (fz); long. sec., USNM 250097, ×50.——*6. Mecynoecia delicatula* (BUSK), rec., 28–30 m, Medit. Sea off Riou Is., Marseille, France; brood chamber with exterior wall (ew) and aperture (bca); long. sec., USNM 250073, ×50.——*7. Hornera* sp., rec., Flinders Is., Vict., Australia; brood chamber showing pattern of interior skeletal wall and aperture in upper right center, centered on fenestrule; exterior view, USNM 250098, ×7.5.——*8. Crisia* sp., rec., low-tide level, Eng. Channel, Roscoff, France; brood chamber with aperture (bca) and exterior walls; long. sec., USNM 250099, ×100.

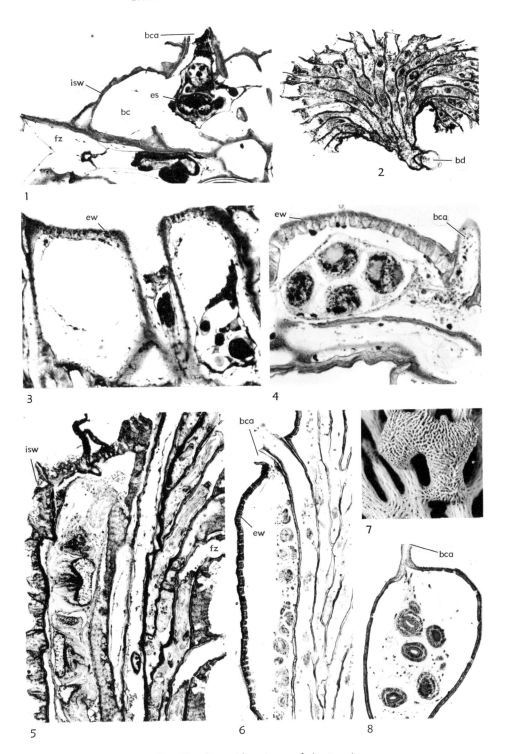

FIG. 52. *(For explanation, see facing page.)*

Most basal discs also have a skeletal layer (Fig. 53,*1,3,7*) calcified by the epidermis from within the disc. The skeletal layer is simple, that is, calcified on its growing edges and inner surfaces only.

The basal disc is generally larger in diameter than the diameter of the distal extension of the ancestrula and the diameters of living chambers of associated feeding zooids. In some Paleozoic species, however, the proximal part of the disc is smaller and may be nearly pointed (Fig. 53,*2*; BOARDMAN & MCKINNEY, 1976, pl. 7, fig. 3; CUMINGS, 1912, pl. 19, fig. 3, 4, 11, 12).

Continued growth of the simple exterior wall of the disc does not complete the skeleton of the ancestrula, but extends the wall laterally to produce the encrusting basal wall of the colony in most taxa (Fig. 25; 26; 53,*3,6,7*). The ancestrula is completed distally by skeletal body walls that are either simple and exterior as in the uniserial corynotrypids (Fig. 31,*1,2*; BOARDMAN & CHEETHAM, 1973, fig. 33A, B), compound and interior (Fig. 25; 53,*1,4,6,7*) or a combination (Fig. 26; 53,*3*). (Compound walls are calcified on edges and both sides and are, therefore, necessarily interior walls that partition existing body cavity.) The few ancestrulae of preserved colonies studied contain a feeding lophophore and gut, which retract down into the basal disc (Fig. 53,*3*; NIELSEN, 1970).

In at least some taxa of the order Fenestrata, the encrusting wall of the basal disc is reportedly not calcified from inside the disc (TAVENER-SMITH, 1969a; GAUTIER, 1972). As reconstructed, a circular flap of ectodermal epithelium (TAVENER-SMITH, 1969a, p. 295) projected from the aperture of the basal disc and folded over so that the flap rested on the exterior cuticle of the outer surface of the disc. A calcified layer was then deposited on the outer surface of the disc by the ectodermal epithelium of the flap.

It should be made clear that in this reconstruction, the hypothesized flap has to be a complete exterior membranous wall enclosing body cavity. It is assumed, therefore, that the folding places the exterior cuticles of the basal disc and flap back to back (questioned by GAUTIER, 1972). The skeletal wall of the disc is here interpreted to be an exterior wall, equivalent to the basal colony wall folded over on top of the basal disc in a lichenoporid (left basal side, Fig. 25). The distal neck of the ancestrula and skeletons of subsequent zooids and extrazooidal structures of fenestrates are interior in origin, surrounded by epidermis and body cavity on all sides.

ASEXUAL GROWTH

The aperture of the basal disc of the ancestrula is covered by an exterior membranous wall consisting of an outermost cuticle and

FIG. 53. Ancestrulae.——*1,2. Orbipora distincta* (EICHWALD); *1,* Kuckers Sh., Ord. Kohlta, Est., basal disc (bd), encrusting colony wall (ecw), primary wedge to right, secondary wedge to left, long. sec., USNM 250100, X30; *2, Echinospherites* Ls., Ord., Reval, Est., exterior view of underside of zoarium showing small, nearly pointed basal disc, USNM 250101, X4.——*3.* Fixed-walled tubuliporate, rec., Popoff Str., Alaska; ancestrula with feeding organs surrounded by membranous sac (ms) and retracted into basal disc (bd), perimetrical attachment organ (pao), distally wall of basal disc connected to encrusting colony wall (ecw), outer walls of ancestrula both exterior (ew) and interior (iw); long. sec., USNM 186542, X150. ——*4,5. Eridotrypa briareus* (NICHOLSON), Ord. (Trenton.); *4,* Cynthiana F., ancestrula (a), notch or fold, which initiates secondary wedge (n₂), direction of growth of secondary wedge (arrow), exterior encrusting wall of secondary wedge (ew), long. sec., USNM 250102, X50; *5,* Catheys Ls., 3.2 km SE. Mt. Pleasant, Tenn., ancestrula (a), notch or fold of secondary wedge (n₂), exterior encrusting wall folded over (ew), direction of growth of secondary wedge (arrow), deep sec. parallel to encrusting colony wall, USNM 250103, X50.——*6–8. Lichenopora* sp., rec., Galapagos Is.; ancestrula (a), basal disc of ancestrula (bd), exterior wall of ancestrula (ewa), notch or fold of primary wedge of zooids (n₁), notch or fold of secondary wedge of zooids (n₂), direction of growth of secondary wedge (arrow), exterior encrusting colony wall (ecw), exterior encrusting colony wall of secondary wedge (ew₂, ecw₂), feeding zooids (fz), polymorphs (pm), vertical walls (vw); *6,* long. sec., X75, *7,* long. sec., X300, *8,* deep sec. parallel to encrusting colony layer, X100, all BMNH specimens.

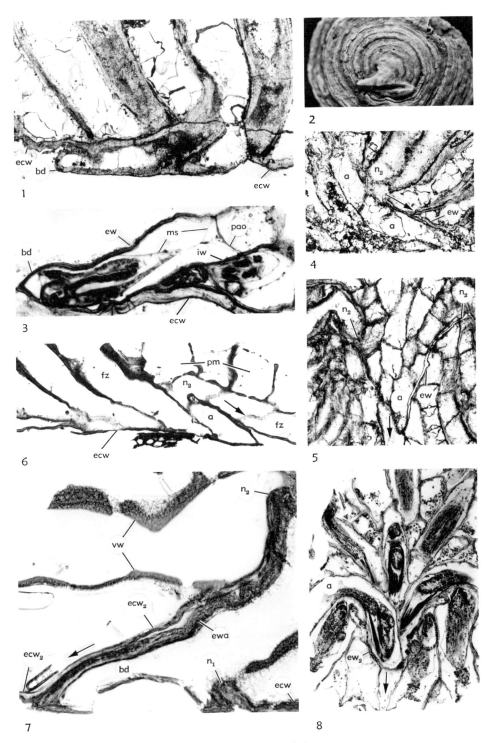

FIG. 53. (*For explanation, see facing page.*)

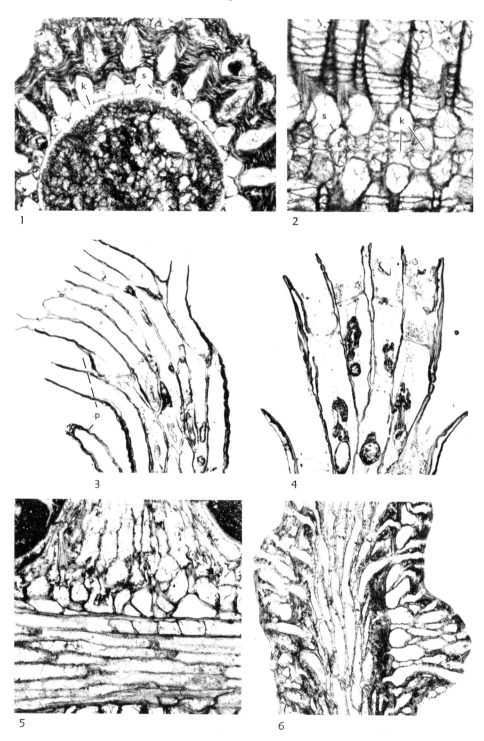

FIG. 54. *(For explanation, see facing page.)*

epidermis. The cuticle is expanded from within itself by multiplying epidermal cells, and this growing cuticle apparently is present in all exterior walls.

Zooidal wall development and confluent budding zones.—In stenolaemate bryozoans, asexual reproduction of zooids (**budding**) begins by the growth of interior vertical walls into existing confluent body cavities. Interior vertical walls of buds are initiated by localized growth produced by infolding into existing body cavity of epidermal cell layers from established skeletal surfaces (Borg, 1926a, p. 322). Skeletal layers of vertical walls (and the entire skeleton of stenolaemate colonies) are, therefore, secreted on outer sides of the epidermis and are exoskeletal throughout. (For an endoskeletal interpretation of vertical wall growth, see Ross, 1976.)

Vertical walls of buds grow from: (1) encrusting basal walls of colonies that are exterior and multizooidal in origin (Fig. 25; 39,5), (2) erect walls of colonies that are exterior and multizooidal in origin (reverse walls of many unilaminate colonies; Fig. 26; 28,5,6), (3) erect walls of colonies that are interior and multizooidal (median walls of bifoliate colonies, Fig. 30,1,3a; possibly reverse walls of some unilaminate colonies), (4) walls of zooids that are interior (dendroid colonies, Fig. 48,4; some unilaminate colonies, Fig. 28,1a,b), (5) extrazooidal parts that are interior (few cystoporates, see Utgaard, this revision), and (6) peristomes of fixed-walled zooids that are exterior (few tubuliporates, Harmelin, 1976, fig. 7).

Most budding is interzooidal, that is, it occurs outside of living chambers of zooids. Buds commonly are centered on growing edges or corners of interior vertical zooidal walls that are necessarily shared by 2 to 4 older supporting zooids. The growing edges of vertical walls of buds and contiguous supporting zooids are grown cooperatively and advance evenly into confluent budding spaces. Therefore, these buds never occupy spaces within living chambers of supporting zooids, regardless of whether the buds are centered on walls or centered on the living chamber of older zooids on encrusting colony walls. In the great majority of taxa, therefore, buds can not be related to single parent zooids.

Intrazooidal budding does occur where buds develop from within established living chambers of single supporting zooids. The budding of subcolonies in some multilaminate stenolaemates (Hillmer, 1971, p. 27, fig. 4, 5, 25, 26) is an example. (For a different concept of intrazooidal budding, see McKinney, 1977b.)

Exterior membranous walls and enclosed confluent body cavities precede budding distally, apparently in all but uniserial stenolaemates. Confluent budding spaces connect body cavities of a few to many existing buds or combinations of buds and zooids. The entire confluent budding zone (apparently the common bud of authors) includes the confluent body cavity, enclosing membranous exterior walls, and any exterior multizooidal basal wall that is present (Fig. 25; 26; 39,5).

Confluent multizooidal budding zones

Fig. 54. Asexual growth.——*1. Ceramophylla vaupeli* (Ulrich), Ord. (Eden.), Brown Street, Cincinnati, Ohio, paralectotype; sinuses (s) and keels (k) in recumbent endozones on basal colony walls of hollow-branched zoarium; transv. sec., USNM 245040, ×30.——*2. Peronopora decipiens* (Rominger), Corryville Mbr., McMillan F., Ord. (Maysvill.), Cincinnati, Ohio, lectotype; sinuses (s) and keels (k) developed irregularly from median wall; transv. sec., UMMP 6676-3, ×50.——*3. Mecynoecia delicatula* (Busk), rec., 35 m, Grand Salaman, Marseille, France; growing tip of branch of fixed-walled colony with feeding zooids opening into confluent budding space, peristomes (p); long. sec., Harmelin Coll., ×50.——*4. Cinctipora elegans* Hutton, rec., 110 m, off Otago Heads, South Is., N.Z.; growing tip of branch of free-walled colony with confluent zooidal budding zone; long. sec., USNM 250064, ×30.——*5. Polycylindricus clausus* Boardman, Centerfield Mbr., Ludlowville F., Dev. (Erian), Paines Cr., Cayuga Lake, N.Y.; secondary branch (projecting upward) grown from exozone of supporting branch without an overgrowing basal encrusting wall; long. sec., USNM 250104, ×20.——*6.* Free-walled tubuliporate, Paleocene, Vincentown, N.J.; secondary branch to right; long. sec., USNM 250105, ×30.

Bryozoa

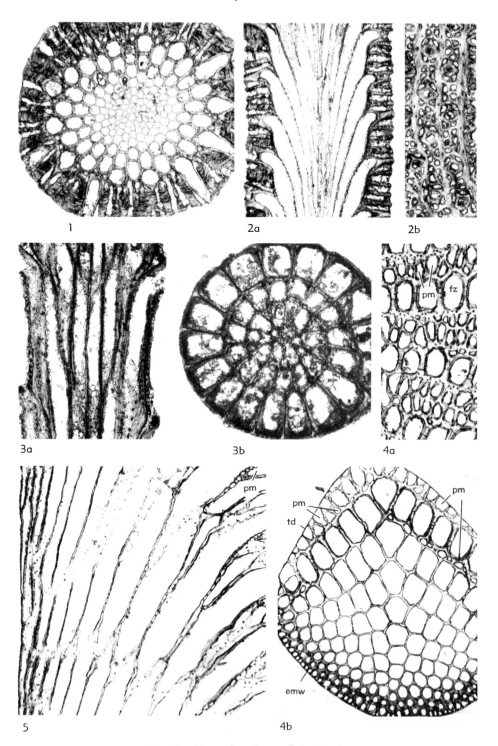

FIG. 55. *(For explanation, see facing page.)*

occur opposite endozones that contain only buds (Fig. 32, 33). The confluent budding space and enclosing exterior walls originate and are at the time of budding outside of zooidal boundaries. As colony growth proceeds and budding zones advance distally, proximal parts of confluent multizooidal budding spaces and enclosing walls become parts of zooids.

Confluent zooidal budding zones occur where buds are interspersed with fully developed zooids (Fig. 54,*3,4*). The budding zone is considered zooidal because the available confluent spaces and enclosing exterior walls are parts of established zooids (Fig. 37) before budding begins. As buds develop, the interspersed zooids share the expanding confluent space with the intervening buds. The relative concentrations of fully developed zooids and buds in distal budding zones can be determined by zooecial patterns in sections cut through more proximal parts of many colonies.

Buds develop in endozones from basal encrusting walls of both free- and fixed-walled colonies and grow into multizooidal body cavities, which are peripherally confluent around margins of colony bases (Fig. 25; 39,*5*; 60,*1–4*).

In free-walled colonies, growing regions of both endozones and exozones are potential budding zones because outer body cavities are confluent over both regions. Budding in exozones generally occurs in confluent zooidal spaces because zooids are fully developed there and all confluent spaces are either parts of zooids or are extrazooidal (Fig. 25). Endozonal budding occurs in multizooidal budding zones in distal ends of erect parts of some taxa of free-walled colonies of unilam-

inate (Fig. 55,*4,5*), bifoliate (Fig. 54,*2*), and dendroid (Fig. 50,*5a,b*; 55,*1,2*) growth habits. Endozonal budding occurs in zooidal confluent budding zones containing interspersed feeding zooids and buds in the distal ends of erect free-walled colonies of some unilaminate (Fig. 30,*2*) and dendroid (Fig. 54,*4*; and possibly Fig. 48,*4*) taxa.

In fixed-walled colonies, confluent budding zones and budding occur only in the most distal regions of colonies, at growing margins and tips. Most budding in fixed-walled colonies, therefore, occurs in endozones and not in exozones. In some taxa of erect fixed-walled colonies, buds are grouped at distal ends and grow into multizooidal budding spaces (Fig. 26; 30,*3a*; 33,*2*; 36,*4a,b*; 41,*5c*; 55,*3a,b*). In other fixed-walled taxa buds are interspersed with established feeding zooids and grow into zooidal budding spaces (Fig. 54,*3*).

Zones of astogenetic change.—The ancestrula and the one to several asexually produced generations of founding zooids commonly differ morphologically from more distally placed zooids. Some of these differences are sequential by generation, are not entirely assignable to ontogeny or polymorphism, and are generally too constant from colony to colony of the same species to be interpreted as microenvironmental in origin. These sequential differences occur as regular developmental features of colonies and, therefore, are assumed to be expressions of astogeny.

The sequential changes of the earliest generations of a colony provide a morphologic transition between the single ancestrula and the complex of zooids and extrazooidal structures that are either repeated or continued

FIG. 55. Zooidal patterns.——*1,2. Petalopora* sp., Cret. (Coniac.), Villedieu, Loire-et-Cher, France; *1,* zooecia arranged radially, grown from multizooidal budding zone around branch axis, transv. sec., USNM 216468, ×30; *2a,b,* polymorph small tubes in exozones, *a,* long. sec., *b,* polymorphs and feeding zooecia of same zoarium cut transversely, tang. sec., both USNM 250106, ×30.——*3a,b. Spirentalophora* sp., Cret. (Coniac.), Villedieu, Loire-et-Cher, France; spiral zoarial pattern, budding at axis into multizooidal budding zone; *a,b,* long., transv. secs., USNM 213321, ×50.——*4,5. Tennysonia* sp., rec., Algoa Bay, S. Afr.; feeding zooecia (fz) budded from exterior multizooidal wall (emw) on reverse side of zoarium, small polymorphs (pm) bud in outer exozone, apertures of polymorphs covered by terminal diaphragms (td); *4a,b,* tang., transv. secs., USNM 216467, ×30; *5,* long. sec., USNM 216466, ×30.

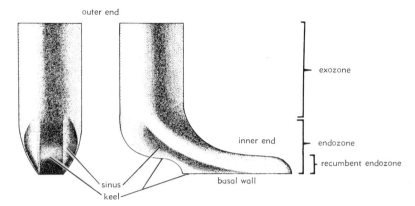

FIG. 56. Zooidal patterns. Idealized drawing of single zooecium from encrusting or bifoliate colony showing generalized shape, including keel and lateral sinuses.

during the growth of colonies. The ancestrula and the one or more transitional generations of a colony are together called the primary zone of astogenetic change (Fig. 21, 27).

In zones of change of many stenolaemates, the exterior wall of the basal disc of the ancestrula grows laterally to become the basal encrusting exterior wall of the colony. Many colonies, initially at least, grow in one general direction from the ancestrula along the substrate, here called the **primary direction of encrusting growth** (Fig. 25, 26, right of disc). The wall of the basal disc develops a small fold (Fig. 53,7) on the primary-growth-direction side, which takes the wall of the disc down to the substrate to be continued laterally as the encrusting colony wall. The distal part of the ancestrula commonly bends toward the primary direction of growth. Zooids of the first encrusting generations bud from the encrusting colony walls into confluent budding zones and zooids of many species display a subparallel orientation to form generally wedge-shaped young colonies of variable proportions (Fig. 52,2), the **primary wedge of encrusting zooids**.

If the downfold of the wall completely encircles the colony (Fig. 25, left side, and 53,6,7), the encrusting colony wall can grow laterally and support progressively younger generations of zooids in all directions from the ancestrula. This encircling exterior wall

provides a basal colony wall for a **secondary wedge of encrusting zooids** growing opposite to the primary direction of growth (Fig. 25, left side; 53,6, right side; CUMINGS, 1912; BOARDMAN, 1971, pl. 3, fig. 4; BOARDMAN & McKINNEY, 1976, pl. 7, fig. 2a,b). Contacts between primary and secondary wedges of zooids produce a typical discordant pattern as seen in deep sections parallel and perpendicular to encrusting colony bases in some stenolaemates of all ages (Fig. 53,*1,4–6,8*).

The number of generations of zooids in both primary and secondary wedges that constitute primary zones of astogenetic change varies in different taxa. Other arrangements of zooids and multizooidal structures in the zone of change have not been described in detail.

Zones of astogenetic repetition.—Zooids commonly develop in repeated patterns and extrazooidal structures are extended to establish colony growth habits distal to zones of change in colonies. These distal parts of colonies are termed zones of astogenetic repetition (Fig. 27). A zone of repetition begins with the first generation of zooids that repeats the morphologies of zooids of the preceding generation. Zooidal patterns and repeated maculae and subcolonies described below are from zones of repetition.

Zooidal patterns.—Zooidal patterns are the three-dimensional shapes and interrela-

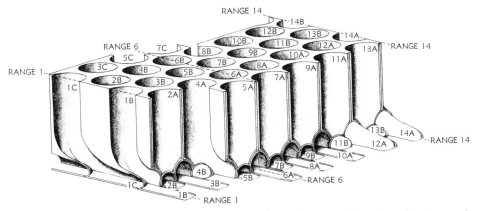

FIG. 57. Zooidal patterns. Idealized cutaway diagram of part of an encrusting colony, showing zooecia similar in shape to the one in Figure 56 arranged in the basic rhombic pattern of zooids in stenolaemate colonies. Recumbent segments of zooecia are long enough in this specimen for adjacent zooecia to overlap in numbered range as seen in longitudinal section. The longitudinal section is cut along the left front side, the transverse section along the right front side.

tionships of zooids within colonies. They are particularly useful in understanding the modes of growth of colonies and in differentiating taxa. Most of the more common growth habits of colonies as described externally can be produced by several different internal patterns of zooids. The common arrangement of zooids in rhombic patterns on colony surfaces in stenolaemates can be produced by a number of different internal zooidal patterns (see discussion of *Petalopora* and *Meliceritites* below). Patterns of zooids and their positional relationships with multizooidal and extrazooidal skeletal structures provide character states that are generally constant enough in occurrence to suggest a high degree of genetic control, and can be expected to produce a more detailed classification.

Factors that are basic to understanding zooidal patterns in three dimensions include: (1) **budding patterns**, that is, shapes of buds and their relative positions on supporting structures; (2) the three-dimensional shapes of zooids during their ontogeny; (3) the manner in which zooids or zooids and adjacent skeletal structures fit together; and (4) the position of depositing epidermis relative to skeletal microstructures.

Three-dimensional regularity of zooidal patterns is indicated by regularity in patterns of oriented two-dimensional sections. It is necessary to convert the two-dimensional patterns to three-dimensional reconstructions to understand fully zooidal patterns and the way zooids fit together to form colonies (for example, see BOARDMAN & MCKINNEY, 1976).

Zooids in many taxa develop **sinus** and **keel** configurations in recumbent endozones (Fig. 56–58; BOARDMAN & UTGAARD, 1966, p. 1083) of encrusting colonies or erect bifoliate colonies. The sinus and keel shape of the zooids allows their narrow recumbent portions (Fig. 54,*1,2*) to fit together in a generally rhombic arrangement and to expand into the full cross-sectional size of zooids in exozones. Variations in this basic pattern can be caused by a number of factors, including differing lengths of recumbent zones, intervening extrazooidal skeleton modifying zooecial shapes, patterns other than rhombic for relative budding positions, and irregular substrates.

In a species of *Tennysonia,* a free-walled tubuliporate (Fig. 55,*4,5*), feeding zooids bud from exterior multizooidal walls on the back side of a unilaminate colony without

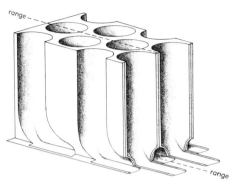

FIG. 58. Zooidal patterns. Idealized cutaway diagram of part of an encrusting colony in which recumbent segments of zooecia are short enough that adjacent zooecia do not overlap in range. Note lack of the mushroom shape caused in transverse section of Figure 57 by zooecial overlap.

developing the sinus and fold. The six-sided zooids fit together in a rhombic pattern and increase in cross-sectional area ontogenetically toward the front of the colony (Fig. 55,*4b*, top). Small polymorphs are budded (pm) near the front surface of the colony in transverse rows, forcing feeding zooids out of the rhombic pattern and into an alternating transverse pattern with the polymorphs (Fig. 55,*4a*). The outermost row of zooids in the transverse section (Fig. 55,*4b*) contains the polymorphs that are covered by terminal diaphragms.

In a species of *Petalopora,* a free-walled tubuliporate (Fig. 55,*1,2*), feeding zooids bud from an axial region with the buds positioned so that the six-sided zooids remain in a rhombic pattern as they extend radially to the surface. The numerous small polymorphs are restricted to the outer exozone.

Spiral zooidal patterns are formed when bud locations occur in a spiral about an axial structure or region. A simple spiral pattern occurs in a species of *Spirentalophora,* a fixed-walled tubuliporate. The buds of feeding zooids are spaced spirally about a linear axis (Fig. 55,*3a,b*) so that the four-sided zooids are aligned radially in transverse section. The outer sides of the zooids at any one level combine to form a continuous outwardly spiraling wall as the zooids develop ontogenetically. Apparently the zooids vary progressively in length because the zooidal apertures are arranged in annular rings at the colony surface, giving little external indication of a spiral budding pattern.

In some species of *Meliceritites* the buds of feeding zooids are arranged spirally about an axial cylinder (Fig. 36,*4a,b*) and are so closely spaced that the transverse view suggests two alternatives. Either the zooids grew radially out to the exozone in a rhombic pattern, or the zooids themselves curved in a clockwise direction part way around the branch axis as they grew. The zooids remain in profile throughout their length in a longitudinal plane through the center of the branch (Fig. 36,*4a*), demonstrating that the zooids are radially arranged in the rhombic pattern that shows on the colony surface. (For another spiral pattern of different origin, see Fig. 48,*6–8*.)

Maculae.—Maculae (monticules of some authors) occur in the exozones of many Paleozoic genera and have been reported in

FIG. 59. Maculae.——*1,2. Constellaria* sp., Catheys F., Ord. (Mohawk.), E. side Harvey Knob, N. of Liberty Pike about 8 km E. of Franklin, Tenn.; *1,* radial or stellate maculae surrounded by feeding zooecia, tang. sec., USNM 250107, ×20.0; *2,* limb of macula cut transversely showing solid to vesicular skeleton and feeding zooecia (fz) on either side, long. sec., USNM 250108, ×20.0.——*3. Constellaria florida prominens* ULRICH, Mount Hope Sh. Mbr., Fairview F., Ord. (Maysvill.), reservoir near Newport, Ky.; star-shaped maculae in relief; external view, USNM 189916, ×1.5.——*4a,b. Amplexopora* sp., Mount Auburn Sh. Mbr., McMillan F., Ord. (Maysvill.), Cincinnati, Ohio; macula in exozone surrounded by feeding zooecia (fz) showing some budded polymorphs of irregular shape; *a,b,* long., tang. secs. of same zoarium, USNM 250109, ×30.0.——*5,6. Crepipora venusta* (ULRICH), Economy F., Ord. (Eden.), river quarries, W. Covington, Ky., paralectotypes; macula of small tabular polymorphs surrounded by feeding zooecia (fz) with lunaria (lu); *5,* tang. sec., USNM 159707, ×30.0; *6,* long. sec., USNM 213295, ×30.0.

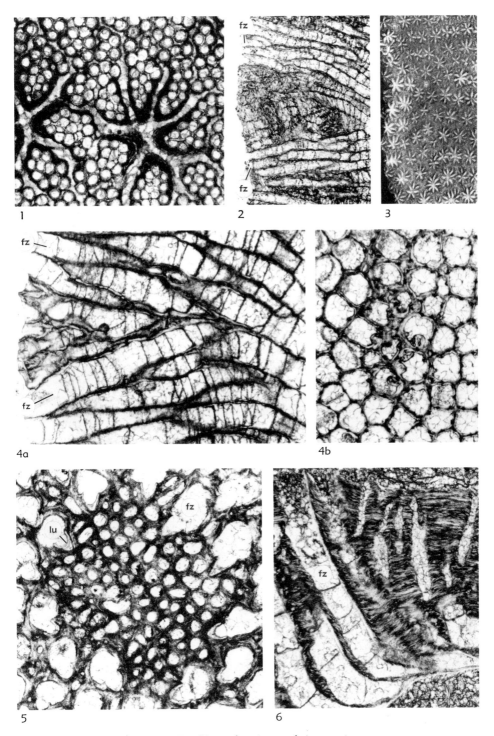

FIG. 59. *(For explanation, see facing page.)*

post-Paleozoic tubuliporates (e.g., TILLIER, 1975; NYE, 1976, pl. 13, fig. 1b, pl. 20, fig. 1b; TAYLOR, 1975). Maculae are generally small equidimensional clusters of polymorphs, or polymorphs in combination with possible feeding zooids, extrazooidal skeleton, or both. They are isolated from each other by areas dominated by assumed feeding zooids and are more or less regularly spaced on colony surfaces (Fig. 59,3). They commonly form prominences, or less commonly their surfaces are flush with or depressed below the colony surface. The word macula is used here instead of monticule because its more general definition better satisfies empirical requirements as expressed on colony surfaces, that is: any of various anatomical structures having the form of a spot differentiated from surrounding tissues.

Some of the common skeletal details that distinguish maculae from surrounding regions of feeding zooids include: differences in size and shape of zooids; increased zooecial wall thicknesses; changing distinctness of zooecial boundaries; different intrazooidal structures such as basal diaphragms and cystiphragms, or differences in their spacing, configuration, or both; increased size and concentration of extrazooidal styles; and central masses of extrazooidal skeleton.

The prominence of maculae on zoarial surfaces attracted attention to them and their possible function or functions early in the study of Paleozoic Bryozoa. A reproductive function was suggested by ULRICH (1890, p. 940), which has been referred to by many subsequent authors. ULRICH compared the large polymorphs occurring in most maculae with the gonozooids of living stenolaemates. Gonozooids are large enough to brood several embryos and therefore much larger than associated feeding zooids. The relative differences in size, however, between feeding zooids and macular polymorphs in many Paleozoic species is less than size differences between feeding zooids and gonozooids of living tubuliporates. Moreover, the character is not constant; some species within genera and even some colonies within species have

maculae, and other congeneric species or conspecific colonies do not (e.g., BOARDMAN & McKINNEY, 1976, p. 60). In some cystoporate genera (UTGAARD, this revision) colonies have both blisterlike skeletal structures on distal ends of some zooecia, which appear to be brood chambers, and maculae with large polymorphs. The reproductive hypothesis for larger polymorphs in maculae of taxa of Paleozoic age should be viewed as speculative until better evidence is available.

Budding of polymorphs occurs in maculae of some taxa. Both maculae and included smaller polymorphs, commonly mesozooecia or exilazooecia (Fig. 59,5,6) are restricted to exozones, so that the smaller polymorphs occurring in maculae are necessarily budded there. In most taxa in which maculae occur, however, budding in maculae produces few polymorphs as large as associated feeding zooids. Relatively narrow exozones typical of erect colonies generally provide little opportunity for budding of the larger polymorphs within maculae (Fig. 59,4a,b). In erect parts of colonies the great majority of both feeding zooids between maculae and larger polymorphs within maculae are budded in endozones at growing tips.

In the larger **massive** and **hemispherical colonies,** endozones can be relatively narrow on basal colony layers and exozones can be many times wider. Exozones can be either uninterrupted throughout most of a colony, or interrupted by endozone-exozone cycles. The cycles are produced by intracolony overgrowths indicated by basal encrusting walls, or rejuvenations in which zooidal chambers are continuous between cycles except for possible basal diaphragms or abandoned chambers. Maculae have more opportunity to contribute polymorphs in the wider exozones of massive and hemispherical colony, and distally these polymorphs can become feeding zooids between maculae in a few species (ANSTEY, PACHUT, & PREZBINDOWSKI, 1976). In massive and hemispherical colonies zooids are budded in varying proportions from basal colony walls, basal walls of overgrowths, other zooids between maculae, and from other zooids within maculae.

There seems to be no morphologic evidence that deregulation of the budding rate of a macula can produce a branch in an erect colony as suggested by ANSTEY, PACHUT, & PREZBINDOWSKI (1976, p. 144). Most branching stenolaemates have uninterrupted endozones from supporting stalks to branches, without intervening maculae at branch bases. A necessary function of endozones is the asexual reproduction of zooids in budding zones at distal ends of colony branches. Budding and more rapid growth of the thinner zooid walls in endozones produce the distal lengthening of branches. Maculae, where present, develop proximal to distal ends in exozones and grow relatively slowly and laterally at right angles to branch length. Certainly maculae are not necessary for branching because many branching species lack maculae.

Branches grown from exozones proximal to growing tips occur in a few species. These secondary branches are generally smaller in diameter at bifurcations than supporting branches and grow at right angles by rejuvenation on supporting branch exozones (Fig. 54,5,6). Two trepostomate species that developed secondary branches (see *Polycylindricus* BOARDMAN, 1960, p. 67) do not have recognizable maculae and the branches arose by rejuvenation from outer surfaces of supporting branch exozones without skeletal interruption of living chambers and with little or no budding. The secondary branches have both endozones and exozones.

See the following section on feeding currents for further discussion of possible functions of maculae.

Feeding currents and subcolonies.—Recent observations of colony-wide feeding currents in several species of stenolaemate suggest that feeding currents produced by ciliated tentacles of zooids were also colony-wide in many fossil species. Colony growth habit and spatial patterns of different kinds of zooids and extrazooidal skeleton on colony surfaces are major factors in the production of colony-wide feeding currents.

As has long been known, feeding currents

of a zooid are incoming toward the mouth and surrounding colony surface. They are produced by motion of cilia on the tentacles. The tentacles themselves are nearly motionless in an expanded feeding position unless struck by larger particles. To reject such particles, one to several tentacles bat the particles out of the incoming current and away from the mouth area. Rejected particles can be bounced from zooid to zooid until they are finally taken beyond the colony. (For detailed discussion of morphology and feeding behavior of bryozoans, see WINSTON, 1978.)

Some basic assumptions can be made relative to the formation of colony-wide feeding currents and to the reconstruction of hypothetical feeding currents for fossil colonies. Surely more assumptions will be suggested as more living colonies of stenolaemates are observed.

1. The prevailing directions of incoming currents of feeding zooids are presumably parallel to the central axes of the outermost lengths of zooidal living chambers. **Tentacle crowns** in recent stenolaemates do not extend far enough beyond skeletal apertures for lophophores to bend independently of zooidal walls. As a result, current directions set up by zooids presumably must parallel their axes. This assumption is more speculative in taxa of Paleozoic age because lengths of extensions of tentacle crowns are unknown.

In contrast, tentacle crowns of cheilostomates can bend in different directions, causing changes in current direction. For example, in a broad unilaminar cheilostomate genus the tentacle crowns of clusters of a few zooids lean away from the centers of the clusters to form excurrent chimneys that permit unopposed outflows of water. No indications of chimneys are reflected in zooidal skeletons (BANTA, MCKINNEY & ZIMMER, 1974).

2. Colonies with broad interrupted surfaces dominated by feeding zooids presumably have some method that permits incoming water to escape from colony surfaces without passing out through actively feeding tentacle crowns and thereby opposing incoming currents. This assumption is supported

partly by observations of different methods employed to release water from colony surfaces in living species, only a few of which are discussed below.

3. Any colony surface area that lacks feeding zooids or in which feeding zooids are not feeding, and which is large enough to be unaffected by surrounding incoming currents, will function as an excurrent chimney because outflow is unopposed.

4. Skeletal apertures of feeding zooids in many taxa are raised by peristomes above the colony surface so that water can escape to colony margins or excurrent chimneys along colony surfaces between peristomes and under tentacle crowns (Fig. 54,3; 61,1,4a).

5. In some taxa, spacing between skeletal apertures of adjacent feeding zooids can be wider than tentacle crowns so that unopposed excurrent space surrounds single zooids. Wider spaces between skeletal apertures may result from sparse budding patterns, the thickening of vertical walls in exozones, intervening extrazooidal skeleton, the growth of frontal walls, diverging peristomes, or presence of interspersed nonfeeding polymorphs. These spacing factors are expressed skeletally but are difficult to evaluate in most fossil colonies because of lack of evidence of diameters of tentacle crowns. It can be generally assumed, however, that lengths of feeding tentacles will be less than axial lengths of their living chambers, because tentacles of living stenolaemates are more or less straight in retracted positions.

6. Colonies of slender branches of one to several feeding zooids at any one level apparently need no special arrangements for water removal because water apparently can flow past branches relatively unimpeded. Unilaminate **fenestrate colonies** are a growth habit modification in which slender branches separated by rectangular open spaces called **fenestrules** are arranged in a reticulate pattern to form broad fronds (Fig. 60,1). In living fenestrate cheilostomes, feeding tentacle crowns pump incoming water through fenestrules and out past the nonzooidal or reverse sides of the fronds. It is assumed that this is also the normal feeding current direction for fenestrate stenolaemates of all ages (e.g., McKINNEY, 1977a).

Most recent species of the free-walled tubuliporate genera *Lichenopora* and *Disporella* are small, circular, convex colonies in which feeding zooids are arranged in radial rows and have long interior-walled peristomes (Fig. 60,2–4). Polymorphs occur between rows of feeding zooids. The polymorphs are without tentacles and form a general zoarial surface (Fig. 25) below peristome apertures and therefore below feeding tentacle crowns (Fig. 60,3, left side). The lower surfaces formed by polymorphs rise toward high central areas consisting of brood chambers or polymorphs, both lacking feeding tentacles.

In this radial growth habit, walls of feed-

FIG. 60. Feeding currents.——*1.* Cystoporatid encrusting reverse side of fenestellid, Road Canyon F., Perm. (Leonard.), 2.4 km N. 19° W. of Hess Ranch House, Hess Canyon Quandrangle, Texas; radial arrangement of feeding zooecia with lunaria budded from encrusting colony wall (ecw); fenestrules (fn) provided passageway for feeding currents through frond of colony; exterior view, USNM 250110, ×10. ——*2. Disporella* sp., rec., Jamaica; feeding zooids arranged radially around large central area of polymorphs lacking tentacles; external view, USNM 250111, ×8.——*3. Disporella* sp., rec., off Riou Is., Marseille, France; polymorphs (pm) on left side of section rise to central area (ca) of colony and form lower zoarial surface at general level of dashed line, feeding zooids (fz) with peristomes (p), encrusting colony wall (ecw); long. sect., USNM 250112, ×30.——*4. Lichenopora* sp., rec., 10–20 m, between Rotones and Caribe Is., Puerto Rico; radially arranged feeding zooids around central brood chamber with large aperture at upper right; external view, USNM 250113, ×15.——*5,6. Prasopora* sp., Ord. (Trenton.), Trenton Falls, N.Y.; *5,* feeding zooecia with cystiphragms (c) surrounding living chambers (lc) radially arranged on sides of zooecia nearest center of macula (m), consisting of smaller mesozooecia, tang. sect., USNM 250114, ×20; *6,* macula (m) in center indicated by smaller, closely tabulated mesozooecia, surrounded by feeding zooecia containing cystiphragms (c) and living chambers (lc) that change in position from center of zooecia at 1 to sides nearest maculae at 2, long. sec. USNM 250115, ×10.

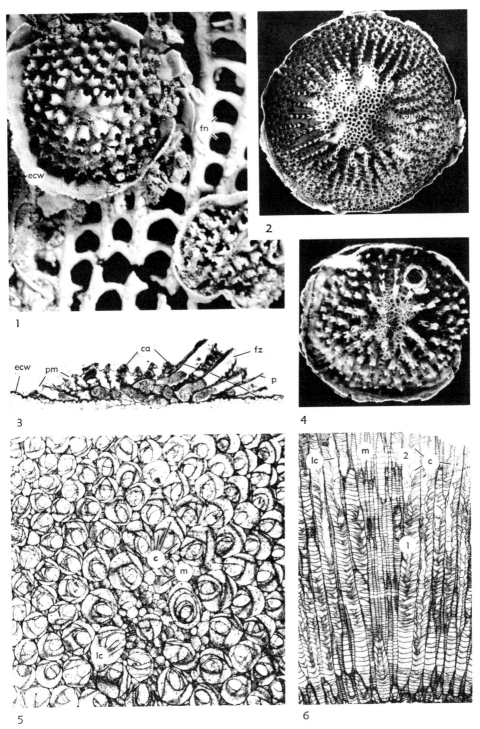

Fig. 60. *(For explanation, see facing page.)*

ing zooids commonly bend away from colony centers distally (Fig. 60,3) so that incoming zooidal feeding currents (Fig. 25) that pass between tentacles and tentacle crowns are directed (assumption 1) along the lower surfaces formed by the polymorphs (assumption 4). These currents are reflected up to colony centers, apparently because of the inward-facing obtuse angles between peristome axes and the lower surfaces. In colony centers excurrent chimneys are formed because there are not feeding zooids to set up incoming currents to oppose outflow (assumption 3). For a contrasting analysis of the origin of feeding currents of *Lichenopora,* see COOK (1977).

Outgoing currents from centers of lichen-oporid colonies are strong, rising several colony thicknesses above the colonies, where any lateral currents of surrounding environments could carry rejected debris away. The colony-wide currents also have the advantage of keeping colony surfaces free of moderate amounts of settling mud in quiet-water environments. Presumably this kind of cooperative action among feeding zooids of a colony is more efficient than zooids acting individually in both food intake and colony cleaning. These may be reasons why the radial growth habit has developed independently many times in stenolaemate history. For examples of comparable colonies of tubuliporate species of Cretaceous age, see BROOD (1972, pl. 45–47, 50).

A cystoporate species of Permian age (Fig. 60,1) has radial surface features comparable to those of recent lichenoporids, suggesting similar colony-wide currents. Several of these colonies occur on the reverse side of a large, erect fenestrate frond. If both lichenoporid and fenestrate colonies were alive at the same time, the feeding currents passing through the fenestrules may have been reversed and captured in the feeding currents of the smaller radial colonies.

Larger colonies in some species of *Lichenopora* develop several radial centers (see Fig. 9,5). Each of these centers presumably develops radially incoming feeding currents

and central excurrent chimneys. Repeated morphologic groupings on colony surfaces of many recent and fossil taxa suggest the concept of subcolonies. **Subcolonies** are groupings of zooids and any extrazooidal structures within colonies, which may or may not be skeletally identifiable, but which carry out most or all of the functions of whole colonies. In many taxa containing subcolonies, the subcolonies develop in exozones of zones of repetition. It is not implied here that subcolonies are necessarily independently budded units.

Among fossil stenolaemates, many maculae apparently were subcolonies. In *Constellaria,* a cystoporate genus of Ordovician age (Fig. 59,1–3), the distinctive radial maculae compare closely with the radial subcolonies of recent species of *Lichenopora.* Species of *Constellaria* range from small circular colonies of one macula to erect branching colonies of many maculae. Some of the presumed feeding zooids of *Constellaria* are radially arranged in the stellate maculae but do not develop isolated peristomes. Thin-walled, closely tabulated mesozooecia or vesicles, both lacking living chamber space, form the interrays. The interrays can be lower than, flush with, or above apertural levels of the feeding zooids (UTGAARD, this revision), so that excurrent chimneys might have been at the center of the macula as in *Lichenopora,* or over the stellate nonfeeding interrays. For examples of comparable maculae of post-Paleozoic age, see HILLMER (1971, pl. 22, fig. 9) and NYE (1976, pl. 13, fig. 1b).

The concept of many types of maculae as subcolonies or centers of subcolonies is suggested in some taxa of Paleozoic age by the radial orientation of eccentrically placed living chambers on sides of feeding zooids either nearest to (Fig. 60,5,6), or farthest from (BOARDMAN & UTGAARD, 1966, p. 1094), centers of the nearest maculae. In monticuliporid trepostomates, cross-sectional areas of living chambers of feeding zooids are considerably reduced from areas of entire zooecia by skeletal cystiphragms. Living chambers are on proximal sides of zooecia and cystiphragms are concentrated on distal sides in early growth stages near endozonal-exozonal boundaries. As maculae developed during ontogeny in a few monticuliporids, living chambers and cystiphragms of some of the zooids both within and surrounding the maculae twisted around zooidal axes so that living chambers were nearest to centers of the nearest maculae (Fig. 60,6). The amount of twisting was variable and controlled, resulting in living chambers of nearby zooids being radially oriented around macular centers in later growth stages. In some species it is possible to divide most feeding zooids into groups surrounding adjacent maculae based on radial orientation of living chambers in later growth stages.

Maculae and surrounding zooids with radially oriented living chambers such as those in the monticuliporid trepostomates are interpreted as subcolonies because that orientation itself suggests a cooperative function. Macular centers in these species generally consist of clustered mesozooecia. The best functional inference presently to be made is that these macular centers resulted in excur-

rent chimneys (assumption 3). The eccentricity of living chambers in some monticuliporids may not have affected feeding currents because in other monticuliporid species, living chambers remained on proximal sides of feeding zooids throughout their ontogeny and no radial orientation developed.

Maculae of many Paleozoic species consist of clusters of larger polymorphs that form prominences above intermacular feeding zooids. The macular polymorphs are larger than adjacent feeding zooids and have larger living chambers. At present there is no evidence that these larger polymorphs lacked tentacles, that they were not extended when surrounding zooids were feeding, or that their cilia created outgoing currents. There seems to be no evidence, therefore, that these maculae formed excurrent chimneys. Nevertheless, these maculae commonly occur in large colonies that should have had some provision for outgoing currents. Observations of large living stenolaemate colonies having closely spaced feeding zooids may suggest methods of forming excurrent chimneys not necessarily reflected in skeletons.

GENETIC AND ENVIRONMENTAL CONTROL, COLONY INTEGRATION, AND CLASSIFICATION

The procedure preferred here for obtaining character states for use in phylogenetic classifications is described in the introduction to this revision in the section on taxonomic character analysis. The goal of character analysis is to obtain states of morphologically independent characters that are largely genetically controlled.

A character should be morphologically independent to the extent that its observable states are not partly determined by states of other characters within the group of taxa being classified. Independent characters can be derived from morphologic units ranging organizationally from single cells to entire colonies. Such characters are determined to be independent only by comparisons among potentially homologous morphologic struc-

tures. These structures are generally similar in mode of growth and most have some functions in common. For example, frontal walls of gymnolaemates and stenolaemates are potentially homologous. They are exterior in origin in both classes. At class level the flexibility of frontal walls in tentacle protrusion is an independent character whose states separate the two classes, flexible in gymnolaemates and inflexible in stenolaemates.

Dependent characters are not considered in the classification. These are of at least two types, redundant and ambiguous. Ambiguous characters can produce equivocal results because they combine states of two or more characters, which can vary independently from specimen to specimen. For example, a commonly cited character of Paleozoic stenolae-

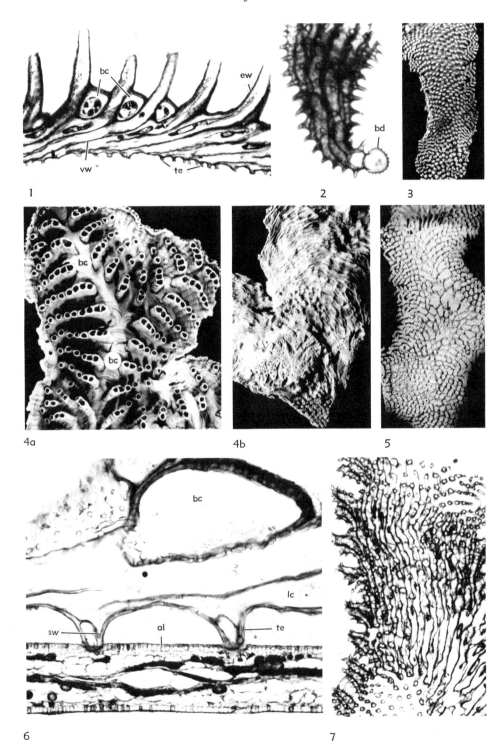

Fig. 61. *(For explanation, see facing page.)*

mates is the number of zooecia of feeding zooids in a standard area or length. Several independent characters of different morphologic units are combined in these counts: the diameter of zooecial chambers, thickness of zooecial walls, and dimensions of any intervening polymorphs or extrazooidal skeleton. The same counts can be obtained from colonies with large living chambers and thin walls and colonies with small living chambers, thick walls, and intervening polymorphs or extrazooidal skeleton. If such a count is presented as the major statement of zooecial size in a description, it could be misleading. If presented as measures of zooecial spacing, however, such counts would be independent character states.

Redundant characters are those whose states are necessarily determined by other characters. For example, in stenolaemates the presence of frontal walls necessarily determines that orificial walls are fixed (attached to the frontal wall) and that confluent outer body cavity between fully formed zooids is absent. Within stenolaemates, therefore, two of the three characters are redundant. Among all three classes of Bryozoa, however, not all orificial walls are fixed to frontal walls (phylactolaemates) and confluent body cavity is present because vertical walls are incomplete or lacking (phylactolaemates). At the class level, therefore, the three characters are independent.

Genetic control of taxonomic characters is expressed to the extent that their observable states correlate with genetic differences among colonies (Boardman, Cheetham, & Cook, this revision). Some character states vary within colonies. These are largely controlled by ontogeny, astogeny, polymorphism, and microenvironment, and are increasingly recognizable in stenolaemates as study techniques improve. Character states that vary within colonies are subject to intracolony analysis and are presented as stages of series or as limits of variation in order to be in a form that can express possible genetic control. Other character states appear to be uniform within colonies and intracolony analysis is not necessary. Both variable and uniform character states of colonies apparently can express different proportions of genetic and environmental control.

Degrees of environmental control of taxonomic characters are expressed by morphologic differences of character states in response to environmental differences within essentially constant gene pools. Environmentally controlled character states, therefore, should correlate closely with environmental differences, although such correlations do not necessarily rule out taxonomically significant degrees of genetic control.

Morphologic limits of environmentally controlled states of characters are presumably genetic, and so these states have taxonomic significance as expressed by their limits. Further division of a taxon, however, based on environmentally controlled states of characters within those limits would have no genetic significance and, therefore, no validity in phylogenetic classifications.

Environmentally controlled character states that are colony-wide or community-wide are

FIG. 61. Microenvironmental modification.——*1–5. Tubulipora anderssoni* Borg, rec., 12 m, Bay of Islands, N.Z.; *1,* tubular extensions (te) in encrusting colony wall, interior vertical walls (vw) and exterior frontal walls (ew) of feeding zooids; extrazooidal brood chamber (bc) between rows of feeding zooids, long. sec., USNM 250116, ×30; *2,* proximal end of colony showing small spines at base of basal disc (bd) and lateral positions of some tubular extensions, USNM 250117, ×60; *3,* underside of colony with evenly distributed tubular extensions, USNM 250118, ×15; *4a,b,* views of *a,* upper side of colony showing distribution of feeding zooids and elongated brood chamber (bc), *b,* underside of same colony with tubular extensions in proximal region only, both USNM 250119, ×15; *5,* underside of colony with tubular extensions unevenly spaced, USNM 250120, ×15.——*6. Tubulipora* sp., rec., intertidal, Leigh Cove, N.Z.; tubular extensions (te) with skeletal wall (sw) projecting into soft algal substrate (al), living chambers of feeding zooids (lc), brood chamber (bc); long. sec., USNM 250121, ×150.——*7. T. anderssoni,* same data as *1;* pattern of tubular extensions from within colony; sec. parallel to encrusting colony layer, USNM 250122, ×30.

most difficult to recognize because they could equally well be genetically controlled. Experimentation with the same living stenolaemate colonies in different natural environments is the obvious approach to distinguishing the effects of changes of environment on character states. A presently available but less satisfactory alternative is the series of indirect assumptions concerning both genetically and environmentally controlled character states, which are listed under the section on intercolony analysis (BOARDMAN, CHEETHAM, & COOK, this revision). The assumptions are useful both as a means of making interpretations and of indicating new questions to be investigated.

ENVIRONMENTALLY CONTROLLED CHARACTERS

Environmentally controlled modifications of parts of colonies are termed microenvironmental and are caused by local environmental differences within a colony. The morphologic limits of microenvironmentally controlled states of a character are set by the constant genetic makeup of the colony and those limits can be valid parts of taxonomic descriptions. Intermediate morphologies or virtually the entire range of morphologic variation can be displayed within a single colony. Colonies exhibiting microenvironmental differences can aid in distinguishing environmentally controlled morphologic states that might be uniform in other colonies of the species and therefore more difficult to recognize (assumption 3, BOARDMAN, CHEETHAM, & COOK, this revision).

Some microenvironmental modifications demonstrate how colonies repair themselves after environmental accidents or reveal something about the environment itself. Many modifications are the results of fortuitous accidents or occurrences and are so trivial that their description adds nothing to the concepts of taxa in phylogenetic classifications.

Modifications involving exterior walls.— An example of a microenvironmentally controlled modification as an aid to the recog- nition of colony-wide environmental differences (assumption 3) can be inferred in the concept of the species *Tubulipora andersonni* BORG, 1926a. Colonies of this and some other species of *Tubulipora,* which grow on kelp and other soft algae, develop tubular extensions of exterior encrusting colony walls (Fig. 61). These function as basal attachments to the algae, leaving their impressions on the algal surfaces when colonies are removed (Fig. 61,6). The ends of the tubes are only partly calcified. The spaces within the tubes of *T. andersonni* are confluent with body cavities of living chambers of feeding zooids (Fig. 61,1) that bud from the encrusting colony walls. The tubes, therefore, are apparently not polymorphs (kenozooids) as suggested by BORG (1944, p. 46), and as they appear to be externally. In other species of *Tubulipora* (Fig. 61,6) skeletal walls are present between the tubes and feeding zooids, but not enough material is available to determine whether the tubes are entirely sealed off.

Within *T. anderssoni,* BORG (1944, p. 46) also included colonies that grow on hard substrates and that have comparable morphology except for the lack of basal tubes. Recently collected colonies from New Zealand presumably belong to the same species and have their encrusting surfaces either partly (Fig. 61,4b) or entirely (Fig. 61,3,5) covered with basal tubes. The converse of assumption 2 apparently applies here, that exterior walls grown adjacent to the environment can reflect greater degrees of environmental modification than interior walls protected by the body cavity of the colony.

Tube distribution is apparently microenvironmentally controlled in colonies with tube development restricted to parts of encrusting walls. These intermediate states within colonies support BORG's interpretation that the presence or absence of tubes under entire encrusting surfaces of particular colonies is environmentally controlled within a broad species concept. If so, the limits of tube distribution can be considered to be a genetically controlled taxonomic character state of that species and partial distribution that is microenvironmentally controlled in conspecific colonies is a valid part of the species description.

Colonies of many free-walled taxa of Paleozoic age are especially susceptible to interruptions of growth of localized groups of zooids. The interruptions appear to be fortuitous because of irregularities in the position and numbers of zooids in the localized groups. Repair of these growth interruptions is generally by the development of **intracolony overgrowths**. The overgrowths (Fig. 27; 36,*1*) originated from adjacent surviving zooids and were initiated by simple basal encrusting walls, which presumably were exterior and had exterior cuticles.

One can only speculate about causes of Paleozoic growth interruptions. Rupture of exterior membranous colony walls is commonly indicated by debris-filled living chambers of the overgrown zooecia (Fig. 62,*5,6*). Accidental rupture of the membranous walls, therefore, might have been a cause of zooids being killed in parts of colonies. These interruptions and repairs primarily involving exterior walls are so common in some species occurring in calcareous mudstones and shales that it is difficult to find wider uninterrupted exozones of advanced growth stages for description and illustration. Descriptions of these overgrowths in some species could possibly establish genetically controlled limits to some of their environmentally controlled character states.

In some post-Paleozoic stenolaemates, cyclic intracolony overgrowths apparently are the normal colony growth pattern (Fig. 62,*7*; HILLMER, 1971). Thus, a mode of colony growth that started as a means of injury repair may have evolved into a more genetically controlled growth habit not initiated by fortuitous environmental factors. In a few taxa, a number of overgrowths can start simultaneously on a colony surface and develop subcolonies, so that each cycle of intracolony overgrowth consists of adjacent subcolonies (HILLMER, 1971, p. 27, pl. 11, 12).

Intracolony overgrowths form subsequent, more distal zones of astogenetic change and repetition (Fig. 27). These subsequent zones of change, which are produced asexually, lack ancestrulae.

Another indication of accidental rupture of exterior membranous colony walls is the presence of obviously foreign organisms within free-walled colonies. Tubuliporate colonies commonly react by growing simple exterior skeletal walls around the foreign organisms (Fig. 62,*1,4*), presumably to contain their advance and to protect surrounding living zooids. The protective exterior skeletal wall can conform to the most minute patterns on surfaces of foreign bodies to provide an apparently tight seal (Fig. 62,*1*). This kind of fortuitous microenvironmental interruption is useful in demonstrating methods of colony repair but adds little to taxonomic concepts.

Exterior frontal walls serve to complete the skeletal living chambers of the zooids because of their outermost positions in fixed-walled colonies. In that role frontal walls necessarily compensate for minor irregularities of size and shape of supporting vertical walls in order to establish apertures in more or less regular external patterns.

For example, within one colony of *Heteropora pacifica* BORG, 1933, p. 317, the most common frontal walls of feeding zooids are exterior-walled peristomes formed by outward extensions of thin zooecial linings from interior vertical walls (Fig. 34,*1a*). In adjacent polymorphs (also *1a*), thick terminal calcified diaphragms containing closely spaced pseudopores form a second kind of skeletal exterior wall. The two kinds join in a few feeding zooids of the colony to form frontal walls (Fig. 34,*1c*, upper zooecium). In a fourth zooecium near the growing tip of the branch (Fig. 34,*1b*) the vertical wall makes a smaller angle with the colony surface than those of most of the other zooecia so that the thicker exterior wall necessarily forms a longer frontal wall in order to complete the living chamber before growing the presumed peristome. Inclusion of these largely microenvironmental variations in the species description seems both valid and useful and conceivably could establish genetically controlled limits of variation.

Modifications involving interior walls.— Examples of microenvironmental modifications of interior walls of colonies (Fig. 62,2) generally seem to be either less common or less obvious than examples for exterior walls. If true, this tentative generalization supports assumption 2, that structures grown within body cavities are more sheltered from some kinds of environmental interferences than are exterior walls. For example, the interior vertical walls of *Tubulipora andersonni,* described above, are apparently not affected by the presence or absence of basal tubes in exterior encrusting walls. Likewise, interior vertical walls of the colony of *Heteropora pacifica* (above) show less variation in construction than exterior frontal walls, except perhaps for the obvious angle difference of the vertical walls in the zooecium (Fig. 34,1b).

Body-cavity protection (assumption 2) apparently can be overcome by environmental changes of short duration relative to colony life, which affect either interior or exterior structures, or both (assumption 5). For example, the erect part of the skeleton of a bifoliate trepostome colony is of interior origin. Zooecia of one side of one of these colonies (Fig. 62,3, right side) are shorter than on the other, the exozonal walls are thicker, and the cystiphragms and diaphragms are more closely spaced. Some directional micro-

environmental factors must have caused these differences. Although morphologic differences within colonies are rarely so pronounced, theoretically these different states could be produced by different colony-wide environments and a thin-walled population of this species might well be conspecific with a thick-walled population from another environment (assumption 3).

In many erect forms of Paleozoic age, colony branches are extended by a series of growth cycles of interior vertical walls of zooids (BOARDMAN, 1960, p. 38). A cycle starts with the establishment of exozones around growing tips, followed by resorption of the outermost segments of zooecia in the exozones leaving behind traces of exozonal position of that cycle, followed by rejuvenation and growth of thin endozonal walls, followed again by growth of exozones at the new growing tips. In a large colony many of these growth cycles combine to form a branch. Distances between remnants of growing tips commonly vary from cycle to cycle within a branch (Fig. 62,6) or from branch to branch within the same colony. Skeletal walls in both endozones and exozones are interior in origin so that body-cavity protection (assumption 2) is again overcome by environmental changes of short duration (assumption 5).

Modifications involving colony growth habit.—The most comprehensive taxonomic study of fixed-walled tubuliporates relative

---

FIG. 62. Microenvironmental modification.——*1.* Heteroporid tubuliporate, rec., Neah Bay, Wash.; colony with exterior skeletal wall (ew) fitted precisely to minute pattern of echinoderm spine, interior vertical walls (iw) of feeding zooecia; long. sec., USNM 250123, ×30.——*2. Orbignyella* sp., Bellevue Ls. Mbr., McMillan F., Ord. (Maysvill.), Cincinnati, Ohio; region of zoarium apparently injured during life and partly filled with cystiphragms (c) to reestablish living chamber (lc); transv. sec., USNM 167689, ×50.——*3. Peronopora decipiens* (ROMINGER), Corryville Sh. Mbr., McMillan F., Ord. (Maysvill.), quarry at Dent, W. of Cincinnati, Ohio; walls thicker and cystiphragms more closely spaced on narrower exozone to right than in exozone to left; long. sec., USNM 250124, ×20.——*4. Densipora corrugata* MAC-GILLIVRAY, rec., 5-m wave-cut platform, Western Port Bay, W. end Phillip Is., Australia; protective exterior wall (ew) around foreign growth, interior vertical walls (iw); long. sec., USNM 250125, ×100.——*5. Atactotoechus fruticosus* (HALL), Windom Mbr., Moscow F., Dev. (Erian), Kashong Cr., Seneca Lake, N.Y.; living chambers filled with terrigenous material under overgrowth (arrow); long. sec., USNM 133941, ×2.——*6. Leptotrypella asterica* BOARDMAN, Kashong Mbr., Moscow F., Dev. (Erian), Little Beards Cr., Leicester, N.Y., paratype; living chambers filled with terrigenous material under overgrowth (arrow), remnant of growing tips, cycles 1 to 6; long. sec., USNM 133895, ×5.——*7. Atagma macroporum* (HAMM), Cret. (Maastricht.), S. of Mons, Belg.; remnants of cyclic growing tips and related overgrowths; long. sec., USNM 186564, ×7.

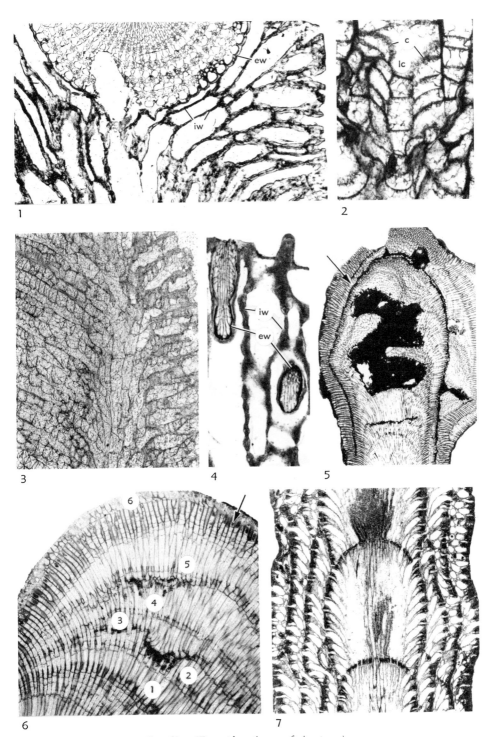

F_{IG}. 62. *(For explanation, see facing page.)*

to environments (HARMELIN, 1976) indicates that colony growth habits of many species are environmentally controlled (assumption 4). Character states derived from exterior frontal walls, such as wall thickness and peristome length and diameter, are correlated with changes in growth habit and therefore those states are interpreted to be environmentally controlled. Within the same species, microstructure of frontal walls, including the density and size of pseudopores, is relatively constant in different environments, so some character states of exposed exterior walls can be assumed to be more nearly genetically controlled (assumption 1). Skeletal structures of HARMELIN's species such as hemisepta, hemiphragms, and some mural spines are grown within zooidal body cavities and are relatively constant in occurrence in different environments. These are interpreted here to be largely genetically controlled (assumption 2).

Summary.—Many characters of exterior structures appear to be largely environmentally controlled and many characters of interior structures appear to be largely genetically controlled. Just the reverse can be true, however, for other characters. From the examples above, body-cavity protection (assumption 2) seems to cause some reduction in microenvironmental and environmental modifications. Some variations of characters apparently caused by environmental changes of short duration (assumption 5), especially those reflecting amounts or rates of growth, can occur within colonies. There seems to be no universally reliable group of indirect approaches to the recognition of all environmental modifications. Reasonable approximations can be achieved for some characters, however, resulting in improvements in attempts at phylogenetic classifications.

GENETICALLY CONTROLLED CHARACTERS

A number of taxonomic characters have been used in the classification of stenolaemates of Paleozoic age. Although generally unexpressed, it apparently has been assumed that these characters were largely genetically controlled because their states, or the patterns of their changing states, were relatively constant through significant intervals of geologic time (assumption 1). Longer lasting character states generally have been evaluated at higher taxonomic levels and more rapidly changing character states tend to be used at lower taxonomic levels.

Microstructural patterns of skeletal layers of interior vertical walls are a major source of taxonomic characters inferred to be genetically controlled. Wide experience by many workers with thousands of stenolaemate specimens of Paleozoic age has produced many different patterns of microstructure in vertical walls (see discussion above). Microstructural patterns are distributed within colonies, among colonies, and among taxa with such high degrees of constancy (assumption 1) that their genetic control has generally been assumed. As a result, different aspects of microstructure have been used in the classification of Paleozoic forms at most hierarchical levels. Microstructure of vertical walls has the added advantage of being present in all specimens except those that are modified diagenetically. The nature of zooecial boundaries within vertical walls is correlated to different degrees with wall microstructure and is also assumed to be largely genetically controlled. Body-cavity protection (assumption 2) is assumed to be a factor in genetic control of vertical zooidal walls.

In post-Paleozoic stenolaemates the microstructures of both interior vertical walls and exterior frontal walls give promise of comparable usefulness in classifications. Sectioning has not been a standard part of the study of post-Paleozoic stenolaemates, the Tubuliporata, however, and no significant amount of information exists in the literature on the taxonomic characters of their vertical walls. We have sectioned approximately two hundred kinds of post-Paleozoic tubuliporates, including both fossil and modern species. This preliminary survey reveals a wider range of microstructural patterns in vertical walls (e.g., Fig. 32,*1–4*; 33,*1–3*;

42,*1,5,6*) than has been discovered in Paleozoic forms, indicating later evolutionary developments.

Laminae of adjacent zooecia form patterns that are convex outward (Fig. 29,*1,3*) in Paleozoic taxa, indicating that surfaces of the laminae were approximate growth surfaces. Many post-Paleozoic species have laminae with that same orientation (e.g., Fig. 31,7; 32,5; 50,5; 55,*1,2*; NYE, 1976, pl. 15, 36, 40, 45). The similarity of orientation and generally comparable microstructures of vertical walls of Paleozoic and many post-Paleozoic taxa suggest the possibility of phylogenetic relationships between the two groups (BOARDMAN, 1973, 1975). (For contrasting interpretations, see BROOD, 1976.)

In many other post-Paleozoic species, including both fixed-walled taxa and free-walled taxa, the direction of inclination of laminae of compound vertical walls is reversed (BOARDMAN & TOWE, 1966, p. 2; BOARDMAN & CHEETHAM, 1969, p. 211) from convex outward to convex inward (Fig. 29,2; 33,*2,3*; 42,*5,6*). This reversal necessarily places the laminae at high angles to growing surfaces, requiring edgewise growth of all laminae simultaneously as vertical walls are extended.

The geometric perfection of patterns of vertical zooidal wall arrangements in endozones of many taxa, especially if they remain unchanged in communities having different environments, suggest that zooidal patterns (see above) can be genetically controlled. Most stenolaemates have less regular zooidal patterns; however, it is possible that genetic control, suggested by regularity of zooidal patterns in some taxa, is just as strong in taxa with less regular patterns. All zooidal patterns should be described in detail in taxonomy until more direct evidence of genetic and environmental control is available.

The presence of basal and lateral skeletal structures that project into zooidal body cavities, such as diaphragms, cystiphragms, hemiphragms, hemisepta, and mural spines, generally has been assumed to be genetically controlled, judging from their use in classification. They have been given approximately

the same taxonomic weight as vertical wall microstructure in many taxa, possibly because they are attached to vertical walls.

Enough differences in the distribution of projecting skeletal structures and vertical wall microstructure have been recognized to suggest that projecting structures should be independently evaluated in different taxa. In some cryptostome taxa, hemisepta occur in virtually all zooecia of feeding zooids and are apparently genetically controlled. In other taxa, however, hemisepta occur in some zooecia and not in others in the same zoarium. This irregular intrazoarial distribution could be interpreted as an indication of polymorphism. It seems best interpreted as the result of microenvironmental control, however, because of a general lack of other observable morphologic differences between the two kinds of zooecia.

Variation in the distribution of hemisepta within colonies is comparable in Paleozoic species and in the few post-Paleozoic species that have them. One species (HARMELIN, 1976) apparently has hemisepta in all feeding zooids and another species of Cretaceous age (Fig. 50,5) lacks them in many zooids. This variation suggests that their presence is subject to significant degrees of environmental control at lower taxonomic levels (assumption 3). The variation also illustrates the assumption that proportions of genetic and environmental control of a potential taxonomic character may differ in different taxa (assumption 6).

Cystiphragms are the single monothetic character defining the family Monticuliporidae NICHOLSON, 1881, in the order Trepostomata (see BASSLER, 1953, p. G94). Cystiphragms are generally present in all assumed feeding zooecia in the zoaria of most included genera, so their presence can be considered to be genetically controlled (assumption 1), although they can vary at least microenvironmentally in spacing and thickness (assumption 3).

The problem of noncorrelation of occurrences of apparently genetically controlled character states is illustrated by the Monti-

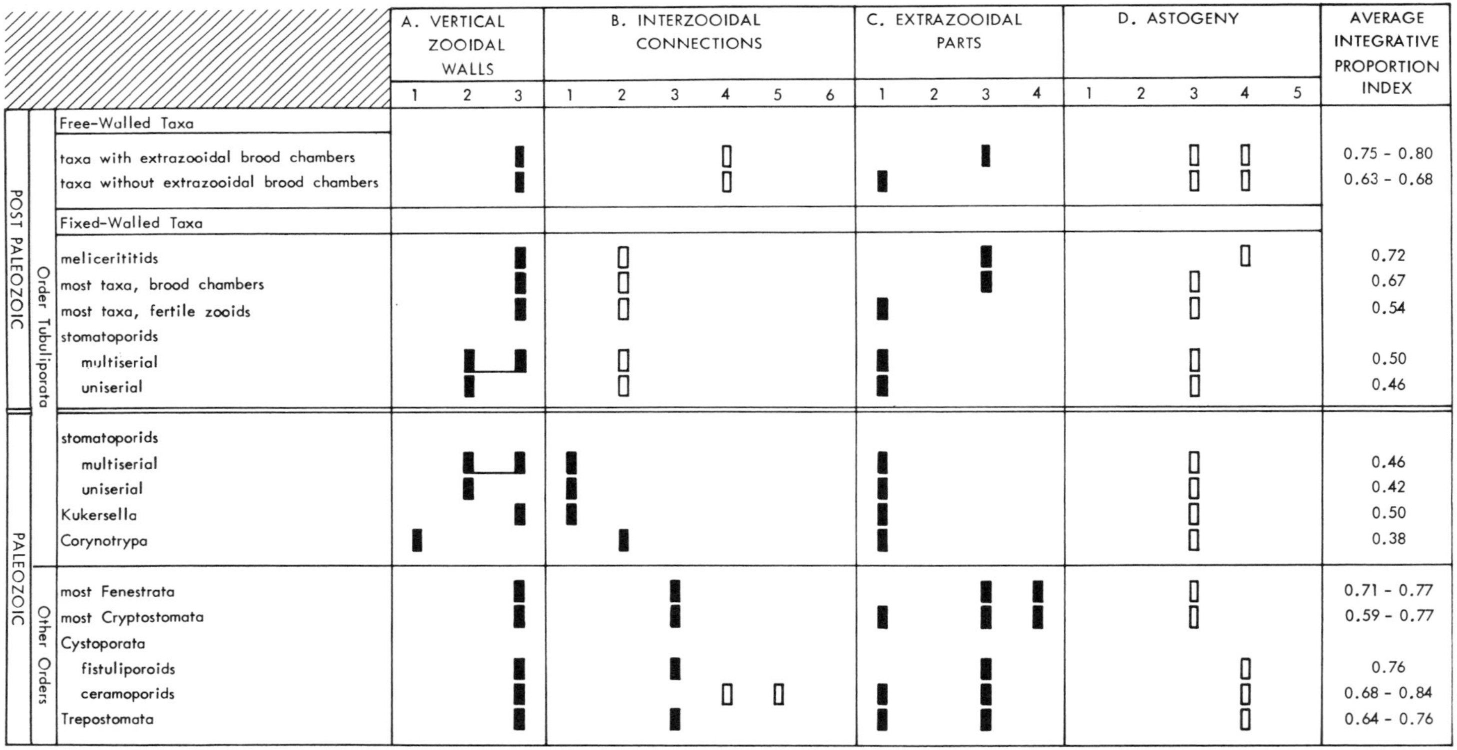

FIG. 63. Colony integration. Summary of integrative states of stenolaemate Bryozoa directly or indirectly discussed in text. Numbered states for each character, A through D, are described by Boardman, Cheetham, and Cook (colony control of function and morphology, this revision). Higher numbers of states and of average integrative proportion indexes indicate higher degrees of integration. Vertical position of groupings of bryozoans within Paleozoic and post-Paleozoic intervals have no relative time significance. The chart merely shows that Paleozoic tubuliporates are not as highly integrated as most post-Paleozoic tubuliporates. Observed states are indicated by black rectangles; inferred states are indicated by open rectangles. Two rectangles of same grouping joined by horizontal line indicate both states in same colony.

	A. VERTICAL ZOOIDAL WALLS	B. INTERZOOIDAL CONNECTIONS	C. EXTRAZOOIDAL PARTS	D. ASTOGENY	AVERAGE INTEGRATIVE PROPORTION INDEX
	1 2 3	1 2 3 4 5 6	1 2 3 4	1 2 3 4 5	
POST PALEOZOIC — Order Tubuliporata					
Free-Walled Taxa					
taxa with extrazooidal brood chambers					0.75 – 0.80
taxa without extrazooidal brood chambers					0.63 – 0.68
Fixed-Walled Taxa					
meliceritids					0.72
most taxa, brood chambers					0.67
most taxa, fertile zooids					0.54
stomatoporids multiserial					0.50
uniserial					0.46
PALEOZOIC — Other Orders					
stomatoporids multiserial					0.46
uniserial					0.42
Kukersella					0.50
Corynotrypa					0.38
most Fenestrata					0.71 – 0.77
most Cryptostomata					0.59 – 0.77
Cystoporata					0.76
fistuliporoids					0.68 – 0.84
ceramoporids					0.64 – 0.76
Trepostomata					

culiporidae. Several different vertical wall microstructures and other presumably genetically controlled morphologic differences occur with the cystiphragms in the Monticuliporidae. Certainly, a family with a single diagnostic character is suspect. Noncorrelation of the states of presumably long-lasting characters thought to be genetically controlled suggests that more natural family groupings might be achieved by using all of the available characters in a polythetic approach.

The taxonomic application, especially in higher categories, of the presence or absence of frontal walls and the resulting concepts of free, fixed, or combined orificial walls seems unpredictable until detailed study of colony interiors is carried out on a significant number of genera. The first division of the stenolaemates into fixed- or free-walled groups as suggested by BORG (1944, p. 18) should be tested because comparable vertical wall structures (compare Fig. 29 with 32,*2* and 33,*3* with 42,*5,6*) and different methods of forming frontal walls (contrast Fig. 33 and 34) suggest the possibility of several independent origins of free- and fixed-walled taxa. If true, BORG's monothetic grouping is polyphyletic.

COLONY INTEGRATION AND GENETIC CONTROL

The concept of the integration of colonies is based on morphologic and associated functional characteristics that occur in colonies and not in solitary animals. It assumes that feeding zooids of bryozoan colonies are more nearly comparable to solitary animals than are whole colonies. Degrees of integration of colonies depend on the extent to which zooids in combination with any extrazooidal parts differ morphologically from solitary animals. States of characters of colonies ranging from nonintegrated to highly integrated provide the basis for the integration series presently recognized (see section on colony control of function and morphology, BOARDMAN, CHEETHAM, & COOK, this revision).

A corollary to the assumption of body-cav-

ity protection of structures of interior origin (assumption 2) states that many integrated structures are grown within the protection of the body cavity and so are relatively sheltered from the environment. Therefore, they can display character states more nearly reflecting genetic control. For example, vertical body walls of zooids in most stenolaemates are integrated structures because they are interior body walls grown cooperatively by adjacent zooids within the body cavity. Similarly, body-cavity connections among zooids through and around vertical walls are integrated features. Neither interior cooperatively grown body walls nor body-cavity connections are possible between solitary animals.

To the extent that integrated structures are interior in origin the two concepts of integration and body-cavity protection are overlapping. Either one or both might be a source of genetically controlled characters. The concept of integrated structures, however, extends beyond wholly interior structures to include structures that are at least partly exterior in origin. For example, basal encrusting colony walls are multizooidal in origin and therefore express a degree of integration although they are exterior walls. Covering walls of many extrazooidal brood chambers are exterior walls but express a degree of integration because extrazooidal structures are not possible in solitary animals.

The concept of integration becomes important to the classification of bryozoans if integrated characters as a group provide a measure of genetic control. A significant proportion of genetic control would be indicated by an apparent development of and selection for integrated structures and associated functions during the evolutionary history of bryozoans. The earliest taxa of the Cheilostomata exhibit low degrees of integration, which increase progressively through time in major evolving stocks of the order (CHEETHAM & COOK, this revision). The stenolaemates are less well known and comparable detail is not available, especially concerning polymorphs (Fig. 63).

Paleozoic tubuliporates have the lowest

integration indices among the stenolaemates and are unique to the phylum because they are fixed-walled colonies with calcified frontal walls and apparently no communication pores in interior vertical walls. Once the zooidal walls were calcified, therefore, no interzooidal connections existed and except for being physically connected the zooids lived like solitary animals. The few Paleozoic tubuliporates known produced small colonies suggesting a minimum of success.

The great majority of post-Paleozoic tubuliporates evolved communication pores in interior vertical walls. Fixed-walled taxa, therefore, had presumed interzooidal connections and were more highly integrated in that character than fixed-walled taxa of Paleozoic age.

Free-walled stenolaemates of Paleozoic age apparently all had interzooidal connections through confluent outer body cavity around ends of vertical walls. They were more highly integrated, therefore, than the few fixed-walled species of the same age.

Free-walled post-Paleozoic tubuliporates were more highly integrated in interzooidal communication than free-walled Paleozoic stenolaemates because, in addition to confluent outer body cavities, they developed communication pores (only the few ceramoporids had communication pores in the Paleozoic). It is possible that some free-walled Paleozoic stocks continued into the post-Paleozoic. If so, stenolaemates evolved toward more means of interzooidal communication and higher integration indices through time.

The phylogenetic relationships of post-Paleozoic free-walled taxa with fixed-walled taxa of equivalent ages can not be inferred convincingly because of lack of evidence to date, so no claim is made here that one or several stocks of fixed-walled forms evolved communication pores and free walls (see Brood, 1976) resulting in increasing interzooidal communication and integration.

The most highly integrated free-walled stenolaemates are the ceramoporids (Fig. 63). They were highly integrated partly because they had communication pores in vertical walls when they first appeared in the Ordovician. They apparently became extinct in the Devonian (Utgaard, this revision), and communication pores of post-Paleozoic stenolaemates were evolved independently. The other orders presently considered to be restricted to the Paleozoic also were highly integrated when they first appeared. Perhaps the tubuliporates are the only stenolaemate order that has its earlier fossil record available so that patterns of integration can be studied throughout its existence.

The few functional interpretations available of integrated characters suggest that there is increasing functional cooperation among zooids and extrazooidal parts of colonies as degrees of morphologic integration increase. Functional cooperation of the kinds that should prove advantageous to colony survival presumably would be selected for over long periods of time. If future work indicates that integrated structures increased in number and degree of integration with time, many of their character states can be inferred to have been selected for in the evolutionary process and many integrated characters can be assumed to be genetically controlled.

As now understood, steps in the integration series (Boardman, Cheetham, & Cook, this revision) for stenolaemates (Fig. 63) express long-lasting character states and associated functions that define generalized evolutionary stages of development in taxa of the higher categories. Long-lasting character states suggest genetic control (assumption 1), whatever the underlying reasons.

Steps in the integration series, however, are only a few of the many character states derived from integrated structures. Many others are relatively short-lived. Unfortunately, it does not seem possible to assume that all characters which can be derived from integrated structures are largely genetically controlled. Examples described above of states of integrated structures interpreted to be environmentally controlled include: (1) the distribution of tubes in encrusting walls of multizooidal origin within colonies of *Tubulipora andersonni*; (2) the variable lengths of growth of vertical walls in endozones between cyclical, abandoned, branch tips within colonies; and (3) the variable thickness of vertical walls in the exozones within colonies of many taxa.

Examples of integrated structures having character states that apparently are either genetically or environmentally controlled suggest that it is too early to predict the ultimate importance of the concept of colony integration as an independent source for genetically controlled characters in the classification of stenolaemates.

GENERAL FEATURES OF THE CLASS GYMNOLAEMATA

By A. H. Cheetham and P. L. Cook

[Smithsonian Institution, Washington, D.C.; British Museum (Natural History), London]

The Gymnolaemata are here considered to be one of three classes of the phylum Bryozoa. Distinguishing characteristics of the class are given by Boardman, Cheetham, and Cook in this revision (p. 26).

The Gymnolaemata include a great diversity of morphologies, ranging from simple uncalcified and partly calcified genera to elaborately integrated soft-bodied and complexly calcified genera. Among living Bryozoa, the Gymnolaemata are the dominant class in abundance and number of species, and the only class with representatives that live in fresh, brackish, and marine waters. The fossil record of the class extends more that 400 million years, beginning in the Late Ordovician; however, the record is sparse before the Late Cretaceous, approximately 100 million years ago. Proliferation of the Gymnolaemata beginning in the Late Cretaceous coincided with the decline in the Stenolaemata (Voigt, 1972b; Boardman, this revision), the only other bryozoan class with a significant fossil record. Numerical dominance in marine environments was achieved by the Gymnolaemata toward the close of the Cretaceous and has increased through the Cenozoic.

The class Gymnolaemata comprises two orders, the Cheilostomata and the Ctenostomata. Most fossil evidence of gymnolaemate history has been produced by the Cheilostomata, which have body walls with continuous calcareous layers that can be readily preserved and from which the morphology of soft parts can generally be interpreted. The body walls of Ctenostomata have only scattered or no calcareous parts, and fossils confidently assigned to this order occur sporadically as external molds.

Fossil Cheilostomata are abundant in many calcareous marine deposits of late Mesozoic and Cenozoic age from throughout the world. In some Upper Cretaceous limestones in Europe, and in some limestones, calcareous sands, and calcareous clays of Tertiary and Quaternary age in Europe, North America, and Australia, cheilostomates are the most abundant remains of megascopic invertebrates. Some cheilostomates having microscopic colonies outnumber even Foraminifera of similar size in some deposits. Similar high abundances of Cheilostomata have recently been reported in cores taken by the Deep Sea Drilling Project in the Atlantic, Pacific, and Indian oceans from deposits of Paleocene to Pleistocene age (Cheetham & Håkansson, 1972; Wass & Yoo, 1975; Cheetham, 1975a; Labracherie & Sigal, 1975). The oldest deposits from which cheilostomates have been reported are of Late Jurassic age (Pohowsky, 1973).

Fossils that have been confidently assigned to the Ctenostomata are much rarer than fossil cheilostomates and are distributed sporadically in marine deposits of Paleozoic, Mesozoic, and Cenozoic age. All Paleozoic and many younger fossils that have been closely compared with living ctenostomates are shell-penetrating forms. **Borings** made by these ctenostomates in calcareous substrates are molds reflecting the external morphology of zooids and the budding patterns of colonies and are comparable to those of living shell-penetrating representatives of the order (Voigt & Soule, 1973; Pohowsky, 1974).

The only fossils of nonpenetrating ctenostomates comparable in morphologic detail to borings of shell-penetrating species are external molds produced by overgrowth of the soft-bodied colonies by such shelled organisms as oysters (Voigt, 1966, 1968, 1971a). Nonpenetrating ctenostomates are known from deposits as old as Middle Jurassic (Voigt, pers. commun., 1976). Other fossils of earlier Mesozoic and Paleozoic age, which historically have been interpreted as nonpenetrating ctenostomates, seem not to be comparable in morphology with living representatives of the order or in mode of preservation with younger fossils and so remain problematical. One Jurassic genus, *Vinelloidea,* previously assigned to the Ctenostomata, has recently been demonstrated to belong to the Foraminifera (VOIGT, 1973).

The abundance and wide distribution of fossil Gymnolaemata are equaled by those of living representatives of the class. Gymnolaemates have been reported from the Arctic to the Antarctic and from freshwater lakes and streams to the abyssal depths of the oceans. A number of gymnolaemate species are important components of fouling communities in fresh, brackish, and marine habitats, and many of these species are cosmopolitan. Many nonfouling gymnolaemate species also have wide geographic distributions. Circumtropical distributions of shallow water species and tropical submergence of shallow to deepwater species have been reported (distributions summarized by CHEETHAM, 1972; LAGAAIJ & COOK, 1973, and references listed therein).

Even though many more living than fossil Ctenostomata are known, the number of living species of Cheilostomata apparently far exceeds that of Ctenostomata. Living species of Cheilostomata are found in brackish to marine water, some in water of variable salinity. The great majority is limited to marine water of shelf depth. Ctenostomates are found in fresh as well as brackish and marine water. Marine representatives of both orders have been found at abyssal depths (SCHOPF, 1969b; D'HONDT, 1975), but only cheilostomates have been reported from depths exceeding 5,000 meters.

Marine Gymnolaemata seem to be most abundant and diversified where available firm substrates and low turbidity and turbulence permit encrusting and erect growth. Less favorable conditions, such as those in intertidal zones, commonly permit habitation by some species with encrusting or flexible growth forms, some of which may be highly specialized in modes of growth. The most specialized growth forms appear to be the free-living, partly mobile colonies of some cheilostomate and ctenostomate species adapted for life on or in unstable seafloor sediments. (For a variety of cheilostomate growth forms, see Fig. 13–15.)

The variety of simple to specialized growth forms in differing combinations with a high diversity of zooidal and, where present, extrazooidal morphologies limits the number of character states shared by all members of the Cheilostomata and Ctenostomata. The few shared states recognized are related to orientation of zooid walls and to the soft parts (see Table 1). Even this small number of states has become recognized only gradually during the long history of gymnolaemate studies.

Different combinations of states of numerous morphologic characters, inferred to reflect independently more genetic than environmental control, provide a rich basis for classification of the two orders. Although a greatly increasing amount of detailed information on morphology and functions of living gymnolaemates and their closely similar fossil relatives has become available during the past 100 years, attempts to generalize about modes of growth and to base classifications on monothetic hierarchies of drastically limited numbers of key characters have produced much instability in taxonomy and conflicting interpretations of phylogenetic relationships. As modern studies confirm and extend the diversity of modes of growth and functions of living gymnolaemates suggested by some earlier workers, a new polythetic basis is being developed to evaluate the multitude of fossil and living genera now included in the class. The detail in which many morphologic features known in diverse groups of living gymnolaemates can be recognized in fossil representatives of the class suggests that comparisons based on all available morphologic characters can be closely approached. By testing such comparisons against the stratigraphic record of the class, a fuller understanding of the evolutionary history of this major group of Bryozoa should be achieved.

Acknowledgments.—We are indebted to W. C. BANTA, R. S. BOARDMAN, E. HÅKANSSON, and G. LUTAUD for technical reviews of the manuscript; and to J. S. RYLAND, L. SILÉN, E. VOIGT, and numerous other colleagues for valuable comments and suggestions during its preparation. Text figures were drawn by J. SANNER, who also provided assistance in compiling literature references and preparing photographs. D. A. DEAN and P. J. CHIMONIDES prepared sections of specimens. W. R. BROWN and M. J. MANN prepared scanning electron micrographs. Specimens and locality data were made available by H. V. ANDERSEN, E. C. HADERLIE, J. B. C. JACKSON, H. T. LOEBLICH, and E. R. LONG. Financial support was provided by the Smithsonian Research Foundation (Grants 427206 and 430005).

HISTORICAL REVIEW

The abundance and wide distribution of Gymnolaemata in modern seas and in late Mesozoic and Cenozoic marine sediments assured that members of this class were available for even the earliest studies of Bryozoa. Among the five living Mediterranean species of Bryozoa catalogued and illustrated (as Pori) nearly 400 years ago by IMPERATO (1599), four are now recognized as members of the gymnolaemate order Cheilostomata and one as a member of the stenolaemate order Tubuliporata (=Cyclostomata of BUSK). Of the eight species of Bryozoa included (as Zoophita) in the work of BASSI (1757) on Pliocene invertebrates of Italy, reportedly the first publication in which fossil bryozoans were described and illustrated, seven are now assigned to the Cheilostomata and one to the Tubuliporata (ANNOSCIA, 1968).

In North America, the first Bryozoa to be reported (as Polypi) were three species from the Paleocene of New Jersey (MORTON, 1829, 1834) and four species from the Eocene of Alabama (LEA, 1833), all but one of which are now assigned to the Cheilostomata.

Pioneer observations of morphology and functions of living Bryozoa during the late eighteenth and early nineteenth centuries were made largely on marine species now assigned to the Gymnolaemata. ELLIS's studies establishing the animal nature of Bryozoa, synthesized in a major work (1755c), included many cheilostomates and a few ctenostomates. GRANT's (1827) detailed observations of the arrangement and movement of tentacular cilia were made on cheilostomates. The classic demonstrations of anatomical differences between bryozoans and coelenterates (AUDOUIN & MILNE-EDWARDS, 1828; THOMPSON, 1830) were based on gymnolaemates. LISTER (1834) and FARRE (1837) provided further detailed descriptions and illustrations of lophophores, retractor and parietal muscles, and other organs, together with observations on their functions, in several species of cheilostomates and ctenostomates. The independent establishment by THOMPSON (1830) and EHRENBERG (1831) of the phylum as now recognized was based on studies of Ctenostomata.

Freshwater Gymnolaemata, comprising a few geographically widespread living genera, are all now assigned to the Ctenostomata. They apparently went unnoticed until nearly 100 years after the first freshwater Bryozoa (members of the class Phylactolaemata) were described by TREMBLEY (1744). Since their discovery by EHRENBERG (1831), freshwater ctenostomates have commonly been included in studies of freshwater Bryozoa. Indeed, ALLMAN's establishment (1856) of the Gymnolaemata and Phylactolaemata as orders of Bryozoa was based on his anatomical comparisons of freshwater genera belonging to both groups.

By the middle of the nineteenth century enough was known about the morphology of Bryozoa for BUSK (1852) to establish Cheilostomata, Ctenostomata, and Cyclostomata (called Tubuliporata in this revision) as suborders of living marine Bryozoa (Table 2), partly paralleling taxa above the family level previously recognized by JOHNSTON (1847). ALLMAN (1956) placed BUSK's suborders, together with freshwater ctenostomates (suborder Paludicellea of ALLMAN) and freshwater entoprocts (suborder Urnatellea of ALLMAN), in the Gymnolaemata (Table 2). BUSK (1859) followed ALLMAN in considering the freshwater ctenostomates to be a suborder of the Gymnolaemata separate from the Ctenostomata, but did not include entoprocts in the Gymnolaemata (see BOARDMAN, CHEETHAM, & COOK, this revision). It was not until late in the nineteenth century that freshwater gymnolaemates were assigned to the Ctenostomata (KRAEPELIN, 1887) and in the twentieth century that the Tubuliporata (=Cyclostomata of BUSK) were removed from the Gymnolaemata (BORG, 1926a).

D'ORBIGNY (1851–1854), in his large monograph of the post-Paleozoic Bryozoa of France, proposed a different classification based principally on study of fossil species but also including numerous living species. Most genera now assigned to the Cheilostomata he placed in an order Bryozoaires cellulinés (1851, p. 23), and a few genera of cheilostomates were placed with the tubuliporates in an order Bryozoaires centrifuginés (1853, p. 585). Each of D'ORBIGNY's orders was divided into suborders on colony forms (1852, p. 318; 1853, p. 591). This classification gained little following, even among paleontologists. GABB and HORN (1862) employed the D'ORBIGNY classification in monographing the fossil Cenozoic Bryozoa of the United States, but BUSK's suborders have been adopted throughout subsequent paleontologic literature.

Fossil species were assigned to the Cheilostomata soon after the suborder was established (BUSK, 1859). As early as 1851, REUSS arranged his descriptions of numerous Tertiary species of Bryozoa so that the species now assigned to the Cheilostomata all preceded those now assigned to the Tubuliporata (=Cyclostomata of BUSK). By 1864, REUSS employed BUSK's subordinal names for this arrangement.

Fossils now assigned to the Ctenostomata were first described and illustrated near the middle of the nineteenth century (D'ORBIGNY, 1839; FISCHER, 1866). However, these species were not distinguished from cheilostomates, and definite assignment of fossil species to the Ctenostomata apparently was not made until late in the nineteenth century (ULRICH, 1890). The anatomy of living shell-penetrating ctenostomates, on which interpretation of much of the fossil material of Ctenostomata depends, remained virtually unknown until nearly the middle of the twentieth century (MARCUS, 1938b).

Most paleontologists have assumed that the morphology and functions of whole zooids and colonies can be inferred from the study of fossil gymnolaemates and by comparison with living species. Only a few paleontologists (for example, BRYDONE, 1929, p. 5–6) have thought that skeletal evidence is generally insufficient for making such inferences and have advocated separate classifications for fossil and living taxa. Recently, it has been proposed that ctenostomates known only from their borings should be classified as ichnotaxa (BOEKSCHOTEN, 1970; BROMLEY, 1970; HÄNTZSCHEL, 1975), but bryozoan workers contend that such borings preserve sufficient evidence of zooid morphology and budding patterns to be compared with living shell-penetrating taxa (VOIGT & SOULE, 1973; POHOWSKY, 1974). No major classification of the Gymnolaemata has been proposed for fossil species alone.

BUSK's subordinal classification emphasized zooid morphology and thus stimulated more detailed observation of both living and fossil gymnolaemates. At lower levels BUSK (1859, 1884, 1886) continued to rely upon colony form and zooid arrangement, but late nineteenth century and early twentieth century workers produced much information on morphology, modes of growth, and functions of zooids with which the classification continued to be refined.

SMITT (1865, p. 115; 1866, p. 496; 1867, p. 279) raised BUSK's suborders to ordinal

TABLE 2. *Major Classifications of the Class Gymnolaemata above the Superfamily Level.*

(Boldface indicates a taxon now wholly included; italic, a taxon now partly included; subtaxa of now-excluded taxa are omitted. Author and date are footnoted for taxa the earliest reference to which is not shown, as are some usages of other authors. Correlations are approximate and informal.)

BUSK (1852)	ALLMAN (1856)	SMITT (1865–1868)	JULLIEN (1888a)	GREGORY (1893)
Order *Polyzoa Infundibulata*[a]	Order *Gymnolaemata*	Tribe *Infundibulata*		Subclass *Gymnolaemata*
Suborder **Cheilostomata**	Suborder **Cheilostomata**	Order **Cheilostomata**	Order *Cheilostomata*	Order **Cheilostomata**
Multiserialaria			Suborder *Diplodermata*[d]	Suborder **Stolonata**[i]
Inarticulata		Suborder **Flustrina**	Tribe *Monopesiata*[e]	
				Suborder **Cellularina**
		Suborder **Cellularina**		Suborder **Athyriata**
Articulata			Tribe **Opesiulata**	
Uniserialaria			Tribe **Anopesiata**	
		Suborder **Escharina**[b]	Suborder *Monodermata*[d]	
			Tribe **Inovicellata**	Suborder **Schizothyriata**
		Suborder **Celleporina**[b,c]	Tribe **Subovicellata**[f]	Suborder **Holothyriata**
			Tribe **Superovicellata**[f]	
	Paludicellea		Order **Paludicellea**	
Ctenostomata	Suborder **Ctenostomata**	Order **Ctenostomata**	Order **Ctenostomata**	
			Suborder **Halcyonellina**	
			Suborder **Urticularia**	
			Tribe **Orthonemida**[h]	
			Tribe **Campylonemida**[h]	
Suborder **Cyclostomata**	Suborder **Cyclostomata**	Order **Cyclostomata**		Order **Cyclostomata**
	Suborder **Urnatellea**			

[a] GERVAIS, 1837.
[b] Used as family names by EHRENBERG, 1839.
[c] JOHNSTON, 1847.
[d] JULLIEN, 1881.
[e] = *Opesiata*.
[f] JULLIEN, 1882.
[g] see HARMER column.
[h] HINCKS, 1880.
[i] BUSK, 1884.

TABLE 2.   *(Continued from preceding page.)*

LEVINSEN (1909)	HARMER (1915–1957)	SILÉN (1942)	BASSLER (1953)	RYLAND (1970)
	Tribe *Gymnolaemata*[k]	Order *Gymnolaemata*	Class *Gymnolaemata*	Class **Gymnolaemata**
Order **Cheilostomata**	Order **Cheilostomata**[l]	Suborder **Cheilo-Ctenostomata**[p]	Order **Cheilostomata**	Order **Cheilostomata**
Suborder **Anasca**	Suborder **Anasca**		Suborder **Anasca**	Suborder **Anasca**
Division **Malacostega**[l]	Division Inovicellata	Section Inovicellata	Division Inovicellata	
	Division Malacostega	Section Protocheilostomata	Division Malacostega	
		Section Membranidea		
		Division Scrupariina		
		Division Malacostega		
	Division Cellularina	Division Cellularina	Division Cellularina	
		Section Cryptocystidea		
Division **Coilostega**[l]	Division Coelostega	Division Coelostega	Division Coilostega	
Division **Pseudostega**	Division Pseudostega	Division Pseudostega	Division Pseudostega	
	Division Cribrimorpha[m]		Division Cribrimorpha	Suborder **Cribrimorpha**
Suborder **Ascophora**	Suborder **Ascophora**	Section Spinocystidea	Suborder **Ascophora**	Suborder **Gymnocystidea**
	Division Ascophora imperfecta	Section Gymnocystidea		Suborder **Ascophora**
	Division Ascophora vera			
	Order **Ctenostomata**[l]	Section Carnosa	Order **Ctenostomata**	Order **Ctenostomata**
	Group Paludicellea	Division Paludicellea	Suborder Paludicellea	Suborder Carnosa
	Group Carnosa[n]	Division Halcyonellea[b,x]	Suborder Carnosa	
	Group Vesicularina[c]	Section Stolonifera	Suborder Stolonifera	Suborder **Stolonifera**
	Group Stolonifera[o]	Division Vesicularina	Suborder Vesicularina	
		Division Valkerina		
Order **Cyclostomata**[l]	Order **Cyclostomata**[l]	Suborder **Cyclostomata**	Order Cryptostomata[1]	
			Order Cyclostomata	
			Order Trepostomata[1]	

[l] LEVINSEN, 1902.
[k] Order in 1915.
[l] Suborder in 1915.

[m] Following informal usage of LANG, 1916.
[n] GRAY, 1841.
[o] EHLERS, 1876.

[p] Following informal usage of BORG, 1926.
[q] VINE, 1884.
[r] ULRICH, 1882.

rank and within the living Cheilostomata established a series of suborders based upon the assumption that ontogenetic and astogenetic gradients recapitulate phylogeny (SMITT, 1868; transl. SCHOPF & BASSETT, 1973). SMITT's suborders (Table 2) ranged from simple, slightly calcified cheilostomates, compared by him to the Ctenostomata, to increasingly complexly calcified cheilostomates. Some of SMITT's suborders were readily adopted for fossil species (KOSCHINSKY, 1885). Recognition of a broad evolutionary trend of increasingly complex calcification among fossil cheilostomates led GREGORY (1893) to propose a series of suborders (Table 2) for both living and fossil species partly paralleling those of SMITT. As more detailed understanding of modes of growth and functions emerged around the turn of the century, however, the SMITT and GREGORY classifications were soon superseded.

Early histologic studies by NITSCHE (1869, 1871), VIGELIUS (1884), OSTROUMOV (1886a, b), DAVENPORT (1891), and others provided detailed evidence of the arrangement of cellular and noncellular layers of body walls and of the structure of interzooidal communications in a number of cheilostomates and ctenostomates. These studies are the foundation for modern understanding of modes of growth in the Gymnolaemata, but emphasis was on taxa in which body walls are uncalcified or only slightly calcified and zooids are relatively simple in morphology. Information on more complex taxa was gained more slowly.

As modes of growth of more complex gymnolaemates were studied, attention was directed to modifications of the frontal structure of zooids, and especially to the hydrostatic system for everting the lophophore. The morphology and function of the hydrostatic system in the Ctenostomata and simple, lightly calcified Cheilostomata had been known at least from the time of FARRE (1837). As frontal structures of more complexly calcified cheilostomates were compared with those of simple gymnolaemates, new characters became available not only for classification within the Cheilostomata but also to

establish basic morphologic similarities between cheilostomates and ctenostomates. The diversity of morphologies in the Cheilostomata, however, makes these relationships complex.

Around the turn of the century, it was realized that the hydrostatic function in many of the more complexly calcified cheilostomates is performed by an inner compensating sac or ascus, instead of the exposed flexible frontal wall to which parietal muscles are attached in ctenostomates and simple cheilostomates (see Morphology and Mode of Growth, below). The concept of the ascus is generally attributed to JULLIEN (1888b,c), who did not, however, distinguish its method of operation. Further, JULLIEN applied this and other morphologic concepts heterogeneously in his taxonomic studies and derived a classification (1888a, p. 7) that bears little resemblance to twentieth century classifications based upon his discoveries (Table 2). JULLIEN's suborders were employed by CANU (1900) in revising D'ORBIGNY's Cretaceous species of Cheilostomata, but the JULLIEN classification gained little following.

The first detailed evidence of the arrangement of cuticular, calcareous, and cellular layers on the frontal sides of zooids in more complex gymnolaemates was presented in a major work by CALVET (1900) on the comparative histology of species of cheilostomates, ctenostomates, and tubuliporates. Some of JULLIEN's concepts were clarified and refined at the histologic level, but CALVET (1900, p. 166) did not distinguish between different modes of growth of similar frontal structures (see Morphology and Mode of Growth, below). Further, CALVET (1900, p. 278) rejected the concept of the ascus, believing the parietal muscles to be attached to calcareous frontal structures. Despite this denial, CALVET presented evidence for an ascus in at least two genera (1900, p. 168–169; fig. 21, pl. 7, fig. 1).

The first major comparison of modes of growth of structures on the frontal sides of simple to complex gymnolaemate zooids was presented by HARMER (1901, 1902). HAR-

MER recognized JULLIEN's concept of the ascus and presented evidence for two different methods by which it is formed (1902, p. 280–281, 294–295). Each mode of ascus formation was thought by HARMER to correlate with a particular mode of growth of the overlying calcified wall, although he (1902, p. 333) suggested two possibilities for the origin of one wall type. In both developmental types, HARMER recognized parietal muscles that insert on the flexible ascus floor, thereby establishing a morphologic comparison with the hydrostatic system of nonascus-bearing gymnolaemates.

HARMER suggested that differences in mode of ascus formation provide a basis for classification within the Cheilostomata (1902, p. 294) but did not then propose formal taxa. Perhaps because HARMER did not formalize his ideas, some were countered almost immediately. LEVINSEN (1902, p. 4) accepted one concept (HARMER, 1902, p. 280–281) but considered this mode of growth to apply to all ascus-bearing genera. LEVINSEN thus seems to have rejected the other concept (HARMER, 1902, p. 294–295), although his later description of one species (LEVINSEN, 1909, p. 18, 33) agrees with HARMER's in some respects. An entirely different, but in many ways unclear concept (see BANTA, 1970, p. 50) was thought by OSTROUMOV (1903) to apply to ascus-bearing taxa.

Attempts to generalize and simplify ideas on development of gymnolaemate frontal structures obscured the important point made by HARMER that features such as the ascus can develop differently in major groups of Gymnolaemata. This point was ignored until 40 years later, when SILÉN (1942) developed a classification of largely new groupings within a combined cheilostomate-ctenostomate taxon (Table 2). Even then, it was assumed that some features, such as parietal muscles and the membranous walls on which they insert, are developmentally homologous throughout these groups (SILÉN, 1942, p. 44).

As a consequence of his attempt to generalize development of certain morphologic features, LEVINSEN proposed a classification (Table 2) in which all ascus-bearing cheilostomates were assigned to one taxon (Camarostega LEVINSEN, 1902; suborder Ascophora LEVINSEN, 1909) and all cheilostomates lacking an ascus to another (suborder Anasca LEVINSEN, 1909). LEVINSEN regarded some lightly calcified anascans as providing a link between the Cheilostomata and Ctenostomata (1909, p. 92, 95) but did not propose a taxonomic revision to reflect this link. Relationships stated or implied in the LEVINSEN classification have been widely accepted by twentieth century workers on fossil and living gymnolaemates.

The LEVINSEN classification, like its nineteenth century predecessors, relied at higher taxonomic levels on the monothetic use of a few morphologic characters. Most discussions of the basis of classification in both the nineteenth and twentieth centuries have concerned the characters selected for monothetic arrangements at each taxonomic level (see HINCKS, 1887, 1890). LEVINSEN, however, recognized with WATERS (1913, p. 460) that a character too variable for taxonomic use in some taxa can be relatively consistent in others. LEVINSEN therefore avoided a strict monothetic adherence to a hierarchy of characters below subordinal level. Indeed his diagnoses of some taxa, for example the Coilostega (LEVINSEN, 1909, p. 161), are quite polythetic.

CANU and BASSLER (1917, 1920, and later works), in a widely used modification of the LEVINSEN classification, returned to more consistently monothetic arrangements in both the Cheilostomata and Ctenostomata, with no close relationship suggested between the two orders. This classification was used with some modifications in the first edition of this *Treatise* (BASSLER, 1953; see Table 2). Diagnoses at all hierarchic levels became severely abbreviated. The hierarchic arrangement of characters was attempted in correlation with essential functions. However, observations on which the functional significance of some characters can be interpreted were not available to CANU and BASSLER, and the ideas of NITSCHE, CALVET, HARMER, and others on modes of growth were not taken fully into account in the hierarchy of functions. Although CANU and BASSLER established numerous taxa at familial and lower levels, their higher level taxa, such as the cheilostomate division Hexapogona, have been little used.

In his large monograph of living Bryozoa of Indonesia, HARMER (1915–1957) synthesized a classification (Table 2) incorporating many features of the Levinsen classification, some of Harmer's earlier ideas, and some new revisions. LEVINSEN's cheilostomate suborders were retained, and HARMER (1926, p. 187) suggested that lightly calcified anascans gave rise independently to two groups of Ctenostomata, the Stolonifera and Carnosa. However, HARMER did not propose taxonomic revisions to reflect this inferred diphyly, or the suggested close phylogenetic relationship between cheilostomates and ctenostomates. Some taxa, such as the Cellularina reintroduced by HARMER (1926), were emended on at least a partly polythetic basis. Other taxa, such as HARMER's divisions of the Ascophora, however, are monothetically based. Unfortunately, HARMER's concepts of ascophoran divisions remained incomplete when he died in 1950 (HASTINGS in HARMER, 1957), and polythetic and phylogenetic evaluation of these groupings is only beginning.

Despite a growing realization that the Cheilostomata and Ctenostomata have certain strong similarities in zooid morphology and mode of growth apparently not shared with other bryozoan orders, the monothetic basis of the Gymnolaemata (ALLMAN, 1856) to include stenolaemate bryozoans continued to be followed by early twentieth century workers. Fundamental works on embryology, larval morphology, and metamorphosis by BARROIS (1877, 1882), REPIACHOFF (1880), VIGELIUS (1886, 1888), KRAEPELIN (1892), HARMER (1893), BRAEM (1897), and CALVET (1900) further emphasized resemblance between living cheilostomates and ctenostomates. Eventually, study of living Tubuliporata (=Cyclostomata of BUSK) by BORG (1926a) revealed striking contrasts with

cheilostomates and ctenostomates and led him to remove the Tubuliporata from the Gymnolaemata (see BOARDMAN, this revision). However, BORG held the traditional view that the Cryptostomata are closely related to the Cheilostomata and left both taxa, together with the Ctenostomata, in the emended Gymnolaemata. An extreme application of this view was BASSLER's (1935) assignment of a Paleozoic cryptostomate genus to the Cheilostomata. It has only been in the last few years that the stenolaemate characters of the Cryptostomata have been recognized and this order removed from the Gymnolaemata (see BOARDMAN, this revision).

MARCUS (1938a) and SILÉN (1942) proposed different means of formalizing the similarities between Cheilostomata and Ctenostomata while retaining the older concept of Gymnolaemata to include the Tubuliporata.

MARCUS (1938a, p. 116) established an order Eurystomata to include suborders Cheilostomata and Ctenostomata. His concept of the Eurystomata was based on embryologic similarities between the Ctenostomata and both anascan and ascophoran Cheilostomata (1938a, p. 123), and morphologic similarities including a generally wide orifice relative to the size of the zooid (1938a, p. 116).

SILÉN (1942, p. 3) went a step farther than MARCUS, by rejecting the concepts of Cheilostomata and Ctenostomata altogether and merging their component taxa in a suborder, which he named Cheilo-Ctenostomata following an informal usage of BORG (1926a, p. 482). Later, SILÉN (1944a, p. 98; and subsequent papers) followed BORG in removing the Tubuliporata from the Gymnolaemata, which SILÉN then regarded as an order including only cheilostomates and ctenostomates. In this later revision, SILÉN continued to reject Cheilostomata and Ctenostomata as taxa.

SILÉN's concept of the Gymnolaemata (=Cheilo-Ctenostomata) and its component taxa (Table 2) was based on a series of phylogenetic inferences from the morphology of living genera, and on the morphology of the feeding apparatus, which ". . .does not show any differences of importance but is surprisingly monotonous throughout the two groups" (SILÉN, 1942, p. 2). SILÉN (1942, p. 52–58) assigned all ctenostomates to two groups, the Stolonifera and Carnosa, which he inferred to have evolved separately, a conclusion similar to that of HARMER (1926). The hypothetical gymnolaemate ancestor of these two groups was inferred by SILÉN to be similar morphologically to a living genus for which he proposed the taxon (section) Protocheilostomata. The Protocheilostomata were regarded by SILÉN as the central gymnolaemate stock leading to five major taxa (sections) of cheilostomates (see Table 2). These include three for anascans and two for ascophorans, although LEVINSEN's suborders were also rejected in the SILÉN classification. SILÉN's ascophoran taxa were based on modes of growth of the calcified wall overlying the ascus. However, SILÉN (1942, p. 43–44) considered the ascus to originate the same way in both groups. His concept of ascus formation appears to correlate with that of HARMER's Ascophora imperfecta.

Some aspects of SILÉN's classification have been incorporated in current classifications of the Gymnolaemata (for example, PRENANT & BOBIN, 1966; MAWATARI, 1965; RYLAND, 1970; see Table 2). RYLAND (1970), BANTA (1971), and indeed SILÉN (1942) himself have emphasized the highly tentative state of some groupings established on virtually monothetic criteria. Emendations of SILÉN's major gymnolaemate taxa have included: (1) rearrangement of component genera (SOULE, 1954; SOULE & SOULE, 1969; RYLAND, 1970); (2) recombination of parts of different taxa (assignment of ascophoran genera of the Spinocystidea to the Gymnocystidea by RYLAND, 1970; assignment of some ascophoran genera to the Cryptocystidea by BANTA, 1970, 1971; new groupings of ctenostomates proposed by JEBRAM, 1973a); and (3) reintroduction of taxa apparently excluded from SILÉN's classification (Ascophora as emended by RYLAND, 1970).

Most subsequent workers have not followed SILÉN in rejecting intermediate level taxa between the Gymnolaemata and these major groupings, however. Cheilostomata and Ctenostomata are generally retained as orders following the usage of SMITT more than 100 years ago, even though POHOWSKY (1975) has suggested the possibility that the Cheilostomata as well as the Ctenostomata may be polyphyletic.

In contrast, BANTA (1970, 1971) has elevated the emended Cryptocystidea to ordinal rank within a subclass Cheilostomata. As phylogenetic relationships become better understood, the diversity of morphologies embraced by the Gymnolaemata, especially within the Cheilostomata, may well justify significant increases in the categorical ranks of component taxa. Here, however, the Cheilostomata and Ctenostomata are tentatively retained as taxa of ordinal rank.

Some workers still retain the broader concept of Gymnolaemata to include the stenolaemates, and follow MARCUS in recognizing Eurystomata (=Eurylaemata of MAWATARI, 1965) as an intermediate level taxon. Reasons for not employing this two level classification are presented by BOARDMAN, CHEETHAM, and COOK (this revision).

The concept of the Gymnolaemata followed here is that of SILÉN (1944a), RYLAND (1970), and some other authors. A tentative phylogenetic basis for this concept is given below (Possible Evolutionary Relationships). To suggest taxonomic emendations within the Gymnolaemata or to review the many fundamental works on lower level taxa, principally at superfamilial and familial rank, would obviously be premature before restudy of the approximately 1,000 nominal gymnolaemate genera has been completed. These reviews will appear in subsequent volumes of this revision of the *Treatise*.

A single example will perhaps serve to illustrate the extensive internal rearrangements in classifications of the Gymnolaemata that have been brought about by changing morphologic emphasis in the predominantly monothetic use of characters. The Cribrimorpha, comprising genera with frontal shields composed of fused spinelike **costae** (see Morphology and Mode of Growth, below), are now usually considered to be a suborder of the Cheilostomata (BUGE, 1957; RYLAND, 1970; and others). These genera were placed by LEVINSEN (1909) in the morphologically simplest of his divisions of the suborder Anasca, emphasizing their simple membranous frontal walls underlying costal shields. HARMER (1926) considered the Cribrimorpha to be morphologically the most complex division of the Anasca, forming a link with the Ascophora, because of the structure of their frontal shields. CANU and BASSLER (1920) placed the cribrimorph genera in the Ascophora, and BASSLER (1935) considered the Cribrimorpha to be a division of the Ascophora, emphasizing the ascuslike cavity between frontal wall and frontal shield. SILÉN (1942) included the cribrimorphs with some ascophorans in his section Spinocystidea on the basis of phylogenetic inferences. The current subordinal position of the cribrimorphs thus seems to be a compromise between more extreme assignments. The systematic positions of this and other major taxa of the Gymnolaemata can only become better known through detailed comparisons of component living and fossil genera, considering all available morphologic characters and the distribution of their states in time and space (for example, LARWOOD, 1969).

Uneven progress over the past 125 years in deriving a stable classification of the Gymnolaemata, at the levels of class, orders, suborders, and lower level taxa, has resulted partly from the sheer number of genera to be understood morphologically and distributionally, as well as from repeated changes in the monothetic bases of classification. However, another, human factor also seems to have been involved. Some of the most significant morphologic discoveries have been ignored, rejected, or misrepresented, often to emphasize shifts in monothetic criteria, and so rediscovered decades later. Some misunderstandings have doubtless been encouraged by a confusing manner in which interpretations were expressed, especially if in a new and complex terminology, or by a failure to present sufficient supporting evidence: "these heroic attempts. . .made without facts to bear them out. . .are usually ignored, and so bring their own punishment" (WATERS, 1889, p. 3). However, over the years the prevalence of such rejections, or worse yet misrepresentations, must make many bryozoologists sympathize with SMITT's (1872, p. 246, 247) comment on contemporary misunderstanding of his work: "Thus I could not think that any one should impute to me such a thought. . .such an opinion would be an absurdity."

MORPHOLOGY AND MODE OF GROWTH

Colonies in the Gymnolaemata range from a few zooids in the free-living ctenostomate *Monobryozoon* to estimated tens of millions of zooids in multilaminate encrusting species of such cheilostomates as *Membranipora* and *Schizoporella*. Major parts of colonies in some taxa are extrazooidal. Principal growth directions of zooids and of major parts of most colonies approximately coincide. Zooids within colonies are commonly polymorphic. **Autozooids** (zooids having protrusible lophophores, some with feeding ability and others without) have orificial walls consisting of one or more movable folds, the outer sides

of which are continuous with an elongated frontal wall (Fig. 64). When closed, the orificial wall generally lies subparallel to the frontal wall and to the principal direction of zooid growth. Part or all of the frontal wall, or an infolded sac derived from it, is flexible by means of attached parietal muscles and functions in the **hydrostatic system** for protruding the lophophore. A variety of supportive and protective structures may be associated with the frontal wall. Other supporting zooid walls include lateral walls, and in most taxa basal walls, elongated generally subparallel to the principal direction of zooid

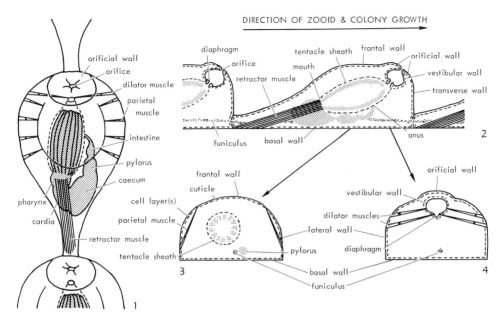

Fig. 64. General features of the class Gymnolaemata. Diagrams of the autozooid of a generalized, uncalcified, encrusting gymnolaemate bryozoan, based on a ctenostome morphologically comparable to the earliest cheilostomates. Body walls of zooid are virtually entirely exterior walls.——*1*. Frontal view, showing retracted feeding organs and muscles through transparent frontal and orificial walls (compare with Fig. 3, 4).——*2*. Median longitudinal section, showing orientation of basal, transverse, frontal, and orificial walls relative to principal growth direction of zooid and colony.——*3*. Transverse section through frontal wall, retracted lophophore, and gut, showing parietal muscles that depress part of frontal wall in lophophore protrusion.——*4*. Transverse section through orificial wall, vestibule, and diaphragm, showing muscles that dilate vestibule and diaphragm in lophophore protrusion.

growth, and transverse walls oriented subperpendicular to the principal growth direction. A plane of bilateral symmetry bisects the orificial, frontal, transverse, and basal walls, but some contained zooid organs as well as some body wall structures may be markedly asymmetrical.

In this section some characters of the Gymnolaemata are considered in expanded form to explain and illustrate some of the great diversity of morphologies in taxa included in the class. To facilitate correlation of this discussion with the distinguishing characteristics of the class as listed by BOARDMAN, CHEETHAM, and COOK (this revision), characters are considered in approximately the same sequence here, but not all are discussed. Throughout this discussion, an attempt is made to emphasize those characters that have

recognizable states in both the Ctenostomata and the Cheilostomata. However, the highly unequal diversity of morphologies in the two orders and their even more unequal representation in the fossil record result in considerable emphasis going to character states, and also some characters, known only in the Cheilostomata. Emphasis on the Cheilostomata is particularly apparent in the sections on calcification, the frontal wall and associated structures, and extrazooidal parts.

CALCIFICATION

The Gymnolaemata apparently comprise the only bryozoan class that includes both uncalcified taxa (Ctenostomata) and calcified taxa (Cheilostomata). In the Cheilostomata, mineral composition and microstructure of

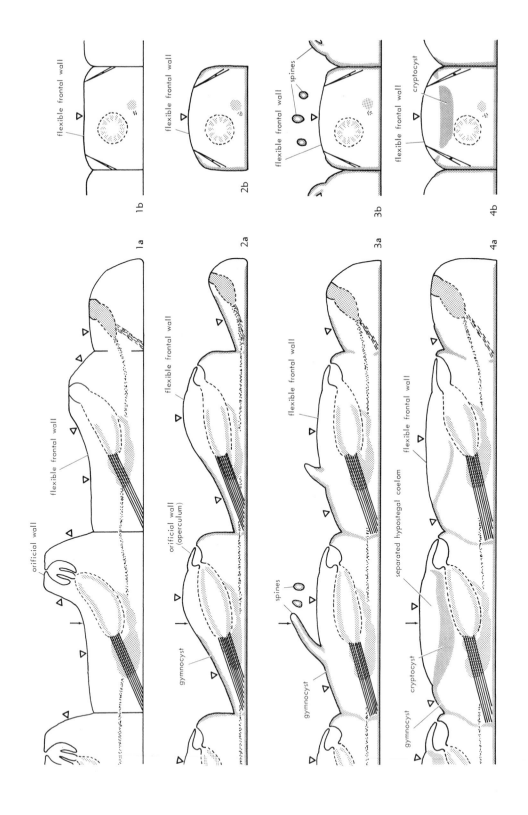

flexible frontal wall

1b

flexible frontal wall

2b

spines

flexible frontal wall

3b

flexible frontal wall

cryptocyst

4b

1a

orificial wall

flexible frontal wall

2a

orificial wall (operculum)

flexible frontal wall

gymnocyst

3a

spines

flexible frontal wall

gymnocyst

4a

separated hypostegal coelom

flexible frontal wall

gymnocyst

cryptocyst

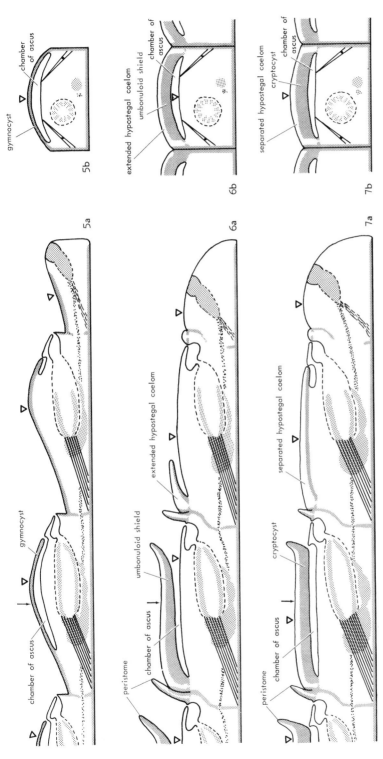

Fig. 65. General features of the class Gymnolaemata. Diagrams of median longitudinal (left) and transverse (right) sections through developing autozooids at and near growing edges of unilaminate (or one layer of bilaminate) simple to complex encrusting or erect colonies. Zooids at more advanced ontogenetic stages lie toward left of longitudinal sections, growing tips at right. Frontal walls of zooids (indicated by open triangles) originate as exposed membranous walls at growing tips and are retained in nearly unmodified form or are modified in various ontogenetic patterns. In all patterns, frontal walls are exterior walls, whether calcified or uncalcified. (Representation as in Fig. 64,2–4, but cellular layers of body walls omitted. Calcareous layers finely stippled. Lophophore and gut more coarsely stippled. Positions of transverse sections indicated by vertical arrows.)——1. Carnose ctenostomate, showing uncalcified body walls and peristomelike frontal-wall prominence on which orificial wall becomes elevated.——2. Simple anascan cheilostomate, showing frontal wall calcified proximally and laterally to form protective shield of (Continued on p. 153.)

skeletal walls also seem to be more variable than in other calcified Bryozoa (Stenolaemata). If the Cheilostomata evolved from the Ctenostomata, as their comparative morphology and stratigraphic records suggest, calcareous skeletons in the Stenolaemata and the Gymnolaemata evolved independently. (See Dzik, 1975, for a contrasting interpretation of separate evolutionary origins of the Ctenostomata and the Cheilostomata from the Stenolaemata, and inferred close relationship of skeletons in cheilostomates and stenolaemates.)

Body walls in the Gymnolaemata consist of cellular layers and more or less stiffened noncellular layers. The great majority of Ctenostomata has body walls stiffened only by cuticular layers (Fig. 64; 65,*1*; 66,*1–3*), but scattered calcareous particles have been reported in the cuticle of one freshwater species (Kraepelin, 1887). With few possible exceptions (Banta, 1975), Cheilostomata have some body walls of zooids and, where present, of extrazooidal parts reinforced with continuous calcareous layers, in addition to the less stiffened cuticular layers. These calcareous layers collectively form the skeleton of a colony (zoarium). Zooid skeletons (zooecia) in the Cheilostomata can include few, thinly calcified walls (Fig. 65,*2*), or most zooid walls can be calcified; some zooid skeletons continue to receive calcareous deposits throughout zooid life (Fig. 65,*6,7*). A variety of ontogenetic patterns of calcification have been described between these two extremes (see section on the frontal wall and associated structures; and Sandberg, this revision).

Calcareous layers of both interior and exterior body walls in the Cheilostomata are all exoskeletal, deposited outside the adjacent epidermal cells on the side away from the body cavity (Fig. 67,*1b*). Epidermal cells of different shapes adjacent to cuticular and skeletal layers have been reported to possess secreting structures (Tavener-Smith & Williams, 1972). No morphologic differences have been observed between epidermal cells adjacent to skeletal layers of different mineral

composition in the same skeleton (Banta, 1971). Skeletal layers of both interior and exterior walls have been reported to lie between noncellular organic sheets and to contain noncellular organic networks continuous in places with these sheets (Banta, 1969). The cuticular nature of outer organic sheets on calcified parts of exterior walls and of the whole sequence of organic sheets on uncalcified parts of exterior walls suggests that the cheilostomate skeleton can be regarded as **intracuticular** (Banta, 1969).

Skeletons in the Cheilostomata are composed of either calcite or aragonite (**monomineralic skeleton**), or a combination (**bimineralic skeleton**). At present, the Cheilostomata are the only order in the phylum in which bimineralic skeletons are known. The generally consistent results obtained in analyzing cheilostomate species from different geographic areas suggest that skeletal composition is closely controlled genetically (Poluzzi & Sartori, 1975). In some bimineralic species, there is evidence that aragonite : calcite ratios may increase in populations living in warmer water (Rucker & Carver, 1969), but the ratio can be strongly affected by ontogenetic gradients within colonies (Cheetham, Rucker, & Carver, 1969; Sandberg, 1971).

More than 150 cheilostomate species have been analyzed (Poluzzi & Sartori, 1975, and references listed therein), over 80 percent from recent specimens only. Of the analyzed species about 50 percent have all skeletal layers composed of calcite, about 40 percent include both calcite and aragonite, and about 10 percent have only aragonite. In bimineralic species in which intracolony distribution of skeletal components has been studied, calcite and aragonite are present in discrete layers. In many of these species, aragonite layers succeed calcite layers ontogenetically in some zooecial walls, whereas other walls in the same zooecium remain entirely calcitic throughout ontogeny (Fig. 67,*1c*; 68,*1d,1e,2*) (Sandberg, 1971). In a few species, zooecial walls have been found to be calcitic and associated with aragonitic extrazooidal skeleton in the same zoarium (GREELEY, 1969; RUCKER & CARVER, 1969).

At higher taxonomic levels, there appears to be less consistency in skeletal composition than within species. The species analyzed are distributed in more than 80 genera, of which only 33 include more than one analyzed species. Of these 33 genera, 11 include only calcitic species and 3 only aragonitic species. Monomineralic skeletons thus appear to be in a slight minority (about 40 percent) among analyzed genera, in contrast to their majority (about 60 percent) among analyzed species. The 19 bimineralic genera analyzed appear to be of two kinds. In 11 genera all species analyzed are bimineralic. The remaining 8 genera include some species of entirely calcitic composition and some of either mixed or entirely aragonitic composition. Examples of all four compositional types of genera are known among both anascan and ascophoran Cheilostomata. Calcite is apparently dominant among anascan species (47 species calcitic, 4 aragonitic, 10 mixed), and the earliest cheilostomates known are morphologically similar to modern anascan species having entirely calcitic skeletons. Bimineralic and aragonitic compositions are more common among ascophoran species, but a large proportion of ascophorans retain calcitic skeletons (36 species calcitic, 20 aragonitic, 41 mixed). The oldest fossil cheilostomates in which aragonite has been reported are of Late Eocene age (GREELEY, 1969; RUCKER & CARVER, 1969). Even diagenetically altered fossil cheilostomate skeletons have been found to contain relic inclusions and textural evidence of their original composition and microstructure (SANDBERG, 1975a), and so it is at least theoretically possible to interpret stratigraphic distribution of skeletal composition in cheilostomate lineages.

Diversity in skeletal composition in the Cheilostomata is paralleled by, but not precisely correlated with, a variability in skeletal microstructure (SANDBERG, 1971, 1973). Calcitic skeletal layers can assume a variety of structures, from laminated subparallel to wall surfaces (Fig. 68,*1a–c*; 69,*1d–f*; 70,*1a–*

FIG. 65. *(Explanation continued from page 151.)*

exterior origin (gymnocyst) and exposed and flexible distally and medially to form hydrostatic membrane to which parietal muscles are attached.——*3.* Complex anascan, showing frontal wall as in *2,* but with flexible part overarched by calcified tubular exterior-walled outpocketings (spines) forming a protective (costal) shield. Spines contain extensions of body cavity of zooid.——*4.* Complex anascan, showing frontal wall as in *2,* but with the flexible part underlain by a frontal shield of interior origin (cryptocyst) protecting feeding organs. The cryptocyst grows between layers of epidermis folded into the body cavity of the zooid, partitioning the cavity into an overlying hypostegal coelom and an underlying principal body cavity. The cryptocyst is thickened ontogenetically by addition of skeleton to its frontal surface by epidermis underlying hypostegal coelom.——*5.* Simple ascophoran cheilostomate, showing frontal wall calcified, except at the proximal margin of the operculum, to form a protective shield of exterior origin (gymnocyst), beneath which the ascus becomes infolded from the proximal margin of operculum. Parietal muscles are attached to the flexible floor of the ascus (compare with the simple anascan in *2*).——*6.* Complex ascophoran cheilostomate, showing the flexible frontal wall, with attached parietal muscles, forming the floor of the ascus overarched by a protective shield of exterior origin (umbonuloid shield). The umbonuloid shield grows on the basal side of a double-walled outfold, within which is the extension of the zooid body cavity (hypostegal coelom), and it is attached to vertical walls by interior wall segments. The umbonuloid shield is thickened ontogenetically by addition of skeleton to its frontal surface by epidermis underlying the extended hypostegal coelom. The calcified layer of peristome is an extension of umbonuloid shield (compare with complex anascan in *3*).——*7.* Complex ascophoran cheilostomate, showing exposed uncalcified frontal wall underlain by cryptocyst grown as in *4,* protecting feeding organs. However, the cryptocyst is underlain by an ascus infolded from the proximal margin of the operculum, as in *5.* Parietal muscles are attached to the floor of the ascus. The cryptocyst is thickened ontogenetically by addition of skeleton to its frontal surface by epidermis underlying separated hypostegal coelom. The calcified layer of the peristome is continuous with the cryptocyst, but of exterior origin, formed after attachment of the rim of the cryptocyst to the membranous frontal wall (compare with complex anascan in *4*).

c,2) to fibrous parallel or transverse to wall surfaces (Fig. 71,*1a–d*). Most aragonitic layers are fibrous either parallel or transverse to wall surfaces (Fig. 67,*1c–e*; 68,*1d,1e*; 72,*1,2*), but more blocky textures have recently been recognized in aragonitic cheilostomates (for further discussion, see SANDBERG, this revision).

BODY WALLS OF AUTOZOOIDS

In living gymnolaemate colonies, some or all zooids can be observed to possess bodywall features associated with protrusible lophophores and thus be recognized as autozooids. In colonies of many taxa all autozooids are capable of feeding at some stages of their ontogeny. In colonies of a few taxa, some autozooids concerned with sexual reproduction, and possibly other functions, remain incapable of feeding. Nonfeeding sexual autozooids have recognizable bodywall differences from feeding autozooids in all but a few species. Colonies in most gymnolaemate taxa also have nonfeeding polymorphs without protrusible lophophores and with distinctive body wall features reflecting this major difference from autozooids.

Body walls expressing morphology by which autozooids can be recognized in the Gymnolaemata are principally the orificial wall defining the orifice through which the lophophore is protruded and the frontal wall and associated structures functioning in the hydrostatic system for protruding the lophophore. In the Cheilostomata, this morphology commonly is reflected in the skeleton. Basal and vertical walls of autozooids may be different from or similar to those of polymorphs and thus are less significant in recognizing the major functional organization of a colony.

Basal walls.—Basal walls generally are present in gymnolaemate autozooids and serve to enclose basal sides of body cavities, to support vertical walls, and to provide attachment for some muscles or organs. Zooids may lack basal walls in some taxa having erect cylindrical colony branches, along the axes of which lateral walls of zooids meet directly to enclose zooids basally (CHEETHAM, 1971, pl. 12, fig. 1–4). Zooids may also lack basal walls in some taxa in which autozooids were budded frontally from hypostegal coeloms of subjacent autozooids and are enclosed basally by frontal structures of subjacent zooids.

Most commonly basal walls of autozooids are exterior walls, which extend the body of the colony (Fig. 64; 65; 68,*1a*). Exterior basal walls may be present in both encrusting and erect parts of colonies. Exterior basal walls of zooids most commonly form the surfaces by which encrusting colonies adhere to other objects (Fig. 69,*1a–c*; 71,*1a–d*; 72,*1,2*) or to overgrown parts of the same colony (Fig. 68,*1a–e*). Encrusting bases of erect colonies and of initial portions of free-living colonies can also adhere to objects by means of exterior basal walls of variable numbers of founding zooids (HÅKANSSON, 1973, pl. 2, fig. 4). Medial surfaces of erect **bilaminate** branches in colonies of many taxa and of subcylindrical branches in colonies of some taxa are formed by exterior basal walls of zooids adherent back to back (Fig. 70,*1a*; 73,*1a,b,2a,c*). Reverse surfaces of unilaminate branches in colonies

---

FIG. 66. Carnose ctenostomates.——*1,2. Elzerina blainvilli* LAMOUROUX, rec., S. Afr.; *1,* Port Alfred, Pondoland, erect branching colony composed of alternating rows of autozooids (az) and kenozooids (kz), BMNH 1922.8.23.1, ×9; *2a,b,* Durban, *a,* autozooids with operculumlike orificial wall (ow), flexible frontal wall (fw), tentacles (te), and retractor muscle (rm), embryos (emb) brooded in diverticulum of tentacle sheath (ts), diaphragm marked by pleated collar (pc), long. sec., *b,* autozooids flanked by kenozooids (kz), parietal muscles (pm) of autozooids originating on cuticular lateral walls (lw) and inserting on frontal wall (fw), transv. sec.; both BMNH 1942.8.6.25, ×120.——*3. Alcyonidium nodosum* O'DONOGHUE & DE WATTEVILLE, rec., S. Afr.; autozooid with orificial wall (ow) slightly elevated on frontal wall (fw) to which parietal muscles (pm) are attached, diaphragm marked by pleated collar (pc), diaphragmatic dilator muscles (dm) originating on cuticular transverse wall (tw); long. sec., BMNH 1942.8.6.1, ×120.

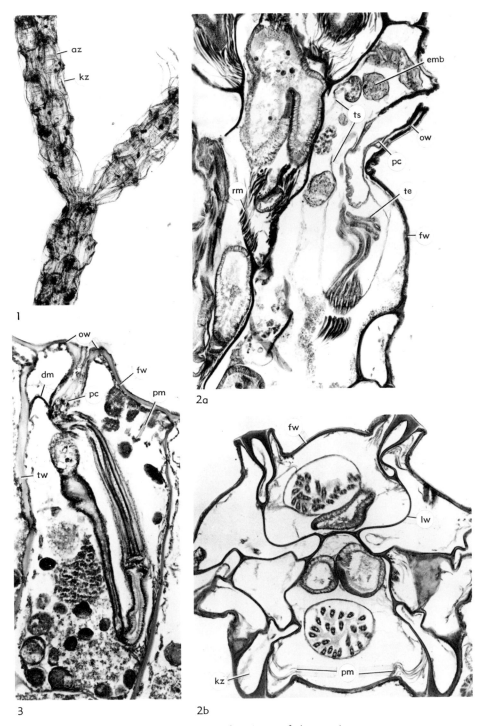

FIG. 66. *(For explanation, see facing page.)*

of some taxa are formed by exterior basal walls of zooids in direct contact with the environment.

Exterior basal walls have at least outermost cuticular layers continuous from zooid to zooid within a budding series (**multizooidal layers**). In the Cheilostomata, exterior basal walls may be calcified (Fig. 68, *1a–e*; 71, *1a–d*) or wholly or partly uncalcified (Fig. 69, *1f*; **basal window** of BANTA, 1968). Some skeletal layers of calcified basal walls are also multizooidal (Fig. 69, *1d*; **basal plate** of BOARDMAN & CHEETHAM, 1969; **basal platform** of SANDBERG, 1971).

Basal walls of zooids in parts of some free-living colonies beyond initial adherent portions (HÅKANSSON, 1973) and in parts of some erect colonies beyond encrusting bases (see Fig. 78, *1a–c*) are interior walls, which partition preexisting body cavity of the colony. In free-living and unilaminate erect colonies (see Fig. 78, *1a–c*), interior basal walls of zooids adjoin interior extrazooidal walls which, together with extrazooidal body cavity and exterior extrazooidal walls, separate the basal walls of zooids from the environment. Interior basal walls of zooids are known only in the Cheilostomata, in which they include some skeletal layers that are continuous among zooids (multizooidal).

Vertical walls.—Vertical walls of autozooids in the Gymnolaemata comprise **lateral** and **transverse walls**, distinguished by orientation relative to the principal growth directions of zooids in most colonies (Fig. 64, 65, 74). Lateral walls give length and, together with transverse walls, depth to the body cavities of zooids. In most taxa lateral and transverse walls are further distinguished by modes of growth. In the Cheilostomata vertical walls include skeletal layers, some but not all of which commonly form a continuous structural unit (zooecial lining of SANDBERG, 1971).

In the great majority of gymnolaemates, both ctenostomates and cheilostomates, lateral walls are exterior walls that extend the body of the colony in **lineal series** of sequentially budded zooids (Fig. 65; 66, *1,2*; 75; 76, *1–4*; 77). Within a lineal series, bounding cuticles and, in the Cheilostomata, some skeletal layers of lateral walls are continuous from zooid to zooid as multizooidal layers. Some multizooidal layers of lateral walls are also continuous with multizooidal layers of basal and frontal walls (Fig. 69, *1f*; 73, *1a–c*). Contiguous lineal series have separate lateral walls (Fig. 69, *1f*; 70, *1,2*; 71, *1c,d*; 72, *2*), although contiguous bounding cuticles apparently can be breached to form interzooidal communication organs (Fig. 70, *2*) (BANTA, 1969) or confluent extrazooidal parts of colonies (Fig. 70, *1b*; see below).

In a few cheilostomates and ctenosto-

FIG. 67.   Ascophoran cheilostomate.——*1a–e. Margaretta cereoides* (ELLIS & SOLANDER), rec., Naples, Italy; *a,* growing tip (gt) with distalmost membranous wall of lineal series nearly intact, distal zooid with walls nearly complete, but outer part of transverse wall (tw) not calcified and operculum not formed, cryptocyst (cry) nearly complete, but without underlying ascus, proximal part of frontal wall (fw) calcified to form gymnocyst (gy), the shape of which reflects future brood chamber to be roofed by peristome of proximal zooid (compare *d*), proximal zooid with operculum (op) and ascus (fa, floor of ascus), but peristome little developed, long. sec.; *b,* detail of cryptocyst of distal zooid with adjacent epidermis on both sides, outer side overlain by hypostegal coelom (hy) and membranous frontal wall (fw), long. sec.; *c,* cryptocyst of more proximal zooid in same segment with thin initial skeletal layer (il) nonstaining in Feigl's solution (presumed calcitic) and thick superficial skeletal layer (sl) staining in Feigl's solution (aragonitic), cuticle of frontal wall (fw) heavier than that forming roof of ascus (ra) immediately adjacent to underside of cryptocyst without intervening epidermis or body cavity, hypostegal coelom (hy) extending into funnel-shaped depression (fd) at base of which is uncalcified spot (un) in initial skeletal layer (compare Fig. 82, *3b*), long. sec.; *d,* brood chamber (bch) floored by gymnocyst (gy) of distal zooid and roofed by outfolded peristome (of) surrounding operculum (op) of maternal zooid, long. view; *e,* ordinary autozooid with heavily reinforced operculum (op) supported circumferentially by skeleton and surrounded by outfolded peristome (of), opening to ascus (oa; fa, floor of ascus) passing through frontal wall (fw) and cryptocyst (cry), long. sec. (for diagram of zooid, see Fig. 7); USNM 242573, *a,d,e,* ×100, *b,c,* ×300.

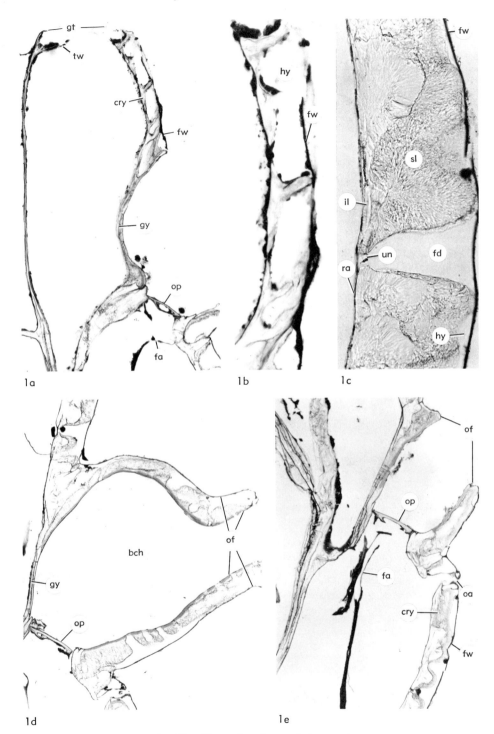

FIG. 67. *(For explanation, see facing page.)*

mates, transverse walls are also largely exterior, formed as extensions of lateral walls enclosing distal ends of zooids (Fig. 64; 65,2; 69,1e; 74,1; 75,1–6,8; 77,1a). Near basal margins of transverse walls, small interior walls extend from inner surfaces to form pore plates of communication organs, separating zooid body cavities within lineal series.

Transverse walls in most cheilostomates, and apparently in most ctenostomates, are developed principally as extensive interior walls, completely partitioning body cavities within lineal series (Fig. 65,1,3–7; 74,2; 75,7). These walls contain pore plates of interzooidal communication organs (Fig. 67,1e; 68,1a,b,e; 71,1a,b; 78,1a,e). Basally, laterally, and frontally, interior transverse walls are attached to inner surfaces of multizooidal layers of exterior walls (Fig. 68,1a–e; 69,1d,e). Parts of these exterior walls can become incorporated in the transverse walls by expansion in a frontal direction. In chei-

lostomates, skeletal layers of adjoining interior transverse walls belonging to contiguous zooids are commonly distinguishable at distinct organic boundaries (Fig. 78,1e), by distinctive skeletal structure (Fig. 71,1a,b), by continuity of laminae with those in basal walls above multizooidal layers (Fig. 69,1d) or by a combination (Fig. 68,1d). Walls on either side of a boundary vary from subequal to markedly unequal in thickness.

In a few taxa, apparently restricted to the Cheilostomata, both lateral and transverse walls develop as interior walls partitioning the colony body cavity within multizooidal budding zones similar to those in the class Stenolaemata (Fig. 74,3; 75,9). Skeletal layers of these walls form a unit continuous with interior basal walls (Fig. 78,1a–c) (HÅKANSSON, 1973). Frontally, interior vertical walls are attached to multizooidal cuticles, although attachment may remain incomplete on some zooidal margins in at

---

Fig. 68. Ascophoran cheilostomates.——*1a–e. Metrarabdotos (Uniavicularium) unguiculatum cookae* CHEETHAM, rec., Ghana, W. Afr.; *a,* distal bud (db) at growing tip of lineal series of encrusting intracolony overgrowth, membranous frontal wall (fw) and calcified basal (bw) and proximal transverse (tw) walls enclosing body cavity of bud, frontal portion of interior transverse wall attached to outer membranous wall to form skeletal rim for orificial wall (ow) of proximal zooid, calcified exterior peristomial wall (now collapsed, original position of inner end indicated by arrow) continued from transverse wall as part of distal bud; *b,* next proximal zooid in same lineal series as *a,* with extensive, thin umbonuloid frontal shield (fs), and overlying hypostegal coelom (hy) and outer membrane, all overarching proximal portion of membranous frontal wall (fw), orificial wall with lightly reinforced operculum (op) complete, but more proximal organs of zooid (dev) in early stage of development; *c,* proximal part of zooid just proximal to distal bud in a lineal series neighboring that in *a* and *b,* with frontal shield (fs) and associated soft parts in an early stage of development, overarching membranous frontal wall (fw); *d,* fully developed zooid just proximal to zooid in *b,* with lophophore fully retracted against proximal transverse wall by retractor muscle (rm), tentacle sheath (ts) attached at outer end to calcified shelflike extension of distal transverse wall beneath operculum (op), distal transverse wall attached at outer end to orificial wall (ow) to form skeletal rim, frontal shield extending over operculum to complete peristome, which has denticles (pd) that check operculum, when open, from closing chamber between frontal shield and membranous frontal wall (fw), frontal shield two-layered, with initial layer (il) of calcite and superficial layer (sl) of aragonite, hypostegal coelom (hy) communicating with principal body cavity of zooid through pore plate (ppl) plugged with cells placed at margin of frontal shield; *e,* fully developed zooid, third proximal to zooid in *c,* with thickened superficial aragonitic layer (sl) of frontal shield, occlusor muscle (om) at lateral margin of calcified distal shelf (see *d*) inserting on operculum (op), funicular strand (fu) attached to cells passing through pore plate (ppl) in transverse wall; all long. secs., USNM 243229, ×100.——*2. Metrarabdotos (Uniavicularium) unguiculatum unguiculatum* CANU & BASSLER, rec., Norseman Sta. 348, off Bahia, Brazil, 50 m; encrusting colony with distal bud (db; compare with *1a*) and autozooids near growing edge; right distal zooid with transverse wall (tw) and frontal shield (fs) with marginal pore plates (ppl), at approximately same stage of development as zooid in *1c,* left central zooid with frontal shield intermediate between those of zooids in *1b* and *1d,* with initial calcitic layer (il) extended to form peristome with denticles (pd), proximal zooids on left and right at comparable stages to zooid in *1e,* with frontal shields and peristomes covered by superficial aragonitic layer (sl), adventitious avicularia (av) with pivotal bars (piv) for mandibles partly or completely developed; frontal view, USNM 243230, ×50.

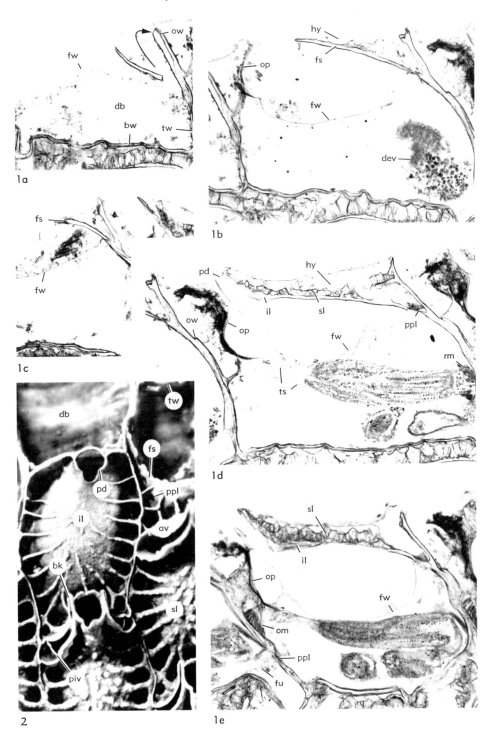

FIG. 68. *(For explanation, see facing page.)*

least one genus (Fig. 78,*1c*).

In a few cheilostomates, autozooids in subsequent zones of astogenetic change and repetition have all exterior vertical walls, which extend the body of the colony in a frontal direction (Fig. 79,*3*; **frontal budding**). In some taxa, these frontally budded zooids originate from hypostegal coeloms of underlying zooids in the primary zone of astogenetic repetition, which have exterior lateral and interior transverse walls oriented with respect to zooidal growth direction as in most other cheilostomates (BANTA, 1972). Vertical walls of some adventitious polymorphs (see below, polymorphism) may be oriented similarly to those of frontally budded autozooids.

Interior vertical walls are grown cooperatively by contiguous zooids, as indicated by microstructure of skeletal (Cheilostomata) or cuticular (Ctenostomata) layers and by complementary configuration (Fig. 74,*3*) (both orders). Configurations of exterior walls of zooids may suggest either autonomous (Fig. 74,*1*; 76,*1–4*; 77,*1,2*) or cooperative growth (Fig. 74,*2*; 80,*2*). Development of interzooidal communication organs in both interior and exterior vertical walls in most taxa also involves cooperative growth.

Frontal walls and associated structures.— As used here, the term frontal wall refers to the outer exterior body wall that bounds the frontal side of a zooid at least in early ontogenetic stages (marked by open triangles in Fig. 65), no matter how modified it may become in later ontogenetic stages. This restricted meaning for the term follows the usage of HARMER (1930, p. 112, 113; 1957, p. 655–657) and SILÉN (1942, p. 5), although several different usages are common in the literature. Frontal walls in the Gymnolaemata support and space orificial walls of autozooids, function directly or indirectly in lophophore protrusion, and in some taxa can be partly calcified to increase colony support and protect retracted zooid organs. Protective and supportive structures in many taxa, however, form a complex of features associated with the frontal wall in addition to any forming parts of the frontal wall itself.

Frontal walls characterize autozooids throughout the Ctenostomata and the Cheilostomata (Fig. 65). In most taxa frontal walls are subparallel to orificial walls and at high angles to vertical walls. In some ctenostomates and a few cheilostomates that have erect tubular autozooids arising from stolonlike bases, frontal walls are at high angles to orificial walls and subparallel to vertical walls. In these taxa, few of which are known as fossils, the distinction between frontal and vertical walls may be arbitrary.

FIG. 69. Ascophoran cheilostomate.——*1a–f. Hippothoa hyalina* (LINNÉ), rec., Cape Cod Bay, U.S. Fish Comm., 1879, 50 m; *a,* growing tip of lineal series with distal bud (db) and autozooid with operculum (op), calcified frontal wall (gy, gymnocyst) and partly formed organs, but no ascus, long. sec.; *b,* next proximal, feeding autozooid with fully formed organs and ascus reaching nearly to proximal end (oa, opening of ascus; fa, floor of ascus), frontal buds (fb) with exterior basal walls (bw) present on both feeding autozooids, long. sec.; *c,* still more proximal, feeding autozooids with partly (fb) and fully developed, frontally budded maternal autozooids, fully developed maternal zooid with ascus (fa, floor of ascus; oa, opening to ascus), brood chamber (bch) enclosed by part of maternal zooid outfolded (of) from distal wall, upper side of brood chamber roof protecting embryo (emb) calcified but with uncalcified spots (un), long. sec.; *d,* junction of basal (bw) and transverse (tw) walls of contiguous zooids in lineal series (distal to top), initial skeletal layers (il) of basal wall continuous between zooids (multizooidal), long. sec.; *e,* distal bud (distal to right) with cuticular and skeletal layers of frontal wall (gy, gymnocyst) continuous with layers of transverse wall of proximal zooid, long. sec.; *f,* laterally contiguous and frontally budded (fb) zooids with bounding cuticles and skeletal layers continuous from basal to lateral to frontal walls (gy, gymnocyst), skeletal layers pinching out medially in basal wall of zooid to right (bw), some basal and lateral wall laminae continuing into interior wall partitioning pore chamber (pch) from principal body cavity (coel) of zooid, transv. sec.; all USNM 242568, *a–c,* ×100, *d,f,* ×800, *e,* ×300.——*2. H. hyalina,* New England coast; encrusting colony with feeding autozooids (az), frontal buds (fb), female autozooids with brood chambers (bch), and male autozooids; frontal view, USNM 242569, ×50.

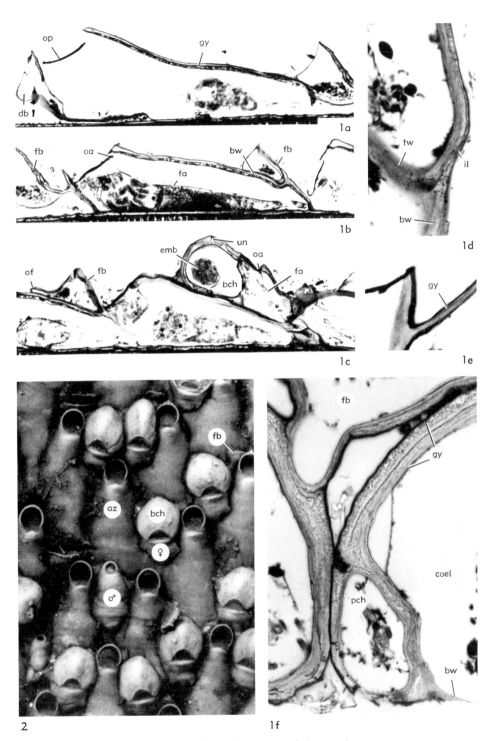

F<small>IG</small>. 69. *(For explanation, see facing page.)*

In the great majority of gymnolaemates, which have some or all vertical walls of zooids developed as exterior walls, frontal walls originate as membranous walls at growing tips of lineal budding series (Fig. 65; 68, *1a*). Laterally, frontal walls in these taxa are continuous in part with exterior vertical walls. Proximally, frontal walls are initially continuous with frontal walls of contiguous zooids within the same lineal series. Attachment of interior components of transverse walls or pore plates transforms the initially multizooidal frontal wall into part of a zooid.

In the few Cheilostomata known to have all interior vertical walls, frontal walls originate as membranous walls in multizooidal budding zones. Developing frontal walls in these taxa are continuous with those of contiguous zooids both laterally and proximally. Attachment of developing interior vertical walls transforms multizooidal frontal walls into parts of zooids. In most of these taxa, both lateral and transverse walls become attached, but lateral walls can remain unattached and parts of body cavities confluent laterally (Fig. 78, *1a–c*).

In the Ctenostomata (Fig. 65, *1*; 66, *2, 3*) and a few Cheilostomata, frontal walls remain entirely flexible and exposed throughout zooid life to function in the hydrostatic system. In most Cheilostomata, frontal walls become modified ontogenetically by calcification, by addition of overlying or underlying calcified structures, or by a combination of these processes (Fig. 65, *2–7*). For the resulting diverse protective and supportive skeletal structures, the general descriptive term **frontal shield** of HARMER (1902, p. 282) is employed here.

Growth of simple to complex frontal shields is partly correlated with slight or extensive changes in the hydrostatic system, conventionally forming the basis for arranging the Cheilostomata in two major morphologic groups (see Table 2). In most taxa generally assigned to the anascan group, the flexible hydrostatic membrane remains largely to partly exposed (Fig. 65, *2–4*). In the ascophoran group, the flexible hydrostatic membrane is overlain by a continuous protective cover (Fig. 65, *5–7*).

Frontal shields in the Cheilostomata comprise skeletal layers of either exterior or inte-

FIG. 70. *Ascophoran cheilostomate.——1a–c. Metrarabdotos (Biavicularium) tenue tenue* (BUSK), rec., Caroline Sta. 68, off NE. coast of Puerto Rico, 20 m; *a,* ordinary autozooid just proximal to growing edge of erect bilaminate colony, basal (bc) and lateral (lc) cuticles of calcified exterior walls forming boundaries with zooids in adjacent lineal series, membranous exterior frontal wall (fw) attached by parietal muscles (pm) to lateral walls, overarched by umbonuloid frontal shield (fs) with overlying hypostegal coelom (hy) and outer membranous wall (compare with Fig. 68, *1a–c*), communication between principal body cavity of zooid and hypostegal coelom through pore plate (ppl), section at midlength of zooid; *b,* part of same colony about 2.5 cm proximal to *a,* ordinary autozooids and adventitious avicularia (av) occluded by extrazooidal skeleton (exs), initial calcitic layer of frontal shield (il) overlain by superficial layer (sl), also calcitic, in turn succeeded without interruption by calcitic extrazooidal skeleton, extrazooidal skeletal layers continuous from zooid to zooid, terminating lateral walls (lc, lateral cuticle) so that hypostegal coelom (hy) confluent around circumference of branch; *c,* part of same colony between *a* and *b,* with ordinary autozooids and adventitious avicularia (av), frontal shield of autozooid with superficial layer (sl) within cuticular boundaries; all transv. secs., USNM 243231, ×100.——*2. M. (B.) t. tenue,* same data as *1*; brooding autozooid about 0.5 cm from growing edge of erect bilaminate colony, embryo (emb) contained in chamber (bch) outside body cavity of colony, surrounded by inner membranous wall (im) and calcified frontal shield (fs), uncalcified spots (un) in frontal shield of brood chamber open into hypostegal coelom (hy); transv. sec., USNM 243232, ×100.——*3a,b. M. (B.) t. tenue,* same data as *1*; *a,* part of erect bilaminate colony about 1 cm from growing edge, with ordinary and brooding autozooids (bch, brood chamber) and two forms of adventitious avicularia, smaller avicularia (av) similar to those in *1b* and *1c* and with simple pointed mandibles, larger avicularia with rounded, bilobed mandibles (md), cuticular boundaries (lc, lateral cuticle) discernible between some but not all zooids; *b,* some of same zooids as in *a* with outer membranous walls and avicularian mandibles removed, pointed and bilobed beaks (bk) conforming in shape to mandibles, both types of avicularia with complete pivotal bars (piv) for mandibles; both frontal views, USNM 243233, ×50.

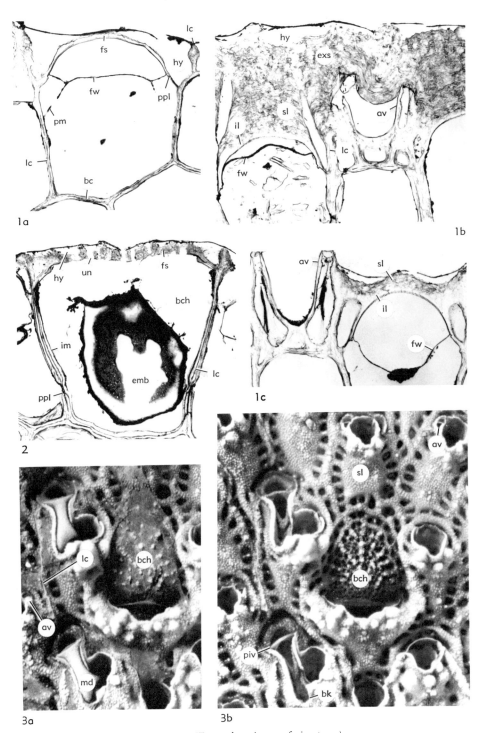

FIG. 70. *(For explanation, see facing page.)*

rior walls. In some taxa (for example, Fig. 65,4) a frontal shield can combine both exterior and interior elements. The ultrastructural characteristics of exterior and interior frontal shield elements are discussed and illustrated by SANDBERG (this revision).

The simplest type of frontal shield is part of the exterior frontal wall itself (**gymnocyst** of HARMER, 1930, p. 113). As the membranous frontal wall develops at a colony growing tip, calcification follows just proximally to produce a gymnocyst extending from the proximal margin of a zooid varying distances distally (Fig. 65,2–5). Gymnocysts of similar appearance are found in both anascans (Fig. 65,2–4; 72,4; 76,1–4; 77,1,2; 80; 81,1,2,4) and ascophorans (Fig. 65,5; 67,1a,d,e; 69). Prominent transverse growth banding is commonly evident on outer surfaces of gymnocystal shields (Fig. 69,2; 76,1,2) (SANDBERG, 1976, pl. 2, fig. 1, 2). Relationship of the gymnocyst to the hydrostatic membrane, however, is different in the two groups (see below).

More complex types of exterior frontal shields are also known in both anascans and ascophorans. These shields differ in the two groups, not only in relation to the hydrostatic system, but also in morphology. In both groups complex exterior frontal shields are

parts of structural features of zooids extending into the environment to overarch the preexisting, more or less completed, flexible part of the frontal wall. These overarching extensions consist of body wall and a contained body cavity, at least initially confluent with the principal body cavity of the zooid.

In anascans, an overarching tubular outpocketing (**spine**) or series of outpocketings, each consisting of exterior body wall with contained coelom, can form a discontinuous cover (**costal shield**) over the flexible frontal wall (Fig. 65,3). Exterior walls of spines can be entirely calcified, or contain uncalcified spots, or be calcified except in a ring where attached to the frontal wall of the supporting zooid. Body cavities of spines can be broadly confluent with that of the supporting zooid (Fig. 71,1c,d) or have openings into the zooidal coelom constricted by body wall (SILÉN, 1942a), in that case being difficult to distinguish from some kinds of polymorphs. In fossils, unfused spines are rarely preserved intact, but **spine bases** are commonly recognizable where they emanate from continuous skeletal structures (Fig. 77,2). In some taxa, spines can be fused at medial ends and intermittently along lengths to produce a more nearly continuous costal shield (**cribrimorph** structure; Fig. 71,1–3). Fused or

FIG. 71.   Cribrimorph cheilostomates.——*1,2. Figularia figularis* (JOHNSTON), rec., Medit.; *1a–d*, Oran, Alg., 100 m; *a*, maternal autozooid with heavily reinforced operculum (op) continuous with membranous frontal wall (fw; collapsed proximally); overarching costal shield (cs) composed of internally thickened spinelike costae, brood chamber (bch) floored by gymnocyst (gy) and roofed by part of costal shield (cs) of distal autozooid, long. sec., ×100; *b*, communication organ in thinned portion (ppl, pore plate) of transverse wall (distal to left), initial granular layer (il) marking boundary between zooids in lineal series and approximately reaching bounding cuticle of basal wall (bc), long. sec., ×300; *c*, brood chamber (bch) floored by gymnocyst (gy) and roofed by costal shield (cs) with narrow central cavities (ccc) opening into body cavity (coel) of autozooid distal to maternal zooid, transv. sec., ×100; *d*, contacting lateral walls of zooids in contiguous lineal series, bounding cuticles (lc) continuous with bounding cuticle of basal wall (bc), narrow central cavities of costae (ccc) opening into body cavities of zooids, all skeletal layers nonstaining in Feigl's solution (presumed calcitic), transv. sec., ×300, all USNM 242565; *2*, Naples, Italy, encrusting colony with autozooids having costal shield (cs) margined by gymnocyst (gy) and interzooidal avicularium with complete pivotal bar (piv) for mandible, autozooids with condyles (cd) for hinging operculum, frontal view, USNM 242566, ×50.——*3. Figularia figularis* (JOHNSTON)?, rec., specimen labeled Albatross Sta. D3987, presumably from Hawaiian Is., 100 m; encrusting colony with maternal and nonmaternal autozooids having dimorphic opercula (op) and interzooidal avicularia having elongate mandibles (md) and smaller membranous postmandibular area (pmd), costal shields of autozooids with openings (ofc) between fused costae and uncalcified spots (un) near peripheral ends of costae (covering cuticle broken in proximal zooid), covering of brood chamber (bch) part of costal shield of autozooid distal to maternal zooid; frontal view, USNM 242567, ×50.

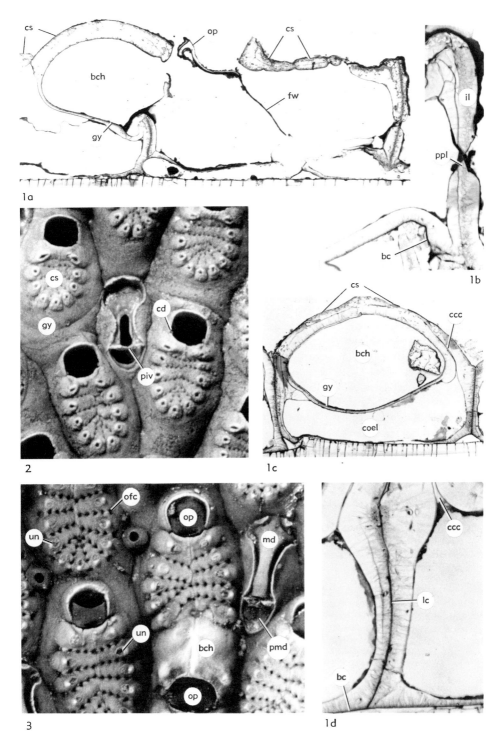

FIG. 71. *(For explanation, see facing page.)*

unfused spines in these anascans emanate from a marginal gymnocyst of variable extent, with which their skeletal layers are structurally continuous (Fig. 71,*1c,d*) (TAVENER-SMITH & WILLIAMS, 1972, p. 111).

In ascophorans, a double-walled exterior outfold with contained coelom can overarch the flexible frontal wall from its proximal and lateral margins (Fig. 65,6; 68,*1b–e*; 70,*1a–c*). The overarching outfold isolates the frontal wall laterally and proximally from the vertical walls of the zooid. The body wall on the basal side of the outfold, facing the membranous frontal wall, is calcified to form an exterior frontal shield (**umbonuloid shield** of HARMER, 1902, p. 332), the underside of which can show prominent growth banding (SANDBERG, 1976, pl. 1, fig. 2–4; this revision). An umbonuloid shield is attached laterally and proximally to vertical walls of the zooid by calcified interior wall segments (SANDBERG, 1976; this revision) forming pore plates of communication organs (Fig. 68,*1b,d,e*) or more extensive walls (SANDBERG, 1976, pl. 1, fig. 2). Body wall on the exposed frontal side of the overarching outfold remains uncalcified (Fig. 68,*1b–e*).

Frontal shields also develop as parts of interior walls that grow into and partition body cavities of zooids in both anascans (Fig. 65,*4*) and ascophorans (Fig. 65,*7*). These frontal shields (**cryptocysts** of JULLIEN, 1881, p. 274; HARMER, 1902, p. 331, 333; BANTA, 1970, p. 39) underlie and approximately parallel preexisting, membranous parts of frontal walls, which bear varying relationships to the hydrostatic system (see below). In anascans, cryptocysts vary from narrow proximal and lateral calcareous shelves (Fig. 80,*3*) to calcareous walls approximately coextensive with flexible parts of frontal walls (Fig. 72,*1,3,4*). In ascophorans, cryptocysts are all approximately coextensive with uncalcified parts of frontal walls (Fig. 67,*1a*; 73,*1a,c*; 78,*1a*). In both anascans (Fig. 72,*1–3*; 81,*3*) and ascophorans (Fig. 78,*1a,c*) cryptocysts can be attached directly to vertical walls laterally and proximally. In many anascans (Fig. 80,*3*) cryptocysts are attached to marginal gymnocysts of varying extent. Some ascophorans (Fig. 67,*1a*) also can have gymnocysts to which cryptocysts are attached proximally.

Different types of frontal shields in anascan and ascophoran cheilostomates differ in potential for ontogenetically increasing in

FIG. 72. Anascan cheilostomate.——*1–4. Monoporella nodulifera* (HINCKS), rec., Jolo Light, Jolo, Philip., 40 m; *1*, Albatross Sta. D5142, maternal autozooid with heavily reinforced operculum (op) attached to flexible frontal wall (fw) overlying hypostegal coelom (hy) and cryptocyst (cry), cryptocyst with membranous attachment to frontal wall just proximal to operculum, distal part of cryptocyst continuous with inner calcified part of transverse wall (tw) and subparallel to outer membranous part of transverse wall, which faces brood chamber (bch), brood chamber enclosed by parts of distal zooid, floored by proximal gymnocyst (gy) and roofed by outfold (of) originating at junction of gymnocyst, cryptocyst (cry), and membranous frontal wall (fw), lower side of outfold calcified, its initial skeletal layer (il) continuous with gymnocyst and superficial layer (sl) continuous with cryptocyst, all of which stain in Feigl's solution (aragonitic), zooids communicating through pore chambers (pch), long. sec., USNM 242561, ×100; *2*, Albatross Sta. D5142, cluster of polymorphic autozooids forming brooding structure; brood chamber (bch) roofed by outfold (of) from distal zooid through openings in which spines (sp) on distal margin of maternal zooid (mz) protrude, membranous frontal walls of laterally adjacent zooids (lz) fitted into lateral openings (lo) of brood chamber, lateral cuticles (lc) separating zooids, are continuous with basal cuticle (bc), transv. sec., USNM 242562, ×100; *3*, Albatross Sta. D5142, encrusting colony with cluster of polymorphic autozooids forming brooding structure, brood chamber (bch) part of zooid distal to maternal zooid (mz), from which spines (sp) project through brood-chamber roof, cryptocysts of laterally adjacent zooids with shapes reflecting lateral openings (lo) of brood chamber, cryptocysts of all polymorphic autozooids with distolateral openings for parietal muscles (opm), frontal view, USNM 242563, ×50; *4*, Albatross Sta. D5137, self-encrusting part of colony with growing edge having brood chamber in early stage of development, distal bud (db) with gymnocyst (gy) to form floor of brood chamber, gymnocysts lacking in other zooids, lateral opening of brood chamber reflected in shape of cryptocyst of lateral zooid with opening for parietal muscle (opm) deeply set, zooids communicating through pore chambers (pch); frontal view, USNM 242564, ×50.

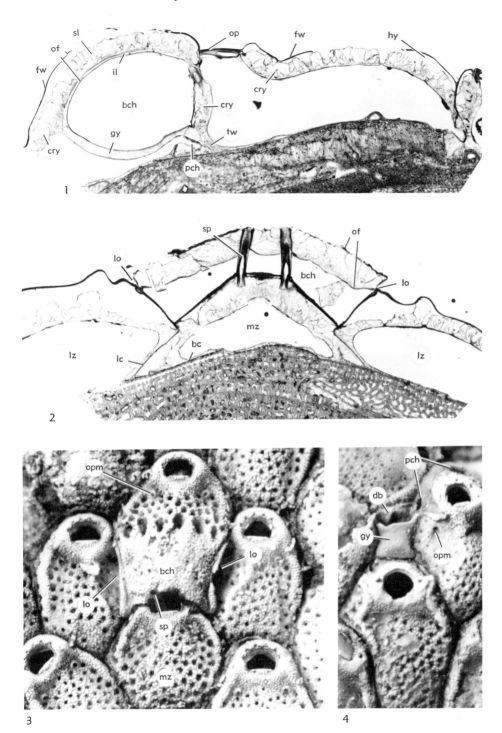

FIG. 72. *(For explanation, see facing page.)*

complexity. Depending on the nature of soft parts overlying their frontal surfaces, some kinds of frontal shields undergo little ontogenetic change except at or near growing tips of colonies, and others continue to undergo extensive changes far proximal to growing tips. With respect to this potential, some exterior frontal shields are similar to interior frontal shields, even though differing in their initial mode of growth.

Gymnocysts and costal frontal shields are covered frontally only by contiguous outermost cuticle (Fig. 69,1e; 71,1a,c). Cuticles and most or all calcareous layers are continuous with those of vertical walls to which the shields are attached (Fig. 69,1f; 71,1c). Like the calcareous layers of vertical walls, these frontal shields cease to be deposited relatively early in zooid life and characteristically remain relatively thin.

Cryptocysts and umbonuloid frontal shields are overlain frontally by cellular layers and intervening body cavity, with outermost cuticle (Fig. 67,1b,c; 68,1b–e; 70,1a–c; 73,1a,c; 78,1a–c). The body cavity overlying a cryptocyst is a separated part of the original body cavity of a zooid (Fig. 65,4,7), and it is to this structure that the term hypostegia or **hypostegal coelom** was originally applied by Jullien (1881, p. 276). The latter term has been broadened, however, to include an extension of the original body cavity of a zooid overlying an umbonuloid frontal shield (Banta, 1970, p. 39; Tavener-Smith & Williams, 1972, p. 110) (see Fig. 65,6). Earlier, Calvet (1900, p. 166) also regarded

this cavity in umbonuloid ascophorans as a hypostegal coelom, but termed the underlying frontal shield a cryptocyst, without distinguishing its mode of growth.

In most anascans and ascophorans having hypostegal coeloms, zooid body cavities from which hypostegal coeloms are derived (by ingrowth of cryptocysts or by outfolding of body wall) are completely separated from those of other zooids (Fig. 65,4,6,7). In later ontogenetic stages, hypostegal coeloms in some ascophoran genera may coalesce to form extrazooidal parts (see below) and in other ascophoran genera may expand to become frontal buds (see below). In one ascophoran (Fig. 78,1c) hypostegal coeloms overlying cryptocysts are confluent laterally throughout ontogeny.

Cryptocysts and umbonuloid frontal shields have initial layers that are continuous with some skeletal layers in vertical walls, marginal gymnocysts, or interior wall segments attached to vertical walls (Fig. 68,1b–e). Initial layers of anascan and some ascophoran cryptocysts clearly show deposition on both basal and frontal sides (Tavener-Smith & Williams, 1970, 1972; Banta, 1970, 1971; Sandberg, 1973), but in ascophorans deposition on the basal surface is soon cut off by development of the ascus (see below). The thin initial layer of some ascophoran cryptocysts shows little evidence of basal deposition (Fig. 67,1a–e; 73,1c; 78,1a). Initial layers of umbonuloid shields, which are of exterior origin, are deposited from the frontal side only (Fig. 68,1b,c). (For further discus-

---

Fig. 73.   Ascophoran cheilostomate. *Margaretta cereoides* (Ellis & Solander), rec., Naples, Italy.——*1a–c.* Walls; *a,* fully developed autozooid about 0.5 cm proximal to growing tip of branch, with membranous frontal wall (fw) intact and overlying completed cryptocyst (cry); floor of underlying ascus (fa) complete but broken; zooid contacting three others at branch axis along its basal (bw) and lateral (lw) walls, and two others outward from axis along its lateral walls, transv. sec., ×100; *b,* detail of basal wall–lateral wall junctions at axis of branch near growing tip, zooid contact along cuticles of basal (bc) and lateral (lc) walls, transv. sec., ×300; *c,* detail of frontal wall–lateral wall junctions in proximal part of zooid near growing tip of branch, cryptocyst (cry) fully developed but proximal to end of ascus, cuticles of lateral walls (lc) continuous with outer cuticular layer of frontal wall (fw), transv. sec., ×300; all USNM 249641.——*2a–c.* Walls; *a,* detail of basal wall–lateral wall junctions at axis of branch near growing tip, with cuticles of basal (bc) and lateral (lc) zooidal walls, gold-plated polished etched transv. sec., SEM, ×1,000; *b,* lateral walls of two contacting zooids, same section as *a,* ×1,000; *c,* detail of *a,* ×6,600; all USNM 249642.

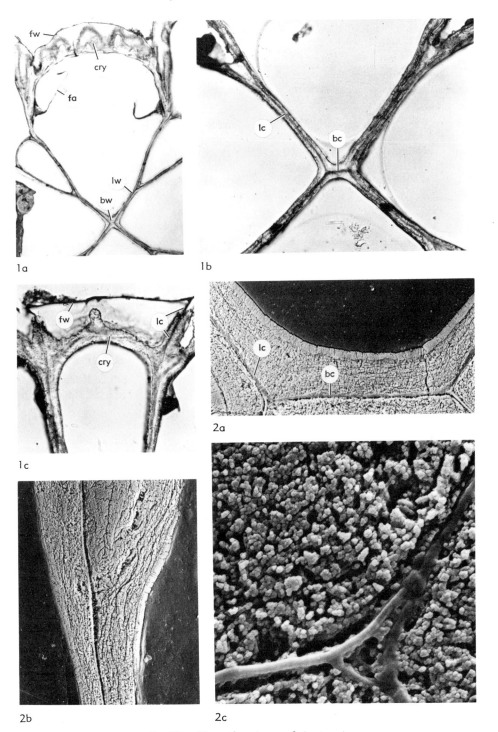

FIG. 73. *(For explanation, see facing page.)*

sion, see SANDBERG, this revision.)

Cellular layers and the hypostegal coelom overlying the frontal sides of cryptocysts and umbonuloid shields allow continued accretion of calcareous deposits on the frontal sides of zooids (Fig. 67,*1a–e*; 68,*1d,e,2*; 70,*1b,c*; 82,*3a,c*), in many species long after deposition in other zooid walls has ceased. Resulting **superficial layers** of these shields can be many times as thick as their **initial layers** and can differ in microstructure (Fig. 70,*1b,1c*) and mineral composition (Fig. 67,*1c*; 68,*1d,e*) (see SANDBERG, this revision). The morphology of cryptocysts and umbonuloid shields can become correspondingly complex, with markedly differing appearance in proximal and distal parts of a colony. With the formation of an ascus, fully formed zooids having cryptocysts and umbonuloid shields can become almost identical in appearance (Fig. 65,*6,7*) (COOK, 1973b).

In ascophorans having gymnocysts and cryptocysts part of the membranous frontal wall becomes infolded beneath the shield after initial calcification is completed. Infolding can occur at the proximal margin of the orificial wall (Fig. 65,*5,7*; 69,*1a,b*; 78,*1a*) or proximal to the orificial wall on the frontal wall (Fig. 67,*1e*). Infolding forms an exterior-walled, flexible-floored sac, the **ascus** (Fig. 65,*5,7*), which opens to the exterior to function in the hydrostatic system. In most species examined, the cuticular roof of the ascus is subjacent to the calcareous frontal shield (Fig. 67,*1c,e*) or possibly lacking (TAVENER-SMITH & WILLIAMS, 1970), the intervening cellular layers apparently having migrated with the proximally advancing edge of the developing ascus. In a few species, the roof of the ascus

is separated wholly (COOK, 1975) or in part (Fig. 78,*1a*) from the frontal shield by cellular layers and intervening body cavity.

In cheilostomates having umbonuloid frontal shields, the flexible frontal wall floors a chamber nearly identical in topology and analogous in function to that enclosed by an infolded ascus, even though formed by overarching (Fig. 65,*6*; 68,*1d,e*; 70,*1a,c*). This structure conventionally is included in the concept of the ascus, and cheilostomates possessing it are regarded as ascophorans but not necessarily as members of the Ascophora (see Table 2). The corresponding space between the flexible frontal wall and the costal shield in some anascans (Fig. 65,*3*) and cribrimorphs (Fig. 71,*1a*) is not generally regarded as an ascus chamber, even though formed in much the same way as the chamber in umbonuloid cheilostomates.

Orificial walls.—In all gymnolaemates, autozooids have orificial walls at or near distal ends of frontal walls (Fig. 65). Outer sides of orificial walls are continuous with frontal walls, from which they become differentiated during ontogeny by infolding of the vestibular wall and distal migration in the growing bud (see LUTAUD, this revision). Inner sides of orificial walls are continuous with vestibular walls (Fig. 66,*2a*).

In most ctenostomates and a few cheilostomates, an orificial wall consists of a radial series of body-wall folds, or a single continuous ringlike fold (Fig. 64). When closed, the orifice is slitlike or puckered and contained within the margins of the orificial wall. In the overwhelming majority of cheilostomates and in some ctenostomates (Fig. 66,*2a*), an orificial wall consists principally

FIG. 74. Cheilostomate vertical walls. Diagrams of sections through vertical walls of developing zooids at and near colony growing edges. Outermost cuticles are represented by solid lines, calcareous layers are stippled, and cellular layers and other soft parts are omitted.——*1.* Uniserial cheilostomate having virtually all exterior vertical walls except for pore plates of communication organs (compare with Fig. 65,*2,5*). ——*2.* Multiserial cheilostomate having interior transverse and exterior lateral walls, and lateral as well as transverse pore plates. Each zooid of a laterally contiguous pair has a separate bounding cuticle shown as a single line (compare with Fig. 65,*3,4,6,7*).——*3.* Multiserial cheilostomate having entirely interior vertical walls.

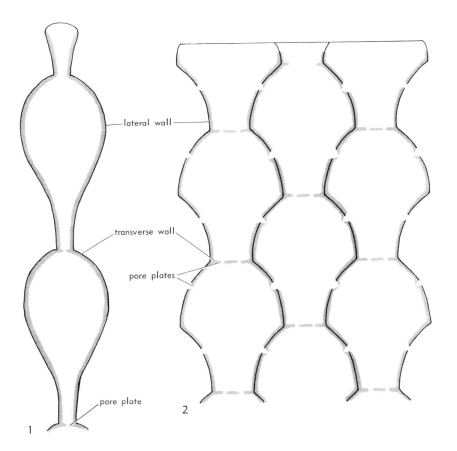

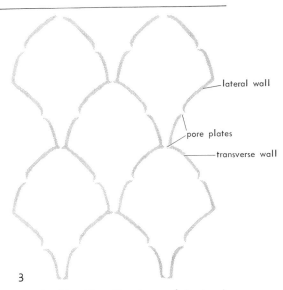

FIG. 74. *(For explanation, see facing page.)*

of a distally directed flaplike fold. When closed, the orifice in these taxa is a crescentic slit defined by the distal and lateral margins of the orificial wall. A weaker, opposing distal flap of wall can be present in a few ctenostomates (Fig. 66,*2a*) and cheilostomates (Fig. 68,*1b,d*),

In most cheilostomates and a few ctenostomates, the distally directed flap is stiffened peripherally or over its whole outer (and in some, inner) surface to form an operculum (Fig. 67,*1d,e*; 68,*1b–e*; 71,*1a*; 72,*1a*; 78,*1a*). Opercula are calcified in a few cheilostomates, but in the great majority stiffening is entirely cuticular. Traces (**opercular scars**) of originally cuticular opercula are known in some fossil cheilostomates in autozooids that lost functioning lophophores with development of calcareous **frontal closures** (Fig. 76,*3*), which are seemingly analogous to terminal diaphragms in fixed-walled members of the class Stenolaemata. Preserved calcareous opercula have been reported in two Cretaceous genera assigned to the Cheilostomata (VOIGT, 1974; TURNER, 1975).

In Ctenostomata orificial walls are supported entirely by membranous frontal walls or membranous vertical and frontal walls. Commonly, orificial walls are elevated above the frontal surface at outer ends of more or less elongate peristomelike extensions of frontal wall (Fig. 65,*1*; 66,*3*). When the lophophore is everted, the diaphragm, which bears a pleated membranous **collar** (Fig. 66,*2a,3*), is exposed at the frontal surface.

In Cheilostomata orificial walls also can be supported entirely by membranous frontal walls (Fig. 80,*2*). In most taxa, however, distal and lateral margins of the operculum or the distal unstiffened part of the orificial wall (Fig. 68,*1b,d,e*) are supported by a skeletal rim generally corresponding in form to the orifice. This skeletal rim comprises the frontal edge of a calcified transverse wall (Fig. 67,*1a,d,e*; 69,*1a*; 78, *1a*), calcified parts of the frontal wall (Fig. 72,*1*), or a combination (Fig. 76,*1–3*). In most cheilostomates the orificial wall is attached proximally to membranous frontal wall (Fig. 68,*1b,d,e*; 71,*1a*; 72,*1a*) or to the floor of an infolded ascus (Fig. 69,*1b*; 78,*1a*), and the skeletal rim of the orifice is thus incomplete (Fig. 65,*2–7*). In those having the opening of the ascus removed from the orificial wall (Fig. 67,*1e*), the skeletal rim is completed proximally by the margin of the frontal shield, and apparently then is analogous to skeletal apertures in fixed-walled members of the class Stenolaemata.

Peristomes are commonly developed in ascophoran Cheilostomata as tubular outfolds of body wall and contained coelom, which together surround the operculum at their inner ends (Fig. 67,*1e*) (BANTA, 1970). Peristomial skeleton is part of the exterior body wall facing inward around the operculum. Proximally and laterally, peristomial skeleton is continuous with the frontal shield and commonly is included in frontal accretion of superficial skeletal layers (Fig. 67,*1d,e*; 68,*1d,e,2*). Distally, peristomial skeleton can be part of an exterior body wall of a distal zooid (Fig. 68,*1d,e,2*) (see SANDBERG, this

---

FIG. 75.   Cheilostomate vertical walls. Diagrams of sections through vertical walls of zooids in early astogenetic stages of encrusting colonies developed from single (a = ancestrula) or multiple (p) primary zooids. Budded generations of zooids are numbered, bud origins are indicated by arrows, outermost cuticles are represented by solid lines, and calcareous layers are stippled.——*1–6*. Uniserial cheilostomate, showing distal budding from ancestrula (*1,2*) to produce single lineal series (*); distal and distolateral budding from ancestrula (*3*) and from budded zooid (*4*) to produce branched lineal series; and other budding sites on the ancestrula (*5*) and budded zooid (*6*) found in some colonies.——*7*. Multiserial cheilostomate with combination of exterior and interior vertical walls and combination of budding directions similar to those in *8*.——*8*. Multiserial cheilostomate with virtually all exterior vertical walls, showing combination of distal, distolateral, and proximolateral budding, and zooids produced by fusion of buds.——*9*. Multiserial cheilostomate with all interior vertical walls, showing circumferential multizooidal budding zone.

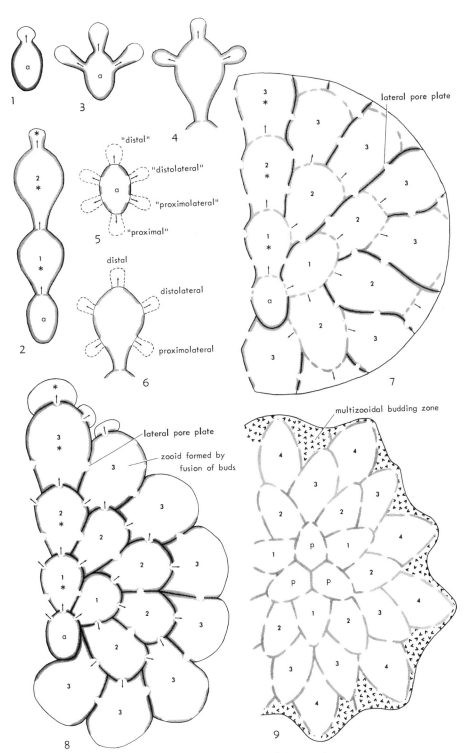

FIG. 75. *(For explanation, see facing page.)*

revision). In some ascophorans, the opening of the ascus is proximal to the peristome (Fig. 67,*1d,e*; 82,*3b,c*). In others, the ascus opens within the peristome into a proximal peristomial channel (Fig. 68,*2*) or a separate opening on the proximal side of the peristome (Fig. 83,*1–3*).

Various elevated structures around orificial walls in both anascan and ascophoran cheilostomates can be similar in appearance to peristomes but be produced by closely spaced spines or adventitious avicularia.

BODY CAVITIES OF AUTOZOOIDS AND CONTAINED ORGANS

The perigastric or **principal body cavity** of a gymnolaemate autozooid is generally enclosed by basal, vertical, and orificial walls, and frontal wall, cryptocyst (and adjacent inner cellular layer), or floor of the ascus (Fig. 65). This body cavity varies markedly in shape in both the Ctenostomata and the Cheilostomata from box- or saclike to cylindrical. Whatever its shape, the principal body cavity of an autozooid in most taxa tends to remain relatively fixed in its dimensions after completion of the vertical walls early in zooid ontogeny. Early completion of the cavity provides a relatively constant position for retracted organs through any later changes in zooid morphology, such as those associated with the frontal wall (Fig. 68,*1a–e*).

The principal body cavity is occupied almost fully by retracted organs and muscles, except in autozooids that have degenerated.

These contained structures include (Fig. 66,*2,3*; 68,*1d,e*; 69,*1b,c*): protrusible lophophore, with or without feeding capability; functional or rudimentary alimentary tract; muscles concerned with protrusion and retraction of lophophore; funicular strands; parts of communication organs; and, in some zooids, structures concerned with sexual reproduction. In the Cheilostomata, some of these structures are reflected directly or indirectly in the skeleton.

Protruded, the lophophore characteristically extends far beyond the orifice, carrying the tentacle crown on an elongate neck (Fig. 4). This **lophophore neck** is formed by the everted tentacle sheath, turned inside out to produce a flexible structure capable of individual or cooperative movement to concentrate exhalant currents away from feeding lophophores (BANTA, MCKINNEY, & ZIMMER, 1974; COOK, 1977). Cooperative current production by groups of zooids may or may not be reflected in skeletons in the Cheilostomata. The presence of elongate tubular peristomes in some presumably restricts movement of the lophophore neck because the orifice remains at the inner end of the peristome.

Lophophore protrusion involves contraction of parietal muscles to depress the hydrostatic membrane of the autozooid and cause pressure in the principal body cavity (Fig. 3, 4). Bilaterally arranged parietal muscles (Fig. 66,*2b*; 70,*1a*) traverse the cavity in one to several pairs, or rarely are arranged unilaterally in highly asymmetrical zooids in a few

FIG. 76.   Anascan cheilostomates.——*1,2. Pyriporopsis?* *catenularia* (FLEMING), rec., Brit. Is.; *1a,b,* Plymouth, Eng.; *a,* encrusting colony with uniserially arranged, predominantly distally and laterally budded autozooids, several injuries repaired by growth of distally and "proximally" budded autozooids (rz); *b,* autozooids and distal bud (db) at growing tip of lineal series, with uncalcified spots on lateral walls (un) opening into pore chambers (pch); both frontal views, USNM 242555, ×30, ×60; *2,* Hastings, Eng.; autozooids, proximal one with frontal and orificial walls completely calcified to form frontal closure preserving traces (scars) of operculum (op) and parietal muscle insertions (pm); frontal view, USNM 242556, ×50.——*3,4. Pyriporopsis? texana* (THOMAS & LARWOOD), Fort Worth F., Cret. (Alb.), Fort Worth, Texas; *3,* encrusting colony with distally budded, uniserially arranged autozooids, one zooid with frontal closure preserving trace (scar) of operculum (op), frontal view, USNM 216139, ×30; *4,* autozooids with uncalcified spots on lateral walls (un); frontal view, USNM 216138, ×50.——*5. P.? catenularia,* same data as *1,2* except exact locality unknown; proximal region of encrusting colony with presumed primary zooids attached by proximal extremities; frontal view, BMNH 1847.9.18.107, Johnston Coll., ×30.

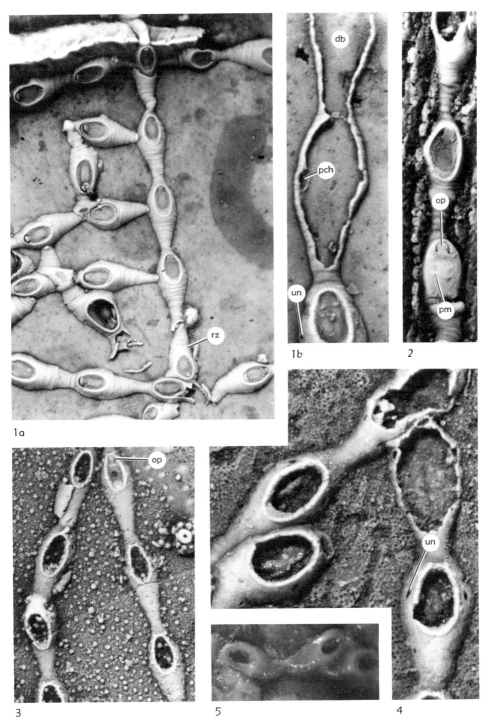

Fig. 76. *(For explanation, see facing page.)*

anascan genera. Parietals originate on lateral walls or on lateral margins of the basal wall and insert on the flexible part of the frontal wall or on the floor of the ascus. In cheilostomates having extensive cryptocysts and no ascus, parietals commonly pass from the principal body cavity into the hypostegal coelom through pores or notches in the lateral margins of the cryptocyst (Fig. 81,*3b*). In one genus, LEVINSEN (1909, p. 162) reported parietals to originate on the frontal side of the cryptocyst. Calcified frontal closures preserving traces of opercula can also have traces of parietal insertions (Fig. 76,*2*).

Parietal muscles may develop in different groups of living gymnolaemate genera at different ontogenetic stages relative to formation of other muscles (SOULE, 1954) or to calcification of a frontal shield, where present. In all Gymnolaemata, parietals develop before the lophophore can be protruded. In cheilostomates having an infolded ascus, parietal muscles grow to insert on the ascus floor as it develops proximally beneath the frontal shield (HARMER, 1902). In those having an overarched frontal wall, parietal muscles may develop before the frontal shield has formed.

During lophophore protrusion, the diaphragm at the outer end of the tentacle sheath is dilated by radially or bilaterally arranged muscles (Fig. 66,*3*) that originate on vertical walls. In some ctenostomates additional radially or bilaterally arranged dilators insert on the vestibular wall (Fig. 64). In some cheilostomates a pair of muscles in series with the parietals is attached to the proximal margin of the operculum to form **divaricator muscles** for opening the operculum. Evidence of

dilator and opercular divaricator muscles has not been reported in fossil cheilostomates. MEDD (1964) inferred that depressions on the inside of basal walls of avicularia of some Upper Cretaceous cheilostomates are scars of mandibular divaricators.

Lophophore retraction is accomplished by contraction of the retractor muscle, as in the other bryozoan classes. In a cheilostomate genus, the retractor muscle has been measured to have one of the fastest contraction rates known in animals (THORPE, SHELTON, & LAVERACK, 1975a). The origin of the retractor muscle can be on the proximal part of the basal wall or on the proximal transverse wall (Fig. 68,*1d*). No traces of retractor muscles have been reported in fossil cheilostomates.

As the lophophore is retracted, the operculum in cheilostomates closes, generally by contraction of a pair of opercular **occlusor muscles**. Opercular occlusors extend from lateral walls or the proximal side of the distal transverse wall to insert on the proximobasal side of the operculum (Fig. 3; 4; 68,*1e*). Various skeletal expressions of occlusor attachments have been reported (HARMER, 1926; MEDD, 1964; CHEETHAM, 1968).

In living Gymnolaemata, connections between principal body cavities of fully developed zooids and between zooids and extrazooidal parts are limited to **interzooidal communication organs** (Fig. 67,*1e*; 68,*1e*; 70,*2*; 71,*1b*). Even in ascophoran cheilostomates in which hypostegal coeloms remain confluent laterally, principal body cavities of zooids communicate with each other and with their hypostegal coeloms only by means of communication organs (Fig. 78,*1a*).

FIG. 77.   Anascan cheilostomate.——*1,2. Allantopora irregularis* (GABB & HORN), Vincentown F., Paleoc., Noxontown Millpond, Del.; *1a,b,* primary zone of astogenetic change of encrusting colony with uniserially arranged, distally and laterally budded zooids; *a,* ancestrula (an) produced bud distally only, size and shape of zooids change from ancestrula through successive generations of budded zooids in zone of change (db, distal bud; lb, lateral bud), frontal view, ×30, *b,* ancestrula with extensive proximal gymnocyst (gy), frontal view, ×50, both USNM 242557; *2a–c,* autozooids in zone of astogenetic repetition, all have spine bases (sp) ringing inner margin of gymnocyst, some have distal brood chambers (bch) preserved in various states of completeness; *a–c,* all frontal views, USNM 242558, ×50.

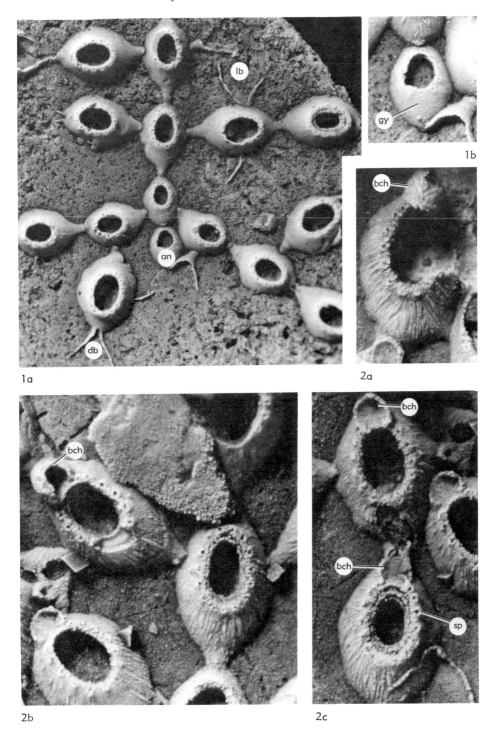

FIG. 77. *(For explanation, see facing page.)*

Throughout the Gymnolaemata, communication organs therefore appear to form the only means of transporting nutrients from feeding autozooids to nonfeeding polymorphs and extrazooidal parts, except possibly those at growing tips of colonies. BOBIN (1964, 1971) presented direct biochemical evidence that nutrients are transferred through cells that make up communication organs.

A **communication organ** consists of a complex of interdigitating cell types together with a cuticular or calcareous pore plate bearing one or more communication pores (BOBIN & PRENANT, 1968; BANTA, 1969; BOBIN, 1971; GORDON, 1975). Cells of special form extend through communication pores to provide the actual interzooidal connection. Communication organs occur on open expanses of walls or on parts of walls partly enclosed within **pore chambers** (Fig. 69,1f; 76,1b; 80,1). Interzooidal communication organs occur in vertical walls of zooids, whether interior or exterior, and can also be present in basal walls and frontal shields. Development of communication organs in preexisting exterior walls that are in contact involves cooperative dissolution of bounding cuticles (BANTA, 1969). Communication organs similar to those connecting zooids occur intrazooidally in cryptocysts of ascophorans (BANTA, 1970, 1971) and around margins of umbonuloid shields of ascophorans (Fig. 68,1b,d), to connect the hypostegal coelom with the principal body cavity.

POLYMORPHISM

In the overwhelming majority of living species in both the Cheilostomata and the Ctenostomata, zooids within a colony may differ discontinuously in morphology and function at the same stages of ontogeny and in the same asexual generations. This polymorphism is most commonly reflected in skeletons in the Cheilostomata and therefore is generally recognizable in fossil species. A few examples of **soft-part polymorphism** without apparent skeletal expression have been reported in living cheilostomates (for example, GORDON, 1968). These include sexual dimorphism of lophophores, differences in vestibule structure for brooding embryos, and differences in tentacle length for producing exhalant water currents, all of which are correlated with skeletal differences in some other species. In some examples, soft-part polymorphs apparently alternate within the same body cavity during degeneration-regeneration cycles.

Polymorphs in the Gymnolaemata include autozooids, which differ from ordinary feeding autozooids in size, shape, tentacle num-

FIG. 78.   Ascophoran cheilostomate.——*1a–e. Euthyrisella obtecta* (HINCKS), rec., Queensl., Australia; *a,* autozooids and adjacent extrazooidal parts of colony (exp), autozooids with heavily reinforced dimorphic opercula (op), extensive hypostegal coeloms (hy), membranous frontal walls (fw), and cryptocyst (cry) underlain by ascus (fa, floor of ascus; ra, roof of ascus) opening (oa) at proximal margin of operculum, ascus roof in contact with cryptocyst except at distal end, where small body cavity intervenes, skeletal layers of zooid and extrazooidal walls very thin throughout colony, organic sheets (os) form boundaries between basal walls of zooid (bw) and inner wall of extrazooidal parts, membranous basal wall (bm) of extrazooidal parts attached to calcified inner wall by membranous filaments (arrows) that may be calcified at inner ends, long. sec., ×100; *b,* growing tip (gt) with outer membrane intact but shriveled, interior walled zooecia and extrazooidal skeleton fragmented but entirely within colony body cavity, proximal zooid with ascus (fa, floor of ascus), ascus lacking in distal zooid, long. peel, ×50; *c,* autozooids with calcified lateral walls (lw) not reaching membranous frontal walls so that hypostegal coeloms (hy) are confluent, frontal wall (fw) attached to cryptocyst (cry) by filaments (arrow) similar to those in extrazooidal parts, injured membranous basal wall (bm) of extrazooidal parts replaced inwardly by a second membrane with foreign particles in intervening space, transv. sec., ×100; *d,* erect colony with autozooids having continuous membranous frontal walls and dimorphic opercula (op), frontal view, ×50; *e,* communication organ (ppl, pore plate) in transverse walls of contiguous zooids (distal to left), organic sheet (os) marking boundary between zooids, floor of ascus (fa) reaching to transverse wall which is continuous with cryptocyst (cry), long. sec., ×300; all USNM 242577.

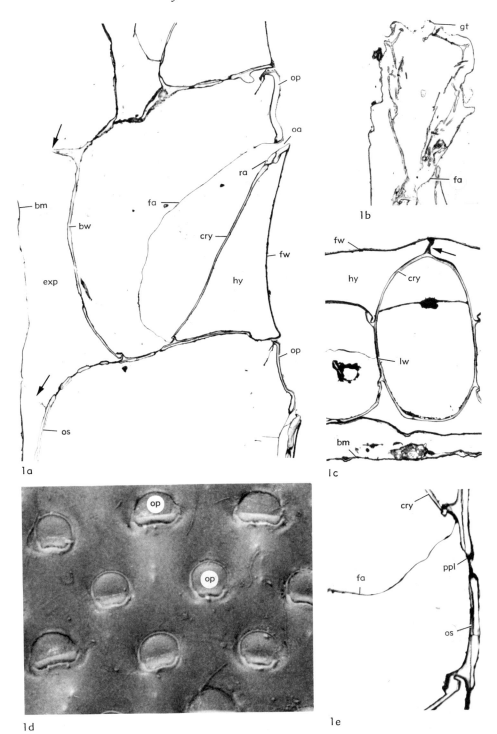

FIG. 78. *(For explanation, see facing page.)*

ber, and other features, but retain protrusible lophophores with or without feeding capability; and **heterozooids**, which have non-protrusible or no lophophores (and therefore no apparent feeding capability), different or no musculature, and specialized organs present or lacking. Different combinations of polymorphic autozooids and heterozooids can differ so in appearance that were they not in the same colony, they might be placed in different taxa (heteromorphy of Voigt, 1975) (Fig. 84,*1*). Autozooidal and heterozooidal polymorphs lacking feeding ability presumably are nourished through interzooidal communication organs that connect them directly or indirectly to feeding autozooids.

Polymorphs may communicate with just one other zooid (adventitious polymorphs) (Fig. 68,*2*; 70,*1b,c,3*; 79,*2,3*; 82,*1,2*; 83,*2,3*; 84,*1–3*), in the extreme form being almost a structural appendage of that zooid. Polymorphs may also be intercalated within budding series and communicate with two or more zooids or with extrazooidal parts of colonies. **Interzooidal polymorphs** (Fig. 71,*2,3*; 81,*1,2*) are intercalated in spaces smaller than those occupied by ordinary feeding autozooids. **Vicarious polymorphs** (Fig. 79,*3*; 81,*3,4*) are intercalated in spaces subequal to or larger than those occupied by ordinary feeding autozooids. Interzooidal and vicarious polymorphs may be arranged among ordinary feeding autozooids either regularly or seemingly at random. Regularly arranged polymorphs may occur at isolated positions among ordinary feeding autozooids or in clusters. Clusters of polymorphs may be restricted to one part of a colony, such as a basal stalk, or may recur throughout a colony. A cluster can consist of one kind of polymorph (Fig. 82,*3a*) or a variety of polymorphs (Fig. 70,*3*; 72,*1–4*) either keyed to a single function (such as reproduction, brooding of embryos, support of the colony, or connection of other zooids) or serving a broad spectrum of functions (including, for example, feeding and defense with other functions).

Diversity in morphology of polymorphs throughout the Gymnolaemata is at least as great as that in ordinary feeding autozooids. However, taxa having autozooids of quite different appearance can have similar heterozooids (compare Fig. 70,*3*; 71,*2*; 79,*3*).

Other than ordinary feeding autozooids, the only kind of zooid that is present virtually throughout the class is the kenozooid (Fig. 66,*1,2b*; 85,*3*). Kenozooids in the Gymnolaemata have body walls enclosing body cavities containing funicular strands and parts of communication organs, but empty of alimentary canal and, in most, of musculature. Kenozooids therefore are all heterozooids apparently incapable of feeding. Basal and vertical walls of kenozooids, and frontal walls of some (including presence of parietal muscles), are comparable in some characters with those of autozooids in the same colony. The function of these parietal muscles must be different from that of parietals in autozooids, which act to protrude the lophophore. Kenozooids lack orifices and orificial walls, and the structures associated with frontal walls, such as cryptocysts, may also be quite different from those of autozooids or lacking.

Adventitious kenozooids much smaller than, and placed in consistent positions upon autozooids can be difficult to distinguish from zooidal structures such as spines. This difficulty is increased if, as suggested by Silén (1942), wall constrictions at spine bases correspond to pore plates of communication organs. Frontal structures of some ascophorans containing hypostegal coeloms that communicate with principal body cavities of autozooids only by means of communication organs are distinguishable from kenozooids, only by their possession of a part (frontal wall) of the supporting functional autozooid. Vicarious kenozooids larger and less regular in shape than autozooids can be difficult to distinguish from some kinds of extrazooidal parts. Distinction of some basic morphologic and functional units in the Gymnolaemata, because of this morphologic continuity throughout a colony, is unavoidably arbitrary.

Most species in the Cheilostomata possess

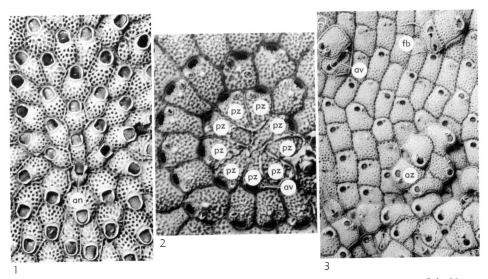

FIG. 79. Ascophoran cheilostomates.——*1. Cryptosula pallasiana* (MOLL), rec., Monterey, Cal., 30 cm below lowest tide; primary zone of astogenetic change of encrusting colony with multiserial budding throughout, size and other characters of autozooids change from ancestrula (an) through successively budded generations, zooid proximal to ancestrula belongs to third asexual generation (compare with Fig. 75,7); frontal view, USNM 242581, X17.——*2,3. Stylopoma spongites* (PALLAS), rec.; *2,* Atl., Fowey Light, 24 km S. of Miami, Fla., 80 m; small encrusting colony with primary zone of astogenetic change beginning with central group of nine primary autozooids (pz) smaller than those of succeeding budded generations, primary zooids supporting adventitious avicularia (av) like those in succeeding generations; frontal view, USNM 242583, X34; *3,* Discovery Bay, Jamaica, West Bull no. 1, 30 m; encrusting colony with autozooids (az) and vicarious avicularia (av) of discontinuous secondary zone of astogenetic change budded frontally from hypostegal coeloms of autozooids (fb, frontal bud) in primary zone of astogenetic repetition, frontally budded autozooids less regular in shape and orientation than those in primary zone of repetition, small adventitious avicularia present on frontal shields of autozooids in both zones, vicarious avicularia also present in primary zone of repetition, not shown; frontal view, USNM 242582, X17.

another kind of polymorph, the **avicularium**, which itself can occur in two or more distinct forms within a colony (Fig. 70,3; 79,2). Avicularia are zooids in which the equivalent of the orificial wall, the **mandible,** is relatively larger and more intricately reinforced than orificial walls (opercula) of ordinary feeding autozooids (Fig. 70,3a; 71,3; 81,3a; 84,3). The mandible is opened and closed by greatly augmented divaricator and occlusor muscles (Fig. 70,1c). In some living cheilostomates, avicularia can be autozooids with feeding organs, but much more commonly are heterozooids with only a nonprotrusible rudiment of lophophore and nondigesting rudiment of alimentary canal. Movement of the mandible is apparently at least partly independent of feeding and in

some species has been inferred to play a role in cleaning (COOK, 1963; GREELEY, 1967) and defense (for example, KAUFMANN,1971, and references therein). Vertical and basal walls of avicularia tend to resemble those of ordinary feeding autozooids, but in some species may be elongated to form stalks that attach the avicularia to other zooids (for example, HASTINGS, 1943). The skeletal rim supporting the free tip and lateral edges of the mandible, the **beak,** may (Fig. 70,3; 71,3; 84,3) or may not (Fig. 81,3) closely approximate the mandible in shape. A partial or complete rim may form the **condyles** or **pivotal bar** on which the fixed edge of the mandible is hinged (Fig. 68,2; 70,3b; 71,2; 79,2,3; 81,3b; 83,2,3; 84,1–3). The frontal wall is relatively smaller than that of ordinary feeding autozooids, typically forming only a small membranous **postmandibular area** (Fig. 71,*3*; 84,*3*) on which the mandibular divaricator muscles are inserted.

Polymorphism associated with sexual reproduction is highly diverse in the Gymnolaemata. Sex cells are produced by zooids, sperm most commonly on funicular strands and eggs on parts of the body wall within the principal body cavity. Sexes may be combined within single zooids, but not necessarily at the same time. There may be a distinct tendency for zooids to be male at earlier stages and female at later ones (SILÉN, 1966), without skeletal expression of sex change. In many species production of eggs is limited to zooids associated with brooding structures with a great diversity of polymorphic expression. Both sperm- and egg-producing zooids are autozooids with protrusible lophophores that may or may not be capable of feeding (RYLAND, 1976, and literature cited therein). In some species **sexual zooids** are distinct feeding polymorphs (Fig. 70,*3*; COOK, 1973a). In others, they are nonfeeding polymorphs, with or without skeletal expression of their functional specialization (Fig. 69,*2*) (MARCUS, 1938a; GORDON, 1968; COOK, 1968c).

EXTRAZOOIDAL PARTS OF COLONIES

In most Gymnolaemata parts of colonies proximal to growing tips or margins consist entirely of morphologically distinguishable zooids of one or more morphologic kinds. At growing tips zooids originate as buds or as parts of multizooidal budding zones that have some (multizooidal) body wall layers continuous with those of other zooids. These walls of multizooidal origin become parts of zooids early in ontogeny, through the completion of the bounding walls of zooids. In a few taxa, apparently limited to the Cheilostomata, major parts of colonies commonly many times as large as autozooids are extrazooidal, with continuous body walls enclosing unpartitioned body cavity devoid of feeding and reproductive organs and musculature, although probably transversed by funicular strands (LUTAUD, pers. commun., 1976). Extrazooidal parts can be restricted to more proximal regions of colonies or can extend from proximal regions to growing tips. Once developed, extrazooidal parts lie outside boundaries of zooids throughout the life of the colony. Extrazooidal body cavities are connected to body cavities of zooids by communication organs similar to those connecting zooids to each other. It is through these connections that extrazooidal tissues apparently are nourished.

Some structures interpreted as extrazooidal parts in cheilostomates may intergrade morphologically with some kinds of polymorphic zooids. It is also possible that structures interpreted as polymorphs in ctenostomates (such as masses of rootlets or basal

FIG. 80. Anascan cheilostomates.——*1. Wilbertopora mutabilis* CHEETHAM, Grayson F., Cret. (Cenoman.), Roanoke, Texas; growing edge of encrusting multiserial colony with staggered lineal series (db, distal buds); autozooids with pore chambers (pch) and some with partly developed brood chambers (bch) distal to maternal zooids; frontal view, USNM 216141, ×50.——*2. Aplousina gigantea* CANU & BASSLER, rec., Bogue Sound, Beaufort, N. Car., 6 m; encrusting colony with apparently coordinated lineal series forming smooth growing edge (ge; db, distal buds); autozooids have membranous frontal walls margined by narrow gymnocysts (gy) and continuous distally with lightly reinforced operculum (op); frontal view, USNM 242559, ×50.——*3. W. mutabilis,* holotype, same data as *1* except Fort Worth F., (Alb.), Krum; primary zone of astogenetic change of encrusting colony; ancestrula (an) produced buds distally and distolaterally to initiate multiserial arrangement evident throughout colony, size of zooids increasing from ancestrula through successive generations of budded zooids in zone of change; zooecia with narrow cryptocysts (cry) attached to marginal gymnocysts (gy); frontal view, LSU 4500, ×50.——*4. W. mutabilis,* same data as *1* except Pottsboro; primary zone of astogenetic change of encrusting colony; ancestrula (an) produced bud distally only, in initially uniserial arrangement, lateral and distal budding in following generations resulted in multiserial arrangement throughout remainder of colony; some zooecia with frontal closures preserving trace (scar) of operculum (op); frontal view, USNM 216140, ×50.

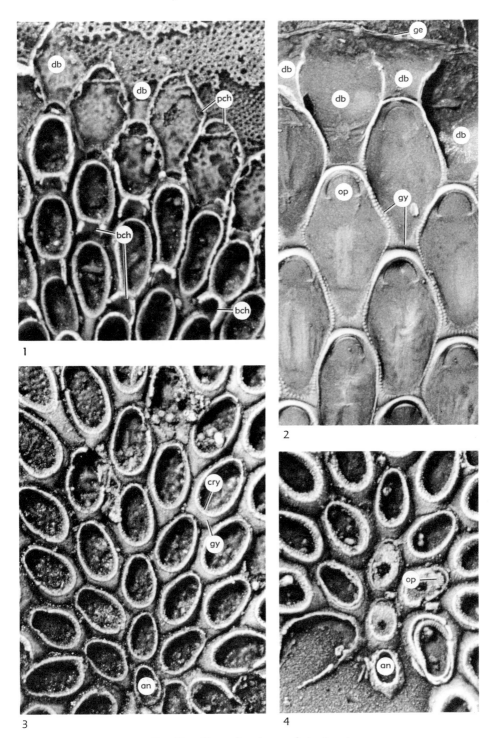

Fɪɢ. 80. *(For explanation, see facing page.)*

stalks) may be extrazooidal, as suggested by HARMER (1915, p. 61).

Extrazooidal parts limited to proximal regions of colonies are found in some ascophorans having autozooids with hypostegal coeloms overlying frontal shields of either exterior (umbonuloid; Fig. 70,*1–3*) or interior (cryptocystal) origin. Extrazooidal parts in these taxa are apparently formed by coalescence of hypostegal coeloms and associated body walls of contiguous preexisting zooids. The first step in this process is apparently dissolution of cuticles at frontal margins of vertical walls; this dissolution of vertical wall cuticles seems similar to that occurring in the formation of some communication organs (BANTA, 1969). Ontogenetically thickened frontal shields with zooid boundaries marked on their frontal surfaces by bounding cuticles (Fig. 70,*1a*; 83,*3*) then are succeeded without interruption by calcareous layers that are continuous across zooid boundaries (Fig. 70,*1b*; 83,*2*). The next step is overgrowth, by these continuous layers, of orificial walls and other structures such as adventitious avicularia (Fig. 70,*1b*); similar overgrowth by zooidal skeleton has been reported in the formation of frontally budded zooids from hypostegal coeloms (BANTA, 1972, supraopercular space). Coalesced extrazooidal coeloms continue to communicate with some underlying zooidal body cavities through communication organs originally filling marginal openings in frontal shields. At proximal ends of colonies, these openings may also become covered with extrazooidal skeleton, but more distal ones apparently remain functional. The extrazooidal coelom is apparently confluent throughout, so that its more proximal parts can continue to be nourished through connection of its distal parts with feeding zooids.

Extrazooidal skeleton produced by coalesced body walls originally bounding zooidal hypostegal coeloms is especially prominent in ascophorans having erect colonies (Fig. 13,*1*; 83,*1*). These deposits are thickest at the most proximal ends of erect colonies, where they cover the ontogenetically oldest zooids. As growing tips of a colony advance distally, extrazooidal skeleton not only thickens at the proximal end of the colony, but also encroaches distally over zooidal frontal shields, as more zooidal hypostegal coeloms and associated body walls become coalesced. The colony thus can be strengthened as it grows (CHEETHAM, 1971), but at the expense of feeding and some other functional abilities earlier possessed by its more proximal zooids. In anascans and some ascophorans having erect colonies and zooidal frontal shields overlain by hypostegal coeloms, skeletal thickening can occur entirely within zooid boundaries (Fig. 82,*3a,c*)

FIG. 81. Anascan cheilostomates.——*1,2. Wilbertopora mutabilis* CHEETHAM, Cret., Texas; *1,* Grayson F., (Cenoman.), Salado, encrusting colony with autozooids and interzooidal avicularia budded distally and distolaterally, most autozooids provided with brood chambers (bch) that are parts of zooids distal to maternal zooids (see Fig. 80,*1*), avicularia with pointed beaks (bk) and condyles (cd) for hinging mandible, frontal view, USNM 216143, ×50; *2,* Kiamichi F., (Alb.), Fort Worth, encrusting colony with ordinary autozooids and interzooidal aviculariumlike polymorph (av) budded distolaterally, frontal view, USNM 216142, ×50.——*3a,b. Smittipora levinseni* (CANU & BASSLER), rec., Atl., 33°41.6′ N., 76°42.4′ W., 70–87 m; *a,* encrusting colony with autozooids and vicarious avicularia having membranous frontal walls, opercula (op), and mandibles (md) intact; membranous mandibles with strongly reinforced central axes are in open (right) and closed (left) positions; postmandibular walls of avicularia (pmd) are similar to frontal walls of autozooids; *b,* autozooids with membranes removed, showing extensive cryptocysts notched (opm) for parietal muscles; small brood chambers (bch) roofed by skeleton continuous with cryptocysts of zooids distal to maternal zooids; vicarious avicularium, budded distolaterally, divided by pivotal condyles (cd) into rounded mandibular part, much shorter than mandible, and postmandibular part; both frontal views, USNM 242560, ×50.——*4. W. mutabilis,* same data as *1,2,* and Fort Worth F., (Alb.), Fort Worth; encrusting colony with ordinary autozooids and vicarious avicularia budded distolaterally; avicularia with rounded beaks (bk) and condyles (cd) for mandible; frontal view, USNM 186572, ×50.

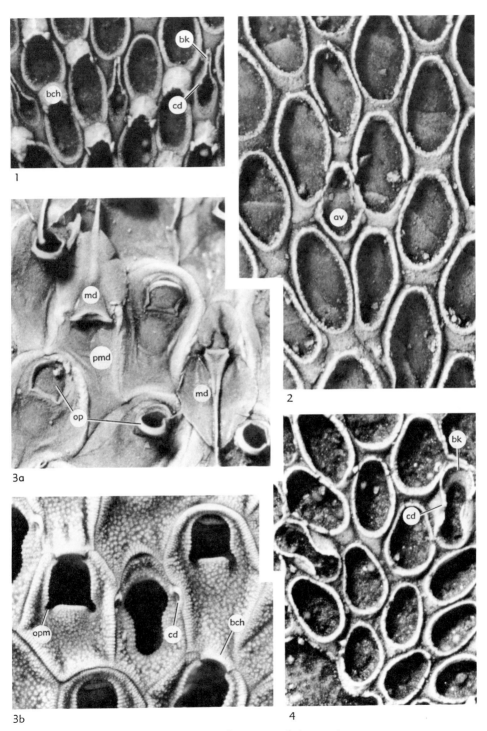

FIG. 81. *(For explanation, see facing page.)*

(Cheetham, 1971, pl. 8, 9), without extrazooidal coalescence. In some of these colonies, proximal zooids also apparently lost their feeding function with overgrowth by zooidal skeleton of their orificial walls (Fig. 82,3c). Presumably communication with feeding zooids then can be maintained through underlying principal body cavities of zooids.

Extrazooidal parts developed at growing tips or margins of colonies concurrently with budding of zooids are known in a few anascans and ascophorans having all interior vertical walls (Harmer, 1902; Håkansson, 1973). These structures form one side of free-living and unilaminate erect colonies. It was for this type of structure that Harmer (1901, p. 16) proposed the term extrazooecial. Calcareous layers of extrazooidal walls in these taxa are parts of interior walls shared with basal walls of contiguous zooids (Fig. 78,1a–c). Communication between extrazooidal body cavity and principal body cavities of zooids is through communication organs in interior basal walls of zooids and through confluence with body cavities of developing zooids at the growing tip or edge of the colony (Fig. 78,1b).

BROODING AND LARVAE

Embryos are brooded in the great majority of Cheilostomata and in most Ctenostomata. In two genera of Ctenostomata brooding is reportedly within the body cavity (Harmer, 1915; Ryland, 1970), as in the classes Phylactolaemata and Stenolaemata, but in other brooding gymnolaemates, embryos are held topologically outside the body cavity of the colony within water-filled brood chambers (Fig. 66,2a; 69,1c; 70,2) partly enclosed by the body walls of one or more kinds of polymorphs. In cheilostomates that brood, this function is generally reflected in the skeleton even though walls enclosing brood chambers are not invariably calcified.

Body walls enclosing brood chambers in the Gymnolaemata most commonly are parts of zooids but can comprise, together with contained coelom and parts of interzoidal communication organs, a whole zooid (polymorph). In the Ctenostomata and many Cheilostomata, the enclosing walls are apparently entirely part of the **maternal zooid** that deposits eggs in the brood chamber. In many other Cheilostomata the enclosing walls are parts of one or more zooids distal or distolateral to the maternal zooid. If part of a maternal zooid, enclosing walls can lie internally, as for example a diverticulum of the vestibule or tentacle sheath (Fig. 66,2a), or can extend distally from the zooid, as for example a double-walled outfold from the distal transverse wall. The outer surface of such outfolded enclosing walls may be exposed at the surface of the colony (Fig. 67,1d; 69,1c,2; 70,3; 77,2a–c; 82,3b), or hidden beneath the surface of the distal zooid (Fig. 81,3b). If distal to the maternal zooid, a brood chamber can be enclosed by body walls of a kenozooid (Woollacott & Zim-

Fig. 82. Anascan and ascophoran cheilostomates.——1,2. *Setosellina* aff. *S. folini* (Jullien), rec., Gulf of Mexico, 28°51′ N., 88°18′ W., Albatross Sta. D2385, 1,500 m; *1*, free-living colony with proximal autozooid apparently broken from preexisting colony, autozooids products of left distolateral budding, each with a distally budded adventitious avicularium (av); setiform mandibles of avicularia, which pivoted on small condyles (cd), missing, frontal view, USNM 242570, ×75; *2*, free-living colony with right distolaterally budded autozooids, uncalcified spots (un) present on left lateral walls, frontal view, USNM 242571, ×75.——*3a–c. Margaretta cereoides* (Ellis & Solander), rec., Naples, Italy; *a*, distal segment of erect, jointed colony with growing tip, ordinary and maternal (bch, brood chambers) autozooids (forming proximal cluster) with relatively thin frontal shields and dimorphic peristomes; *b*, detail of same segment with ordinary and maternal autozooids having dimorphic peristomes and distinct cuticular boundaries (lc), frontal shields with numerous funnel-shaped depressions (fd) similar in appearance to opening to ascus (oa) (compare Fig. 67,1c,e); *c*, proximal segment of same colony with frontal shields of autozooids greatly thickened, funnel-shaped depressions nearly filled, and peristomes sealing off underlying opercula, cuticular boundaries (lc) and opening to ascus (oa) still distinct; all frontal views, USNM 242572, *a,c*, ×30, *b*, ×50.

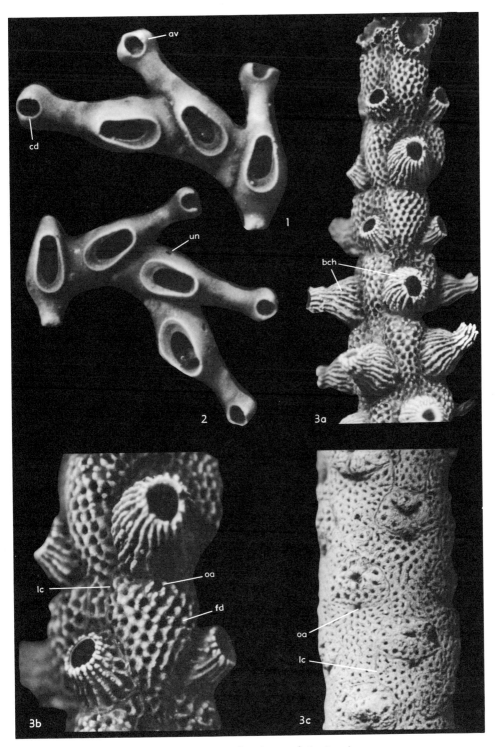

FIG. 82. *(For explanation, see facing page.)*

MER, 1972a) or by exposed or hidden parts of one or more autozooids or heterozooids (Fig. 71,*1a,c,3*; 72,*1–4*). In the Cheilostomata, varying combinations of exposed and hidden walls enclosing brood chambers can be calcified, and especially for brood chambers that have some calcareous enclosing walls the term **ovicell** is commonly used (see RYLAND, 1976, for alternative usage). Whatever its origin, a brood chamber opens near the orifice of a maternal zooid that commonly but not invariably differs in size, shape, or both from nonmaternal ordinary feeding autozooids, and therefore commonly is a polymorph (Fig. 70,*3*). Brood chambers formed by clusters of polymorphic zooids can have multiple openings (Fig. 72,*1–4*). Clusters of brood chambers around the orifice of a maternal zooid have been reported in two species, but only one chamber at a time in such clusters has been observed to be occupied by an embryo (POWELL, 1970).

Maternal zooids must at some ontogenetic stage be female autozooids, provided with protrusible lophophores through which eggs are extruded into the brood chamber. Except in a few living species, maternal lophophores bear tentacles and most such zooids appear to be capable of feeding. An egg produced on the body wall of the maternal zooid makes its way through the body cavity to the lophophore. Fertilization has not been observed in brooding gymnolaemates (RYLAND, 1976), but, as in nonbrooding genera, sperm has been reported to be released through tips of tentacles of male or hermaphrodite zooids (SILÉN, 1966, 1972; RYLAND, 1976, and literature cited therein). Thus, a mechanism for interzooidal or intercolony fertilization appears to be common if not universal in the Gymnolaemata (RYLAND, 1976). Once fertilized, an egg is extruded by the maternal zooid into the brood chamber through a pore in the wall of the lophophore below and between the distal pair of tentacles. After deposition of a fertilized egg in the brood chamber, the maternal lophophore may degenerate.

Except in a few living species, only one egg undergoes embryonic development in a brood chamber at a time, but additional eggs may occupy the same brood chamber sequentially. Embryonic fission is unknown in the Gymnolaemata. In a number of anascan and ascophoran cheilostomates, embryos have been observed to increase in size during development (RYLAND, 1976, and literature cited therein). In one such species, in which embryos may increase tenfold in diameter, evidence has been presented that nutrients are transferred to the developing embryo through a membranous outfold of the maternal zooid occupying the opening of the brood chamber (MARCUS, 1938a, p. 120; WOOLLACOTT & ZIMMER, 1972a,b, 1975). Membranous walls of maternal zooids occupy openings of or face into brood chambers in other species (Fig. 72,*1*), thus providing possible mechanisms for nutrient transfer. In still other species, membranous walls appear to be lacking (Fig. 67,*1d*), and the developing embryo may be physiologically isolated from its maternal zooid. In species in which there is no embryonic size increase, brooded embryos apparently subsist on yolk in the egg (**lecithotrophic development** of RYLAND, 1976). Apparently, both lecithotrophic development and nourishment of brooded embryos can occur within a genus (RYLAND, 1976).

FIG. 83. Ascophoran cheilostomate.——*1–3. Tessaradoma boreale* (BUSK), rec.; *1,* near Georges Bank, small erect colony thinly calcified near growing tips (gt) of branches, thickly calcified with nearly occluded peristomes near encrusting base (eb), USNM 242574, ×20; *2,* Caribb. Sea, 15°24′40″ N., 63°31′30″ W., Albatross Sta. D2117, 1,350 m, thickly calcified proximal part of colony with zooid boundaries covered by extrazooidal skeleton within which peristomes and adventitious avicularia (av) are immersed, USNM 242575, ×50; *3,* Albatross Sta. D2117, thinly calcified more distal part of colony with distal zooids having cuticular boundaries (lc) exposed distally and peristomes and adventitious avicularia (av) not immersed, USNM 242576, ×50.

FIG. 83. *(For explanation, see facing page.)*

Larvae produced by most brooding gymnolaemates lack a digestive tract and after their release from brood chambers continue to subsist entirely on nutrients provided by maternal zooids before or during development (RYLAND, 1970, and references cited therein). These larvae are naked and have variable but relatively short motile stages before metamorphosis.

Even though brooding is widespread in the Gymnolaemata, nonbrooding species are known among both the Cheilostomata and the Ctenostomata. The ctenostomate genera *Alcyonidium* and *Flustrellidra* include both brooding and nonbrooding species. Commonly, fossil cheilostomates that lack evidence of brooding are morphologically similar to living species that do not brood; however, no skeletal evidence of production of nonbrooded larvae is known in either living or fossil gymnolaemates. The presence of nonbrooded larvae in fossil species, therefore, is inferential.

Nonbrooding gymnolaemates release fertilized eggs, commonly in great numbers, directly into the water through a pore at the end of an elongate **intertentacular organ** (absent in all but a few brooding gymnolaemates). The intertentacular organ is on the distal side of the lophophore beneath the tentacle bases in the same position as the pore through which eggs are extruded in brooding species. Lophophores provided with intertentacular organs for releasing eggs apparently can alternate in degeneration-regeneration

cycles with lophophores lacking these organs, and no skeletal expression of this soft-part dimorphism is known.

Nonbrooded embryos undergo extensive development after release, and most of their lengthy motile stage is passed as larvae (Fig. 85,4) with fully functional digestive tracts (planktotrophic) and in most, a bivalved cuticular shell. Both digestive tract and shell are lost in metamorphosis. **Planktotrophic larvae** of gymnolaemates have generally been termed **cyphonautes,** because they were originally described under this name as a genus of planktonic animals.

In a few species larvae developed from brooded embryos have digestive tracts and other morphologic features, including bivalved shells in some, that are similar to those of planktotrophic larvae. The digestive tracts are not functional, however, so that these larvae are not planktotrophic even though included within the concept of cyphonautes.

In one freshwater ctenostomate, larvae have been reported to contain much yolk and to lack a digestive tract, even though not brooded (BRAEM, 1896).

ASTOGENY

The Gymnolaemata apparently include the widest variety of astogenetic patterns known in the phylum. The number of primary zooids formed by metamorphosis of a larva, the presence or absence of a primary zone of asto-

---

FIG. 84.   Ascophoran cheilostomates.——*1a,b. Hippopetraliella marginata* (CANU & BASSLER), rec., Gulf of Mexico, 28°45′ N., 85°02′ W., Albatross Sta. D2405, 60 m; *a,* repaired part of loosely encrusting colony with autozooids having wider orifices and smaller adventitious avicularia with pointed beaks (bk) placed in distolateral corners of frontal shields; *b,* uninjured part of same colony, about same distance from growing edge, with autozooids having narrower orifices and larger adventitious avicularia with rounded beaks (bk) placed nearer middle of lateral margins of frontal shield, avicularia with complete bars (piv) for hinging mandible; both frontal views, USNM 242578, ×50.——*2,3. Petraliella bisinuata* (SMITT), rec., Gulf of Mexico; *2,* 28°45′ N., 85°02′ W., Albatross Sta. D2405, 60 m, loosely encrusting colony with autozooids and adventitious avicularia communicating through frontal shield with underlying principal body cavity of zooid, pivotal bar (piv) separating mandibular (bk) and postmandibular regions of avicularium, frontal view, USNM 242579, ×75; *3,* 22°18′ N., 87°04′ W., Albatross Sta. D2365, 50 m, autozooids and adventitious avicularia with membranous frontal walls, opercula (op), and mandibles intact, mandible (md) hinged to pivotal bar, behind which is postmandibular membranous area for attachment of divaricator muscles (pmd), frontal view, USNM 242580, ×50.

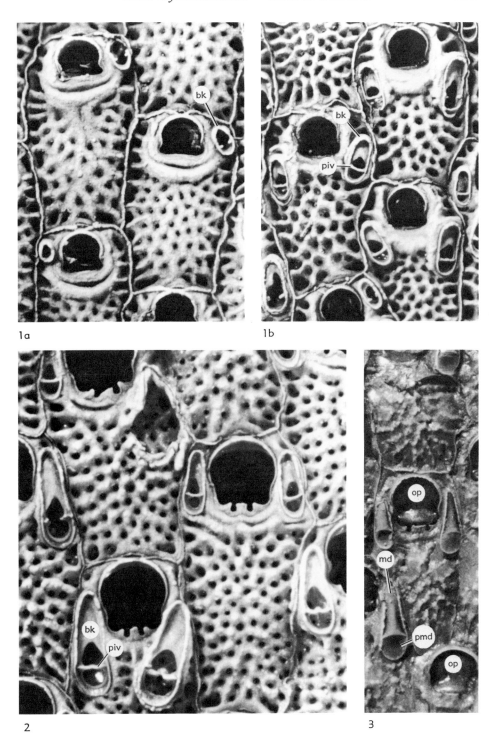

FIG. 84. *(For explanation, see facing page.)*

genetic change, the magnitude and generational duration of astogenetic differences in zooid morphology, and the presence or absence of subsequent zones of astogenetic change and repetition—all can differ between taxa, and some can differ within species.

The soft-bodied sac formed by extensive reorganization of larval tissues becomes the body wall of one or more primary zooids. Usually a single primary zooid (ancestrula) is formed (Fig. 75,1–8; 77,1; 79,1; 80,3,4), but in some cheilostomates two or more primary zooids are partitioned simultaneously by interior walls (Fig. 75,9; 79,2) (EITAN, 1972; COOK, 1973a; HÅKANSSON, 1973). Localized, broad or circumferential swelling of the outer wall of the sac (Fig. 75,1–3,5,7–9) is followed by ingrowth of interior walls or pore plates to cut off the primary zooids from buds.

Primary zooids, whether multiple or a single ancestrula, most commonly are smaller and morphologically simpler than autozooids subsequently produced by budding in the same colony (Fig. 77,1; 79,1). Most have basal, lateral, and distal transverse walls similar to those of succeeding autozooids. The proximal end of an ancestrula commonly includes a more extensive exterior component than those of succeeding zooids (Fig. 75,1–3,7,8; 79,1). Orificial and frontal walls of an ancestrula commonly differ at least in proportions from those of succeeding zooids, but can also differ in structure. Some ascophoran species, for example, have an ancestrula with frontal structure like that of anascan autozooids. In living species, an ancestrula typically has feeding and alimentary organs, developed by infolding of exterior walls, but lacks sex cells. In a few genera of both Cteno-stomata and Cheilostomata, the ancestrula is a kenozooid (HARMER, 1926; RYLAND, 1976).

In a few morphologically simple cheilostomates and ctenostomates, a zone of astogenetic change is apparently lacking, with the primary zooid or zooids having the same morphology as subsequently budded zooids. In most gymnolaemates, primary zooids initiate a primary zone of astogenetic change that extends through one to several asexual generations of zooids of intermediate morphology and ends with a generation of repeatable morphology (Fig. 77,1a; 79,1; 80,3).

In a few morphologically complex cheilostomates a zone of astogenetic repetition is apparently lacking, with zooids continuing to show generational changes throughout colony life (COOK & LAGAAIJ, 1976). In most cheilostomates and ctenostomates primary zones of astogenetic repetition typically consist of numerous generations of one or more kinds of zooids.

In some species of both Cheilostomata and Ctenostomata **subsequent zones of astogenetic change** and **repetition** are developed (BOARDMAN, CHEETHAM, & COOK, 1970). These subsequent zones can be distal or frontal (Fig. 79,3) to zooids in primary zones of repetition. Subsequent astogenetic zones may provide renewed growth or a different form of concurrent growth in some colonies and restrict or end further growth in others (COOK & LAGAAIJ, 1976). In ascophorans having cryptocysts or umbonuloid frontal shields, frontally budded subsequent zones of astogenetic repetition can produce massive nodular multilaminate growth from initially encrusting colonies (Fig. 13,3,4).

FIG. 85. Carnose and stoloniferous ctenostomates, cheilostomate cyphonautes larva.——*1,2. Arachnidium clavatum* HINCKS, rec., Eng.; *1*, Northumberland, encrusting colony with uniserially arranged, distally and laterally budded autozooids, irregular anastomoses (ana) between lineal series common, frontal view, BMNH 1913.7.10.3, ×16; *2*, locality unknown, proximal region of encrusting colony with presumed primary zooids (pz) attached by proximal extremities, frontal view, BMNH 1898.5.7.182, Norman Coll., ×18.——*3. Terebripora* sp., rec., Bay of Santos, Brazil; polyester cast of boring in shell with autozooids (az) connected by stolonlike kenozooids (kz); oblique basal view, ×21 (photograph courtesy R. A. Pohowsky).——*4. Electra pilosa* (LINNÉ), rec., River Crouch, Essex, Eng.; cyphonautes larva; right lateral view, ×215.

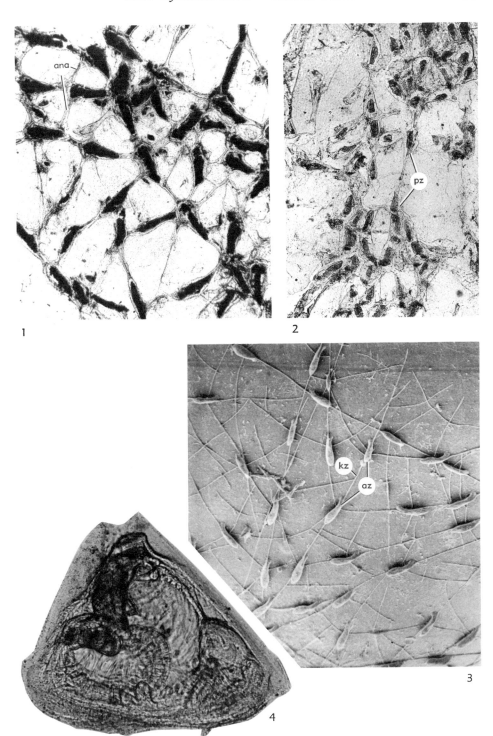

FIG. 85. *(For explanation, see facing page.)*

Some colonies in a few species of both freshwater and marine ctenostomates (JEBRAM, 1975) are produced asexually from encapsulated resistant resting bodies (hibernacula) that develop by inswelling or outswelling of the body walls of parent colonies. Similar asexual reproductive bodies have been reported in a marine cheilostomate (SIMMA-KRIEG, 1969). These colonies may form attached to or detached from the dead parent colony, and unlike colonies produced by fragmentation, have been noted to begin with zones of astogenetic change.

BUDDING

Zooids in the Gymnolaemata typically are budded at distal ends of lineal series (Fig. 74,*1,2*; 75,*1–8*), each bounded basally, laterally, and frontally by exterior walls of multizooidal origin. Buds originate by outswelling of these multizooidal walls (Fig. 68,*1a,2*; 76,*1b*; 77,*1a*; 80,*1,2*). As or after a bud swells, ingrowth of an interior wall separates the newly developing zooid body cavity from that of its proximal asexual parent (Fig. 68,*1a*; 75,*1,2*). Further lengthening of the bud is followed by ingrowth of a second, more distal interior wall that separates the now developed zooid body cavity from that of the next distal bud in the series (stage approximating that of zooid in Fig. 68,*1b*). Facing portions of the two interior walls are the transverse walls, or parts of the transverse walls, of the zooid, and their completion transforms exterior multizooidal walls to basal, lateral, and frontal zooidal walls. Further growth of walls and organs of the zooid takes place from or within these zooidal walls (Fig. 68,*1b,d,e*).

In the Cheilostomata, outswelling to initiate budding occurs on uncalcified parts of the body wall, most commonly as **distal** and **distolateral buds** on distal and distolateral sides, respectively, of vertical walls of parent zooids (Fig. 68,*2*; 72,*3,4*; 75,*1–4*; 76,*1–4*; 77,*2b,c*; 80,*1,2*; 81,*1–3*). **Proximolateral buds** are less common (Fig. 77,*1a*; 79,*1*; 80,*3,4*), and **proximal buds** arising from

ends of zooids appear to be limited to repair of broken zooids (Fig. 76,*1a*) and to **periancestrular budding** in a few species (possibly the one shown in Fig. 76,*5*; note that periancestrular budding indicated in Fig. 75,*7,8* does not include "proximal" budding).

Budding in the Cheilostomata can also be initiated on basal walls of zooids in erect, unilaminate colonies and on frontal walls and associated structures in both anascans and ascophorans (Fig. 68,*2*; 69,*1b,c,f,2*; 70,*1b,c,3*; 79,*3*; 83,*2,3*; 84,*1–3*). Frontal buds most commonly produce such adventitious polymorphs as avicularia, communicating only with the underlying parent zooid, but can also produce ordinary feeding autozooids that communicate with each other through vertical walls (Fig. 79,*3*) (POUYET, 1971; BANTA, 1972). Adventitious polymorphs can also originate distally from vertical walls of parent zooids (Fig. 82,*1,2*).

In some genera of Ctenostomata and anascan and ascophoran Cheilostomata, most zooids arise as single buds (**uniserial budding**) at tips of lineal series that remain mostly separated from each other laterally (Fig. 75,*1,2*; 76,*1–4*; 77,*1,2*). In the great majority of Cheilostomata, zooids arise by **multiserial budding** so that lineal series are in contact along exterior vertical zooidal walls breached by communication organs (Fig. 75,*7,8*). In multiserially budded colonies, zooids can form by fusion of two or more buds emanating from different asexual parent zooids (Fig. 75,*7*) (GORDON, 1971a, b). The interzooidal communication organs breaching exterior vertical zooidal walls in multiserially budded colonies have also been regarded by some workers (SILÉN, 1944b; BANTA, 1969) as buds fused with zooids, and their formation involves much the same process as bud fusion.

In some cheilostomates, buds can become multizooidal by a lag in formation of interior walls leaving two or more zooid lengths of each lineal series unpartitioned. The relative lengths of such multizooidal buds (*Grossknospen* of NITSCHE, 1871; *bourgeons géants* of LUTAUD, 1961), however, may be controlled more by environmental conditions than by genetic differences (LUTAUD, 1961; this revision).

In the few genera of cheilostomates presently known to have all interior vertical walls (Fig. 74,3), zooids are budded in multizooidal budding zones (Fig. 75,9) with body cavity confluent laterally around the colony periphery (HÅKANSSON, 1973) or at distal ends of colony branches (Fig. 78,1b). These budding zones are similar to those in the class Stenolaemata. Relationships between asexual parent and descendant zooids are less distinct in these colonies, and lineal budding series are not recognizable. However, ontogenetic gradients in zooid morphology proximally from growing tips are discernible (Fig. 78,1a,b), as in colonies with lineal series.

Budding in the Ctenostomata, in which the uncalcified walls are apparently predominantly exterior, could be expected to be more flexible than that in the calcified Cheilostomata. However, budding sites in the Ctenostomata tend to be similar in position to those of cheilostomates having similar growth forms (BANTA, 1975). In some major groups of the Ctenostomata, autozooids are budded only from kenozooids (Fig. 85,3).

POSSIBLE EVOLUTIONARY RELATIONSHIPS

Similarities in morphology and mode of growth among living representatives of the Ctenostomata and the Cheilostomata have long been regarded as evidence of a close phylogenetic relationship between the two orders. The following similarities form a major basis for the modern concept of the class Gymnolaemata, and include features expressed in development both of larvae and of colonies (see summary and discussion by BANTA, 1975). (1) The only nonbrooded larvae known in the Bryozoa are found in the Ctenostomata and the Cheilostomata (cyphonautes larvae). (2) Brooded larvae in the two orders "...seem impossible to distinguish... unless the adult is known" (BANTA, 1975, p. 574). (3) Embryological development in both orders leading to both brooded and nonbrooded larvae proceeds similarly (MARCUS, 1938a; RYLAND, 1970) and is less "aberrant" than in the other bryozoan classes (ZIMMER, 1973). (4) Autozooids in both orders have parietal muscles traversing the coelom to insert on flexible body walls to form the hydrostatic system for protruding the lophophore. (5) Reinforced, distally directed orificial wall flaps form opercula or operculumlike structures in some, but not all genera in each order. (6) Where present, opercula or operculumlike structures are closed by paired occlusor muscles in series with parietals. (7) Interzooidal communication organs form similar complexes of cells and noncellular structures in the two orders (BOBIN, 1964, 1971; BOBIN & PRENANT, 1968; BANTA, 1969, 1975; GORDON, 1975). (8) Budding in both orders commonly is in lineal series between which communication organs are formed in exterior walls in all but a few genera.

In addition, certain features that apparently are present in one order but not in the other can vary markedly in expression where present (BANTA, 1975). For example, the pleated membranous collar on the diaphragm of ctenostome autozooids varies among genera, from rudimentary to prominent. Continuous calcareous layers in body walls of cheilostomates vary from a few lightly calcified zooecial walls to extensive heavily calcified zooecial walls and extrazooidal skeleton. Characteristic cheilostome polymorphs, such as avicularia, are absent in many cheilostome genera of diverse morphologies.

The variable expression in the Gymnolaemata of numerous shared as well as unshared features suggests that some shared features could have evolved independently in the Ctenostomata and the Cheilostomata.

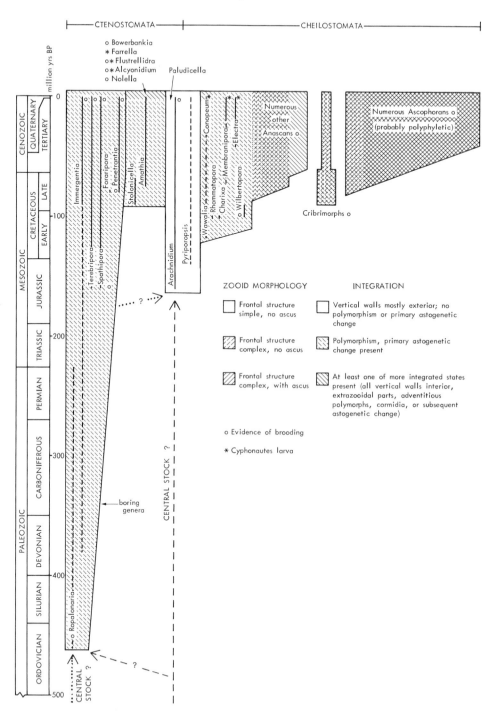

FIG. 86.   (*For explanation, see facing page.*)

Cyphonautes larvae have been reported in ctenostomate genera that, on the basis of morphologies of zooids and colonies, are considered not to be closely related (BANTA, 1975, p. 574). Two ctenostomate genera include species having brooded and cyphonautes larvae. Cheilostomate genera in which cyphonautes larvae have been found are generally similar morphologically, but also show much morphologic similarity, except in reproductive structures, to some brooding genera. Morphology associated with brooding is variable in both orders.

If features considered to be characteristic of the Gymnolaemata, such as cyphonautes larvae, brooding, polymorphism, or interzooidal communication through exterior walls, could have evolved convergently, then the Cheilostomata and the Ctenostomata might have entirely separate evolutionary origins (DZIK, 1975). The question of whether the two orders should form a higher level taxon (class Gymnolaemata) in a phylogenetic classification cannot be answered by comparing morphology of living representatives alone. Evolutionary trends in the morphology of each order through time must be considered in order to suggest how the two orders might be phylogenetically linked.

Phylogenetic inference in the Gymnolaemata is hampered by the sporadic fossil record of the Ctenostomata, inadequate knowledge of distributions of more complex morphologies in the Cheilostomata, and low correlation between characters in both orders. A more precise delineation of major evolutionary stocks within the Cheilostomata can be

attempted after restudy of the nearly 1,000 described nominal genera now assigned to the order is completed. Current understanding of early gymnolaemate morphology and of its apparent relationships to morphology of later gymnolaemates of both orders provides a starting point to suggest a tentative evolutionary basis for the class Gymnolaemata. This understanding has recently been increased by discoveries of new material and modern interpretations of modes of growth.

Phylogenetic relationships of the Cheilostomata and the Ctenostomata to the Stenolaemata or the Phylactolaemata are no less important to an evolutionary concept of the Gymnolaemata, but are more speculative. Significant overlaps occur in mode of growth between the Stenolaemata and some genera of the Cheilostomata (for example, cupuladriids, *Euthyrisella*) in which zooids have all interior vertical walls that grow into confluent body cavities in multizooidal budding zones. These groups of cheilostomate genera, however, seem to have appeared too late in gymnolaemate history (Late Cretaceous to Cenozoic) to provide a phylogenetic link between the Cheilostomata as a whole and representatives of the class Stenolaemata. Moreover, cheilostomates having this mode of growth have zooid morphologies and other characters closely comparable to those of different groups of anascans ranging from simple to complex, to which the cheilostomates seem to be phylogenetically related. Similarities in zooid shape and degree of "frontal" calcification once thought to imply a close phylogenetic relationship between cheilo-

FIG. 86. Possible evolutionary relationships among commonly recognized major morphologic groups of gymnolaemate bryozoans. Groups to right, under Cheilostomata, include many more genera than those to left, under Ctenostomata. Groups of genera are marked with two patterns, representing zooid morphology and integration level. The simplest state of each set of characters is indicated by absence of pattern, the most complex state by solid cross-hatching. Simple ctenostomes and cheilostomes are considered to have zooids with simple frontal structure and low degrees of integration (absence of both patterns). Other ctenostomes also are considered to have zooids with simple frontal structure, but can vary in integration from simple to complex (single pattern). Other cheilostomes vary from simple to complex in both zooid morphology and integration (intersecting patterns). Ranges of a few critical genera, discussed in text, are plotted, including all reported fossil genera confidently assigned to the Ctenostomata. Dotted and dashed arrows indicate two hypotheses of evolution in the Ctenostomata, discussed in text.

stomates and fenestellid cryptostomates (ULRICH, 1890; BASSLER, 1911) are now interpreted as a heterochronous convergence (TAVENER-SMITH, 1971).

Overlaps with the Phylactolaemata, such as lack of calcification in almost all ctenostomates and development of resistant resting bodies by some freshwater ctenostomates and some marine ctenostomates and cheilostomates, seem more difficult to evaluate because of the scarcity or lack of a fossil record in these groups of genera. Furthermore, JEBRAM (1973b) has suggested that early stenolaemate as well as early gymnolaemate and phylactolaemate stocks may have been uncalcified and thus may not be preserved in the fossil record. If this hypothesis is correct, phylogenetic relationships among the three bryozoan classes may remain speculative, unless exceptionally preserved material is eventually discovered.

Even within the Gymnolaemata, in which fossil evidence is available for some uncalcified as well as calcified taxa, study of the phylogenetic significance of such features as presence or absence of larval brooding, monomorphism or polymorphism of zooids, and different budding sites and directions has been based mostly on comparative morphology and development of living representatives of the class. Assumptions that certain states of these features are primitive and others derived are only beginning to be checked against fossil morphology (BANTA, 1975). Some genera that have been considered to link the Cheilostomata and the Ctenostomata, or the Gymnolaemata to other classes, either are not represented in the fossil record at all or have not been found in Mesozoic and older deposits, from which evidence of early gymnolaemate history must come to be convincing (Fig. 86).

The broad outlines of evolutionary relationships in the Gymnolaemata tentatively suggested below emphasize the rich fossil record of the Cheilostomata. As presently understood, evolutionary trends within the Cheilostomata support the inferred close phylogenetic relationship with the Cteno-

stomata. Even though much less adequate than that of the Cheilostomata, the fossil record of the Ctenostomata also supports this inferred relationship, and the two records thus provide some evolutionary basis for the modern concept of the Gymnolaemata. The ctenostomate record, however, seems inadequate to provide a choice between alternative hypotheses of evolutionary trends within that order (Fig. 86), and thus seems to shed little light on the origin of the Gymnolaemata. Major improvements in our understanding of early gymnolaemate history probably will require new discoveries and interpretations of more Paleozoic and early Mesozoic Ctenostomata.

EVOLUTIONARY TRENDS IN CHEILOSTOMATA

Phylogenetic inference in the Cheilostomata begins conveniently with the observation that the group of species having the oldest reported occurrence in the fossil record also has the simplest combination of morphologic features apparent in the order (Fig. 86). This group of species seems to be referable to a single genus, *Pyriporopsis,* and to include *P. portlandensis* POHOWSKY from the Upper Jurassic of England, one or more species of intermediate age, and the living North Atlantic species *P.? catenularia* (FLEMING). In Lower Cretaceous deposits a few genera in addition to *Pyriporopsis* have been reported. These genera are slightly more complex in morphology but, like *Pyriporopsis,* are comparable to some living species. In Upper Cretaceous and younger deposits, an increasing diversity of simple to complex morphologies leads to the numerous groups of living species of Cheilostomata.

Morphologic similarities between fossil and living species in each of the major groups of Cheilostomata permit a high degree of biologic interpretation of the morphology of the order. No major group of Cheilostomata, above the family level, appears to have become extinct. Most morphologic features found in fossil cheilostomates can be studied in living colonies. However, some important questions in the biology of early cheilostomates, such as presence or absence of brooding and functions of different types of avicularia, still remain to be answered, at least in part through further studies of living representatives of their groups.

Earliest cheilostomates.—Morphologic simplicity of Jurassic to recent *Pyriporopsis* is expressed by a low level of integration of zooids in colonies and by lack of structural complication of zooids, particularly in features associated with the frontal wall and hydrostatic system.

Almost entirely exterior-walled zooids in *Pyriporopsis* are budded for the most part uniserially in series that branch irregularly to form encrusting colonies (Fig. 76,*1–5*). Basal walls of zooids, which may be either calcified or uncalcified in the same colony, adhere directly to the substrate, with no tendency to be partly immersed in calcareous substrates as are some stratigraphically younger cheilostomates with similar colony forms (for example, *Electra, Hippothoa*).

Zooids have only slight contact along vertical walls. Within lineal series, contact is through pore plates or pore chambers at the narrowed proximal extremities of zooids. Lateral contacts are irregular and less frequent in the generally open areas between lineal series. Even though the calcified lateral walls of zooids have uncalcified gaps opening into pore chambers (Fig. 76,*1b,4*), these gaps do not match where zooids contact laterally (POHOWSKY, 1973; BANTA, 1975). Interzooidal communication thus appears limited to zooids within lineal series. Uncalcified spots in lateral walls appear to serve only as incipient budding sites (BANTA, 1975).

Frontal walls in *Pyriporopsis* include calcified and flexible portions. An extensive gymnocyst margins a simple, flexible hydrostatic membrane proximally and laterally (Fig. 76,*1,2*). This membrane, commonly preserved in both fossil and modern colonies by formation of frontal closures (Fig. 76,*2,3*), was apparently entirely exposed and unprotected in fossils as it is in modern colonies

and demonstrates the simple anascan structure of the genus. Frontal closures preserve traces (scars) of bilateral series of parietal muscle insertions and the simple flaplike operculum reinforced only on its distal and lateral margins (Fig. 76,*2*) (POHOWSKY, 1973, fig. 1). Cretaceous and living *Pyriporopsis* have narrow cryptocysts within the margins of the gymnocyst, and the Cretaceous species has a pair of minute spine bases flanking the orifices of some zooids. Both spines and cryptocyst are lacking in Jurassic *Pyriporopsis* (POHOWSKY, 1973).

Zooids in *Pyriporopsis* apparently are entirely monomorphic, at least skeletally and in the morphology of the hydrostatic membrane and operculum. This apparent monomorphism and the presence of structures reflecting protrusible lophophores (Fig. 76,*2,3*) suggest that all fully developed zooids in *Pyriporopsis* colonies, except when lophophores and associated organs were degenerate, were able to feed. All fully developed zooids also may have been able to produce sex products, but there is no direct evidence for this known from either living or fossil colonies. It is not known whether living *Pyriporopsis* broods embryos or releases them directly. Modern species of *Electra, Conopeum*, and *Membranipora*, which have similar zooid morphology and only slightly higher levels of integration (Fig. 86), all produce nonbrooded cyphonautes larvae. However, such other genera as *Allantopora* (Fig. 77) appear equally similar to *Pyriporopsis,* except for having skeletally reinforced brood chambers. Genera that are known to brood embryos without apparent skeletal expression of this function, such as *Steginoporella* (COOK, 1964), are morphologically much less similar to *Pyriporopsis*. It therefore seems likely that *Pyriporopsis* is not a brooder.

Astogenetic differences in zooid morphology also appear to be lacking in *Pyriporopsis*. Differences in zooid size and shape reported in Jurassic colonies (POHOWSKY, 1973) appear to be gradational within generations and related to different budding sites. Primary zooids have not been recognized in fossil *Py-*

riporopsis, but a few modern colonies have been found with proximal ends intact. In most fossil and modern colonies, the frequency of regenerative budding, commonly from proximal ends of broken zooids (Fig. 76,*1a*), obscures the proximal region. In intact colonies a pair of proximal zooids, having the same size and shape as those of succeeding generations, are joined by their narrowed "proximal" extremities (Fig. 76,*5*). Whether one zooid is the ancestrula from which the other budded "proximally" (see Fig. 75,*5*) or both grew simultaneously at opposite poles of a postlarval sac has not been determined. Both "proximal" budding (*Conopeum*) and simultaneous differentiation of twinned primary zooids (*Membranipora*) are known in living cheilostomates with generally similar zooid morphology. However, these genera all have primary zones of astogenetic change in which zooids show progressive generational increases in size and changes in other morphologic characters.

Other early cheilostomates.—All four or five other genera that have been reported from the Lower Cretaceous (Dzik, 1975; Larwood, 1975) show one or more increases in morphologic complexity over *Pyriporopsis* (Fig. 86). The fewest changes are evident in *Rhammatopora* and *Wawalia* and the most in *Wilbertopora.* These changes are not strictly progressive, however, but rather show the beginnings of a mosaic evolutionary pattern that typifies Upper Cretaceous and stratigraphically younger cheilostomates (Boardman & Cheetham, 1973).

All genera known from the Lower Cretaceous retained an encrusting growth form generally similar to that in *Pyriporopsis.* Cheilostomates with erect and other specialized colony forms are found in Upper Cretaceous and younger deposits. In the Lower Cretaceous *Rhammatopora,* uniserial budding of zooids was also retained, but other genera are characterized by multiserially budded zooids. In *Wawalia* and most colonies of *Wilbertopora* (Fig. 80,*3*) budding produced multiserial arrangements throughout, beginning at the ancestrula. In some colonies of *Wilbertopora* (Fig. 80,*4*), one or more

generations of zooids initially budded uniserially, and these were followed by generations of zooids arranged like those in fully multiserial budded colonies (Cheetham, 1975b). Some modern species of *Conopeum* and *Electra* also show this pattern. Winston (1976) found that uniserial or multiserial budding in cultured colonies of *Conopeum* can be controlled by varying the kind of food. Variation in arrangements of zooids in *Wilbertopora* may also have been environmentally controlled and related to the low degree of integration, especially in the largely exterior vertical walls of zooids, in this genus.

Multiserial budding represents an advance in integration in that growth of adjacent lineal series is more or less coordinated and thus apparently less autonomous than uniserial growth. Gaps in calcified lateral walls match pore plates or pore chambers in laterally adjacent zooids (Fig. 75,*7,8*; 80,*1*) to provide interzooidal communication between lineal series. Such lateral communications occur in *Wawalia* (Dzik, 1975) and in *Wilbertopora* (Banta, 1975). Growing edges preserved in some *Wilbertopora* colonies (Fig. 80,*1*) show that adjacent lineal series were slightly staggered, suggesting less coordination of growth than in many stratigraphically younger cheilostomates that have smooth growing edges (Fig. 75,*7*; 80,*2*).

Within lineal series, multiserially budded zooids are also more extensively in contact than uniserially budded ones. Increased contact in Lower Cretaceous multiserial cheilostomates is generally produced by widening of proximal extremities of zooids, a shape change that is also common in multiserial parts of predominantly uniserial colonies. In *Wilbertopora* widening of proximal extremities of zooids was achieved by folding back the exterior vertical wall upon itself without greatly increasing the amount of interior wall (Fig. 75,*6*) or changing most of the zooidal outline from the elongated, distally inflated shape common in uniserial colonies (Cheetham & Lorenz, 1976). In this respect *Wilbertopora* remained significantly less integrated than stratigraphically younger multiserial cheilostomates in which broad intraseries contact is along extensive interior transverse walls (Fig. 75,7), generally to produce more squat, uninflated zooid outlines.

Frontal and orificial walls of zooids in these Lower Cretaceous cheilostomates appear to be only slightly different from those in coeval *Pyriporopsis*. In *Rhammatopora* and *Charixa* a row of spine bases rings the inner margin of the gymnocyst. Spines presumably protected the hydrostatic membrane and the orifice, as in living genera such as *Callopora*. A few spine bases, in addition to the pair flanking the orifice, occur in some specimens of *Wilbertopora*. More extensive cryptocysts are evident in the Lower Cretaceous species compared by LARWOOD (1975) to *Conopeum*, in *Wawalia*, and in *Wilbertopora* (Fig. 80,3). There is no evidence, however, of fused spines, cryptocysts extensive enough to reflect passage of parietal muscles, or an ascus in any Lower Cretaceous species. Evidence of a simple frontal wall and operculum similar to those of *Pyriporopsis* has been reported in *Rhammatopora*, *Wawalia*, and *Wilbertopora*.

Polymorphism has been recognized in *Rhammatopora* and *Wilbertopora*. In *Rhammatopora* polymorphs are limited to kenozooids that occur sporadically between autozooids in the uniserial colonies (THOMAS & LARWOOD, 1960). In *Wilbertopora*, colonies with varying combinations of polymorphs that can be interpreted as kenozooids, avicularia, and zooids with brood chambers, together with ordinary autozooids, occur in the same populations as colonies in which zooids were apparently monomorphic (CHEETHAM, 1975b). Structures interpreted as avicularia and brood chambers (Fig. 81,1,2) have been reported from the earliest known *Wilbertopora* populations and thus could have evolved approximately simultaneously in this genus. However, broken brood-chamberlike structures have also been reported in a poorly preserved multiserial anascan that is slightly older stratigraphically (PITT, 1976).

Avicularia in *Wilbertopora* are all interzooidal or vicarious and follow a graded sequence of increasing morphologic difference from ordinary autozooids (Fig.

81,1,2,4). The most differentiated avicularia (Fig. 81,1) are found only in the stratigraphically youngest *Wilbertopora* populations, which also include colonies having less differentiated or no avicularia. The similarity in shape of the less differentiated avicularia (Fig. 81,2,4) to ordinary autozooids in the same colonies suggests that these avicularia may have had feeding organs, as in such living genera as *Crassimarginatella*. It seems unlikely that the most differentiated avicularia had feeding organs because of diminished width of the orificial wall (mandibular) area relative to the frontal wall (postmandibular and gymnocystal) area (Fig. 81,1). More highly differentiated avicularia of adventitious position, which are common in Upper Cretaceous and Cenozoic cheilostomates, have not been found in Lower Cretaceous genera (BOARDMAN & CHEETHAM, 1973).

The polymorphism evident in Lower Cretaceous cheilostomates, especially in *Wilbertopora*, can be inferred to represent at least some separation of functions and therefore a significant advance in the level of integration over the earliest cheilostomates, which appear to have been monomorphic. The apparent variability in polymorphism (presence or absence within a colony, degree of morphologic differentiation of polymorphs, and number of kinds of polymorphs) within *Wilbertopora* populations again suggests that integration was less rigidly controlled than in most stratigraphically younger cheilostomates (CHEETHAM, 1975b).

Astogenetic differences in zooid morphology are commonly preserved in *Wilbertopora*, and similar astogenetic differences have been reported in *Wawalia* (DZIK, 1975). With some variation in arrangement (Fig. 80,3,4), an ancestrula, smaller than but otherwise similar in morphology to distal zooids, is followed by a few generations of distally and generally distolaterally budded zooids of gradually increasing size (CHEETHAM & LORENZ, 1976). Numerous following generations of ordinary autozooids, and commonly polymorphs, form the primary zones of astogenetic repetition. The morphologic

difference between the ancestrula and autozooids of repeated morphology is small compared to that in many stratigraphically younger cheilostomates probably because of the
low level of morphologic complexity of zooids
in *Wilbertopora*. The level of integration
through astogeny shown by *Wilbertopora* thus
seems similar to that shown by many, perhaps even the majority of stratigraphically
younger cheilostomates.

In summary, the stratigraphic sequence of
increasing morphologic complexity among
Lower Cretaceous cheilostomates seems to be:
(1) development of cryptocysts in autozooids
and of primary zones of astogenetic change
and multiserial budding of zooids in colonies,
with concomitant establishment of interzooidal communication through exterior walls
of zooids in adjacent lineal series (*Wawalia*);
(2) development of spines on gymnocysts of
autozooids and differentiation of kenozooids
(*Rhammatopora, Charixa*); and (3) development of brood chambers and differentiation of avicularia (*Wilbertopora*). The Early
Cretaceous record of the Cheilostomata is
probably not well enough known, however,
to attach much significance to the exact order
of appearance of new morphologic features
in this sequence. The possibility of brood
chambers in the poorly preserved multiserial
anascan slightly older than *Rhammatopora,
Charixa,* and *Wilbertopora* (PITT, 1976)
already suggests that revisions in this sequence
will be forthcoming as further studies are
made. It does seem apparent even from this
tentative sequence that autozooidal frontal
structure and colony integration increased
approximately simultaneously and at least
partly independently in the early evolution
of the Cheilostomata. For example, gymnocystal spines and cryptocysts are present both
in better integrated genera such as *Wilbertopora* and in poorly integrated ones such as
Rhammatopora.

*Mosaic evolution in younger cheilosto-
mates.*—The many hundreds of genera of
Cheilostomata known from deposits of Late
Cretaceous and Cenozoic age display a range
of morphologic differences markedly increased
over that shown by Early Cretaceous repre-

sentatives of the order. This diversification
involved progressive appearances of major
groups of genera having autozooids with more
complex frontal structure, colonies with
higher states of integration, or both (Fig. 86).

At least some changes in zooid morphology and colony integration in the Cheilostomata appear to be functionally linked to
evolution of more specialized growth habits
(CHEETHAM, 1971; BOARDMAN & CHEETHAM,
1973). In contrast to the exclusively encrusting habit of Jurassic and Early Cretaceous
cheilostomates, younger representatives of the
order exhibit an increasing variety of growth
habits, eventually to include: (1) encrusting
colonies of unilaminate, multilaminate, and
loosely attached form; (2) erect colonies of
rigid, flexible, jointed, and fenestrate form;
and (3) free-living colonies of discoid and
conical form. (See Fig. 13–15 for growth
habits in living representatives of the Cheilostomata.) The earliest evidence of rigidly
erect, jointed erect, and free-living colonies
in the Cheilostomata has been found in Upper
Cretaceous deposits (VOIGT, 1959, 1972b).
These and other specialized growth habits
numerically dominate fossil and living Cenozoic marine bryozoan assemblages (STACH,
1936; CHEETHAM, 1963; LAGAAIJ & GAUTIER,
1965; COOK, 1968b; LABRACHERIE, 1973; see
SCHOPF, 1969a, for a review). However, the
simpler growth habits also continue to be
represented in many assemblages and even to
dominate some of them.

For approximately 100 years, frontal
structure of autozooids conventionally has
been regarded as providing the most significant morphologic characters for phylogenetic interpretation of the Cheilostomata. This
assumption has been inadequately tested on
a polythetic basis against the fossil record;
however, available evidence continues to suggest that increasing complexity of frontal
structure is the apparent evolutionary trend,
with the most obvious sequence of intermediate morphologies in the Cheilostomata (Fig.
86). Considered against the trend in frontal
structure, characters derived from colony
growth form and levels of integration form
patterns suggesting uneven rates of evolution

or parallel or convergent trends in the several major evolutionary stocks within the order.

Characters expressing growth habit seem particularly to have been subject to parallel or convergent evolution. The most highly specialized growth habits, such as jointed erect and free-living colonies, are found in groups of genera ranging from simple anascans *(Nellia, Cupuladria)* to complex ascophorans *(Margaretta, Mamillopora)*. Numerous examples of simple encrusting to more specialized growth habits are known within the same genus, also in groups ranging from simple anascans *(Membranipora)* to complex ascophorans *(Metrarabdotos)*. Observed environmental plasticity of growth habits within species, and even within some colonies (COOK, 1968a), further suggests that some similarities in colony form among otherwise morphologically distinct genera may be induced directly by the environment (STACH, 1936).

The generally increasing level of integration evident in the stratigraphic record of the Cheilostomata appears to have proceeded at uneven rates (Fig. 86), partly but not entirely correlated with specialization in colony form. For example, both encrusting and erect species of *Metrarabdotos* and *Schizoporella* have similar high levels of integration in their combination of interior and exterior vertical zooid walls, transverse and lateral communication organs, brooding autozooids, and adventitious avicularia. Erect species of *Metrarabdotos* have extensive extrazooidal skeleton, which is only partly or not developed in the encrusting species, and thus a higher level of integration. However, encrusting species of *Schizoporella* have subsequent zones of astogenetic change and repetition not found in erect species of this genus, and thus are the more highly integrated.

Some integrative characters reached peak states in groups of cheilostomate genera having increasingly different types of frontal structure and either high or low levels of other integrative characters. Some peak states occur in genera so different in other morphologic characters that convergence in integrative characters seems highly probable. Conver-

gence seems especially probable in integrative characters with states associated with differences in environment. For example, species possessing avicularia in stable environments can lack them under unstable conditions of salinity or temperature (SCHOPF, 1973). In colonies that are either uniserial or multiserial under the influence of different foods (WINSTON, 1976), it seems likely that integrative characters of zooid walls and interzooidal communication may suffer direct environmental modification.

Detailed review of the combinations of states of integrative and frontal characters can be made only when all the genera now assigned to the Cheilostomata have been restudied. The following examples are intended to show a few of the extreme combinations that have been reported previously (as reviewed by BOARDMAN & CHEETHAM, 1973), or are illustrated in this section.

Genera having extensive interior vertical walls include anascans *(Cellaria,* BANTA, 1968; SANDBERG, 1971; cupuladriids, HÅKANSSON, 1973) and ascophorans *(Euthyrisella,* Fig. 78; HARMER, 1902; *Myriapora, Mamillopora,* and conescharellinids, SANDBERG, 1973), with erect colonies of jointed, flexible, or rigid form and free-living colonies. The erect *Euthyrisella* and free-living cupuladriids are further integrated in having extrazooidal parts formed concurrently with budding of zooids. Extrazooidal parts are apparently absent in other genera in this group. Some cupuladriids are even more highly integrated through the presence of subsequent zones of astogenetic change and repetition (BOARDMAN, CHEETHAM, & COOK, 1970). Some genera with erect or free-living habit *(Myriapora, Mamillopora,* conescharellinids) have highly specialized polymorphs (avicularia) adventitious upon autozooids or in clustered arrangements. Others also erect or free-living *(Cellaria,* cupuladriids) have interzooidal or vicarious avicularia in irregular or regular, nonclustered arrangements. Still others *(Euthyrisella)* lack highly specialized polymorphs but have dimorphic autozooids in apparently random intermixtures.

TABLE 3. *Comparison of Some Morphologic Characters of Early Gymnolaemata.*

Character	Paleozoic–Early Mesozoic (Boring) Ctenostomes	Cretaceous Amathia, Stolonicella	Jurassic–Early Cretaceous Arachnidium	Jurassic Pyriporopsis	Early Cretaceous Cheilostomes
Form of colony	Totally immersed in calcareous substrates	Encrusting to erect, not immersed	Encrusting, not immersed	Encrusting, not immersed	Encrusting, not immersed
Budding and communication	Uniserial, lateral fusions regular	Uniserial, lateral fusions irregular or absent	Uniserial, lateral fusions irregular	Uniserial, no lateral fusions	Uniserial-multiserial, lateral fusions generally regular
Vertical zooid walls	Virtually all exterior, uncalcified	Virtually all exterior, uncalcified	Virtually all exterior, uncalcified	Virtually all exterior, calcified	With significant interior components, calcified
Frontal zooid walls	Entirely flexible but protected by immersion in substrate	Entirely flexible and exposed	Entirely flexible and exposed	Flexible portion exposed, but reduced by rigid gymnocyst	Flexible portion largely exposed, but commonly protected by spines or underlain by cryptocyst, and reduced by gymnocyst
Orificial walls	Probably radially disposed, unreinforced folds[a]	Ringlike or radially disposed, unreinforced folds	Ringlike, unreinforced fold	Operculum reinforced distally and laterally	Operculum reinforced distally and laterally
Polymorphism	Generally present, integral to budding pattern	Present, integral to budding pattern	Apparently absent	Apparently absent	Generally present
Brooding	Generally present	Probably present	Probably present	Possibly absent	Generally present
Astogenetic change	Primary zone present	Probably present	Probably absent	Probably absent	Primary zone generally present

[a] Excludes Cretaceous *Penetrantia.*

Highly specialized adventitious and clustered interzooidal or vicarious polymorphs are commonly found among the numerous cheilostomate genera that retained extensive exterior vertical walls. These genera include anascans (*Monoporella*, Fig. 72; *Setosellina*, Fig. 82,*1*,*2*) and ascophorans (*Hippothoa*, Fig. 69; *Tessaradoma*, Fig. 83; *Hippopetraliella*, Fig. 84,*1*; *Petraliella*, Fig. 84,*2*,*3*; *Stylopoma*, Fig. 79,*2*,*3*; *Metrarabdotos*, Fig. 68,*2*; 70,*1b,c,3*) with a wide variety of growth habits. The specialized adventitious or clustered polymorphs include brooding and other sexual zooids (for example, *Monoporella*, *Hippothoa*, and *Metrarabdotos*) and avicularia. Some ascophoran genera in this group develop extrazooidal parts through coalescence of parts of zooids (*Tessaradoma*, *Metrarabdotos*), and others have subsequent zones of astogenetic change and repetition formed by frontal budding from hypostegal coeloms (*Stylopoma*). Some anascans in this group can also have subsequent astogenetic zones formed by distal budding (*Nellia*; *Poricellaria*, Boardman, Cheetham, & Cook, 1970).

A great diversity of Late Cretaceous, Tertiary, and living genera include species that have frontal structures of moderate to high complexity but have not reached peak states of any integrative characters considered here. These genera even include relatively complex ascophorans (*Margaretta*, Fig. 67; 73; 82,*3*; *Cryptosula*, Fig. 79,*1*) with both specialized and simpler growth habits.

Flexibility of different integrative morphologic features in combination with different zooidal frontal structures may well have provided the broad adaptability in growth habit evident in late Mesozoic and Cenozoic Cheilostomata, and consequently assured the increasing evolutionary success of the order (Boardman & Cheetham, 1973). Despite the great numbers of elaborately integrated and morphologically complex species present in modern faunas, however, even the simplest morphology, as represented by *Pyriporopsis* and similar forms, continues to have its niche in present seas.

POSSIBLE LINKS BETWEEN CHEILOSTOMATA AND CTENOSTOMATA

There has been no convincing evidence reported of the existence of calcified Cheilostomata before Late Jurassic time. The earlier Mesozoic and Paleozoic fossil record of the uncalcified Ctenostomata, however fragmentary, provides strong evidence that representatives of this order considerably preceded the earliest cheilostomates in time (Fig. 86).

Present understanding of gymnolaemate morphology makes it appropriate to seek the ancestry of the Cheilostomata among Ctenostomata approximately coeval with and similar in morphology to early, simple, *Pyriporopsis*-like cheilostomates (Banta, 1975). Three groups of ctenostomate genera have been reported from Mesozoic or earlier deposits (Fig. 86): genera that penetrate calcareous substrates (boring genera), stoloniferous nonboring genera (*Amathia*, *Stolonicella*), and a carnose genus (*Arachnidium*). These genera show different degrees of morphologic similarity to *Pyriporopsis* (Table 3).

Similarities between *Pyriporopsis* and some simple uniserial stenolaemates of Paleozoic age (corynotrypids) led Dzik (1975) to propose that the Cheilostomata and the Ctenostomata each separately evolved from the Stenolaemata. This hypothesis requires that basic features shared by zooids throughout the Cheilostomata and Ctenostomata—such as flexible frontal walls or their derivatives, parietal muscles, and the folded structure of the orificial wall, together with negative features such as the absence of a membranous sac—all evolved convergently. These convergences would be in addition to those that possibly produced cyphonautes larvae, extracoelomic brooding, or polymorphism in the two gymnolaemate orders.

Nonboring carnose ctenostomates.—Although lacking calcification and possessing typical ctenostomate features such as unreinforced orificial walls, *Arachnidium* is closely similar in morphology to *Pyriporopsis* (Table 3). As in other **carnosans**, autozooids bud directly from other autozooids. Predominantly uniserial colonies lack apparent zones of astogenetic change and begin with a pair of proximally opposing zooids (Fig. 85,2). Irregular tubular extensions connect some zooids in neighboring lineal series (Fig. 85,1). Zooids are monomorphic and similar in shape to those of *Pyriporopsis*. Living species brood embryos in diverticula of vestibules of otherwise unmodified zooids. Although *Arachnidium* is marine, hibernacula have been reported in one species (JEBRAM, 1975).

Arachnidium thus appears to be more specialized reproductively and slightly more advanced in integration than *Pyriporopsis*, even though occurring in slightly older deposits (Middle Jurassic; VOIGT, pers. commun., 1976). Simpler ctenostomates might have existed before the earliest cheilostomates, but there is as yet no fossil evidence. The morphologic similarities are enough, however, to make a close phylogenetic relationship between *Arachnidium* and *Pyriporopsis* likely.

Other carnosans, none known as fossils, display differing but higher levels of integration. Genera such as the freshwater *Paludicella* are similar to *Arachnidium*. At the upper end of the scale are genera such as *Flustrellidra, Elzerina,* and *Alcyonidium* (Fig. 66,1–3) with clustered arrangements of autozooids and kenozooids and some other features paralleling those of advanced cheilostomates (Fig. 86). The reproductive features of these genera display a pattern seemingly best interpreted as the result of convergence.

Nonboring stoloniferous ctenostomates.—Colonies of stoloniferous ctenostomates are comparable in levels of integration to most complex carnose genera. Autozooids in **stoloniferans** are budded entirely from kenozooids. Budding patterns typically include lineal series of kenozooids forming stalks or encrusting networks from which regularly grouped clusters of autozooids arise. This highly organized budding pattern seems to exclude stoloniferous genera from consideration as a possible link to early cheilostomates.

Boring ctenostomates.—Even though boring ctenostomate genera have a long fossil record preceding the earliest known cheilostomates (Fig. 86), their morphology and mode of life suggest that they did not include the direct ancestors of the Cheilostomata.

A few modern boring genera penetrate noncalcareous substrates, apparently by mechanical means (SOULE & SOULE, 1969). None of these genera is known from fossils. Colonies of fossil boring genera are completely immersed in calcareous substrates. Most of these genera have living representatives (VOIGT & SOULE, 1973) found exclusively, or nearly so, in calcareous substrates. Growth of colonies in calcareous substrates is accomplished by some chemical means of penetration not well understood (SOULE & SOULE, 1969, p. 801). SILÉN (1947) presented chemical evidence that in *Penetrantia* dissolution of mollusk shell may be accomplished by secretion of phosphoric acid. In some cheilostomates (*Electra, Hippothoa*) basal walls of zooids in encrusting colonies may be immersed in calcareous substrates to produce pits, which in some respects seem comparable to ctenostomate borings (PINTER MORRIS, 1975). However, there is no evidence that the earliest cheilostomates or their modern representatives produced such pits.

Within calcareous substrates, zooids of boring ctenostomates are connected in lineal series and laterally by a complex system of elongate, anastomosing tubes to form colonies with relatively widely spaced autozooidal orifices (Fig. 85,3). In all but one genus (*Immergentia*) the connecting tubes are kenozooids separated from autozooids by pore plates so that the autozooids themselves are widely separated. This arrangement is similar to that in some nonboring stoloniferans, to which most boring genera are considered to be related.

Polymorphs in addition to connective kenozooids have been reported in a number of Paleozoic and Mesozoic genera (VOIGT & SOULE, 1973; POHOWSKY, 1974, 1975; RICHARDS, 1974). In a Cretaceous species, these polymorphs have been compared in shape and position to brooding autozooids in living species of the boring genus *Penetrantia*. Living species of other boring genera all brood embryos without apparent modification of autozooidal size or shape.

The complex budding patterns and polymorphism of boring ctenostomates thus represent a significantly higher level of integration than that reached by early cheilostomates. Even though one boring genus, *Penetrantia*, has features such as opercula and associated musculature similar to those in the Cheilostomata (SOULE & SOULE, 1975), it shares the high level of integration of other boring genera. Moreover, reinforced flaplike orificial walls even more similar to the opercula of early cheilostomates also occur in other groups of ctenostomates (for example, *Elzerina*; Fig. 66,2a). If *Penetrantia* should be assigned to the Cheilostomata (SOULE & SOULE, 1969), its ctenostomate features probably indicate convergence (possibly through adoption of the boring mode of life), rather than a phylogenetic link between the orders.

Summary.—Even though other groups of ctenostomates also occur in deposits older than those containing earliest (Late Jurassic) *Pyriporopsis*, Middle Jurassic to Early Cretaceous *Arachnidium* is most comparable morphologically to early cheilostomates. Simple *Arachnidium*-like ctenostomates therefore seem likely to have been the mid-Mesozoic ancestors of the Cheilostomata and to provide a phylogenetic basis for the class Gymnolaemata.

NATURE OF EARLY CTENOSTOMATA

Evolutionary relationships of simple *Arachnidium*-like ctenostomates both to more highly integrated boring and nonboring genera of the Ctenostomata, and to representatives of other bryozoan classes, are much more difficult to infer from available evidence. Critical to such an inference is whether nonboring ctenostomates existed during Paleozoic and early Mesozoic time and, if so, whether they were as highly integrated as Paleozoic and early Mesozoic boring genera or possessed a low level of integration comparable to that of *Arachnidium*. Problemat-

ical Paleozoic fossils historically interpreted as nonboring ctenostomates have not yielded morphologic evidence that permits comparison with living ctenostomates (DZIK, 1975). The nature of early Ctenostomata thus remains speculative, with only the few boring ctenostomate genera providing stratigraphic evidence for the early history of the group.

If nonboring ctenostomates of the *Arachnidium* type did not evolve until mid-Mesozoic time, as the sporadic Paleozoic record of Ctenostomata suggests, the central gymnolaemate stock would likely have lain among relatively highly integrated forms of boring and perhaps nonboring habit (dotted arrows on left side of Fig. 86). Evolution of *Arachnidium*-like ctenostomates then would have involved a decrease in integration through loss of polymorphism and simplification of astogeny and budding patterns. Such a decrease would be in contrast to prevailing evolutionary trends toward higher levels of integration in the Cheilostomata.

Conversely, if trends increasing integration could be assumed to have characterized the class Gymnolaemata as a whole, then simple *Arachnidium*-like ctenostomates would have existed throughout much of Paleozoic and early Mesozoic time as the central gymnolaemate stock (dashed arrows, center of Fig. 86). Ctenostomates within this hypothetical central stock should have been similar in some morphologic features to the ancestors of the Gymnolaemata.

Although the ancestry of the Gymnolaemata must now be the most speculative inference of all, the morphology of the class as a whole is slightly more similar to that of the Stenolaemata than to that of the Phylactolaemata (Table 1), even allowing for convergence in some modes of growth. Some uniserial stenolaemates of early Paleozoic age (corynotrypids) are comparable, especially in level of integration, to gymnolaemates of the *Arachnidium-Pyriporopsis* type (BOARDMAN & CHEETHAM, 1973, p. 144; DZIK, 1975; BANTA, 1975; see BOARDMAN, this revision). In contrast, no close comparison between boring ctenostomates and any group of stenolaemates seems to have been suggested.

AUTOZOOID MORPHOGENESIS IN ANASCAN CHEILOSTOMATES

By Geneviève Lutaud

[Laboratoire Cytologie, Université de Paris VI]

Modes of growth and subdivision of initial buds of zooidal series, as well as ontogenetic folds of the undifferentiated wall of the bud, are fundamental manifestations of the diversification of species in Bryozoa. It is necessary therefore to coordinate structural observations on the temporal evolution of zooid shape and of skeletal deposits with biological observations on the underlying cellular layers and their capacity for proliferation and organization. The zooecium is not a simple tegumental protection for the feeding organ, or polypide. It is the persistent and physiologically active organ of the entire functional zooid.

Early anatomists, notably Braem, Calvet, Claparède, Nitsche, Seeliger, and Smitt, established in the late nineteenth and early twentieth centuries the biological details of the phylum. These are: the community of the body wall within a colony, which results from a continuous process of asexual reproduction by budding and implies an incomplete anatomical and physiological autonomy of zooids; and internal budding and periodic renewal of the polypide from the parietal layers of the zooecial compartment, which implies that the digestive epithelium in the adult does not derive directly from the larval endoderm, but from a secondary invagination of zooecial epithelium.

In adult zooids of any shape and functional adaptation, the bryozoan wall includes a pavemental epithelium externally covered by its cuticular and skeletal secretions (Fig. 87), and an inner peritoneal lining limiting the body cavity and including several cellular categories. In stenolaemates and cheilostomates, the superficial cuticle is reinforced by an underlying deposit of calcium carbonate within an organic matrix (Fig. 87,2). Undifferentiated epithelium is columnar in the bud

and restricted areas of tissue proliferation in the adult wall. Both epithelium and peritoneum are present and mitotically active in the bud wall.

Confusion in terminology arose from use of the terms ectocyst and endocyst with different meanings in early descriptions of zooecial wall structure. According to different authors, ectocyst may mean either cuticle only, or include epithelium, or epidermis, and its cuticular and skeletal protection. Endocyst has been used to mean both cellular layers or only the peritoneum. More recent authors have preferred the terms ectoderm and mesoderm to designate epithelium and peritoneum. Although this is justified by the organogenetic potential of the two layers in the bud, ectoderm and mesoderm are embryologic terms that cannot be directly applied to budding and adult tissues before the precise relationship between these tissues and larval layers throughout metamorphosis is established. The general term mesenchyme for a comprehensive designation of subepithelian tissues is simply descriptive of their destiny during morphogenesis, and more appropriate than mesoderm. Here, cellular layers of the wall are designated by the terms epithelium and peritoneum, which account for their cytological character, function, and relative situation in the bud, zooecial wall, and polypide.

The bryozoan wall has a propensity to proliferate whenever space is free and energy is supplied. Primary buds around the ancestrula arise as hollow outward expansions of the parietal layers from distal and lateral areas in the ancestrular wall, which locally retain undifferentiated characters. In gymnolaemates, buds grow in a linear direction and by the development of lateral areas of proliferation that may or may not be able to expand depending on specific budding patterns, physiological and trophic regulations, and intrinsic or incidental obstacles.

Fundamental phylogenetic options based on evolution of zooid shape and colony construction will not be discussed here. However, for a better understanding of the basic process of proliferation, which will be described for Anasca, it is noted that colony construction is regulated by specific differences in relative intensity of distal and lateral budding, and by rhythms of the transverse and longitudinal subdivisions of buds. In the simplest colonial pattern of such ctenostomates as *Arachnidium,* or of such uniserial Anasca as *Pyropora,* new zooids are formed one after another in divergent series from equal distal and lateral buds borne by successive zooids. In Stolonifera, the distal portion of the **stolon,** or stolonal bud, grows in a rapid linear progression while lateral buds are formed with a specific periodicity. Lateral buds develop into autozooids, which are separated from the stolon by a basal septum. Other transverse septa separate segments along the stolon. Division of the stolon at the **growing tip** leads to branching. In Carnosa and some Anasca, multiserial colonies are built when new zooidal series formed from the longitudinal division of the bud are kept together by reciprocal pressure and by adherence of the cuticular and skeletal layers of adjacent series. Lateral proliferation is then inhibited, or restricted to the formation of rows of heterozooids, kenozooids, and pore chambers. Thus, a phylogenetic and morphogenetic difference is apparent between longitudinal and transverse partitions. According to SILÉN (1944a), a unique peripheral evagination, or "common bud," would have first appeared around a solitary ancestral zooid. Then, this "common bud" would have been subdivided by peripheral indentations of the "exterior wall," as a consequence of the formation of several polypides when space became sufficient for their development. Transverse septa, or "interior walls," would have secondarily separated successive zooids along zooidal lines. Longitudinal partitions

are now universally interpreted as the contiguous lateral walls of adjacent zooidal series growing together. Transverse partitions are formed from an invaginated fold of the parietal cellular layers, in the middle of which a skeletal lamina is secreted.

Two principal modes of colonial construction occur among encrusting cheilostomates (HARMER, 1931; BOARDMAN & CHEETHAM, 1969). In the simplest colonial pattern, linear series of zooids in concordant or alternate rows are regularly produced, first from peripheral buds around the ancestrula, then by growth of distal buds of linear series at the periphery of the colony. The formation of lateral buds is inhibited. With increase in surface area and circumference of the growing colony, buds tend to enlarge until their normal width is reestablished by longitudinal subdivision (LUTAUD, 1961). In some species, the longitudinal subdivision occurs earlier and young peripheral zooids bear two distal buds. In species of quincuncial or spiral pattern, every new zooid is formed between two preceding zooids from an axillary bud, which may be either a dominant lateral bud or a distal bud of distorted orientation. The colonial pattern is often complicated by partial development of distal and lateral buds that build an intercalary range of pore chambers around the anterior portion of every fully developed zooid. Then, new zooids are formed from distal or lateral buds arising from distal or lateral pore chambers (*Fenestrulina,* GORDON, 1971a,b). Only the simple mode of **lineal growth** will be taken into account in the following description of the budding process in Anasca.

In bilaminate and encrusting cheilostomates of lineal growth mode, new zooids are formed from the proximal portion of the bud, which is separated from the proliferating distal portion by formation of a new transverse septum. The proximal portion absorbed during the formation of every new zooid varies in length according to the speed of proliferation and to specific zooecial dimensions. The rhythm of transverse divisions depends on both genetic regulation and the abundance of metabolites transmitted by preceding feeding zooids and accumulated in the parietal tissues of the bud. It is a general rule in Anasca that rapid colony growth, with increase in number of feeding units, leads to an increase in length of buds. Growth, being proportional to the number of cells participating in mitosis, is intensified in long buds (LUTAUD, 1961). In slowly growing species, in young colonies, or in unfavorable conditions, buds are not much longer than the average size of a zooid. Except for the tip, they are almost entirely absorbed in the formation of every successive zooid. Rest periods while metabolites are consumed by organogenesis may interrupt proliferation, and budding then is discontinuous. In large colonies, when nutrition and climate are good, proliferation becomes so rapid that the formation of transverse partitions and the organization of newly formed zooids are delayed in comparison to the progression of buds along the substrate. This growth acceleration reaches an exceptional potential in large colonies of *Membranipora membranacea* (LINNÉ), which cover many square feet of kelp frond. In large tongue-shaped colonies, a thick margin of **giant buds** is progressively developed in a dominant growth direction. Several rows of incomplete zooids showing the successive phases of organogenesis extend behind the growing margin. Moreover, the frontal wall without a gymnocyst is simple and transparent. The systematic position of the group, near the divergence of the orders Ctenostomata and Cheilostomata, indicates that this species offers the best possibility to observe basic organizational processes of cellular wall layers before generic diversification introduces parietal superstructures. These are the reasons for choosing this particular species for a study of autozooid morphogenesis in Anasca.

BUD PROLIFERATION IN MEMBRANIPORA MEMBRANACEA

EPITHELIUM AND SECRETION OF CUTICLE

Sagittal sections through a bud of the growing margin in *Membranipora membranacea* show decreasing thickness of the epithelium from tip to proximal septum. Epithelial cells are columnar and high at the tip, as in the bud of other cheilostomates, and become progressively lower in the median region of the bud; epithelium becomes abruptly flat and pavemental in the clearing proximal region, which will be absorbed during formation of a new zooid. At equal distance from the tip, epithelium is thicker on the basal wall than on the frontal wall.

Normally, parietal epithelium in invertebrates is one-layered with a determinate polarity in the orientation of its secretory activity, and with the ability to secrete an external cuticular coating.

Cytological features of columnar epithelial cells at the tip of the bud indicate their intense secretory activity and their participation in the construction of the cuticle (LUTAUD, 1961; TAVENER-SMITH & WILLIAMS, 1972). Density of the cytoplasm and its affinity for standard histological dyes correspond to the development of granular **endoplasmic reticulum**. Mitochondria are abundant around a median nucleus with large multiple nucleoli. These are the normal characters of any embryonic epithelium; however, this cytological aspect in *M. membranacea* corresponds to a relatively stable region of the bud (see Fig. 88,2). In live and preserved specimens, the cytoplasm of the columnar apical cells clears abruptly a short distance beneath the fragile cuticular coating already protecting the tip of the bud (Fig. 88,1). The loose cytoplasmic web of the external pole of the cells beneath the cuticle contains granular secretions and a vesicle of diffuse substances. An important **Golgi apparatus** lies next to this vesicle. Positive reactions to such histochemical tests as the PAS, controlled by the reversible acetylation reaction, indicate that mucopolysaccharides are dominant in the subcuticular secretions and in the internal layer of the cuticle. However, the secretory activity of the cells is diversified. Part of the granular secretions, intermixed with diffuse secretions in the external pole of the cells, shows affinities for stains of proteins. Concomitant protein and polysaccharide secretions, produced by undifferentiated columnar stages of the epithelium at the tip of the bud, are consistent with the hypothesis that the glycoprotein frame of the cuticle is built at this level (see Fig. 88,2). Supple cuticular coating would be later hardened by one of the tanning processes that are known to occur in the superficial organic pellicle of the exoskeleton in other invertebrates.

TAVENER-SMITH and WILLIAMS (1972) studied the structure of the wall by transmission and scanning electron microscopy in the adult and in the bud of *M. membranacea* and of a few other Anasca. According to their observations, the cuticle, which they called "periostracum," is externally bounded by a "triple-unit membrane" consisting of an electron-light layer between two dense layers. The "triple-unit membrane" is internally reinforced by a thicker fibrillar formation.

The cuticular coating cannot be confused with a basal limit of the epithelium. Although no differentiated membrane separates the parietal epithelium from the peritoneal lining, there is no doubt that the basal pole of the cells is their internal extremity, in direct contact with the underlying peritoneal tissues, in the adult as well as in the bud. The implications of this fundamental orientation of the polarity of epithelium must be taken into account when interpreting the superposition of calcified layers in the skeleton of higher cheilostomates. A reversal of the orientation of activity of the epithelial cells would be the adaptation of their external border to an absorption function, as is the case in the digestive tract and in the tentacle sheath.

HYMAN (1958), using the chitosan test of CAMPBELL, found evidence of glycosaminic components of chitin in the organic substrate of the exoskeleton of several cheilostomates and ctenostomates. SCHNEIDER (1963) estimated that chitin represented approximately 10 percent of the exoskeleton in *Bugula,* considering together the cuticle and the organic matrix of the calcified deposits. JEUNIAUX (1963, 1971), using a precise method of enzymatic digestion by chitinolases, confirmed the presence of chitin in the cuticle and in the matrix of various Anasca, at the rate of 3 to 6 percent of the organic material; in cheilostomates, 1.6 percent of this would be free chitin, and the rest would be combined with a glycoprotein substrate.

SUBTERMINAL GROWTH OF THE BUD

Cinematographic observations showed the feeble adhesion of the columnar apical cells to the thin cuticular membrane at the tip of the bud. SCHNEIDER (1958), in a cinematographic study of the phototropic orientation of growth of the autozooidal bud in *Bugula,* observed that the positive response beneath the cuticle was due to displacement of apical cells toward the light source. In *M. membranacea* cultured on glass slides, the progression of the bud, gliding forward along the smooth experimental substrate, is accompanied by a slow but perpetual horizontal oscillation of the columnar epithelial cells at the tip (LUTAUD & PAINLEVÉ, 1961). This movement stirs permanently the fluid secretions of the external poles of cells beneath the cuticle.

Colored markers of vital dye have been applied on the frontal wall of the giant bud of *M. membranacea,* at various levels between the tip and the proximal partition. Change of the marks during growth shows that bud elongation is preapical and that the apex does not proliferate as a **blastema**, where cellular multiplications would be localized and from which new cells would be added to preceding tissues (LUTAUD, 1961). Marks applied at the tip remain concentrated in place. Marks applied in median and proximal portions of the bud are dispersed both by cellular mul-

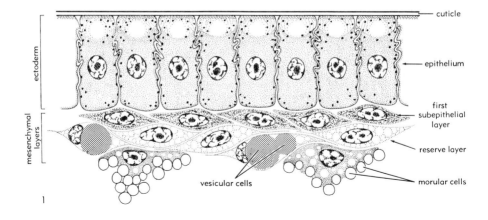

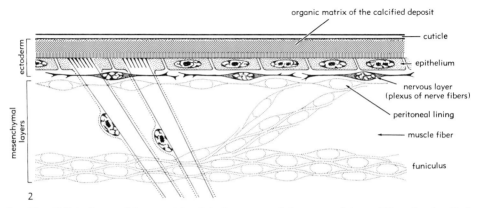

Fig. 87.   Cellular layers of the zooecial wall in Anasca.——*1*. Structure of the undifferentiated wall of
a bud.——*2*. Organization of parietal tissues in the wall of an adult zooid.

tiplication and by pavemental spreading of the epithelium. Analysis of the distribution of mitoses by precise counts shows that cellular multiplication occurs in the epithelium along the entire length of the bud. However, mitotic activity is maximal in the median region for the frontal wall, and in the anterior half of the bud for the basal wall. It is significantly minimal among the columnar apical cells, which participate to a lesser extent in bud elongation. This means that the tip of the bud is pushed forward by proliferation of the preceding regions and by general spreading of parietal tissues in the proximal region. The apical cells that Tavener-Smith and Williams called "archaetype cells" show a remarkable stability of their undifferen-

tiated character and corresponding secretory features. This preapical mode of growth implies a permanent stretching of the preexisting cuticular membrane at the tip of the bud where the secretion of the primary glycoprotein frame of cuticle is presumed to take place. The precise process of cuticle extension at the tip of the bud, under the pressure of growing subjacent tissues, is unknown. According to Tavener-Smith and Williams (1972) ". . .the existing central apical zone of periostracum is gradually pushed aside as newly secreted material displaces it, either physically or by longitudinal impregnation of an adjustable protein-chitin fabric that has not yet polymerized. . . ."

In the giant bud of *M. membranacea,* the

proliferating region behind the apex is extensive. The mitotically active zone, between the tip and the proximal septum, is more restricted in shorter buds of lateral zooidal series diverging from the dominant direction of growth, or in normal buds of other species. The maintenance or appearance of a group of columnar epithelial cells actively secreting glycoprotein substances characterizes any region in the wall capable of proliferation or temporary dedifferentiation. A localized group of columnar cells is formed at the growing tip of spines, in healing areas after a wound, around the ancestrula at the origin of initial buds, at the origin of communication chambers, and, in Stolonifera, at the origin of lateral autozooidal buds.

EVOLUTION OF EPITHELIAL SECRETIONS DURING DIFFERENTIATION

In *M. membranacea,* the progressive lowering of epithelium, from the columnar stages at the tip of the bud to a steady pavemental state in the wall of the adult zooid, is accompanied by a reduction of granular endoplasmic reticulum, by a reduction of the length of mitochondria, and by migration of the Golgi apparatus toward the basal pole (LUTAUD, 1961). Mucopolysaccharides and protein granules are still actively produced by the differentiating epithelium. However, these cytological modifications correspond to an evolution in the nature or proportions of organic substances that first reinforce the primary cuticular membrane, then are deposited on the inner surface of the cuticle and form the organic substrate of the skeleton in calcified regions. Secretion of this organic matrix and concomitant deposition of calcium carbonate persist in the pavemental epithelium of the adult, and the skeleton is reinforced in young adult zooids.

Organic matrix of calcified deposits always remains after cautious decalcification. The matrix shows the histochemical affinities of mucopolysaccharides. Observed with the transmission electron microscope on ultrathin sections through lateral walls in *M. membranacea,* and through the frontal gymnocyst or basal and lateral walls in *Electra pilosa* (LINNÉ), the matrix appears as a thick fibrillar formation lying beneath the internal fibrillar layer of cuticle. The matrix itself consists of two unequal layers differing in the density and orientation of their **fibrillation**: the thickest, next to the cuticle, shows a looser web and would correspond to a primary deposit of calcium carbonate; the internal layer of the matrix next to the epithelium may correspond to newly secreted material. According to TAVENER-SMITH and WILLIAMS, this stratification of matrix indicates that two successive phases occur in the deposit and crystallization of calcite.

In study of *M. membranacea* by polarized light, calcite crystals in lateral walls are first detected in the proximal region of buds, in front of the first transverse partition. In *Bugula,* calcification proceeds on the basal and lateral walls by continuous growth of a calcified lamina, and later extends to the frontal wall to form the gymnocyst (CALVET, 1900). According to SCHNEIDER (1963), who did not discriminate cuticle and matrix, organic fibers and calcite crystals grow together by preapical construction behind the group of columnar apical cells. The frontal wall of a newly formed zooid undergoes invagination of the polypide and development of the tentacle sheath (see Fig. 90,3). Of course, a coherent shield of calcite cannot solidify in the frontal wall while the underlying cellular layers are still undergoing morphogenetic movements, and the extension of calcification to the frontal wall is normally delayed. Consolidation of a calcified layer requires mechanical stabilization of the epithelium.

Without entering into a fundamental discussion of skeletal evolution, and of the significance of the superposition of calcified layers in the frontal wall of Ascophora, an open question of bryozoan biology is how calcium carbonate is produced at the cellular level. Modern cytochemical techniques that are now used in the study of animal secretion of calcium carbonate in other phyla have not been

214 Bryozoa

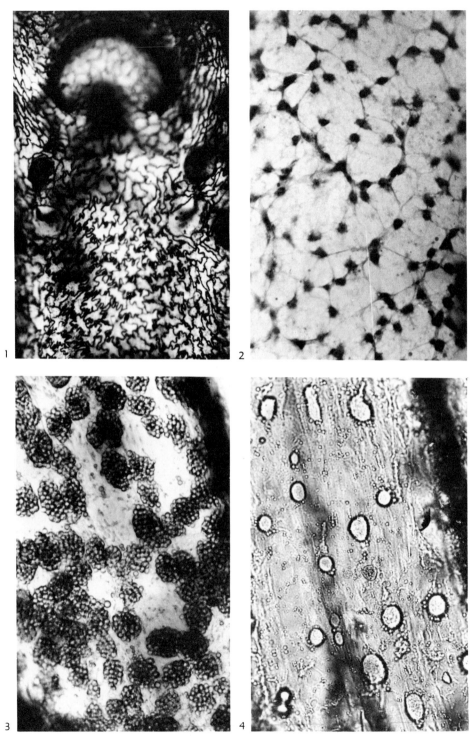

FIG. 88. (For explanation, see facing page.)

applied to Bryozoa (VOVELLE, 1972). The secretory or eliminative process in cellular metabolism, which releases calcium carbonate, is unknown. It has not been established whether organic substances of the matrix and calcite were simultaneously produced and interwoven, or whether ionic calcium carbonate impregnates a preformed organic frame and then precipitates. The evolution in epithelial metabolism within gymnolaemates, which induces calcification in Anasca, then its reinforcement in Ascophora, is entirely unknown.

SUBEPITHELIAL CELLULAR LAYERS

In the adult zooid, peritoneum lining the inner surface of the epithelium in basal, lateral, and frontal walls is a thin network of stellate cells (Fig. 87,2; 88,2). Diffuse endings of parietal funicular strands could be intermixed with the peritoneal network. The peritoneum of the wall includes various cellular categories, among which are mucocytes presumed to liberate acid mucopolysaccharides into the body cavity, and different cells carrying protein granules, granular glycogen, globular glycoprotein inclusions, or lipid droplets (CALVET, 1900; LUTAUD, 1961; BOBIN & PRENANT, 1972). Two kinds of predominant cells, attached to the peritoneal network and to funicular strands, occur in all ectoprocts. These are cells occupied by a voluminous vesicular inclusion, called **vesicular cells** (Fig. 87,1; 88,4), and cells filled with a cluster of refringent spherules, called **morular cells** (Fig. 87,1; 88,3). Amoeboid **phagocytes** are are also liberated into the body cavity (BOBIN & PRENANT, 1957, 1972).

Sections through a bud of *M. membranacea* show that a thick lining of undifferentiated tissue lies beneath the epithelium (Fig. 87,1), extending from the tip to the clearing proximal region, where it is dissociated into longitudinal strands. Subepithelial tissues in the bud are composed of two distinct superposed layers (LUTAUD, 1961; TAVENER-SMITH & WILLIAMS, 1972). An external, first subepithelial layer (Fig. 87,1), lying against the epithelium in cellular membrane to membrane contact, is composed of spindle-shaped cells poor in inclusions. An internal reserve layer (Fig. 87,1) is thicker, especially in the basal wall. It is multistratified and composed of large vacuolated cells carrying chains of lipid droplets and glycoprotein inclusions of various sizes, which tend to concentrate into large globular vesicles.

Vesicular cells, or "vesicular leucocytes" of CALVET, are simply distended cells occupied by a voluminous vesicle showing a positive reaction to the PAS test (Fig. 87,1). This vesicle results from the confluence of smaller glycoprotein droplets in the reserve layer. Vesicular cells are dispersed along peritoneal strands in the clearing proximal region of the bud. They are usually abundant in the basal and lateral walls of newly formed zooids (Fig. 88,4). They are partly consumed during development of the polypide. They appear in adult zooids under high nutrient conditions. Vesicular cells have specific shapes and are commonly subdivided.

Morular cells (Fig. 87,1) are quite different in structure and significance. At the inner surface of the reserve layer, protruding cells of irregular shape are formed. Their dense cytoplasm is progressively invaded by growing vacuoles. Condensation of the vacuolar contents forms spherules that protrude at the periphery of the cells and that deplete the cytoplasm. Finally, a small residual area of cytoplasm, including a distorted nucleus, remains against the cluster of spherules retained within a cytoplasmic film. Morular

FIG. 88. Parietal tissues in *Membranipora membranacea* (LINNÉ).——*1.* Epithelium in the frontal wall of an adult autozooid; silver impregnation, ×150.——*2.* Peritoneum in the frontal wall of an adult autozooid; decalcified whole mount, stained with hematoxylin, ×250.——*3.* Morular cells in the wall of a bud; live specimen, ×200.——*4.* Vesicular cells in the basal wall of a newly formed zooid, live specimen, ×200.

cells are probably liberated into the body cavity at the end of their cytological development. They have not been found in mitotic division, and are probably formed from divisions of other elements in the reserve layer. CALVET (1900) interpreted morular cells as coelomocytes, which he called "leucocytes spherulaires"; however, the nature of the spherules and their function are unknown. The term morular cells, used by BOBIN and PRENANT (1957) because of their shape, seems preferable to leucocyte, which has precise physiological implications. Spherules are refractory to most usual histochemical stains, including PAS. They might be sclerotized proteins, comparable to pigments. Morular cells are numerous in the bud (Fig. 88,3). They are still produced in the wall of the adult zooid. They may be related in some way to metabolism of parietal tissues, particularly active during proliferation. Shape and refringence of the spherules are specific characters.

CALVET (1900) presumed that mesenchymal cells were produced at the tip of the bud from divisions of the columnar apical cells. Superficial observation of live or preserved specimens might give the impression that mesenchymal cells detach from the epithelium at the tip of the bud. However, this interpretation implies that a parietal sheet is formed in the bud during asexual reproduction, after metamorphosis of the larva, and is contradicted by more recent observations. Sections through the bud of *M. membranacea* show that the two subepithelial layers are already present at the tip of the bud, and that mitoses occur in subepithelial tissues from the tip to the proximal partition (LUTAUD, 1961). It seems more probable that epithelium and peritoneum both participate in evagination of initial buds around the ancestrula, and proliferate concomitantly further.

FORMATION OF INTERZOOIDAL WALLS

LONGITUDINAL DIVISION OF THE BUD

Parallel buds of the growing margin in *M. membranacea* grow rapidly in a linear progression while successive zooids are individualized from their proximal extremity by the formation of new **transverse partitions** at regular intervals (Fig. 89,1). However, with increase in colony size, longitudinal divisions occur occasionally in certain enlarged buds in favorable locations, particularly in rounded margins of the colony (Fig. 89,2).

A **longitudinal partition** begins at the tip of the bud as a median notch (Fig. 89,3,4). Cinematographic observation showed that the initial indentation was preceded by a local disturbance in regularity of the apical epithelial cells when the width of the bud exceeded an average dimension (LUTAUD & PAINLEVÉ, 1961). Colored marks applied on the initial notch remained concentrated as the tips of newly formed buds issued from the

longitudinal division. Marks applied along more developed partitions were dispersed in the same way as marks of similar level on the frontal wall of undivided buds (LUTAUD, 1961). This means that lateral double walls do not grow from the tip of the bud to the proximal septum, but elongate distally from their origin by the parallel growth of the two new buds that they separate. This is confirmed by the presence of two contiguous layers of cuticle in the middle of the calcified skeleton, with sand, bacteria, and dirt particles enclosed between.

The consequence of this mode of construction is that interzooidal communications are secondarily pierced in lateral walls (SILÉN, 1942b; LUTAUD, 1961; BANTA, 1969; BOBIN, 1977). In *M. membranacea,* zooidal rows generally alternate. Two pairs of lateral communications are formed in the clearing proximal region of the bud, a little in front of every newly formed transverse partition. The formation of a pore plate is prepared by a

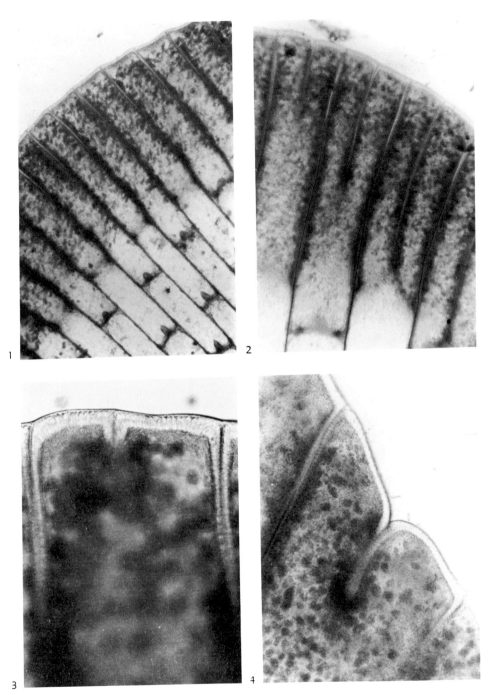

FIG. 89. Longitudinal division of the bud in *Membranipora membranacea*; all live specimens.——*1.* Buds in the growing margin of the colony, ×25.——*2.* Formation of longitudinal partitions, ×40.—— *3.* Initiation of a longitudinal partition at the tip of the bud, ×100.——*4.* Further stage in development of a longitudinal partition, ×80.

unilateral lenticular concentration of epithelial cells, slightly bulging into the wall of the
adjacent bud. The opposite wall immediately
reacts by a coinciding epithelial thickening.
According to Banta, in *Watersipora* the
intercalary cuticle is then dissolved in the
middle of the double epithelial thickening.
A complete pore plate is later secreted in the
perforated area during differentiation of
pedunculate cells of an organ called a **rosette**,
which obstructs every pore in the adult (see
Fig. 98,*1–3*). Silén (1944b) analyzed the
alternation of communications chambers, or
septulae, in several anascans of quincuncial
colonial pattern (*Electra, Flustra,* and *Callopora*). According to his interpretation, communication chambers would have the significance of lateral buds stopped in their
development by the presence and tissue reaction of the adjacent obstructing wall. Development of normal autozooidal buds from
septulae, after accidental or experimental
destruction of adjacent zooids, is often
observed in *Electra,* and has been recorded
in various other cheilostomates. Alternation
of communication chambers, originating on
opposite sides of the common double wall of
two zooidal series, might induce an alternation in the orientation of rosettes across pores
and, thus, alternation of the direction of lateral exchange from a zooid on one side to the
next on the other side.

AUTOZOOID INDIVIDUALIZATION

In *M. membranacea,* the length of buds in
the growing margin of medium-sized colonies is 2 to 5 mm. Daily progression under
experimental conditions on the substrate is
approximatively twice the length of the buds
(Lutaud, 1961). As the average length of a
zooid is between 0.8 and 1.2 mm, a new
transverse partition separates a new zooidal
compartment every 4 to 6 hours.

Calvet working with *Bugula,* and earlier
authors working with other eurystomes,
described the formation of a transverse partition between a new zooid and the distal
bud. The partition proceeded from an annu-

lar invagination of the cellular layers of the
wall, closing like an iris diaphragm. Accel-
erated cinematography showed that the
beginning of the septal invagination in *M.
membranacea* coincides with a maximum
contrast of density in the wall between the
distal proliferating portion and the proximal
clearing portion of the bud (Lutaud &
Painlevé, 1961). A localized disruption in
thickness of parietal tissues may have a part
in the initiation of the partition. The initial
annular fold is asymmetrical and begins on
the basal and lateral walls, later extending to
the frontal wall. This slight asymmetry in
dynamic development of the transverse partition in an encrusting anascan is related to
the unequal thickness of the basal and frontal
walls in the bud, to their divergent organogenetic evolution in the new zooidal compartment, and to the concomitant formation
of a polypide on the distal side of the closing
partition (Fig. 90). According to recent
observations on skeletal growth (Boardman
& Cheetham, 1969), this asymmetrical
development of the transverse wall is more
pronounced in higher cheilostomates, in which
the partition grows from the basal to the
frontal wall.

Interzooidal communications in a transverse wall are formed during closure of the
annular septal fold. Pores, either irregular in
their distribution or grouped in pore plates,
are maintained through the epithelial layers
and median skeletal deposit of the closing
partition when peritoneal tissues, grouped in
the center, are intersected by the epithelial
fold. In *M. membranacea,* peritoneal strands
are grouped in the center of the closing partition in two bundles from which the main
funicular branches are formed. Rosette cells
are differentiated from elements of the funicular strands surrounded by epithelium
(Lutaud, 1961). According to Bobin
(1958a,b), in Stolonifera, undifferentiated
mesenchyme and accumulated mucoid substances first obstruct the central hole of the
growing septum, which separates the autozooidal bud from the stolon. Then, special
cells differentiate unilaterally and insert

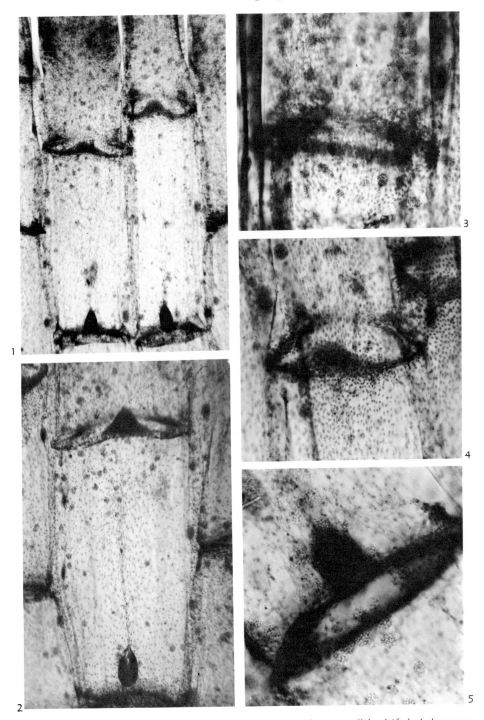

Fɪɢ. 90.   Individualization of the autozooid in *Membranipora membranacea*; all decalcified whole mounts, stained with hematoxylin.——*1.* Formation of transverse partitions at the rear of the buds, ×50.——*2.* Separation of a new zooid, ×100.——*3.* Early stage of formation of the transverse partition, ×125. ——*4.* Formation of the polypidean bud, ×125.——*5.* Closure of the transverse partition, ×150.

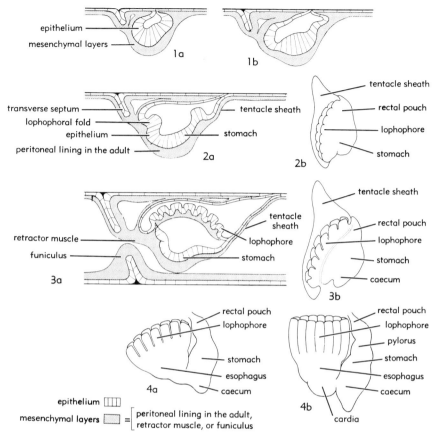

FIG. 91.   Development of the polypide in *Membranipora membranacea*, an encrusting Anasca; based on histological sections and whole mounts.——*1a,b*. Early formative stages of a polypidean vesicle, in section. ——*2a,b*. Formation of the atrial bag, lophophoral fold, and digestive pouch; *a*, in section, *b*, in profile. ——*3a,b*. Differentiation of the tentacle sheath, tentacles, and retractor muscle; *a*, in section, *b*, in profile. ——*4a,b*. Allometric development of the lophophore and subdivisions of the digestive tract, both in profile.

pedunculate prolongations between the epithelial cells of the septal fold, thus impeding locally the secretion of cuticle. In cheilostomates, the manner in which peritoneal strands are invested by the epithelial fold during closure of the partition, and correlatively the number and distribution of pores or pore plates, are specific characters. In some anascans, a single funicular bundle is formed, attached to a central pore plate. In *M. membranacea,* two main funicular strands are attached to two pore plates. In *Electra,* funicular strands are spread across the partition through a range of single pores.

Histological and ultrathin sections show that transverse walls, like longitudinal walls, include two opposite sequences of epithelium and peritoneal lining on either side of the median skeleton. However, structural differences in the skeletal deposits may correspond to morphogenetic differences in the moment and modalities of the initiation of transverse and lateral partitions. Recent observations show that cuticle is lacking in the middle of the transverse wall, and that the calcareous layer is homogeneous, at least in certain species. This might be related to the occurrence of transverse and lateral partitions at different stages of secretory evolution of the epithelium: a lateral double wall begins at the level of columnar stages in the epithelium where secretion of cuticle is presumed to be particularly active, and grows forward between the tips of adjacent buds. Transverse partitions are formed later and grow inward, at the beginning of differentiation of parietal tissues and shortly before calcification.

FORMATION AND LOCATION OF THE FIRST POLYPIDIAN BUD

At the beginning of septal invagination, a cluster of epithelial cells rapidly condenses at its frontal edge and distal side (Fig. 90,*3,4*). This is the **polypidian bud** of the next autozooid invaginating into the body cavity with the internal edge of the closing partition. The initial epithelial cluster, surrounded by the subepithelial layers, quickly increases in volume by cellular multiplication. Then a central lumen is formed by cavitation with the concentric alignment of the epithelial cells (SOULE, 1954; LUTAUD, 1959a). By this time, the polypidian bud has become a double-layered **polypidian vesicle**. An internal epithelium is oriented toward the central cavity and is enclosed by a thickened mesenchymal envelope where lipid and glycoprotein droplets accumulate (Fig. 91,*1*). A constriction separates the epithelial vesicle from the parietal epithelium. However, the polypidian vesicle remains attached to the internal edge of the contiguous partition by the continuity of its mesenchymal envelope with the parietal peritoneal lining.

Simultaneous formation of the transverse partition and the first polypidian bud is fundamental in gymnolaemates. The origin of regenerated polypidian buds during cyclic renewals of the polypide is not precisely established. A regenerated polypidian vesicle is usually found next to a brown body.

Experimental dissociation of the polypidian bud from the concomitant partition has been attempted in *M. membranacea* to understand in detail the determinism of their coinciding formations (LUTAUD, 1961). A reversed orientation of bud proliferation is obtained by removing all recently formed zooids behind the growing margin of the colony. If sufficiently rich in reserves, isolated buds resume growth and formation of transverse divisions after healing of the wound. When an incision is made behind proximal bud partitions, growth goes on in the initial direction. When an incision is made in front of proximal partitions, newly formed buds of reversed orientation are regenerated from the cut, on the proximal side of the next partition. A single polypidian bud is borne on one or the other side of this partition separating the operated bud from the proximal regenerated bud. Of course, traumatism is important. One or several partitions may abort, and a giant zooidal compartment is formed. This does not impede the formation of polypidian buds at regular intervals. Such monstrous zooids are occupied by two or three successive polypides of similar or opposite orientations, each with a normal aperture and tentacle sheath. The retractor muscle of polypides formed without a partition is inserted on a lateral wall. The formation of a polypide depends first on the available space, as suggested by SILÉN (1944a). The formation of a partition and of a polypidian bud occurs at the same moment as differentiation of parietal tissues. The orientation of the polypide is the immediate consequence of the orientation of bud growth.

AUTOZOOID ORGANIZATION

EARLY DEVELOPMENTAL STAGES OF THE POLYPIDE

Development of the polypide has been precisely studied by CALVET (1900) in *Bugula simplex* HINCKS, and by HERWIG (1913) in *Alcyonidium gelatinosum* (LINNÉ). These classical descriptions have been corroborated

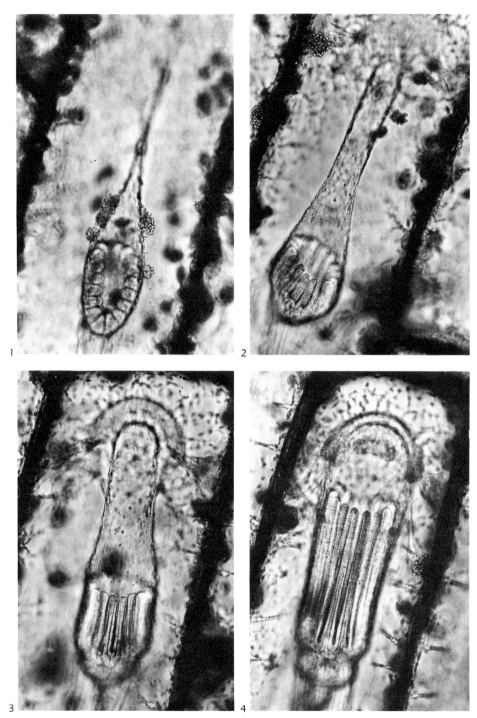

Fig. 92.   Development of the polypide in *Membranipora membranacea*; all live specimens, ×125.——
1. Early stages in differentiation of the lophophore and tentacle sheath.——*2.* Development of the tentacle
sheath and orientation of the lophophore.——*3,4.* Elongation of the tentacles and formation of the
operculum.

by Soule (1954) in Carnosa and Stolonifera, and by Lutaud (1959a,b, 1961; Lutaud & Painlevé, 1961) in *Membranipora membranacea* by histological and cinematographic observations.

In *M. membranacea*, under experimental conditions, complete development of the polypide, tentacle sheath, and aperture requires approximately two days. The first step in organization of the polypidian vesicle is the development of a central lumen with the rapid multiplication and concentric orientation of internal epithelial cells (Fig. 91,*1a*). The next step is the asymmetrical development of the polypidian vesicle, still attached to the frontal wall and to the contiguous transverse partition; the superior, or frontal, region tends to spread while the bottom, or dorsal, region thickens (Fig. 91,*1b*). This is a determinant morphogenetic stage initiating the differentiation of an **atrial bag** from which the tentacle sheath is formed, and of a dorsal pouch from which the digestive tract is formed. Very soon, a slight constriction delimits more clearly the two unequal regions differing in the height of the epithelium, and subdivides the central cavity into a lophophoral atrium and a digestive lumen.

A protuberance next appears at the limit of the atrial and digestive regions. This is the **lophophoral fold** into which the peritoneal layers of the polypidian vesicle penetrate (Fig. 91,*2a*). Meanwhile, the digestive pouch is unequally subdivided by a new constriction into a distal rectal pouch and a stomach pouch (Fig. 91,*2b*). The atrial region extends into a conical bag arising from the base of the lophophoral fold. This atrial bag elongates unilaterally toward the distal end of the zooid along a median tract induced on the frontal wall by invagination of the polypidian bud (Fig. 90,*2*).

The slightly oblong polypidian vesicle has then acquired the shape of a coffee bean, with the furrow of the intestinal lumen opening between the symmetrical pads of the lophophoral fold (Fig. 92,*1*). The tentacle sheath, lophophore, and digestive tract are already clearly delimited. The polypidian vesicle is

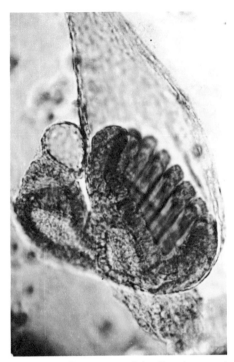

Fig. 93. Differentiation of the digestive tract in *Membranipora membranacea*; live specimen, ×350.

now enclosed in a bag becoming the tentacle sheath, which derives from the frontal portion of the polypidian vesicle and is ontogenetically a polypidian organ. The mesenchymal layers of the polypidian vesicle follow the epithelium in all its successive folds and constrictions; their continuity with the peritoneal lining of the wall is never interrupted during development of the polypide.

DEVELOPMENT OF LOPHOPHORE AND DIGESTIVE TRACT

The next period is characterized by two simultaneous morphogenetic movements, migration of the polypidian vesicle toward the center of the zooecium and orientation of the lophophore toward the apertural area (Fig. 92,*2*). Tentacular stubs are separated by regular slits between secondary folds in the lophophoral protuberance, and then elongate within the atrial bag (Fig. 91,*3*; 92,*3,4*). The

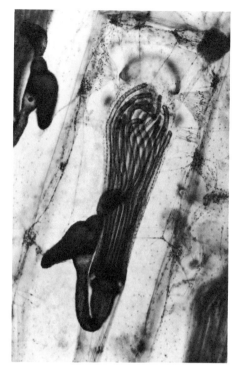

Fig. 94. The adult polypide in *Membranipora membranacea*; decalcified whole mount stained with hematoxylin, ×100.

peritoneal layers, infiltrated into the lophophoral fold, penetrate further into the internal interstice of each tentacular stub. Meanwhile, successive constrictions delimit the different organs of the digestive tract (Fig. 91,4; 93).

In an adult cheilostomate, the successive regions of the digestive tract, from the mouth at the base of the lophophore to the anus opening through the tentacle sheath into the **tentacular atrium**, are the **esophagus**, **cardia**, stomach, **pylorus**, and **rectal pouch** (Fig. 91,4; 94). The esophagus is sometimes mentioned as the **pharynx**, although the term pharynx is usually restricted to the transitional area of the oral constriction. The esophagus is characterized by a vacuolated myoepithelium, which contracts strongly during ingestion (MATRICON, 1973). The esophagus opens into a curved cardial tube, itself opening into the stomach. In certain

ctenostomates, an additional **gizzard** is differentiated from the stomach portion of the cardia. The stomach is prolonged by a blind **caecum**, in which food remains for some time, and opens into a ciliated pylorus. In the pylorus, the remnants of digestion are agglutinated with mucins into a whirling **stylet** by vibratile cilia, before being expelled into the rectal pouch.

The first constriction in the dorsal pouch of the polypidian vesicle separates the rectal pouch from the stomach (Fig. 91,2b). The thick mesenchymal connection that persists between the polypidian vesicle and the contiguous partition, and from which the great retractor muscle of the polypide is formed, retains the posterior portion of the stomach pouch, which elongates into a posterior caecal prolongation (Fig. 91,3b). Meanwhile the esophagus bulges slightly at the base of the lophophore (Fig. 91,4a). Then, the transitional area between esophagus and stomach elongates into a cardial tube, while the pylorus is differentiated from the subrectal portion of the stomach (Fig. 91,4b). Thus, the early subdivisions of the digestive pouch are the esophagus, rectum, and stomach, which have different cytological characters in the adult. The caecum, cardia, and pylorus are localized parts of the stomach pouch. All these subdivisions occur early during development of the polypidian vesicle, while the lophophore develops into a low tentacle crown, and are completed when the young polypide reaches its definitive position in the center of the zooecium. Its further development consists simply of allometric growth of the different organs to their final shape and proportions (Fig. 93, 94).

TENTACLE SHEATH AND FORMATION OF THE APERTURE

The zooidal aperture is secondarily pierced as a result of tension exerted on the frontal wall by development of the tentacle sheath of the first polypide. The aperture is lacking when the polypidian bud aborts, and two apertures are formed in abnormal zooids with

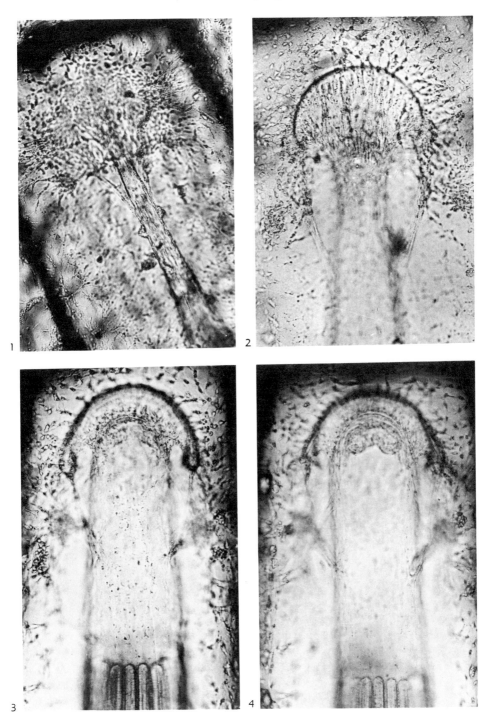

F<small>IG</small>. 95. Formation of the aperture in *Membranipora membranacea*; all live specimens, ×150.——*1. Orientation of parietal tissues around the top of the embryonic tentacle sheath.——2. Junction of the tentacle sheath with the frontal wall and secretion of the edge of the operculum.——3. Differentiation of the vestibule.——4. Differentiation of the diaphgram.*

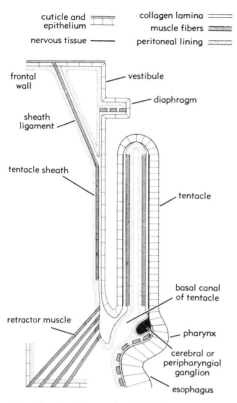

cuticle and ⊤⊤⊤⊤⊤ epithelium

nervous tissue ────

collagen lamina ════

muscle fibers ▨▨▨▨

peritoneal lining ⌇⌇⌇⌇

frontal wall

vestibule

diaphragm

sheath ligament

tentacle sheath

tentacle

basal canal of tentacle

retractor muscle

pharynx

cerebral or peripharyngial ganglion

esophagus

FIG. 96. Continuity of cellular layers in the wall and polypide of adult cheilostomates.

twin polypides.

Displacement of the top of the atrial bag along the median tract, which prolongs it on the frontal wall, is the mechanical consequence of invagination and growth of the polypidian bud. It is not known whether elements from the cellular layers of the frontal wall are absorbed in the growing sheath. Organization of the vestibule, between the brim of the aperture and the muscular diaphragm that closes the tentacular atrium at the top of the sheath, is complex. It has long been presumed that the vestibule, between the operculum and diaphragm, derived from invagination of the frontal wall in the apertural area as a result of traction exerted by the embryonic sheath.

Cinematographic observation of live specimens in *M. membranacea,* confirmed by stained whole mounts, shows that a special sutural process occurs between the conical top of the growing sheath and a subapertural epithelial thickening. The first step in organization of the aperture is the concentric alignment of epithelial cells around the top of the sheath in the distal region of the frontal wall (Fig. 95,1). Then this semicircular area, delimiting the shape of the future operculum, thickens by cellular concentration. A refringent line, corresponding to a localized hypersecretion of cuticle on the surface of the opercular area, appears at the periphery (Fig. 95,2). Cuticle is reinforced at the edge. Meanwhile, the conical top of the growing sheath adheres to the subopercular epithelial pad. The suture proceeds from the center to the corners of the operculum while the sheath enlarges with the elongation of the tentacles (Fig. 95,3). At the end, a double epithelial ring, from which the diaphragm is formed, appears at the precise level of the suture (Fig. 95,4). In the adult, the diaphragm consists of two upper and lower fans of epithelial folds enclosing peritoneal cells, mucocytes, and the fibers of a sphincter muscle. By the time that operculum, vestibule, and diaphragm are completed, the adult polypide is already active and striving for protrusion. A slit appears along the hardened edge of the operculum, which bursts open under repeated pressure from the lophophore (LUTAUD & PAINLEVÉ, 1961).

The relative positions of epithelium, musculature, and peritoneum in polypidian organs result from the ontogenetic continuity of parietal and polypidian layers (Fig. 96). The wall of the tentacle sheath includes the complete sequence of epithelium on the atrial side, muscle fibers, and peritoneal lining on the coelomic side. At its junction with the base of the lophophore, the epithelium of the tentacle sheath is in continuity with the aboral cellular rows of the tentacular epithelium; the oral rows of the tentacular epithelium are in continuity with the digestive epithelium in the pharyngeal area. In the polypide and in the tentacle sheath, the epithelium is supported by an elastic lamina of collagen that does not exist in the zooecial wall. The collagen lamina is not a basal membrane of the epithelium, for collagen is presumed to be of mesenchymal origin in invertebrates. The collagen lamina is reinforced at the insertion of the great retractor muscle on the base of the lophophore and inside tentacles, where it forms an elastic tube limiting the internal tentacular canal. The peritoneal lining forms a continuous envelope in the tentacle sheath and digestive tract, and joins the reticular peritoneal lining of the wall at the aperture. Annular and longitudinal muscle fibers in the digestive tract, and longitudinal muscle fibers in the tentacle sheath, lie on the coelomic side of the collagen lamina. Muscle fibers are imbedded in the peritoneal lining.

In tentacles, musculature and peritoneal lining penetrate into the collagen tube (Fig. 96). Peritoneal tissues fill the internal space, except for a narrow central lacuna. Muscle fibers lie against the collagen in two oral and aboral groups, in prolongation of muscle fibers of the digestive tract and tentacle sheath. The internal lacunae of all tentacles open at the base into a circumoral lacuna called the **basal canal** of the lophophore. The tentacular and basal lacunae derive from the initial space of the lophophoral fold of the polypidian vesicle, and are enclosed within the peritoneal lining. They are prolongations of the body cavity into the lophophore. The question arises whether the basal canal of the lophophore is closed in the adult, or freely communicates with the body cavity. Communication occurs at least during breeding periods for the passage of eggs and spermatocytes, which are formed in the body cavity and liberated into the tentacular atrium by means of the lophophoral canals.

In the tentacle sheath, tentacles, and digestive tract, the epithelium has lost cuticle and acquired **microvilli** on the external border of cells, which indicate a potential absorption function. It acquires also vibratile cilia in specialized regions of the tentacles and digestive tract. A fundamental function of epithelium in Bryozoa is the secretion of mucoid substances, which form the substrate of the cuti-

FIG. 97. Funiculus in the autozooid of *Electra pilosa* LINNÉ; decalcified whole mount stained with hemalun, ×90.

cle and matrix in the exoskeleton. In the vestibule, epithelium is still protected by a supple cuticular coating. In polypidian organs, the epithelium liberates mucopolysaccharides on the outer surface of the tentacles and into the lumen of the digestive tract. These function in prey capture, protection of tentacles, and digestion. The tentacle sheath is more than a tissue connection between the polypide and the wall. Because of its absorbent or secretory potential, it may have important functions in the physiology of the entire zooid, particularly in respiratory or excretory exchanges between seawater and the coelomic cavity.

FORMATION AND FUNCTION OF THE PERITONEAL-FUNICULAR SYSTEM

Two functionally and topographically distinct tissues derive from differentiation of the two subepithelial layers of the bud wall, the peritoneal-funicular system and the musculature.

In adult zooids, a **funiculus** comprises thick funicular strands (Fig. 87,2), which extend across the body cavity and join the digestive tract to every interzooidal communication in the transverse and lateral walls. In Stolonifera, the funiculus is a simple axial strand in stolons, with branches to the basal septum of autozooids; in autozooids, it joins the basal septum to the stomach and caecum. In multiserial cheilostomes, funicular strands are multiple and ramified (Fig. 97). The main funicular ramifications, attached to pore plates, extend from proximal to distal partitions. They lie between the digestive tract and basal wall, and wrap the stomach, caecum, and pylorus along the way. Divergent branches join every pore plate in lateral walls. In adjacent zooids, correspondent ramifications attach to the other side of the pore plates. Thus, the funicular system extends throughout the colony across interzooidal pores (Fig. 98,1,2). Funicular ramifications are present in heterozooids and kenozooids. Parietal funicular strands extend and ramify in the wall of the zooid. In encrusting Anasca, particular parietal strands of funiculus run in vertical walls at the periphery of the zooid, from one pore plate to the next. Funicular ramifications in the wall and across the body cavity persist during cyclic renewals of the polypide.

Funicular strands are made of spindle-shaped cells (Fig. 98,4) of feeble cohesion, free to diverge and join crossing or adjacent strands. At their junction with the polypide, they simply fuse with the peritoneal lining of the digestive tract, of similar nature and origin. Funicular cells are characterized by coiled formations of granular endoplasmic reticulum, which indicate an intense synthesis activity. They carry lipid droplets and diffuse glycoprotein substances. In *M. membranacea,* the peripheral parietal strand, at the base of vertical walls, is so charged with diffuse reserves that it becomes a canal with the formation of a central lacuna filled with glycoprotein material (LUTAUD, 1961). This phenomenon of accumulation occurs in other species in rich nutrient conditions, or at certain periods of the life cycle.

The funicular system is presumed to transmit metabolites from the digestive tract to the wall and from one zooid to another through interzooidal pores. The rosettes (Fig. 98,3), which obstruct every pore, consist of a group of dumbbell-shaped cells. The nucleated portion of these special cells is on one side of the pore plate and extends into a narrow pedunculate prolongation; the cell extends through the pore and swells on the other side into an anucleated blister in the adjacent zooid (BANTA, 1969; BOBIN, 1977; GORDON, 1975). In *Membranipora, Electra,* and *Watersipora,* two special cells occupy every pore. In Stolonifera, a single rosette of several special cells occupies the central perforation of each stolonal or autozooidal partition. Rosettes are surrounded on both sides by a semicircular row of limiting cells by which funicular strands are attached to pore plates. Within this cellular boundary, diffuse glycoprotein material accumulates in a lacuna around the nucleated portion of the special cells. These special cells would absorb metabolites by the microvillous border of their nucleated portion, which is presumed to be on the transmitting side of the pore plate. Metabolites are released on the anucleated side of the special cells. According to BOBIN (1958a,b), orientation of the special cells may be reversed when the direction of need for

---

FIG. 98.   Structure of funiculus and interzooidal communications in *Electra pilosa.*——*1.* Funicular strands and their junction with pore plates in a lateral wall; decalcified whole mount stained with hematoxylin, ×200.——*2.* Pore plate in lateral wall; decalcified whole mount stained with hemalun, ×350. ——*3.* Rosette cells through a range of pores in a transverse partition; silver impregnation, ×500. ——*4.* Structure of funicular tissue; histological section stained with hematoxylin, ×500.

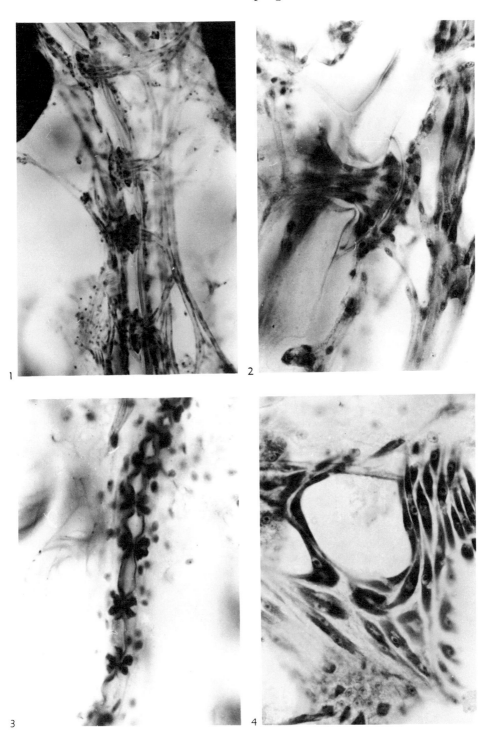

FIG. 98. *(For explanation, see facing page.)*

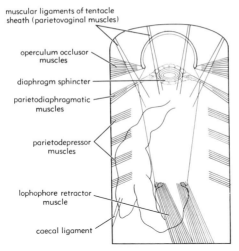

muscular ligaments of tentacle sheath (parietovaginal muscles)

operculum occlusor muscles

diaphragm sphincter

parietodiaphragmatic muscles

parietodepressor muscles

lophophore retractor muscle

caecal ligament

FIG. 99. General external muscle pattern in encrusting cheilostomates.

energy changes between adjacent zooids.

Funicular strands are formed from the reserve layer of the undifferentiated wall of buds (Fig. 87,2). In *M. membranacea,* the cells of the two subepithelial layers are segregated into separate strands in the clearing proximal region of the bud, which is absorbed during formation of a new zooid. Different thickness of the basal and frontal walls in the bud results in a different organization of subepithelial tissues in basal and frontal walls of the zooid. Segregation of peritoneal strands is multidirectional in the frontal wall under combined developmental tensions of the tentacle sheath and aperture; the peritoneal lining becomes reticular. On the basal wall, the thicker reserve layer is dissociated into bundles of anastomosing, longitudinal, funicular strands, which are partly detached from the wall by partitions. The junction of the digestive tract and funiculus occurs early, proceeding either by simple adhesion of the bottom of the polypidian vesicle to the underlying funicular bundles or by fusion of the dorsal funicular bundles with peritoneal strands in the mesenchymal connection that attaches the polypidian vesicle to the center of the proximal partition. The precise destiny of the first subepithelial layer and its contribution to the

peritoneal network of the wall are not clearly established; however, reserve cells, vesicular cells, and morular cells, originating from the reserve layer, are attached to the peritoneal lining of the adult wall.

FORMATION OF MUSCULATURE

CALVET (1900) and earlier authors described how fibers of the retractor muscle, inserted on the proximal partition and at the base of the lophophore, are formed from **myocytes** in the mesenchymal connection that persists between the polypidian vesicle and the partition. Myocytes are stretched and separated from peritoneal-funicular strands during growth of the polypidian vesicle toward the center of the zooecium. According to CALVET, a muscle fiber is formed from two associated myocytes. The retractor muscle is contractile early in development, and the young polypide is capable of sudden retractions before differentiation and elongation of the muscle fibers are completed.

The general muscle pattern in gymnolaemates includes **external muscles** inserted on transverse or lateral walls and polypidian muscles formed within the polypidian vesicle. External muscles include the retractor muscle of the polypide and the parietal muscles (Fig. 99). The parietal muscles include the **parietodepressor** and the **apertural muscles.** In Anasca, the parietodepressor muscles are inserted on lateral walls and on the flexible frontal membrane, at regular intervals around the opesia. In Ascophora, they are inserted on lateral walls and on the ascus beneath the calcified shield of the frontal wall. In Stolonifera, they are inserted on the lateral and abanal sides of the tubular autozooid. Their contraction exerts a pressure on the polypide and incites the protrusion of the lophophore. There are two pairs of apertural muscles. One is the occlusor muscles of the operculum in cheilostomates or of the collar in ctenostomates. The other pair is the **parietodiaphragmatic muscles,** which insert on lateral walls and on the diaphragm at the junction of the tentacle sheath and vestibule.

In *M. membranacea* and *E. pilosa,* fibers of the retractor, parietodepressor, and apertural muscles are not striated; however, in the avicularia and vibracula of other species, homologues of the occlusor muscles of the operculum, which animate the mandible or seta, are striated.

Parietodepressor muscles originate at the periphery of the opesia from small groups of myocytes at the corner of the lateral and frontal walls. Myocytes are stretched and detached from the wall by development of the tentacle sheath and aperture. Apertural muscles are formed from similar groups of myocytes at the level of the apertural area. Occlusor muscles of the operculum are formed from a distal pair of thick mesenchymal bridges stretched between lateral walls and the base of the opercular area. Parietodiaphragmatic muscles are formed from minor groups attached at the junction of the conical top of the embryonic sheath with the frontal wall; their frontal insertion is later drawn in during development of the vestibule.

The polypidian muscles adhere to the collagen lamina along their entire length. In adults, the esophagus is surrounded by an almost continuous layer of large annular muscle fibers. In the pharynx, muscle fibers form a sphincter around the mouth (Fig. 99). Thinner annular muscles, overcrossed by longitudinal fibers, surround other subdivisions of the digestive tract.

In the tentacle sheath, the muscular layer consists of parallel longitudinal fibers arising at some distance from the base of the lophophore, and of a few annular fibers grouped in the sphincter of the diaphragm. Longitudinal muscle fibers of the tentacle sheath are collected below the diaphragm into suspending ligaments attached at the base of the distal transverse partition, and on the frontal wall near the aperture. Ligaments of the tentacle sheath have been designated by CALVET (1900) as **parietovaginal muscles.** In ligaments, muscle fibers are imbedded in collagen within a tubular peritoneal envelope; epithelium is lacking (Fig. 96). Ligaments are formed from early mesenchymal anastomoses between the top of the embryonic sheath and the wall. Their contraction lifts the polypide toward the aperture during protrusion, and completes the action of the parietodepressor muscles. Another ligament of identical structure links the caecum to the nearest lateral wall and retains the digestive tract during protrusion.

In *M. membranacea* and *E. pilosa,* annular muscle fibers of the esophagus and intracellular myofibrils of the esophageal epithelium are striated (MATRICON, 1973). Internal muscles of the tentacles are also striated. Longitudinal and annular muscle fibers in the tentacle sheath, sheath ligaments, and diaphragm are smooth.

The precise origin of muscle fibers during differentiation of parietal and polypidian organs is not established; however, in all polypidian organs, the position of muscles between the epithelium and peritoneal lining suggests that myocytes are formed from the first subepithelial layer of undifferentiated mesenchyme of the bud.

NERVOUS COORDINATION OF PARIETAL AND POLYPIDIAN ORGANS

The tentacle sheath is the substrate of important **peripheral nerves** that arise from the cerebral ganglion at the base of the lophophore and serve the aperture and zooecial wall. Motor and nonmotor nerve endings, or nerve cells, are found in extensive or restricted dispersion in the free, external, zooidal wall. Two coexistent pathways of parietal innervation, of different degree of differentiation, occur in Bryozoa. They are either clearly separate in their topographical pattern or intermixed in their connectives to or from the cerebral center. The first consists of motor and sensory endings borne by parietal branches of the great mixed nerves of the tentacle sheath, which exist in all gymnolaemates. The second is a nerve net, or plexus, of more primitive character, which has been found at present in the wall of phylactolaemates and gymnolaemates. A similar plexus has been found by HILTON (1923) in the body wall of entoprocts.

CEREBRAL CENTER AND INNERVATION OF THE POLYPIDE

The **cerebral ganglion** lies in the oral constriction between the base of the lophophore and the esophagus on the anal side of the polypide. An annular ganglionic belt, called the **peripharyngeal ganglion**, lies between the basal canal of the lophophore and the epithelium of the pharynx. The cerebral ganglion and its circumoral prolongation, as well as lophophoral and visceral nerves, are basiepithelial and lie between the epithelium and the collagen lamina (Fig. 96). The cerebral and peripharyngeal ganglia are formed early in the polypidian vesicle from a secondary fold of reversed orientation at the base of the lophophoral fold.

In *Electra,* the cerebral ganglion includes 40 to 50 cells, among which are neurons of different kinds, secretory cells, and investing nonnervous elements (LUTAUD, 1977). Neurons are arranged in a fixed pattern around a deep core of intermixed fibers and intracerebral connectives. Arrangement of the cerebral cells is constant in *Electra*; however, specific variations occur in different cheilostomate families. Nevertheless, three areas always remain distinct: (1) a central aggregate with topographical potential for general cerebral coordination; (2) the distal brim where chains of neurons in the peripharyngeal ganglion are initiated and sensory nerves from the lophophore are received; and (3) symmetrical proximal clusters including giant neurons from which the main peripheral nerves arise.

Two pairs of sensory and motor nerves along every tentacle arise at regular intervals from the peripharyngeal ganglion (Fig.

100,2). Twin nerves arise from branched intertentacular stems on either side, and converge to run along the oral edge of every tentacle. These are presumed to be sensory nerves to which sensory cells in the tentacular epithelium would be sporadically attached (MARCUS, 1926). They run beneath two rows of monociliated epithelial cells along the oral edge of the tentacle, which are presumed to have a tactile function (LUTAUD, 1973). Another pair of median oral and median dorsal nerves arise in the axis of the tentacle. Although they lie on the epithelial side of the collagen lamina, their pathway coincides with the position of the internal tentacular muscles, and they are presumed to be either motor or mixed.

The digestive tract is served by a median dorsal visceral nerve along the esophagus, and by a pair of branched lateral visceral nerves arising from a small group of ganglionic cells below the cerebral ganglion (Fig. 100,2). Short connectives and anastomoses link the visceral stems to nervous strands of the peripharyngeal ganglion, and provide a plausible pathway for coordination of lophophore activity and of contractions and peristaltic waves of the digestive tract.

INNERVATION OF THE APERTURE AND FRONTAL WALL

In gymnolaemates, two pairs of peripheral nerves arise from the proximal cellular clusters of the cerebral ganglion, and emerge together through lateral openings. They first diverge, then meet again and fuse on their way toward the aperture along the tentacle sheath (Fig. 100,1). Equivalent peripheral nerves, of slightly different pathway, exist in phylactolaemates. The peripheral nerves run in the tentacle sheath on the peritoneal side of the collagen lamina, imbedded in the peritoneal lining. They are, on either side, a thick fibrillous strand directly joining the aperture and a thin three-branched motor nerve, called the **trifid nerve** (BRONSTEIN, 1937). The three branches of this motor nerve are: (1) a branch around the pharynx to the insertion of the

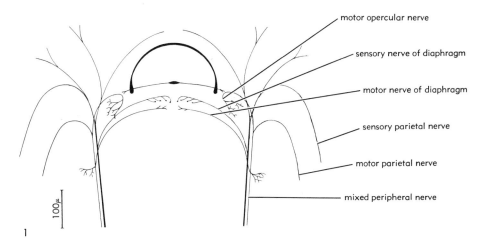

motor opercular nerve

sensory nerve of diaphragm

motor nerve of diaphragm

sensory parietal nerve

motor parietal nerve

mixed peripheral nerve

100μ

1

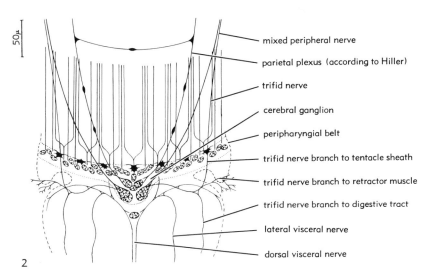

mixed peripheral nerve

parietal plexus (according to Hiller)

trifid nerve

cerebral ganglion

peripharyngial belt

trifid nerve branch to tentacle sheath

trifid nerve branch to retractor muscle

trifid nerve branch to digestive tract

lateral visceral nerve

dorsal visceral nerve

50μ

2

FIG. 100.   Main nervous pathways in anascan cheilostomates.——*1.* Parietal and apertural branches of main peripheral nerves; frontal view.——*2.* Cerebral ganglion and innervation of the polypide, dorsal view.

retractor muscle; (2) a visceral branch bending down to the esophagus; and (3) an axial branch bending up along the sheath and joining the **direct nerve**. Below the junction, an annular ramification of the axial branch surrounds the tentacle sheath at the level of the basal extremities of the longitudinal muscle fibers of the tentacle sheath and sheath ligaments.

In Anasca, the great **mixed nerves**, formed by conjunction of the direct and trifid nerves on either side, ramify at the top of the sheath into three couples of motor and sensory branches (Fig. 100,*1*). A motor ramification innervates first the parietodiaphragmatic muscle, then joins the sphincter in the diaphragm; a corresponding sensory ramification develops arborizations in the upper folds of the diaphragm. Parietal ramifications join the zooecial wall next to the aperture. At this level, an opercular motor branch to the occlusor muscles of the operculum diverges, while a transverse nonmotor ramification follows the hinge of the operculum. Twin proximal ramifications around the opesia, or motor parietal nerves, run through frontal insertions of all parietodepressor muscles. In Electridae, parallel nonmotor branches around the opesia expand into diffuse fibers with terminal knobs or cells at the base of every marginal spine (LUTAUD, 1977). Distal fibers with peculiar cellular endings of undetermined function join the base of the distal partitions. Nerve strands in the zooecial wall run between the epithelium and the peritoneal lining.

Thus, the main nerves of the tentacle sheath carry motor impulses to all muscles working together during lophophore protrusion and retraction. They are also the probable pathways of an unelaborate perception of the environment at the level of the free external wall and at the entry of the tentacular atrium. However, it is not established whether superficial endings of nonmotor parietal nerves are nerve cells or epithelial receptors.

HYPOTHETICAL PATHWAYS OF COLLECTIVE INTERZOOIDAL INFORMATION

The observations of GERWERZHAGEN (1913) and MARCUS (1926, 1934) brought evidence of the existence of a nerve net in the body wall of ectoprocts. In phylactolaemates, both GERWERZHAGEN in *Cristatella* and MARCUS in *Lophopus* observed a network of large multipolar cells, selectively stained by the vital methylene-blue dye after ERLICH, in the tentacle sheath and external wall. MARCUS found a similar plexus in the wall of tubular zooids of the stoloniferan ctenostomate *Farella repens* (FARRE), spreading from a nonmotor parietal ramification of the main tentacle sheath nerves. This parietal nerve net has been recently observed again in the autozooids and stolons of *Bowerbankia gracilis* LEYDI (LUTAUD, 1974). The large meshes of the net-

work, which is probably continuous all over the colony across interzooidal pores, are brightly stained by methylene blue in orthochromatic tones and cannot be confused with the underlying peritoneal network of quite different appearance and affinity for the stain. According to BRONSTEIN (1937) a similar plexus would exist in the external wall of all gymnolaemates. However, in the encrusting carnosan *Alcyonidium polyoum*, as well as in *Electra pilosa* and other malacostegans, this unorganized network is replaced in the frontal wall by a more differentiated set of nerve fibers spreading from the sensory branch of the parietal nerve, with cellular endings around the orifice and at the periphery of the frontal area (LUTAUD, 1981).

Another form of methylene-blue positive network coexisting with the superficial developments of the parietal nerve in the frontal wall was discovered by HILLER (1939) in *Electra pilosa* (confirmed by LUTAUD, 1969). In Electridae and other encrusting anascans, methylene-blue staining reveals a linear chain of bipolar cells running at the base of the interzooidal partitions (Fig. 101). This internal pathway around the basal wall is linked to prominent cells in the central cellular cluster of the cerebral ganglion by twin connectives of similar structure running on the dorsal side of the tentacle sheath to join the peripheral plexus filament at the distal corners of the zooid. Short transverse branches penetrate into every septular chamber on its concave side. Interzooidal bonds, which are made of a modified plexus cell replacing a rosette across one pore of every lateral pore plate and certain pores in transverse partitions, periodically link the parallel filaments running on both sides of the common partitions of adjacent zooids (LUTAUD, 1979). Ultrastructural investigation confirms the nervous nature and cellular structure of the network. No homologous nervous pathway in a similar location was found in the carnosan *Alcyonidium polyoum*.

Thus, the differentiation of two separate pathways of parietal innervation is observed in encrusting cheilostomates. One is a group

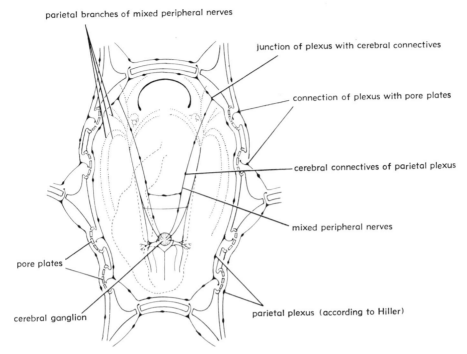

parietal branches of mixed peripheral nerves

junction of plexus with cerebral connectives

connection of plexus with pore plates

cerebral connectives of parietal plexus

mixed peripheral nerves

parietal plexus (according to Hiller)

pore plates

cerebral ganglion

FIG. 101. Pathways of parietal innervation in *Electra*; dorsal view.

of nerve fibers with sensory endings in the frontal area, which informs the nervous center of each polypide about external variations. The other is a dorsal pathway of plexus structure along interzooidal partitions. By its colonial continuity across certain pores in communication organs and its connection with the cerebral centers of all the polypides, the plexus is the most probable pathway for a primitive collective communication among groups of zooids within a colony. The interzooidal transmission of experimental stimuli in all directions along zooidal rows was recently demonstrated in malacostegans by electrophysiological experimentation (THORPE,

SHELTON, & LAVERACK, 1975b). However, the colonial plexus cannot be interpreted from its cytological features as a colonial nerve, which would imply oriented impulses from one zooid to the next and the incomplete autonomy of the cerebral center of a polypide. Colonial coordination should rather be understood as a simple consequence of the morphogenetical continuity of cellular layers of the wall in bryozoans, and as response to a diffuse collective perception of the general activity of polypides, which might be reinforced or counterbalanced by individual perceptions of the medium at the level of the frontal wall.

CONCLUSIONS

The complete succession of soft tissues in the wall of the adult zooid, from exoskeleton to body cavity, is epithelium, nervous layer (either diffuse peripheral endings or plexus),

musculature, and peritoneal-funicular strands. Muscles and funiculus are partly detached from the wall by formation of partitions and by development of the polypide.

Except for exoskeletal layers secreted by epithelium, a similar succession is present in all organs of the polypide as a result of its formation from an investigation of soft layers of the wall. The body cavity and its prolongations into the lophophore, with the function of a coelom, are enclosed by mesenchymal tissues.

Except for musculature that is differentiated after individualization of the autozooid, all layers of the wall contribute to functional coordination of the colony. Continuity of epithelium is not interrupted during the formation of partitions. Coalescence of exoskeletal layers in transverse and double lateral walls maintains the mechanical cohesion of zooidal units. Nervous coordination of the colony might be possible by the ontogenetic continuity of a parietal nerve net, closely associated with epithelium, which is already present in the wall of the bud. Interzooidal continuity of the funiculus is the consequence of the initial continuity of subepithelial tissues in the bud, maintained or secondarily reestablished during formation of septal folds and during longitudinal division of the bud.

The peritoneal-funicular system, with a trophic function, does not have an equivalent in other invertebrates. Constant interaction of its different organs regulates the equilibrium between individual and colonial energy needs. The funiculus carries excess metabolites from feeding autozooids toward regions in the colony where the claim for energy is maximal, such as the growing margin, rows of active heterozooids, or rows of autozooids during renewal of their polypides. Orientation of the filtering apparatus in communication pores is related to directions of budding and to changing needs of zooids. Funicular pathways are the means of the colonial community of reserves. The transit of metabolites is not the only function of the peritoneal-funicular system. The peritoneal lining is also presumed to have the function of wall nutrition and of storage of reserves, in the adult zooid as well as in the bud. Intensity of proliferation and secretory activity of

the epithelium, and therefore the metabolism of calcification, depend on the availability of these reserves. The peritoneal lining and funicular tissues also contribute to secretion and equilibrium of the coelomic medium by such specialized elements as mucocytes, phagocytes, and morular cells.

The zooecial wall contributes to physiological regulation of the entire zooid by its secretions toward external and internal media, and by storage and utilization of reserves. There are presumptions that it has also a function in respiration, either through the flexible frontal membrane in Anasca, or through frontal pores, or through uncalcified developments of the frontal wall in Ascophora. It is also the durable organ of the entire zooid. Persistence of its organogenetic potential allows renewal of the polypide when it becomes poisoned by waste products of its own digestive function. When the polypide degenerates, resorption of the tentacle sheath interrupts the continuity of parietal and polypidian tissues at the level of the aperture. The retractor and parietodiaphragmatic muscles are destroyed; the opercular occlusor and parietodepressor muscles remain. Funicular ramifications surround the brown body formed by encysted remnants of the degenerated polypide; they are later reconnected with a new polypidian vesicle. The early stages of a regenerating polypidian vesicle, and the restoration of the junction of a new tentacle sheath with the preexisting aperture and opercular muscles, have not been closely investigated. The superficial plexus of the ctenostomates, and the internal plexus of *Electra,* persist during renewal of the polypide; however, their connectives to the cerebral ganglion are broken. Plexus, nerve endings, or epithelial receptors in the wall are secondarily reconnected with the new peripheral nerves arising from the cerebral ganglion of the regenerated polypide.

Individuality of the autozooid is maintained by the occurrence of a polypide and by the presence of a central nervous system. Although it is not necessary to the survival and nutrition of the zooecium, the polypide is more than a feeding organ, for it includes the cerebral center, which coordinates the functional activity of zooecial and polypidian organs, and controls all relations with the environment. Nervous extensions are everywhere present in the wall and in the polypide with the same ubiquitous dispersion as in any other animal. The zooecial wall is not inert, and parietal innervation is controlled by the cerebral ganglion. The plexus structure of part of the parietal innervation in certain families, and the absence of any protective sheath around nerve bundles, are primitive characters that may have some phylogenetic value. The parietal plexus, when it exists, shows a tendency to form linear pathways related to shape of the zooid and to pattern of colonial construction. Perception is elementary, by means of dispersed or ordered nerve endings or epithelial receptors in the wall and lophophore. Although rudimentary in its organization, the cerebral center shows delimited districts of innervation that are found in distinct ganglionic lobes in other lophophorates. Potential for integration of individual perceptions of the environment may overcome, compensate, or reinforce interzooidal information, and allow autonomous behavior of the zooid.

Anatomical pathways of the functional unity of the autozooid have been described here in Anasca with their ontogenetic rela-tionships. These structures are fundamental in the gymnolaemates, with minor specific variations; and, based on the soft parts, the group appears homogeneous. Fundamental characters now used in classification are: (1) laws of colony construction; (2) presence or absence of calcification with correlative modification of the chitinous protection of the aperture, collar (in ctenostomates), or operculum (in cheilostomates); and (3) degree of calcification of external walls, with displacement of parietal muscle insertions on a compensative internal fold of epithelium when calcification extends to the frontal wall. Less apparent variations in soft parts are observed in different orders, families, and genera, which could be used for systematic differentiation. The most evident are the subdivisions of the stomach and the number of tentacles. An increasing complexity of the funiculus and variations in direction of the transit of reserves may be determinant in evolution of the budding mode. From Stolonifera to higher cheilostomates, an increasing complexity in structure of the cerebral ganglion and in nervous ramifications is also noticed. In ctenostomates, and in different families of cheilostomates, evolution of cellular categories of mesenchyme, particularly of protein inclusions, is observed. Modifications in epithelial metabolism are at the origin of skeletal evolution.

ULTRASTRUCTURE AND SKELETAL DEVELOPMENT IN CHEILOSTOMATE BRYOZOA

By Philip A. Sandberg

[Department of Geology, University of Illinois, Urbana]

Calcification of cheilostomate bryozoan skeletons is often rapid, so that intermediate morphologies can be difficult to observe, even in living colonies. Most of the ontogenetic gradient is in the fragile, distal colony margin, which is most susceptible to breakage, abrasion, or corrosion. Banta (1970, p. 52–53) and Cook (1973b, p. 259) noted that it is difficult to determine ontogenetic differences in fossil or dry material lacking cuticle or epidermal layers, or even in dissected mature colonies. Nevertheless, the data available in fossil cheilostomates are the skeletons, without soft tissues, but with diverse, distinctive morphologic features and ultrastructural details. Even many modern species are known only from dry material. Morphologic features, mainly of the frontal wall, have been of major importance in taxonomic and functional studies of cheilostomates.

Bryozoan skeletons, like mollusk and brachiopod shells, grow by continuous additive calcification. Earlier ontogenetic states are preserved in each zooecium, but are mostly covered by later skeletal increments. As noted by Banta (1970), major developmental differences may be obscured by the similar mature morphology of such ontogenetically different groups as **umbonuloids** (ascophorans with the frontal shield formed by calcification on the lower side of an **epifrontal fold**) and **lepralioids** (ascophorans with the frontal shield formed as a cryptocyst). Despite morphological similarities even at the microscopic structural level in the mature state, bryozoan skeletons developed by different ontogenetic modes should show recognizable differences at the ultrastructural level. In this discussion I briefly characterize skeletal ultrastructure among cheilostomates and consider to what degree ultrastructure can provide data for ontogenetic reconstructions and for taxonomic-phylogenetic inferences. First, however, it is advisable to consider whether studies of skeletal ultrastructure have actually accomplished their stated objectives. Following are some major reasons for ultrastructure studies and at least a preliminary evaluation of the usefulness of such studies.

ULTRASTRUCTURE AS A TAXONOMIC KEY

In a paleontologic parallel of the reductionist philosophy championed by some biologists, one might engage in ultrastructure studies in the hope that the very fine skeletal structure will afford a highly refined criterion for determining taxonomic relationships. Certainly, some organisms do produce skeletal crystals that, individually or in aggregate structural units, are distinctive. Most notable among these are the monocrystalline skeletal elements of echinoderms and the crossed-lamellar structure of mollusks. These, and various other distinctive arrangements of skeletal crystals, are indeed usable for taxonomic recognition, even in very fine fragments of skeletal debris (Majewske, 1969; Hay, Wise, & Stieglitz, 1970; Stieglitz, 1972). However, only fairly high-level taxonomic differentiation is possible, and, more importantly, not all organisms produce skeletal ultrastructures that are so distinctive or diagnostic, even at the class or phylum level. Indeed, small fragments of spherulitic aragonite produced by scleractinian corals, cheilostomate bryozoans, the modern cephalopod *Nautilus,* some codiacean algae, or even by inorganic, submarine cementation can be very difficult to distinguish (Sandberg, 1975a). It is usually the size and morphologic features of such ultrastructurally similar fragments that are diagnostic.

As the skeletal ultrastructure of bryozoans and diverse other groups has become better understood, it has become evident that ultrastructures are largely not taxonomically dependent. Rather they reflect the degree and nature of the biological interference that the organism exerts upon the calcification process. This vital effect is manifested in such things as the composition of the fluid medium from which skeletal precipitation occurs and the amount and distribution of organic matrix. Depending on the nature of that interference, the properties (for example, mineralogy, crystal morphology, cation makeup, stable carbon and oxygen isotope composition) of the resulting skeletal carbonates may resemble or differ by varying degrees from equivalent properties of actually or potentially coprecipitated inorganic carbonates (SANDBERG, 1975a). On the basis of ultrastructural and mineralogical studies, it appears that biological interference with calcification may vary greatly both topographically or ontogenetically within skeletons of a single taxon, but also may show great similarities even among skeletons of diverse phyla. In addition to the common spherulitic structures discussed above, for example, much morphologic similarity exists among lamellar or "nacreous" skeletal units of cheilostomate bryozoans, brachiopods, and bivalves in their development of similar **screw-dislocation** structures (Fig. 102,*4,5*; 103,*3,4*; WILLIAMS, 1971b; WADA, 1972). It should be noted that, despite the morphological resemblance, the lamellar crystals are calcite in the bryozoans and brachiopods, but aragonite in the bivalves, presumably reflecting differences in biological interference.

Contrary to earlier hopes, ultrastructure is not a panacea for resolving questions of taxonomic relationships. Used cautiously within well-defined taxonomic units, ultrastructure can be taxonomically useful. However, because of the relationship between ultrastructure and both the general functional-structural properties of skeletons and the variable degree to which different organisms interfere with calcification, ultrastructure is not a broadly applicable, general taxonomic criterion.

ULTRASTRUCTURE AND SKELETAL GROWTH MODES AND SUCCESSIONS

Despite these taxonomic limitations, ultrastructure can be quite useful in understanding modes of skeletal development and ontogenetic or even phylogenetic changes in the nature of the skeletal material deposited. The basic patterns of skeletal growth in cheilostomates and their relationship to skeletal growth in other groups, notably mollusks, will be treated in greater detail later. For the moment, it is sufficient to say that distinctive ultrastructures and ultrastructural successions afford considerable information on growth of cheilostomate skeletons, the location of cuticles, and the structural-functional role of the skeleton in various groups. Successful interpretation of some fairly complex examples of cheilostomate skeletal growth has been possible on the basis of ultrastructure, even in fossil material. Such interpretations would previously have been available only from suitably prepared histologic samples or from observation of growth in living colonies. The implications for bryozoan paleontology are obvious.

The ultrastructural feature most promising for growth mode interpretations was called "parallel fibrous" ultrastructure (SANDBERG, 1971) (see later discussion of exterior wall recognition). That term (parallel fibrous) has been used in various earlier papers, but leaves the reader with the question "parallel to what?" The growth to which that term has been applied is a form of two-dimensional **spherulitic ultrastructure** of the type discussed by BRYAN (1941) and BRYAN and HILL (1941). For the sake of clarity and to minimize proliferation of new terms, their descriptive term **planar spherulitic ultrastructure** will be used in this discussion. In cheilostomate bryozoans, this ultrastructure is the first calcification against the cuticle in exterior walls. It is fundamentally a two-

Bryozoa

dimensional spherulitic growth. An equivalent ultrastructure occurs on the undersides of dissepiments of scleractinian and tabulate corals. Significantly, in those latter groups it is also useful in selecting among various theories of dissepiment growth (WELLS, 1969). CHEETHAM (1971) related zooid structure to colony form, particularly with reference to calcification of frontal walls and its relationship to structural support of the colony. Knowledge of ultrastructural successions and of distributions of individual ultrastructures, particularly the planar spherulitic ultrastructure, can provide a clearer picture of development of the fundamental skeletal "box" of the zooecium and its later calcareous embellishments, and their relationship to such things as zooidal function, colonial morphology and stress distribution, and physiological changes during zooid ontogeny.

ULTRASTRUCTURE AND THE INTERPRETATION OF ANCIENT SKELETAL REMAINS

The skeletons we observe as fossils are the products of not only the biological interference that the organisms exerted on the calcification process, but also the vagaries of postmortem diagenetic effects. If we are to use ultrastructure for interpretation of fossil skeletons and their genesis, we must have a better understanding of the extent and nature of recrystallization than has been evident in much paleontologic literature. Some excursions into the areas of carbonate petrography and geochemistry are required for proper appreciation of the processes involved (for discussion of problems and reference to much of earlier literature, see BERNER, 1971; LIPPMANN, 1973; MILLIMAN, 1974; BATHURST, 1975; SANDBERG, 1975a,b).

It appears that, in the paleontological literature, "recrystallization" has been sometimes overworked and sometimes underestimated. The reasons for these two extremes are quite distinct. Overuse of inferred recrystallization to explain observed textures in fossil skeletons has resulted from too strict an adherence to analogy with modern inferred relatives and their skeletal products. For example, one argument that had rather wide acceptance earlier was—"The rugose corals, an extinct group, have lamellar or spherulitic calcite skeletons; these must have originally been spherulitic aragonite, the exclusive skeletal material of the modern relatives, the scleractinian corals." In interpretation of the original skeletal makeup of extinct taxa, the sort of analogy described above becomes increasingly less reliable as the taxonomic distance between the two groups increases. Furthermore, as discussed elsewhere (SANDBERG, 1975a) and briefly reviewed below, there are certain predictable patterns of diagenetic behaviors for the various skeletal carbonate phases. Knowledge of these patterns can be used to support or, as in the case of the rugose corals, refute inferences of original state of fossil skeletons.

Since the time of ROSE (1859) and SORBY (1863, 1879) it has been recognized that carbonate skeletons or skeletal parts of differing mineralogies and cation composition have varying susceptibilities to textural disruption by diagenesis. Solid-state processes have

FIG. 102. Growth surfaces of interior walls.——*1,2. Metrarabdotos tenue* (BUSK), rec., Caroline Sta. 68 off NE. Puerto Rico; *1,* frontal exterior, growth surface, etched, note numerous superimposed layers accreting laterally simultaneously in a manner analogous to gastropod nacre, ×8,000 (bar = 1 μm); *2,* detail of lower left region of *1,* ×16,000 (bar = 1 μm); both USNM 209434.——*3,4. Tremogasterina robusta* (HINCKS), rec., Perim Is., Aden, Red Sea; basal interiors, distal toward top of photograph; note rhombic crystal shapes, accretionary banding, and occasional screw dislocations; both ×5,000 (bar = 2 μm), BMNH 1966.2.24.1.——*5. Arachnopusia unicornis* (HUTTON), rec., N.Z.; basal interior; screw dislocations more common than in *3,4* and accretionary banding virtually absent; ×2,000 (bar = 5 μm), BMNH 1886.6.8.4–5.——*6. Labioporella calypsonis* COOK, rec., Konakrey, Senegal; frontal exterior of cryptocyst near distal edge; note overlapping flat crystals and numerous, minute screw dislocations; ×4,900 (bar = 2 μm), BMNH 1964.9.2.31.

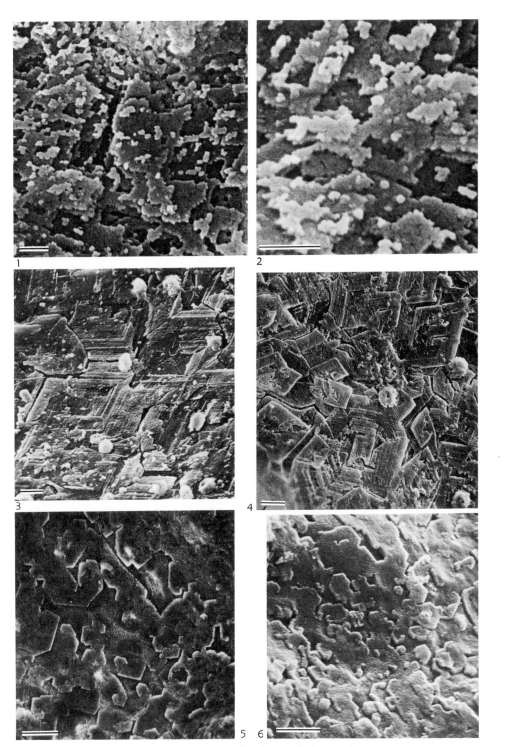

FIG. 102. (For explanation, see facing page.)

sometimes been invoked to explain observed textural states of fossils. Although these may well be influential in cation exchange, they are not generally regarded as significant in producing textural changes in shells in comparison to the temperature-pressure region in which diagenesis occurs (FYFE & BISCHOFF, 1965; BATHURST, 1975).

Over the past century, numerous workers have studied fossil carbonate skeletons of known or inferred original composition. Compilation of their observations allows a few generalizations about preservation or disruption of skeletal detail that are independent of taxonomy.

Aragonite skeletons.—Fossil skeletons or skeletal parts that were originally aragonite, unless protected by such unusual, generally impermeable deposits as the Buckhorn Asphalt and Kendrick Shale (Pennsylvanian) (STEHLI, 1956; HALLAM & O'HARA, 1962), will undergo solution removal (leaving a mold) or else transformation to calcite by microscale solution-redeposition, but with profound disruption of original crystal texture (SORBY, 1879; HUDSON, 1962; BATHURST, 1975). In microscale replacement, the new calcite crystals are several orders of magnitude larger than the original skeletal aragonite crystals and may be traversed by fine relics of the organic sheets occurring at growth surfaces or ultrastructural unit boundaries. Unfortunately, those organic relics have been misconstrued by some as indi-

cations of preservation of fine skeletal detail, despite the transformation to calcite. This is a good example of the underestimation of the effect of recrystallization. Although some few crystallites of original aragonite may occur as inclusions in the replacement calcite (SANDBERG, SCHNEIDERMANN, & WUNDER, 1973; SANDBERG, 1975a,b), it should be emphasized that these are only scattered relics and that the main mass of such shells has been drastically altered to coarse calcite. Otherwise, preserved aragonite shells, which are not especially common in older rocks, will retain their original ultrastructure, sometimes in a chalky state, because of partial solution removal.

Calcite skeletons.—Fossil skeletons that were calcite in their original state commonly show little if any change in texture, at least at the light-microscopic level. However, varying amounts of $MgCO_3$ may exist in solid solution in the skeletal $CaCO_3$. High-Mg calcite is metastable relative to low-Mg calcite and, in diagenetic environments, alters by microscale solution-redeposition or by surface exchange or solid-state diffusion processes to produce calcite and an Mg^{++} enriched solution. Various workers have observed that this alteration occurs without textural disruption at the light-microscopic level (SANDERS & FRIEDMAN, 1967; LAND, MACKENZIE, & GOULD, 1967; PURDY, 1968) and in some cases even at the electron microscopic level (TOWE & HEMLEBEN, 1976), but

---

FIG. 103. Frontal walls and organic matrices.——*1. Petraliella crassocirca?* (CANU & BASSLER), rec., Albatross Sta. D4880, off Japan; frontal shield element between two tremopores; note that initial calcite frontal (C) has concentric lamellae around a more massive core; lamellae later overlain, on frontal side only, by superficial spherulitic aragonite (A); etched transv. sec., USNM 209443, ×1,680 (bar = 5 μm).——*2. Watersipora subovoides* (D'ORBIGNY), rec., locality unknown; single bar in cryptocyst; note concentric laminations reflecting secretion on all sides of frontal shield; heavily etched transv. sec., BMNH 1970.6.1.32, ×3,500 (bar = 2 μm).——*3,4. Labioporella calypsonis* COOK, rec., Konakrey, Senegal; *3,* detail of etched cryptocyst, note abundant screw dislocations of thin rhombic crystals, ×2,000 (bar = 5 μm); *4,* detail of another region of cryptocyst surface, ×5,000 (bar = 2 μm); both BMNH 1964.7.2.31.——*5. Adeona* sp., rec., locality unknown; basal-lateral wall junction; aragonite forms short, broad laths between very numerous, close-spaced organic sheets; etched transv. sec., BMNH 1934.2.10.20, ×9,000 (bar = 1 μm).——*6. Melicerita obliqua* (THORNELY), rec., Antarctic; zooecial lining layers, very heavily etched, zooecial interior toward top; massive, spherulitic zooecial lining with some accretionary banding, and organic sheets, shown in this photograph, very closely spaced; BMNH 1967.2.8.119, ×9,500 (bar = 1 μm).

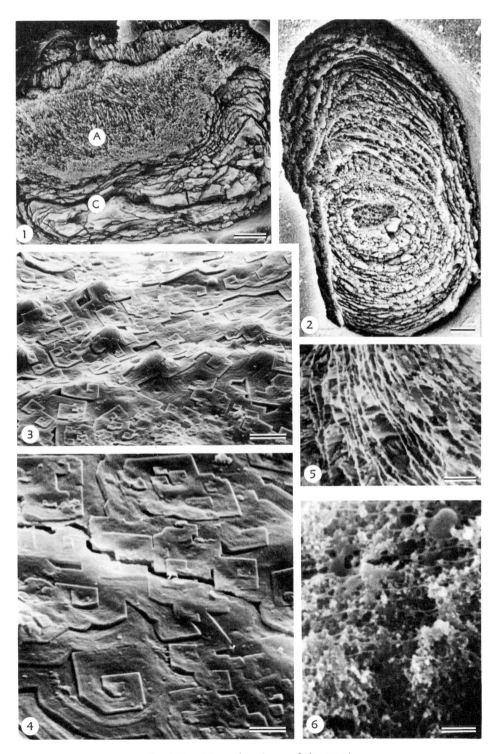

FIG. 103. *(For explanation, see facing page.)*

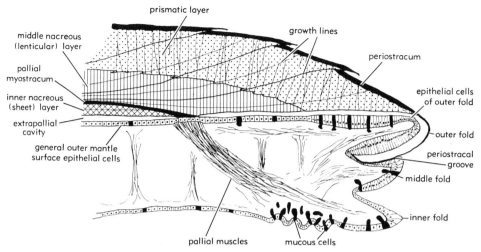

Fig. 104.   Skeletal growth. Radial section of the shell and mantle at the valve margin in the bivalve *Anodonta cygnea* Bosc (after Taylor, Kennedy, & Hall, 1969).

others (BANNER & WOOD, 1964; SCHNEIDER-MANN, 1970) have found an apparent relationship between original $MgCO_3$ content and diagenetic textural disruption. More recently, SEM work has shown that significant textural reorganization not clearly evident in the light microscope can occur during Mg^{++} loss from even low-Mg calcite (bryozoan skeletons; SANDBERG, 1975a).

From the foregoing we can expect, in the fossil record of bryozoans (and other groups as well), that originally calcitic skeletons will preserve much of the original texture—more so with lesser original $MgCO_3$ content. At least in many ancient stenolaemate bryozoans, textural preservation appears excellent. That textural retention is most likely related to an initially low $MgCO_3$ content, like that found in skeletons of modern stenolaemates (cyclostomates) (SCHOPF & MANHEIM, 1967). The textural disruption commonly observed in skeletons of some stenolaemates, such as *Nicholsonella,* may well be a function of higher original $MgCO_3$ content.

Statements about the degree of resemblance between observed state (texture, mineralogy, cation, and stable isotope composition) of a fossil skeleton and its original state are thus dependent on knowledge of first,

stability of the various skeleton-forming carbonates in diagenetic environments, and second, the nature of the skeletal products of modern forms (or rare, unusually well-preserved fossil forms) most closely related to the fossils of interest. This latter application of "biological uniformitarianism" (BEER-BOWER, 1960) is much more reliable than a uniformitarian comparison between nonskeletal carbonates in modern sediments and similar nonskeletal carbonates in ancient limestones. This is because the composition of those nonskeletal carbonates will closely reflect the physical-chemical conditions of the general environment and will vary with temporal changes in those conditions. In contrast, biological interference with calcification and the nonequilibrium stability of biological systems (PRIGOGINE, NICOLIS, & BABLOYANTZ, 1972) act as buffers, tending to minimize the influence that any temporal changes in external environmental conditions would have on the resulting skeletal carbonates.

Terminology.—Although qualifying adjectives (calcareous, skeletal) have generally been used in reference to walls in this discussion, it should perhaps be further emphasized that this discussion deals almost

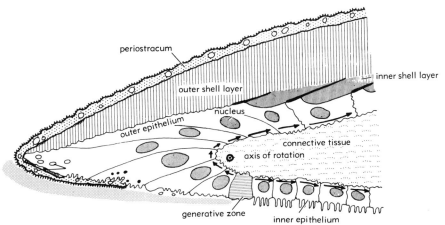

FIG. 105. Skeletal growth. Conveyor-belt model of shell growth, as exemplified by the brachiopod
Notosaria (after Williams, 1971a).

exclusively with skeletal walls. In order to
underscore this skeletal emphasis, HARMER's
(1902) term "frontal shield" has been used
to differentiate calcified walls in the frontal
region from the original membranous frontal
wall (see Fig. 65).

BANTA (1968) discussed the interpretation
of skeletal walls in cheilostomates as intracuticulate. It should be emphasized that
statements on cuticulate or noncuticulate
walls in this discussion refer not to that possible intracuticulate condition, but rather to
the presence or absence of an outermost
boundary cuticle (BANTA, 1968, p. 498).

COMPARISON OF SKELETAL GROWTH IN CHEILOSTOMATES, BRACHIOPODS, AND MOLLUSKS

General models of growth can be most
useful in understanding developmental patterns, both ontogenetically and phylogenetically, in skeletons of diverse organisms.
Growth models can be derived from observation of such features as skeleton-tissue
relationships, the distribution of organic partitions or bounding layers (e.g., cuticle, periostracum) and ultrastructural successions in
the skeletons, the orientation of skeletal
growth lines and surfaces, or by a more theoretical or function-analysis approach (particularly significant with extinct organisms).

The growth model proposed for cheilostomate bryozoans by SILÉN (1944a,b) is no
longer fully satisfactory, largely because of
the great diversity of morphologies and
observed or inferred skeletal-epithelial relationships subsequently discovered among
members of that group. Nevertheless, the
nature of cheilostomate skeletal development
can be compared and contrasted with skeletal
growth in groups for which well-established
growth models exist, such as the brachiopods
and the mollusks, especially bivalves. TAV-
ENER-SMITH and WILLIAMS (1972) implied
some similarities, in details of fine skeletal
structure, between bryozoans and brachiopods. However, if one compares the higher
level skeletal development of bryozoans (particularly cheilostomates) and brachiopods-
mollusks, certain major distinctions and dis-
similarities emerge.

Brachiopod-molluscan growth model.—The
fundamentally open-ended, one-sided growth
of brachiopod and bivalve shells involves
dimensional expansion and lateral displacement of the skeletal accretion surface along

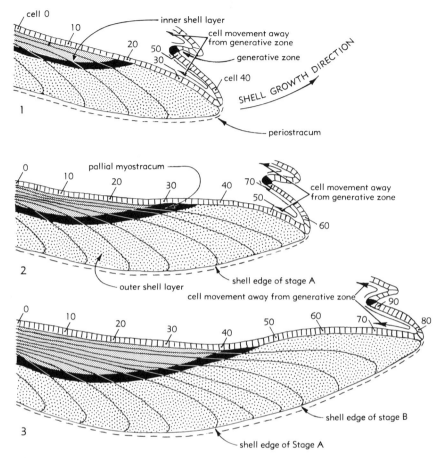

F<small>IG</small>. 106.   Skeletal growth. "Road-paver" model of shell growth in mollusks and brachiopods. Note the fixed position of each cell relative to the shell surface and the continual distal generation of new cell "pavement" (by the "paving machine," the generative zone) and the later development of subjacent shell. Note the change in secretory function of the numbered cells as the valve-mantle margin and generative zone move away distally. Growth lines marking the positions of the shell edge at earlier stages are indicated in *2* and *3*.

a growth spiral (Y<small>ONGE</small>, 1953; R<small>AUP</small>, 1966). That surface is divided into zones whose skeletal products are distinctive and whose positions relative to one another are, with few exceptions, fixed. Dominance and areal extent of any given ultrastructural unit may vary, such as the variations in thickness and distribution of the myostracal layer in the mussel *Mytilus* as described by D<small>ODD</small> (1963). However, such variations are merely topological distortions of skeletal units in a fixed succession.

Secretion begins on the outer surface, at a

periostracum, and the shell is thickened by successive accretions added medially (toward the mantle cavity and body) across the entire inner surface of the shell in more or less uniform, concentric zones (Fig. 104). The nature of mantle formation and the successive secretory regimes through which a mantle epithelial cell passes have been portrayed by a "conveyor-belt" model (W<small>ILLIAMS</small>, 1968). This model is commonly represented by a radial section through the shell and mantle showing the positional and physiological change in cells as shell growth progressively alters their loca-

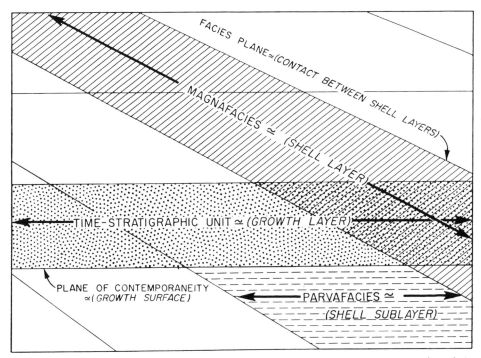

Fɪɢ. 107. Skeletal growth. Diagrammatic comparison of shell-unit and growth-increment boundaries with facies concepts of Cᴀꜱᴛᴇʀ (1934) (after McClintock, 1967).

tion relative to the shell margin (Fig. 105). Along any radial sector line, cells may be expected to pass through all secretory regimes present proximally (adapically), with the obvious exceptions of spatially limited adductor attachments and, in some groups, discontinuous pallial attachment. If it is a conveyor belt, it is an odd one with one end attached and the other continuously generating new belt and moving forward. The conveyor-belt analogy is perhaps not a good one, because the cells, once in position adjacent to the shell, are each "nailed down," that is they do not move laterally relative to that shell. Rather they undergo a series of physiological changes in secretory function dependent on their positive relative to the mantle-generative zone.

The mantle system would be better portrayed as a road-paving system in which the moving "paving machine" (the mantle-generative zone) produces a fixed "pavement" of cells as it moves distally (Fig. 106). Even that analogy is limited because the cells, unlike the passive macadam pavement formed by a real road paver, are active producers of new subjacent layers (the shell), migrate vertically relative to the shell surface as new shell is secreted, and change in function as the "paving machine" of the generative zone moves away distally.

MᴀᴄCʟɪɴᴛᴏᴄᴋ (1967) pointed out the analogy between molluscan skeletal ultrastructural units with their "outcrop bands" on the inner shell surface and Cᴀꜱᴛᴇʀ's (1934) stratigraphic concepts of magnafacies and parvafacies, respectively (Fig. 107; see also Wᴇꜱᴛʙʀᴏᴇᴋ, 1967). The ultrastructural types present on the interior of the shell are thus analogous to laterally adjoining, temporally equivalent depositional environments. Vertical successions of ultrastructures through the shell reflect the lateral (distal) shift of those "environments" in a microscopic corollary of

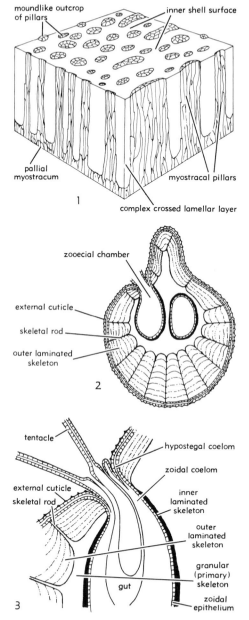

FIG. 108. Skeletal growth. Comparison of myostracal pillars in the bivalve *Chama rubea* REEVE (*1*) (after Taylor, Kennedy, & Hall, 1969) with the rods of "primary" skeleton extending through the lamellar "secondary" skeleton in a fenestellid stenolaemate bryozoan (*2,3*) (after Tavener-Smith, 1969).

a transgressive sequence. Attempts at demonstration of "correlation" or, more correctly, "lithic equivalence" (i.e., not necessarily contemporaneous deposition but simply "same unit") among shells of different bivalves using the myostracal layer (prismatic aragonite formed at the line of mantle attachment) as a datum plane have not been totally successful (TAYLOR, KENNEDY, & HALL, 1969, p. 9). This is because, in some forms, numerous myostracal layers are interleaved with other shell units, and, in some other forms, mantle attachment is absent or secondary.

Any given skeletal unit is obviously not contemporaneous throughout, but the generally clear growth lines allow determination of earlier instantaneous growth surfaces and of time-equivalent skeletal deposits, whether in brachiopods or in mollusks (Fig. 104, 106, 107).

In the molluscan-brachiopod model there is a single open-faced secretory surface whose dimensions are continually expanding (even in successive chambers of cephalopods). The resulting individual skeleton produced at that secretory surface is composed of skeletal layers that are essentially continuous sheets or wedges from their distal exposure at the growth surface back toward the proximal region of the larval shell. There are some discontinuous units, such as the myostracal pillars in *Chama* (TAYLOR, KENNEDY, & HALL, 1969). As indicated in Figure 108, those pillars are analogous at least in appearance to the rods of the primary layer extending through the laminated secondary layer of fenestellid stenolaemate bryozoans (TAVENER-SMITH, 1969).

Parenthetically, it should be noted that the one-sided, external shell model given here for mollusks does certainly have some unusual, if relatively rare exceptions or modifications. Notable ones includes extreme mantle extension and envelopment of the shell by secondary deposits (as in the gastropods *Cypraea* and *Calyptrophorus*) and the development of an internal shell in belemnites. The chambered shells of ectocochlear cephalopods, which are external and one-sided, are the closest molluscan approximation of the compartmentalization of the colonial skeleton in bryozoans.

Cheilostomate bryozoan skeleton.—The bryozoan skeleton (like shells of brachiopods and bivalves) is epidermal in origin (i.e., an exoskeleton), regardless of the often complex, infolded topological distortions of that epidermis into the colonial coelom (BOARDMAN & CHEETHAM, 1973, p. 124). The "standard" growth pattern (see later definition and discussion) among cheilostomates is, like the brachiopod-bivalve model, one-sided. That is, calcification occurs in one direction from an exterior wall with a bounding organic layer (periostracum in bivalves-brachiopods, cuticle in cheilostomates). However, in cheilostomates and other bryozoans, that exterior wall surface may comprise only the basal side of the ancestrula.

One may reasonably expect significant differences between skeletons of solitary and colonial animals, the most obvious being some degree of compartmentalization of the colonial skeleton, delineating the individuals. The distinctiveness of that delineation varies greatly among cheilostomates. In fact, the significance of boundaries between individuals in a bryozoan colony is a subject of some controversy, relating primarily to the degree of integration or of separateness in functioning of individuals. Some work, such as the recent study of response to mechanical stimuli by THORPE (1975), suggests a high degree of integration of individuals. Even when superficial calcification of the frontal shield occurs without breakdown of intercalary cuticles, the amount of such calcification is closely determined by the position of the individual zooid. A striking example of this occurs in adeonids, in which **dendritic thickenings** extend over the zoarium (Fig. 109,2). Along those axial dendritic zones, normal zooids are overlain by a thick sequence of heavily calcified kenozooids quite distinct from the outer, laterally adjacent zooids. Despite the extreme thickening, the lateral intercalary cuticles persist throughout (Fig. 109,1).

In the cheilostomates, the general colonial growth field, the equivalent of the valve interior in the brachiopod-mollusk model, is subdivided into numerous, repeated skeletal compartments, the zooecia. There is a clear developmental gradient of morphological change from the zooecial buds at the leading edge of the colony through the heavily calcified, sometimes occluded zooecia in the proximal region. However, the zooecial unit dimensions are fixed early, and, except for small variations produced in zones of astogenetic change or as a result of crowding, similar sized units (sometimes polymorphic) are repeated as the colony growth continues.

Once the fundamental calcareous box or structural framework of each zooid is formed, the coelomic volume tends to decrease as skeletal secretion continues for a time on the interior of that box (Fig. 110,2). Although brachiopod and mollusk shells also grow by secretion inward, continuing marginal expansion of the open-ended skeletal enclosure more than compensates in living space for the inwardly growing skeleton (Fig. 110,1).

Recognition of contemporaneous skeletal deposits.—The ultrastructurally distinctive shell layers that compose the skeletons of mollusks and brachiopods are all present in the postlarval shell, and the secretory regimes in which they are each produced are displaced laterally as the shell grows. This lateral displacement is generally rapid relative to the rate of thickening of the skeletal unit formed in any of those regimes, and the boundaries between resulting shell layers tend to be at relatively low angles to the secretory surface. Nevertheless, distinctive growth lines allow recognition of contemporaneous parts of the different molluscan shell layers (Fig. 104).

For several reasons, such easy demonstration of contemporaneity of deposition is not generally possible for parts of cheilostomate bryozoan skeletons. It should be emphasized here that, because of compartmentalization of the developmental gradient from the colony margin inward, the major problem is determination of contemporaneity of skeletal

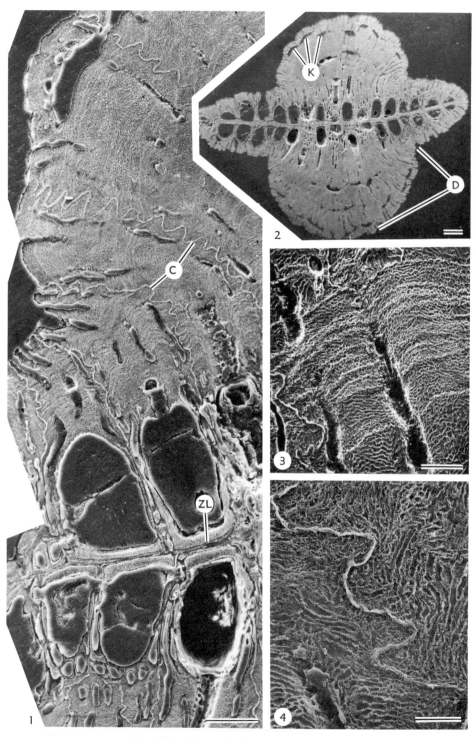

FIG. 109.   *(For explanation, see facing page.)*

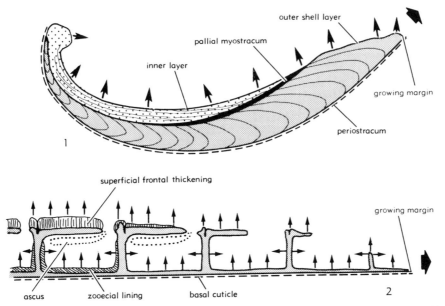

FIG. 110. Skeletal growth. Diagrams of skeletal growth fields in a bivalve (*1*) and a cryptocystidean (lepralioid) cheilostomate bryozoan (*2*). Broad arrows in each diagram indicate the direction of expansion (marginal growth) of the skeletal field. The diagram of the cheilostomate shows only the very distal edge of the field in order to allow sufficient magnification to show wall thickening directions (narrow arrows). In the cheilostomate, the skeletal growth of the underside of the frontal is terminated at the time of ascus formation. Because the frontal-thickening and zooecial-lining deposits are discontinuous, demonstration of contemporaneous growth increments in those deposits is difficult.

parts within and among individual zooecia at different positions along that gradient. In cheilostomates, skeletal deposition rates are often quite rapid, and physiological changes resulting in difference in secretory function (hence different skeletal ultrastructures) tend to move rapidly over the secretory epithelium of any given wall. Therefore, boundaries between ultrastructural types may sometimes

be nearly equivalent to growth surfaces. Also, except for the planar spherulitic ultrastructure on exterior walls (Fig. 111, 112) or the distally oriented spherulites of interior-walled aragonitic cheilostomates (Fig. 113, *1–3*; SANDBERG, 1973, figs. 3, 4), skeletal accretion is not reflected by growth banding at an angle to the zooecial wall. Therefore, growth increments may be quite clear in ultrastruc-

FIG. 109. Frontal thickening.——*1,2. Adeona* sp., locality unknown; *1,* transverse section, etched, of zoarial branch at junction between axial thickening and lateral "normal" zooecia, sinuous lateral intercalary cuticles (C) extend all the way through the frontal despite extreme thickening of frontal shields of the occluded axial zooecia, frontal composed of numerous organic-bounded units of spherulitic aragonite (compare Fig. 114,*2–5*), note the zooecial linings (ZL) deposited only in the lower part of each zooecial interior, ×140 (bar = 100 µm); *2,* lower magnification view of specimen in *1,* note extreme thickening of frontal shields of the axial zooecia in the dendritic thickening (D) as well as blisterlike kenozooid chambers (K) in that frontal thickening, ×25 (bar = 200 µm); both BMNH 1920.12.10.1.——*3. Adeonella atlantica* BUSK, rec., Nightingale Is., near St. Helena, S. Atl.; frontal shield; note accretionary layers and organic-bounded aragonite units; transv. sec., BMNH 1887.12.9.725, ×220 (bar = 50 µm).—— *4. Adeona* sp., rec., locality unknown; detail of upper end of *1*; note sinuous intercalary cuticle and organic boundaries of aragonite units; ×400 (bar = 20 µm).

tures that have crystal orientations perpendicular to the growth surface (Fig. 113,4; 114,1,3). However, if there is only lamellar structure (Fig. 115,2; 116,1,2), then the location of growth increments (contemporaneously grown surfaces) is equivocal.

A further problem in demonstration of contemporaneity in bryozoan skeletons is that, unlike the condition in molluscan shells, some ultrastructurally distinctive skeletal units (e.g., superficial frontal layers, **zooecial linings**) are not present initially in a cheilostomate zooecium. Furthermore, in mollusks an epithelial cell lying along a radial expansion vector passing through an adductor scar will pass through all or essentially all skeletal secretory regions. In contrast, in cheilostomate secretory epithelia each cell's function and possible skeletal products are limited by its location within the zooid. In calcitic or bimineralic species, only the calcitic framework portion of the skeleton is continuous along basal surfaces for all zooecia and, as upward projections, into the lateral, transverse, and frontal walls of the zooecia. Other skeletal units, whether calcite or aragonite, occur as localized, discontinuous deposits, most commonly on the frontal exterior or on

---

Fɪɢ. 111.   Growth surfaces of exterior walls.——*1. Petraliella bisinuata* (Sᴍɪᴛᴛ), rec., Albatross Sta. D2405, Gulf of Mexico; basal wall, distal indicated by arrow; note distally radiating fans of acicular calcite crystals and intermittent zones of very strong and very subdued accretionary banding; etched exterior, USNM 209448, ×1,100 (bar = 10 μm).——*2. Posterula sarsi* (Sᴍɪᴛᴛ), rec., Gulf of St. Lawrence; surface of calcified inner layer of ovicell just distal to orifice of fertile zooid, distal toward top; curved line near the bottom (arrow) marks line of emergence of ovicell as an exterior-walled lobe; BMNH 1911.10.1.1360A–B, ×1,150 (bar = 10 μm).——*3. Megapora ringens* (Bᴜsᴋ), rec., Shetland Is.; basal wall; planar spherulitic ultrastructure in rosettes radiating from scattered sites of initial calcification; etched exterior, BMNH 1911.10.1.630, ×1,180 (bar = 10 μm).——*4. Onychocella angulosa?* (Rᴇᴜss), rec., locality unknown; basal wall; see comments on *3*; exterior, BMNH 1911.10.1.140, ×1,850 (bar = 5 μm).——*5. Arachnopusia unicornis* (Hᴜᴛᴛᴏɴ), rec., N.Z.; detail of planar spherulitic ultrastructure on exterior surface of inner calcified layer of ovicell; arrow indicates distal direction; BMNH 1886.6.8.4,5, ×4,900 (bar = 2 μm).

Fɪɢ. 112.   *(See p. 254.)*

Fɪɢ. 113 (p. 255).   Spherulitic structure in aragonitic and calcitic walls.——*1,3,4. Mamillopora cupula* (Sᴍɪᴛᴛ), rec., Gulf of Panama; *1,* detail of a single aragonite spherulite, note initial poorly etched core (organic rich?), lateral compromise boundaries between adjacent spherulites, distally expanding acicular crystals, and transverse accretionary banding, ×3,200 (bar = 2 μm); all USNM 184151; *3,* lateral wall, distal to right, spherulitic aragonite radiating from rows of rather evenly spaced centers of calcification to produce pattern very like trabecular structure of aragonite in scleractinian corals, etched oblique long. sec., ×1,325 (bar = 5 μm); *4,* distal wall, distal to left, calcification of transverse wall, in contrast to that in most cheilostomates, one-sided, with aragonite spherulites beginning at proximal side and growing distally, long. sec., ×1,300 (bar = 5 μm);——*2. Flabellopora arculifera* (Cᴀɴᴜ & Bᴀssʟᴇʀ), rec., Albatross Sta. D5315, Philip.; zooecial lining projecting basally from frontal shield just inside orifice; note numerous, very finely spaced accretionary bands and coarse spherulitic calcite; frontal direction toward bottom of photograph, etched transv. sec., USNM 209437, ×2,550 (bar = 5 μm).

Fɪɢ. 114.   *(See p. 256.)*

Fɪɢ. 115 (p. 257).   Lamellar ultrastructure.——*1–3. Metrarabdotos tenue* (Bᴜsᴋ), rec., Caroline Sta. 68, off NE. Puerto Rico; *1,* basal interior, large rhombic crystals with accretionary bands and incipient screw dislocation, ×6,100 (bar = 2 μm); *2,* frontal wall, showing only a few of the very numerous calcite lamellae that make up superficial frontal thickening (growth surface of this ultrastructure shown in Fig. 102,1,2), long. sec., ×4,200 (bar = 2 μm); *3,* detail of *2,* note subunits in crystals of lamellae, ×16,800 (bar = 1 μm); all USNM 209434.——*4–6. Labioporella calypsonis* Cᴏᴏᴋ, rec., Konakrey, Senegal; *4,* frontal shield interior, fractured at very low oblique angle to surface, subunits of large rhombic-hexagonal crystals that make up lamellae emphasized by heavy etching of this specimen, ×5,100 (bar = 2 μm); *5,* another part of same wall surface, ×10,200 (bar = 1 μm); *6,* detail of *5,* ×20,400 (bar = 0.5 μm); all BMNH 1964.9.2.31.

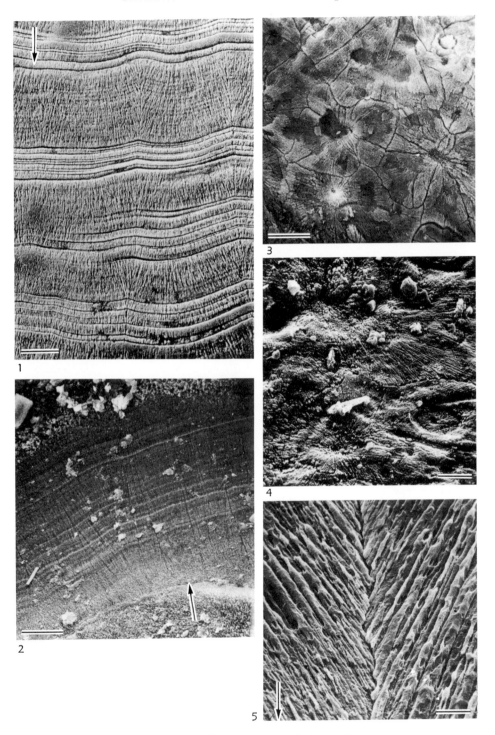

FIG. 111. *(For explanation, see facing page.)*

Bryozoa

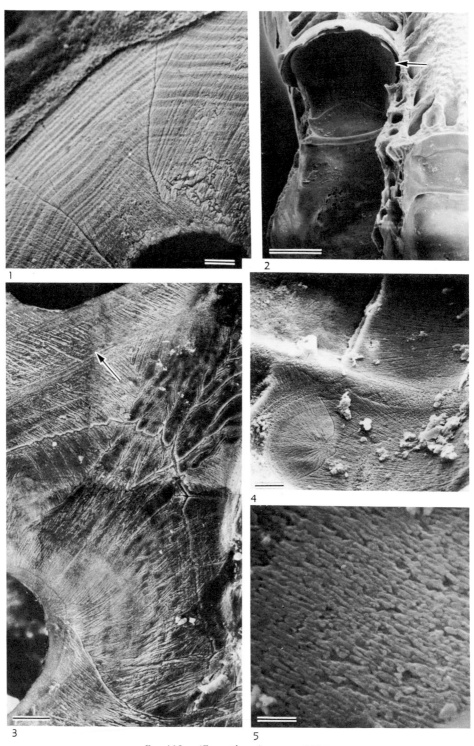

Fig. 112. *(For explanation, see p. 259.)*

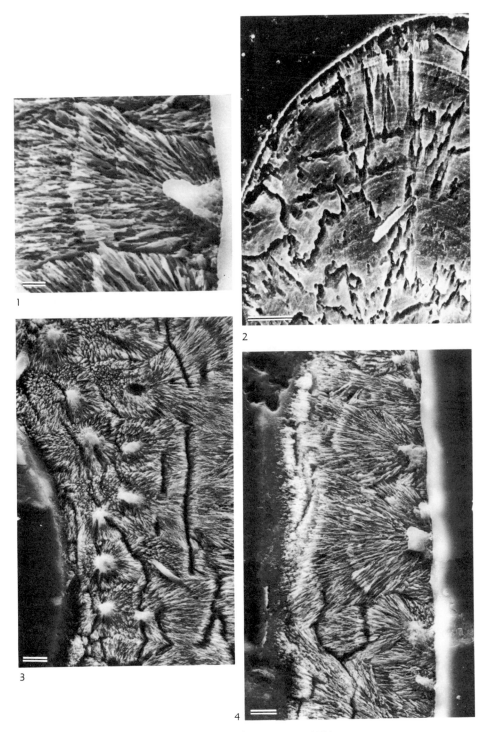

FIG. 113. *(For explanation, see p. 252.)*

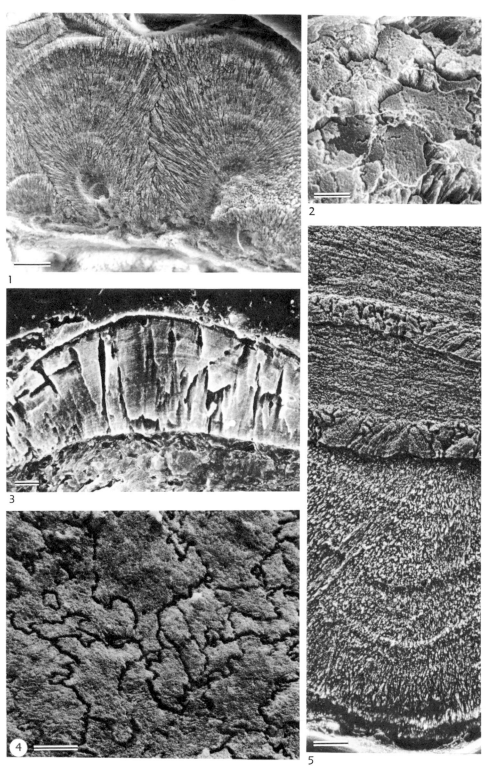

F𝗜𝗚. 114. *(For explanation, see p. 259.)*

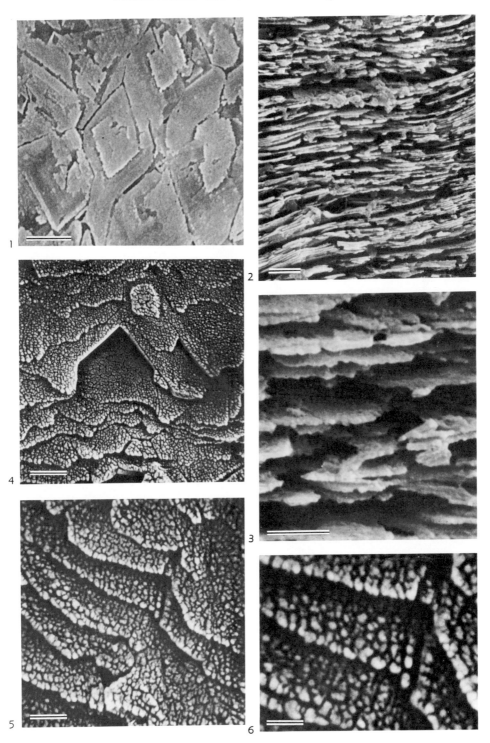

FIG. 115. *(For explanation, see p. 252.)*

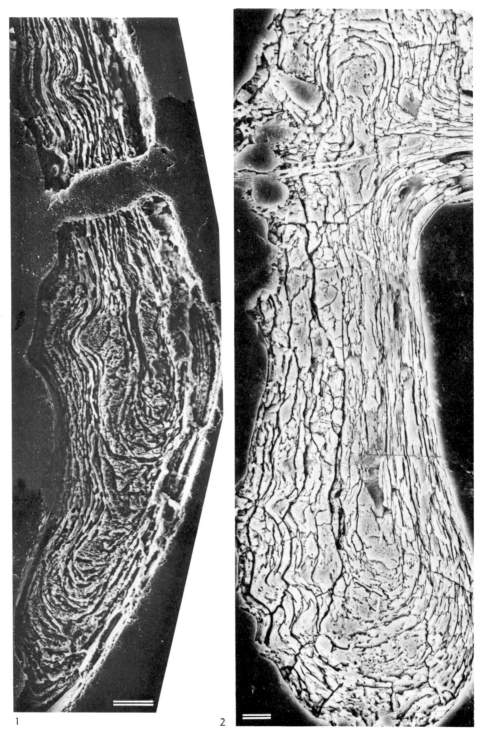

1   2

Fig. 116.   *(For explanation, see facing page.)*

the zooecial interior below the compensation device, less commonly on basal exteriors (of interior-walled forms). Totally aragonitic skeletons tend to be ultrastructurally uniform (acicular crystals in spherulitic arrays) throughout, although differences in spherulite orientations can indicate differences in direction of wall growth (Fig. 113, *1–3*; 117, *1–4*; 118, *1–3*).

Because different ultrastructural types have different expressions of growth increments, and because of discontinuity of skeletal units between and even within zooecia (e.g., units present only on frontal exterior, only on zooecial interior), it has not yet been possible to correlate the products of an instant of skeletal deposition in different parts of a colony or even among the various parts of a single zooecium. Tagging with radioisotopes (^{45}Ca or ^{14}C) and use of sectioning and microautoradiography should provide the data needed for such correlation.

In some cheilostomates (e.g., *Metrarab-dotos*) it appears there may be some simul-

---

Fig. 112 (p. 254). Exterior walls in ovicells and peristomes.——*1,2. Posterula sarsi* (Smitt), rec., Gulf of St. Lawrence; *1*, upper exterior surface of outer calcified layer of ovicell with planar spherulitic ultrastructure growing back proximally (toward bottom) and medially, superficial frontal thickening layers of distal zooecium encroaching over ovicell in upper left, ×1,575 (bar = 5 μm); *2*, frontal view of an ovicelled zooecium with frontal wall and part of the ovicell broken away, arrow marking lumen between inner and outer calcified layers of ovicell, ×140 (bar = 100 μm); both BMNH 1911.10.1.1360A,B.—— *3. Eurystomella bilabiata* (Hincks), rec., Pacific Grove, Cal.; basal exterior of two adjacent zooids; stripes of planar spherulitic ultrastructure grow in toward central basal window of each zooecium; lateral junction between zooecia marked by arrow; BMNH 1964.1.2.1, ×1,500 (bar = 10 μm).——*4. Megapora ringens* (Busk), rec., Shetland Is.; planar spherulitic ultrastructure near proximal edge of inner calcified layer of ovicell, distal toward left; BMNH 1911.10.1.630, ×780 (bar = 10 μm).——*5. Reteporella myriozoides* Busk, rec., Challenger Sta. 148, Possession Is., Indian O.; etched surface of a peristome, distal toward upper left; crystals of planar spherulitic ultrastructure very nearly parallel and less elongate than in some other species; BMNH 1887.12.9.516, ×1,050 (bar = 10 μm).

Fig. 113. *(See p. 255.)*

Fig. 114 (see p. 256). Spherulitic structure in aragonitic and calcitic walls.——*1. Hippoporidra sene-gambiensis* Cook, rec., Konakrey, Senegal; frontal shield; spherulitic arrays of acicular aragonite start at scattered centers and meet at roughly planar compromise boundaries; etched long. sec., BMNH 1970.8.10.24, ×1,000 (bar = 10 μm).——*2. Adeona* sp., rec., locality unknown; etched section perpendicular to growth direction of spherulitic arrays of an aragonite wall; note organic membranes; BMNH 1934.2.10.20, ×4,850 (bar = 20 μm).——*3. Flabellopora arculifera* (Canu & Bassler), rec., Albatross Sta. D5315, Philip., frontal shield; crudely laminated initial calcite wall covered, on basal side, by a zooecial lining of spherulitic calcite with many, close-spaced accretionary bands; frontal surface toward bottom, etched transv. sec., USNM 209437, ×3,120 (bar = 20 μm).——*4. Micropora* sp., rec., Albatross Sta. D2856; lower (basal) surface of a cryptocyst constructed of spherulitic calcite; spherulitic arrays separated by convoluted interlocking boundaries; USNM 209438, ×2,450 (bar = 50 μm).——*5. Tub-iporella magnirostris* (MacGillivray), rec., Port Philip Head, Australia; lower part of a basal wall of an interior-walled form; crudely laminated calcite of basal wall interlayered near its lower limit with two spherulitic calcite layers, the second of which is followed by a spherulitic aragonite superficial layer on the basal exterior surface; etched transv. sec., BMNH 1927.8.4.24, ×975 (bar = 10 μm).

Fig. 115. *(See p. 257.)*

Fig. 116. Lamellar walls.——*1. Labioporella calypsonis* Cook, rec., Konakrey, Senegal; distal part of cryptocyst, distal toward bottom; note continuity of layers around distal end of frontal as well as in more massive central portions of wall; the somewhat more massive skeletal layers near middle of wall composed of fine, transverse, lathlike subunits; etched long. sec., BMNH 1964.7.2.31, ×2,200 (bar = 5 μm). ——*2. Membranipora grandicella* (Canu & Bassler), rec., Albatross Sta. D5315, Philip.; cryptocyst, distal toward bottom (see Fig. 122,2 for lower magnification view); note continuity of layers out of upper end of transverse vertical wall and around distal end of cryptocyst; thin, central poorly laminated portion was well developed distally before inception of concentric lamellae; cuticle incorporated into calcified wall extends well down below frontal surface above vertical transverse wall; etched long. sec., USNM 209441, ×750 (bar = 10 μm).

taneous edgewise growth (BOARDMAN & TOWE, 1966) of numerous lamellae in a narrow zone at the very distal growing edge of the skeleton. This is suggested by the occurrence of multiple lamellae at a rather thick but apparently unbroken edge of a well-preserved modern colony of *Metrarabdotos*. More commonly, distal ends of skeletal walls, in growing colonial margins of most cheilostomates studied, feather out to quite thin edges. The progressive, very broad, distally thinning zones over which lamellae accrete are evident in longitudinal sections of some embedded colonies (Fig. 119,*3,4*).

In the most broad sense, skeletal lamellae grow at their edges. New lamellae often arise as screw dislocations (Fig. 103,*3,4*), as seed crystals scattered on a narrow to broad zone or step (Fig. 102,*1–3*), or a combination of the two (Fig. 102,*4,6*; 115,*1*). Seed crystals grow at their edges until they impinge on adjacent crystals (Fig. 102,*3,4*) of the same lamellar "step" and form a solid layer, the "tread" of the lamellar step (Fig. 102,*3*). In some forms, lamellae arise as distally growing, superimposed steps formed by sheets of flat blades (Fig. 119,*1,2*). The width of the lamellar tread may be narrow relative to the total zooecial skeletal width (as in the bladed structure) or may extend over most of the secretory surface that is producing the lamellae (as in the laterally accreting, seed crystal structure). What has been called edgewise growth is effectively a more extreme form of **lamellar growth** than the latter. In it, numerous lamellar steps with very narrow "treads" are crowded into a narrow growth zone. The different parts of lamellae in **lamellar ultrastructure** are of different ages; the magnitude of the difference relates to the steepness of the lamellar "staircase." Thus edgewise growth in bryozoan skeletons is more like the narrow-zone development of the nacreous layer in gastropods (WISE, 1970), which has a very much shorter "tread" than do the broad nacre "steps" in bivalves (WADA, 1972; WISE, 1969).

Types of lamellar growth are clearly affected by the shape of crystals that compose the layers, i.e., equant crystals growing from numerous scattered centers or elongate and bladelike crystals advancing at their distal ends. One should much more reasonably expect the type of edgewise growth discussed by BOARDMAN and TOWE (1966) in forms with bladelike crystals (Fig. 119,*1,2*), than in those where there is seeding of scattered equant crystals over a large area (Fig. 102,*1–4*), as in bivalve nacre.

The planar spherulitic ultrastructure, which

FIG. 117.   Spherulitic aragonitic walls.——*1–5. Cleidochasma porcellanum* (BUSK), rec., Albatross Sta. 2405, Gulf of Mexico; *1,* frontal shield, spherulitic aragonite of two orientations meeting along an irregular boundary to right of center, several marked accretionary bands occurring near upper frontal surface, etched transv. sec., ×2,200 (bar = 5 μm); *2,* detail of *1,* ×5,400 (bar = 2 μm); *3,* low magnification view of same frontal shield, note tendency for greater wall thickening marginally, ×1,100 (bar = 10 μm); *4,* frontal exterior view, ×50 (bar = 200 μm); *5,* etched vertical section along sutured lateral zooecial boundary, interfingering of zooecial walls as well as accretionary banding of frontally growing aragonite spherulites well shown, ×550 (bar = 20 μm); all USNM 209439.

FIG. 118.   *(See p. 262.)*

FIG. 119 (see p. 263).   Foliated and lamellar ultrastructure.——*1,2. Tessarodoma boreale* (BUSK), rec., Shetland Is.; *1,* frontal shield exterior growth surface, distal toward top; lathlike growth similar to foliated structure in bivalves and some structures shown in cyclostomates by BROOD (1972), ×3,150 (bar = 2 μm); *2,* detail of adjacent wall, ×7,900 (bar = 1 μm); both BMNH 1911.10.1.841.——*3,4. Schizoporella errata* (WATERS), rec., Gharadaqa, Red Sea; *3,* distal margin of colony, note extent (more than two zooecial lengths) of basal wall produced by multizooidal bud distal of last transverse wall, long. sec., ×50 (bar = 200 μm); *4,* detail of *3,* distal indicated by arrow, note that any lamination traced distally comes closer to basal exterior surface, effectively paralleling growth surface of distally thinning multizooidal zone shown in *3,* near top of wall, plastic has pulled away, disrupting some carbonate and organic layers, ×900 (bar = 10 μm); both BMNH 1937.9.28.18.

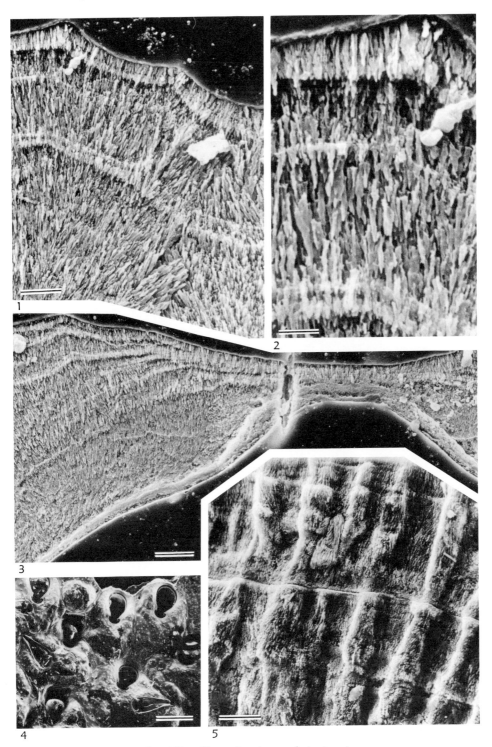

FIG. 117. *(For explanation, see facing page.)*

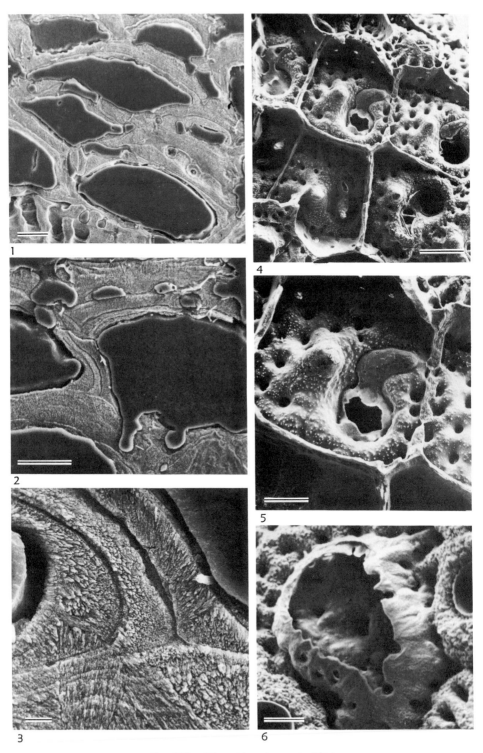

FIG. 118. *(For explanation, see p. 265.)*

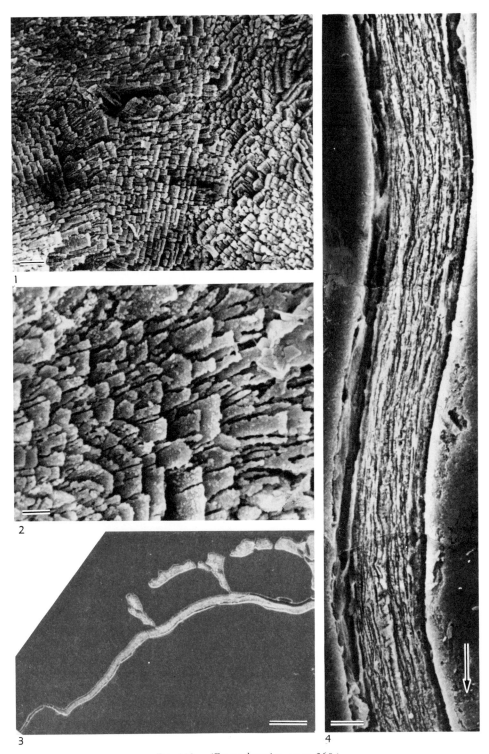

FIG. 119.   *(For explanation, see p. 260.)*

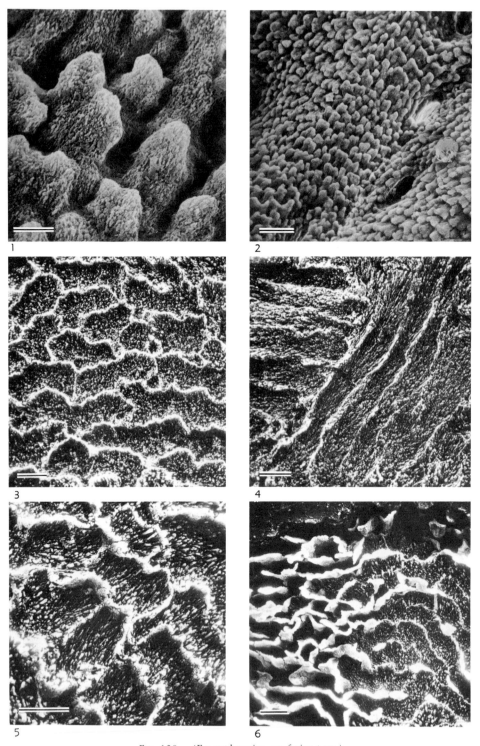

Fig. 120. *(For explanation, see facing page.)*

is the initial calcification of exterior walls against cuticle (SANDBERG, 1971), gives clear evidence, in its **accretionary banding**, of the difference in time of deposition of the various parts of that thin initial skeletal layer.

Bryozoan workers have long noted the inception of deposition (along the proximal and lateral margins of each zooecium) of a superficial frontal calcification, which then progresses distally and toward the zooecial midline.

In an allusion to a temporal succession of skeletal units, the terms "primary," "secondary," and sometimes such higher order terms as "tertiary" have been used in the literature to designate particular ultrastructural units. Even in the relatively simpler system of stenolaemate skeletons the temporal connotations of "primary" and "secondary" are misleading. Although "primary" skeleton may be deposited at the growing tip before any "secondary" skeleton (HINDS, 1975), skeletal rods of "primary" material are deposited contemporaneously throughout the entire time of "secondary" material deposition in some forms (TAVENER-SMITH, 1969, 1973; GAUTIER, 1973) (Fig. 108).

In the cheilostomates, meaningful appli-cation of such terms with temporal connotations as "primary" and "secondary" becomes even more difficult because of the greater diversity and complexity of skeletal subunits and growth modes in cheilostomate skeletons. For example, cheilostomate skeletons may be either all calcite, all aragonite, or bimineralic. Furthermore, not only are there two distinct mineralogies present, but also a diversity of ultrastructural types. This ultrastructural variety is particularly true for calcite, although recent work (SANDBERG, 1976 and unpublished) indicates greater ultrastructural diversity exists for aragonite than was previously thought. Also, the skeleton of any given species is commonly made up of three ultrastructural types, sometimes four, five, or perhaps more types. An opposite problem is that, in some aragonitic forms, except for a poorly developed outer planar spherulitic layer, the entire skeleton is constructed of a single ultrastructure.

Cheilostomate skeletons may, like stenolaemate skeletons, be produced by any one of a broad spectrum of growth modes ranging from only interior walls (except for ancestrular attachment surface) to all exterior walls (except for pore plates), with a variety of

---

FIG. 118 (see p. 262). Frontal budding.——*1–3. Hippoporidra senegambiensis* (CARTER), rec., S. of Tema, Ghana; *1,* numerous superimposed, frontally budded zooecia, transv. sec., ×170 (bar = 50 μm); *2,* detail of adjacent area, note absence of lateral cuticles, ×300 (bar = 20 μm); *3,* detail of *2,* note blisterlike nature of new frontal zooecium and spherulitic aragonite. ×1,475 (bar = 5 μm); all BMNH 1970.8.10.24.——*4,5. Schizoporella floridana* (OSBURN), rec., W. coast of Fla.; *4,* frontal view of region with developing, frontally budded zooecia, note all vertical walls are doubled and overlie vertical walls of lower zooecia, ×60 (bar = 100 μm); *5,* detail of *4,* note occluded orifice, ×120 (bar = 100 μm); both USNM 184158. ——*6. Porella compressa* (SOWERBY), rec., Sound of Mull, Scot.; frontal view of developing, frontally budded zooecium; note thin walls and ultrastructural difference from adjacent superficial calcification of older zooecia; BMNH 1888.6.9.45, ×120 (bar = 100 μm).

FIG. 119. (See p. 263.)

FIG. 120. Compartmentalized spherulitic aragonite walls.——*1,2. Adeonellopsis distoma* (BUSK), rec., Madeira; *1,* outer growth surface of frontal shield, etched, wall constructed of numerous parallel, fingerlike projections of spherulitic aragonite, ×2,250 (bar = 5 μm); *2,* lower magnification view of area of *1,* note reverse orientation of aragonite lobes in lower right, ×370 (bar = 200 μm); both BMNH 1911.10.1.927. ——*3,4. Adeonella atlantica* BUSK, rec., Nightingale Is., near St. Helena, S. Atl.; *3,* frontal shield, showing aragonite lobes with organic envelopes, note shape similarity to pillow lava, etched transv. sec., ×1,675 (bar = 5 μm); *4,* etched section parallel to long axis of some aragonite lobes in frontal shield, compare orientation of individual aragonite needles to those in *3,* ×1,675 (bar = 5 μm); both BMNH 1887.12. 9.725.——*5,6. Adeona* sp., rec., locality unknown; *5,* aragonite lobes in frontal shield, etched transv. sec., ×2,825 (bar = 5 μm); *6,* outer portion of frontal shield, note heavy organic partitions between aragonite lobes, etched transv. sec., ×1,400 (bar = 5 μm); both BMNH 1920.12.10.1.

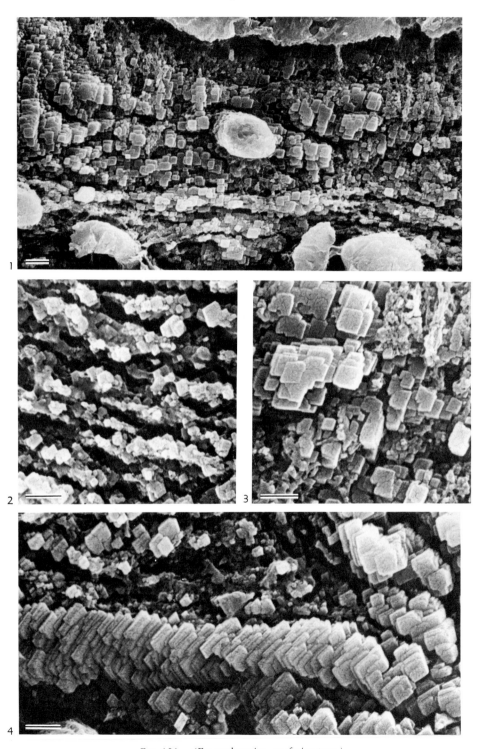

FIG. 121. *(For explanation, see facing page.)*

intermediate modes.

The skeletal succession of some cheilostomates may include repetitions of one or more ultrastructural units. This does not refer to the commonly observed identity of ultrastructure in morphologically equivalent parts in successive zooecia. Rather it is the recurrence, within the wall of a single zooecium, of an ultrastructural type already present in the previously deposited succession in that same wall (Fig. 114,5). This recurrence is comparable to the repetition of myostracal layers in some bivalves (TAYLOR, KENNEDY, & HALL, 1969).

The mineralogic and ultrastructural diversity tabulated above reflects the greater range of variations in skeletal makeup known among cheilostomates than among stenolaemates.

The distinctive ultrastructural-mineralogic units of any individual cheilostomate skeleton are clearly sequential. There is little problem in any part of the skeleton in determining a "local stratigraphic section" in that skeleton and applying a sequential terminology (such as "primary," "secondary," "tertiary") to the observed units in that local section. However, great difficulty and uncertainty exists in making any inference of equivalence (correlation) of skeletal units among skeletons of different cheilostomes. Is "secondary" in one form with four ultrastructural units the same as "secondary" in a form with only three? Note that this is a separate issue from the question of contemporaneity of deposition in different parts of a zooecium or zoarium.

One could apply the terms "primary" and "secondary," respectively, to the initial structural framework of the zooecium, and the elaborations or thickenings added to it. This usage is similar to that of CHEETHAM, RUCKER, and CARVER (1969), who used the terms primary and "superficial." Except possibly for this structural approach, the use of such terms as primary, secondary, tertiary should be avoided, especially for designation of individual, ultrastructurally distinctive units. Such terms imply the existence, among skeletons of diverse taxa, of a sort of stratigraphic equivalence of units that simply cannot be demonstrated.

MINERALOGY AND ULTRASTRUCTURAL TYPES IN CHEILOSTOMATE SKELETONS

MINERALOGY

Skeletons of cheilostomate bryozoans are composed of calcite, aragonite, or both (LOWENSTAM, 1954; SCHOPF & MANHEIM, 1967; RUCKER, 1967; RUCKER & CARVER, 1969; SANDBERG, 1971; POLUZZI & SARTORI, 1973, 1975). Calcite skeletons of living cheilostomates contain more $MgCO_3$ (3 to 12 mole percent, mean about 8 mole percent and most species between 6 and 9 mole percent; see POLUZZI & SARTORI, 1973, 1975; SCHOPF & MANHEIM, 1967; LOWENSTAM, 1963, 1964b) than do skeletons of living cyclostomates (all calcitic). Cheilostomate skeletons (or skeletal parts) composed of aragonite contain little $MgCO_3$, but their Sr/Ca ratios are at or near that of seawater (LOWENSTAM, 1964a,b; SCHOPF & MANHEIM, 1967; DODD, 1967).

In organisms with bimineralic carbonate skeletons, including cheilostomates, the two

---

FIG. 121. Dendritic calcite structures.——*1–4. Umbonula ovicellata* (HASTINGS), rec., Gairloch, NW. Scot.; *1,* frontal shield, distal to left, elliptical and curved features in center and lower edge are plastic fillings of endolithic algal borings, crystals quite uniformly rhombic, etched long. sec., ×1,000 (bar = 7.5 μm); *2,* detail of another area of section, note organic matrix sheets, ×4,000 (bar = 2.5 μm); *3,* detail of *1,* ×4,000 (bar = 2.5 μm); *4,* detail of another area of section, crystals in lower region forming an elongate dendritic array, ×4,000 (bar = 2.5 μm), all BMNH 1963.3.6.8.

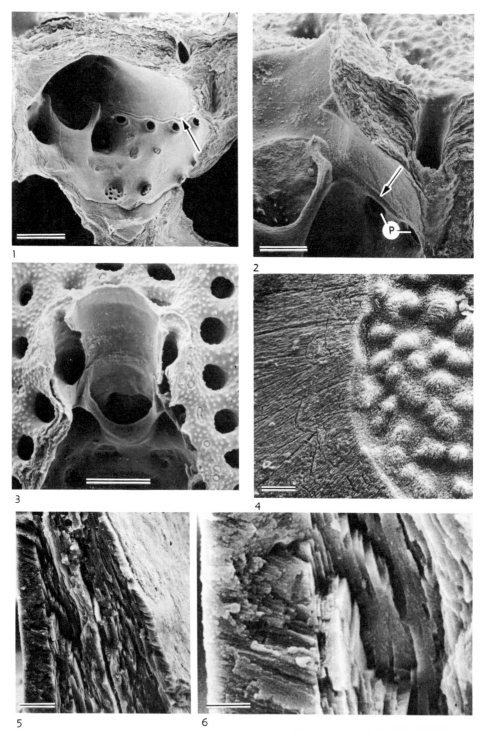

FIG. 122. *(For explanation, see facing page.)*

CaCO₃ polymorphs form discrete "microarchitecturally separate elements" (LOWENSTAM, 1954). In bimineralic cheilostomates the basic structural box of the zooecial skeleton is calcite; aragonite is added as elaborations or reinforcements whose position and degree of development vary considerably among taxa. Commonly, aragonite occurs as superficial layers on the frontal exterior surface only (see Fig. 125,*1*) (CHEETHAM, RUCKER, & CARVER, 1969; SANDBERG & others, 1969), even in **cryptocystideans**. Neither bimineralic nor aragonitic **gymnocystideans** (*sensu* BANTA, 1970; SANDBERG, 1976) have been encountered. In some zoarial forms in which basal skeletal walls beyond the ancestrula are interior walls (some **lunulitiform colonies**, GREELEY, 1969; HÅKANSSON, unpublished; some **petraliiform colonies**, SANDBERG, 1976), aragonite may also occur or even only occur as superficial basal thickenings. Only one example of aragonitic zooecial lining in a bimineralic species has been reported (SANDBERG, 1976). In bimineralic adeonids and adeonellids, calcite may be clearly present in the basal and lateral walls, probably (but not yet certainly observed) in the lower part of the frontal; aragonite makes up the vast bulk of the frontal and even lateral and transverse walls of those bimineralic skeletons.

ULTRASTRUCTURAL TYPES

The predominant ultrastructural types in cheilostomate skeletons can be broadly characterized as lamellar and spherulitic. The crystal morphologies are largely mineralogically controlled. However, in contrast to the situation found in bivalves, in which most ultrastructures are aragonite (KOBAYASHI, 1969, 1971), calcite forms most ultrastructures in cheilostomate bryozoan skeletons.

The individual CaCO₃ crystals in cheilostomate skeletons have a wide range of observed morphologies. These include such distinctive types as very elongate needle-shaped or lath-shaped crystals elongate in the *c*-axis direction and thin rhombic or hexagonal crystals flattened in a plane perpendicular to the *c*-axis. Those latter, planar crystals commonly make up the lamellar skeleton units and may exhibit spiral growth steps (screw dislocations) (Fig. 102,*4–6*; 103,*3,4*) as a result of lattice defects (WISE & DEVILLIERS, 1971; WADA, 1972). TAVENER-SMITH and WILLIAMS (1972, fig. 106, 139) noted such spiral growth, but appear to have too broadly applied the term "spiral growth" to include various arrangements of minute polycrystalline arrays (TAVENER-SMITH & WILLIAMS, 1972, fig. 31, 54).

There is some uncertainty (as discussed by

FIG. 122. Interior-exterior wall boundaries and calcite wall ultrastructures.——*1–3. Metrarabdotos tenue* (BUSK), rec., Caroline Sta. 68, off NE. Puerto Rico; *1*, transverse fracture section; scalloped double line (arrow) along right lateral wall marks attachment of membranous frontal, exterior epifrontal wall above, areolar pores below, and rather tubular oral shelf distally (Y-shaped in this view); also note multiporous septulae and numerous layers of frontal wall, ×135 (bar = 100 μm); *2*, detail of fractured section of frontal, line of attachment of membranous frontal (arrow) extending above areolar pores (P) and around distal margin of oral shelf; above that line and continuing out peristome, wall bears planar spherulitic ultrastructure, ×260 (bar = 50 μm); *3*, frontal view of zooecium with frontal broken out, longitudinal stripes of planar spherulitic ultrastructure faintly visible on distal wall of peristome, ×175 (bar = 100 μm); all USNM 209434.——*4. Megapora ringens* (BUSK), rec., Shetland Is.; detail of kenozooid in frontal view; exterior wall (gymnocyst) with planar spherulitic ultrastructure to left, interior wall (cryptocyst) with tuberculate thickening to right; curved line between marking outer edge of both membranous frontal and hypostegal coelom; BMNH 1911.10.1.630, ×950 (bar = 10 μm).——*5,6. Watersipora subovoidea* (D'ORBIGNY), rec., locality unknown; *5*, fractured transverse section, unetched, of lateral walls of adjacent zooecia; walls composed of mainly lamellar calcite, but with zooecial linings of massive, apparently spherulitic calcite, ×1,900 (bar = 5 μm); *6*, detail of *5*, ×6,000 (bar = 2 μm); both BMNH 1970.6.1.32.

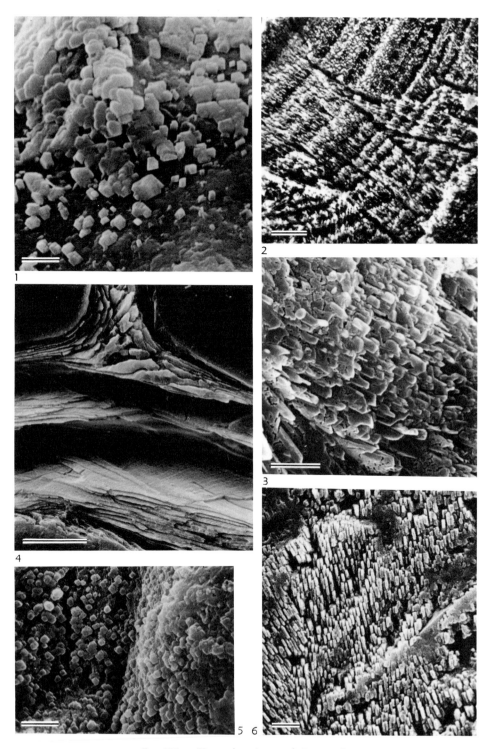

FIG. 123.   *(For explanation, see facing page.)*

TOWE & CIFELLI, 1967, p. 744–745; TOWE, 1972, p. 2–4) as to what constitutes "a crystal." Certainly, in cheilostomate skeletons there are some "crystals" that, after strong etching, appear to be composed of many aligned "subcrystal units" (Fig. 115).

The shapes of individual skeletal crystals in bryozoans (as well as other organisms) are related to mineralogy, amount and distribution of organic matrix, rate of carbonate deposition, and other factors. There is a general similarity in those controls that transcends even phylum boundaries. For example, very similar spherulitic arrays of acicular aragonite occur in bryozoans, mollusks, scleractinian corals, and sclerosponges. In contrast, the effects of some factors may vary among taxa. Broad rhombic lamellar crystals with screw dislocations are only aragonite in gastropods and bivalves and only calcite in bryozoans and brachiopods, despite great morphologic similarity of all those crystals.

Organic matrix can be quite abundant in cheilostomate skeletal carbonate, whether as distinct intercrystalline sheets, separating and surrounding individual crystals or regions of crystals (Fig. 103,2; 109,3,4; 114,2; 120,3–6), or as a more diffuse intracrystalline network visible only after extensive etching (Fig. 103,6). In some instances, as in the more tabular aragonite (Fig. 103,5), the distribution of intercrystalline organic matrix seems strongly to affect the carbonate crystal shape.

In earlier published polarized-light studies of cheilostomate skeletons and in my limited number of such observations, the *c*-axis orientation of skeletal crystals has generally been easy to determine. The *c*-axes are usually aligned parallel to the wall surface in longitudinal stripes of planar spherulitic ultrastructure and perpendicular to the wall surface in most lamellar or **transverse spherulitic ultrastructures.** A detailed comparative study of optic orientation and crystal morphology and arrangement would be most beneficial for a clearer understanding of skeletal structure, particularly of the seemingly irregular or homogeneous units.

Individual crystals of some cheilostomate skeletal units may be quite striking (Fig. 119,1,2; 121) but nevertheless may resemble crystals in skeletons of other phyla. Therefore, the main value that skeletal ultrastructure may have in ontogenetic or phylogenetic reconstructions is not in the individual crystals, but rather in the aggregate units of crystals or successions of units. Examples of these aggregate units are the planar or lenticular lamellae, the planar spherulitic fans, and the conical or palisade spherulitic arrays. There are a number of less clearly organized (or at least less clearly understood) aggregates of crystals. These ultrastructures are poorly ordered arrays, usually involving minute, equant crystals for which preferred orientation individually or in aggregate is not evident. Similar ultrastructure in bivalves has been referred to as "homogeneous." Al-

FIG. 123. Ultrastructures of wall surfaces and sections.——*1. Membranipora grandicella* (CANU & BASSLER), rec., Albatross Sta. D5315, Philip.; frontal shield (cryptocyst) exterior, distal toward top, USNM 209441, ×5,250 (bar = 2 μm).——*2.* Sertellid sp., rec., locality unknown; calcite zooecial lining layers on proximal side of basal-transverse wall junction; competitive, interfering growth of spherulitic calcite arrays clearly shown; etched long. sec., BMNH 1892.1.28.112, ×5,100 (bar = 2 μm).——*3. Tubiporella magnirostris* (MACGILLIVRAY), rec., Port Phillip, Vict., Australia; detail of basal exterior surface with lobes of acicular aragonite crystals, crystals here the more massive laths; BMNH 1887.6.27.1, ×6,800 (bar = 2 μm).——*4. Micropora* sp., rec., Albatross Sta. D2856; basal-transverse wall junction, distal toward right; note broad lathlike crytals in basal wall, separated because of embedding-plastic shrinkage; USNM 209442, ×300 (bar = 20 μm).——*5. Reteporella myriozoides* BUSK, rec., Challenger Sta. 148, Possession Is., SW. Indian O.; frontal shield exterior surface; rhombic calcite crystals seeded over entire wall surface, forming layers by lateral accretion; BMNH 1887.12.9.516, ×5,250 (bar = 2 μm).——*6. Ogivalia gothica* (BUSK), rec., Challenger, Prince Edward Is.; frontal shield interior surface, etched; calcite crystals here stubby rods, but in some nearby wall areas more flattened and rhombic; BMNH 1887.12.9.358, ×3,800 (bar = 2 μm).

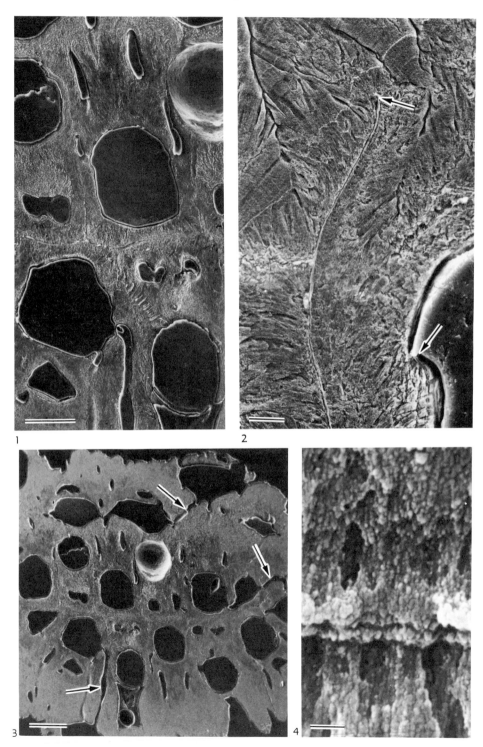

FIG. 124. *(For explanation, see facing page.)*

though some of the molluscan (and bryozoan) ultrastructures may approach a truly homogeneous texture, the term really reflects not a distinct ultrastructural type, but rather a complex of as yet poorly understood fine textures that have been beyond the resolution limit of the light microscope.

It is probably best not to perpetuate, as general categories, the "parallel" and "transverse" groups proposed earlier (SANDBERG, 1971). The difference between lamellar ("parallel") and spherulitic ("transverse") ultrastructures can be totally a function of organic sheet development with no change in crystal morphology or orientation. This was particularly well shown by MUTVEI (1972) for the cephalopod *Nautilus*. Furthermore, even in the "parallel" lamellar ultrastructure, the tabular crystals may sometimes be composed of "transverse" lath or needlelike subcrystals (Fig. 116,*1*; 122,*5,6*; ERBEN, 1974).

It is important to recognize that much of the distinction between the ultrastructural categories previously proposed for cheilostomate skeletons is a matter of degree of dominance of textural details either parallel to or perpendicular to the wall surface. With the exception of the planar spherulitic layer and skeletons of some interior-walled forms, the skeletal units are accreting over broad areas parallel to the wall surface. Textural features that are oriented parallel to the wall include discrete tabular or lenticular units separated by diffuse or distinct organic sheets (lamellar ultrastructure) and accretionary banding (most commonly seen as distal increments on crystals growing generally *trans-*

verse to the wall). The main textural component perpendicular to the wall surface is that of crystals elongate in that growth direction. Those acicular or bladed crystals most commonly are arranged in spherulitic bundles transverse to the wall. Massive units common as zooecial linings provide an example of a possible arbitrariness in the terminology. Those massive units tend to be composed of spherulitic ultrastructure, but have accretionary banding. Depending on which of those components is most evident, the massive unit could be called either "parallel" or "transverse." Consequently those terms, although useful for orientational and descriptive purposes, should not be used as names of specific distinctive ultrastructural groups.

INDIVIDUAL CRYSTALS AND CRYSTAL AGGREGATES IN CALCITE AND ARAGONITE

Aragonite in cheilostomate skeletons was earlier (SANDBERG, 1971) known only as acicular crystals, mainly in transverse, usually spherulitic arrays (Fig. 114,*1*; 117,*1–3,5*; SANDBERG, 1971, pl. 3, fig. 1–8; pl. 4, fig. 1–3). Such spherulitic aragonite arrays comprise the entire skeleton (above an aragonite planar spherulitic layer) in some cheilostomates. Recent SEM study has shown that more blocky aragonite can occur if there are closely spaced organic sheets (Fig. 103,*5*) transverse to the direction of spherulite growth. Furthermore, acicular aragonite commonly has been found as planar spherulitic ultrastructure against cuticle in exterior walls (SANDBERG, 1971, pl. 3, fig. 3).

FIG. 124. Frontal loss of intercalary cuticles.——*1–4. Porella compressa* (SOWERBY), rec., Sound of Mull, Scot.; *1,* intercalary cuticles extend along basal surfaces of primogenial layer and up between lateral calcified walls to only slightly above upper end of interior zooecial cavity, overlying, frontally budded zooecia lack cuticle and are irregularly arranged, etched transv. sec., ×140 (bar = 100 μm); *2,* detail of a frontal-lateral junction, intercalary cuticle overgrown (upper arrow) by extrazooidal frontal thickening with distinct accretionary banding and spherulitic structure, shelf on inner lateral wall surface (lower arrow) marks upper end of maximum interior thickening of lateral wall and perhaps also lower edge of ascus, etched transv. sec., ×925 (bar = 10 μm); *3,* lower magnification view of area of *1,* note continuation of areolar pores through frontal thickening to frontally budded zooecia (arrows), ×55 (bar = 200 μm); *4,* detail of spherulitic structure in *2,* radiating arrays composed of minute aligned calcite crystals, ×18,500 (bar = 0.5 μm); all BMNH 1888.6.9.45.

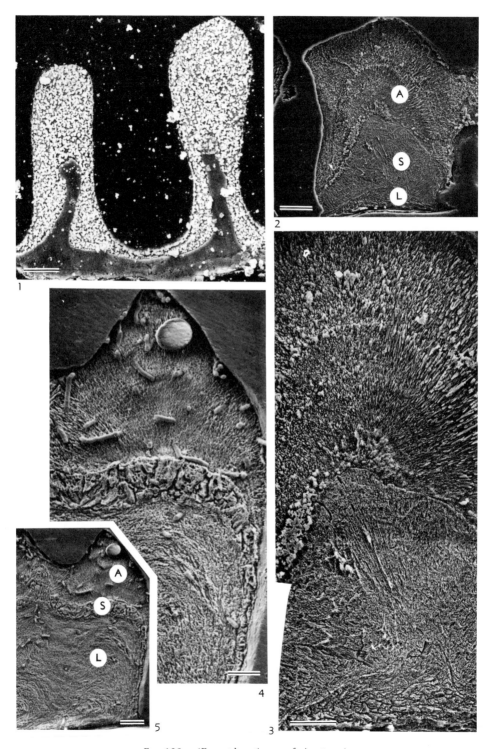

FIG. 125. *(For explanation, see facing page.)*

Calcite in cheilostomate skeletons is much more diverse in crystal morphology, but most ultrastructures are basically lamellar or spherulitic (note comparable groupings in bivalve shells as discussed by TAYLOR, 1973). Individual crystal morphologies and the aggregate arrangements for calcite in cheilostomate skeletons include:

1. Rhombic or hexagonal crystals, flattened in a plane perpendicular to the *c*-axis, are commonly quite large and often show screw dislocations (Fig. 102; 115,*1*). These crystals make up much of the lamellar ultrastructural units. Marginal growth of the individual crystals may leave distinctive accretionary banding (Fig. 102,*3*). Closely stacked or dendritic arrays of this general crystal type make up some of the lenticular or crudely laminated skeletal units, as well as some of the irregular, massive units (Fig. 121,*1–4*; 123,*1,5*).

2. In a few cheilostomates, broad calcite lamellae either show no clear individual crystal units (Fig. 123,*4*; not studied in polarized light) or have large, usually thin, bladelike crystals with indications of accretion at the end of the blade (Fig. 119,*1,2*). The latter crystals are similar to some bladelike crystals in stenolaemate bryozoan and brachiopod skeletons and represent a form of edgewise growth.

3. Some irregular massive or lenticular skeletal units are made up of a mixture of the flattened rhombic crystals discussed above and small truncated, rod-shaped crystals. These rods are parallel within individual arrays, but occur in diverging arrays that suggest a form of spherulitic growth (Fig. 123,*3,6*).

4. In the planar spherulitic layers occurring against cuticle, calcite crystals are most commonly acicular, rarely flattened laths (Fig. 111), and generally arranged in wedge- or fan-shaped arrays. In those arrays, individual crystals may stand out clearly or be subordinate in clarity to larger, usually triangular or trapezoidal aggregates of crystals (Fig. 111,*1,2*; 112,*1,3*; 122,*4*). These generalities of crystal orientation and appearance hold whether the calcification that produced the planar spherulitic ultrastructure occurred at an advancing linear front or radially from scattered centers of crystallization (Fig. 111,*3,4*).

5. It was noted above that spherulitic arrays transverse to the wall are the most common skeletal unit in aragonite. Similar spherulitic units are also quite common in calcite, but are made of crystals with a far greater variety of individual shapes. Acicular crystals of calcite comparable in size and shape to those that commonly form the planar spherulitic layer on exterior walls also make up some of the transverse spherulitic units (Fig. 123,*2*). In addition, such spherulitic calcite units may be made up of needlelike columns of very minute, equant crystallites (Fig. 124,*4*). In other cheilostomates the spherulitic arrays are composed of massive, crudely conical calcite masses with crenulate, interlocking boundaries (Fig. 114,*2–5*; 125,*4*). Such arrays are almost certainly the source of the pattern called "cell-mosaic" by LEVINSEN (1909) and SANDBERG (1971) and

FIG. 125. Bimineralic walls.——*1. Metrarabdotos unguiculatum* (CANU & BASSLER), rec., Albatross Sta. D2405, Gulf of Mexico; frontal shield, treated with Feigl solution; precipitate (Ag and MnO$_2$) selectively formed on superficial aragonite, leaving lower calcite unit etched but unstained; long. sec., USNM 184156, ×900 (bar = 10 μm).——*2,3. Pentapora foliacea* (ELLIS & SOLANDER), rec., Cornwall; *2*, frontal shield, initial, calcite portion composed of a lower, laminated unit (L) and an upper, spherulitic unit (S), which is, in turn, surmounted by a superficial spherulitic aragonite unit (A), long. sec., ×450 (bar = 20 μm); *3*, detail of *2*, ×1,250 (bar = 10 μm); both BMNH 1911.10.1.1561.——*4,5. Tubiporella magnirostris* (MACGILLIVRAY), rec., Port Philip Head, Australia; *4*, frontal shield, upper portion of wall penetrated by numerous borings (now plastic-filled), etched transv. sec., ×850 (bar = 10 μm); *5*, lower magnification view of same wall; lower, poorly laminated calcite portion (L) overlain by thin spherulitic calcite unit (S) and thicker, superficial aragonite unit (A), ×340 (bar = 20 μm); both BMNH 1927.8.4.24.

Bryozoa

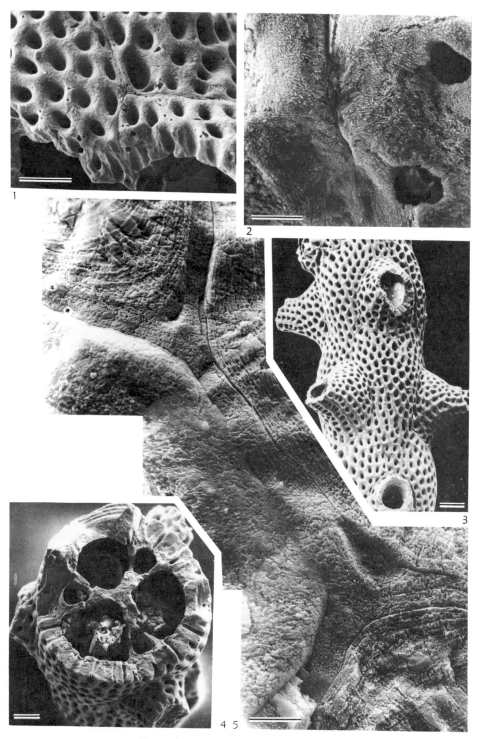

Fig. 126. *(For explanation, see facing page.)*

not actual, single-cell secretory zones (see also SANDBERG, 1976). In some such skeletal units the calcite appears to be in quite large crystals; etching reveals no substructure other than accretionary banding (Fig. 113,*4*; 114,*3*). In others, the etching shows a very fine, granular or acicular substructure that may be related to the arrangement of intracrystalline organic matrix. In some massive layers there occur calcite crystals that are arrayed as laths at an angle to the wall surface, sometimes in roughly conical groupings (Fig. 122,*6*; SANDBERG, 1971, pl. 2, fig. 9). These crystals are clearly flattened and appear to be growing by terminal extension toward the adjacent coelomic space with some intercalation of new crystals between in the conical arrays.

Spherulitic calcite commonly occurs in zooecial linings (Fig. 114,*2,3*; 122,*5,6*; 126,*5*) as well as superficial thickenings, especially in frontal shields. In those frontal shields, the spherulitic calcite is often followed by spherulitic aragonite (Fig. 125,*2–5*). In some cheilostomates, spherulitic aragonite alone may compose the superficial thickening of the frontal shield (Fig. 103,*1*; SANDBERG, 1971, pl. 4, fig. 1, 2).

Organic matrix occurs most commonly as intercrystalline networks, sheaths around individual crystals, or as bounding sheets at the outer surface of exterior walls or between some ultrastructural units. In some few groups there may be developed an intermediate level of organic matrix as envelopes compartmentalizing regions of an ultrastructural unit. For example, in adeonids the transverse spherulitic aragonite, which makes up most of the skeleton, is subdivided into long, fingerlike units by tubular organic sheaths. Within each sheath, numerous, minute aragonite needles are arranged with their long axes generally parallel to the long axis of the enclosing organic tube (Fig. 120,*3–5*). These organic-walled, fingerlike skeletal units appear to originate by distal prolongation of the numerous lobelike skeletal projections on the frontal exterior surface (Fig. 120,*1,2*).

CORRELATION OF ULTRASTRUCTURE AND SKELETAL GROWTH MODES

ULTRASTRUCTURAL RECOGNITION OF EXTERIOR AND INTERIOR SKELETAL WALLS

The recognition, on the basis of skeletal features, of ontogenetic development patterns and major taxonomic groups among cheilostomates depends heavily on the ability to differentiate calcified interior and exterior walls (SILÉN, 1944a,b), especially in the frontal region. Combining the various definitions of SILÉN (1944b, p. 436), BANTA (1970, p. 39), BOARDMAN and CHEETHAM (1973, p. 131), and BOARDMAN, CHEETHAM, and COOK (this revision): **exterior skeletal walls** are those walls which calcify against cuticle and which occur in body walls that (in their precalcified, membranous state) expanded the

FIG. 126. Frontal intercalation of cuticle.——*1–5. Margaretta tenuis* HARMER, rec., Albatross Sta. D5134, Philip.; *1,* fractured frontal shield, etched, distal toward top, note lines of cuticle intercalation in lateral (running vertically through figure) and transverse positions, ×140 (bar = 100 μm); *2,* detail of *1,* intersection between line of lateral cuticle intercalation and transverse fracture, showing that cuticle is a near-surface phenomenon in frontal shield and does not extend down into lateral walls, ×1,400 (bar = 10 μm); *3,* exterior view of internode, distal toward top, note superficial cuticulate boundaries both proximal and distal to each peristome, as well as laterally, ×35 (bar = 200 μm); *4,* transverse fractured section, etched, of a zoarium, ×70 (bar = 100 μm); *5,* detail of *4,* axis of zoarial segment at triradiate junction (lower right), figure shows massive zooecial lining deposits of four zooecia surrounding thinner, initial skeletal layers, significantly, these initial layers are not subdivided by intercalary cuticles, ×1,400 (bar = 10 μm); all USNM 209446.

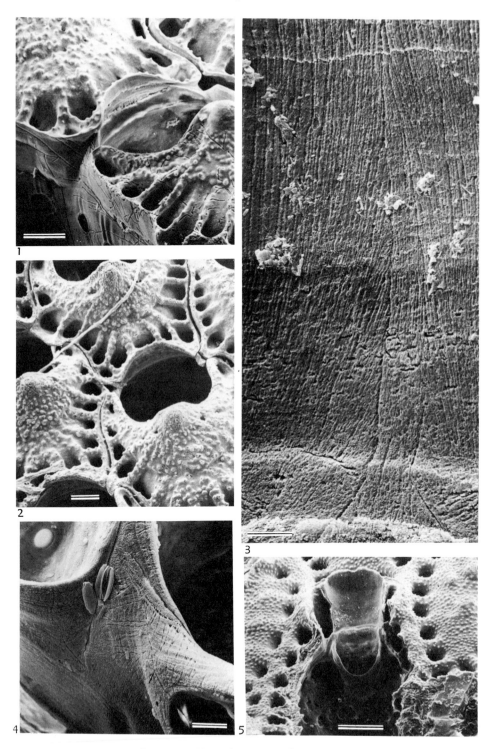

FIG. 127.   *(For explanation, see facing page.)*

coelomic volume of the colony. **Interior skeletal walls**, in the same terms, are walls that grow off the inner surface of exterior skeletal walls (or other interior skeletal walls) by apposition and partition preexisting coelomic volume of the colony. In the absence of observations on preserved distal colony edges or on living colony growth (obviously not possible with fossils), such differentiation must be based on some skeletal record of the growth mode. In the case of contiguous exterior walls (e.g., lateral walls of two adjacent lineal series) the presence of doubled intercalary cuticles (in dead modern material) or of a sharp central break between the two walls (in fossils) is indicative of exterior walls. However, that criterion does not work for single, noncontiguous exterior walls, such as those in the frontal region. The ultrastructure of interior and exterior skeletal walls is distinctive and provides an excellent supplement to morphological criteria for wall differentiation. Superimposed lamellae or accretionary bands on both sides of a cheilostomate skeletal wall reflect the presence of secretory epithelia on both sides and therefore the origin of that wall as an interior wall. In contrast, initial calcification adjacent to cuticle in an exterior wall is an array of planar spherulites (Fig. 111; 112,*1,3–5*; 127,*3,4*; 128). Such planar spherulitic ultrastructure (see discussion in SANDBERG, 1971, 1973, 1976) is produced by calcification at a linear front advancing over a surface (the exterior cuticle). The spherulitic ultrastructure is visible on exterior walls of cheilostomates because it is left exposed on the outer surface of the skeletal wall by the one-sided skeletal growth away from cuticle. Planar spherulitic ultrastructure characterizes cuticulate exterior walls in skeletons of not only cheilostomate but also cyclostomate bryozoans (SÖDERQVIST, 1968; TAVENER-SMITH & WILLIAMS, 1972; BROOD, 1973). Similar ultrastructure has been found on the underside of coral tabulae and dissepiments (WELLS, 1969; BARNES, 1970; SORAUF, 1971, 1974), which, although not cuticulate, are formed by one-sided growth with calcification advancing on a linear front.

The existence of a distinctive ultrastructure on exterior walls in cheilostomate skeletons is of great significance in the interpretation of genesis of frontal shields and other walls, ovicells, spines, and other calcified features. Its presence or absence can usually be determined easily by SEM study of surfaces of specimens freed from sedimentary matrix. However, techniques for recognition of that planar spherulitic ultrastructure in sections must be developed before it could be of general use for study of solidly embedded cheilostomates or of ancient stenolaemates, which commonly occur in dense crystalline limestone.

Similar ultrastructure may occur in stenolaemates on the undersides of diaphragms, which calcify from one side only. However, because of the difference in zooecial shape and growth mode in cheilostomates, such one-sided, later ontogenetic, proximal partitions evidently do not occur. Even in stenolaemates, the interior wall nature of the dia-

FIG. 127. Cormidial apertures, exterior walls in peristomes, and transverse walls.——*1,2. Umbonula ovicellata* (HASTINGS), rec., Gairloch, NW. Scot.; *1,* oblique frontal view of zooecial aperture, zooecial row along left removed, exposed lateral walls showing planar spherulitic ultrastructure and numerous borings (algal?), note formation of peristome by distal zooid, ×120 (bar = 100 μm); *2,* frontal view, note sides of secondary orifice formed in part by lobes of superficial calcification by zooids of adjacent lateral rows, ×80 (bar = 100 μm); both BMNH 1963.3.6.8.——*3. Metrarabdotos tenue* (BUSK), rec., Caroline Sta. 68, off NE. Puerto Rico; detail of distal wall of peristome with stripes of planar spherulitic ultrastructure oriented distally (toward top); USNM 209434, ×2,400 (bar = 5 μm).——*4. U. ovicellata,* same data as *1;* detail of transverse-lateral wall junction, distal toward upper left; line of cuticle incorporation into upper transverse wall partly obscured by diatoms; planar spherulitic ultrastructure showing direction of growth of various wall regions; BMNH 1936.3.6.8, ×460 (bar = 20 μm).——*5. M. tenue,* lower magnification view of specimen in *3;* distal peristome wall is the cuticle-bounded proximal end of frontal shield of distal zooecium; ×120 (bar = 100 μm).

Bryozoa

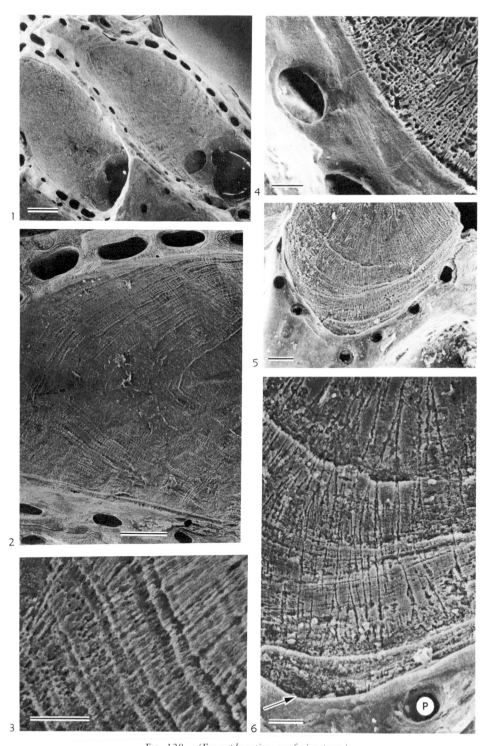

Fɪɢ. 128. *(For explanation, see facing page.)*

phragms is discernible by study of topological relationships between a diaphragm and surrounding walls.

When planar spherulitic ultrastructure is composed of longitudinal stripes, the accretionary lineations often form a scalloped curve (Fig. 111,*1*; SANDBERG, 1971, pl. 2, fig. 1, 2, 4). On flat or convex surfaces the scalloped curve is convex in the distal growth direction. The situation may be different on such concave surfaces as the undersides of umbonuloid frontal shields (Fig. 128,*2,6*). Because the growth front on such concave walls is continually decreasing in radius of curvature as it advances toward the zooecial midline, there is a tendency toward distal narrowing of the arrays of planar spherulitic ultrastructure, and sometimes even distally concave scalloped growth lines. Growth front curvature, together with the distal bifurcation and expansion of spherulitic arrays, may be used to determine the direction of growth, even in small areas (or even fragments). The ability to recognize growth direction is significant in the reconstruction of earlier ontogenetic stages from mature zooecia. For example, the planar spherulitic ultrastructure in the region of the upper transverse wall and the secondary orifice of umbonuloids lends itself very well to a chronicling of the ontogeny of calcified structures in that region (including orifices that are **cormidial**, i.e., the joint product of more than one individual in the colony) (Fig. 122,*1–3*; 127).

It must be realized that, even in the simplest exterior wall construction, any calcified wall that might form is not expanding the coelomic volume. Rather it develops in a calcification zone that lags slightly behind the front of cuticle intussusception, where the coelomic expansion is occurring (see SCHNEIDER, 1957, 1963). Nevertheless, it is the intimate association of carbonate skeleton and the outermost bounding cuticle that demonstrates the exterior wall origin of such skeletal walls in cheilostomates. As discussed above, such exterior walls characteristically have, as their outermost carbonate unit, planar spherulitic ultrastructure. From that ultrastructure (with a few cautions mentioned below) we can recognize the exterior walls.

It should be emphasized that the cuticle-calcified wall association referred to here is the one developed penecontemporaneously at the distally advancing margin of wall growth. It does not refer to the possible later contact of cuticle with an already calcified wall, such as apparently occurs on the undersides of some frontal shields during ascus invagination.

Exterior walls commonly give rise to interior walls by apposition, but that first wall can never become an interior wall. It will always have an exterior cuticulate surface on one side, even if the wall is subsequently overgrown by skeletal layers of another part of the colony, for example, ovicells. However, it is relatively common for individual skeletal walls to have both interior and exterior wall portions. This is not surprising, because the operational difference between those two major wall types is deposition in

---

FIG. 128. Frontal exterior walls.——*1–3. Posterula sarsi* (SMITT), rec., Gulf of St. Lawrence; *1*, basal view of several zooecia, etched, with basal walls broken away, distal toward lower right, ×80 (bar = 100 μm); *2*, detail of upper zooecium in *1*, distal toward right, note marginal areolae, line of membranous frontal attachment, and accretionary banding and distally oriented crystals of planar spherulitic ultrastructure on basal (exterior) surface of epifrontal shield, ×260 (bar = 50 μm); *3*, detail of *2*, ×1,650 (bar = 10 μm); BMNH 1911.10.1.1360A,B.——*4. Umbonula ovicellata* (HASTINGS), rec., Liverpool Bay, Eng.; detail of areolar pores, line of membranous frontal attachment and, in upper right, stripes of planar spherulitic ultrastructure, etched; BMNH 1936.12.30.380, ×425 (bar = 20 μm).——*5, 6. Metrarabdotos tenue* (BUSK), rec., Caroline Sta. 68, off NE. Puerto Rico; *5*, basal view of a zooecium, etched, with basal wall broken away, distal toward upper right, ×130 (bar = 50 μm); *6*, detail of *5*, planar spherulitic ultrastructure begins immediately at line of membranous frontal attachment (arrow), above areolar pore (P), note numerous instances of competitive growth interference and accretionary bands of varying intensity, ×500 (bar = 20 μm); both USNM 209434.

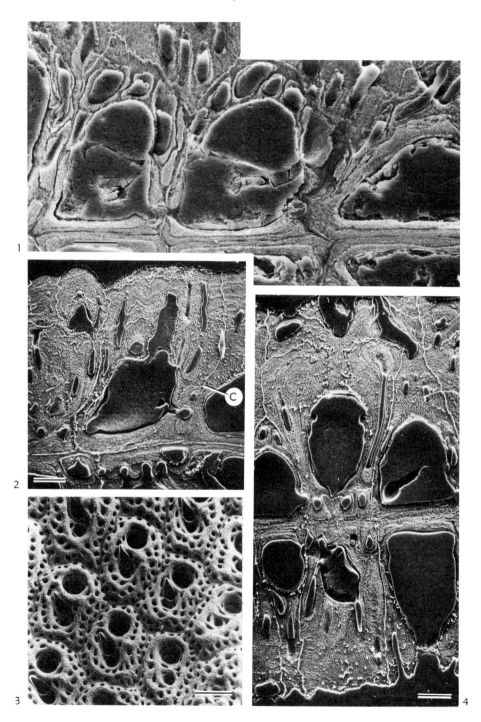

FIG. 129.   *(For explanation, see facing page.)*

contact, or not in contact, with cuticle. Various growth patterns or positions of wall development are naturally in contact (or not) with cuticle, and a single wall may go through both stages. This occurs most commonly in the transverse vertical wall. The lower central part, often most of the wall, is an interior wall, grown by apposition off the inner surface of the basal wall. The upper part of the wall impinges on the frontal cuticle and continues upward growth, incorporating cuticle and producing a doubled exterior wall region with intercalary cuticle (Fig. 129,2). Most commonly that upper exterior wall part of the transverse wall is short relative to the total height of the wall (Fig. 116,2). However, it increases as frontal thickening occurs, and in some forms, such as the adeonids, comprises the bulk of the transverse wall. In those adeonids, the interior wall portion of the transverse wall is narrow relative to the total zooecium. As a result, near the lateral walls, intercalary cuticle extends through the transverse wall from frontal to basal surfaces (Fig. 129).

Several examples of ontogenetic transitions from interior to exterior walls are known in frontal walls. The frontal shield of an umbonuloid ascophoran originates as a cryptocystal extension (thus an interior wall) off the inner surface of the lateral and transverse walls below the membranous frontal wall (Fig. 130). The figure construction of TAVENER-SMITH and WILLIAMS (1970, fig. 35) is accurate, although they called the initial, marginal part of the umbonuloid frontal shield a gymnocystal wall, as did COOK (1973b). As that initially cryptocystal wall grows upward and medially, it ultimately meets the membranous frontal wall, attaches to the outermost frontal cuticle and extends to produce an epifrontal fold with cuticle and skeleton on its lower surface and a hypostegal coelom (which is extending the colonial coelomic space) above. In that process, the umbonuloid frontal shield becomes an exterior wall, calcifying (with planar spherulitic ultrastructure) against cuticle (Fig. 122,1; 128).

When extreme frontal thickening of cryptocystidean frontal shields occurs, the colonial coelom expands upward and calcification against cuticle occurs marginally (laterally, proximally, distally). The central portion of the frontal thickening was deposited below an upwardly advancing hypostegal coelom. However, at the vertical boundaries of the zooecium, the carbonate was secreted in contact with the cuticle just below an upward-moving front of cuticle intussusception (see BANTA, 1972, fig. 3). This exterior-walled growth pattern is analogous to the distal extension of exterior basal and vertical walls at the colony margins in the "standard" cheilostomate pattern. Adding zooecial cavities in this upwardly growing skeletal succession would produce the frontally-budded lineal series of zooecia discussed by BANTA (1972) (Fig. 118,4,5). In some cheilostomates frontal budding occurs as thin-walled, blisterlike zooecia without extensive frontal thickening (Fig. 118,6).

"STANDARD" GROWTH MODE FOR CHEILOSTOMATE SKELETONS

Most discussions of skeletal wall genesis in cheilostomates have implicitly or explicitly

---

FIG. 129. Frontal thickening.——*1–4. Adeona* sp., rec., locality unknown; *1,* lower part of three zooecia, distal to right, cuticles in transverse walls come down to join basal wall cuticles, note also zooecial lining deposits, especially on basal and lower vertical walls, etched long. sec., ×220 (bar = 50 μm); *2,* zooecium, distal to right, note considerable frontal wall thickening and that intercalary cuticle in vertical wall proximal to zooecium extends to basal wall, but in vertical wall distal to that zooecium, below point C, does not, etched long. sec., ×85 (bar = 100 μm); *3,* frontal exterior view of part of a colony, distal toward top, transverse wall boundaries with intercalary cuticle like that of lateral wall boundaries, ×50 (bar = 200 μm); *4,* colony branch, etched transv. sec., ×90 (bar = 100 μm); all BMNH 1920.12.10.1.

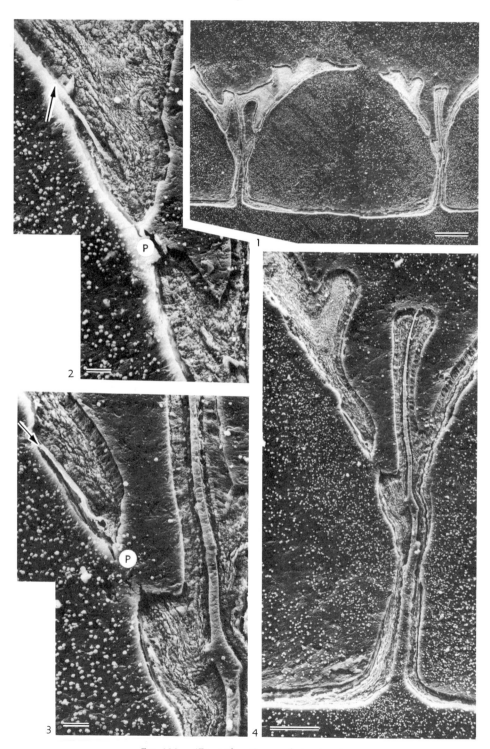

FIG. 130.   *(For explanation, see facing page.)*

made use of what may be called the "standard" pattern of cheilostomate skeletal development. This pattern is the most common one among cheilostomates and is characterized by predominantly exterior-walled skeletal growth. That is, the calcified walls are cuticulate (exterior walls) laterally, basally, and frequently frontally, but non-cuticulate (interior walls) distally (in the transverse wall) and in some frontal shields (see SILÉN, 1938, 1944a,b; BANTA, 1969, 1970; RYLAND, 1970).

However common it may be, this "standard" pattern clearly does not fit all cheilostomates. Recent studies have shown that it lies near the middle of a broad spectrum of cheilostomate skeletal growth modes which ranges from types with nearly all exterior walls to types with nearly all interior walls. It has been noted (BOARDMAN & CHEETHAM, 1973, p. 163) that interior walls in cheilostomates may be limited to pore plates between zooecia (e.g., *Pyripora* and similar uniserial forms, THOMAS & LARWOOD, 1956, 1960). At the other end of the spectrum, there appear to be more forms with interior-walled skeletons than was earlier realized. BANTA (1968, 1969, 1970, 1972), HÅKANSSON (1973), and SANDBERG (1973, 1976) have pointed out diverse cheilostomates in which interior walls comprise most or all of the zoarium except for the basal surface of the ancestrula or the multiple primary zooids (HÅKANSSON, 1973; MATURO, 1973). This interior-walled growth is particularly common in discoidal or linguiform zoaria (cupuladriids, conescharellinids, mamilloporids: HÅKANSSON, 1973; SANDBERG, 1973), some petraliiform zoaria

(SANDBERG, 1976), and some erect forms (cellariids and *Myriapora*: BANTA, 1968, SANDBERG, 1973; *Euthyrisella,* HARMER, 1902). The earlier inclusion of the sertellid *Triphyllozoon* as an example of this growth mode (SANDBERG, 1973, p. 308) appears to be an error based on misinterpretation of the cuticle distribution.

Skeleton construction in the majority of cheilostomates appears to follow the intermediate, "standard" pattern. Those "standard" cheilostomates may be grouped on the basis of relative spatial arrangement of features in the frontal region. Those features include the cuticle, secretory epithelia, initially and subsequently calcified portions of the wall, hydrostatic mechanism, and coelomic spaces.

Knowledge of cuticle distribution in vertical walls of cheilostomate skeletons is thus important in deducing wall type and growth mode. One might expect simple inspection of the frontal surface of the zooecia to produce this information. However, because cheilostomates often deposit superficial thickenings of the frontal, that is often not the case, especially where zooecia in ontogenetically earlier stages are not preserved. It is common for secondary thickening to take the form of an extrazooidal, colonial calcification in which the cuticles that exist in vertical walls (usually only in lateral walls) become detached from the frontal cuticle and are buried beneath the resulting extrazooidal wall (Fig. 124,1,2). Inspection of frontal exteriors of such forms would fail to reveal the exterior wall nature of the vertical (usually lateral) walls.

FIG. 130.   Cross sections of exterior frontal walls.——*1–4. Umbonula ovicellata* (HASTINGS), rec., Gairloch, NW. Scot.; *1,* zooecium, adjacent zooecia partially separated along lateral wall intercalary cuticles, embedding plastic filling intervening space, areolar pore on right open (and plastic-filled) all the way to frontal surface in plane of section, etched transv. sec., ×90 (bar = 100 μm); *2,* detail of region of areolar pore (P) of left adjacent zooecium in *1,* note incorporation of cuticle (arrow) into calcareous wall, above arrow lower surface of wall bears planar spherulitic ultrastructure, ×680 (bar = 10 μm); *3,* detail of right areolar pore of central zooecium in *1,* see comments on *2,* ×680 (bar = 10 μm); *4,* intermediate magnification view of right zooecial boundary in *1,* note continuous cuticulate boundary between vertical walls and thin but distinctive superficial frontal layer wrapping around upper frontal surface, ×260 (bar = 50 μm); all BMNH 1963.3.6.8.

Not only may cuticles be buried, but intercalary cuticles that do not extend down through the skeleton may be added near the frontal surface. Most commonly an upward growing interior transverse wall attaches to the frontal cuticle and produces a transverse intercalary cuticle during subsequent frontal thickening. This development also results from frontal thickening, but, in this case, the frontal cuticle has become embedded in the frontal skeletal wall. The frontal surfaces of such forms suggest the presence of cuticles in lateral and even transverse vertical walls (Fig. 126,3). However, as vertical sections of some such specimens show (Fig. 126,1,2), the vertical walls lack cuticle except at the frontal surface. Results of this study, using *Margaretta tenuis* HARMER, showed superficial cuticulate zooecial boundaries at the zoarial surface but an absence of intercalary cuticles in lateral walls within the zoarium (Fig. 126,1,5). However, observations of other species of that same genus (CHEETHAM & COOK, this revision; CHEETHAM, unpublished) revealed well-developed or intermittent intercalary cuticles throughout the lateral walls. Thus, at present it appears that one should not ascribe much taxonomic value to this character.

GENERAL FEATURES OF THE CLASS PHYLACTOLAEMATA

By TIMOTHY S. WOOD

[Department of Biological Sciences, Wright State University, Dayton, Ohio]

Within the large and diverse group of animals known as Bryozoa there occur several dozen species whose unique morphology, development, and ecology indicate a long independent evolutionary history. In 1856, ALLMAN established for these species the distinct class Phylactolaemata (*phylasso*, guard + *laimos*, throat) named for the small liplike lobe of tissue overhanging the mouth. Easily recognized by the horseshoe-shaped lophophore in all but one genus, the phylactolaemates are also distinguished by an exclusively freshwater habitat, a relatively large polypide (Fig. 131), a muscular body wall, free encapsulated buds (statoblasts) and an unusual ciliated colony progenitor, which develops from the zygote.

Though easily overlooked, phylactolaemate colonies are often dominant among organisms attached to substrates under water. They occur in nearly every clean body of fresh water where there exists suitable submerged substrate of wood, stone, vegetation or firm synthetic material. Before the practice of sand filtration in public waterworks, enormous quantities of these colonies chronically clogged the water mains of such cities as Boston, Hamburg, and Rotterdam (WHIPPLE, 1910; KRAEPELIN, 1886; DEVRIES, 1890). Most species, however, occur in shallow bodies of standing water, and it is not unusual to find colonies of four or five different species inhabiting the same small pond. Both *Plumatella emarginata* ALLMAN and *Fredericella sultana* (BLUMENBACH) may flourish in flowing water (BUSHNELL, 1966), although neither has obvious adaptations to a lotic habitat. With few exceptions, phylactolaemates grow within a temperature range of 15° to 26°C. A record high temperature of 37°C was recorded by BUSHNELL (1966) for living colonies of *Plumatella repens* (LINNÉ) and *P. fruticosa* ALLMAN at a shallow lake margin. Only *Fredericella sultana* is perennial in temperate latitudes, surviving under ice at temperatures close to freezing. Toxicity bioassays and field observations indicate a sensitivity of many species to low concentrations of certain industrial and domestic pollutants (BUSHNELL, 1974). ROGICK and

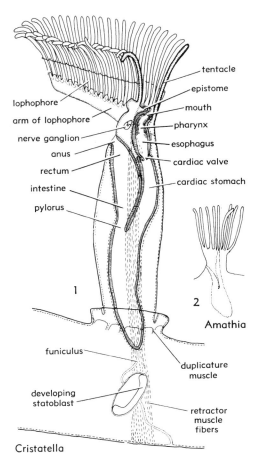

Cristatella

FIG. 131. Phylactolaemate zooid morphology. ——*1*. Transverse section of zooid of *Cristatella mucedo* CUVIER (after Brien, 1960).——*2*. Zooid of the ctenostomate *Amathia convoluta* LAMOUROUX drawn to the same scale as *1*.

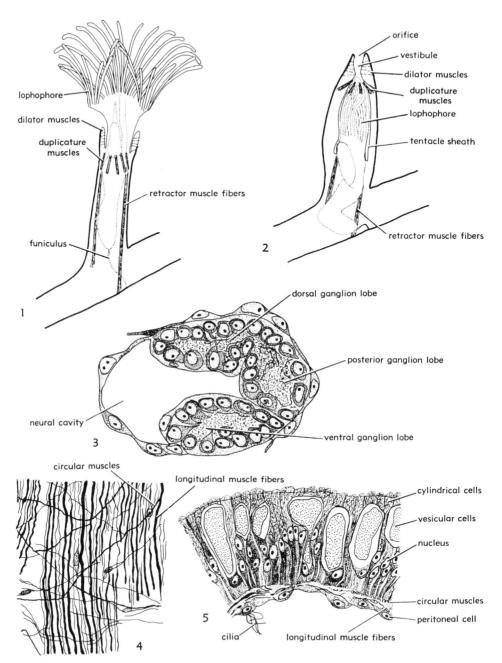

Fig. 132.   Phylactolaemate zooid anatomy.——*1*. Intercoelomic muscles of *Plumatella casmiana* Oka with polypide in feeding position.——*2*. Intercoelomic muscles of *P. casmiana* with polypide in retracted position.——*3*. Sagittal section of nerve ganglion in *Lophopus crystallinus* (Pallas) (after Marcus, 1934). ——*4*. Surface view of zooecial muscles beneath epithelium in *Lophopodella carteri* (Hyatt) (after Rogick, 1937).——*5*. Cross section of portion of zooecium in *Cristatella mucedo* Cuvier (after Brien, 1960).

Brown (1942), nevertheless, collected *Plumatella repens* from a Puerto Rican stream

contaminated with livestock wastes.

ZOOID MORPHOLOGY

In a schematic sense the phylactolaemate colony is a vessel of coelomic fluid in which are suspended many independently moving organ systems performing major physiological functions. Each active unit, known as a polypide, communicates directly to the colony wall through muscle fibers, a funiculus, and a common peritoneum. The polypide and its adjacent colony wall are customarily combined in the term "zooid," defined as the individual member of a colony. Such a unitary concept, however, is awkward when applied to the phylactolaemate Bryozoa where septa are infrequent in many species and the colony may be little more than a sac of communal polypides. For lack of specific identity between a polypide and a section of colony wall it is useful to distinguish these parts and to use "zooid" only in reference to an individual in a more abstract sense.

Colony wall.—The phylactolaemate colony wall is a histologically complex structure composed of well-defined tissue layers beneath an externally secreted integument (Fig. 132,5). Although details may vary considerably among species and even in different areas of the same colony, the basic pattern may be generalized. In *Pectinatella* the nonliving outer material is a gelatinous deposit consisting largely of water, but according to Kraepelin (1887) also containing some protein, chitin, and other organic materials. The dendritic colonies of *Plumatella* and *Fredericella,* however, develop a firm cuticle composed mainly of chitin (Hyman, 1958). Prior to chitin secretion, young zooids usually have a sticky exterior which allows them to adhere to the substrate or each other and to collect a thin crust of particles from the ambient water. The presence of a slightly raised longitudinal keel has some diagnostic value in species identification.

Beneath this nonliving material lies a single **epithelial layer** consisting of two cell types (Fig. 132,5). The columnar cylindrical cells form a uniform surface and are apparently involved in secretion of the outermost material (Brien, 1953). The larger vesicular cells contain fatty deposits that led Marcus (1934) to suggest a role in food storage. At the anterior budding region of the zooids, a distinct layer of undifferentiated cells underlies the epithelium. These are apparently totipotent for either cylindrical or vesicular cells, and transitional forms have been described by Brien (1960).

An interesting feature of the colony wall is the presence of thin **circular** and **longitudinal muscle layers** below the epithelium (Fig. 132,4,5). Circular muscle fibers, presumably derived from peritoneum, are able to execute limited orienting movements of the zooid.

The innermost tissue is a thin peritoneum bearing scattered tracts of cilia, particularly in the anterior portions of the zooids. The cilia beat continuously, driving coelomic fluid in random eddies among the polypides. This coelom was long thought to be separated from the two coelomic spaces of the episteme and lophophore. Together these were considered respectively homologous to the **metacoel, protocoel** and **mesocoel** of other lophophorates, and were named accordingly; however, Brien (1960) believes that all three cavities are continuous and can be characterized histologically only by their ciliation. The distinguishing terminology remains tentatively in use.

Polypide.—The polypides of a phylactolaemate colony are basically **monomorphic**. Each is autonomous with a lophophore of ciliated tentacles, a recurved digestive tract, and a single funiculus joining the gut caecum to nearby peritoneum (Fig. 131,*1*; 133,*4*). In addition, a dorsal nerve ganglion and mus-

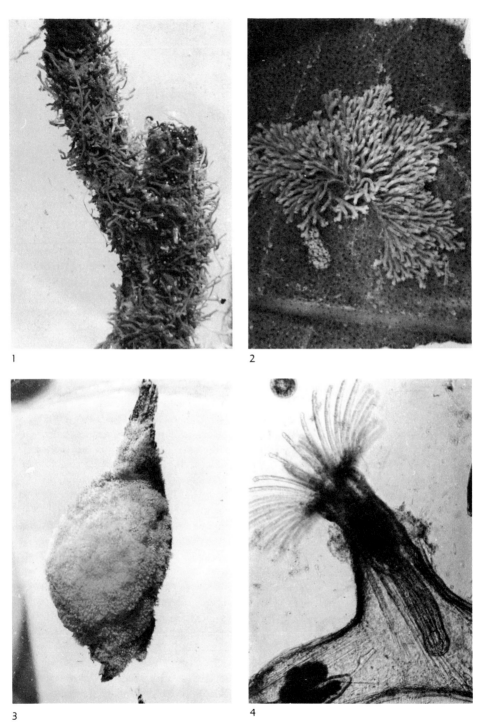

1

2

3

4

FIG. 133. *(For explanation, see facing page.)*

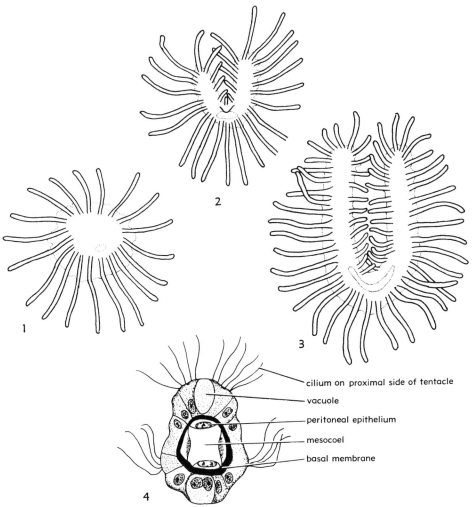

cilium on proximal side of tentacle

vacuole

peritoneal epithelium

mesocoel

basal membrane

Fig. 134.   Structure of the phylactolaemate lophophore.——*1. Fredericella sultana* (BLUMENBACH), anterior view of the lophophore showing circular conformation.——*2. Plumatella casmiana* OKA, anterior view of the lophophore showing moderate dorsal inflection.——*3. Pectinatella magnifica* (LEIDY), anterior view of the lophophore showing pronounced dorsal inflection.——*4.* Tentacle cross section in *Lophopodella carteri* (HYATT) (after Rogick, 1937).

culature is associated with movements of the polypide.

Lophophore.—In species of *Fredericella* the lophophore is reminiscent of Gymnolaemata,

a small bell-shaped structure formed by 20 or so tentacles arranged in a circle around the mouth. In all other species, however, the ring of tentacles is inflected dorsally to produce

FIG. 133.   Phylactolaemate colony form.——*1.* Colony of *Fredericella sultana* (BLUMENBACH) growing on a submerged twig, ×3.0.——*2.* Colony of *Plumatella casmiana* OKA growing on underside of floating leaf of *Nelumbo lutea*, ×3.0.——*3.* Colony of *Pectinatella magnifica* (LEIDY) from the underside of a floating log, ×0.5.——*4.* Polypide of *P. casmiana* showing lophophore, gut caecum, and retractor muscles; the funiculus is clearly visible extending from the polypide on the left, ×50.0.

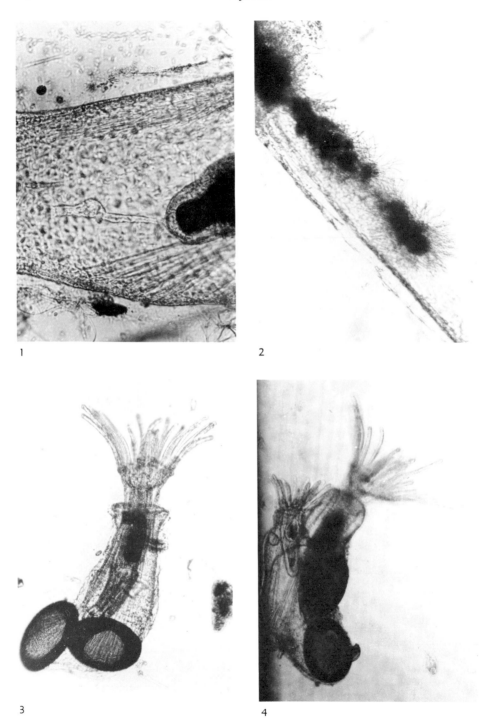

1

2

3

4

Fig. 135. *(For explanation, see facing page.)*

the characteristic horseshoe shape with two arms projecting freely on either side of the mouth. Species with the deepest inflection carry the greatest number of tentacles (Fig. 134,*1,3*). On *Pectinatella gelatinosa* OKA polypides, over 100 tentacles have been reported (TORIUMI, 1956). Tentacles of the outer series are longer than those of the inner, and a membrane connects all of them near the base. The projection of tentacles on both sides of each arm creates a narrow central groove in which food particles are collected and passed along to the mouth. Thus, a horseshoe-shaped lophophore operates differently from one of circular design and the two may differ in function and efficiency.

In all species the continuous mesocoel of the lophophore extends for the length of each hollow tentacle. Tentacles are roughly triangular in cross section (Fig. 134,*4*), and bear one medial and two lateral tracts of cilia (MARCUS, 1934; ROGICK, 1937). Stiff hairlike projections extending laterally between tentacles are easily seen in living specimens but have not been described from prepared sections. Longitudinal muscle fibers and several tentacular nerves allow tentacles to respond individually to impinging particles. Zooids of *Plumatella emarginata* have been observed to bring the tips of tentacles together repeatedly and trap protozoa near the mouth region. In *Plumatella casmiana* OKA the tentacles of individual zooids occasionally maintain a curious rhythmic flicking movement of several pulses per second (VIGANÒ, 1968).

Digestive tract.—The recurved digestive tract varies only slightly from the gymnolaemate plan. Unique to the Phylactolaemata is a triangular flap of tissue known as an epistome, which overhangs the mouth (Fig. 131,*1*). By means of muscle fibers within its coelomic interior, the epistome moves about

actively, and although it never closes the mouth it can alter the shape of the mouth opening. Most likely the function of this structure is chemosensory.

The mouth is a **stomodaeal cavity** that leads to a strongly ciliated vestibule, the pharynx, in which particulate food is collected and tumbled about (Fig. 131,*1*). A nonciliated esophagus opens periodically to receive a cluster of particles and push it through the cardiac valve into the stomach. Slow peristaltic contractions originating at the caecum move slowly along the **cardiac stomach** and thoroughly mix the food. A little at a time, food is eased through an unciliated pylorus into the so-called intestine, where it is packed into a dense mass and expelled through the anus as a fecal pellet. Although MARCUS (1926) testified to pH gradients in various parts of the digestive tract, the observations of living rotifers in fecal pellets of *Lophopodella* indicate suprisingly mild—or at least selective—enzymatic activity (ROGICK, 1938). Phylactolaemates are known to ingest quantities of bacteria, but the possibility of intracellular digestion of these and other minute organic particles has never been seriously explored.

Funiculus.—A single funiculus spans the metacoel from the stomach caecum to a certain point on the body-wall peritoneum, according to species (Fig. 131,*1*; 135,*1*). It is a tubular strand of tissue incorporating small muscle fibers, and it is the major site of spermatogenesis and asexual production of statoblasts. These critical roles will be discussed later in some detail.

Intracoelomic muscles.—Retraction and protrusion of the polypide is effected by coordinated action of several distinct sets of muscles (Fig. 132,*1,2*). Most conspicuous of these are two bundles of retractor fibers originating

FIG. 135. Phylactolaemate reproduction.——*1.* Funiculus extending from the gut caecum of *Plumatella casmiana* OKA showing the earliest stage of statoblast formation, ×150.——*2.* Sperm developing along the funiculus in *Fredericella sultana* (BLUMENBACH), ×250.——*3.* Ancestrula of *P. casmiana* recently emerged from a statoblast; the two statoblast valves are clearly evident, ×50.——*4.* Parietal budding in a young colony of *P. casmiana*; note the new duplicate bud to the left of the smaller polypide, ×50.

posteriorly on the colony wall, extending laterally along the polypide, and inserting at various points from the esophagus to the lophophore. Sudden contraction of these muscles jerks the polypide into the colony interior, carrying with it a thin membranous portion of the zooecial tip that turns inward to become a tentacle sheath (Fig. 132,2). Anteriorly the sheath opening is constricted by a sphincter, beyond which lies a small chamber called the vestibule. Prior to polypide eversion the sphincter relaxes and small muscle fibers dilate the vestibule. Bundles of duplicature muscle fibers, radiating from the tentacle sheath to the colony wall, slowly contract against coelomic pressure, widening the space through which the lophophore must pass. Almost simultaneously the retractor muscles relax and allow the polypide to emerge, pushed by the pressure of coelomic fluid. The tentacle sheath everts and the duplicature muscles now relax and become taut, serving as fixator ligaments to halt the polypide's outward progression. The lophophore opens, cilia beat, and feeding resumes.

Nervous system.—Every polypide in a colony has a nerve ganglion located dorsally in the mesocoel of the lophophore between the mouth and the anus. It is essentially a vesicle delimited by a thin nucleated membrane enclosing large dorsal, ventral, and posterior ganglionic lobes. Each lobe has a central fibrillar region and a periphery of ganglion cells (Fig. 132,3). A large tract from each side of the ganglion passes dorsally into the nearest lophophore arm, accompanied for the proximal third of its length by an extension of the neural vesicle. The tracts bifurcate into right and left branches to innervate internal and external rows of the tentacle sheath, and then branch out as a plexus between the epidermal and muscular layers of the colony wall. Other nerves from the ganglion provide a network of presumably bipolar cells along the entire digestive tract. The episome is well innervated, supporting the suggestion of a sensory function. Specialized sensory cells occur on the tentacles, intertentacular membrane, episome, and in the unsclerotized epithelium at the zooid tip.

The most detailed accounts of the phylactolaemate nervous system are those of Gerwerzhagen (1913) and Marcus (1934), working with *Cristatella* and *Lophopus,* respectively. There are yet many aspects to be clarified, including innervation of retractor muscles and the question of interzooidal nervous communication.

PARIETAL BUDDING OF ZOOIDS

Bryozoan colonies grow in size by the addition of new zooids, and colony morphology is to a large extent determined by patterns of sequential budding. Among most gymnolaemate species the budding process generally begins with the formation of a septum across the parental zooid, creating an additional small sac in which the new polypide is to develop. Phylactolaemates, however, like the living cyclostomes, reverse this sequence: the new polypide appears first and gradually draws away from the parental zooid as the colony wall elongates or enlarges.

A long succession of investigators have observed and interpreted the budding process in *Cristatella* (Davenport, 1890), *Pec-* *tinatella* (Oka, 1891), *Plumatella* and other genera (e.g., Kraepelin, 1887, 1892; Brien, 1936, 1953). The **primordium** originates from a cluster of dedifferentiated epithelial cells on the ventral body wall of a parental zooid (Fig. 136). Their mass bulges into the metacoel, pushing ahead of it a thin covering of peritoneum. A central cavity appears and from it develop two narrow dorsal and ventral invaginations that elongate, and eventually converge and fuse to form a continuous U-shaped tube (Fig. 136). This becomes the digestive tract, with the future cardiac valve at the point of fusion. Oka (1891) described a somewhat modified series of events for *Pectinatella gelatinosa,* but the effect is the same.

Meanwhile, as the developing bud elongates, a narrow strand of peritoneum separates from the ventral side, remaining attached to the colony wall at one end and to the distal part of the bud at the other (Fig. 136). Eventually this strand develops a hollow interior and becomes the funiculus. A third invagination now pinches off from the original central cavity, forming a small vesicle that becomes the nerve ganglion lying close to the pharynx. From points behind and in front of the ventral mouth opening, small fingerlike projections appear and extend laterally as tentacles of the two arms of a lophophore. This places the anus beyond the dorsal row of tentacles and orients the mouth squarely between the lateral arms. When the bud is fully formed, an orifice is created by a rupture in the body wall, and the diminutive polypide protrudes and begins feeding immediately.

Painstaking observation by BRIEN (1953) revealed a fascinating hierarchy of three bud primordia occurring on every mature zooid (Fig. 136). The so-called **main bud** is largest of the three and is always the first in line of succession. Close beside it ventrally lies a minute **duplicate bud** (Fig. 135,4), and on the dorsal side toward the parental zooid is a small **adventitious bud.** As the main bud develops into a new polypide the following adjustments are made: the duplicate bud becomes a main bud to the new polypide, the adventitious bud becomes a main bud to the parental polypide, and new duplicate and adventitious primordia appear in appropriate new locations.

The combination of stimuli required to initiate bud development is unknown. Occasionally the zooids of laboratory-reared colonies, while appearing perfectly healthy, will cease budding and eventually die without being replaced, even though good bud primordia are present. In other cases a colony may suddenly enter a growth phase in which new generations of zooids develop every day

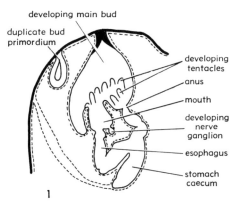

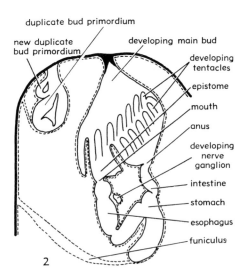

Fig. 136. Late stages of phylactolaemate zooid budding (after Brien, 1960).——*1.* Developing main bud with small saclike duplicate bud.——*2.* Further development and appearance of new duplicate bud primordium.

for several days. In *Plumatella casmiana* it has been shown that in old colonies a main bud primordium may apparently be stimulated to develop by the death of its parental zooid (WOOD, 1973). These aspects of polypide budding deserve further study.

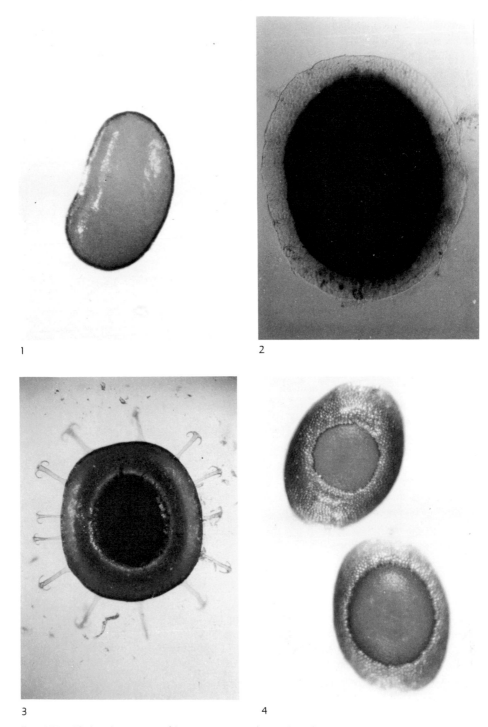

1

2

3   4

FIG. 137.   Phylactolaemate statoblasts.——*1.* Piptoblast of *Fredericella sultana* (BLUMENBACH), ×100. ——*2.* Sessoblast of *Plumatella emarginata* ALLMAN, ×180.——*3.* Floatoblast of *Pectinatella magnifica* (LEIDY), ×50.——*4.* Dorsal (upper) and ventral (lower) sides of floatoblasts of *Plumatella repens* (LINNÉ), ×65.

STATOBLASTS

The temperate freshwater habitat has an inconstant environment, fluctuating in temperature, pH, dissolved oxygen, nutrients, turbidity, water level, and in other chemical and physical conditions. For all but a few stream-tolerant organisms there is seldom a water route by which individuals can escape suboptimal conditions of a pond or lake and disperse to other locations. It is not surprising, therefore, that most invertebrate species living in fresh water have in their life histories a dormant resistant stage that may serve both as a disseminule and as a mechanism for surviving periods of unfavorable conditions. Certain adult rotifers, nematodes, and tardigrades can withstand prolonged dehydration (CROWE, 1971). Thick-walled cryptobiotic eggs occur among many aschelminths and crustaceans, and protozoan cysts are common. Sponges and bryozoans, the two groups of exclusively colonial organisms in fresh waters, both produce highly resistant structures by asexual processes unknown among their marine relatives. The sponge gemmule is an accumulation of food-filled amoebocytes enclosed in a spherical thick-walled capsule (LEVEAUX, 1939). The bryozoan statoblast is a discoid envelope of chitin containing large yolky cells and an organized germinal tissue capable of becoming a single polypide ancestrula for a new colony (Fig. 135,*3,4;* 137). Statoblasts can endure severe environmental stress and will survive freezing in both dried and undried conditions. ODA (1959) was able to germinate statoblasts of *Lophopodella carteri* (HYATT) that had been dried for over six years.

The development of statoblasts has been traced by many workers including KRAEPELIN (1892), BRAEM (1890), OKA (1891), and more recently by BRIEN (1954). The important role of the funiculus begins soon after its initial appearance alongside the developing bud. Dedifferentiated cells of epithelial origin migrate from the parental zooecium into the tubular funiculus, forming a loose axial strand. As the funiculus elongates it

shifts its distal position away from the developing bud. The axial cells slowly proliferate near the concave side facing the polypide bud, while a few muscle fibers appear opposite them. This sets the stage for statoblast production, which may follow immediately, but often occurs some time later or not at all. Among rapidly growing colonies of *Plumatella repens,* however, statoblasts begin to form on the funiculi of developing polypides that have not yet emerged from the colony interior.

The environmental or physiological conditions favoring statoblast production are unknown. The first sign of activity is a small bulge to one side of the funiculus where axial cells arrange to form a vesicle, and yolk-filled funicular cells accumulate on its proximal side (Fig. 135,*1;* 138,*1,2*). As the cell mass mushrooms away from the side of the funiculus it remains covered with a thin layer

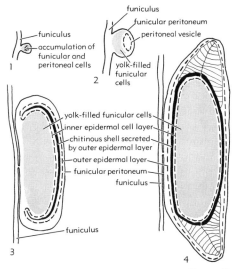

FIG. 138. Statoblast development in *Plumatella fungosa* (PALLAS) (based on Brien, 1954).——*1.* An accumulation of yolk-filled funicular cells surrounding peritoneal cells.——*2.* Appearance of vesicle lined with peritoneal epithelium.——*3.* Radial growth of peritoneal vesicle around yolky mass.——*4.* Formation of external chitinous structures.

of funicular peritoneum. The vesicle enlarges, flattens to a two-layered disc and spreads out along the surface to enclose the large accumulation of yolky cells (Fig. 138,3). The inner epidermal layer now is destined to give rise to a new polypide. The outer cells become columnar and begin secreting a chitinous protective shell on all sides. Those cells along the margin may become particularly large and surround themselves with thin walls of chitin (Fig. 138,4). They then lyse and are replaced by a gas, giving the statoblast a peripheral area that provides buoyancy. The completed capsule has a marginal suture along which two halves will separate when the new polypide is ready to emerge (see BUSHNELL & RAO, 1974, for excellent scanning micrographs of statoblasts). When fully formed the statoblast is released from its peritoneal envelope on the funiculus and remains free in the metacoel. In certain species it may be discharged through a temporary pore of a living zooid (MARCUS, 1941; VIGANÒ, 1968), although usually statoblasts are released upon disintegration of the colony. The number of statoblasts produced by a single polypide varies according to species. BUSHNELL (1966) reported as many as twenty per polypide of *Plumatella repens,* whereas *Pectinatella, Cristatella,* and *Lophopus* typically form only one. Where multiple statoblasts occur they arise in close succession in a proximodistal gradient along the funiculus.

As a rule, statoblasts do not germinate immediately, but enter a dormant or quiescent state, lasting from several days to many months. The major studies of statoblast dormancy are somewhat contradictory regarding the factors that trigger germination (BROWN, 1933; ODA, 1959; MUKAI, 1974). Variability is apparently introduced by differential ages of the statoblasts, their specific developmental histories, and exposure to varying regimes of temperature, light, moisture, and water chemistry. In an excellent review of this subject, BUSHNELL and RAO (1974) suggested that considerable species differences exist and that much experimental work has yet to be done.

Gross morphological features of the statoblast are often important for diagnosis of phylactolaemate species. Recent scanning micrographs by BUSHNELL and RAO (1974) and WIEBACH (1975) show excellent surface details on statoblasts of a few species. Those statoblasts with a peripheral pneumatic **annulus** are produced in all genera but *Fredericella,* and are called **floatoblasts** (Fig. 137,3,4). In genera such as *Pectinatella* and *Cristatella* these are equipped with marginal hooks (Fig. 137,3), which seem to suggest dispersal by catching onto bird plumage. BROWN (1933), however, is probably correct in his belief that they serve more to prevent the washing away of dormant statoblasts from favorable substrates. Hooks are absent from the floatoblasts of *Plumatella* species (Fig. 137,4), but the holdfast function is retained by a second type of statoblast called a **sessoblast** (Fig. 137,2). These are generally larger than *Plumatella* floatoblasts and lack the buoyant annulus. Generally formed simultaneously with floatoblasts, they appear in the zooecial tubes nearest to the substrate and are firmly cemented directly to the substrate along with an underlying portion of the body wall. Long after the colony has disintegrated these sessoblasts remain attached, appear in linear patterns of small black dots on rocks or submerged logs. Curiously, the sessoblast seems to form directly against the colony wall rather than the funiculus, but despite the careful attention given to every other aspect of phylactolaemate development, sessoblast origins remain obscure.

A remarkable species is *Plumatella casmiana,* which produces at least three different morphological types of floatoblasts in addition to the sessoblast (WIEBACH, 1963). One of these, called a **leptoblast,** bypasses diapause and may complete polypide development while still within the parental colony. Upon release through a vestibular pore, the leptoblast germinates almost immediately (VIGANÒ, 1968).

In *Fredericella,* generally considered the most primitive of all phylactolaemates (LACOURT, 1968), statoblasts have neither hooks nor a buoyant annulus, nor are they cemented to a subtrate (Fig. 137,*1*). They may, in fact, never be released at all but instead held firmly within the narrow tubular zooecium. To distinguish these structures from the cemented sessoblasts with which they are so often confused, EVELINE MARCUS (1955) proposed the name **piptoblast.** Since it is never liberated, the piptoblast can serve only the function of maintaining a population through suboptimal conditions. It is fragmented portions of the upright zooecial branches that serve as disseminules in this species (WOOD, 1973).

SEXUAL REPRODUCTION

Sexuality is an enigma in the Phylactolaemata, for it appears to have little real function. As a means of reproduction it is vastly out-performed by the asexual development of statoblast colonies. Its potential for genetic recombination is blocked by the apparent habit of self-fertilization (BRAEM, 1897; MARCUS, 1934). Nevertheless, sexual activity has been observed in all major species, occurring at various seasons of the year in colonies both large and small. If sexuality is a vestigial process in the Phylactolaemata there is at least no evidence that it faces negative selection. Published information, however, is scanty and incomplete, and further investigation of the process is definitely needed.

Colonies are **monoecious,** producing both eggs and sperm, although typically only a few of the zooids in a colony participate in gametogenesis. Sperm develop in clusters from peritoneal tissue on the funiculus (Fig. 135,*2*) or, in the case of *Cristatella,* on mesodermal strands of tissue spanning the metacoel (BRAEM, 1890). They differ from the sperm of Gymnolaemata, having a shorter head, a helical mass of mitochondrial material in the middle region, and more cytoplasm in the tail (FRANZÉN, 1970). They are released into the coelomic fluid and apparently never leave the colony. Eggs arise from a short invagination of peritoneum between the parental polypide and its adventitious bud. The invagination, constituting an ovary, becomes somewhat pedunculate and typically contains 20 to 40 eggs in various stages of maturity (Fig. 139,*1*). Only one egg among these is ever fertilized, the rest detaching from the zooid wall and eventually disintegrating. Meanwhile, an invagination of elongated cells from the zooecial wall occurs beside the ovary opposite the adventitious bud. This grows to become an embryo sac, involving all tissue

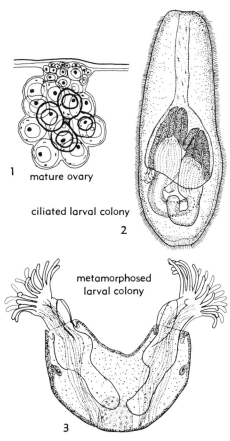

1 mature ovary

ciliated larval colony
2

metamorphosed
larval colony

3

FIG. 139. Phylactolaemate sexual reproduction. ——*1.* Mature ovary of *Plumatella fungosa* (PALLAS) (after Brien, 1960).——*2.* Ciliated larval colony of *P. fungosa* (after Brien, 1954).——*3.* Metamorphosed larval colony of *P. fungosa* (after Brien, 1960).

layers from the metacoel to the colony exterior. The fertilized egg will not undergo further changes until it enters the embryo sac. Such entry has not been witnessed.

In the embryological events that follow, the zygote undergoes **holoblastic cleavage** and forms an elongate **blastula** whose cells become arranged in two distinct layers. At one pole of the embryo 2 to 4 small polypides develop in a fashion similar to parietal budding, and a fold of body wall grows up as a **mantle** from the middle of the embryo nearly to cover the new polypides. A mass of cilia completely cover the embryo colony, and a rupture of the embryo sac releases the entire structure to the ambient water.

The so-called larva (Fig. 139,2) has at its leading aboral pole an accumulation of nervous tissue described by Marcus (1926), and from this end it probes potential substrates for a period up to 24 hours. Preliminary experiments by Hubschman (1970) with *Pectinatella* larvae indicate an importance of particle size in substrate selection and a distinct preference for natural over manufactured surfaces. Upon contact with a suitable substrate, the larva attaches with a glandular secretion from the aboral pole, the mantle fold pulls back, and the new polypides emerge and begin feeding (Fig. 139,3).

COLONY MORPHOLOGY

Despite the small number of phylactolaemate species, there is an impressive variety in colony morphology within the class, ranging from strongly tubular to essentially globular. The massive gelatinous colonies of *Pectinatella magnifica* (Leidy) have been reported with diameters as large as 0.6 m (Geiser, 1937), with many thousands of polypides crowded together over the surface (Fig. 133,3). By contrast, *Fredericella sultana* (Blumenbach) often exists as a stringy tangle of tubules and widely separated polypides (Fig. 133,1). Other species may have been very flattened colonies closely adhering to the substrate in a crustose mat (Fig. 133,2). In every case, zooids throughout a colony are essentially identical in morphology and in the manner in which they form new buds, and nearly all polypides arise from positions ventral to the parental zooid. Any slight differences in morphology between the ancestrula and subsequently budded zooids are generally temporary and are almost certainly environmentally induced. Also, while ancestrula tissues may be expected to contain initially more yolky food reserves than those of subsequent zooids, there is nothing more to suggest an astogenetic gradient. However, we can recognize at least four conditions influencing morphology of the colony as a whole: differential interzooidal growth, varied time interval between successive buds, directional orientation of buds, and the density of zooids.

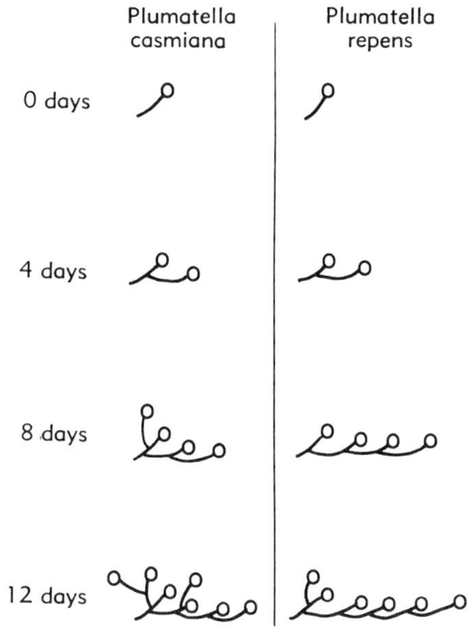

Fig. 140. Phylactolaemate budding patterns; comparison of *Plumatella repens* (Linné) and *P. casmiana* Oka, based on mean data from 265 zooids in natural populations occurring together (after Wood, 1973).

Differential interzooidal growth.—Elongation of tubular branches draws new polypides away from the parental zooids, resulting in very open, dendritic colonies. In *Fredericella sultana* this growth is so pronounced that the branches cannot maintain continuous contact with the substrate throughout their length, and hence they tend to be largely free. Luxuriant colonies of this species form dense spongy tufts several centimeters thick, which may occur on the surface of lake sediments away from any solid substrate. By contrast, compact tubular parts of certain *Hyalinella* colonies have so little linear growth that several polypides may all seem to emerge from a slightly enlarged portion of the metacoel.

Varied time interval between successive buds.—The polypides of *Plumatella repens* and *P. casmiana* are morphologically very similar, yet colonies of the former are usually open and reptant while those of the latter are often dense and compact. Wood (1973) considered this dichotomy to be largely the result of different time intervals between successive bud production. Observations of 265 colonies showed a mean lapse of 7.7 days between the appearance of first and second buds of a zooid in *P. repens*. In *Plumatella casmiana* the interval is only 3.7 days. Moreover, the first bud in *Plumatella repens* emerges at a mean zooid age of 2.7 days, whereas *P. casmiana* zooids are generally 3.5 days old before their first bud is feeding. The effect of these temporal differences is a dense colony in one species and a more open or reptant one in the other (Fig. 140). There is some evidence that the compactness of *Plumatella casmiana* provides some protection from damage by midge larvae.

Directional orientation of buds.—In the families Plumatellidae and Fredericellidae, where zooids are mainly tubular, new buds arise directly ventral to the parental zooid. Whether they eventually bend to the left or right appears largely a matter of chance. A significant departure from this randomness is shown by the gelatinous colonies of Pectinatellidae and Cristatellidae, in which suc-

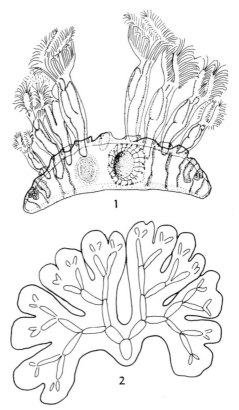

Fig. 141. Directional orientation of phylactolaemate buds.——*1.* Transverse section of *Cristatella mucedo* Cuvier showing progression of young to old zooids from lateral edges toward the midline (after Brien, 1954).——*2.* Schematic surface view of *Lophopus crystallinus* (Pallas) colony showing fan shape (after Marcus, 1934).

cessive buds generally arise on alternate sides of the ventral sagittal plane. The effect is a colony whose shape is specific and predetermined according to species, such as the fan shape of *Lophopus* and the linear configuration of *Cristatella*.

Colony growth in *Lophopus crystallinus* (Pallas) has been detailed by Brien (1954). The colonies are small, soft and transparent, seldom having more than 12 polypides. All polypides are oriented in the same direction, and all share a common saclike body cavity. The colony ancestrula, emerging from a statoblast, produces two daughter zooids in succession to the right and left of the ventral sagittal plane. Each of these in turn buds two zooids, and the process repeats itself as the colony spreads into a fan shape (Fig. 141,*1*). Before long, lobulations develop and the sinuses between them deepen to fragment the colony into smaller sections.

In *Cristatella mucedo* CUVIER the statoblast ancestrula gives rise to as many as five daughter zooids, both lateral and medial to the ventral sagittal plane (KRAEPELIN, 1887). These in turn produce zooids until the colony is heart-shaped with its cleft on the dorsal side of the first zooid. Budding is most active at the two upper lobes, and these gradually spread apart to become poles of an elongated colony. At this point the oldest zooids occupy a medial position in the colony and the youngest occur along the lateral edges (Fig. 141,*2*). Between these on both sides is a budding zone in which the ventral sides of all polypides face the periphery and new buds orient to the parental zooid exactly as in *Lophopus*. As a row of new zooids forms along the lateral edges, the older medial zooids become senescent and are resorbed into the colony, so that while the colony may grow longer it does not become any wider.

Density of zooids.—In most species each zooid may produce 2, 3 and occasionally as many as 5 daughter zooids. With no predetermined limits to growth it often happens that zooids normally adherent to the substrate exhaust the available two-dimensional space and can only grow vertically from the surface. This often occurs around small sticks and submerged branches where space is limited. A similar situation is faced by young colonies developing simultaneously from a dense accumulation of statoblasts. The result is a thick spongelike mat of contiguous vertical zooids that may give the colony surface a honeycomb appearance (BUSHNELL & WOOD, 1971). In *Plumatella fungosa* (PALLAS) this growth form is typical of the species.

PHYLOGENETIC CONSIDERATIONS

In 1741 when ABRAHAM TREMBLEY discovered the first known phylactolaemate colony, he assumed it to be closely related to colonial hydroids. The systematics of freshwater bryozoans has been a challenge to investigators ever since. With uniform polypide morphology and considerable phenotypic plasticity, the phylactolaemates offer few morphological features for species identification. Key diagnostic characters, based entirely on external anatomy, presently include statoblast dimensions, tentacle number, and certain details of the colony wall. All of these have been shown to be quite variable within a single population. Numerous taxonomic schemes have been proposed for phylactolaemate Bryozoa, notably those of ALLMAN (1856), JULLIEN (1885), KRAEPELIN (1887), VANGEL (1894), ROGICK (1935), TORIUMI (1956), and LACOURT (1968). Only TORIUMI has clarified the status of certain species by the laboratory rearing of colonies, and this approach deserves further attention.

It can hardly escape notice that, when properly arranged, phylactolaemate species exhibit three simultaneous morphological trends. A general decrease in interzooidal distance is accompanied by increased complexity of statoblasts and a rise in the number of tentacles borne by each lophophore. This is not to imply a monophyletic lineage, however; LACOURT (1968) proposed, in fact, a rather complex systematic scheme with at least three major divisions. Nevertheless, beginning with the Fredericellidae, the morphological gradients from simple to complex are distinct and provide support for the following assumptions:

1. Phylactolaemate evolution has brought about increased confluence among zooids, to the point that polypides are grouped together in open coelomic chambers. At the same time, chitinous sheathing exterior to the colony wall (*Fredericella, Plumatella*) has been replaced

by a gelatinous material of variable thickness (*Hyalinella, Pectinatella, Lophopodella*).

2. The simple unadorned statoblast of *Fredericella* must be considered primitive. Buoyant and cemented statoblasts represent a progressive step in *Plumatella* species, succeeded by the multifunctional spinous structures of *Pectinatella, Cristatella,* and *Lophopodella*.

3. An increase in the number of tentacles on the lophophore apparently confers certain advantages, perhaps in feeding or gaseous exchange, and is interpreted as representing phylogenetic advancement. Providing space for additional tentacles requires that the lophophore become deflected inwardly. Thus the horseshoe shape, often incorrectly cited as a vestige of primitive bilateralness, is barely evident in the Fredericellidae, but attains its greatest development in the more recently evolved Pectinatellidae and Cristatellidae (Fig. 134,*1*).

Phylactolaemate relations with other bryozoan groups are by no means obvious, although it is reasonable to suppose that the class represents an ancient lineage with an origin early in bryozoan evolution. The relatively large monomorphic zooids, the retention of three body regions (protosome, mesosome, metasome), and the muscular colony wall all point to this conclusion. Several authors have suggested that ancestral phylactolaemates preceded the gymnolaemates on the basis of morphological similarities between certain members of the former group and the phoronids (e.g., CORI, 1941; MARCUS, 1958; HYMAN, 1959; DAWYDOFF & GRASSE, 1959; BRIEN, 1960). However, NIELSEN (1971) described morphogenic inconsistencies that he believed to make such a close relationship unlikely. Further suggestive evidence for the precedence of phylactolaemate Bryozoa was offered by JEBRAM (1973b) in the observation that all species bud in an oral direction. This can be most easily explained by assuming the origin of phylactolaemates from a sessile or semisessile ancestor, development of an orally directed budding pattern, and subsequent reversal of budding polarity. This would permit an erect serial type of growth, conferring certain advantages to the colony which helped realize the spectacular adaptive radiation in gymnolaemate species, while phylactolaemates remained at a phylogenetic dead-end.

With the curious habitat isolation between phylactolaemate and gymnolaemate Bryozoa, both groups have become highly modified. One is left with the rather safe hypothesis that both groups share with the extinct Trepostomata a common Precambrian ancestor (BUGE, 1952). The virtual lack of fossil information on such ancient geneology warrants little further speculation on this point.

GLOSSARY OF MORPHOLOGICAL TERMS

Compiled by R. S. Boardman and A. H. Cheetham

[Smithsonian Institution, Washington, D.C.]

Terms and definitions are as used in this volume and include variant usages by volume authors. This glossary is not complete for the phylum. Additional terms and definitions will be included in volumes to follow as revisionary work proceeds.

This glossary does not distinguish recommended terms or their usages, for it is not meant to be authoritative. We do not believe that terms or their definitions should be fixed. Morphologic concepts are progressive approximations of full biologic understanding. Therefore, definitions should be constantly revised as knowledge of biologic relationships increases. Unthinking acceptance of a definition can lead to failure to ask critical questions, and progress is retarded. As understanding improves, concepts are modified or discarded and new concepts added. Terms applied to those concepts are more numerous than the concepts themselves and even more subject to change.

Definitions followed by one or more authors' names or by figure numbers in parentheses are as specifically used or cited by those authors in this volume. Definitions not followed by authors' names or figure numbers are as used or cited in the papers by Boardman, Cheetham, and Cook. Synonyms are those cited by authors in this volume or used by them as defined terms.

For some terms in this glossary, there are additional usages common in the literature but not included by any author in this volume. An example is the use of aperture for the opening in the skeleton of an anascan cheilostomate occupied by the membranous part of the zooidal frontal wall.

abandoned chamber. In stenolaemates, abandoned part of zooidal chamber generally sealed off by basal diaphragm (Fig. 142, 146).

acanthopore. Synonym of style or stylet in stenolaemates.

acanthorod. Synonym of style or stylet in stenolaemates.

acanthostyle. (a) In stenolaemates, a type of stylet; core a well-defined, smooth rod of nonlaminated calcite, sheath laminae usually strongly deflected toward zoarial surface, and sheath lamellar bundle wide. Acanthostyles usually larger than paurostyles (Fig. 219,9; 270,1) (Blake). (b) Rod-shaped calcite structure in zooecial walls or in extrazooidal vesicular tissue; core of hyaline calcite, may have sheath of cone-in-cone laminae if located in laminated wall or tissue; protrudes at zoarial surface as spine (Fig. 248,3b) (Utgaard). (c) Synonym of style or stylet in stenolaemates.

accretionary banding. Banding transverse to direction of growth of skeletal wall, of part of wall, or of individual crystal, resulting from addition of distinct growth increments (Fig. 102,3; 109,3; 111,4).

adventitious bud. In phylactolaemates, small bud primordium on dorsal side of main bud toward parental polypide; becomes main bud to parental polypide as original main bud develops into new polypide (Wood).

adventitious polymorph. In gymnolaemates, polymorph that communicates with just one other zooid; generally smaller than, and in extreme form almost structural appendage of, that zooid (Fig. 84,1–3).

aktinotostyle. Type of stylet in cryptostomates; core constructed of laminae that are medially arched toward zoarial surface, laterally deflected to form spines; contains scattered nonlaminated granules and, rarely, a continuous nonlaminated rod may be present in core; sheath laminae weakly to strongly directed toward zoarial surface; sheath lamellar bundle narrow (Fig. 219,7; 270,3).

anascan. Cheilostomate in which autozooids have hydrostatic system including largely to partly exposed flexible part of frontal wall and, consequently, no ascus (Fig. 72,1–4).

anastomosing colony. Branching erect colony in which branches join and rebranch to form open network (Fig. 15,1) (Utgaard).

ancestrula. (a) Zooid formed by metamorphosis of larva to found colony in stenolaemates and most gymnolaemates; generally differs in size and other morphologic characters from other zooids in colony; compare primary zooid. (b) In phylactolaemates, first zooid formed from a statoblast to found colony (Wood).

annulus. Outer epidermal layer of statoblast of phylactolaemates that encircles protective capsule containing germinal mass; can be air-filled and with or without marginal hooks, causing statoblast to float (floatoblast), or can form an adhesive encrusting layer (sessoblast) (WOOD).

apertural muscle. One of either of two pairs of muscles, occlusor muscles of operculum or diaphragmatic dilator muscles, of cheilostomate autozooid (LUTAUD).

aperture. In stenolaemates, terminal skeletal opening of zooid (Fig. 25, 26).

articulate colony. In stenolaemates, erect colony with jointed branches; node or joint consists of noncalcified, thick cuticular material in life (UTGAARD).

ascophoran. Cheilostomate in which autozooids have hydrostatic system including ascus beneath continuous protective frontal shield developed as cryptocyst, gymnocyst, or umbonuloid shield (Fig. 68,*1,2*; 69,*1,2*).

ascus. Exterior-walled, flexible-floored sac beneath frontal shield of autozooid in ascophoran cheilostomates; encloses water-filled chamber opening at or near orifice to function in hydrostatic system; derived by infolding of part of frontal wall beneath gymnocyst or cryptocyst, or by overarching of frontal wall by umbonuloid shield and associated structures (Fig. 68,*1d,e*; 69,*1b,c*; 78,*1a*).

astogenetic differences. Differences in morphology unique to zooids comprising asexual generation and therefore restricted to zones of astogenetic change in colony.

astogeny. Course of development of sequence of asexual generations of zooids and any extrazooidal parts which together form colony.

atrial bag. Part of polypidian vesicle attached to frontal wall of developing zooid, from which tentacle sheath is formed by slight constriction separating it from digestive lumen (Fig. 91,*2*).

autozooid. (a) Zooid having at some stages of ontogeny protrusible lophophore, with or without feeding ability. (b) Usual, common zooid containing feeding organs in colony; capable of carrying out all life functions in monomorphic colony (UTGAARD; BLAKE; KARKLINS).

autozooidal polymorph. Autozooid differing from ordinary feeding zooids in size, shape, tentacle number, or other feature, which may or may not be reflected in any skeletal parts present, but retaining protrusible lophophore with or without feeding ability (Fig. 69,*1c*).

avicularium. In cheilostomates, autozooidal or more commonly heterozooidal polymorph having equivalent of orificial wall relatively larger and more intricately reinforced than those of ordinary feeding autozooids, to form mandible (Fig. 71,*2,3*; 81,*3*).

axial bundle. In Rhabdomesina, cluster of axial zooecia differentiated as distinct axial structure (Fig. 283).

axial zooecium. In Rhabdomesina, elongate polymorph that parallels zoarial axis for part or all of its length. Those that diverge from axis become

typical autozooecia (BLAKE).

basal attachment. In Ptilodictyina, proximal part of zoarium consisting of encrusting base adnate to substrate and connecting segment that develops distally into zoarium (Fig. 223,*2*).

basal bud. In gymnolaemates, bud arising from basal wall of parent zooid, as on reverse surface of unilaminate erect colony branch and in some uniserial erect colony branches.

basal canal. Circumoral lacuna of lophophore into which internal lacunae of all tentacles open (Fig. 96).

basal diaphragm. In stenolaemates, diaphragm that acts as floor of living chamber (Fig. 2, 37).

basal disc. In stenolaemates, encrusting proximalmost part of ancestrula; direct development of metamorphosis of the larva (Fig. 25, 26).

basal layer. Synonym of skeletal part of encrusting colony wall in stenolaemates.

basal plate. Synonym of basal platform in cheilostomates.

basal platform. Multizooidal skeletal layers of basal zooidal walls in cheilostomates.

basal window. Uncalcified subcentral portion of exterior basal zooidal wall in encrusting cheilostomate colony (Fig. 69,*1d*).

basal zooecium. Small polymorph on basal surface of some Ceramoporidae (Fig. 145).

basal zooidal wall. In stenolaemates and gymnolaemates, exterior or interior zooidal supporting wall, opposite and generally parallel to orificial wall; can be absent in zooids budded above encrusting base in some erect and some multilaminate encrusting colonies.

beak. Pointed, rounded, or lobate skeletal rim on which mandible of cheilostomate avicularium occludes and to which it may or may not conform in length and shape (Fig. 70,*3b*; 84,*1,2*).

bifoliate colony. In stenolaemates, erect colony in which two layers of zooids bud back to back from interior multizooidal median wall (Fig. 30,*1,3a,4*).

bilaminate colony. In cheilostomates, colony with erect branches consisting of two layers of zooids with separate but adjacent, commonly exterior basal walls; flexible or rigid, depending on degree of calcification (Fig. 70,*1a*).

bimineralic skeleton. Cheilostomate zoarium or zooecium having some layers composed of calcite and others of aragonite (Fig. 67,*1c*; 68,*1e*).

biological interference. Effect exerted by organism on mineralogy, crystal morphology, and other properties of its skeleton, which make those properties different from equivalent properties of actually or potentially coprecipitated inorganic carbonates (SANDBERG).

bisexual. Zooid or colony that produces both male and female gametes.

blastema. Undifferentiated part from which organ develops or tissues proliferate (LUTAUD).

blastula. Single-layered embryonic stage produced by cleavage of zygote (STEEN, 1971).

body cavity. Space enclosed by zooidal, multizooidal, or extrazooidal walls containing zooidal organs or other structures suspended in body fluid.

body wall. (a) Wall enclosing the body cavity of a colony and its parts, including zooids, parts of zooids, multizooidal parts, and any extrazooidal parts; consists of inner cellular peritoneum, outer cellular epidermis, and outermost noncellular layers, including cuticular, gelatinous, or skeletal material, or a combination; in phylactolaemates, includes layers of longitudinal and circular muscles between epidermis and peritoneum. (b) Wall of zooid or bud, consisting of inner cellular peritoneum, outer cellular epithelium, and at least in exterior walls outermost cuticle with or without underlying skeleton; in fully developed zooids, with nerve layer (diffuse peripheral endings or plexus) between epidermis and peritoneum (Fig. 89).

boring. External mold of ctenostomate colony immersed in calcareous substrate; produced by chemical penetration during colony growth (Fig. 85,3).

branch midrib. Protruding, central, compound range wall in center of branch in some bifoliate fistuliporines (Fig. 210,2c).

brood chamber. (a) In stenolaemates, zooidal or extrazooidal coelomic chamber in which eggs develop into larvae (Fig. 52). (b) In most gymnolaemates, water-filled space partly enclosed by body walls of one or more polymorphs, within which embryos are held during development, generally one at a time, topologically outside body cavity of colony (Fig. 69,1c; 70,2).

brown body. (a) In stenolaemates and gymnolaemates, encapsulated mass of degenerating cells from lophophore, gut, some muscles, and some other nonskeletal parts of zooid varying in different groups; either retained in zooidal body cavity or expelled after regeneration of feeding and digestive organs (see Fig. 40,3b). (b) Synonym of brown deposit (KARKLINS).

brown deposit. In stenolaemates, granular deposit of iron oxide or pyrite presumably representing fossilized remains of organic material which were either functional organs or brown bodies of degenerated states in life (Fig. 40,1,5; 43,3; 46,1,4a).

bud. (a) In stenolaemates and gymnolaemates, newly developing, asexually produced zooid, initiated as body walls. (b) In phylactolaemates, newly developing, asexually produced zooid, initiated as statoblast or polypide (WOOD).

budding. Asexual reproduction of zooids.

budding pattern. In stenolaemates, shapes of buds and their relative positions on supporting structures.

CaCO₃ polymorph. Either of two forms of CaCO₃,

calcite or aragonite, which constitute cheilostomate skeletons (SANDBERG).

caecum. Blind prolongation of stomach portion of digestive tract in which food remains for some time (Fig. 91; 95,4).

canaliculus. Large style in Actinotrypidae that inflects autozooecial wall, producing ridge in zooecial chamber parallel to zooecial length, each with a septumlike appearance (Fig. 194, 195).

cardia. Curved tubular part of digestive tract into which esophagus opens; in some ctenostomates (and one cheilostomate) differentiated into gizzard and stomach portions (Fig. 91; 95,4).

cardiac stomach. In phylactolaemates, tubular part of stomach between cardiac valve and caecum (Fig. 131,1).

carina. Protruding median ridge on surface of zoarium of Goniocladiidae formed by protruding vertical mesotheca (UTGAARD).

carnosan. Ctenostomate in which autozooids bud directly from other autozooids, or alternate in groups with groups of kenozooids (Fig. 66,1–3).

celluliferous side of colony. Synonym for frontal side in stenolaemates.

cerebral ganglion. Nerve center lying in oral constriction between base of lophophore and esophagus on anal side of polypide (Fig. 96; 100,2).

circular muscle layer. Outer of two thin muscle layers in body wall of phylactolaemates between peritoneum and epithelium (Fig. 132,4,5).

closure. Synonym of frontal closure in cheilostomates.

coelom. Body cavity lined with peritoneum.

collar. Pleated membranous structure attached to diaphragm of ctenostomate zooid; contained within vestibule when lophophore is retracted and exposed at frontal surface when lophophore is everted (Fig. 66,2a,3).

colony. Morphologic and functional unit that interacts with the environment as a complete organism, consisting of one or more kinds of physically connected zooids, multizooidal parts, and in some colonies extrazooidal parts, all assumed to be genetically uniform.

colony control. Process influencing growth and functions of zooids to make then differ morphologically and functionally from solitary animals because of membership in colony.

colony wall. In phylactolaemates, body wall composed of outer noncellular cuticle or gelatinous layer, epithelial layer, longitudinal and circular muscle layers, and inner peritoneum (WOOD).

columnar epithelium. Mitotically active epithelium of body wall of bud or of more restricted area of proliferation, capable of secreting cuticle (Fig. 87,1).

common bud. Synonym for confluent budding zone in stenolaemates.

communication organ. Complex of interdigitating

cell types, together with cuticular or calcareous pore plate, which form exclusive means of communication between principal body cavities of fully developed gymnolaemate autozooids, between parts of some zooids, and between zooids and extrazooidal parts (Fig. 68,*1d,e*).

communication pore. (a) In stenolaemates, pore in interior wall through which physiological communication among zooids or between zooids and extrazooidal parts is assumed (Fig. 35,*4*; 46,*3*). (b) In gymnolaemates, single or one of several minute pores in pore plate traversed by cells of communication organ.

compensating sac. Synonym of ascus in cheilostomates.

compound skeletal wall. Skeletal wall calcified on edges and both sides, therefore necessarily an interior wall. Most vertical walls in stenolaemates.

condyle. One of pair of bilaterally arranged skeletal protuberances on which operculum of autozooid or mandible of avicularium is hinged in some cheilostomates; in asymmetrical avicularia of some cheilostomates can be single (Fig. 81,*3b*).

confluent budding zone. In stenolaemates, coelomic budding space and enclosing exterior walls connecting body cavities of a few to many buds or combinations of buds and zooids (Fig. 25, 26).

confluent multizooidal budding zone. In stenolaemates, confluent budding zone that originates outside of zooidal boundaries opposite endozone and which contains only buds at distal ends or edges of colony (Fig. 25, 26).

confluent zooidal budding zone. In stenolaemates, confluent budding zone that originates within outer coelomic space of zooids opposite exozone, or in some taxa opposite distal endozone (Fig. 54,*3,4*).

connecting segment. In Ptilodictyina, part of zoarial attachment between encrusting base and regularly developed distal part of zoarium (Fig. 223,*2*).

core. In stenolaemates, one of two structural elements forming stylets; formed either of laminated or nonlaminated skeletal material or a combination of both; generally separated from sheath laminae by growth discontinuity (BLAKE).

cormidial orifice. In cheilostomates, skeletal support for zooidal orifice which is joint product of more than one zooid (Fig. 122,*1–3*; 127).

cortex. In stenolaemates, main portion of zooecial wall adjacent to zooecial boundary (UTGAARD).

costa. One of usually paired spines fused medially and commonly intermittently laterally to form costal shield of cribrimorph cheilostomate zooid.

costal shield. Discontinuous frontal shield or part of frontal shield of cheilostomate zooid, formed by unfused or intermittently fused spines overarching uncalcified part of frontal wall (Fig. 71,*1–3*).

cribrate colony. In stenolaemates, sheetlike or frondose colony with flattened, anastomosing branches separated by fenestrules (UTGAARD).

cribrimorph. Cheilostomate with autozooids having costal shields composed wholly or in part of spines fused medially, and most commonly intermittently along lengths (Fig. 71,*1–3*).

cryptocyst. Continuous frontal shield or part of frontal shield of cheilostomate zooid, formed by calcification of interior wall grown into body cavity subparallel to and beneath frontal wall; completely calcified or with uncalcified spots covered by cuticle or plugged with cellular and noncellular material; in anascans, commonly with lateral notches or openings for passage of parietal muscles; in ascophorans, with marginal or submarginal communication organs connected to underlying principal body cavity of zooid (Fig. 67,*1a–e*; 72,*1,3*).

cryptocystidean. Anascan or ascophoran cheilostomate with autozooids having frontal shields (cryptocysts) formed by calcification of interior body walls grown into body cavities subparallel to and beneath frontal walls (SANDBERG).

cuticle. Noncellular organic outer layer of body wall secreted by columnar epithelium of bud (Fig. 87); composed of mucopolysaccharides in glycoproteinic frame, hardened by a tanning process.

cyphonautes larva. In gymnolaemates, ciliated larva with bivalved cuticular shell; most commonly planktotrophic, but in some cheilostomates having nonfunctional digestive tract (Fig. 85,*4*).

cystiphragm. In stenolaemates, lateral skeletal partition extending from zooecial wall into chamber and curved inward to form cyst or collar that extends partly or entirely around zooidal chamber (Fig. 30,*1*; 46,*7*).

cystoidal diaphragm. In stenolaemates, transverse skeletal structure formed by two diaphragms in contact only part way across zooecial chamber to form an enclosed compartment between them (Fig. 264).

cystopore. Synonym of vesicular tissue in stenolaemates.

dactylethra. In stenolaemates, defined originally as an aborted, shortened polymorph; interpreted here to be a degenerated feeding zooid closed by terminal diaphragm (Fig. 48,*6–8*).

dendrite. Short, usually branched process of nerve cell that conducts impulses to cell body (STEEN, 1971).

dendritic thickening. In some erect bilaminate cheilostomates, extreme skeletal thickening along axes of colony branches formed by thickened frontal shields of axial autozooids and overlying kenozooids (Fig. 109,*2*).

dendroid colony. In stenolaemates, erect branching colony with branches circular in cross section and most zooids budded from vertical walls of other zooids.

diaphragm. (a) In stenolaemates, membranous or skeletal partition that extends transversely across entire zooidal chamber (Fig. 31,5; 36,1). (b) In gymnolaemate autozooid, muscular ring of body wall forming attachment between inner end of vestibular wall and outer end of tentacle sheath; commonly connected to vertical walls of zooid by diaphragmatic dilator muscles (Fig. 66,3).

diaphragmatic dilator muscle. One of generally bilaterally paired muscles that traverse body cavity of gymnolaemate autozooid to insert on muscular diaphragm at connection between vestibular wall and tentacle sheath (Fig. 66,3).

digestive epithelium. Cellular lining of digestive tract derived from secondary invagination of epithelium of body wall (LUTAUD).

dilator muscle. One of commonly multiple, radially or bilaterally arranged muscles that traverse body cavity of gymnolaemate autozooid to insert on diaphragm (diaphragmatic dilator muscle) or vestibular wall (vestibular dilator muscle) for dilation during lophophore protrusion.

direct nerve. One of two nerve strands composing each of twin peripheral nerves following tentacle sheath toward orifice and frontal wall.

distal bud. In gymnolaemates, bud arising from distal side of vertical wall of parent zooid to continue growth in principal growth direction of parent, as in most encrusting and erect colonies (Fig. 76,1b; 77,1a; 80,1,2).

distal direction. Principal direction of growth of colony or of major part of colony, away from founding zooid or zooids (ancestrula, multiple primary zooids, statoblast ancestrula, or preexisting colony fragment); can be subparallel or subperpendicular to principal growth directions of zooids.

distal hemiseptum. In stenolaemates, hemiseptum projecting from distal zooidal wall or mesotheca.

distolateral bud. In gymnolaemates, bud arising from distolateral side of vertical wall of parent zooid to initiate growth in direction slightly diverging from principal growth direction of parent, as in most encrusting and erect colonies (Fig. 75,1a).

divaricator muscle. One of pair of muscles that traverse body cavity of cheilostomate avicularium to insert near fixed margin of mandible, and of some cheilostomate autozooids to insert near fixed margin of operculum, both of which are opened by their action.

double-walled colony. Synonym of free-walled colony in stenolaemates.

duplicate bud. In phylactolaemates, minute bud primordium lying close beside main bud ventrally; becomes main bud to new polypide (Fig. 135,4).

duplicature muscle fiber. In phylactolaemates, one of bundles of muscle fibers that widen anterior end of tentacle sheath through which lophophore passes during protrusion, and serve as fixator ligaments for protruded polypide (Fig. 132,1,2).

ectocyst. Variously used to correspond to cuticular layer of body wall or also to include epidermis, cuticle, and skeleton (LUTAUD).

ectoderm. Embryological term sometimes applied to epidermis in bryozoans (Fig. 87).

edgewise growth. Skeletal growth in which calcification of walls occurs by simultaneous addition of calcite to edges of crystals at growing ends of walls; wall laminae may be at any angle to growth lines (Fig. 29,1,2).

embryonic fission. In tubuliporates, asexual division of primary embryo into secondary, and in some species, tertiary embryos, presumably all with the same genetic makeup.

encrusting colony. (a) Colony in which most zooids are attached to substrate by their basal walls. (b) In gymnolaemates, colony in which each autozooid of unilaminate colony or of basal layer of multilaminate colony is attached to substrate by all of its basal wall (tightly encrusting), or by protruding parts of its basal wall or kenozooids budded from its basal wall (loosely encrusting).

encrusting wall of colony. In stenolaemates, basal wall of colony adjacent to substrate (Fig. 25; 26; 28; 30,5a,b).

endocyst. Variously used to include both epidermis and peritoneum, or peritoneum alone (LUTAUD).

endoplasmic reticulum. Organelle consisting of fine, branching, anastomosing tubules, spaces, or isolated vesicles present in cytoplasm of most cells (STEEN, 1971).

endozone. In stenolaemates, inner parts of zooids of a colony, characterized by one or a combination of growth directions at low angles to colony growth direction or colony surface, thin vertical walls, and relative scarcity of intrazooidal skeletal structures (Fig. 10, 11).

entosaccal cavity. In stenolaemates, that part of zooidal body cavity within membranous sac (Fig. 2).

epidermis. Epithelium of body wall; secretes cuticle and, in stenolaemates and cheilostomates, underlying deposit of calcium carbonate (skeleton) within organic matrix (Fig. 2) (LUTAUD).

epifrontal fold. Double-walled fold of exterior body wall and contained body cavity overarching membranous frontal wall in umbonuloid cheilostomates (SANDBERG).

episome. Small, movable, liplike lobe of tissue and contained coelom overhanging the mouth of a phylactolaemate zooid (Fig. 131,1).

epithelial layer. In phylactolaemates, single layer consisting of two cell types, columnar cells that secrete outermost noncellular material of colony wall, and vesicular cells containing fatty deposits (WOOD).

epithelium. Outer cellular layer of zooid body wall (epidermis) and internal cellular layer lining lumen of alimentary tract (digestive epithelium) (Fig. 87; 88,*1*).

erect colony. Colony that extends into water from relatively small encrusting base or rootlets.

esophagus. (a) In phylactolaemates, nonciliated part of digestive tract between pharynx and cardiac valve (Fig. 131). (b) Used, in part, as synonym of pharynx (LUTAUD).

eustegal epithelium. In free-walled stenolaemates, epithelium that secretes exterior cuticle (Fig. 142, 143).

excurrent chimney. Localized current created by the feeding action of adjacent zooids which carries excess water and any rejected particles away from colony surface (Fig. 25).

exilazooid, exilazooecium. In stenolaemates, generally small polymorph originating in outer endozone or exozone between feeding zooids with few or no basal diaphragms so that living chamber space is available for possible organs (Fig. 48,*5*).

exosaccal cavity. In stenolaemates, that part of zooidal body cavity between membranous sac and body wall (Fig. 2).

exozone. In stenolaemates, outer parts of zooids of colony, characterized by one or more combinations of growth directions at high angles to colony growth directions or colony surfaces, thick vertical walls, and concentrations of intrazooidal skeletal structures (Fig. 10, 11).

explanate colony. Erect, sheetlike or frondose colony, in some with lobate extensions (KARKLINS).

exterior skeletal wall. In cheilostomates, skeletal wall that calcifies against cuticle and occurs in a body wall that, in its precalcified, membranous state, expanded coelomic volume of colony (SANDBERG).

exterior wall. Body wall that extends body of zooid and of colony; includes outermost cuticular or gelatinous layer (Fig. 1).

external muscle. Muscle, such as retractor or parietal, which extends across body cavity from body wall to lophophore or digestive tract, or to other body wall (Fig. 99).

extrazooidal part. Protective or supportive colony structure which, once developed, remains outside zooidal boundaries throughout the life of a colony; in phylactolaemates, the exterior colony body walls and adjacent body cavity transitional with exterior zooidal vertical walls and body cavities.

extrazooidal skeleton. In cheilostomates, skeletal layers of extrazooidal body walls produced by coalescence of body walls originally bounding hypostegal coeloms of zooids or formed concurrently at growing extremities with budding of zooids (Fig. 70,*1b*).

feeding zooid. A zooid that at some ontogenetic stage(s) possesses a protrusible lophophore, a digestive tract, muscles, a nervous system, and funicular strands capable of functioning to provide nourishment to itself and to any connected nonfeeding zooid or other nonfeeding part of colony; may include some or all zooids within a colony.

Feigl's solution staining. Mineralogical staining technique by which location of aragonite within cheilostomate skeleton can be recognized by selective precipitation of silver and MnO_2 on aragonite (Fig. 125,*1*).

fenestrate colony. Erect colony in which branches form a reticulate pattern (Fig. 15,*1,3*; 60,*1*).

fenestrule. One of the open spaces between branches of fenestrate colonies (Fig. 60,*1*).

fibrillation. Arrangement of myofilaments in muscle fibers (LUTAUD).

firmatopore. Type of kenozooecium consisting of slender, proximally directed tubule on reverse side of zoarium in tubuliporates (BASSLER, 1953).

fixed-walled colony. In stenolaemates, colony in which orificial walls of feeding zooids are fixed directly to apertures so that confluent outer body cavities between zooids are eliminated (Fig. 26).

flask-shaped chamber. In stenolaemates, chamber defined by skeletal funnel cystiphragm within zooidal living chamber (Fig. 46,*6,7*).

flexibly erect colony. In gymnolaemates, erect colony in which zooids and any extrazooidal parts present are uncalcified (ctenostomates) or lightly calcified (some cheilostomates), thus permitting extensive motion in moving water (Fig. 13,*2*; 66,*1*).

floatoblast. Statoblast with peripheral pneumatic annulus, having or lacking marginal hooks (Fig. 137,*3,4*).

fragmentation. Asexual reproduction of colony by direct growth from zooid or group of zooids broken from preexisting colony (compare hibernaculum, statoblast).

free-living colony. Cheilostomate or ctenostomate colony without general attachment to substrate; commonly partly mobile on or in unstable seabottom sediments by means of specialized polymorphs (Fig. 14,*4*).

free-walled colony. In stenolaemates, colony that is loosely covered by membranous exterior walls not attached at apertures of feeding zooids so that confluent outer body cavities connecting zooids are produced (Fig. 25).

frondose colony. In stenolaemates, erect colony with branches flattened into leaflike shapes and zooids budded from vertical walls of other zooids (UTGAARD).

frontal budding. In gymnolaemates, budding from frontal wall or associated structure, such as hypostegal coelom of parent zooid, to produce autozooids in some multilaminate encrusting colonies and some free-living colonies; or to produce adventitious polymorphs in many kinds of colonies (Fig. 69,*1b*,*2*; 79,*2*).

frontal closure. In cheilostomates, calcified frontal and orificial walls of autozooid that were membranous when lophophore was functional, but became permanently sealed; commonly retains traces (scars) of cuticular operculum and parietal muscle insertions (Fig. 76,*2*,*3*; 80,*4*).

frontal membrane. Flexible, membranous part of frontal wall of cheilostomate autozooid (LUTAUD).

frontal shield. Protective and supportive skeletal structure on frontal side of retracted organs of cheilostomate autozooid, grown as part of frontal wall or as part of exterior body wall overlying, or interior body wall underlying frontal wall (Fig. 65,*2*–*7*).

frontal side of colony. In stenolaemates and gymnolaemates, side of unilaminate colony that contains orifices of feeding zooids (Fig. 28, left sides of *1b* and 6; 76,*1*–*5*; 78, right side of *1a*).

frontal structure. In gymnolaemate autozooid, relationship of frontal wall and, where present, of frontal shield to hydrostatic system.

frontal wall. (a) In fixed-walled stenolaemates and all gymnolaemates, an exterior zooidal wall attached to and wholly or partly supporting the orificial wall; provides front side to zooid more extensive than orificial wall alone (Fig. 1, 4). (b) In gymnolaemates, bounds frontal side of zooid at least in early ontogenetic stages, but commonly is modified by partly calcified supportive and protective structures in cheilostomates (Fig. 65,*1*–*7*).

funicular strand. Cellular tissue traversing the body cavities of zooids, buds, and extrazooidal parts of gymnolaemate colonies to connect feeding organs and communication organs to body walls; produces sperm in male or hermaphrodite autozooids (Fig. 68,*1e*).

funiculus. (a) System of strands of spindle-shaped cells that are continuous with peritoneum of digestive tract and body wall, extend across body cavity and along body wall between pore plates, are attached to special club-shaped cells through communication pores of pore plates in body walls, and thus extend from zooid to zooid throughout colony (Fig. 4; 87,*2*) (LUTAUD). (b) In phylactolaemates, tubular strand of tissue incorporating small muscle fibers spanning metacoel from caecum to peritoneum of colony wall (Fig. 142,*1*) (WOOD).

funnel cystiphragm. In stenolaemates, skeletal structure within zooidal living chamber which defines flask-shaped or funnel-shaped chamber, interpreted to be walls of intrazooidal polymorph (UTGAARD) or calcified parts of membranous sac or orificial-vestibular wall of feeding zooid (BOARDMAN) (Fig. 46,*6*,*7*).

funnel-shaped chamber. Synonym of flask-shaped chamber in stenolaemates.

fused-wall colony. Synonym of fixed-walled colony in stenolaemates.

giant bud. In gymnolaemates, unpartitioned distal end of lineal series two or more zooid lengths in extent, formed by lag in formation of interior transverse walls relative to growth of exterior walls of multizooidal origin (LUTAUD).

gizzard. In some ctenostomates (and one cheilostomate), spheroidal to elongate inner portion of cardia with epithelial surface supporting few to many, pointed or rounded plates or teeth.

glycoprotein. One of group of protein-carbohydrate compounds, such as mucin (STEEN, 1971).

Golgi apparatus. Organelle, well developed in cytoplasm of secretory cells, consisting of a set of flat formations of endoplasmic reticulum (LUTAUD).

gonozooecium. In stenolaemates, inflated polymorph that provides brood chamber in which eggs develop into larvae (Fig. 52,*8*).

granular microstructure. In cystoporates, skeletal microstructure characterized by subquadrate crystallites; generally dark-colored in thin section (UTGAARD).

granular-prismatic microstructure. In cystoporates, skeletal microstructure characterized by blocky to prismatic crystallites elongated perpendicular to epithelium that secreted skeleton; generally light-colored in thin section (UTGAARD).

growing tip. Proliferating distal extremity of colony, colony branch, or lineal series of zooids, characterized by columnar, mitotically active epithelium and undifferentiated peritoneal layers (Fig. 89,*1*,*4*).

growth habit. General form or shape in which a colony grows, and its relationship to the substrate; examples are a unilaminate encrusting colony or a conical free-living colony.

gymnocyst. Continuous frontal shield or part of frontal shield of cheilostomate zooid, formed by calcification of exterior frontal wall; completely calcified or with uncalcified, cuticle-covered spots (Fig. 69,*1*,*2*; 71,*1*,*2*).

gymnocystidean. Ascophoran cheilostomate with autozooids having gymnocysts as their frontal shields (SANDBERG).

hemiphragms. In stenolaemates, shelflike skeletal projections in zooidal living chamber, which alternate in ontogenetic series from opposite sides of zooecia; hemiphragms in any one zooid commonly comparable in morphology (see Fig. 40,*5*).

hemisepta. In stenolaemates, shelflike skeletal projections in zooidal living chambers, generally on proximal walls or in one or two pairs in alternate positions on proximal and distal sides of zooecia. Proximal and distal hemisepta commonly different in morphology in zoaria of Paleozoic age (Fig. 32,*1*; 267,*1*).

hemispherical colony. In stenolaemates, colony of approximately hemispherical shape in which zooids bud from encrusting colony wall and vertical walls of other zooids, and in some taxa from intracolony overgrowths.

heterostyle. Type of stylet in cryptostomes; core of lenses of nonlaminated calcite separated by bands of laminae continuous with sheath laminae; sheath laminae weakly to strongly directed toward zoarial surface; sheath lamellar bundle narrow (Fig. 219,5; 270,2).

heterozooid. In gymnolaemates, a polymorph with nonprotrusible or no lophophore, and therefore no apparent feeding ability, musculature different from that of autozooids or lacking, and specialized organs present or lacking (Fig. 70,1c,3).

hibernaculum. Encapsulated bud in some gymnolaemates, with fusiform to irregular stiffened cuticular cover containing yolklike material and partly developed feeding and digestive organs capable of germinating to produce first zooid of new colony, either attached to or detached from dead parent colony; formed as inswellings or outswellings of body wall of parent zooid.

hollow ramose colony. In stenolaemates, erect branching colony in which zooids bud from cylindrical axial colony walls (Fig. 36,4b; 54,1).

holoblastic cleavage. Mitotic division of zygote to form blastula consisting of cells approximately equal in size (STEEN, 1971).

hydrostatic system. System for protruding lophophore in gymnolaemate autozooid, consisting of flexible part of frontal wall, or infolded sac derived from it, and attached parietal muscles.

hypostegal coelom. (a) Part of body cavity of cheilostomate zooid separated from principal body cavity by ingrowth of body wall to form cryptocyst, or extended from principal body cavity enclosed in double-walled outfold to form umbonuloid shield; remains confluent with principal body cavity or is connected to it only by communication organs (Fig. 67,1b; 68,1b,d; 71,1). (b) Synonym of outer coelomic space in freewalled stenolaemates.

hypostegal epithelium. In free-walled stenolaemates, epithelium that secretes extrazooidal skeleton (Fig. 143).

hypostegia. Synonym of hypostegal coelom in cheilostomates.

immature region. Synonym for endozone in stenolaemates.

inferior hemiseptum. Synonym of distal hemiseptum in stenolaemates (KARKLINS).

initial layer of skeleton. Layer of cryptocyst or umbonuloid shield in cheilostomate zooid first deposited by proliferating epidermal cells, commonly of different microstructure or mineral composition from superficial skeletal layers (Fig. 67,1c; 68,1d).

inner epithelium. In free-walled stenolaemates, epithelium that secretes skeleton, including both zooidal epithelium, which secretes zooidal skel-

etal walls, and hypostegal epithelium, which secretes extrazooidal skeleton (Fig. 142, 143).

integration. Extent to which zooids in combination with any extrazooidal parts differ morphologically from solitary animals because of colony control of growth and functions.

intercalary cuticle. Cuticle composed of outermost layers of lateral walls of contiguous lineal series of zooids in gymnolaemate colony (Fig. 116,1).

interior skeletal wall. In cheilostomates, skeletal wall that grows off inner surface of exterior skeletal wall or other interior skeletal wall by apposition and partitions preexisting coelomic volume of colony (SANDBERG).

interior wall. Body wall that partitions preexisting body cavity into zooids, parts of zooids, or extrazooidal parts; may or may not include cuticular or gelatinous layer (Fig. 1).

interray. Area between rays of monticular zooecia in star-shaped monticules, generally composed of extrazooidal vesicular tissue in cystoporates (UTGAARD).

intertentacular organ. In some gymnolaemates, elongate protuberance of body wall on distal side of lophophore beneath tentacle bases bearing terminal pore through which fertilized eggs are released to develop generally into planktotrophic larvae.

interzooidal budding. In stenolaemates, budding that occurs outside of living chambers of zooids, so that one bud cannot be related to single parent zooid.

interzooidal communication organ. In gymnolaemates, communication organ that connects one zooid to another.

interzooidal growth. In phylactolaemates, growth of colony wall between newly budded polypides and parental polypides (WOOD).

interzooidal polymorph. In gymnolaemates, polymorph intercalated in budding series to communicate with two or more zooids, in space smaller than those occupied by ordinary feeding zooids (Fig. 71,2,3).

intracoelomic muscle. Synonym of external muscle.

intracolony overgrowth. Overgrowth of encrusting zooids onto colony surface, initiated from adjacent surviving zooids (Fig. 36,1).

intracuticular skeleton. In cheilostomates, skeletal layers that lie between noncellular organic sheets or within noncellular organic networks continuous with cuticles of uncalcified exterior walls.

intrazooidal budding. In stenolaemates, budding that occurs within the living chamber of a single parent zooid.

intrazooidal communication organ. In cheilostomates, communication organ that connects hypostegal coelom to principal body cavity of the same zooid.

intrazooidal polymorphism. In stenolaemates, sequential development of two different kinds of zooids in same living chamber (Fig. 49,6,7).

intrinsic body-wall muscles. Circular and longitudinal muscle layers in body walls of phylactolaemates.

jointed erect colony. In cheilostomates, erect colony in which zooids and any extrazooidal parts present are well calcified except at more or less regular intervals along branch lengths, thus permitting motion in moving water (Fig. 14,1).

keel. (a) In stenolaemates, flat median portion of zooid wall between sinuses in recumbent part of endozone (Fig. 56). (b) Synonym of carina in some stenolaemates. (c) In phylactolaemates, a longitudinal medial ridge extending along the recumbent tubular portions of a colony (Wood).

kenozooid. (a) In stenolaemates, any polymorph lacking lophophore and gut, muscles, and orifice. (b) In gymnolaemates, polymorph lacking orificial wall or its equivalent, lophophore, alimentary canal, and, in most, muscles (Fig. 66,1,2).

lamellar growth. Skeletal growth involving many parallel to subparallel layers or lamellae. A lamella grows either by advancement of bladelike crystals at the distal end or by marginal increase and impingement of scattered seed crystals on a broad-to-narrow zone or step to form solid layer; different parts of each lamella are of different ages (Sandberg).

lamellar ultrastructure. In cheilostomates, broad group of skeletal ultrastructures consisting of planar or lenticular aggregates of commonly tabular crystals of calcite forming layered units (lamellae) oriented parallel to wall surfaces; aggregates separated by diffuse or distinct organic sheets (Fig. 115,2).

lanceolate colony. In Ptilodictyina, erect, unbranched, bifoliate colony with proximally tapering zoarial segment (Fig. 242,1e).

larva. (a) Sexually produced, motile, ciliated immature individual from which most colonies of stenolaemates and gymnolaemates are developed by metamorphosis and growth; in stenolaemates, the larva is incapable of feeding, and is developed by fission of brooded embryos; in gymnolaemates, the larva is either capable of feeding (planktotrophic) and generally developed without brooding, or incapable of feeding (lecithotrophic) and developed from a brooded embryo. (b) In phylactolaemates, a brief motile phase composed of one or more fully-developed polypides enclosed in a ciliated mantle, the product of sexual reproduction (Wood).

lateral skeletal projections. Skeletal structures in living chambers of stenolaemates that occupy positions opposite feeding organs; including hemisepta, hemiphragms, ring septa, mural spines, and skeletal cystiphragms.

lateral wall. One of pair of vertical walls of gymnolaemate zooid, elongated generally subparallel to principal direction of zooid growth to give length, and together with transverse wall, depth to body cavity of zooid; most commonly developed as exterior wall extending body of colony in series of lineally budded zooids; in cheilostomates, includes skeletal layers (Fig. 70,1–3).

lecithotrophic development. In gymnolaemates, production by brooding of naked ciliated larva lacking digestive tract and subsisting entirely on nutrient supplied by maternal zooid; larva has variable but short motile stage before metamorphosis.

lepralioid. Ascophoran cheilostomate in which autozooids have frontal shields formed as cryptocysts (Sandberg).

leptoblast. Floatoblast that germinates almost immediately after release from parent colony (Wood).

ligament. Muscle fibers embedded in collagen with tubular peritoneal envelope (Lutaud).

lineal growth. Formation of zooidal line by successive development of new zooids from proximal portion of bud by growth of transverse partitions separating zooids from proliferating distal portion of bud (Fig. 89,1).

lineal series. In gymnolaemates, single line of connected zooids sequentially related by direct asexual descent; bounded basally, laterally, and frontally by exterior walls of multizooidal origin, through which communication organs generally are formed to connect with zooids in adjacent lineal series (Fig. 76,1b,2,3; 77,2; 80,2).

lipid. Organic compound insoluble in water but soluble in organic solvents, and upon hydrolysis generally yielding fatty acids (Steen, 1971).

living chamber. In stenolaemates, outermost part of zooidal body cavity in which major organs are housed when lophophore is retracted (Fig. 37).

longitudinal direction. Direction parallel to colony growth direction.

longitudinal muscle layer. Inner of two thin muscle layers in body wall of phylactolaemates between peritoneum and epithelium (Fig. 132,4,5).

longitudinal partition. In gymnolaemates, common double wall consisting of contiguous lateral walls of adjacent zooidal series growing together and kept together by reciprocal pressure and adherence of cuticular and skeletal layers; formed by peripheral indentation of exterior wall at growing tip (Fig. 89,2–4).

longitudinal ridge. Short, vertical plate perpendicular to mesotheca in some bifoliate Fistuliporina; a multizooidal skeletal structure (Utgaard).

longitudinal section. (a) In stenolaemates, section oriented so that zooids are cut parallel to their entire length. (b) In gymnolaemates, section oriented so that zooids are cut parallel to length and perpendicular to width.

longitudinal wall. In Ptilodictyina, compound skeletal wall between laterally adjacent zooecia that is structurally continuous for variable distances in general colony growth direction (Fig. 228).

lophophoral fold. Part of polypidian vesicle from which lophophore is formed, by development at constriction between atrial bag and digestive lumen and infiltration of peritoneal layers (Fig. 91,2).

lophophore. Part of the body wall beginning at inner end of vestibule and ending at mouth, including tentacle sheath and tentacles; comprises the feeding organ of a feeding zooid and a specialized organ of some nonfeeding polymorphs.

lophophore neck. Elongate movable cylindrical structure formed by everted tentacle sheath carrying tentacle crown far beyond orifice of gymnolaemate autozooid.

lunarial core. In cystoporates, one central hyaline projection or several subcylindrical spinelike hyaline projections in the lunarium, which serve as centers of growth of the lunarial deposit (UTGAARD).

lunarial deposit. Synonym of lunarium in stenolaemates.

lunarium. In cystoporates, microstructurally distinct or thicker part of autozooecium or large monticular zooecium; on proximal or lateral side of zooecium and projecting above zooecial aperture or peristome as a hood; commonly with shorter transverse radius of curvature than remainder of zooecial wall (Fig. 144).

lunulitiform colony. In cheilostomates, free-living colony of discoidal to conical shape (SANDBERG).

macula. In stenolaemates, cluster of a few polymorphs, extrazooidal skeleton, or a combination; clusters more or less regularly spaced among feeding zooids, commonly forming prominences, less commonly flush or depressed areas on colony surfaces (Fig. 59).

main bud. Largest of three bud primordia occurring on every mature zooid in phylactolaemates; first to form new polypide (Fig. 135,4).

mandible. Orificial wall equivalent to avicularium of cheilostomates, opened and closed by greatly augmented divaricator and occlusor muscles (Fig. 70,1c; 71,3; 81,3a).

mantle. Ciliated fold of colony wall nearly covering one to four small polypides of sexually produced colony progenitor in phylactolaemates; lost after release from parent colony and settlement (WOOD).

marginal zooecium. In Ptilodictyina, zooecium of polymorph at lateral margins of bifoliate zoarium, commonly without endozone (Fig. 230,2).

massive colony. In stenolaemates, colony of irregular shape in which zooids bud from the encrusting colony wall, from vertical walls of other zooids, and in some taxa, from intracolony overgrowths.

maternal zooid. In gymnolaemates, autozooid with

or without feeding ability which extrudes eggs, generally one at a time, into brood chamber through pore in lophophore wall below and between distal pair of tentacles (Fig. 72,1–4).

mature region. Synonym for exozone in stenolaemates.

median granular zone. In Ptilodictyina, middle layer of mesotheca with granular microstructure (Fig. 227–229).

median lamina. Synonym for median wall in stenolaemates.

median rod. In Ptilodictyina, long rodlike extrazooidal skeletal structure oriented longitudinally in median granular zone of mesotheca (Fig. 227, 231, 235).

median tubule. Synonym of median rod in stenolaemates.

median tubuli. In stenolaemates, aligned pustules or mural lacunae in laminated skeleton (BLAKE).

median wall. In stenolaemates, erect colony wall parallel to colony growth direction, interior and multizooidal, from which zooids bud back-to-back to form bifoliate colony (Fig. 30,1,3a,4).

membranous sac. In stenolaemates, membrane that surrounds digestive and reproductive system of zooid, dividing body cavity into two parts, the entosaccal cavity within sac, and the exosaccal cavity between sac and zooidal body wall (Fig. 2).

mesenchyme. All tissues derived from embryonic mesoderm, including connective tissues, parietal peritoneal network, funiculus, and muscles (LUTAUD).

mesocoel. Body cavity of second division of deuterostome body; assumed to correspond to cavity within and at base of tentacles in Bryozoa (STEEN, 1971).

mesoderm. Embryological term sometimes applied to peritoneum in bryozoans.

mesopore. Synonym of mesozooecium in stenolaemates (see mesozooid).

mesotheca. Synonym of median wall in stenolaemates.

mesozooid, mesozooecia. In Paleozoic stenolaemates, space-filling polymorph in exozone between feeding zooecia; closely tabulated out to distal end so that no room available for functional organs (Fig. 42,3).

metacoel. Body cavity of third division of deuterostome body; assumed to correspond to principal body cavity of zooid in Bryozoa (STEEN, 1971).

metamorphosis. An extensive external and internal reorganization of a larva to produce a founding zooid (ancestrula) or multiple founding zooids (primary zooids) of most stenolaemate and gymnolaemate colonies.

metapore. In Rhabdomesina, slender tubular opening in exozonal wall, oriented approximately perpendicular to zoarial surface (Fig. 262, 286). Metapores generally originate at base of exozone;

with diaphragms in few species.

microenvironmental variation. Differences within colony in morphology of zooids or extrazooidal parts, which cannot be inferred to express ontogeny, astogeny, or polymorphism; may be irregular or gradational and related to crowding, irregularities in substrate, encrustation, turbulence, breakage, boring, or sedimentation.

microvilli. Minute cylindrical processes forming striated or brush borders of epithelium (LUTAUD).

midray partition. Compound vertical wall along center of monticular ray or cluster of zooecia in cystoporates; may be multizooidal, extrazooidal, or both (Fig. 171,*1c*).

minutopore. Synonym of mural tubula in cystoporates.

mitochondrion. One of minute spherical, rod-shaped or filamentous organelles present in all cells and of primary importance in metabolic activities (STEEN, 1971).

mixed nerve. Nerve formed by conjunction of motor and sensory fibers (LUTAUD).

monila. In stenolaemates, concentric thickening of zooecial wall; resulting in beadlike appearance in longitudinal or transverse section (UTGAARD).

monoecious. Hermaphrodite; producing both female and male sex cells, as colonies and some zooids in Bryozoa (STEEN, 1971).

monomineralic skeleton. Cheilostome zoarium having all skeleton present composed exclusively of either calcite or aragonite (Fig. 70,*1b*; 72).

monomorphic colony. Colony in which one kind of zooid occurs in the zone of astogenetic repetition.

monomorphic polypides. Independent organ systems of one morphologic kind throughout zone of astogenetic repetition in a phylactolaemate colony (WOOD).

monomorphic zooids. Zooids of one morphologic kind throughout zone of astogenetic repetition in a gymnolaemate colony.

monticule. In stenolaemates, generally applied to cluster of polymorphs which makes a prominence on colony surface; also synonym of macula in stenolaemates.

morular cell. Cell filled with cluster of refringent spherules, found in peritoneal network and in funicular strands (Fig. 87,*1*; 88,*3*).

mucopolysaccharide. One of a series of complex organic compounds consisting of mixtures of glycoproteins and polysaccharides (STEEN, 1971).

multifoliate colony. In stenolaemates, erect colony with more than three mesothecae radiating from colony or branch center, each mesotheca supporting feeding zooids in bifoliate pattern (UTGAARD).

multilaminate colony. In cheilostomates, encrusting, generally nodular colony, commonly with irregular erect protuberances, consisting of two or more superposed layers of zooids produced by frontal budding, intracolony overgrowth, or a combination (Fig. 13,*3,4*; 79,*2*).

multiserial budding. In gymnolaemates, budding in which lineal series are regularly and most commonly continuously in contact, zooids in adjacent series are regularly connected by communication organs through exterior walls, and adjacent series form more or less coordinated growing edge for major part of colony (Fig. 80,*1,2*).

multizooidal bud. Synonym of giant bud in gymnolaemates.

multizooidal budding zone. In cheilostomates, distal region of colony with laterally confluent body cavity, within which all vertical walls of zooids arise as interior walls to partition zooid body cavities from each other.

multizooidal layer. Noncellular, cuticular or skeletal layer of body wall continuous from zooid to zooid and into buds or budding zones in gymnolaemate colonies (Fig. 69,*1d*).

multizooidal part. Part of a colony, such as continuous wall layers of zooids, buds, or budding zones, which is grown outside existing zooidal boundaries but becomes part of zooids as colony develops.

mural lacuna. Synonym of pustule in stenolaemates.

mural spine. In stenolaemates, small skeletal spine extending into zooidal chamber from skeletal wall or diaphragm (Fig. 41,*1–4*).

mural style. In Ptilodictyina, small rodlike skeletal structure consisting of superposed flexed segments of zoarial laminae; rarely containing discontinuous minute core; may project as minute spine above zoarial surface (Fig. 227, 235).

mural tubula. In cystoporates, small calcite rod in wall cortex, generally perpendicular to wall and zooecial boundary (UTGAARD).

muscle layer. In phylactolaemates, one of two adjacent layers of muscles, longitudinal and circular, lying between epithelial and peritoneal layers of colony wall to function in lophophore protrusion (WOOD).

myocyte. Embryonic cell of mesodermal origin that develops into a muscle fiber.

myoepithelial cell. Contractile ectodermal cell with intracellular striated muscles (LUTAUD).

nanozooid. In tubuliporates, polymorph with single tentacle, muscular system, reduced alimentary canal, and membranous sac (Fig. 49,*5–7,9*).

nematopore. In tubuliporates, slender tubular kenozooecium opening on reverse side of zoarium with tubes directed in obliquely distal direction (BASSLER, 1953).

noncelluliferous side of colony. Synonym for reverse side of colony in stenolaemates.

obverse side of colony. Synonym for frontal side of colony in stenolaemates.

occlusor muscle. One of pair of bilaterally arranged muscles, in series with parietals, which traverse body cavity of gymnolaemate zooid to insert on operculum or mandible and function in closing (Fig.68,*1e*).

ontogenetic variation. Differences in morphology of zooids or extrazooidal parts arising from changes during course of zooidal or extrazooidal development; recognizable in stenolaemate or gymnolaemate colony as increases in size or complexity among zooids or extrazooidal parts along proximal gradient from growing extremities toward founding zooid or zooids.

opercular scar. Trace of cuticular operculum preserved in frontal closure of cheilostomate autozooid.

operculum. (a) Presumably hinged, skeletal covering of zooecial aperture in melicerititid tubuliporates (Fig. 36,*2–4*). (b) In gymnolaemates, distally directed, flaplike fold of orificial wall, reinforced by cuticular or calcified margins, axes, or general surface, which by means of attached occlusor muscles closes orifice when lophophore is retracted (Fig. 66,*2a*; 68,*1d*; 72,*1*).

opesia. (a) Opening defined by inner margin of cryptocyst, serving as passageway for lophophore in some anascan cheilostomes. (b) Membranous area of frontal wall defined by inner margin of cryptocyst (LUTAUD).

orifice. Porelike or puckered opening within, or slitlike opening on margin of orificial wall, through which the lophophore is protruded and retracted (Fig. 2).

orificial wall. (a) Exterior, terminal or subterminal zooidal wall that bears or defines orifice and is attached through orifice to the vestibular wall; it may be attached to or free from supporting zooidal walls (Fig. 2). (b) In stenolaemates, a single, membranous, exterior, generally terminal body wall that covers the skeletal aperture and includes a simple circular orifice through which the tentacles protrude (Fig. 25, 26). (c) In gymnolaemates, a body wall that defines or contains the orifice through which the lophophore of an autozooid is protruded; commonly a single flaplike fold, reinforced to form operculum, at or near distal end of a frontal wall with which it is structurally and developmentally continuous; in most cheilostomes, synonymous with operculum (Fig. 66,*2a*,*3*).

outer coelomic space. In free-walled stenolaemates, coelomic space between outer skeletal surface and exterior membranous wall (Fig. 25).

ovicell. (a) In cheilostomes, structure consisting of body walls, some or all of which are calcified, enclosing brood chamber; commonly placed at or near distal end of maternal zooid (Fig. 72,*1–3*). (b) Synonym of gonozooid in stenolaemates.

parallel fibrous ultrastructure. Synonym of planar spherulitic ultrastructure in cheilostomes (SANDBERG).

parietal muscle. (a) One of commonly multiple,

usually bilaterally paired muscles that traverse body cavity of gymnolaemate zooid to insert on flexible part of frontal wall or floor of ascus, generally to function in hydrostatic system (Fig. 66,*2b*; 70,*1a*). (b) One of two sets of external muscles in anascan cheilostomate zooid (LUTAUD).

parietodepressor muscle. Parietal muscle (b) originating on lateral wall and inserting on flexible frontal wall, and therefore a synonym of parietal muscle (a) in gymnolaemates (Fig. 99).

parietodiaphragmatic muscle. Parietal muscle (b) originating on lateral wall and inserting on diaphragm, and therefore a synonym of diaphragmatic dilator muscle in gymnolaemates (Fig. 99).

parietovaginal muscle. One of muscular ligaments extending from muscle fibers of tentacle sheath to base of distal transverse wall of gymnolaemate autozooid (Fig. 99).

PAS test. Cytological technique by which location of polysaccharides within cell can be determined (STEEN, 1971).

paurostyle. Type of stylet in cryptostomates; core irregular, may be weakly differentiated rod of nonlaminated material; sheath laminae weakly deflected toward zoarial surface; sheath lamellar bundle narrow. Paurostyles usually smaller than acanthostyles (Fig. 219,*4*; 270,*1*).

periancestrular budding. In gymnolaemates, budding to produce zooids surrounding ancestrula, either radially from ancestrula, or more commonly by wrapping of distolaterally and proximolaterally budded lineal series around proximal end of ancestrula (Fig. 75,*5*,*7*; 79,*1*).

perigastric cavity. Synonym of principal body cavity of zooid.

perimetrical attachment organ. In stenolaemates, circular, collarlike membrane, attached at inner perimeter to tentacle sheath, at outer perimeter both to outer end of membranous sac and to skeletal body wall (Fig. 43,*1*; 45,*2–4*,*5b*).

peripharyngeal ganglion. Prolongation of cerebral ganglion around oral orifice, lying between basal canal of lophophore and epithelium of pharynx (Fig. 96; 100,*2*).

peripheral nerve. Any nerve serving extrapolypidian organs and wall (Fig. 100, 101).

peristome. (a) In stenolaemates, an outward tubular extension or rim of zooidal body wall beyond general surface of colony; either extension of interior vertical wall in free-walled colony (Fig. 39,*2*), or exterior frontal wall in fixed-walled colony (Fig. 28,*6*; 54,*3*). (b) In ascophoran cheilostomes, tubular outfold of body wall and contained body cavity together surrounding operculum and orifice at inner end, with calcified wall of exterior origin facing inward around orifice; can be produced entirely by one zooid or have components from adjacent zooids; opening of ascus can be inside or outside peristome (Fig. 67,*1d*,*e*; 68,*1d*,*2*; 82,*3a*,*b*).

peritoneum. (a) Inner cellular layer of body wall lining body cavity in both bud and fully developed zooid, continuing into tentacles and around digestive tract, and consisting of various cellular categories (Fig. 88,2). (b) In phylactolaemates, thin innermost layer of colony wall, bearing scattered tracts of cilia that drive coelomic fluid among polypides (WOOD).

petraliiform colony. In cheilostomates, encrusting unilaminate colony loosely attached by protruding parts of basal walls of autozooids or by basally budded kenozooids (SANDBERG).

phagocyte. Cell having ability to engulf particles (STEEN, 1971).

pharynx. Strongly ciliated part of digestive tract into which mouth opens (Fig. 2; 4; 96; 131,1).

pinnate growth habit. In stenolaemates, erect colony in which lateral branches grow in same plane from opposite sides of main axial branch.

piptoblast. Statoblast lacking both annulus and marginal hooks, often adhering to the colony wall by small keel-like projections on the basal valve, not released from parent colony (Fig. 137,1).

pivotal bar. Complete skeletal rim on which fixed edge of mandible is hinged in some cheilostomate avicularia (Fig. 71,2; 84,2).

planar spherulitic ultrastructure. Skeletal ultrastructure consisting of essentially two-dimensional, wedge- or fan-shaped arrays of acicular or rarely flattened laths of calcite or aragonite, formed in cheilostomates as first calcification against cuticle in exterior walls (Fig. 111).

planktotrophic larva. In gymnolaemates, ciliated larva generally produced without brooding, possessing functional digestive tract, and having lengthy motile phase before metamorphosis.

pleated collar. See collar.

polyembryony. Synonym of embryonic fission in tubuliporates.

polymorph. In stenolaemates and gymnolaemates, a zooid that differs distinctly in morphology and function from ordinary feeding zooids at same stage of ontogeny and in same asexual generation within a colony; may be a feeding or nonfeeding zooid specialized to perform sexual, supportive, connective, cleaning, defensive, or other functions; minimally includes body cavity and enclosing body walls.

polymorphic colony. Colony with more than one kind of zooid in zone of repetition (UTGAARD).

polymorphism. Repeated, discontinuous variation in morphology of zooids in colony; may be recognized in many stenolaemate and gymnolaemate colonies in the same generation of a zone of astogenetic change or among any zooids at the same ontogenetic stage in a zone of astogenetic repetition.

polypide. (a) Feeding organ of zooid, internally budded and periodically renewed from cellular layers of body wall; includes lophophore and digestive tract (pharynx, esophagus, cardia, stomach and caecum, pylorus, rectum), tentacle sheath, and cerebral ganglion (Fig. 91). (b) Major organs of autozooid contained in membranous sac of tubuliporate bryozoan (UTGAARD). (c) In phylactolaemates, independently moving organ system performing major physiological functions, suspended with other polypides in common vessel of coelomic fluid (Fig. 133,4).

polypidian bud. Newly developing digestive tract and feeding organs, originating as cluster of epithelial cells on distal side of growing transverse partition to invaginate into body cavity, together with surrounding subepithelial layers, to form first polypide of developing zooid (Fig. 90,4).

polypidian vesicle. Double-layered polypidian bud, with central cavity lined by undifferentiated internal epithelium formed early in development of polypide (Fig. 92,1).

polysaccharide. One of a group of complex carbohydrates, which upon hydrolysis yields more than two molecules of simple sugars (STEEN, 1971).

pore chamber. Part of body cavity of gymnolaemate zooid partly separated by interior wall continuous with portion of zooidal wall containing communication organ (Fig. 69,1f; 76,1b).

pore plate. Part of communication organ in gymnolaemates formed as thin calcareous or cuticular part of body wall of zooid or extrazooidal part, bearing one or more minute pores through which cells of special form project; grown as interior wall, but can be continuous with, and provide communication through either interior or exterior walls (Fig. 68,1d,e).

postmandibular area. Membranous part of frontal wall equivalent of cheilostomate avicularium, on which mandibular divaricator muscles insert; commonly separated from beak by partial or complete skeletal rim on which fixed edge of mandible is hinged (Fig. 71,3; 81,3a).

primary bud. One of buds arising as hollow outward expansions of cellular layers from distal and lateral areas of body walls of ancestrula (LUTAUD).

primary direction of encrusting growth. In stenolaemates, general direction along substrate of encrusting growth of ancestrula and first generations of colony (Fig. 25, 26, right of disc; also see Fig. 52,2).

primary wedge of encrusting zooids. In stenolaemates, ancestrula and first generations of colony that all grow in same general direction along substrate (Fig. 52,2).

primary zone of astogenetic change. Zone of astogenetic change forming proximal part of colony, beginning with founding zooid or zooids (ancestrula, statoblast ancestrula, or multiple primary zooids), commonly continuing distally through a few generations, and followed distally by primary zone of astogenetic repetition (see zone of astogenetic change).

primary zone of astogenetic repetition. Zone of astogenetic repetition following primary zone of astogenetic change distally and commonly consisting of numerous generations of zooids (see zone of astogenetic repetition).

primary zooid. Ancestrula, or one of two or more simultaneously partitioned zooids formed after metamorphosis of larva to found colony in some cheilostomes; commonly smaller and otherwise morphologically different from subsequently budded zooids (Fig. 79,4).

primordium. First accumulation of cells comprising identifiable beginning of developing organ or structure (STEEN, 1971) (compare blastema).

principal body cavity. In gymnolaemate autozooid, body cavity generally enclosed by basal, vertical, and orificial walls, and frontal wall, cryptocyst (and adjacent inner cellular layer), or ascus floor; occupied almost fully by retracted organs and muscles, except in degenerated stages (Fig. 66,3; 68,1b; 69,1b).

protocoel. Body cavity of first, most anterior of three divisions of deuterostome body; assumed to correspond to cavity of epistome in phylactolaemates (STEEN, 1971).

proximal bud. In gymnolaemates, bud arising from proximal side of vertical wall of parent zooid to initiate growth in direction opposite to principal growth direction of parent, as in repair of injury (Fig. 76,1a).

proximal direction. Direction opposite to distal, toward founding zooid or zooids of colony.

proximal hemiseptum. In stenolaemates, hemiseptum projecting from proximal zooidal wall.

proximolateral bud. In gymnolaemates, bud arising from proximolateral side of vertical wall of parent zooid to initiate growth in direction greatly diverging from principal growth direction of parent, as in parts of encrusting colonies (Fig. 75,5,6).

pseudocoel. Body cavity lined at least in part by epidermis.

pseudopore. Pore that penetrates all or part of skeletal layer but not cuticle in many exterior walls (Fig. 26; 35,4).

pustule. In stenolaemates, small equidimensional skeletal structure consisting of crinkled segments of skeletal laminae (Fig. 246,1a).

pylorus. Ciliated part of digestive tract into which stomach portion of cardia opens and in which remnants of digestion are agglutinated with mucins into a whirling mass (Fig. 2; 95,1).

ramose colony. Synonym of dendroid colony in stenolaemates.

range of zooids. Zooids aligned in direction of colony growth.

range partition. In Ptilodictyina, elongate structure of extrazooidal stereom between zooecial ranges (Fig. 229,2; 236,1,2).

range wall. In cystoporates, wall parallel to colony growth direction between ranges of zooids; discontinuous or relatively continuous; extrazooidal or partly extrazooidal and partly multizooidal in origin (Fig. 209; 210,1).

ray. In cystoporates, cluster of monticular zooecia radiating from center of star-shaped monticule (Fig. 171,1c).

rectal pouch. Part of digestive tract into which pylorus opens, and which ends at anus (Fig. 91).

rectum. In phylactolaemates, so-called intestine in which fecal pellets are formed and passed through anus (Fig. 131,1).

regenerative budding. In cheilostomes, budding from within zooecial walls of broken zooid (Fig. 76,1a).

retractor muscle. (a) One or more bundles of muscle fibers originating on basal or vertical zooidal walls or on colony wall, and inserting on base of lophophore and pharyngeal or cardiac regions of digestive tract; retracts tentacles and introverts tentacle sheath (Fig. 2–4). (b) In phylactolaemates, two bundles of muscle fibers originating on colony wall and inserting on polypide at various points from esophagus to lophophore (Fig. 131,1; 133,4).

reverse side of colony. In stenolaemates, back side of erect unilaminate colony; side opposite to that on which feeding zooids open (Fig. 28,6, right side of colony).

rigidly erect colony. In cheilostomes, erect colony in which zooids and any extrazooidal parts present are well calcified, generally increasingly so toward proximal encrusting base, thus permitting little motion in moving water (Fig. 13,1; 14,2; 83,1).

ring septum. In stenolaemates, centrally perforated skeletal diaphragm in zooidal living chamber (Fig. 40,2).

rosette. Cellular apparatus of communication organs of the funicular system, made of club-shaped cells across pores (LUTAUD).

sagittal section. Median longitudinal section in gymnolaemates (LUTAUD).

screw dislocation. Spiral growth steps induced by lattice defects in thin rhombic or hexagonal crystals that make up lamellar skeletal unit; in cheilostomes, thus far seen only in calcite skeletons (Fig. 102,4–6; 103,3,4).

secondary direction of encrusting growth. In stenolaemates, general direction of growth of wedge of zooids along substrate opposite to that of ancestrula and first generations (see left side of Fig. 25).

secondary wedge of encrusting zooids. In stenolaemates, wedge of zooids that buds from downfold of encrusting colony wall resting on upper surface of primary wedge, and that grows in general direction opposite to primary direction of colony growth (see Fig. 25, left side).

septula. Synonym of communication organ.

septum. (a) In stenolaemates, newly-formed compound, interior, body wall of bud (Utgaard). (b) Synonym of canaliculus in Actinotrypidae. (c) Synonym of interior wall (Lutaud).

sessoblast. Statoblast cemented through colony wall to substrate, usually with rudimentary annulus, but lacking marginal hooks or spines (Fig. 137,2).

sexual zooid. In gymnolaemates, autozooid in which eggs, sperm, or both are developed, with or without skeletal expression of this function in cheilostomates; can have or lack feeding ability (Fig. 69,1c,2).

sheath laminae. In stenolaemates, one of the two structural elements forming stylets; sheath laminae concentrically enclose the core of a stylet and are directed toward the zoarial surface. Sheath laminae are continuous with those of the remainder of zoarium, differing only in orientation (Blake).

simple skeletal wall. In stenolaemates, skeletal wall calcified on edges and one side only, either exterior or interior.

simple-walled colony. Synonym of fixed-walled colony in stenolaemates.

single-walled colony. Synonym of fixed-walled colony in stenolaemates.

sinus. In stenolaemates, groove on either side of keel in zooid wall in recumbent part of endozone, which accommodates inner end of next younger zooid in rhombic zooidal arrangement (Fig. 56).

skeleton. In stenolaemates and cheilostomates, calcareous layers of body wall and any connected calcareous structures deposited by epidermis on its external side opposite peritoneum and body cavity, and therefore exoskeletal throughout zoarium.

soft-part polymorph. Cheilostomate zooid differing from ordinary feeding zooids in having sexual features, membranous structures for brooding embryos, or elongate tentacles for producing exhalant water currents, which are apparently not reflected in differences in skeletal parts.

solid ramose colony. Synonym of dendroid colony in stenolaemates.

spherulitic ultrastructure. In cheilostomates, group of skeletal ultrastructures consisting essentially of either two-dimensional wedge- or fan-shaped arrays oriented parallel to wall surfaces (planar spherulitic ultrastructure), or three-dimensional conical or palisade arrays oriented transverse to wall surfaces (transverse spherulitic ultrastructure); arrays of acicular to bladelike or blocky calcite or aragonite crystals (Fig. 111; 113,1–3).

spine. In cheilostomates, tubular to flattened outpocketing of calcified exterior body wall and contained body cavity, commonly in groups overarching uncalcified part of frontal wall of autozooid to form costal shield, or margining orificial wall distally and laterally to form peristomelike structure (Fig. 72,2,3).

spine base. In cheilostomates, collarlike skeletal remnant of attached end of spine (Fig. 77,2).

statoblast. Free encapsulated bud in discoid envelope of chitin, with large yolky cells and organized germinal tissue capable of giving rise to polypide to start most phylactolaemate colonies; formed on funiculus of parent zooid by migration of epithelial cells (Fig. 135,3).

statoblast ancestrula. First zooid produced by germination of statoblast to found new phylactolaemate colony (Fig. 135,3).

stereom. In stenolaemates, extrazooidal skeletal deposits, consisting of either dense skeleton or vesicle roof skeleton (Fig. 201,1b; 205).

stolon. In stoloniferous ctenostomates, tubular kenozooids or extensions of autozooids from which autozooids are budded.

stoloniferan. Ctenostomate in which one or more autozooids are budded from a single kenozooid generally of elongate tubular form (Fig. 85,3).

stomodaeal cavity. Anterior part of gut lined with ectoderm infolded to form mouth (Steen, 1971).

striae. In Ptilodictyina, small skeletal ridges consisting of tightly arched skeletal laminae projecting above general zoarial surface (Fig. 224,2a).

style. (a) In stenolaemates, general term for rodlike skeletal structure approximately parallel to adjacent zooecia, which forms spinose projection on zoarial surface (Fig. 51,1–8). (b) Synonym of stylet.

stylet. (a) In stenolaemates, any member of class of rodlike skeletal structures, oriented approximately perpendicular to zoarial surface and parallel to zooecia (Blake). (b) Synonym of style.

subcolony. In stenolaemates, grouping within colony of zooids and any extrazooidal structures, which may or may not be skeletally identifiable, but which carries on most or all functions of whole colony (Fig. 59,1–3).

subsequent zone of astogenetic change. Zone of astogenetic change following primary or subsequent zone of astogenetic repetition distally; develops asexually from zone of astogenetic repetition and therefore lacks ancestrula (see zone of astogenetic change).

subsequent zone of astogenetic repetition. Zone of astogenetic repetition following subsequent zone of astogenetic change distally (see zone of astogenetic repetition).

superficial layer of skeleton. One of commonly multiple layers successively deposited on frontal side of advancing initial skeletal layer of cryptocyst or umbonuloid shield in cheilostomate zooid, commonly increasing thickness of frontal shield several-fold; commonly different in microstructure or mineral composition from initial layer (Fig. 67,1c; 68,1d).

superior hemiseptum. Synonym of proximal hemiseptum in stenolaemates (KARKLINS).

supporting walls. Body walls of zooids that support orificial walls; includes basal walls, vertical walls (lateral and transverse walls of gymnolaemates), and frontal walls.

tangential section. In stenolaemates, section just under surface of colony oriented so that zooids are cut at approximate right angles near outer ends.

tentacle. One of a row of tubular extensions of body wall and contained body cavity that surrounds the mouth in a circular or bilobed pattern; in feeding zooids, ciliated to produce water currents that concentrate food particles near mouth (Fig. 2, 4).

tentacle crown. Tentacles of a zooid in expanded feeding position.

tentacle sheath. Part of body wall that is introverted to enclose tentacles in their retracted position and everted to support tentacles in their protruded position; boundary with vestibular wall is generally the sphincter muscle, forming diaphragm in gymnolaemates (Fig. 2–4).

tentacular atrium. Cavity enclosed by retracted tentacle sheath, containing tentacles (Fig. 92,2).

tergopore. Type of kenozooecium on reverse side of zoarium, with polygonal aperture in some tubuliporates (BASSLER, 1953).

terminal diaphragm. In stenolaemates, membranous or calcified diaphragm near zooecial aperture that seals living chamber from surrounding environment. Calcified terminal diaphragms are either exterior (tubuliporates and ceramoporines) or interior (most Paleozoic stenolaemates) (see Fig. 27; 34; 42; 43,3).

transverse partition. Interior wall separating successive zooids in zooidal lines, formed from invaginated fold of cellular layers in middle of which skeletal lamina is secreted (Fig. 90,1).

transverse section. (a) In stenolaemates, section oriented so that recumbent or inner ends of zooids are cut transversely. (b) In gymnolaemates, section oriented so that zooids are cut parallel to width and perpendicular to length.

transverse spherulitic ultrastructure. In cheilostomates, skeletal ultrastructure consisting of three-dimensional conical or palisade arrays of acicular to bladelike or blocky calcite or aragonite crystals oriented transverse to wall surfaces (Fig. 113,1–3).

transverse wall. One of pair of vertical walls of gymnolaemate zooid, oriented generally subperpendicular to principal direction of zooid growth; together with lateral walls gives depth to body cavity of zooid; most commonly developed at least in part as interior wall completely separating body cavities of zooids within lineal series; in cheilostomates, includes skeletal layers (Fig. 68,1,2).

trifid nerve. Three-branched peripheral motor nerve, with branches to insertion of retractor muscle, esophagus, and along tentacle sheath to direct nerve (Fig. 100,2).

trifoliate colony. In stenolaemates, erect colony with three mesothecae radiating from colony or branch center, each supporting feeding zooids in bifoliate pattern (UTGAARD).

tunnel. Elevated, branched anastomosing ridge on colony surface in Rhinoporidae; a curved roof covers branched tunnel-like space that possibly was brood space (Fig. 192).

umbonuloid. Ascophoran cheilostomate in which autozooids have frontal shields formed by calcification on basal side of epifrontal fold (umbonuloid shield) (SANDBERG).

umbonuloid shield. Continuous frontal shield or part of frontal shield of cheilostomate zooid, formed by calcification of inner wall of exterior double-walled fold and contained body cavity, overarching flexible part of frontal wall from its proximal and lateral margins to face flexible part of frontal wall; attached to vertical walls by interior wall segments pierced by pores of marginal communication organs connecting hypostegal coelom to underlying principal body cavity of zooid, and in some by additional uncalcified, cuticle-covered openings (Fig. 68,1,2; 70,1,3).

unilaminate colony. Encrusting or erect colony consisting of a single layer of zooids opening in approximately the same direction.

uniserial budding. In gymnolaemates, budding in which lineal series rarely and irregularly come in contact, communication organs are absent or rare between zooids in adjacent series, and each lineal series forms more or less independent growing tip of colony (Fig. 76,1–5; 77,1,2).

uniserial colony. In stenolaemates, encrusting colony in which zooids bud in single row in direct parent-descendant relationship (Fig. 31,1,2).

unisexual. Zooid or colony that produces either male or female gametes but not both.

vertical plate. In cystoporates, platelike compound wall, generally parallel to colony growth direction; commonly extrazooidal but may be in part multizooidal (Fig. 196,2).

vertical wall. (a) One of zooidal supporting walls that is entirely or in part at high angle to basal and orificial walls, giving depth, length, or both to zooidal body cavity; can be exterior, interior, or a combination, and if interior, complete or incomplete (Fig. 1, 10, 11). (b) In gymnolaemates, a lateral or transverse wall of zooid.

vesicle. In Fistuliporina, blisterlike, boxlike, or less commonly tubelike element of extrazooidal vesicular tissue bounded by calcified walls and roof; space in vesicle presumably contained no soft tissue (UTGAARD).

vesicle roof. In Fistuliporina, flat or curved skeletal component of vesicle on distal or frontal side of a vesicle; simple interior wall (UTGAARD).

vesicle wall. Straight to curved lateral sides of vesicle, generally simple, interior wall, may be compound in a few genera (UTGAARD).

vesicular cell. Cell occupied by voluminous vesicular inclusion, found in peritoneal network and in funicular strands, and partly consumed in development of digestive tract and feeding organs (Fig. 87,*1*; 88,*4*).

vesicular tissue. Extrazooidal skeletal structures in Fistuliporina composed of adjacent and superimposed vesicles (Fig. 143).

vestibular dilator muscle. One of commonly multiple, radially arranged muscles that traverse body cavity of gymnolaemate autozooid to insert on vestibular wall.

vestibular wall. That part of the body wall surrounding the vestibule and connecting tentacle sheath to orificial wall (Fig. 2, 3).

vestibule. (a) Variable space through which lophophore passes in protruding and retracting (Fig. 2; 3; 132,*2*). (b) In cryptostomates, that part of zooecial chamber between aperture and either hemisepta or boundary between exozone and endozone.

vibraculum. Type of avicularium in cheilostomates, with mandible elongated beyond beak and commonly slung between asymmetrical condyles (LUTAUD).

vicarious polymorph. In gymnolaemates, polymorph intercalated in budding series to communicate with two or more zooids, in space subequal to or larger than those occupied by ordinary feeding zooids (Fig. 81,*3,4*).

zoarium. In stenolaemates and cheilostomates, the skeleton of a colony, consisting of zooecia together with any connected multizooidal and extrazooidal skeleton.

zone of astogenetic change. Part of colony in which zooids show morphologic differences from generation to generation in more or less uniform progression distally, ending with pattern capable of endless repetition of one or more kinds of zooids.

zone of astogenetic repetition. Part of colony in which zooids show one or more repeated morphologies from generation to generation distally in pattern capable of endless repetition.

zooecial compartment. Body cavity of zooid (LUTAUD).

zooecial lining. (a) In stenolaemates, distinct skeletal layer lining zooidal chamber, generally laminated, with laminae parallel to chamber surfaces. (b) In cheilostomates, skeletal layer structurally continuous around inner surface of vertical and commonly basal walls of zooid.

zooecial wall. (a) Skeletal wall of zooid. (b) Body wall of zooid including skeletal layers and underlying soft cellular layers (LUTAUD).

zooecium. (a) In stenolaemates and cheilostomates, the skeleton of a zooid, consisting of calcareous layers of zooidal walls and any connected intrazooidal calcareous structures. (b) In phylactolaemates, consisting of any nonliving secreted parts of the body wall (WOOD).

zooid. (a) One of the physically connected, asexually replicated morphologic units which, together with multizooidal parts and any extrazooidal parts present, compose a colony; it may separately perform major colony functions with systems of organs or other internally organized structures, much like a solitary animal, or it may be a polymorph consisting minimally of body cavity and enclosing body walls. (b) In phylactolaemates, polypide and its adjacent colony wall (Fig. 133,*4*).

zooidal autonomy. Extent to which zooids are comparable morphologically to solitary animals.

zooidal bend. In stenolaemates, region of zooid where growth direction turns outward to colony surface; in outer endozone or inner exozone (UTGAARD).

zooidal boundary. (a) Outermost extent of body walls of zooid. (b) In stenolaemates, boundary generally referred to along vertical walls between zooid and contiguous zooids or contiguous extrazooidal structures; most commonly indicated by abutting laminae from contiguous walls, organic-rich partitions, granular zones, or centers of bilateral symmetry where boundaries not indicated microstructurally (Fig. 2). (c) In gymnolaemates, boundary between zooid and contiguous zooids, extrazooidal parts, or the environment, especially along vertical walls; most commonly marked by combination of outermost cuticles of contiguous exterior lateral walls and parts of transverse walls and microstructural differences or centers of symmetry in interior parts of transverse walls.

zooidal control. Process influencing growth and functions of zooids to make them comparable morphologically and functionally to solitary animals in spite of membership in colony.

zooidal pattern. In stenolaemates, three-dimensional shapes and interrelationships of zooids within colony.

THE ORDERS CYSTOPORATA AND CRYPTOSTOMATA

OUTLINE OF CLASSIFICATION

The following outline of the orders Cystoporata and Cryptostomata summarizes taxonomic relationships, geologic occurrence, and numbers of recognized genera and subgenera in each suprageneric group. A single number refers to genera; where two numbers are given, the second indicates subgenera in addition to the nominate subgenus.

Order Cystoporata, 90. *Ord.-Perm.*
Suborder Ceramoporina, 10. *M.Ord.-L.Dev.*
  Ceramoporidae, 10. *M.Ord.-L.Dev.*
Suborder Fistuliporina, 80. *Ord.-Perm.*
  Anolotichiidae, 7. *Ord., M.Dev.*
  Xenotrypidae, 2. *?L.Ord., M.Ord.-M.Sil.*
  Constellariidae, 2. *M.Ord.-U.Ord., ?L.Sil.*
  Fistuliporidae, 29. *Sil.-Perm.*
  Rhinoporidae, 2. *L.Sil.-M.Sil.*
  Botrylloporidae, 1. *M.Dev.*
  Actinotrypidae, 3. *L.Miss., Perm.*
  Hexagonellidae, 14. *L.Dev.-U.Perm.*
  Cystodictyonidae, 11. *M.Dev.-L.Perm.*
  Etherellidae, 2. *Perm.*
  Goniocladiidae, 7. *Miss.-Perm.*

Order Cryptostomata, 78; 2. *Ord.-Perm.*
Suborder Ptilodictyina, 38. *Ord.-Perm.*
  Ptilodictyidae, 8. *M.Ord.-L.Dev.*
  Escharoporidae, 6. *M.Ord.-L.Sil.*
  Intraporidae, 2. *M.Dev.-U.Dev.*
  Phragmopheridae, 1. *U.Carb.*
  Rhinidictyidae, 10. *L.Ord.-M.Sil.*
  Stictoporellidae, 3. *L.Ord.-M.Sil.*
  Virgatellidae, 2. *M.Ord.*
  Family Uncertain, 6.
Suborder Uncertain, 1.
Suborder Rhabdomesima, 39; 2. *Ord.-Perm.*
  Arthrostylidae, 17. *L.Ord.-L.Perm.*
  Rhabdomesidae, 7. *U.Sil.-U.Perm.*
  Rhomboporidae, 6; 1. *?U.Dev., L.Miss.-U.Perm.*
  Bactroporidae, 1. *M.Dev.*
  Nikiforovellidae, 4. *?L.Dev., M.Dev.-U.Perm.*
  Hyphasmoporidae, 3; 1. *L.Carb.-U.Perm.*
  Family Uncertain, 1.

RANGES OF TAXA

The stratigraphic distribution of orders, superfamilies, and families of Bryozoa recognized in this volume of the *Treatise* is indicated graphically in Table 4, which follows (compiled by JACK D. KEIM).

TABLE 4. *Stratigraphic Distribution of the Cystoporata and Cryptostomata*

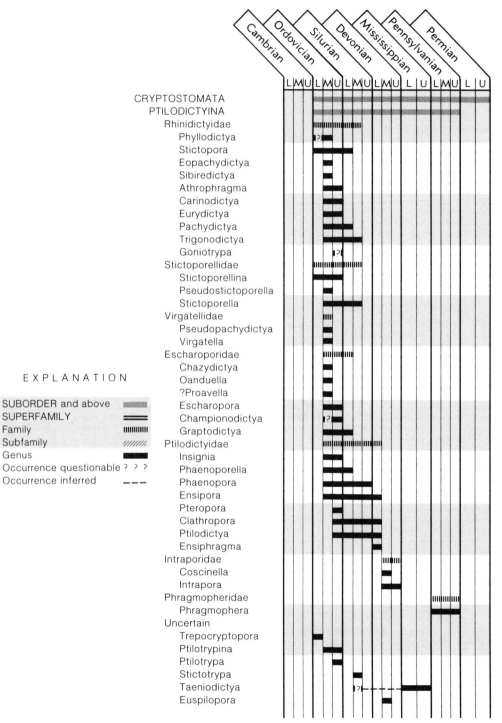

TABLE 4. (*Continued.*)

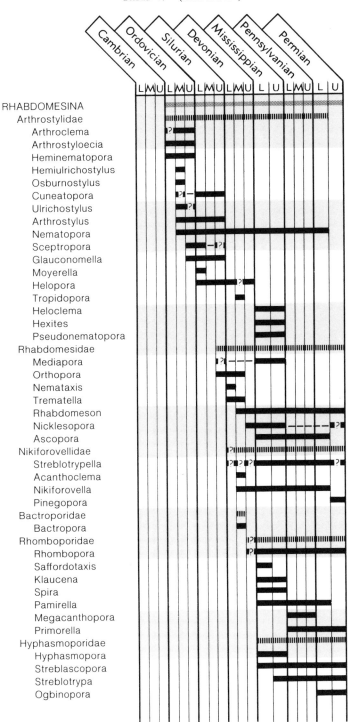

TABLE 4. (*Continued.*)

Taxon	Cambrian L	M	U	Ordovician L	M	U	Silurian L	M	U	Devonian L	M	U	Mississippian L	M	U	Pennsylvanian L	U	Permian L	U
Uncertain																			
Petaloporella																█			
UNCERTAIN																			
Heliotrypa																		█	
CYSTOPORATA			▓	▓	▓	▓	▓	▓	▓	▓	▓	▓	▓	▓	▓	▓	▓	▓	▓
FISTULIPORINA			▓	▓	▓	▓	▓	▓	▓	▓	▓	▓	▓	▓	▓	▓	▓	▓	▓
Xenotrypidae				?	█	█	█	█	█										
Xenotrypa				?	█	█													
Hennigopora						█	█	█											
Anolotichiidae			█	█	█	█	–	–	–	█									
?Lamtshinopora				█															
Profistulipora				█															
Bythotrypa				█															
Scenellopora				█															
Anolotichia				█															
Crassaluna				█															
Altshedata													█						
Constellariidae			█	█	█	?													
Revalopora			█																
Constellaria			█	█	?														
Rhinoporidae					█	█													
Rhinopora					█														
Lichenalia					█														
Fistuliporidae					█	█	█	█	█	█	█	█	█	█	█	█	█	█	
Fistuliporella					█	█	█	█	█										
Fistulipora					█	█	█	█	█	█	█	█	█	█	█	█	█	█	█
Diamesopora						█													
?Pholidopora							█												
Duncanoclema							█												
Fistuliramus							█												
Coelocaulis							█												
Buskopora							█												
Favicella							█												
Fistuliphragma							█												
Fistuliporidra							█												
Lichenotrypa							█												
Odontotrypa							█												
Pileotrypa							█												
Pinacotrypa							█												
Selenopora							█												
Kasakhstanella							█												
Cyclotrypa							█	?											
Canutrypa							█												
Cystiramus							█												
Eofistulotrypa							█												
Cliotrypa								█											
Strotopora								█											

T ABLE 4. (*Continued.*)

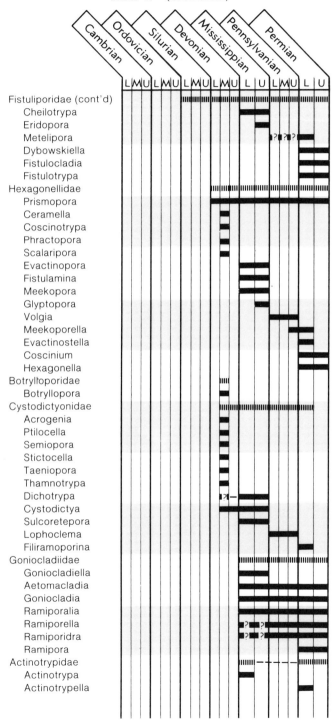

TABLE 4.   (*Continued.*)

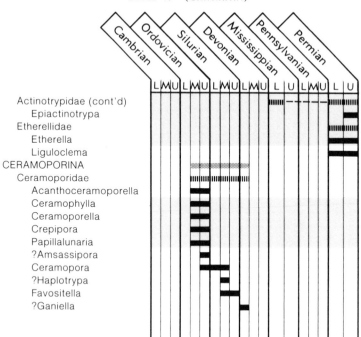

PALEOBIOLOGY AND TAXONOMY OF THE ORDER CYSTOPORATA

By JOHN UTGAARD

[Southern Illinois University at Carbondale]

GENERAL MORPHOLOGY

Cystoporates are extinct, marine, double-walled bryozoans displaying a wide variety of growth habits and belonging to the class Stenolaemata. They typically have long conical or tubular autozooecia with basal diaphragms, although some have short autozooecia without diaphragms. The growth direction of the autozooecia changes and diverges from the growth direction of the colony, producing an endozone and an exozone. Basal layers, where observed, are of simple construction (Fig. 142) and are exterior walls that presumably were covered by an outermost layer of cuticle. Interior vertical zooecial walls are compound. However, in many genera of the suborder Fistuliporina, particularly in the Fistuliporidae, the lateral and distal sides of an autozooecium are composed of superimposed vertical parts of extrazooidal vesicular tissue (vesicle walls), so that part of the autozooid is bounded by a simple interior wall (Fig. 143, 153). This is a rare, if not unique, feature in double-walled bryozoans.

Another unusual feature, found in most Fistuliporina, is the partial to complete isolation by extrazooidal vesicular tissue of new autozooecia budded on either the basal layer or mesotheca. The presence of these new autozooids, isolated from their neighbors, suggests colony-wide control of budding rather than direct parent-daughter autozooecial budding. In many fistuliporines, new autozooecia are budded on top of extrazooidal vesicular tissue in the exozone.

Cystoporates in the suborder Fistuliporina are characterized by large amounts of extrazooidal vesicular tissue (=cystopores) and stereom. Vesicular tissue is almost invariably composed of simple interior skeletal deposits (Fig. 143) secreted only from the upper or outer side. Available evidence suggests that vesicular tissue housed no viable soft parts and served as a buttress between isolated or partly isolated zooecia. Stereom is fairly dense skeletal material produced by essentially continuous deposition or contiguous deposit of vesicle roofs without intervening vesicle walls or chambers.

Most genera in the Cystoporata have a lunarium, which projects above the general zoarial surface (Fig. 144) and above the rim or the peristome, if present, of the autozooecial orifice. The lunarium consists of a microstructurally distinct or thicker deposit developed throughout the exozone of the autozooecium. It is located on the proximal side of each autozooecium or rotated to the left or right lateral side (see Fig. 205). Lunaria are known in such post-Paleozoic, double-walled tubuliporates as *Lichenopora*, in which the membranous sac occupies the proximal half of the living chamber, next to the lunarium.

Most genera of the suborder Ceramoporina evidently had two means of interzooidal communication, via coelomic fluid in the hypostegal coelom, as in other double-walled tubular bryozoans, and via communication pores in the compound skeletal zooecial walls (Fig. 142). Pores in the zooecial walls apparently are restricted to ceramoporines among Paleozoic tubular bryozoans. Contrary to many published reports, I have seen no undoubted communication pores in members of the suborder Fistuliporina. Evidently their only means of interzooidal communication was via coelomic fluid in the hypostegal coelom.

Some genera of the Fistuliporina, partic-

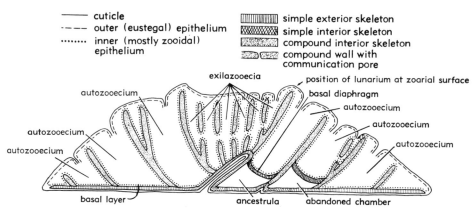

FIG. 142. Cystoporate morphology. Longitudinal section through hypothetical double-walled ceramoporine (after Utgaard, 1973). Simple skeletal walls include the basal layer and basal diaphragms in
autozooecia. The basal layer folds back upon itself on the left side of the ancestrula. Some exilazooecia
(right-center) are shown with a terminal-vestibular membrane, as if they had an extrusible polypide;
others (left-center) are shown with an imperforate terminal membrane. Communication pores are shown
in exilazooecial walls. Autozooecial walls also have communication pores, but the section is through the
lunarial deposit on the proximal side of each zooecium, and this deposit is imperforate.

ularly the cystodictyonids, etherellids and
goniocladiids, are monomorphic. Many fistuliporines have polymorphic colonies and
most are dimorphic, with larger monticular
zooecia in addition to the normal autozooecia. A few fistuliporines have intermonticular
autozooecia with expanded subspherical
outer ends termed **gonozooecia**. Their morphology and development in a colony suggests that these zooids were polymorphs, possibly involved with production of eggs and
brooding of embryos. One monomorphic
goniocladiid has a subspherical expansion in
the vesicular tissue that possibly served as a

brood space. Most genera of the Ceramoporina are trimorphic and have normal autozooecia, large monticular autozooecia, and
exilazooecia, which are small, tubular zooids
(formerly called mesopores or cystopores)
developed in the exozone (Fig. 142). One
ceramoporine has, in addition, **basal zooecia**
(Fig. 145) and displays the greatest degree
of polymorphism in the Cystoporata. Funnel
cystiphragms and flask-shaped chambers in
many Cystoporata suggest that intraautozooecial polymorphism may be widespread
in the order.

AUTOZOOIDS

RECOGNITION OF AUTOZOOIDS
IN THE CYSTOPORATA

Autozooids are the normal individuals in
a colony (BORG, 1926a, p. 188), which perform all the usual body functions (RYLAND,
1970, p. 29). At one or more stages in their
ontogeny, autozooids have a protrusible
lophophore (BOARDMAN, 1971, p. 2). Using
the definition of zooid as an individual member of a colony, minimally consisting of body
wall enclosing a coelom and connected by the

body wall to other members of a colony
(BOARDMAN, CHEETHAM, & COOK, this revision), and evidence from microstructure,
budding, and colony construction, several
kinds of zooecia in the Cystoporata could
have been autozooecia.

In the Fistuliporina, the choice is narrowed
to one or usually no more than two kinds of
zooecia. Some fistuliporines are monomorphic in both intermonticular and monticular
areas; all zooecia can be considered to be
autozooecia that housed feeding organs.

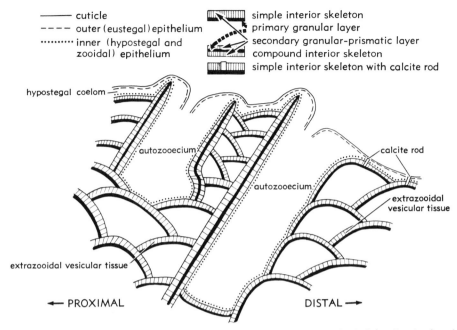

Fig. 143. Cystoporate morphology. Longitudinal section through hypothetical fistuliporine based on observed sections but showing double-walled construction (after Utgaard, 1973). The autozooecium on the left was formed by septa produced on old vesicular tissue by folding of the inner (hypostegal) epithelium. It has a peristome and compound walls. The autozooecium on the right has a compound wall on the lunarian side, which is proximal (left side in figure), but the distal side is composed of superimposed vertical simple walls of extrazooidal vesicular tissue. It has no peristome and the inner, zooidal epithelium curves up and out to continue as the inner (hypostegal) epithelium below the hypostegal coelom. Walls and roofs of the vesicles or vesicular tissue can be composed of the inner granular primary layer alone or the primary layer and a secondary granular-prismatic layer. Outer cuticle is shown attached to the calcite rod of small acanthostyles or tubuli in the vesicle roofs.

Most fistuliporines have intermonticular autozooecia that are slightly smaller than the otherwise similar monticular zooecia (see Fig. 180, *1b*); both types probably housed feeding organs. The intermonticular autozooecia are considered to be the common kind of autozooecia and the monticular zooids to be polymorphs (UTGAARD, 1973, p. 324). A few fistuliporines have zooecia with expanded outer ends on the colony surface; these are probably autozooecia modified to serve a brooding function.

The polymorphic colonies of the Ceramoporina have as many as four different kinds of zooecia. All are possible autozooecia. Only two of these types are common to all ceramoporines: the large intermonticular

zooecia and the slightly larger monticular zooecia. The intermonticular zooecia probably housed feeding organs because they compare in size, number, and position with autozooecia of monomorphic fistuliporines. As with the Fistuliporina, the larger monticular zooecia probably also housed autozooids. Smaller zooecia on the frontal or nonbasal surface of some ceramoporine colonies possibly are another kind of autozooecium, a kind of polymorph that is called an exilazooecium. Rare basal zooecia, found so far only in some colonies of *Ceramopora*, were probably not autozooecia.

Inferred autozooecia in the Cystoporata are comparable in relative size, shape, position of origin, extent of living chamber,

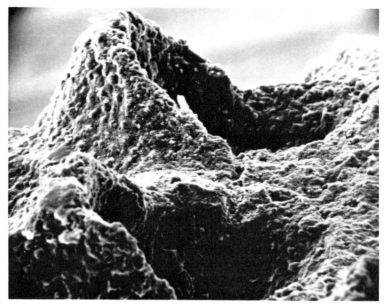

Fig. 144.   Cystoporate morphology. *Cheilotrypa hispida* ULRICH, Glen Dean Ls., Miss. Sloans Valley, Ky. Lunarium on the proximal side of an autozooecium; note the small protrusions on the proximal side (left), which are the surface expression of minute calcite rods in tubuli in the thick lunarial deposit. Scanning-electron photograph, SIUC 3001, ×260.

intrazooecial structures, distribution in the colony, and similarity to probable ancestrula with inferred autozooecia in the Trepostomata (BOARDMAN, 1971) and autozooecia in the Tubuliporata (BORG, 1926a, 1933).

LUNARIA

Most genera in the Cystoporata have a lunarium on the proximal side of each autozooecium and on each large monticular zooecium. Lunaria are not present in exilazooecia, basal zooecia, or in extrazooidal vesicular tissue. The lunarium projects above the general surface of the zoarium (Fig. 144) and above the rim or peristome of the autozooecial orifice in unworn specimens. In most genera, zooecia are radially arranged around monticular centers with lunaria on the sides of zooecia nearest monticular centers (see Fig. 183,*1c,d*). In some bifoliate forms, lunaria rotate to the right- or left-lateral side of autozooecia in the exozone (see Fig. 205,*1b*).

The lunarium, cut transversely in tangential thin sections or acetate peels, generally has a shorter radius of curvature than the remainder of the zooecial orifice (see Fig. 159,*1b,c,e*). In some genera the ends of lunaria project into zooecial cavities (see Fig. 174,*1e,f*; 180,*2a,b*) and greatly modify shapes of skeletal living chambers. In other cystoporates the radius of curvature of the lunarium is approximately the same as that of the distal side of the zooecium, regardless of whether the aperture is elongated in the proximal-distal direction (see Fig. 158,*1c*), or is approximately circular. In some, the lunarium is small or spinelike (see Fig. 183,*2a*; 188,*1b*).

Lunaria appear in early ontogenetic stages of autozooecia and are distinct skeletal structures that generally can be seen in longitudinal and transverse thin section as well as in tangential section. Lunarial deposits in many genera are continuous, extending from outer endozones or inner exozones to zoarial sur-

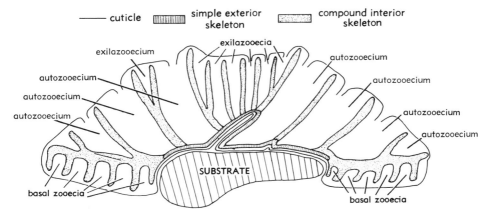

FIG. 145. Cystoporate morphology. Longitudinal section through hypothetical double-walled *Ceramopora* with a celluliferous base (after Utgaard, 1973). Epithelia are omitted. The celluliferous base contains relatively short, narrow, diaphragmless polymorphs called basal zooecia. Probably lacking polypides, they opened into the hypostegal coelom on the free basal margins of the colony, beyond the encrusted substrate. Their walls are compound. Only the basal layer in the ancestrula and adjacent to the encrusted substrate is a simple wall.

faces (see Fig. 158,*1a,e*; 160,*1e,i*). In a few genera, such as *Anolotichia* (see Fig. 164), the structure of lunarial deposits changes markedly near zoarial surfaces. In most cystoporates, the structure of the lunarium is relatively uniform throughout its length. Lunaria commonly increase in size in the exozone in the growth direction of the autozooecium. Thus, in cystoporates with lunaria and exozones, the lunaria are developed and visible externally in well-preserved specimens. Microstructures and ultrastructures of lunarial deposits are not uniform within the order Cystoporata or even within some families.

BORG (1965) and UTGAARD (1968a) reported lunaria in post-Paleozoic hornerids and lichenoporids. BOARDMAN (1971) described the position of the membranous sac in a lichenoporid as being on the proximal (lunarial) side.

AUTOZOOECIAL LIVING CHAMBERS

In modern tubuliporates the living chamber is that part of a zooid lined by zooidal epithelium, and it houses the functional organs of the zooid, if any are present. Skeletal remains of living chambers in autozooids of cystoporates as well as trepostomates (BOARDMAN, 1971, p. 5) can be studied in unworn specimens or beneath protective overgrowths. Estimating the minimum extent of a living chamber by trying to trace "time lines" of skeletal deposition from a basal diaphragm into and along the autozooecial wall to the zooecial boundary is difficult in trepostomates (BOARDMAN, 1971, p. 5), and is virtually impossible in cystoporates. In the Fistuliporina the wall laminae are obscure or walls have a **granular** and **granular-prismatic microstructure**. In the Ceramoporina, this method could be successful, but wall laminae apparently were deposited in bundles in the form of partial cylinders and not as complete cylinders lining the zooecium, as was usual in trepostomates. Minimum length of a living chamber can be estimated from studying relatively unabraded specimens, but the best estimate comes from studying living chambers preserved beneath overgrowths (BOARDMAN, 1971, p. 5).

Autozooecial living chambers in the Cystoporata.—Nearly all living chambers in the cystoporates are either modified cone-cylinder shapes with the smaller, modified cone-shaped inner end on the basal layer or mesotheca and the larger, cylindrical portion in the exozone; or cylinders with a nearly flat basal diaphragm. The length of these living chambers ranges from approximately two autozooecial diameters to about seven. In *Botryllopora*, which has comparatively narrow living chambers, they may be as much as nine times as long as wide. Most living chambers in the cystoporates are approximately three to five times as long as wide.

Cross-sectional shapes of inner ends of living chambers of the modified cone-cylinder type may be hemispherical, mushroom-shaped, subcircular, triangular, or subtriangular. Many bifoliate fistuliporines have right- and left-handed autozooecia that join mesothecae at teardrop- or club-shaped contacts (see Fig. 205,*1f*; 207,*1e*). Cross-sectional shapes of living chambers in exozones may be circular to elliptical. These basic shapes may be modified depending upon the radius of curvature of the lunarium, whether or not the ends of the lunarium project into the autozooecial cavity, or the presence of canaliculi, large septalike styles that inflect autozooecia (see Fig. 194,*1d,e*; 195,*1c,2b*). In many cystoporates the lunarium encloses part of a round to elliptical cylinder-shaped space on the side of and paralleling a larger round to elliptical space enclosed by the remainder of the autozooecial walls (see Fig. 159,*1c*; 180,*2a–c*). Possible polypide remnants in this smaller cylinder suggest that the lunarium formed a groove in which the polypide was located.

In a few constellariids, the basal structures of living chambers may be curved or cystoidal diaphragms or a combination of cystoidal diaphragms and flat diaphragms. In these forms, the living chamber generally has a bisected funnel shape and a smaller cross-sectional area in its inner portion and is cylindrical in its longer, outer portion. The deepest part of the living chamber is on the proximal side of the autozooecium, next to the lunarium if one is present.

In many bifoliate and a few encrusting species of fistuliporines, a proximal hemiseptum partly divides the living chamber into a modified conical inner portion and a cylindrical outer portion. Species of *Strotopora, Cliotrypa* and *Fistuliphragma* have alternating hemiphragms, which are triangular and platelike or curved spines, and which protruded into the living chamber.

Basal diaphragms and abandoned chambers.—As an autozooecium grows and the living chamber reaches a certain length (which is not constant in a colony), a new basal diaphragm is formed in many cystoporates. Other cystoporates with relatively short autozooecia do not have basal diaphragms. Lengthening of the outer end of the autozooecia and, especially, formation of a new basal structure, are probably related to the degeneration-regeneration cycle in an autozooid. As in the trepostomates (BOARDMAN, 1971, p. 18), spacing of basal diaphragms in many cystoporates is such that **abandoned chambers** between successive basal diaphragms are usually much shorter than the living chamber in the same autozooecium. Length of the abandoned chambers in the cystoporates usually ranges from less than one to slightly more than three autozooecial diameters, and generally is less than two autozooecial diameters.

Formation of a new basal diaphragm probably involved proliferation of a new epithelium and peritoneum from the lateral walls of the zooid across the zooecium at a level closer to the surface of the colony than the preceding basal diaphragm. Brown bodies, some other cellular material, and coelomic fluids probably were left behind in the abandoned chamber. If the basal zooidal epithelium next to the basal diaphragm were drawn intact to a new position farther out in the autozooid, brown bodies or fossilized brown deposits would not be found in abandoned chambers, but they commonly are. In the cystoporates, basal structures are simple-walled in construction, being deposited by epithelium on the outer side. Except in some ceramoporines, where an abandoned chamber may be connected to the living chamber of an adjacent autozooecium or exilazooecium through communication pores in zooecial walls, living tissue was probably not present in abandoned chambers. Brown deposits encapsuled with membrane, diffuse brown deposits, and rare membranous linings found in abandoned chambers are probably the remnants of brown bodies, cellular material, coelomic fluid and membrane left behind in the abandoned chamber when a new basal zooidal epithelium was proliferated and a new basal skeletal diaphragm was formed.

In the cystoporates, as in the trepostomates (BOARDMAN, 1971, p. 5), the new living chamber consisted of most of the old living chamber, minus the abandoned segment, and new space where new zooecial walls were secreted at the outer end of the autozooecium while the polypide was degenerated.

Terminal and subterminal diaphragms.— Among Paleozoic bryozoans, subterminal diaphragms, with reverse curvature indicating deposition from the inner side, are known only in some ceramoporines (UTGAARD, 1968b, p. 1445–1446). They are common in autozooecia and some exilazooecia, especially in species of *Ceramoporella*. It is possible that some reversed subterminal diaphragms are compound, being calcified by the zooidal epithelium on the inside and the **inner (hypostegal) epithelium** on the outside (Fig. 146), but most appear to have been calcified from the inside only.

BORG (1933) reported terminal and subterminal diaphragms in autozooecia and kenozooecia in heteroporid tubuliporates. Considerable variation exists, within and among genera of heteroporids, in the abundance of terminal and subterminal diaphragms. NYE (1968, p. 112) reported pore-bearing terminal and subterminal diaphragms and imperforate intermediate diaphragms in some post-Paleozoic tubuliporates. Both the terminal and intermediate diaphragms have laminae flexed toward the inner end of the autozooecium where the diaphragm joins the zooecial walls, indicating at least partial deposition from the inner side.

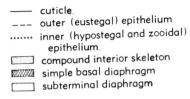

— cuticle.
– – – outer (eustegal) epithelium
······· inner (hypostegal and zooidal) epithelium
▨ compound interior skeleton
▨ simple basal diaphragm
☐ subterminal diaphragm

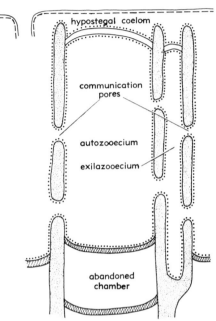

FIG. 146. Cystoporate subterminal diaphragms. Idealized diagram of living chambers in an autozooecium and an exilazooecium in the ceramoporine *Ceramoporella*, which are closed by a subterminal diaphragm (after Utgaard, 1973). The subterminal diaphragms clearly are mainly or entirely deposited from their inner side. The inner (hypostegal) epithelium below the hypostegal coelom could have had a part in secreting the subterminal diaphragms from the outer side. The abandoned chamber, between successive basal diaphragms, contained no viable tissue whereas chambers below the subterminal diaphragms contained at least a living epithelium, a peritoneum, and coelomic fluid by virtue of their communication with adjacent zooids via communication pores.

Pores have not been observed in the terminal and subterminal diaphragms in ceramoporines.

Terminal and subterminal diaphragms have been observed in more than a dozen genera of the Fistuliporina. They are of simple construction (see Fig. 186,*1e*), but, unlike those in the ceramoporines, they were deposited by an inner or hypostegal epithelium below the hypostegal coelom on the outer side of the diaphragm. Pores have not been observed in these diaphragms in fistuliporines. Some membranous structures previously referred to as pellicles or opercula in the fistuliporines may be terminal or subterminal diaphragms (UTGAARD, 1973).

BORG (1933, p. 320) thought that the formation of terminal diaphragms was a part of the degeneration-regeneration cycle. It seems likely that the polypide would degenerate in a ceramoporine autozooid when the terminal or subterminal diaphragm was formed, especially if the subterminal diaphragm were formed in the region of the vestibule. It is possible that the inner part of a polypide could lie dormant, nourished through communication pores. The entire zooid probably would be abandoned to degenerate or decompose in a fistuliporine below a terminal or subterminal diaphragm, if the diaphragm lacked a pore. The presence of possible polypide remnants in large "abandoned" chambers (capped living chambers) below terminal diaphragms supports the inference that they underwent decomposition without complete degeneration to form a brown body.

Living chambers capped with terminal or subterminal diaphragms are much longer than normal abandoned chambers in the same colony and even in the same zooecium, indicating that terminal or subterminal diaphragms do not, as a matter of course, become the next basal diaphragm when a polypide is regenerated, as ULRICH (1890, p. 315–316) thought. It is possible that terminal diaphragms were produced during degeneration and were resorbed during regeneration in the degeneration-regeneration cycle. BORG (1933, p. 298, 302–303) reported resorption of the subterminal diaphragms in parts of the colonies of the tubuliporate *Heteropora*. However, the presence of some long, abandoned chambers, including those containing flask-shaped chambers,

funnel cystiphragms, and partial funnel cystiphragms (see below), suggest possible capping by terminal diaphragms rather than the usual basal diaphragms. The relative rarity of obvious terminal diaphragms and longer-than-normal abandoned chambers suggests that they were not consistently part of a normal degeneration-regeneration cycle. In addition, membranous remnants possibly representing sacs of undegenerated polypides have been found in several living chambers capped by subterminal diaphragms.

Perforated apertural structures (opercula) have been reported in several genera of cystoporates. These need further study.

LATERAL STRUCTURES

Cystoporates display relatively few lateral intrazooecial structures as compared to some other stenolaemate bryozoans. Cystoporates lack skeletal cystiphragms. Only a few genera (see Fig. 176,*1a,d*; 183,*1e*; 191,*1c*) have spine- to platelike hemiphragms that alternate across the autozooecium. Formation of hemiphragms may have been related to a degeneration-regeneration cycle. Some of the bifoliate fistuliporines have a proximal or proximolateral hemiseptum (see Fig. 205,*1e*) at the zooecial bend. One genus, *Prismopora,* has a recurved distal hemiseptum at the zooecial bend region. A few fistuliporines have short, hyaline mural spines in longitudinal and horizontal rows in the exozone.

Funnel cystiphragms and partial funnel cystiphragms, lateral structures found in few cystoporates, are discussed below.

Hollow spherical cysts.—Hollow spherical calcareous cysts have been observed in autozooecia in nine species of cystoporates. BASSLER (1911, p. 86, 90) and UTGAARD (1968b, p. 1449) reported such structures from the Ordovician ceramoporine *Crepipora incrassata* BASSLER, and ULRICH (1890, p. 318) reported similar structures in a species of *Fistulipora* from the Devonian of New York. In addition, a survey of the thin-section collection at the U.S. National Museum of Natural History revealed similar cysts in the Ordovician ceramoporine *Crepipora venusta* (ULRICH), the Ordovician anolotichiid *Anolotichia ponderosa* ULRICH, an unidentified Devonian species of *Fistulipora*, an unidentified Permian fistuliporid, the Devonian fistuliporid *Cyclotrypa communis* (ULRICH), the Mississippian hexagonellid *Glyptopora elegans* (PROUT), and the Devonian cystodictyonid *Dichotrypa foliata* ULRICH.

The cysts are generally circular in cross section, although those in *Dichotrypa foliata* and the unidentified Devonian species of *Fistulipora* are ovate to elongate-ovate in cross section. The microstructure of the cysts is granular in those species with granular-prismatic walls and laminated in those species with laminated walls. They are generally attached to the autozooecial walls or basal diaphragms, or both, but a few appear to "float" in the autozooecial cavity if the plane of the section misses the site of attachment. The wall of the cysts generally is 0.01 mm thick or slightly less. The cysts generally range from 0.10 to 0.20 mm in diameter, though some of the elongate-oval ones are up to 0.30 mm in maximum dimension. In *Crepipora venusta,* several abandoned chambers contain clusters of from 3 to 9 smaller calcareous cysts, each about 0.05 mm in diameter. Their walls are fused, and several cysts are aggregated into a grapelike mass.

Some of the cysts are apparently empty but most contain a relatively small amount of brown granular material. The small cysts in grapelike clusters in *Crepipora venusta* are nearly filled with brown residue.

Most hollow cysts are in abandoned chambers. One hollow cyst has been observed near the distal end of a living chamber that is capped by a terminal diaphragm (*Dichotrypa foliata*), one is possibly in a living chamber (*Anolotichia ponderosa*) and one is in an open living chamber (*Glyptopora elegans*). In an unidentified fistuliporid from the Permian, the cyst is in an abandoned chamber immediately below an autozooecial living chamber

that is modified by half of a funnel cystiphragm. Hollow cysts apparently can occur anywhere within an autozooecial living chamber.

ULRICH (1890, p. 318) suggested that the cysts in the tubuliporate *Ceriocava ramosa* D'ORBIGNY (ULRICH, p. 318, fig. 7e, f) from the Cretaceous of France and in the unidentified species of *Fistulipora* from the Devonian of New York were homologous with "true cystiphragms" in monticuliporid trepostomates. However, as CUMINGS and GALLOWAY (1915, p. 351–354, and fig. 17–22) determined, cystiphragms do not contain brown residue and their major functions seem to be to limit the size and impart a generalized shape to the autozooecial living chamber (CUMINGS & GALLOWAY, 1915, p. 354–355; BOARDMAN, 1971, p. 12). Unlike the relatively rare, subspherical to spherical hollow cysts, cystiphragms in monticuliporids generally occur in ontogenetic series and provide a living chamber that was relatively constant in shape during ontogeny (BOARDMAN, 1971, p. 12). The rare hollow cysts generally occupy half to four-fifths of the diameter of a small part of a living chamber and greatly modify the shape of that part of the living chamber. In addition, these cysts generally contain brown residue, and in this respect differ from cystiphragms.

BASSLER (1911, p. 86) called the cysts in *Crepipora incrassata* "rounded, ovicell-like structures" and further suggested (p. 90) that they ". . .would seem to bear most resemblance to the ovicells of the cyclostomatous bryozoans." The presence of a similar, hollow, calcareous cyst in an autozooecium of a modern species of *Hornera* that also has gonozooecia and brown bodies suggests that these structures in Paleozoic cystoporates were neither ovicells nor encysted brown bodies.

BOARDMAN (1960, 1971) and DUNAEVA (1968) reported similar, hollow cysts in the Devonian and Carboniferous trepostomates *Leptotrypella, Aisenvergia,* and *Volnovachia.* The hollow spheres in *Volnovachia,* illustrated by DUNAEVA (1968), are smaller

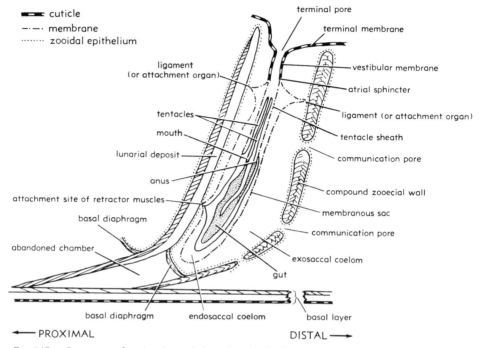

cuticle
— · — · membrane
········· zooidal epithelium

terminal pore

terminal membrane

ligament
(or attachment organ)

vestibular membrane

atrial sphincter

tentacles

ligament (or attachment organ)

mouth

tentacle sheath

lunarial deposit

communication pore

anus

compound zooecial wall

attachment site of retractor muscles

membranous sac

basal diaphragm

communication pore

abandoned chamber

exosaccal coelom

gut

basal diaphragm   endosaccal coelom   basal layer

◄─── PROXIMAL   DISTAL ───►

FIG. 147. Cystoporate functional morphology. Longitudinal section through a hypothetical model of a ceramoporine autozooid based on soft tissue of a recent lichenoporid tubuliporate (after Utgaard, 1973). Eustegal epithelium, peritoneum, and muscles that widen the vestibule are omitted from the drawing.

and more numerous than most of the cysts in the cystoporates. They are about the same size, though in larger numbers, than the cysts in *Crepipora venusta*. DUNAEVA (1968) suggested that the small spheres might be eggs. BOARDMAN (1971) suggested that they are encysted foreign bodies or perhaps even encysted brown bodies.

The size, shape, location, rarity, and associated brown material of hollow spherical cysts suggest that they were: (1) not homologous with monticuliporid cystiphragms that modified the size and shape of normal autozooecial living chambers; (2) not encysted brown bodies formed from the degeneration of a polypide; (3) not ovicells; (4) not eggs (at least not the cysts found in the cystoporates). One of BOARDMAN's suggestions (1971) seems to be the most likely proposed so far: that they are encysted foreign bodies. It seems most likely that the rare cysts were formed in autozooecial living chambers by an

inpocketing of the zooidal epithelium that ranged in shape from hemispherical to spherical with one or more narrow skeletal connections to the autozooecial wall or basal diaphragm. Such an inpocketing could be the result of a local pathological stimulus or presence of a foreign body. The brown residue in most of the cysts suggests a pathologic stimulus or biological foreign body.

INTERPRETIVE FUNCTIONAL MORPHOLOGY OF CYSTOPORATE AUTOZOOIDS

Recent tubuliporates as a model.—Similarities in mode of growth led BORG (1926b, p. 596; 1965, p. 3) and BOARDMAN (in BOARDMAN & CHEETHAM, 1969; 1971, p. 6–7) to look for a growth model for Paleozoic tubular bryozoans in the post-Paleozoic double-walled tubuliporates. The evidence for a double-walled colony construction in tubular bryozoans and BORG's order Stenolaemata, which included the tubuliporates, treposto- mates and cystoporates, led BOARDMAN (1971, p. 6) to recent tubuliporates as a log- ical first approximation for a model for zooid form and function in tubular bryozoans. A generalized autozooid, based on that of a lichenoporid tubuliporate is used here (Fig. 147) as a model for autozooids in the Cys- toporata.

Evidence from Paleozoic cystoporates.—In addition to evidence from budding locations, microstructure, ultrastructure, and growth of the skeleton, the preservation of HCl-resis- tant organic matter in living chambers below overgrowths or terminal diaphragms gives some indication of the nature and extent of zooidal tissues. The nature of this preserved organic material suggests certain character- istics that cystoporate autozooids had in com- mon with autozooids of trepostomates and tubuliporates. One of these is the degenera- tion-regeneration cycles.

Polypide remnants in cystoporates.—A very few fossilized cystoporate autozooecia contain long, tubular, brown deposits that range from interrupted patches of brown granular material to fairly complete tubular membranes (UTGAARD, 1973). These are found in living chambers below overgrowths or terminal or subterminal diaphragms, and probably represent polypide remnants rather than brown bodies from a degenerated zooid. In abandoned chambers of several zoaria con- taining the long, tubular brown deposits are compact brown deposits, more likely rem- nants of brown bodies (UTGAARD, 1973). The long, tubular, membraneous deposits suggest that at least those cystoporates had autozooids with a membranous sac.

A few of these tubular deposits occupy almost the whole width of the living cham- ber; most occupy only a part, some less than half, of the living chamber width (UTGAARD, 1973). Similar polypide and membranous sac placement is common in recent tubulipo- rates. Some of these organic remnants are closer to the proximal or lunarial side, at least toward their outer end, a situation found in

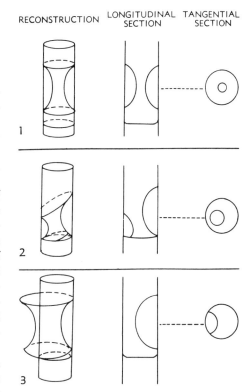

RECONSTRUCTION LONGITUDINAL TANGENTIAL
SECTION SECTION

FIG. 148. Cystoporate flask-shaped chambers. Variations in shape and size of flask-shaped cham- bers as seen in longitudinal and tangential thin sec- tions, produced by different inclinations and posi- tions of the axis of a funnel cystiphragm (after Utgaard, 1973).——*1,* Axis of funnel cystiphragm centered on autozooecial axis and not inclined. ——*2,* Axis of funnel cystiphragm inclined to axis of autozooecium.——*3,* Axis of funnel cysti- phragm parallel to but offset from axis of auto- zooecium.

zooids of recent *Lichenopora* (BOARDMAN, 1971).

Organic linings of autozooecial chambers, possible membranous diaphragms, and enig- matic organic threads and cysts are rare in cystoporates (UTGAARD, 1973) and their pos- sible extent and paleobiological importance are unknown.

FLASK-SHAPED CHAMBERS

The occurrence of complete and partial funnel cystiphragms, which form flask-shaped chambers in the Cystoporata, was summarized by UTGAARD (1973). More variation in shape exists in flask-shaped chambers in the cystoporates (Fig. 148) than in trepostomates. The possibility that they are intraautozooecial polymorphs was discussed by BOARDMAN (1971) and UTGAARD (1973). BOARDMAN and McKINNEY (1976) later suggested that they are calcified partitions reflecting the shapes of organs of the lophophore and gut of the feeding autozooids, calcified during a dormant, nonfunctional stage. If that were so, autozooecial living chambers would be greatly modified by their presence, and the resulting flask-shaped chambers presumably would have been sealed off and abandoned and would not have been occupied by succeeding, normal polypides. In some cystoporates, there is clear evidence that the flask-shaped chambers were occupied by a zooid with a polypide (UTGAARD, 1973, p. 338–339). Funnel cystiphragms are found in ontogenetic series and some of the flask-shaped chambers contain fossilized brown bodies and polypide remnants indicating that they were occupied by a zooid with a polypide, probably a polymorphic one.

ZOOECIAL POLYMORPHS

EXILAZOOECIA

Exilazooecia are found in most genera of the Ceramoporidae but are not developed in the Fistuliporina. The term exilazooecium is a modification of the term "exilapore" (DUNAEVA & MOROZOVA, 1967, p. 87) originally used for mesopores in some Trepostomata that are relatively narrow and long and that lack or have only a few basal diaphragms spaced far apart. By the definition of zooid used in this work, there is little doubt that the exilazooecia in ceramoporines are zooecia of some smaller kind of zooid. Because the soft parts and function may never be known, the descriptive term "exilazooecium" is preferred to terms based on function, tissues, or organs, such as kenozooecium, nanozooecium, and heterozooecium. Exilazooecia in ceramoporines differ from mesozooecia and alveoli. The descriptive term "cystopore" does not fit these structures as well as the term "exilazooecia."

Exilazooecia in the Ceramoporidae.—Exilazooecia in the Ceramoporidae arise by formation of a septum (a compound wall) proximal to a semirecumbent autozooecial wall in the outer endozone or inner exozone (see Fig. 158,*1a*; 159,*1g*). Exilazooecia have never been observed to arise by formation of a compound wall on the basal layer or mesotheca.

Exilazooecia extend from their locus of origin to the zoarial surface. They have a relatively long, narrow, tubular cavity with a rounded cross section. In some genera they are slightly more subangular and larger in cross section in the inner exozone and become progressively more circular and smaller toward the zoarial surface. Exilazooecia can be absent to abundant between autozooecia in intermonticular areas in the ceramoporines, and generally a cluster of exilazooecia forms the center of a monticule. They generally lack intrazooecial skeletal structures but basal diaphragms are present in some relatively long exilazooecia and subterminal diaphragms may be present, generally at the same level as subterminal diaphragms in adjacent autozooecia.

Exilazooecial wall microstructure is similar to that of autozooecia in the Ceramoporidae, that is, with broadly curved laminae and generally a broadly serrated boundary zone. The cortex is thinner and a zooecial lining is thin or absent in the exilazooecia. Exilazooecia do not have lunaria or lunarial deposits (UTGAARD, 1968b, p. 1449). Acanthostyles may be present in exilazooecial walls (see Fig. 157,*1e,g*; 160,*1b*).

In genera of the Ceramoporidae that have communication pores, the pores are present in the exilazooecial walls as well as in the autozooecial walls. Pores may connect the cavities of adjacent autozooecia, adjacent exilazooecia, or contiguous exilazooecia and autozooecia. Communication pores commonly are fewer between exilazooecia than between autozooecia. In one Middle Ordovician *Acanthoceramoporella,* the communication pores are enlarged to huge gaps in the walls, and flat and curved basal diaphragms may extend across several autozooecia and exilazooecia (UTGAARD, 1973, p. 340).

Interpretive functional morphology of exilazooecia.—The presence of compound exilazooecial walls, zooecial linings, basal diaphragms, terminal diaphragms deposited from their inner side, and communication pores in exilazooecia indicate the presence of a secretory zooidal epithelium. Ceramoporines are double-walled bryozoans and coelomic fluid in the exilazooecia could communicate with the (presumably exosaccal) coelomic fluid in adjacent autozooecia and exilazooecia through the hypostegal coelom as well as through communication pores. When subterminal diaphragms without pores were formed, coelomic communication could still take place through communication pores in the zooecial walls (Fig. 146).

Encapsuled brown structures, reminiscent of brown bodies formed from degeneration of a polypide, have been reported (UTGAARD, 1968b, p. 1446) in exilazooecia in *Ceramophylla vaupeli* (ULRICH). These brown structures strongly suggest the presence of live tissue, in addition to the zooidal epithelium. Kenozooids in modern heteroporid tubuliporates (BORG, 1933, p. 362, 368) have a zooidal epithelium, peritoneum, coelomic fluid, and various cells in the coelomic fluid, but they do not degenerate to produce an encapsuled brown body because they do not have a polypide. Nanozooids in recent tubuliporates do have a reduced extrusible polypide and a membranous sac and do degenerate to form brown bodies (BORG, 1926a, p. 234, 236). Thus, some exilazooecia may have contained a modified polypide or other organs.

LARGE MONTICULAR ZOOECIA

The zooecia that immediately surround monticular centers in most cystoporates are slightly larger than common intermonticular zooecia, which are interpreted to have housed the normal feeding autozooids. The larger zooecia have comparable wall structure and thickness, living chamber length, abandoned chamber length, and intrazooecial skeletal structures, including funnel cystiphragms and flask-shaped chambers. The only differences they show with intermonticular autozooecia are a slightly larger diameter and the commonly radial arrangement of the lunarium and proximal side of the living chamber around the monticular center (see Fig. 183,*1c,d*). When they show radial arrangement, the lunarium is on the side nearest the monticular center. Intermonticular autozooecia also display radial arrangement of lunaria in many cystoporates. It seems likely that the larger monticular zooecia housed slightly larger feeding autozooids. The functional significance of the larger monticular autozooids is not yet clear, but it seems likely that they did not serve a reproductive or brooding function. Forms that have inferred gonozooecia also have large monticular zooecia.

BASAL ZOOECIA

Polymorphic basal zooecia are known only in some colonies of *Ceramopora imbricata* HALL (see Fig. 156,*1a*) (UTGAARD, 1969, p. 289). They develop in the free margins of encrusting colonies, beyond the encrusted part of the colony, which has a simple-walled basal layer. The basal polymorphs have relatively short, narrow cavities that are subcircular in cross section and most closely resemble exilazooecia in shape and size. They have compound walls and most likely opened into a basal, centripetal expansion of the hypostegal coelom that continued from the frontal surface of the colony, around the growing margin, to the peripheral part of the base of the colony (Fig. 145). There is no evidence, to date, of a possible modified polypide or membranous sac in these basal polymorphs, but their mode of growth suggests that they were lined by a secretory zooidal epithlium and contained coelomic fluid that was in communication with a hypostegal coelom.

GONOZOOECIA

Intermonticular autozooecia with expanded subspherical outer ends have been known for some time (ULRICH, 1890, p. 383) in a few cystoporates. To date, they have been found only in the Fistuliporidae and only in the genera *Strotopora* and *Cliotrypa*. Spaces in vesicular tissue, interpreted to be gonocysts, have been reported (SHULGA-NESTERENKO, 1933, p. 49).

In *Cliotrypa* and *Strotopora,* an autozooecium of normal living-chamber diameter opens into an enlarged chamber. Several of these gonozooecia are commonly developed at the same level in a colony (see Fig. 176, 191). Colonies that died with gonozooecial expansions at the surface display low, rounded hemispherical blisters, each with a small subcircular pore in no set location on the blister. Many are broken, presumably after death of the colony, and appear as large, hemispherical depressions with elevated rims. In section, the enlarged end of the gonozooecium is seen to cover adjacent extrazooidal vesicular tissue (see Fig. 191,*1b,e*) and, in some cases, adjacent autozooecia. Some adjacent autozooecia adjusted their direction of growth and grew around the expanded part of the gonozooecium (UTGAARD, 1973, fig. 72). Calcified centripetal shelflike structures and curved plates have been reported (UTGAARD, 1973, p. 341) in the enlarged part of some gonozooecia.

Older colonies may contain a zone of abandoned gonozooecial chambers and a younger zone of gonozooecia. Although unproved, these zooecial polymorphs were probably involved with egg production and the brooding of embryos. This conclusion is supported by the zonal, cyclical arrangement, the large chambers connected to normal-sized autozooecia, the pores opening through the surface of the large blisters, and the smaller number of gonozooecia compared to normal autozooecia.

EXTRAZOOIDAL SKELETAL STRUCTURES

Cystoporates in the suborder Fistuliporina are characterized by many extrazooidal skeletal structures, an unusual condition in bryozoans. The more important of these structures are described in the following discussion.

VESICULAR TISSUE AND STEREOM

The most prominent and widespread extrazooidal skeletal structures are vesicular tissue and stereom. Vesicular tissue or vesicles may originate on the budding surfaces (a basal layer or mesotheca), in the endozone, or in the exozone. Vesicles in some fistuliporines partly to completely isolate autozooecia and the distance of isolation may be narrow to wide (see Fig. 167,*1c*; 190,*3b,c*). Vesicle walls and roofs are almost invariably simple interior structures, deposited from the outside or frontal side by an epithelium under the colonial hypostegal coelom. Rarely, vesicle walls are compound; vesicle roofs are always simple. Commonly, the vesicle wall and roof merge into a single curved plate.

Vesicles display a considerable range in size and shape in fistuliporines. A few forms have narrow, long, tubelike vesicles. Many have vesicles with subequal height and width. Most fistuliporines have low, wide, blisterlike vesicles. Walls and roofs may be straight, producing boxlike or polygonal shapes (see Fig. 190,*3*), or curved, producing blisterlike shapes (see Fig. 184,*1d*) with polygonal to subcircular cross sections. Vesicles range from very small, generally less than one-quarter the zooecial diameter, to very large, being several times as wide as the autozooecium. Most are nearly half the diameter of an autozooecium. In many genera, vesicle height decreases outward in the outer endozone and exozone.

Zones of thick vesicle roofs are common within the exozone or at the zoarial surface in many genera, suggesting semicolony-wide control of deposition. In many fistuliporines, particularly those with bifoliate colony construction, vesicle roofs thicken and the space between successive vesicle roofs decreases to create an essentially continuous deposit of stereom. Stereom may occupy the outer exozone (see Fig. 198) or the entire exozone (see Fig. 209).

Available evidence suggests that vesicular tissue and stereom were deposited under colony-wide or semicolony-wide control and served as space filler and buttressing between autozooecia.

VERTICAL PLATES

Extrazooidal skeletal structures that are platelike and lie in planes generally parallel to the growth direction of autozooecia in the exozone are here termed **vertical plates**. They are relatively rare but several distinctive kinds have been observed in some cystoporates belonging to the suborder Fistuliporina.

Midray partitions.—A few genera have, surrounding monticules, radiating clusters or fascicles of autozooecia—rays—separated by **interrays** composed of vesicular tissue. Rays have compound vertical plates, termed **midray partitions**, along the center (see Fig. 171,*1c*; 193,*1a,c*). In some colonies of *Constellaria* (CUTLER, 1973) hyaline material ("yellow tissue") forms a considerable portion of the midray partition. Where bordered by monticular ray zooecia, the midray partitions are, in part, multizooidal structures. In *Revalopora* (see Fig. 172,*1b*) the midray partitions may extend into the center of the monticule, where they are bordered by vesi-

cles rather than by monticular zooecia. Such parts of midray partitions are extrazooidal in origin.

Vertical plates in intermonticular areas.—A few cystoporates have compound vertical plates in the exozone in intermonticular areas. Those in *Hexagonella* (see Fig. 196,*2a*) stand as elevated ridges forming polygons at the zoarial surface, and outline each monticule and its associated intermonticular autozooecia, emphasizing the subcolony aspect of monticules. Portions of some vertical plates border zooecial cavities and may be, in part, multizooidal structures.

Compound range walls.—Genera in the Cystodictyonidae generally have compound **range walls** that separate longitudinal ranges of autozooecia from their locus on the mesotheca to the zoarial surface. In the endozone, they are thin, compound walls that possibly developed early. The dark-colored primary layer in autozooecial walls and vesicles abuts the secondary, light-colored layer of the mesotheca. The primary, dark-colored layer of compound range walls commonly abuts the central, dark-colored primary layer in the mesotheca (see Fig. 209,*1c*), indicating that range walls developed earlier than adjacent autozooecial walls and vesicle walls. A part of the compound range wall may serve as the lateral wall of an autozooecium, and compound range walls can be, in part, multizooidal structures. The range walls may lose their distinctiveness in the exozone, as in *Dichotrypa* (see Fig. 207,*1e*), but commonly are thicker (being called libria), have branched dark zones (valvae) and tubules (see Fig. 210,*1e*), and can protrude on the zoarial surface as longitudinal ribs (see Fig. 210). The central compound range wall, termed the **branch midrib** (see Fig. 210,*2c*), or the central and contiguous lateral range walls (see Fig. 206,*2c*), may be thicker and higher and produce a marked bipartite or tripartite branch symmetry.

TUNNELS

Genera in the small family Rhinoporidae

have elevated, branched to anastomosing ridges on the zoarial surface (see Fig. 192,*1b,c*). They are poorly understood but apparently developed on vesicular tissue, are covered by a rounded roof, and are here called **tunnels**. Their function is not known but they may have been extrazooidal brood spaces.

MULTIZOOIDAL SKELETAL STRUCTURES

Multizooidal skeletal structures in the Cystoporata include basal layers, mesothecae, and low, longitudinal ridges developed on some mesothecae. Some vertical plates (see above) may be, in part, multizooidal.

Basal layers.—Encrusting, hemispherical and massive colonies of cystoporate bryozoans have a basal layer and presumably the initial, encrusting portions of all erect colonies had a basal layer too. The basal layer presumably had an outer, exterior cuticular portion, although none has been found preserved in cystoporates. The basal layer is a simple exterior wall, deposited by an epithelium on its inner side. Microstructure of basal layers differs among cystoporates and is discussed elsewhere.

Mesothecae.—A considerable number of cystoporates have bifoliate, trifoliate, or multifoliate colony construction and have mesothecae of variable thickness and microstructure. Also, mesothecae, though generally planar, may be undulatory, crenulated, or sharply folded.

Like the basal layer, the mesotheca served as the budding surface and as the bottom of the living chamber in most bifoliate, trifoliate, and multifoliate forms. Most of these forms have relatively short autozooecia. Median tubules or calcite rods were observed in the median portion of the mesotheca only in *Glyptopora* and *Aetomacladia*. They are generally absent in cystoporates with a mesotheca.

Vertical longitudinal ridges.—A few genera in the Hexagonellidae and Cystodictyonidae have low, **longitudinal ridges** on the mesotheca. They extend into the proximal part of the autozooecia and into vesicles (see Fig. 199,*1a,b*) and may be developed for a considerable distance, through several autozooecia and vesicles. They evidently were formed within linear folds of the basal epithelium, before vesicle and autozooecial walls were formed, but their function is not known.

SKELETAL MICROSTRUCTURE AND ULTRASTRUCTURE

Many workers have mentioned the "indistinct," "fibrous," "homogenous granular-fibrous," "granular," or "microporous" nature of skeletal walls in the Cystoporata, particularly in the fistuliporines. The cystoporates display a variety of skeletal microstructures that permits interpretive reconstruction of the depositing epithelium. Preliminary studies of the ultrastructure of cystoporates support some of the interpretations based on light-microscope studies, but additional studies of ultrastructure using electron microscopy are needed.

THE CERAMOPORINA

The ceramoporines primarily have a laminated skeletal microstructure with some skeletal elements of dense, light-colored, hyaline calcite. Laminae are here interpreted to have been deposited parallel to the secreting epithelium, like those in trepostomates (BOARDMAN in BOARDMAN & CHEETHAM, 1969, p. 211), and are considered to represent growth surfaces. The irregular discontinuous nature of the laminae, which was noticed in tangential sections by ULRICH

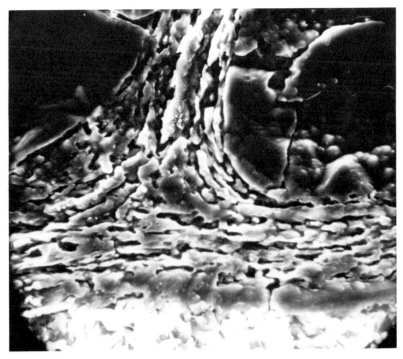

FIG. 149.   Cystoporate skeletal structure. *Ceramophylla vaupeli* (ULRICH), McMicken Mbr., Eden F., U. Ord., Cincinnati, Ohio. Longitudinal section, etched, through basal layer (below) and autozooecial wall in endozone. Note large tabular crystallites in lower portion of basal layer, which has a hyaline appearance in thin sections, and smaller tabular crystallites in laminated upper portion of basal layer; also note tabular crystallites and distinct zooecial boundary in compound, longitudinally laminated autozooecial wall in the endozone. Scanning electron photograph, USNM 159859, ×1,050.

(1890, p. 311), suggests that skeletal material was not uniformly deposited in nested cylinders in the autozooecia. The larger discontinuous laminations are ultralaminae in bundles that feather out against the autozooecial cavity, suggesting that they were secreted now here, now there in an autozooecium. At any one time, new and slightly older laminae lined an autozooecial cavity. Autozooecial and exilazooecial walls are compound, the laminae on either side of the zooecial boundary being secreted by the epithelia of adjacent zooids. In inner endozones, compound walls are thin, have narrow zooecial boundaries, and may be longitudinally laminated. In outer endozones and in exozones, laminae are broadly curved in the growth direction of autozooecia, commonly have an irregularly intertonguing appear-

ance, and zooecial boundaries may be narrow but commonly are wide, serrated zones.

Basal layers in encrusting and hollow ramose zoaria are simple, that is, deposited from only one side, and longitudinally laminate, with laminae parallel to the secreting epithelium. In many ceramoporines the basal layer consists of two parts: a lower or outer primary layer that is light colored and hyaline in appearance, with large tabular crystallites in at least one ceramoporine (*Ceramophylla vaupeli*; Fig. 149); and an upper or inner secondary layer with a distinct longitudinally laminated microstructure and ultrastructure. Presumably the primary layer was bounded by an external cuticle.

Mesothecae in bifoliate species of *Ceramophylla* have a compound, trilayered construction with a central, primary, hyaline

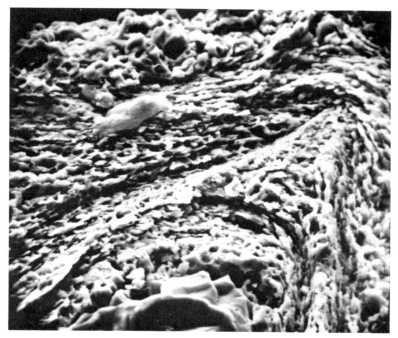

FIG. 150.   Cystoporate skeletal structure. *Ceramoporella flabellata* (ULRICH), Corryville Mbr., Maysville Gr., U. Ord., Jefferson Lake, Ind. Transverse surface, etched, showing discontinuity between small, tabular crystallites in laminated zooecial wall and larger, more subtabular to granular crystallites in lunarial deposit in bottom center of figure. The lunarial deposit is laminated and has a large rib on its proximal side. Scanning electron photograph, SIUC 3002, ×650.

layer and outer, secondary laminated layers like the secondary basal layer of encrusting colonies.

Lunarial deposits in the Ceramoporidae have a dense, light-colored hyaline appearance under a light microscope, but indistinct, distantly spaced laminations are observable in some forms. One to several lighter colored, rod-shaped lunarial cores may extend longitudinally in a lunarial deposit and may extend above the general surface of the lunarium as knobs or spines (UTGAARD, 1968b, p. 1445). Indistinct laminations are concentric around the cores. Preliminary studies of the ultrastructure of the lunarial deposits reveal that some have a more granular, but still recognizably laminate ultrastructure (Fig. 150, 151), with slightly larger crystallites than the wall, and are compound. Some of the hyaline appearance may be due to a preferred orientation of the optical axes of the

crystallites. Part of the lunarial deposit is secreted by zooidal epithelium on the distal side, where a thin, laminated zooecial lining may be later secreted, and part by epithelium on the proximal side of the lunarial deposit. Lunarial cores may be composed of crystallites that are coarser than those in laminated lunarial deposits (HEALY & UTGAARD, 1979, p. 184).

Acanthostyles with a dense, light-colored hyaline core surrounded by thin cone-in-cone laminae and spherical bodies of light-colored, dense hyaline calcite, surrounded by laminated wall material, are known from zooecial walls in two genera of Ceramoporidae (UTGAARD, 1968b, p. 1449, 1452). Their ultrastructure has not been investigated.

Basal diaphragms in zooecia in the Ceramoporina typically are simple. They are deposited by the zooidal epithelium on their upper side, and are laminated, with the laminae running parallel to the depositing epithelium. Terminal and subterminal diaphragms (UTGAARD, 1968b, p. 1445–1446; 1973, p. 328) have a dense hyaline microstructure, and their configuration indicates that they were deposited, primarily at least, from the inside. It is possible that they are compound, and were deposited from inside by the epithelium of a degenerated or resting zooid and from outside by the inner colonial epithelium beneath the hypostegal coelom. The ultrastructure of these diaphragms has not been investigated. The few funnel cystiphragms that have been found in ceramoporines are longitudinally laminated and simple, suggesting deposition by an epithelium lining the flask-shaped chambers or on one side of a membrane.

HEALEY and UTGAARD (1979, p. 185) have shown that communication pores in *Ceramophylla vaupeli* have a lining of laminae smaller than those in the zooecial walls. They indicate in addition that the skeleton of *Ceramophylla vaupeli* consists of ordinary, low-magnesium calcite, and has well-preserved original ultrastructure.

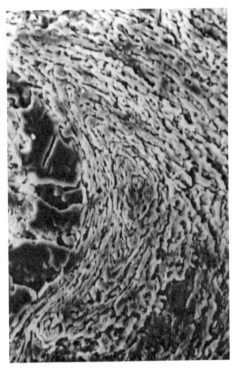

FIG. 151. Cystoporate skeletal structure. *Ceramophylla vaupeli* (ULRICH), McMicken Mbr., Eden F., U. Ord., Cincinnati, Ohio. Tangential section, etched, showing lunarial core with coarser crystallites, laminated lunarial deposit with tabular crystallites, and laminated autozooecial wall with tabular crystallites. Scanning electron photograph, USNM 159859, ×500.

THE FISTULIPORINA

Fistuliporines display laminated, hyaline, granular, and granular-prismatic microstructures. Laminae are interpreted as having been deposited parallel to the secreting epithelium and essentially parallel to growth surfaces. Laminated skeleton is present in some basal layers, mesothecae, autozooecial walls, lunaria, vesicular tissue, stereom, basal diaphragms, terminal diaphragms, and funnel cystiphragms. All of these structures except autozooecial walls, lunaria, and mesothecae are of simple rather than compound construction. Laminated autozooecial walls are generally longitudinally laminate, and the laminae and zooecial boundary are indistinct. **Mural tubulae,** small calcite rods generally perpendicular to the wall surface, may be developed in autozooecial walls.

Hyaline skeletal material is present in the lunarial deposits of some genera, in calcite rods (variously called acanthostyles, "acanthorods," "acanthopores," "minutopores," mural or median tubulae, and septa or canaliculi in the Actinotrypidae), and in calcite masses (yellow tissue in *Constellaria,* midray partitions in *Constellaria*). TAVENER-SMITH (1969b, p. 97; 1973), WILLIAMS (1971a), and HEALEY and UTGAARD (1979) have noted the granular ultrastructure of calcite rods or acanthostyles, which may be composed of large, irregular to rodlike crystallites. TAVENER-SMITH (1969b, p. 97) and WILLIAMS (1971a) have suggested that they served as attachment sites for muscles. They possibly

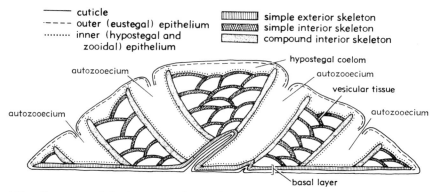

FIG. 152. Cystoporate skeletal structure. Longitudinal section through a hypothetical double-walled fistuliporine (after Utgaard, 1973). The basal layer is a simple exterior wall; all other walls are interior, that is, without an outer bounding membrane. Autozooecia have compound walls and the vesicular tissue is extrazooidal with simple walls secreted by the inner (hypostegal) epithelium immediately below the hypostegal coelom.

anchored the outer membrane of the double-walled colony (UTGAARD, 1973, p. 324).

Some fistuliporines have skeletal structures composed of a thick, primary, dark-colored granular layer. Many have a thin primary layer composed of granular or elongate crystallites and a thicker, secondary, light-colored granular-prismatic layer composed of granular, elongate, or prismatic crystallites with their long axis perpendicular to or nearly perpendicular to the secreting epithelium (HEALY & UTGAARD, 1979). This granular-prismatic microstructure has been interpreted as being secondary, due to recrystallization of the primary, presumably laminated, skeletal crystallites. Basal layers with this type of microstructure have a lower or outer primary layer, which may be thin and composed of fine dark granules. Presumably, a cuticle was secreted at the base of the colony and the primary granular layer was deposited on top of the cuticle (Fig. 152). In many forms, on top of this primary skeletal layer is the thicker secondary layer of granular-prismatic microstructure (Fig. 179,*1a*).

Mesothecae and vertical plates that divide a branch longitudinally or separate rows of autozooecia (compound range walls) are compound: a median primary layer that commonly is thin, dark, and granular is flanked by secondary layers that are lighter colored,

commonly thicker, and are composed of granular-prismatic crystallites.

Autozooecial walls in Fistuliporina with granular microstructure are composed of dark granular material with no evident zooecial boundary. Those with granular-prismatic microstructure have a thin, dark, granular, primary layer containing the zooecial boundary, flanked by secondary, generally thicker, lighter colored, granular-prismatic layers secreted by zooidal epithelium and lining the autozooecial cavities of autozooecia that are in contact. More commonly, where an autozooecium is adjacent to vesicular tissue, the granular-prismatic layer adjacent to the vesicle was secreted by epithelium under the hypostegal coelom, which covered the outer rim of the peristome and extended down and onto the uppermost roof of the interzooidal vesicular tissue (Fig. 143). The secondary layer lining the autozooecial cavity is commonly thicker than the secondary layer formed on the outside of the peristome. The elongate and prismatic crystallites in the secondary layer generally are perpendicular to the zooecial boundary and the epithelium. In *Anolotichia* (UTGAARD, 1968a, p. 1039), elongate and prismatic crystallites fan out toward the zooecial aperture in rings of thickened autozooecial walls, somewhat similar to **monilae** in some trepostomates. The microstructure is reminiscent of orally diverging crystallites in the autozooecial walls in *Lichenopora*. In *Lichenopora,* however, the crystallites apparently are plates composing laminae diverging outwardly. In many fistuliporines the lateral and distal sides (but not the proximal or lunarial side) of some autozooecia lack a peristome and the zooidal epithelium sweeps in a gradual curve onto the last roof of extrazooidal vesicular tissue. In these autozooecia, part of the skeletal wall is made up of the vertical parts (or walls) of overlapping vesicles (Fig. 143). In tangential thin sections, these parts are commonly linear segments, each belonging to one vesicle. In longitudinal thin sections, these parts are serrated (Fig. 153). Such bounding walls are simple. They have a primary, dark, granular layer and, commonly, a secondary, light-colored granular-prismatic layer next to the zooidal epithelium. The secondary layer lining an autozooecial cavity may continue up into the secondary layer of a vesicle two or three vesicles toward the zooecial aperture (Fig. 153), indicating that the secondary layer may have been deposited considerably later than the adjacent primary layer.

Lunarial deposits with granular-prismatic microstructure are of compound construction. The thin, dark, granular, primary layer of the lunarium may continue into the dark, granular, primary layer of the lateral and distal sides of the autozooecial walls (see Fig. 197,*1a*), or may terminate at the ends of the lunarial deposit adjacent to the zooecial cavity. In the latter case (see Fig. 180,*2b,c*), the dark granular primary layer and secondary layers of the lateral parts of the autozooecial wall unconformably abut the secondary proximal layer of the lunarial deposit. Commonly, the secondary layer of the lunarial deposit is relatively thin on the distal side of the lunarium adjacent to the zooidal epithelium, and has a granular-prismatic microstructure. The secondary layer on the proximal side of the lunarial deposit generally is thicker, has a granular-prismatic microstructure, and was lined by epithelium that generally sloped down off the lunarium and con-

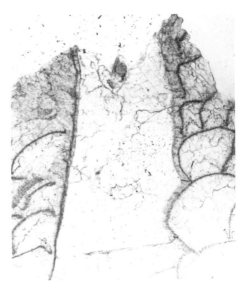

FIG. 153. Cystoporate autozooecial walls. *Meekopora clausa* ULRICH, Glen Dean Ls., Miss., Sloans Valley, Ky. Longitudinal section showing distal side of autozooecium (right) bounded by superimposed vertical vesicle walls. The light-colored, secondary, granular-prismatic layer is thicker in these vesicle walls than in the vesicle roofs to the right. SIUC 3000, ×100.

tinued, below the hypostegal coelom, onto the outermost vesicle roof of the interzooidal vesicular tissue. Where autozooecia were adjacent, the epithelium on the proximal side of a lunarial deposit continued down into the zooidal epithelium of the adjacent autozooid.

In zoaria with a basal layer or mesotheca, new autozooecial walls were deposited as septa, which are unconformable on the basal layer or mesotheca (see Fig. 167,*1a*). The primary layer and the secondary layers, if present, of the autozooecial septa abut the basal structure. Elongate and prismatic crystallites in the secondary layers in the two structures are at right angles.

Extrazooidal vesicular tissue with granular or granular-prismatic microstructure is almost invariably simple. An inner, primary, relatively thin, dark, granular layer is always present and may constitute the entire vesicle. More commonly, it is covered by a thin to conspicuously thicker, light-colored, granular-prismatic layer. The skeleton of vesicular tissue must have been overlain and deposited by epithelium below the hypostegal coelom. In a few forms, for example, in *Lichenotrypa*, some vesicle walls form as septa on older vesicle roofs, and grow upward for a short distance before vesicle roofs are deposited between them. These short segments of vesicle wall are compound, with a central granular primary layer and lateral, secondary granular-prismatic layers.

The primary and secondary layers of vesicles unconformably abut the compound autozooecial walls (usually the secondary layers of autozooecial walls), indicating that the vesicles were deposited after the autozooecial walls. Further, the primary layer and, if present, the secondary layer of superjacent vesicles unconformably abut the upper surface of subjacent vesicles.

WILLIAMS (1971a), TAVENER-SMITH (1973), and TAVENER-SMITH and WILLIAMS (1972) suggested that all bryozoans with a calcified skeleton have a primary skeletal layer consisting of acicular crystals of calcite (rarely aragonite in some gymnolaemate bryozoans) and that all stenolaemates have a secondary layer that invariably consists of carbonate laminae, separated by protein sheets. These laminae consist of tablets commonly perpetuating screw dislocations or overlapping rows of fibers in spiral growth, typically the latter in the extinct orders. TAVENER-SMITH maintained (pers. commun., 1971), that the granular and granular-prismatic microstructures so common in fistuliporines are due to recrystallization. The gradual transition of crystals from the granular-prismatic layer into the obviously secondary sparry-calcite filling of zooecial and vesicle cavities, which is seen in some forms, seems to support the view that these microstructures are produced by recrystallization.

The following facts suggest that the granular and granular-prismatic microstructural types need further study. First, many fistuliporines consistently have granular and granular-prismatic skeletons, including the lunarial deposit (*Hexagonella, Dybowskiel-* *la, Crassaluna*), whereas others consistently have granular and granular-prismatic autozooecial walls and vesicles but hyaline lunarial deposits (*Fistuliphragma, Duncanoclema, Strotopora, Cliotrypa*). If the granular and granular-prismatic microstructure is recrystallized, then in specimens of the latter genera, only a part of the skeleton is recrystallized. Second, *Diamesopora subimbricatum* (HALL) from the Silurian, *Taeniopora exigua* NICHOLSON and *Canutrypa francqana* BASSLER from the Devonian, and *Cheilotrypa hispida* ULRICH from the Mississippian have granular-prismatic microstructure in the basal layer or mesotheca and in autozooecial walls and vesicles in the endozone, but have a laminated microstructure in the exozone. In the generally laminated exozone of *Canutrypa francqana*, distal cystlike structures consistently have a granular-prismatic microstructure. Third, BOARDMAN and CHEETHAM (1973, p. 147) reported that the Silurian tubuliporate *Diploclema sparsum* has interior walls that are well laminated but has simple exterior walls with a columnar structure (a primary granular and a secondary granular-prismatic layer).

It is unlikely that some skeletal layers would be consistently recrystallized and others would not, if in fact they invariably consisted of carbonate laminae, particularly if the layers were of the same composition. The consistency of the change from granular-prismatic microstructure in the endozone to laminated microstructure in the exozone suggests a real ontogenetic change in original skeletal ultrastructure or composition and not vagaries of recrystallization. At this time, it seems premature to assume that all stenolaemates had laminated secondary layers. It is possible that elongate, prismatic, or acicular crystals could have been primary crystallites deposited perpendicular to the secreting epithelium. WARNER and CUFFEY (1973, p. 23) also suggested this possibility. The granular-prismatic microstructure of many fistuliporines resembles that of originally laminated, high-Mg calcite that recrystallized to low-Mg calcite in a cheilostomatous bryozoan

(Sandberg, 1975a). Thus, it possibly was produced by recrystallization of originally laminated calcite. Healey and Utgaard (1979, p. 190–193) found evidence for this in *Cystodictya* where the laminated mesotheca, composed of granular to mostly tabular crystallites of high-Mg calcite arranged in laminae, locally displays granular and granular-prismatic ultrastructure and microstructure, probably as a result of local recrystallization.

Even if the granular and granular-prismatic microstructures were produced by recrystallization, the relationship between the two layers indicates continued thickening of the secondary, granular-prismatic layer after the primary granular layer was formed. The microstructures of different skeletal parts also strongly suggest unconformable relationships. These permit reconstruction of the order of deposition of different skeletal parts and of the changes in configuration of the secreting epithelium.

DOUBLE-WALLED GROWTH MODEL

Borg (1926b) described the double-walled nature of the lichenoporid and hornerid tubuliporates (see free-walled colonies, Boardman, this revision) and suggested (p. 596) that the trepostomates had the same kind of body wall. Borg thought that it would be impossible to demonstrate this positively in fossils. Again, Borg (1926a, p. 482) stated that it was evident that the Trepostomata are more closely related to the double-walled tubuliporates than to the single-walled tubuliporates. Elias and Condra (1957, p. 37–38) alluded to the "sclerenchyma" in fenestrate cryptostomates and in trepostomates as apparently being deposited in the same manner as in *Hornera* and related tubuliporates, that is, by an ectoderm that stretched externally over the whole zoarium. Thus, they surmised that fenestrates had a double wall and proposed the new order Fenestrata, to be included with the orders Cyclostomata (here called Tubuliporata) and Trepostomata in Borg's class Stenolaemata. In a posthumous publication, Borg (1965, p. 3) stressed the relationship of the Fistuliporidae to the Lichenoporidae and the Trepostomata to the Heteroporidae and stated that he had succeeded in showing that they had a covering of soft tissue over the entire colony surface. Tavener-Smith (1968, p. 86, 88, 89; 1969a, p. 291) used the double-walled concept described by Borg as a basis for construction of a double-walled model for fenestellid growth. Boardman

(Boardman & Cheetham, 1969, p. 209, 213) suggested that the double-walled concept of Borg could be extended to most fossil tubular bryozoans (notably the Trepostomata, Cryptostomata and Cystoporata) and later presented (Boardman, 1971, p. 6–7) a more detailed account of the double-walled concept as applied to trepostomates. In addition Boardman (1973) discovered the fossilized remnant of the external cuticle on the outer surface of a colony of a trepostomate from the Ordovician: the most direct proof yet of the double-walled nature of a Paleozoic bryozoan.

THE DOUBLE-WALLED CONCEPT AS A MODEL FOR THE CYSTOPORATA

The presence of new autozooecia budded in the exozone in virtually any part of the colony in many cystoporates, on old autozooecial walls or on extrazooidal vesicular tissue, is strong evidence that cystoporates are double-walled bryozoans with a cuticle surrounding the entire colony. In many fistuliporines, autozooecia are isolated or partially isolated at the basal layer by intervening extrazooidal vesicular tissue. The presence of these new autozooids, isolated from their neighbors, also suggests colony-wide budding control by an outer membrane, rather than direct parent-daughter autozooecial origins. In addition, autozooecial walls in most cystoporates are compound walls, as are interior walls of modern tubuliporates, secreted under an infolding of inner epithelium into a hypostegal coelomic cavity.

The relatively uniform level of the outer surface (exclusive of the basal layer) of cystoporate colonies suggests a colony-wide epithelium and colony-wide control of growth. The only projections are relatively short calcite rods (acanthostyles and tubulae), some vesicle walls, vertical plates, autozooecial peristomes, and lunaria. Projections of similar magnitude are known in modern double-walled lichenoporids and hornerids. Individual, isolated autozooecia do not project significant distances above the general surface of the colony in cystoporates as they do in some single-walled tubuliporates. Growth of interior walls continued at nearly the same rate over the entire frontal surface of the colony.

The nature of the extrazooidal vesicular tissue and stereom in the Fistuliporina strongly suggests a double-walled construction. First, there is no space of relatively constant volume and shape and commonly no space for zooids in the vesicular tissue. In only a few species did vesicle walls form before vesicle roofs. In most species, vesicle walls and roofs are essentially one, curved structural unit. Thus, vesicular tissue is extrazooidal and must have required physiological communication with feeding zooids for nutrients. Second, thick vesicle roofs or stereom at the zoarial surface or in abandoned zones in the exozone indicate deposition from one epithelium on the outside of vesicles. Third, most vesicle walls and all vesicle roofs are simple walls. Some workers have reported pores in vesicle roofs and walls, but such pores are extremely rare. The simple-walled nature of vesicles suggests that they contained no living tissue, at least no secretory epithelium, and were not zooids. For that reason they are not, as Borg (1965) suggested, structures similar to alveoli in lichenoporid tubuliporates. The development of autozooecial walls on the outer side of vesicle roofs in the exozone also indicates that an epithelium existed on the outer side of the vesicles. This epithelium would almost have to be nourished by a coelom that was, in turn, protected from the environment by an outer membrane.

A possible membrane remnant is preserved over the zooecial orifice and extends over the vesicular tissue beneath an overgrowth (Utgaard, 1973, p. 323) on a colony of *Cheilotrypa hispida*.

Utilizing the double-walled concept of growth for the Cystoporata (Fig. 142, 143, 145, 152), it is probable that the inner epithelium (the zooidal and hypostegal epithelia of authors) secreted all of the calcareous skeleton. The outer (eustegal) epithelium secreted only the cuticular cover on the upper surface of the colony (the surface excluding the basal layer), including the terminal-vestibular membranes of the zooids.

ZOARIAL FORM AND LOCUS OF BUDDING

The Cystoporata display a wide variety of zoarial growth habits, including some that are unique as well as nearly all those exhibited by other tubular bryozoans in the orders Trepostomata, Cryptostomata, and Tubuliporata (Boardman & Cheetham, 1969, p. 206). Zoarial growth form in various Paleozoic tubular bryozoans has been discussed by Ulrich (1890, p. 294–296) and Ross (1964b, p. 932–934), among others, and a good summary of growth forms in the Fistuliporidae was presented by Moore and Dudley (1944, p. 248–250, 258–264). Borg (1965), Boardman and Utgaard (1966), and McKinney (1977b) discussed budding and three-dimensional packing of autozooecia in some Paleozoic tubular bryozoans. More work is needed on details of the location and geometry of budding or septa formation, the three-dimensional geometry of autozooecia, zooecial polymorphs and extrazooidal structures, and on the packing of the individual components (colony construction) and their relationship to zoarial growth form. As BOARDMAN (BOARDMAN & CHEETHAM, 1969, p. 216) pointed out, mere reference to the growth form without references to the internal architecture and budding patterns does not distinguish between colonies that may have similar growth habits but significantly different internal construction. For example, the cystoporate genera *Botryllopora, Ceramopora,* and *Fistulipora* and the post-Paleozoic tubuliporate *Lichenopora* can all assume a small, subcircular, discoidal growth form with an encrusting base and one central monticule with autozooecia radiating out from the center of the colony. This superficial resemblance in colony form masks important differences in the location and details of budding of autozooecia and exilazooecia and the geometry and packing of zooecia and any extrazooidal structures. Some information of this nature is available for the Cystoporata, and it is adequate to provide a general summary.

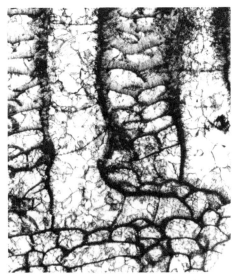

FIG. 154.   Cystoporate autozooecial budding. *Fistulipora waageniana* GIRTY, Wu-shan Ls., Penn., near Daning Xian, China. Longitudinal section showing an autozooecium with a short, recumbent initial portion budded on top of extrazooidal vesicular tissue. USNM 61922, ×50

Encrusting sheetlike zoaria.—In encrusting sheetlike zoaria (see Fig. 157,*1b*; 159,*1a,d,g*; 181,*1a,b*), autozooecia originate by septal formation within a fold of epithelium on top of the calcified basal layer around the growing margin of the colony (Fig. 142, 152). In the Ceramoporina, autozooecia typically are narrower at the basal layer than in the exozone, are hemispherical in cross section at the basal layer, and display a keel and sinus (BOARDMAN & UTGAARD, 1966, fig. 2, 3) in the outer recumbent portion and at the zooecial bend (see Fig. 159,*1a,f*). Exilazooecia originate by septal formation in the outer endozone or inner exozone (see Fig. 158,*1a*; 159,*1a*) so that only autozooecia are in contact with the basal layer. A similar situation, where only autozooecia and not exilazooecia originate at the basal layer, is found in the order Trepostomata. In contrast, in the suborder Fistuliporina of the order Cystoporata, extrazooidal vesicular tissue can originate at the basal layer and partly to completely isolate the autozooecia (see Fig. 181,*1a,b*). Autozooecia may be narrow to full width and may have keels and sinuses. New autozooecia rarely develop except at the basal layer in most encrusting sheetlike zoaria in Cystoporata. In some zoaria of *Ceramopora,* where free margins extend beyond the encrusted substrate (UTGAARD, 1969, p. 289–290), autozooecia originate lateral to the basal layer and above skeletal tissue associated with polymorphs in a celluliferous base (Fig. 145).

Hemispherical and massive zoaria.—In hemispherical and massive zoarial growth forms (see Fig. 161,*1c*; 163,*1a*; 170,*1c*) autozooecia originate at the growing periphery of the colony on the basal layer, as in encrusting sheetlike zoaria. In the Ceramoporina, they may also originate by septal formation on autozooecial walls over the entire upper surface of the colony (see Fig. 161,*1b*), so that new autozooecia are intercalated between older autozooecia in the exozone. The new autozooids reach a maximum diameter in a relatively short distance (two to three autozooecial diameters) above their point of origin. In addition, new autozooecia may also originate by septal formation on top of old extrazooidal vesicular tissue (Fig. 143, 154) in the Fistuliporidae.

Hollow ramose zoaria.—In hollow ramose zoaria in the Ceramoporidae, autozooecia originate by septal formation near the leading edge of the cylindrical basal layer. They generally have a narrow hemispherical outline and a keel and sinus in the zooecial bend region (see Fig. 160,*1g*). As in the encrusting sheetlike zoaria, only autozooecia are in contact with the basal layer. In thicker colonies, new autozooecia may originate between old autozooecia in the exozone, over the entire zoarial surface, by formation of new septa on autozooecial walls.

In Fistuliporidae with hollow ramose growth forms (see Fig. 175,*2c*; 177,*2d*; 179,*1c*), autozooecia originate at the growing edge of an irregular, tubular basal layer by formation of new septa on the basal layer. Unlike other growth forms in the Fistuliporina with a basal layer, autozooecia are not isolated at the basal layer by vesicular tissue, but vesicular tissue does partly isolate the autozooecia at the basal layer in encrusting overgrowths of some hollow ramose forms. In some forms with wide exozones, such as some *Dybowskiella,* new autozooecia may originate in the exozone by septa formation on old vesicular tissue. At the basal layer, autozooecia originate in linear series and generally are arranged rhombically in deep tangential sections in the outer endozone. Preliminary investigation suggests that significant differences in autozooecial geometry and packing exist in different genera. For example, species of *Diamesopora* have hemispherical to subtriangular autozooecial cross sections at the basal layer, with the base of the triangle resting on the basal layer (see Fig. 179,*1c*), whereas in *Cheilotrypa* the autozooecial cross section is mushroom-shaped to subtriangular with a flattened point of the triangle resting on the basal layer (see Fig. 175,*2a*). In many examples of *Cheilotrypa,*

the hollow axial tube displays regularly spaced expansions and contractions (see Fig. 175,*2f,g*) and, in some, the hollow axial tube is present distal to a cylindrical encrusted object, such as a rhomboporoid bryozoan. Such variations in the hollow axial tube and autozooecial shape and packing suggest that there is more taxonomic diversity in hollow ramose growth forms than has previously been suspected and that species should not be uncritically referred to a genus merely because they have a hollow ramose growth form.

Solid ramose zoaria.—In solid ramose growth forms, new autozooecia originate between older autozooecia by formation of new septa on autozooecial walls at the growing tip of the colony. New septa are formed at the growing tip in what becomes the endozone, or axial region of the branch.

In some Ceramoporina (UTGAARD, 1968b, p. 1448) with solid ramose zoaria, rather thick-walled autozooecia reach their full diameter in a distance of about one or two autozooecial diameters. Exilazooecia originate in the inner exozone.

In some solid ramose *Constellaria* (MC-KINNEY, 1975, p. 70, 71; 1977b, p. 323–326), new autozooecia originate in the corners of distally expanding polygons in an irregular to orderly fashion, the most orderly resulting in autozooecia that are triangular in cross section. Autozooecia in the endozone, particularly in *Constellaria,* generally have a larger diameter than they do in the exozone. Autozooecia narrow in the zooecial bend region and many new autozooecia are produced here by septal formation on old autozooecial walls. In addition, extrazooidal vesicular tissue, which may be present or absent in the endozone of ramose zoaria, is formed at the zooecial bend region so that autozooecia may be partially or completely isolated in the exozone by vesicular tissue. Autozooecia generally have a circular cross section in the exozone. In one species from the Baltic, identified as *Constellaria varia* ULRICH by BASSLER (1911, p. 220), endozonal autozooecia are very large and flare toward the zoarial surface. They are crossed by diaphragms that are at nearly the same level in adjacent autozooecia, and in curved zones representing abandoned growing tips. Some new septa in the inner endozone and in the zooecial bend region are formed on these diaphragms (also see McKinney, 1977b, p. 320). Thus, in this form, monticular components, including zooids, are budded in the monticules in the exozone. In addition, one new zooecial septum has been observed forming on a diaphragm in an autozooecium in the exozone of a Middle Ordovician *Constellaria* from Kentucky (Fig. 155).

In Fistuliporidae that have solid ramose growth forms, new autozooecia are produced by septa formation on old autozooecial walls at the growing tip of the colony, which becomes the axial endozone. In the endozone, autozooecia have a cross-sectional shape that commonly displays a keel and sinuses on the distal and lateral sides and a rounded proximal side when the lunarium is developed. McKinney (1975, p. 70; 1977b, p. 320, pl. 8, fig. 2) has determined that in *Canutrypa,* new autozooecia originate in the trough of the keel of the parent autozooecium. Autozooecia are in complete contact or may be isolated by large, elongate blisters of extrazooidal vesicular tissue. New autozooecia can originate in wide exozones by new septa formation on old vesicular tissue.

Slender ramose zoaria of *Fistulocladia* have a cylindrical central endozone composed of narrow, tubelike vesicles with flat vesicle roofs and circular cross section. The central endozone has cyclical zones of stereom. Autozooecia are budded off the flanks of the central cylinder, where they are narrowly isolated by stereom, and are circular in cross section (see Fig. 186,*1c*).

Frondescent zoaria.—In frondescent growth forms—erect, leaflike frondose colonies—in some species of *Ceramoporella* (Utgaard, 1968b, p. 1450–1451), new autozooecia originate at the growing tip, in the endozone, by formation of new septa on autozooecial walls. New autozooecia have a

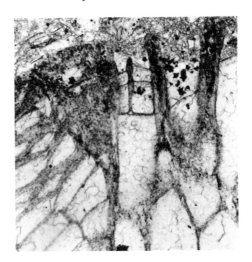

Fig. 155. Cystoporate autozooecial budding. *Constellaria* sp., M. Ord., Mason Co., Ky. Longitudinal section showing new compound zooecial wall produced by septum formation on a basal diaphragm; vesicular tissue to left. USNM 159858, ×100.

long endozonal portion where they parallel older autozooecia, and they reach their normal diameter in a distance of from two to four zooecial diameters. Autozooecia reach their maximum diameter in the inner endozone, where they have a cross-sectional shape that ranges from hemispherical to irregularly polygonal, but commonly with a rounded proximal side where the lunarium is developed. In the outer endozone, autozooecia display a crude rhombic packing, have sinuses and a keel, and a rounded proximal side where the lunarial deposit is situated. The few frondescent Ceramoporidae known so far display a marked reduction in autozooecial diameter from the inner endozone to the exozone and, for the most part, the autozooecia lose their rhombic arrangement in the exozone.

Frondescent Constellariidae are similar to ramose Constellariidae.

Bifoliate, trifoliate, and multifoliate zoaria.—In bifoliate zoaria of *Ganiella* and *Ceramophylla* (see Fig. 158,*1a,d*) in the Ceramoporidae, autozooecia originate at the growing margin of the mesotheca by formation of new septa on the mesotheca. As in the basal layer of encrusting sheetlike and hollow ramose zoaria, only autozooecia are in contact with the mesotheca of these bifoliate zoaria; autozooecia are narrow and hemispherical in cross section and a keel and sinuses are developed in the zooecial bend region.

The Fistuliporina display a wide variety of bifoliate, trifoliate, or multifoliate growth forms. Bifoliate zoaria generally are compressed in the plane of the mesotheca and include branched or unbranched frondose forms, cribrate colonies with anastomosing branches and large, subcircular fenestrules (Fig. 198,*1b*; 212,*2*), and straplike forms, the latter being regularly or irregularly branched or anastomosing, forming an anastomosing colony. One (see Fig. 206,*1b*) has an articulate colony, with flexible joints at dichotomous branchings. Trifoliate forms are narrow with parallel sides and can have regularly or irregularly developed trifoliate branches. Most branching is in the plane of the mesotheca; less commonly frondose bifoliate or narrow trifoliate forms have branches perpendicular to the mesotheca. Multifoliate colonies have a central multifoliate portion with radiating bifoliate branches. New autozooecia arise at the growing edge of the mesotheca by septal formation on the mesotheca.

In Hexagonellidae, autozooecia have relatively long, narrow, recumbent portions and are hemispherical in cross section. They may be partly contiguous, with partial keel and sinus development, or isolated by extrazooidal vesicular tissue. Autozooecia are more commonly contiguous with adjacent autozooecia in longitudinal ranges and may be teardrop- or club-shaped in outline at their contact with the mesotheca. The ranges are not separated by compound range walls in the endozone or exozone.

In the Cystodictyonidae, autozooecia range from partly isolated and almost entirely contiguous to completely isolated by vesicular tissue. Compound range walls, which arise on the mesotheca, separate ranges of autozooids. Autozooecia generally have a narrow proximal end and a club- to teardrop-shaped outline at their contact with the mesotheca (see Fig. 205,*1f*; 207,*1e*). Many genera display right- and left-handed autozooecia, and branches have a plane of bilateral symmetry perpendicular to the mesotheca. Some forms have a pronounced compound vertical plate (librium) in the plane of symmetry.

The Goniocladiidae have cylindrical to laterally compressed branches with a "vertical" mesotheca extending from the center of a reverse side, where it protrudes, forming a ridge or carina, to the center of an obverse side of a branch. Primary branches may have secondary and tertiary branches (generally perpendicular to the plane of the mesotheca), which are paired and laterally or distolaterally directed or alternating and distolaterally directed. Secondary and tertiary branches may fuse to produce reticulate, fenestrate, or pinnate growth forms. New autozooecia arise by septal formation on the mesotheca, at the growing tip of the branch. At the mesotheca, they are partially isolated by vesicular tissue, are hemispherical in cross section, and may have keels and sinuses developed. They curve distally and laterally to open on the rounded to sloping flanks of the obverse surface, in indistinct ranges.

Autozooecia originating on a basal layer or mesotheca in radial or linear series commonly alternate in adjacent ranges so that a basically rhombic packing pattern is achieved. This arrangement is commonly retained into the exozone to produce a rhombic or subrhombic surficial arrangement of autozooecia in large to small areas of a colony.

Little information is available for some genera of Cystoporata and much work is needed on details of autozooecial budding.

Ancestrula and astrogeny.—Virtually no details of the ancestrula, its shape and development, or the early astogenetic development of cystoporate colonies are known. Evidently most encrusting cystoporate zoaria have narrow, subcircular primary zones of astogenetic change (BOARDMAN & CHEETHAM, 1969, p. 208), and the colony consists mostly of the primary zone of astogenetic repetition

(BOARDMAN, CHEETHAM & COOK, 1970, p. 302). Subsequent zones of astogenetic change would be represented, for example, by development of branches in a plane other than the plane of the mesotheca in bifoliate fistuliporines. It is likely that the cystoporates had a funnel-shaped early stage (see Fig. 168,*1d*), like the lichenoporids. The early funnel-shaped stage is covered by later parts of the zoarium. The study by PERRY and HATTIN (1958) provides almost the only quantitative information to date on changes in the size and spacing of autozooecia along ontogenetic gradients in a portion of a large colony.

CLASSIFICATION

Genera here included in the order Cystoporata have commonly been placed in three orders; Tubuliporata (formerly Cyclostomata), Trepostomata, and Cryptostomata. The first edition of this *Treatise* (BASSLER, 1953) and *Fundamentals of Paleontology* (SARYCHEVA, 1960) generally reflect the classifications used in the first half of this century. BASSLER (1953), included the Ceramoporidae, genera now included in the Anolotichiidae, Fistuliporidae (including the Botrylloporidae), Hexagonellidae, and Goniocladiidae in the suborder Ceramoporoidea and order Cyclostomata (now Tubuliporata). The Constellariidae were included in the order Trepostomata. The Sulcoreteporidae (=Cystodictyonidae), Rhinoporidae, and Actinotrypidae were included in the order Cryptostomata. BASSLER further included most of the more obviously fistuliporine bifoliates (that is, those with extensive vesicular tissue) in the Cyclostomata (now Tubuliporata) and the less obviously fistuliporine bifoliates (that is, those with fewer vesicles and more stereom) in the Cryptostomata. The classification in SARYCHEVA (1960) is similar but the Hexagonellidae and Goniocladiidae were included in the Cryptostomata, which thereby contained nearly all bifoliate fistuliporines.

ASTROVA (1964) established the order Cystoporata to include the Ceramoporidae and Dianulitidae in the suborder Ceramoporoidea and the Constellariidae and Fistuliporidae in the suborder Fistuliporoidea of the order Cystoporata. UTGAARD (1968a, p. 1035) suggested that the order Cystoporata should include the Ceramoporidae, Anolotichiidae, Fistuliporidae, Hexagonellidae, and possibly the Lichenoporidae. The latter family of post-Paleozoic tubuliporates shows some features, probably produced by convergence, in common with the Cystoporata but most likely are not living cystoporates. Further increase in the content of the order Cystoporata was suggested (UTGAARD, 1973, p. 319) and the Cystoporata was expanded to include the Ceramoporidae in the suborder Ceramoporoidea and the Constellariidae, Anolotichiidae, Fistuliporidae, Hexagonellidae, Goniocladiidae, Botrylloporidae, Actinotrypidae, and some genera in the Sulcoreteporidae (=Cystodictyonidae) and Rhinoporidae in the suborder Fistuliporoidea. MOROZOVA (1970) proposed the suborder Hexagonelloidea to include the bifoliate cystoporates in the Hexagonellidae, Goniocladiidae, Sulcoreteporidae (=Cystodictyonidae), and Etherellidae.

PRESENT CLASSIFICATION

In this *Treatise,* the order Cystoporata ASTROVA, 1964 is considered to contain two suborders, Ceramoporina BASSLER, 1913 and Fistuliporina ASTROVA, 1964. The Ceramoporina contains only the family Ceramoporidae ULRICH, 1882. The Fistuliporina contains eleven families: Anolotichiidae UTGAARD, 1968a; Xenotrypidae UTGAARD, new family; Constellariidae ULRICH, 1896; Fistuliporidae ULRICH, 1882; Rhinoporidae MILLER, 1889; Botrylloporidae MILLER, 1889; Actinotrypidae SIMPSON, 1897; Hexagonellidae CROCKFORD, 1947; Cystodictyonidae ULRICH, 1884; Etherellidae CROCKFORD, 1957, and Goniocladiidae WAAGEN & PICHL, 1885.

The Cystoporata is a rather heterogenous order. It includes a variety of growth forms and wall microstructures. In many important respects it resembles other orders, notably the Trepostomata and Cryptostomata of the class Stenolaemata. The major, unifying characters of the Cystoporata are the tubular autozooecia and the lunarium in most, but not all genera and families. No single character separates cystoporates from all other Bryozoa and thus, the classification used here is polythetic. Cystoporates differ from other Paleozoic stenolaemates in generally having a lunarium and generally possessing either communication pores, extrazooidal vesicular tissue, or stereom. Autozooecia may be long and tubular with diaphragms, resembling those in most trepostomates, or relatively short and tubular without diaphragms, resembling some cryptostomates.

The suborder Ceramoporina with one family contains forms having most or all of the following features: well-laminated walls, exilazooecia, communication pores, and lunaria. They do not have extrazooidal vesicular tissue or stereom and most do not have acanthostyles.

The suborder Fistuliporina with eleven families contains forms having extrazooidal vesicular tissue or stereom. Most have lunaria, and in most zoarial growth forms the autozooecia are partially to completely isolated at the budding surface by vesicular tissue. They lack communication pores and exilazooecia. Some, such as the Goniocladiidae, are monothetic; many others are polythetic.

Arrangement of families of Fistuliporina in the systematic descriptions is in order of first stratigraphic occurrence and, except for the Etherellidae, does not represent presumed phylogenetic relationships.

STRATIGRAPHIC DISTRIBUTION

This summary of stratigraphic distribution should be used with caution. It is based upon ranges of genera that are given with the systematic descriptions. Some reported occurrences were difficult to evaluate, and ranges used in this discussion are based only on specimens available to me and on available illustrations in which I had reasonable confidence. Cystoporates show good biostratigraphic potential but need further study to increase their usefulness.

Cambrian.—Cambrian fossils reported to be Cystoporata are either unrecognizable or belong to other taxa. No undoubted Cambrian cystoporates are known.

Ordovician.—The Ordovician was the time of origin and flourishing of four families of Cystoporata; the Ceramoporidae, Xenotrypidae, Constellariidae, and Anolotichiidae. Species in all families, but particularly the Ceramoporidae and Constellariidae, are potentially useful in stratigraphic studies of Ordovician rocks.

The oldest known cystoporates, the Lower Ordovician genera *Lamtshinopora* and *Profistulipora,* are members of the family Anolotichiidae and the suborder Fistuliporina. Both are known only from the Soviet Union. A questioned occurrence of *Xenotrypa* has been reported from the Lower Ordovician of the Baltic region.

Middle Ordovician rocks have yielded six described genera of Ceramoporidae, three genera of Anolotichiidae, two genera of Constellariidae, and one Xenotrypidae.

The Ceramoporidae continued to flourish during the Late Ordovician, with seven known genera. Two genera of Anolotichiidae and one constellariid have species reported from Upper Ordovician rocks.

Silurian.—The Silurian apparently was a time of transition for the Cystoporata. The Ceramoporidae were on the decline, with only three known genera. No undoubted

Anolotichiidae has been reported from Silurian rocks and the last genus of the Xenotrypidae and of the Constellariidae are known. The small family Rhinoporidae is known only from the Lower and Middle Silurian. In addition, six genera of the family Fistuliporidae occur in rocks of Silurian age, with two genera reported in the Lower Silurian and four genera in the Middle and Upper Silurian.

Devonian.—The Devonian yields the youngest representative of the Ceramoporidae and of the Anolotichiidae. The greatest generic diversity in the Fistuliporidae, 13 genera in the Middle Devonian, and the occurrence of the sole representative of the Botrylloporidae in the Middle Devonian, are also recorded. The origin and expansion of bifoliate Hexagonellidae and Cystodictyonidae took place during the Devonian, with five hexagonellid genera and seven or eight genera of the Cystodictyonidae being represented in the Middle Devonian. It was the time of greatest diversity for the Cystodictyonidae. The greatest number of cystoporate genera (26 or 27) have been reported from Middle Devonian rocks.

Mississippian.—Rocks of Mississippian age have yielded at least four or five genera of the Fistuliporidae, the earliest representative of the Actinotrypidae, five genera of Hexagonellidae, three genera of Cystodictyonidae and four, possibly six, genera of Goniocladiidae, the oldest representative of the latter family. During the Mississippian, dominance changed from encrusting, massive, and ramose colonies to erect bifoliate, trifoliate, and multifoliate forms.

Pennsylvanian.—The Pennsylvanian has yielded only a few Fistuliporidae: one or two genera. The Hexagonellidae, with three genera, and the Cystodictyonidae, with one genus, declined. The Goniocladiidae have at least three, possibly five, genera from Pennsylvanian rocks.

Permian.—Three new genera of Fistuliporidae increased the total to five genera known from Permian rocks. Two new genera of Actinotrypidae and three new genera of Hexagonellidae evidence a slight revival of those families during Permian time. One Cystodictyonidae is known from the Lower Permian and the two genera in the poorly known family Etherellidae may be cystodictyonids. The Goniocladiidae reached their greatest diversity with six genera known from the Permian. No undoubted post-Paleozoic cystoporates are known.

ACKNOWLEDGMENTS

At Southern Illinois University, JOHN RICHARDSON and FREDDA BURTON did drafting and provided photographic aid; students TERRY GIVENS, JAMES MEACHAM, JOHN POPP, CHARLES PRICE, MAYNARD LITTLE, ROBERT WEBER, NEIL HEALEY and RICHARD MARTIN helped with photographic illustrations or preparing specimens; JUDY MURPHY and LEE DRYER helped with investigations of ultrastructure using the scanning electron microscope; and L. E. HARRIS typed the manuscript. Specimens were lent by R. S. BOARDMAN, ALAN HOROWITZ, MATTHEW NITEKI, E. S. RICHARDSON, W. D. I. ROLFE, EWAN CAMPBELL, R. F. WISE, CURT TEICHERT, A. H. KAMB, R. L. BATTEN, D. B. MACURDA, C. F. KILFOYLE, LOIS KENT, T. E. BOLTON, and G. G. ASTROVA. Photographic prints were kindly furnished by G. G. ASTROVA, A. M. YAROSHINSKAYA, R. V. GORJUNOVA, ALAN HOROWITZ, R. J. CUFFEY, I. P. MOROZOVA, R. E. WASS, V. P. NEKHOROSHEV, and L. NEKHOROSHEVA. ALAN HOROWITZ and R. S. BOARDMAN offered helpful suggestions. The Smithsonian Institution provided space and access to collections during 1972, and Southern Illinois University provided financial assistance.

SYSTEMATIC DESCRIPTIONS FOR THE ORDER CYSTOPORATA

By John Utgaard

[Southern Illinois University at Carbondale]

New photographs of primary types are used when possible. Some primary types are silicified, and new photographs of topotypes are used. Where neither primary types nor topotypes were available for study, specimens of nearly the same age from nearby areas were used. Some photographs were furnished by other paleontologists. For a few genera, drawings based on original published illustrations had to be used. Some reported geologic and geographic ranges cited for genera are difficult to evaluate, and I included only those in which I had reasonable confidence.

Order CYSTOPORATA
Astrova, 1964

[Cystoporata Astrova, 1964, p. 28]

Zoarial growth form variable. Autozooecia tubular; short and lacking diaphragms or long and with diaphragms. Walls laminated, granular, or granular-prismatic. Some (ceramoporines) with communication pores and exilazooecia in exozone. Some (fistuliporines) with extrazooidal vesicular tissue between autozooecia. Most with lunaria. Few with acanthostyles, except in vesicles. *Ord.-Perm.*

Suborder CERAMOPORINA
Bassler, 1913

[*nom. correct.* herein, *pro* Ceramoporoidea Bassler, 1913, p. 326, suborder]

Zoarial growth form encrusting to massive, ramose, frondose, or bifoliate frondose. Monticules low or flush with zoarial surface. Autozooecia originating at basal layer or mesotheca or by interautozooecial budding in growing tip or in exozone. Autozooecia in full contact with one another at basal layer. Diaphragms generally sparse. Walls well laminated; with communication pores or gaps in most genera. Lunaria usually well developed, composed of hyaline calcite. Exilazooecia in exozone in most genera, few to abundant, restricted to monticules in a few forms. *M.Ord.-L.Dev.*

Family CERAMOPORIDAE
Ulrich, 1882

[Ceramoporidae Ulrich, 1882, p. 156] [=Ceramoporellidae Simpson, 1897, p. 481]

Zoaria discoidal, encrusting, hemispherical, massive, ramose, hollow ramose, frondose, or bifoliate frondose. Monticules commonly low or flush with zoarial surface. Lunaria partly to completely radially arranged around monticules in most genera. Basal layer commonly dense calcite in lower half, laminated in upper half. Autozooecia tubelike, small to large; diaphragms generally few or absent. Walls commonly longitudinally laminate in endozone; generally with broadly curved laminae and amalgamate appearance in outer endozone and exozone. Walls in exozone generally thick. Communication pores few to abundant, absent in some genera. Lunaria extending from endozone to zoarial surface. Lunarial deposits commonly dense, light-colored calcite; locally, in many genera, indistinctly laminated. Wall laminae generally absent on distal side of a lunarium. Exilazooecia (cystopores) tubular; developed only in exozone; few to abundant; wall microstructure similar to that of zooecia. Acanthostyles present in some genera. *M.Ord.(Chazy.)-L.Dev.*

Characters of particular importance are well-laminated walls, communication pores, exilazooecia in the exozone, lunaria, and absence of extrazooidal vesicular tissue.

Ceramopora Hall in Silliman, Silliman, & Dana, 1851, p. 400 [*C. imbricata* Hall, 1852, p. 169; SD Hall & Simpson, 1887, p. xviii; Niagaran Gr., M. Sil. (Niag.), Lockport, N.Y., USA]. Zoarium thin discoidal expansions; encrusting, free, or a combination. Zoarium having attached central area with basal layer and free margins with celluliferous base made up of basal zooecia, or basal layer throughout, or free celluliferous base throughout. Small zoaria with one central monticule. Monticules with depressed center; exilazooecia of central cluster larger, more angular than intermonticular exilazooecia. Lunaria in perfect radial arrangement around central monticule; partial to complete radial arrangement around marginal monticules. Autozooecia large, commonly rhombically arranged; cavity ovate to rhomboidal in cross section. Communication pores abundant, most commonly just distal to ends of lunarium. Lunaria small in inner exozone, large at zoarial surface. Diaphrams absent. Exilazooecia few to abundant in intermonticular areas, generally small and subcircular in cross section; lacking diaphrams; with walls commonly oblique to zooecial walls. [Most zoaria of *C. imbricata* have a celluliferous base throughout or in the lateral margins beyond a central encrusting attachment area. Tubular basal zooecia commonly are directed downward or downward and outward; a few are directed downward and toward the center of the zoarium. Thick, poorly laminated walls of basal zooecia have laminae broadly curved and convex toward the base. Configuration of the laminae of basal zooecia indicates that soft tissue was present under the basal zooecia on the underside of the zoarium as well as on the upper surface.] *U.Ord.(Richmond.)-M.Sil.(Niag.)*, N.Am., Eu.——FIG. 156,*1a–c.* *C. imbricata; a,* celluliferous base (below), zooecial walls, lunarial deposits, and no diaphragms; long. sec., lectotype, AMNH 1737-A, ×30; *b,* rhombic arrangement of large ovate autozooecia, large lunarial deposits in proximal (lower) half of zooecia, abundant communication pores, and virtual absence of exilazooecia; tang. sec., lectotype, ×30; *c,* abundant communication pores and small, subcircular exilazooecia in intermonticular area; tang. sec., paralectotype, AMNH 1737-B, ×30.

Acanthoceramoporella UTGAARD, 1968b, p. 1451 [**Ceramoporella granulosa* ULRICH, 1890, p. 466; OD; "Fernvale F.," U. Ord., Wilmington, Ill., USA]. Zoarium encrusting expansions, less commonly hollow ramose. Autozooecia with cavity in exozone elliptical or ovate, less commonly, trilobate in cross section. Thin zooecial lining locally present. Communication pores abundant, large; perpendicular to wall and circular to elliptical in cross section. Local gaps in walls in some species. Beadlike wall segments between closely spaced pores may have a short acanthostyle or subspherical body of poorly laminated, light-colored calcite surrounded by thin layer of laminae proximally and laterally and thicker layer distally. Lunaria small in inner endozone; nearly as large at zooecial bend as at zoarial surface. One to several ill-defined cores in some lunaria.

Locally, distal side of lunarium bordered by thin laminated layer. Diaphragms abundant, straight or curved; a few abutting wall and very few lacking laminations and curving proximally along wall. Exilazooecia irregular in size and shape, partially to completely isolating autozooecia; diaphragms few to abundant, thick to thin. Exilazooecial walls with more communication pores than autozooecial walls. Acanthostyles generally abundant, short to long; core hyaline, large; collar of core-in-cone laminae, thin. Monticules with central cluster of exilazooecia and more acanthostyles and light-colored spherical bodies. [DONALD DEAN, National Museum of Natural History, has noted that a suite of specimens I thought (UTGAARD, 1968b, p. 1452, 1454) were cotypes of *Acanthoceramoporella granulosa* (ULRICH) are topotypes. Also, an old slide figured by ULRICH (1890) is the holotype, not a cotype. Thus, USNM 43227 are thin sections of the holotype and USNM 159715, 159716, and 159717 are topotypes, not the lectotype and paralectotypes.] *M.Ord.(Mohawk.)-U.Ord.(Richmond.)*, E.N.Am., Eu.——FIG. 157,*1a–h.* **A. granulosa* (Ulrich); *a,* monticule (upper left) with partial radial arrangement of lunaria, abundant exilazooecia with many communication pores, and acanthostyles; tang. sec., topotype, USNM 159715, ×27; *b,* lunarial deposits, irregular diaphragms, communication pores, and beadlike wall segments; long. sec., topotype, USNM 159715, ×27; *c,* irregular diaphragms in inner exozone; long. sec., topotype, USNM 159732, ×27; *d,* basal layer (left), abundant diaphragms, and acanthostyles; long. sec., holotype, USNM 43227, ×27; *e,* large monticular zooecia (lower half), lunaria, acanthostyles, and exilazooecia; tang. sec., holotype, ×27; *f,* lunarial deposits, wall laminae, communication pore, and light-colored subspherical body immediately distal to pore (left of center); long. sec., topotype, USNM 159717, ×90; *g,* poorly laminated lunarial deposits, irregular exilazooecia, abundant acanthostyles; tang. sec., topotype, USNM 159717, ×90; *h,* communication pores, spherical, light-colored bodies in beadlike wall segments, and acanthostyle (left of center); transv. sec., topotype, USNM 159716, ×90.

?Amsassipora YAROSHINSKAYA, 1960, p. 394 [**A. simplex*; OD; U. Ord., Altai Mts., USSR]. Zoarium massive with smooth surface, autozooecia with cavity circular to irregularly subangular, variable in size. Walls unevenly thickened, questionably with communication pores. Diaphragms sparse, planar to slightly concave. Lunaria lacking. Acanthostyles at zooecial corners; associated with autozooecial budding. Exilazooecia absent. [Lack of lunaria and of undoubted communication pores as well as presence of acanthostyles associated with budding, which resembles the situation in rhombotrypids,

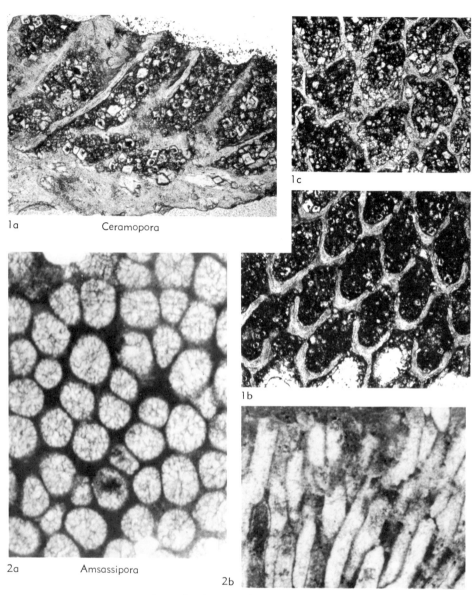

la   Ceramopora

1c

1b

2a   Amsassipora

2b

Fig. 156. Ceramoporidae (p. 358–359).

suggests that this genus may be a trepostomate. It is questionably retained in the Ceramoporidae, where it was placed by Yaroshinskaya and Astrova.} *U.Ord.,* USSR.——Fig. 156,*2a,b.* *A. simplex,* paratype, SNIIGGIMS 951/523a, 523b; *a,* subcircular to subangular autozooecia, acanthostyles at zooecial corners; tang. sec., ×40; *b,* irregular thickness of autozooecial walls, sparse diaphragms, acanthostyles associated with budding loci; long. sec., ×20 (photographs courtesy of G. G. Astrova).

Ceramophylla Ulrich, 1893, p. 331 [**C. frondosa;* OD; Decorah Sh., M. Ord. (Mohawk.), St. Paul, Minn., USA} [=*Coeloclema* Ulrich, Nickles, & Bassler, 1900, p. 24, 211; Bassler, 1953, p. G82, *non* Ulrich, 1883, p. 258; *Coeloclema* Nickles & Bassler, Elias, 1954, p. 53, *non* Ulrich, 1883, p. 258]. Zoarium bifoliate or encrusting expansions or hollow ramose. Monticules small. Mesotheca locally absent in some bifoliate zoaria. Autozooecia with cavity subcircular to circular. In exozone, walls thick, bound-

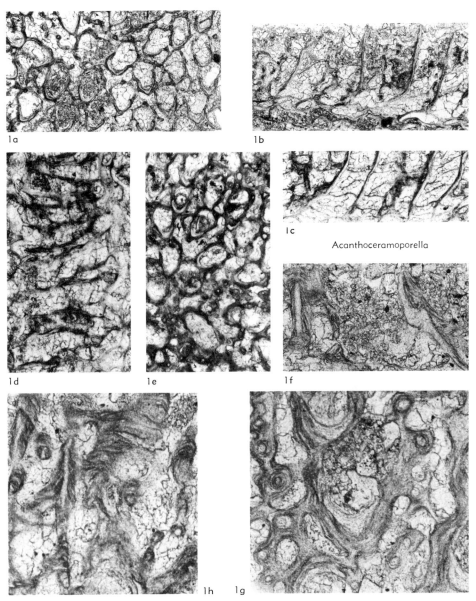

FIG. 157. Ceramoporidae (p. 359).

ary jagged or indistinct. Irregular bundles of wall laminae intertonguing between boundary and cavity, producing mottled appearance in tangential section. Communication pores few, generally in inner exozone. Lunaria small in inner endozone; nearly as large at zooecial bend as at zoarial surface. One to a few cores in many lunaria. Laminae lining distal side of some lunaria. Diaphragms absent. Exilazooecia generally abundant, reduced in diameter toward zoarial surface and in some pinched out or filled with laminated deposit; walls thinner than autozooecial walls; diaphragms absent. Monticular cluster of exilazooecia having thicker walls; monticular center locally subsolid. Lunaria not in radial arrangement or only slightly skewed toward monticular center in autozooecia on flanks of monticule. Zooecia in some slightly larger in monticules. [*Diamesopora vaupeli* ULRICH, 1890, is here reassigned to *Ceramophylla. C. vaupeli* has

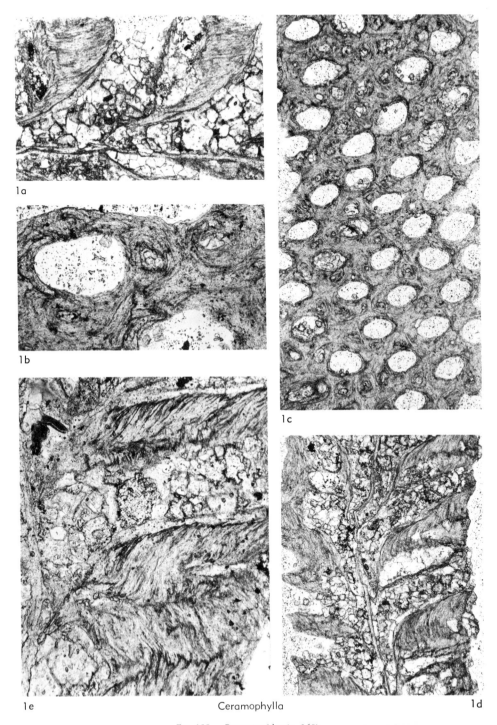

1a

1b

1c

1e

1d

Ceramophylla

Fig. 158.   Ceramoporidae (p. 360).

widely but erroneously been cited as the type species of *Coeloclema* ULRICH (UTGAARD, 1968b, p. 1453). *Coeloclema* ULRICH, 1883 is a fistuliporine with a Silurian type species. *C. vaupeli* and several other species differ from *C. frondosa* mainly by having a hollow ramose growth habit with a basal layer rather than encrusting laminar or bifoliate growth habits with basal layers or mesothecae. Several zoaria of *C. frondosa* have a relatively sharp boundary between clear and muddy cavity fillings in some zooecia, but no remnant of a skeletal diaphragm was observed.] *M.Ord.(Mohawk.)-U.Ord. (Eden.),* E.N.Am., Eu. ——FIG. 158, *1a–e.* **C. frondosa*; *a,* mesotheca, lunarial deposits, wall laminae, and exilazooecia; long. sec., lectotype, USNM 159721, ×50; *b,* poorly defined core in lunarium (left) and irregular wall laminae in mottled amalgamate wall; tang. sec., lectotype, ×100; *c,* thick, irregularly laminated walls and small exilazooecia and lunarial deposits; tang. sec., lectotype, ×30; *d,* mesotheca, thickening of walls at zooecial bend, sparse communication pores, and no diaphragms; long. sec., lectotype, ×30; *e,* intertonguing of laminae in walls, irregular dark zooecial boundary (below), and lunarial deposits; long. sec., paralectotype, USNM 159722, ×100.

Ceramoporella ULRICH, 1882, p. 156 [**C. distincta* ULRICH, 1890, p. 464; SD ULRICH, 1890, p. 380; McMicken Mbr., Eden F., U. Ord. (Eden.), Cincinnati, Ohio, USA] [=*Cheiloporella* ULRICH, 1882, p. 157; *Chiloporella* MILLER, 1889, p. 297, incorrect subsequent spelling; *Ceramporella* CUMINGS & GALLOWAY, 1913, p. 427, incorrect subsequent spelling]. Zoarium encrusting or frondescent. Monticules circular to elongate. Autozooecial cross-sectional area reduced from outer endozone to exozone, particularly in frondose zoaria. Autozooecia moderately small in exozone, cavity ovate to subcircular. Zooecial lining local. Communication pores rare. Lunaria increasing in size distally in endozone; nearly as large at zooecial bend as at zoarial surface; ends locally projecting into cavity; cores commonly one, rarely more, per lunarium; thin laminated layer locally on distal side. Commonly one, rarely more, convex or planar diaphragms of dense, light colored calcite per autozooecium. Most diaphragms abutting wall or curving proximally along wall; a few curving distally along wall. Diaphragms commonly at same level in adjacent zooecia. Exilazooecia partially to completely isolating zooecia; large and subangular in cross section in inner exozone, smaller and more circular distally. Lining commonly lacking. Diaphragms few, similar to zooecial diaphragms. Monticular exilazooecia, commonly of serial origin, clustering on semiprostrate proximal wall of autozooecium. [*Fistulipora flabellata* ULRICH, 1879, p. 28, which was subsequently designated as the type species of *Cheiloporella* ULRICH (1882, p.

257), is consistent with the emended definition of *Ceramoporella. C. flabellata* may have a frondose as well as an encrusting growth habit. Frondose zoaria commonly have longer and larger autozooecia in the endozone than do encrusting zoaria. Some zoaria of *C. flabellata* have thicker walls and autozooecial linings than do those of *C. distincta,* and display only slight or partial radial arrangement of lunaria around the monticules.] *M.Ord.(Mohawk.)-U.Ord.(Richmond.),* E.N. Am.——FIG. 159, *1a–d.* **C. distincta; a,* parts of three encrustations; lunarial deposits and diaphragms at nearly same level; transv. sec., lectotype, USNM 159710, ×30; *b,* radial arrangement of lunaria around monticular center; tang. sec., lectotype, ×30; *c,* large lunaria, thin walls, and abundant exilazooecia in intermonticular area; tang. sec., paralectotype, USNM 159711, ×30; *d,* basal layer (below), lunarial deposit, wall laminae, and exilazooecia in exozone; note light-colored, proximally curved diaphragm (left); long. sec., paralectotype, USNM 159712, ×100. ——FIG. 159, *1e–g. C. flabellata* (ULRICH), U. Ord., Ohio, USA, topotype, USNM 159714; *e,* exilazooecia in monticular center; tang. sec., ×30; *f,* encrusting overgrowths showing sparse diaphragms, lunarial deposits, and secondary overgrowths on right; transv. sec., ×30; *g,* basal layer, lunarial deposit, wall laminae, and remnant of a diaphragm; long. sec., ×50.

Crepipora ULRICH, 1882, p. 157 [**Chaetetes venusta* ULRICH, 1878, p. 93; SM ULRICH, 1882, p. 257; Economy Mbr., Eden F., U. Ord. (Eden.), W. Covington, Ky., USA]. Zoarium encrusting expansions, hollow ramose or solid ramose with conspicuous monticules. Autozooecia with cavity subangular to subcircular, moderately small, commonly rhombically packed. Wall laminae short, irregular; not concentric around living chamber. Boundary irregular, sinuous; forming broad, dark zone. Communication pores abundant in outer endozone and exozone; subcircular in cross section. Proximal surface of pore perpendicular to wall, distal surface commonly oblique, pores with larger diameter at one end; locally, both surfaces parallel and oblique to wall. Lunaria moderately large in endozone of ramose zoaria, moderately small in inner endozone of zoaria with basal layer; lunaria becoming slightly larger in outer exozone, ends in some projecting into autozooecial cavity. Radius of lunarium conspicuously smaller than radius of autozooecium. Central core or cores in many lunaria. Diaphragms thin to thick, irregularly spaced. Exilazooecia never abundant, commonly rare or absent in intermonticular areas; walls commonly thinner than in exozone. Communication pores present but less common in walls between adjacent autozooecia. Diaphragms rare in exilazooecia. Acanthostyles small, few to many in monticular exilazooecial

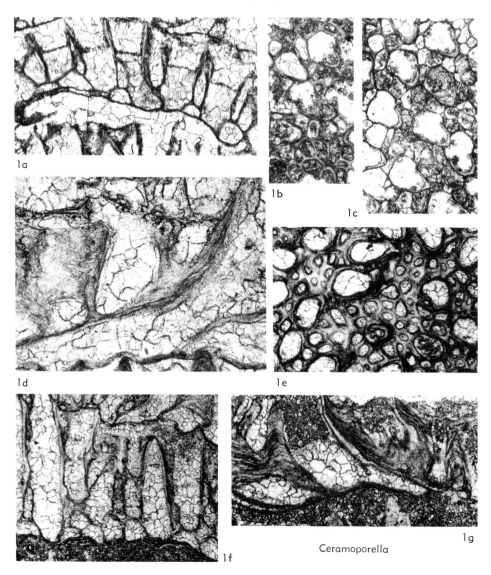

1a

1b

1c

1d

1e

1f

Ceramoporella

1g

FIG. 159. Ceramoporidae (p. 363).

walls, rare, usually lacking in autozooecial walls in intermonticular areas. Monticules having core of small to large, circular to irregularly shaped exilazooecia, acanthostyles, and ring of marginal zooecia slightly larger than intermonticular auto-zooecia. Lunaria in some areas partly radially arranged. Monticular centers rarely subsolid. *M. Ord.(?Chazy., Mohawk.)-U.Ord.(Richmond.),* E.N.Am., Eu.——FIG. 160,*1a–i.* *C. venusta* (ULRICH); *a,* lunarial deposit (left) and exila-zooecia and acanthostyles in monticule; transv. sec., paralectotype, USNM 159707, ×50; *b,* cluster of exilazooecia in monticule and some exi-

lazooecia in intermonticular area; tang. sec., paralectotype, USNM 159707, × 30; *c,* acan-thostyles in exilazooecial walls in monticule; tang. sec., paralectotype, USNM 159708, ×100; *d,* core in lunarial deposit; tang. sec., paralec-totype, USNM 159708, ×100; *e,* basal layer in hollow ramose zoarium and diaphragms in endo-zone and inner exozone; long. sec., lectotype, USNM 159706, ×30; *f,* exilazooecia with few acanthostyles in monticule and partial radial arrangement of lunaria; tang. sec., lectotype, ×50; *g,* basal layer and increase in size of lunarial deposit from endozone to inner exozone; transv.

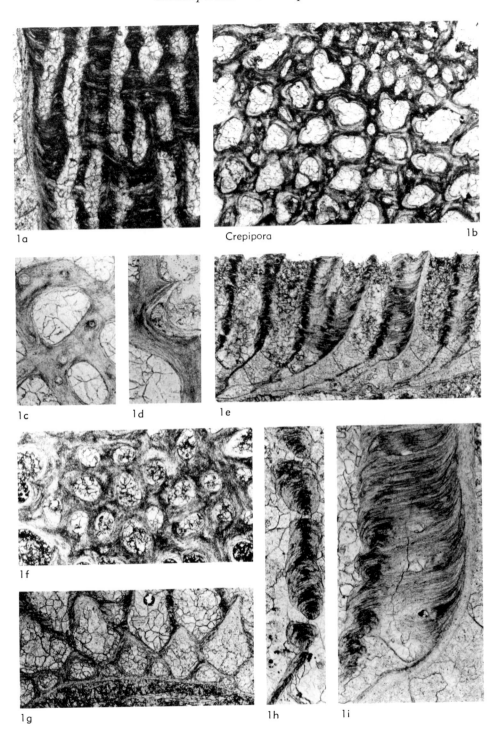

FIG. 160. Ceramoporidae (p. 363).

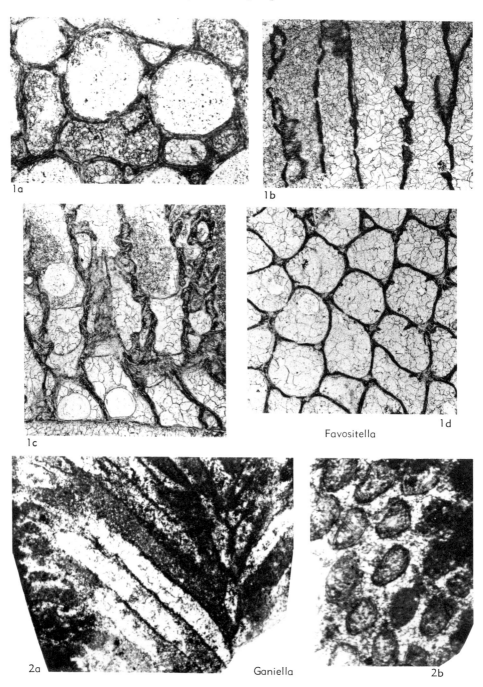

FIG. 161.   Ceramoporidae (p. 366–368).

sec., lectotype, ×50; *h,* wall laminae and communication pores in exozone; long. sec., lectotype, ×100; *i,* lunarial deposit, wall laminae, and communication pores in exozone; long. sec., lec-

totype, ×100.

Favositella ETHERIDGE & FOORD, 1884, p. 472 [*Favosites interpunctus* QUENSTEDT, 1878, p. 10; OD; M. Sil. (?Wenlock.), Dudley, Eng.]

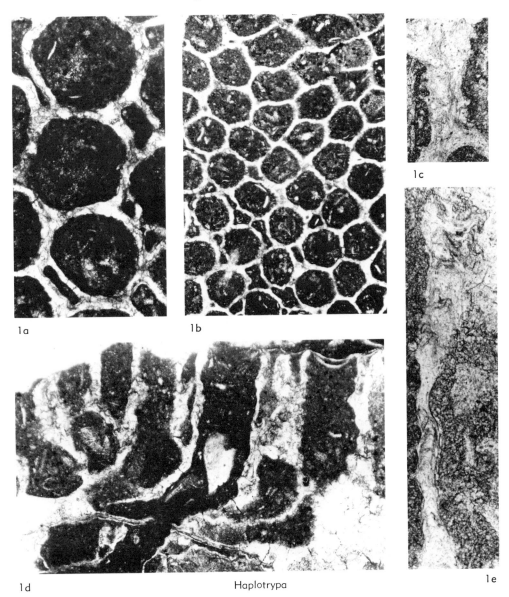

la   lb   lc   ld   Haplotrypa   le

FIG. 162.   Ceramoporidae (p. 369).

[=*Dnestropora* ASTROVA, 1965, p. 130]. Zoarium encrusting to massive, less commonly discoidal or hemispherical; with monticules. Autozooecia large, irregular in size and shape, cavity commonly with subangular to subrounded cross section in exozone. Walls undulatory or crenulated, moderately thick throughout. Locally, wall laminae irregular; with large mural lacunae. Generally short, curved, rod- to platelike protrusions (some resembling incomplete septa having broadly curved laminae) may extend outward (locally inward) at acute angle to zooecial wall, protrusions commonly in contact with one side of autozooecium. Communication pores common, small, subcircular. Diaphragms thin, moderately abundant. Lunaria small and subcircular to crescentic in endozone and inner exozone; increasing in size in outer exozone; small to large at zoarial surface. Laminated lining common on distal side of lunarial deposit. Wall laminae on

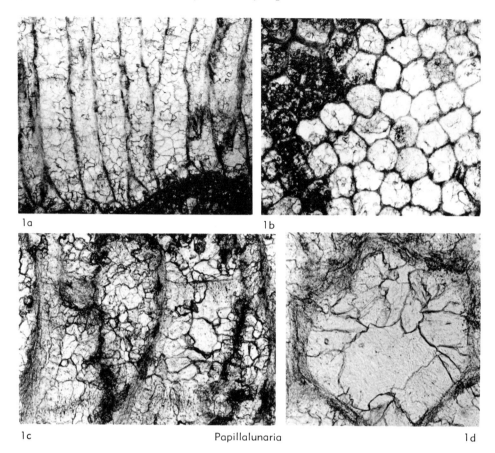

1a   1b

1c   Papillalunaria   1d

FIG. 163.   Ceramoporidae (p. 369).

proximal side of some lunaria, particularly near monticules, with cone-in-cone flexures nearly perpendicular to lunarium or directed slightly outward. Flexures extending proximally from lunarium for short to long distance, locally inflecting autozooecial cavity. Exilazooecia few in intermonticular areas, irregular in shape; diaphragms may be thicker than autozooecial diaphragms; walls irregularly undulatory, with more communication pores and platelike protrusions than in autozooecia. Monticules irregular and ill defined; center of slightly larger monticular zooecia and abundant exilazooecia. [Middle Ordovician species from Estonia that were assigned to *Favositella* by BASSLER (1911) do not fit the emended definitions of *Favositella* or *Bythotrypa*. They resemble *Favositella* more closely but probably represent an undescribed genus. *Dnestropora mirabilis* ASTROVA, 1965, p. 130, the type species by original designation of *Dnestropora* ASTROVA, 1965, p. 130, agrees well with the emended definition of *Favositella*, and *Dnestropora* is considered to be a junior subjec-

tive synonym of *Favositella*.] *M.Sil.-U. Sil.*, E.N.Am., Eu.——FIG. 161,1a–d. **F. interpuncta* (QUENSTEDT); *a,* small, ovate lunarial deposits and abundant exilazooecia; tang. sec., USNM 159729, ×50; *b,* irregular exilazooecium (left), communication pores, and protrusions from walls; long. sec., USNM 159729, ×30; *c,* spheroliths, undulatory walls, communication pores, and irregular exilazooecia; long. sec., USNM 159730, ×30; *d,* small lunarial deposits in some autozooecia (left), large lunarial deposits with proximal flexures (right), and few exilazooecia; tang. sec., USNM 159728, ×30.

?**Ganiella** YAROSHINSKAYA in ASTROVA & YAROSHINSKAYA, 1968, p. 51 [**G. frequens*; OD; L. Dev., Altai Mts., USSR]. Zoarium encrusting or bifoliate, branching. Mesotheca undulatory, thin. Autozooecia with sparse diaphragms; elongate to subcircular in cross section. Wall variable in thickness, questionably with small communication pores; microstructure indistinct, questionably laminated. Lunaria with shorter radius of curvature, indistinct. Questionable exilazooecia in outer exozone. [Indistinct lunaria, lack of vesicular tissue, and possible exilazooecia in the outer exozone suggest that *Ganiella* belongs in the Ceramoporidae, where YAROSHINSKAYA originally placed it; however, the indistinct microstructure and lack of obvious communication pores make the placement of this poorly known genus uncertain.] *L.Dev.*, USSR.——FIG. 161, 2*a,b*. *G. frequens*; *a*, bifoliate zoarium, long autozooecia with no diaphragms, ?exilazooecia in outer exozone (left); long. sec., holotype, (SNIIGGIMS 952/T-20-5, ×40; *b*, elongate autozooecia with lunaria; indistinct microstructure; tang. sec., ?paratype, ×40 (negatives courtesy of A. YAROSHINSKAYA).

?**Haplotrypa** BASSLER, 1936, p. 157 [**H. typica*; OD; Osgood F., M. Sil., Osgood, Ind., USA]. Zoarium encrusting. Autozooecia polygonal to subcircular in cross section. Wall laminae containing mural lacunalike structures and irregular acanthostylelike flexures that are subcircular in cross section. Wall irregular in thickness; irregular spinelike structures projecting into autozooecial cavity, some hooked proximally. Undoubted communication pores not observed. Diaphragms few. Lunaria not observed. Exilazooecia tubelike, subcircular to highly elongate in cross section; few in intermonticular areas. Monticules with more exilazooecia and slightly larger zooecia. [This doubtful genus, known only from the holotype, is questionably placed in the Ceramoporidae.] *M.Sil.*, E.N.Am.——FIG. 162,1*a–e*. **H. typica*, holotype, USNM 92132; *a*, mural lacunalike structures and flexures, no lunarial deposits; tang. sec., ×48; *b*, subangular to subcircular autozooecia and irregular exilazooecia in monticule (lower left); tang. sec., ×19; *c*, irregular flexures, spinelike protuberances, and mural lacunalike structures; tang. sec., ×96; *d*, basal layer, thick and irregular walls, and exilazooecia; long. sec., ×29; *e*, spinelike projections and irregularly flexed laminae (proximal, left); long. sec., ×96.

Papillalunaria UTGAARD, 1969, p. 290 [**Crepipora spatiosa* ULRICH, 1893, p. 323; OD; "Trenton," M. Ord. (Mohawk.), Harrodsburg, Ky., USA]. Zoarium discoidal, hemispherical or irregularly massive; base concave or irregularly convoluted. Autozooecia moderately large, generally angular or subangular in cross section, a few subcircular. Walls generally undulatory, in some crenulated; generally thin, and some with monilalike swellings. Wall laminae indistinct, broadly to sharply curved. Boundary thin, dark, crenulated, and commonly not in center of compound wall, producing integrate appearance, or boundary obscure and walls amalgamate. Communication pores few, commonly absent. Lunaria small in inner exozone, nearly as large as zooecial bend as at zoarial surface. Ends of lunaria projecting into cavity. Cores in some lunaria. Knob- to rodlike

protrusions on proximal side of lunaria in zooecia adjacent to monticules, rarely in intermonticular autozooecia. Laminated layer on proximal side of lunarium thin to absent; where present, laminae flexing around protrusions of lunarial deposit. Diaphragms abundant, thin, tabular, commonly at same level in adjacent autozooecia. Diaphragms laminated or displaying no discernable microstructure. Exilazooecia few. Monticules with few exilazooecia interspersed among autozooecia, monticular autozooecia much larger than intermonticular autozooecia. *M.Ord.-U.Ord.*, E.N.Am.——FIG. 163,1*a–d*. **P. spatisoa* (ULRICH), lectotype, USNM 159723; *a*, thin basal layer, lunarial deposits, slightly moniliform walls, and thin diaphrams; transv. sec., ×20; *b*, larger zooecia in monticle (upper left), small lunarial deposits, and sparse exilazooecia; tang. sec., ×20; *c*, lunarial deposits with protrusions and laminae on proximal (left) side and diaphragms; long. sec., ×50; *d*, lunarial deposit (top) with protuberances on proximal side (toward top), indistinct wall laminations, and thin walls; tang. sec., ×100.

Suborder FISTULIPORINA Astrova, 1964

[*nom. correct.* herein, *pro* Fistuliporoidea ASTROVA, 1964, p. 29, suborder]

Zoarial growth form variable; encrusting, hemispherical, massive, ramose, hollow ramose, bifoliate frondose, bifoliate with narrow straplike branches, bifoliate cribrate, articulate with bifoliate straplike branches, trifoliate, multifoliate, and fenestrate or pinnate with a vertical mesotheca and an obverse and reverse side. Monticules in most genera except some narrow bifoliate forms. Autozooecia short to long; diaphragms generally sparse. Walls laminated, granular or granular-prismatic. Distal and lateral sides of autozooecia bounded by superimposed vesicle walls in many genera. Communication pores absent. Autozooecia budded at basal layer, mesotheca, in endozone, or on vesicular tissue in exozone; partly to widely isolated by vesicles at budding surface. Crescentic to subcircular lunaria in exozone in most genera; granular, granular-prismatic, hyaline, or laminated. Exilazooecia absent. Vesicular tissue (cystopores) in all genera; tubelike with compound walls, like mesozooecia, in a few genera; generally boxlike to blisterlike with simple walls and roofs; granular, granular-prismatic, or laminated. Zones of thick vesicle roofs or stereom in many forms. Gonozooecia in few. Acanthostyles in some forms; most in vesicle roofs or stereom. *Ord.-Perm.*

Family ANOLOTICHIIDAE Utgaard, 1968

[Anolotichiidae UTGAARD, 1968a, p. 1035]

Zoaria encrusting, hemispherical, saucer shaped, massive or irregularly ramose. Monticules regularly spaced, rarely absent or poorly defined. Autozooecia large, tubular, with thin, distantly spaced diaphragms, commonly at same level in adjacent autozooecia. Wall structure indistinct, granular, or having dark granular zooecial boundary and cortex of light-colored layers of irregular rodlike or blocklike crystals or crystal aggregates perpendicular to boundary. Autozooecial walls incomplete; partly composed of superimposed vertical walls of interzooecial vesicles. Lunaria extending at least from outer endozone to zoarial surface, with microstructure similar to that of autozooecial walls but thicker in some. Interzooecial spaces (cystopores) in endozone and exozone; variable and irregular in shape in a zoarium, resembling tubular exilazooecia with compound walls, or, more commonly, tubelike, boxlike, or cystlike vesicles with simple walls; some originating at basal layer, partially isolating zooecia. Microstructure indistinct, granular or granular-prismatic. Monticules with central cluster of interautozooecial spaces. Acanthostyles absent except in *Lamtshinopora* and *Profistulipora*. Undoubted communication pores not observed. *Ord., M.Dev.*

Characters of particular importance are: granular or granular-prismatic wall microstructure; large autozooecia; tubelike, boxlike, or cystlike vesicular tissue that is irregular; lunaria; and absence of acanthostyles and stereom.

Anolotichia ULRICH, 1890, p. 381 [**A. ponderosa*; OD; "Fernvale F.," U. Ord. (Richmond.), Wilmington, Ill., USA]. Zoarium encrusting, massive, or irregularly ramose. Monticules poorly defined or absent. Cross section of autozooecial living chamber angular or subangular in endozone, subangular in exozone where interzooecial spaces few, subcircular where interzooecial spaces abundant. Axes commonly not parallel in adjacent autozooecia. Autozooecial walls undulatory, thin, locally thickened in bands. Transverse bands light colored where wall thinner, commonly at same level in adjacent zooecia. Rod-shaped crystals in thicker walled bands fanning out from subjacent thinner walled band. Two to six clear calcite rods extending for length of lunarium. Rods irregular in diameter and cross-sectional shape, in some with perpendicular extensions connecting rods in same lunarium. Rods enlarged distally, merging in some at zoarial surface to occupy entire thickened lunarium. Interzooecial spaces rare to abundant, angular to subangular, may be subcircular in cross section in exozone. *M.Ord.-U.Ord.*, N.Am.——FIG. 164,1a–f. **A. ponderosa*; *a,* enlarged partially to completely fused rods in lunaria close to zoarial surface; tang. sec., lectotype, USNM 159693, ×50; *b,* autozooecia, light-colored rods in lunaria, and interzooecial spaces; tang. sec., paralectotype, USNM 159696, ×30; *c,* rods in lunarium having lateral projections, some of which extend to adjacent rod; long. sec., lectotype, ×50; *d,* large angular autozooecium, lunarium with rods, and angular interzooecial spaces in endozone; transv. sec., paralectotype, USNM 159697, ×50; *e,* irregular interzooecial spaces and rods in lunaria; long. sec., paralectotype, USNM 159695, ×30; *f,* alternating dark- and light-colored bands in wall, outward fanning of crystal aggregates in thicker walled bands; long. sec., paralectotype, USNM 159694, ×100.

Altshedata MOROZOVA, 1959b, p. 7 [**Fistulipora belgebaschensis* NEKHOROSHEV, 1948, p. 50; OD; *M.Dev.* (Givet.), Altai Mts., Kuznetsk basin, USSR]. Zoarium encrusting or massive. Monticules with large zooecia and more abundant interzooecial spaces. Autozooecial living chamber subangular to subrounded in cross section; ends of lunaria commonly indenting living chamber. Autozooecial walls thick, undulatory; indistinct granular microstructure. Diaphragms closely spaced, oblique, concave, and thin. Interzooecial spaces generally one row of high, narrow blisterlike vesicles, angular in cross section; larger and more abundant in monticules. Lunaria thicker near zoarial surface; projections into zooecial cavity moderately long and thin in inner exozone to short, tapering in outer exozone. *M.Dev.*, USSR. ——FIG. 165,1a–c. **A. belgebaschensis* (NEKHOROSHEV), Kuznetsk basin, USSR, PIN 1204/9.A;b; *a,* oblique, closely spaced diaphragms in autozooecia, narrow interzooecial spaces with one row of vesicles, and undulatory, thick autozooecial walls; long. sec., ×30; *b,* thin diaphragms, thick autozooecial wall, and indistinct granular microstructure of

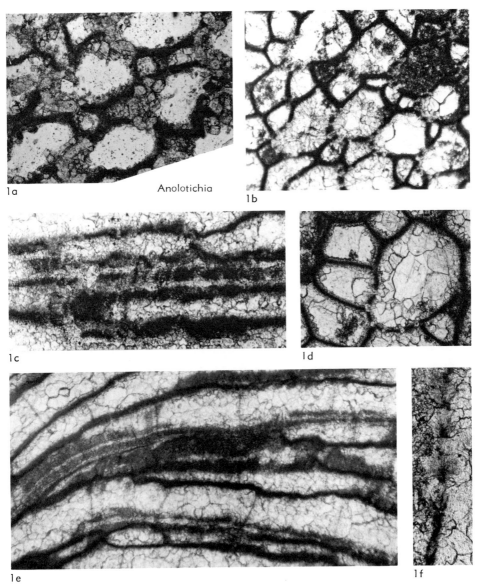

Fig. 164. Anolotichiidae (p. 370).

high, narrow, curved plates of cystose interzooecial spaces; long. sec., ×100; *c,* angular interzooecial spaces (above), some lunaria with ends projecting into zooecial cavity; tang. sec., ×30.

Bythotrypa ULRICH, 1893, p. 324 [**Fistulipora? laxata* ULRICH, 1889, p. 37; OD; "Trenton Gr.," M. Ord., Manit., Can.]. Zoarium encrusting, massive, or hemispherical. Monticules inconspicuous, depressed to slightly elevated. Axes commonly not parallel in adjacent autozooecia in exozone. Walls on distal or lateral sides of autozooecia commonly formed by vertical portions of walls of cystlike interzooecial vesicles; these walls scalloped in longitudinal view and straight in tangential section, producing angular autozooecial cavities. Other zooecia subrounded in cross section. Walls relatively thin throughout, straight to crenulated; granular. Zooecial boundary commonly obscure and walls uniform in appearance. Wall thickenings local with fan-shaped bundles of crystals. Lunaria moderately large in outer endozone, large at zoar-

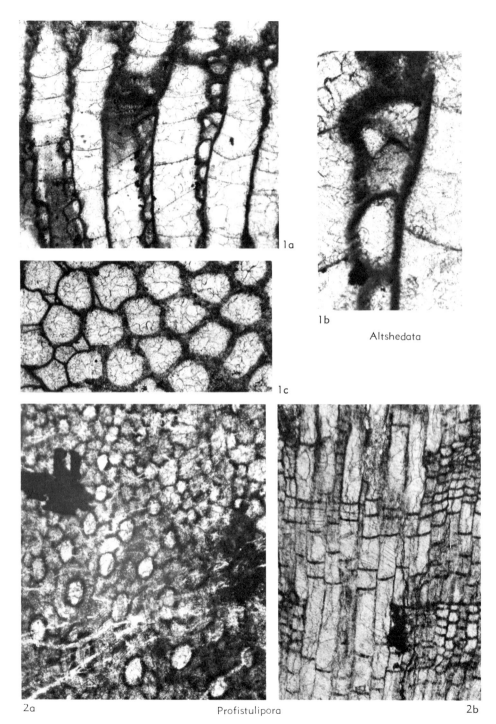

1a

1b

Altshedata

1c

2a

Profistulipora

2b

Fig. 165. Anolotichiidae (p. 376).

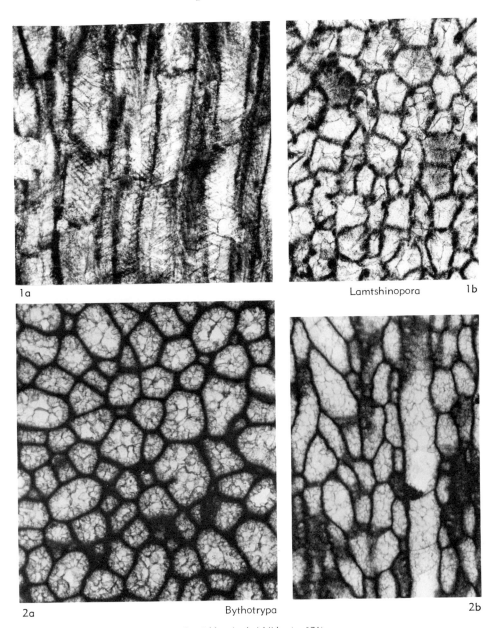

1a

Lamtshinopora 1b

2a Bythotrypa 2b

FIG. 166. Anolotichiidae (p. 371).

ial surface. Vesicular tissue abundant, partially isolating autozooecia; polygonal, boxlike, or cystlike; commonly elongated parallel to axes of autozooecia, locally obliquely, particularly in monticules. Locally top of curved wall of vesicular tissue diaphragmlike. Vesicular tissue in monticular centers commonly more variable in size and shape, obliquely elongated. Lunaria radially arranged around monticular centers.

M.Ord., E.N.Am.——FIG. 166,*2a,b*. **B. laxata* (ULRICH), holotype, USNM 43241; *a*, central cluster of vesicular tissue and autozooecia radially arranged with lunaria nearest monticular center; tang. sec., ×30; *b*, boxlike to obliquely elongated vesicular tissue in monticule (left); long. sec., ×20.

Crassaluna UTGAARD, 1968a, p. 1039 [**Crepipora epidermata* ULRICH, 1890, p. 471; OD; "Fern-

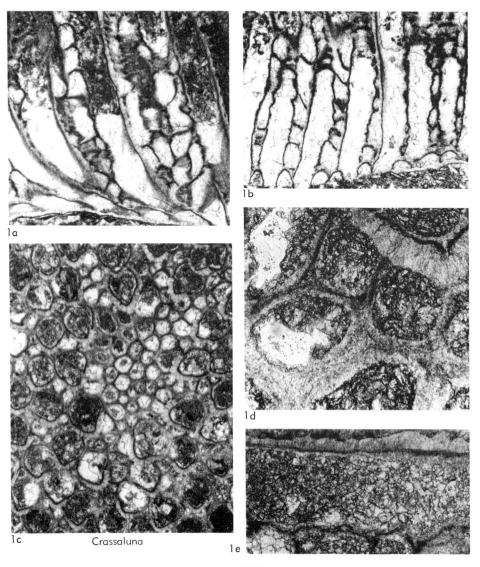

FIG. 167.   Anolotichiidae (p. 373).

vale F.," U. Ord., Wilmington, Ill., USA]. Zoarium thin expansions, flat to convoluted. Monticules small to large, slightly elevated to slightly depressed. Recumbent portion of autozooecia short to long. Axes generally parallel in adjacent autozooecia. Autozooecia angular to subangular in cross section at zooecial bend, subangular to subrounded in outer exozone. Lateral and distal sides commonly composed of overlapping vertical portions of vesicular tissue. Autozooecial walls moderately thick, granular to granular-prismatic. Zones of thicker wall lacking distinct boundary and with crystal aggregates fanning outward, producing minutely spotted amalgamated appearance. Lunaria becoming larger and thicker in outer exozone; proximal side with uneven nodes and ridges; most lunaria with one minute longitudinal corelike structure near center. Walls of adjacent autozooecia and vesicular tissue unconformably abutting proximal side of lunarium. Vesicles variable but generally small, subangular to subrounded in cross section. Monticular centers with irregular tubelike and oblique vesicles surrounded by larger zooecia. Lunaria locally and in part radially arranged around monticules. *U.Ord.,* N.Am.——Fig.

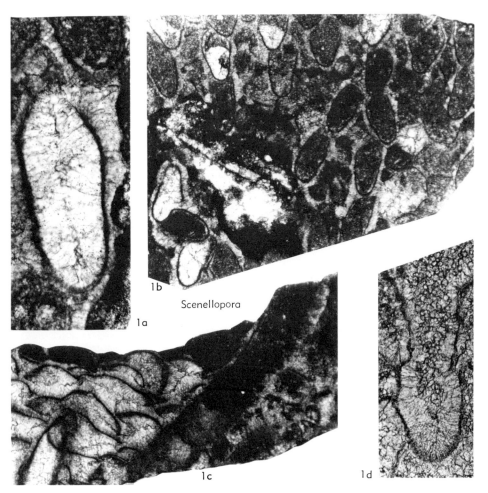

FIG. 168. Anolotichiidae (p. 376).

167,*1a–e*. **C. epidermata* (ULRICH); *a,* overlapping cystlike vesicles forming wall on distal side of autozooecia and serrated proximal sides of lunarial deposits; long. sec., lectotype, USNM 159700, ×20; *b,* basal layer, recumbent sinuses, and variation in shape of vesicular tissue; transv. sec., paralectotype, USNM 159701, ×20; *c,* monticule and partial radial arrangement of zooecia; tang. sec., lectotype, ×20; *d,* lunarium (upper right) with indistinct central corelike structure and thinner walls (center) with dark granular primary layer, and thicker walls (bottom) in vesicular tissue; tang. sec., paralectotype, USNM 159701, ×100; *e,* corelike structure in lunarium (top) and boxlike to cystlike vesicles (bottom); long. sec., paralectotype, USNM 159702, ×50.

?**Lamtshinopora** ASTROVA, 1965, p. 124 [**L. hirsuta*; OD; L. Ord., Vaigach Is., near N. Zemlya,

USSR]. Zoarium encrusting, laminated or massive. Monticules either poorly defined areas of larger zooecia or absent. Cross section of autozooecial living chamber angular to irregularly subangular. Autozooecial walls thick; microstructure granular; light areas in walls may represent pores. Lunaria not readily apparent; poorly developed or absent. Numerous granular acanthostyles in autozooecial walls indenting living chamber, producing pseudoseptate appearance in some autozooecia. Interzooecial spaces apparently tubelike; cross section irregular in size and shape. [Extensive recrystallization makes it difficult to observe communication pores, which apparently are lacking. Similarly, lunariumlike structures are difficult to evaluate. Granular acanthostyles are not present in other Anolotichiidae, and *Lamtshinopora* is placed in this family with reservation.] *L.Ord.,* USSR.——FIG.

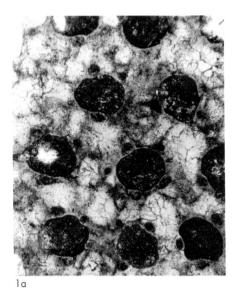

1a

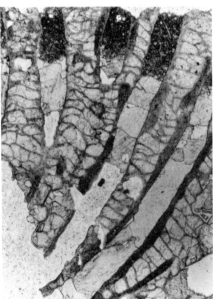

1b Xenotrypa

FIG. 169. Xenotrypidae (p. 377).

166,1a,b. **L. hirsuta*; a, autozooecia with sparse diaphragms and tubelike interzooecial spaces; long. sec., paratype, PIN 1393/382b, ×30; b, irregular interzooecial spaces and angular to subangular autozooecia; tang. sec., paratype, PIN 1393/382a, ×30.

Profistulipora ASTROVA, 1965, p. 144 [**P. arctica*; OD; L.-M. Ord., N. Urals, N. Zemlya, USSR]. Zoarium encrusting to massive, surface smooth. Monticules with slightly larger zooecia, lunaria

not radially arranged. Autozooecia ovate in cross section, widely isolated by interzooecial spaces; diaphragms sparse. Wall granular, some small acanthostyles. Lunaria indistinct, granular. Interzooecial spaces tubular, irregular in height; zones of shorter tubular vesicles in monticules; cross section polygonal; walls and roofs granular; small acanthostyles in walls. *L.Ord.-M.Ord.*, USSR. ——FIG. 165,2a. **P. arctica*, paratype, PIN 1245/57a; monticule with larger zooecia (lower left), ovate autozooecia isolated by moderately large polygonal interzooecial spaces; tang. sec., ×30.——FIG. 165,2b. *P. menneri* ASTROVA, M. Ord., N. Urals, USSR, paratype, PIN 1606/16B; sparse diaphragms in autozooecia, tubelike interzooecial spaces, and zones of lower tubular vesicles in monticule (right); long. sec., ×20.

Scenellopora ULRICH, 1882, p. 150, 158 [**S. radiata*; OD; Trenton Gr., M. Ord., Knoxville, Tenn., USA]. Zoarium small, saucer shaped, commonly with central basal protrusion and one monticule; or thin irregular expansions with several monticules and broadly undulating basal layer. Monticular centers containing large, irregularly cystlike vesicles. One or two sets of fascicles of one or a few irregular rows of autozooecia radiating from margin of depressed monticular center. Fascicles highly elevated and distinct if zooecia tightly packed, low and indistinct if loosely packed. Autozooecial cavity generally oval but distal side may be angular. Walls moderately thin, straight to crenulated. Walls granular-prismatic in holotype; in some, wall granular-prismatic and locally indistinctly laminated. Lunaria moderately large; in most zoaria, some lunaria with one to several light-colored longitudinal rods, locally inflecting autozooecial cavity; rods with widely spaced, curved laminations convex outward; remainder of lunarium granular-prismatic or indistinctly laminated. Lunaria radially arranged and closest to monticular centers. Vesicular tissue irregular in size and shape, with microstructure similar to that of autozooecial walls. *M.Ord., N.Am.*——FIG. 168,1a–d. **S. radiata*; a, lunaria on proximal (bottom) side of autozooecia, subcircular rod near center of upper lunarium; tang. sec., holotype, USNM 43289, ×100; b, irregular fascicles of zooecia radiating from central monticule; tang. sec., holotype, ×30; c, cystlike vesicles in depressed central monticule (left) and zooecia with few diaphragms (right); long. sec., holotype, ×30; d, radial disposition of autozooecia from ancestrula, near base of colony; deep tang. sec., topotype, USNM 159704, ×100.

Family XENOTRYPIDAE
Utgaard, new

Zoaria encrusting, hemispherical, or ramose. Monticules with central cluster of vesicles. Autozooecia large, living chamber inflected by large acanthostyles. Diaphragms few. Wall microstructure indistinct, granular. Lunaria absent. Vesicles large, irregular, blisterlike; large acanthostyles in vesicle walls. *?L.Ord.(Arenig.), M.Ord.-M.Sil.*

Characters of particular importance are: large autozooecia; indistinct, granular, wall microstructure; large, irregular vesicles; large acanthostyles in vesicle walls and inflecting autozooecia; and absence of lunaria.

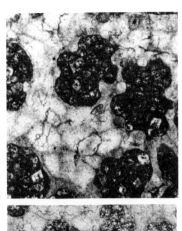

Xenotrypa and *Hennigopora* have previously been included in the Constellariidae but differ from other genera in the family in lacking star-shaped monticules and laminated walls and in having large acanthostyles that inflect autozooecial living chambers. Thus, a new family is established here for these unusual forms. They resemble late Paleozoic Actinotrypidae in having large acanthostyles inflecting the autozooecia but differ from the Actinotrypidae in having large autozooecia, an indistinct granular microstructure, and large acanthostyles in vesicle roofs.

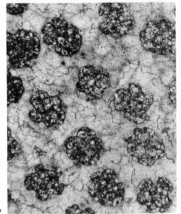

1a

1b

Xenotrypa BASSLER, 1952, p. 381 [*Fistulipora primaeva* BASSLER, 1911, p. 109; OD; Glauconite Ls., M. Ord., B2, Pawlovsk, USSR]. Zoarium high hemispherical. Monticules large, flush; center a cluster of vesicles. Autozooecia large, slightly indented by acanthostyles. Walls indistinctly granular. Vesicles irregular, isolating zooecia; indistinct microstructure. Acanthostyles large, generally in autozooecial walls, some in vesicle walls; centers light to dark in color; projecting as spines. *?L.Ord. (Arenig.), M.Ord.,* Eu.
———FIG. 169,1*a,b.* *X. primaeva* (BASSLER), holotype, USNM 57208; *a,* autozooecia isolated by vesicles, large acanthostyles in zooecial walls and vesicles; tang. sec., ×30; *b,* autozooecia with few diaphragms, irregular vesicles, and large acanthostyles; long. sec., ×20.

Hennigopora BASSLER, 1952, p. 382 [*Callopora florida* HALL, 1852, p. 146; OD; Niagaran Gr., Rochester Sh., M. Sil., Lockport, N.Y., USA]. Zoarium encrusting or ramose, base unattached in some. Monticules slightly elevated, having slightly larger zooecia. Autozooecia large, acanthostyles inflecting living chambers; diaphragms few, straight to oblique. Microstructure indistinct; granular, local indistinct laminations. Lunaria lacking. Vesicles large, subrectangular to cystlike; partially to completely isolating autozooecia in exozone. Acanthostyles few in vesicle walls. *U.Ord.(Richmond.)-M.Sil.,* E.N.Am., Eu.

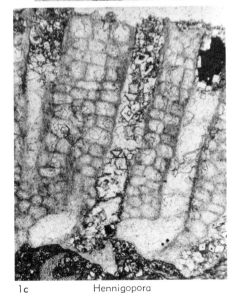

1c   Hennigopora

FIG. 170. Xenotrypidae (p. 377).

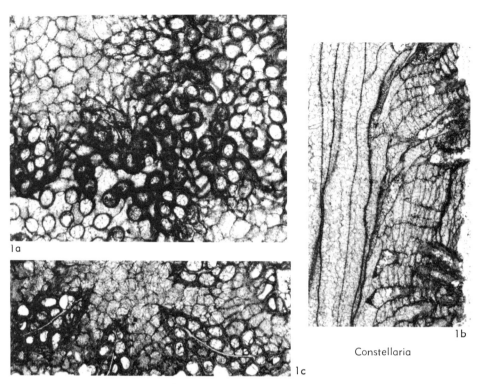

1a

1b

Constellaria

1c

FIG. 171.   Constellariidae (p. 378).

——FIG. 170,*1a–c*. **H. florida* (HALL); *a,* acan-thostyles with light and dark centers in auto-zooecial and vesicle walls; tang. sec., syntype, AMNH 1744/1, ×48; *b,* acanthostyles indent-ing autozooecia, few in vesicles; tang. sec., USNM 159763, ×29; *c,* autozooecia with few diaphragms, acanthostyles (left side), and vesi-cles; long. sec., USNM 159763, ×29.

Family CONSTELLARIIDAE
Ulrich, 1896

[Constellariidae ULRICH, 1896, p. 267] [=Stelliporidae MILLER, 1889, p. 169]

Zoaria encrusting, ramose, or frondose. Monticules star shaped; composed of central cluster and radiating interrays of vesicles (cystopores) between rays of loosely to tightly packed zooecia, with or without midray par-titions. Lunaria present in one genus.[1] Wall structure indistinctly and transversely lami-nated. Acanthostyles small; in zooecial walls

and vesicular tissue. Vesicles quadrate to blisterlike or commonly irregular and vari-able in zoarium. Local zones having thick vesicle roofs. *M.Ord.(Chazy.)-U.Ord. (Rich-mond.), ?L.Sil.*

Characters of particular importance are stellate to substellate monticules, indistinctly laminated wall microstructures, irregular quadrate to blisterlike vesicles, and small acanthostyles.

Constellaria DANA, 1846, p. 537 [**C. constellata* DANA, 1849, pl. 52; SM; ?U. Ord., ?Ohio, USA] [=*Stellipora* HALL, 1847, p. 79]. Zoarium encrusting, ramose, or frondose. Monticules stel-late to subcircular; primary plus secondary rays of zooecia flush or elevated; monticular center and interrays of vesicles depressed, flush, or ele-vated. Autozooecia larger, with irregular polyg-onal cross section in endozone, smaller, with sub-circular cross section in exozone; generally isolated by vesicles in intermonticular areas; lunarium lacking but some autozooecia with thicker proximal wall. Walls indistinctly and transversely laminated; diaphragms few to many, straight to curved. Vesicles boxlike and super-

---

[1] *Lunaferamita* UTGAARD, 1981, described too late for inclusion in this *Treatise*. See J. Paleontol., v. 55, p. 1058–1070. Type species: *L. bassleri* (LOEBLICH); range: M.Ord. (Chazy.), Nev., Okla., Va.

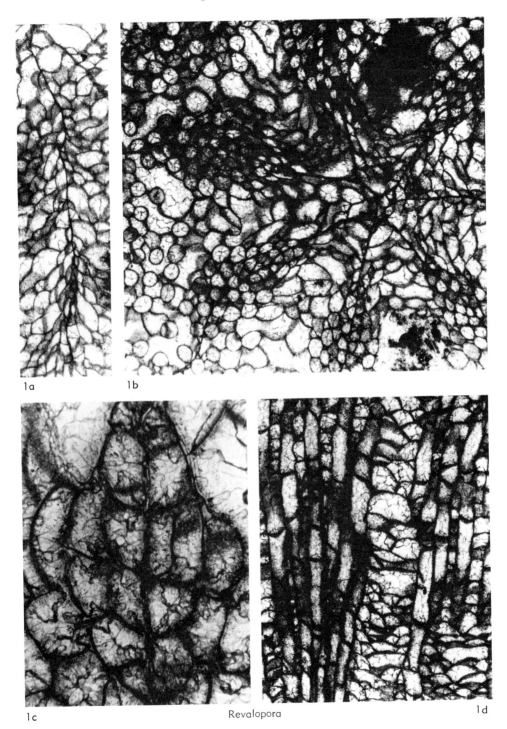

1a   1b

1c   Revalopora   1d

Fig. 172. Constellariidae (p. 380).

imposed (resembling mesozooecia) to irregular blisters; poorly laminated, containing pustules; vesicle roofs thickened in zones. Midray partitions of hyaline calcite; acanthostyles in vesicle and autozooecial walls. [The type species of *Stellipora* (*S. antheloidea* HALL, 1847, p. 79, by monotypy) is a *Constellaria*. The type species of *Constellaria* has been widely but erroneously assumed to be *C. florida* ULRICH, 1882. Whereabouts of the type of *C. constellata* is not known and it is presumed to be lost.] *M.Ord.(Chazy.)-U.Ord.(Richmond.)*, E.N.Am., Eu. [*?L.Sil.*, Sib., see ROSS, 1963a].——FIG. 171, *1a,b. C. florida* ULRICH, Maysville Gr., U. Ord., Ky., USA, lectotype, USNM 159760; *a,* monticular centers (upper left, lower right) of vesicles, rays of zooecia radiating from monticule; tang. sec., ×30; *b,* simple endozone and autozooecia and vesicles in exozone; long. sec., ×30.——Fig. 171, *1c. C. florida prominens* ULRICH, McMicken Mbr., Eden F., U. Ord. (Eden.), Ohio, USA; paratype, USNM 159762; vesicles in monticular center and interrays, ray zooecia, and hyaline midray partitions; tang. sec., ×30.

Revalopora VINASSA DE REGNY, 1921, p. 220 [*Stellipora revalensis* DYBOWSKI, 1877, p. 44; OD; CI or CII, M. Ord., Est.]. Zoarium encrusting. Monticules large, star shaped; rays clusters of generally smaller zooecia; interrays and center depressed, containing vesicles. Autozooecia with straight to curved diaphragms; walls thin, indistinctly laminated. Lunaria not observed. Vesicles large and irregular in monticules and intermonticular areas; walls and roofs poorly laminated; isolating autozooecia in intermonticular areas. Midray partitions laminated, with acanthostyles, extending through vesicles to monticular center and inward in zoarium. *M.Ord.(?Llanvirn., Llandeil.-Caradoc.),* Eu.(Est.).——FIG. 172, *1a-d. *R. revalensis* (DYBOWSKI), Kuckers Sh., Caradoc., C2, near Jewe, Est., USNM 57303; *a,* laminated midray partition and vesicles in monticular center; long. sec., ×20; *b,* monticule, midray partitions, ray zooecia, subcircular intermonticular autozooecia, and large, irregular vesicles; tang. sec., ×20; *c,* poorly laminated walls, midray partition with acanthostyles and subquadrate to subcircular ray zooecia; tang. sec., ×100; *d,* autozooecia with straight to curved diaphragms, superimposed vesicle walls forming part of autozooecial walls, and irregular vesicles; long. sec., ×20.

Family FISTULIPORIDAE
Ulrich, 1882

[Fistuliporidae ULRICH, 1882, p. 156] [=Chilotrypidae SIMPSON, 1897, p. 480; Favicellidae SIMPSON, 1897, p. 556; Fistuliporinidae SIMPSON, 1897, p. 480; Odontotrypidae SIMPSON, 1897, p. 481; Selenoporidae SIMPSON, 1897, p. 557; Cheilotrypidae MOORE & DUDLEY, 1944, p. 266]

Zoaria encrusting, massive, ramose or hollow ramose. Monticules in most genera. Autozooecia partly to completely isolated at budding surface by vesicular tissue. Blister- to boxlike vesicular tissue in exozone in all genera. Walls and vesicular tissue indistinctly laminated or granular or granular-prismatic. Acanthostyles in vesicle walls, roofs, or stereom in most genera. Lunaria in most genera; hyaline or granular-prismatic. *Sil.-Perm.*

Characters of particular importance are: blister- to boxlike vesicles in exozone; acanthostyles in vesicle roofs or stereom; thin, local zones of stereom; and lunaria.

Fistulipora McCOY, 1849, p. 130, *nom. conserv.* ICZN Opinion 459, *non* RAFINESQUE, 1831 [**F. minor*; SD MILNE-EDWARDS & HAIME, 1850, p. lix, *non* CUMINGS, 1906, p. 1293, *ut F. spergensis minor*; L. Carb. Ls., Miss., G. Brit.] [?=*Cucumulites* GURLEY, 1884, p. 2]. Zoarium encrusting or massive, rarely ramose. Monticules elevated or flush, central cluster of vesicles and ring of larger zooecia with lunaria partly to altogether radially arranged, on side of zooecia nearest monticular center. Autozooecia long, tubular, with closely spaced planar or curved diaphragms; partially to completely isolated by vesicular tissue; microstructure granular or granular-prismatic. Lunaria through endozone and exozone, granular or granular-prismatic; radius of curvature shorter than that of autozooecial wall. Vesicles moderately large, angular; thin granular or granular-prismatic walls and roofs; rarely boxlike, commonly polygonal or subquadrate blisters. Local zone of thicker vesicle roofs at zoarial surface. [Many species in this large genus are in need of restudy. Early species tend to have larger, more boxlike vesicles, whereas many late Paleozoic species have smaller, polygonal or cystlike vesicles. *Cucumulites tuberculatus* GURLEY, 1884 is the type species of *Cucumulites* GURLEY, 1884, by original designation. The holotype is silicified and has not been sectioned. Its apparent synonym, *C. tricarinatus* GURLEY, 1884, though mostly silicified and poorly known, has features that are most similar to those of *Fistulipora,* and *Cucumulites* is tentatively placed in synonymy with *Fistulipora.*] *Sil.-Perm.*, worldwide.—— FIG. 173, *1a-d. *F. minor*, paratype, SM 315; *a,* angular to blisterlike vesicles and closely spaced diaphragms in autozooecia in outer exozone, granular to granular-prismatic microstructure; transv. sec., ×30; *b,* lunaria with short radius of curvature and granular microstructure in rounded autozooecia partially isolated by moderately large polygonal vesicles; tang. sec., ×30; *c,* subangular autozooecia with lunaria, partially

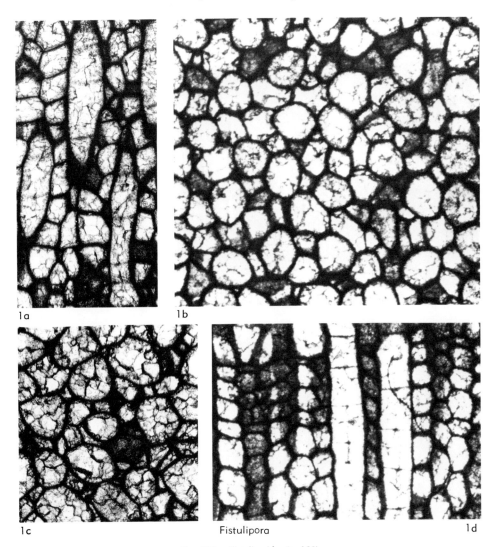

1a   1b

1c   Fistulipora   1d

FIG. 173. Fistuliporidae (p. 380).

isolated by polygonal vesicles in endozone; transv. sec., ×30; *d,* tubular autozooecia with closely spaced planar diaphragms isolated by boxlike to polygonal vesicles; long. sec., ×30 (*a* and *c* are photographs of a section figured by BASSLER, labeled SM 315; *b* and *d* are photographs furnished by A. G. BRIGHTON of presumably the same paratype, SM 315, W. Hopkins Coll., bearing the additional number E5373b).

Buskopora Ulrich, 1886b, p. 22 [**B. dentata;* OD; "U. Helderberg?," M. Dev., Falls of the Ohio, Jeffersonville, Ind., USA]. Zoarium encrusting. Monticules small, elevated clusters of vesicles surrounded by larger zooecia with large, hood-shaped lunaria. Autozooecia with peristomes. Basal layer with dark, granular, primary layer and light-colored, granular-prismatic layer. Autozooecia recumbent for short distance, erect in exozone; diaphragms thick, throughout endozone and exozone. Autozooecia partially to completely isolated by small angular to subrounded vesicles; circular to oval in cross section, deeply indented by bifid lunaria. Walls of autozooecia and lunaria granular-prismatic; dark, granular, central zone of lunarium not continuous with dark, granular boundary of autozooecial wall. Vesicles blister-like, granular-prismatic, local zones of thick vesicle roofs with minute acanthostylelike tubuli. *M.Dev.,* E.USA.——FIG. 174, *1a–g.* **B. dentata;* *a,* topotype, USNM 43273, *b–g,* specimens from Jeffersonville F., Ind.; *a,* peristomes and lunaria projecting into

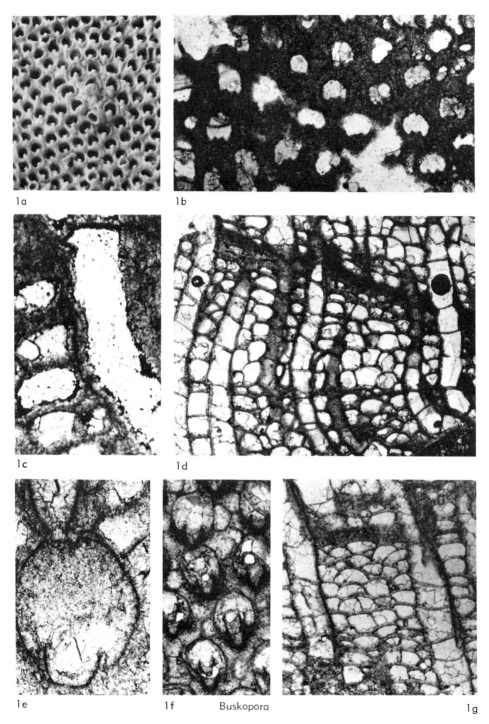

1a

1b

1c

1d

1e

1f   Buskopora

1g

Fig. 174.   Fistuliporidae (p. 381).

subcircular autozooecia like bifid teeth, and large, hoodlike lunaria in monticular zooecia (center), ×10; *b,* subcircular autozooecia indented by bifid lunaria and isolated by small vesicles having thick roofs with minute acanthostylelike structures; tang. sec., SIUC 3003, ×30; *c,* compound autozooecial wall with dark, granular, boundary zone and granularprismatic cortex; thicker, granular-prismatic lunarium on proximal side (right), and granular-prismatic vesicles (left); note simple, granular-prismatic subterminal diaphragm in autozooecium (above); long. sec., SIUC 3004, ×100; *d,* tubular autozooecia with planar diaphragms isolated by boxlike to blisterlike vesicles with two zones of thicker roofed vesicles; long. sec., SIUC 3004, ×30; *e,* granular-prismatic microstructure of autozooecial wall and projecting lunarium (below), light-colored corelike structure in lunarium (above) and granular vesicle walls (upper right); tang. sec., SIUC 3005, ×100; *f,* autozooecia narrowly isolated by small, angular vesicles and indented by lunaria; tang. sec., SIUC 3005, ×30; *g,* autozooecia with thick planar diaphragms and projecting end of lunarium and blisterlike vesicular tissue with zone of thick vesicle roofs containing minute acanthostylelike structures, transv. sec., SIUC 3006, ×30.

Canutrypa BASSLER, 1952, p. 382 [**C. francqana;* OD; U. Dev., Ferques, France]. Zoarium ramose. Low monticules with central clusters of vesicles and ring of slightly larger zooecia. Autozooecia long, tubular; few diaphragms; proximal side rounded, distal side angular or with keel and sinuses in endozone. Wall granular-prismatic to indistinctly laminated. Gentle curve from endozone into exozone where walls thicken, containing minute tubuli perpendicular to boundary. Thin, straight or curved or partial diaphragms in inner exozone. One hemicylindrical cystlike structure with axis perpendicular to zooecial axis in many autozooecia in exozone; granular-prismatic microstructure. Lunaria in exozone; dark boundary and thick, light-colored distal deposit, poorly laminated, with minute tubuli perpendicular to boundary. Vesicular tissue in exozone; partially isolating autozooecia; small subquadrate blisters, subangular to subrounded in cross section; indistinctly laminated or granular-prismatic. Vesicle roofs thickening into solid stereom with minute tubules in outer exozone. *U.Dev., Eu.*——FIG. 175,*1a–e.* **C. francqana; a,* thick lunarial deposits with tubuli perpendicular to zooecial walls and stereom with tubuli; tang. sec., paratype, USNM 113984, ×24; *b,* thin-walled contiguous autozooecia with diaphragms in endozone (left) and vesicular tissue from zooecial bend to zoarial surface, where stereom developed; long. sec., paratype, USNM 113984, ×24; *c,* thin-walled contiguous autozooecia with rounded proximal side and angular

distal side, some with keel and sinus; transv. sec., paratype, USNM 113982-1, ×16; *d,* subcircular autozooecia with thick, light-colored lunarial deposit (below) and nearly straight wall of hemicylindrical cyst; vesicle roofs thickened to form stereom; tang. sec., holotype, USNM 116417, ×40; *e,* autozooecia with closely spaced and straight to curved and incomplete diaphragms in inner exozone, hemicylindrical cystlike structure on distal wall, thick and light-colored lunarial deposit on proximal side, and cystlike vesicular tissue in exozone; long. sec., holotype, ×24.

Cheilotrypa ULRICH, 1884, p. 49 [**C. hispida;* OD; Glen Dean F., U. Miss. (Chester.), Sloans Valley, Ky., USA) [=*Chilotrypa* MILLER, 1889, p. 297, incorrect subsequent spelling]. Zoarium slender, hollow ramose or solid ramose, less commonly encrusting. Monticules small, with central cluster of vesicles, flush with surface; few surrounding zooecia conspicuously larger. Autozooecia narrow in endozone with narrower distal side. In exozone, autozooecia with ovate cavity, wider at proximal end. Walls with dark, granular, boundary zone and granular-prismatic cortex with tubuli. Lining local, laminated on distal side in outer exozone. Diaphragms sparse. Vesicles small, subrounded; autozooecia narrowly isolated; small blisters in inner exozone, granular-prismatic vesicle roofs merging to form stereom in outer exozone; tubuli in stereom. Lunaria in exozone, on wider proximal side, dark granular zone merging with granular zooecial boundary; thick, proximal granular-prismatic layer with tubuli perpendicular to boundary. *Miss., N.Am.* ——FIG. 175,*2a–h.* **C. hispida; a,* hollow ramose zoarium with narrow autozooecia in endozone and vesicular tissue and stereom in exozone; transv. sec., paralectotype, USNM 159754, ×40; *b,* lunaria at wide end of autozooecia (below) and stereom between autozooecia; tang. sec., paralectotype, USNM 159759, ×24; *c,* hollow ramose zoarium, subtriangular autozooecia becoming wider outward in endozone; transv. sec., paralectotype, USNM 159759, ×24; *d,* lunaria with dark zone continuous into zooecial boundary and proximal (below dark zone), thick granular-prismatic zone, small subrounded vesicles (above) narrowly isolating autozooecia; tang. sec., lectotype, USNM 159757, ×40; *e,* granular-prismatic walls, questionable distal lining (above) in autozooecium on right, and stereom with tubuli; tang. sec., paralectotype, USNM 159758, ×80; *f,* hollow axis with constrictions, thin wall resembling basal layer, and thick walls and vesicular tissue in exozone; long. sec., lectotype, ×24; *g,* irregular hollow center (below), thin-walled autozooecia in endozone becoming wider in outer endozone and exozone, vesicles thickening into stereom in exozone; long. sec., paralectotype, USNM 159753, ×24; *h,* stereom isolating

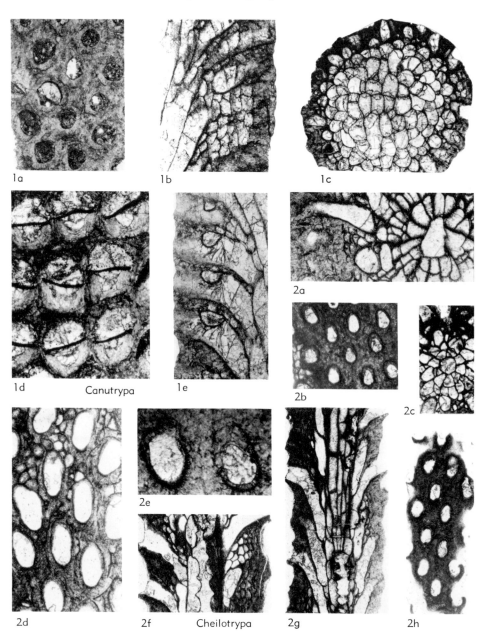

FIG. 175. Fistuliporidae (p. 383).

elongate autozooecia; tang. sec., paralectotype, USNM 159753, ×24.

Cliotrypa ULRICH & BASSLER IN BASSLER, 1929, p. 49 [*C. ramosa* ULRICH & BASSLER IN BASSLER, 1936, p. 160; SM; New Providence F., L. Miss., Kings Mt., Ky., USA]. Zoarium ramose, encrusting overgrowths. Monticules flush; center stereom, ring of slightly larger autozooecia with lunaria radially arranged. Autozooecia in endozone narrow, with proximal side rounded and distal side subangular to sinus-and-keel shaped; cavity subcircular in exozone. Walls transversely laminated; dark boundary, tubuli in cortex perpendicular to boundary. Lunaria in outer endozone and exozone; hyaline with one or two proximal projections. Diaphragms thin, sparse.

Hemiphragms proximally curved, triangular to spinelike; alternating on proximal and distal walls in outer endozone and in exozone. Gonozooecia, as bulbous swellings of autozooecia, in exozone. Vesicles in outer endozone and exozone; blisterlike cysts decreasing in height outward; stereom in most of exozone. Walls and roofs poorly and longitudinally laminated. Acanthostyles in vesicle walls and roofs. *L.Miss.*, N.Am.——Fig. 176,*1a–f*. **C. ramosa*; *a,* small, thin-walled autozooecia in endozone, rounded proximal sides and angular distal sides, some with keel and sinus, vesicles and stereom in exozone, autozooecia and bulbous gonozooecial swellings in exozone, hemiphragms; transv. sec., holotype, USNM 92133, ×20; *b,* large gonozooecia, smaller autozooecia with hyaline lunaria isolated by stereom with abundant acanthostyles; tang. sec., holotype, ×20; *c,* monticular center (right) of stereom with acanthostyles, radial arrangement of hyaline lunaria, and two large gonozooecia (left); tang. sec., holotype, ×20; *d,* narrow autozooecia with diaphragms in endozone (left), vesicles decreasing in height outwardly in exozone, and autozooecia with hemiphragms, long. sec., topotype, USNM 159799, ×30; *e,* poorly laminated vesicle walls and roofs with acanthostyles; transv. sec., topotype, USNM 159800, ×100; *f,* laminated autozooecial walls with tubules (top, center), lunaria with short radius of curvature, and subrounded vesicles; tang. sec., topotype, USNM 159801, ×100.

Coelocaulis HALL & SIMPSON, 1887, p. xvi [**Callopora venusta* HALL, 1874, p. 101; OD; New Scotland F., L. Dev. (Helderberg.), Clarksville, N.Y., USA]. Zoarium ramose, hollow. Monticules absent. Autozooecia contiguous at basal layer, with narrow keel and sinus in outer endozone. Autozooecia oblique to zoarial surface, cavity ovate; walls poorly laminated; diaphragms few; lunaria absent. Vesicular tissue in outer endozone and exozone; blisterlike cysts; vesicle roofs locally thicker; laminated. Vesicles elongate in cross section, partially isolating autozooecia. [*Coelocaulis* is similar in most respects to *Diamesopora* HALL, 1852. Apparent lack of well-defined lunaria and tubules in the laminated walls may result from silicification of the type specimens of *Coelocaulis venusta*.] *L.Dev.*, USA.——Fig. 177,*2a–d*. **C. venusta* (HALL), holotype, NYSM 635; *a,* ovate autozooecia and protruding vesicle walls, ×10; *b,* poorly laminated basal layer (right), walls, and vesicles, planar diaphragm in one autozooecium; long. sec., ×30; *c,* autozooecia isolated by elongate vesicles and vesicle roofs in exozone; tang. sec., ×50; *d,* hollow ramose zoarium; autozooecia narrow at basal layer, keel and sinus in outer endozone, autozooecia larger and isolated by vesicular tissue in exozone, transv. sec., ×20.

Cyclotrypa ULRICH, 1896, p. 269 [**Fistulipora communis* ULRICH, 1890, p. 476; SD NICKLES & BASSLER, 1900, p. 219; Cedar Valley F., M. Dev., Buffalo, Iowa, USA]. Zoarium encrusting; multiple overgrowths common. Monticules low; central cluster of vesicles and ring of larger autozooecia. Basal layer with granular lower part and granular-prismatic upper part. Autozooecia recumbent for short distance; hemispherical in cross section and achieving full width at basal layer; isolated by vesicular tissue; walls with granular boundary zone, granular-prismatic cortex generally thicker outside boundary. Autozooecia erect in exozone, circular in cross section, with thin diaphragms. Lunarium absent. Vesicular tissue granular-prismatic, low blisterlike cysts decreasing in height outward, commonly in repeated cycles, capped by zone of thicker vesicle roofs. Acanthostyles in vesicle walls and some in vesicle roofs. *M.Dev.-U.Dev.,* ?*L.Miss.,* N.Am., Eu., USSR.——Fig. 177,*1a–d*. **C. communis* (ULRICH); *a,* basal layer with granular lower zone and granular-prismatic upper layer, autozooecium with hemispherical cross section in endozone, blisterlike vesicles with roofs thicker at surface, and acanthostyles light-colored; transv. sec., lectotype, USNM 159802, ×50; *b,* large vesicles in monticule, circular autozooecia isolated by medium-sized subangular to subrounded vesicles; tang. sec., lectotype, ×30; *c,* thick outer granular-prismatic layer of autozooecial walls, granular-prismatic vesicle walls; tang. sec., paralectotype, USNM 159803, ×100; *d,* long tubular autozooecia isolated by vesicular tissue, several cycles of outward decreasing of vesicle height and increasing thickness of vesicle roofs; long. sec., paralectotype, USNM 159804, ×30.

Cystiramus MOROZOVA, 1959a, p. 79 [**C. kondomensis*; OD; Vassinskie Beds, U. Dev. (*Frasn.*), Kondoma River, Kuznetsk basin, USSR]. Zoarium ramose, bifurcating. Monticules not reported but small areas of larger zooecia are present. Endozone with short zooecia, rounded proximal side and angular distal side, cyclically budded in hemispherical zones on a curved wall; wall granular. Autozooecia in exozone isolated by vesicular tissue; wall thickening rapidly in inner exozone; boundary granular and granular-prismatic cortex thick with some distally diverging tubuli; diaphragms in inner exozone. Lunaria in exozone, dark boundary zone continuing into lateral autozooecial walls; proximal, light-colored, granular-prismatic layer thickest, with some perpendicular tubuli. Vesicles in exozone; granular-prismatic; subquadrate blisters thickened to stereom through most of exozone; ?tubuli few. *U.Dev.,* USSR.——Fig. 178,*1a–d*. **C. kondomensis*; *a,* endozonal zooecia with curved proximal side and subangular distal side, autozooecia with diaphragms isolated by vesicular tissue in exozone; transv. sec., paratype, PIN

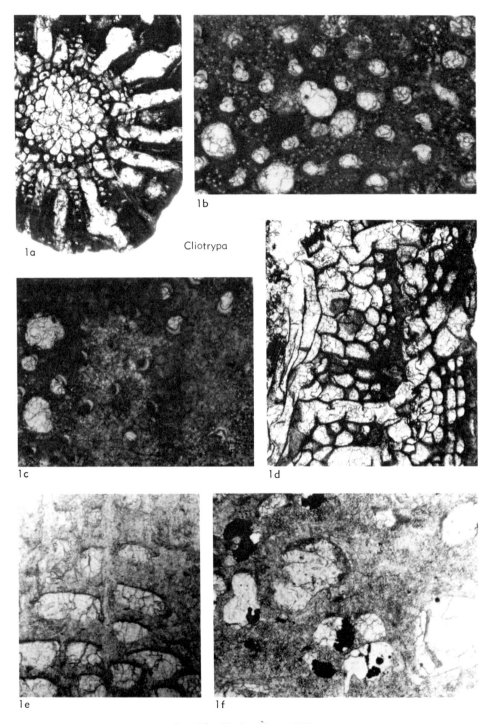

Cliotrypa

1a

1b

1c

1d

1e

1f

Fig. 176.   Fistuliporidae (p. 384).

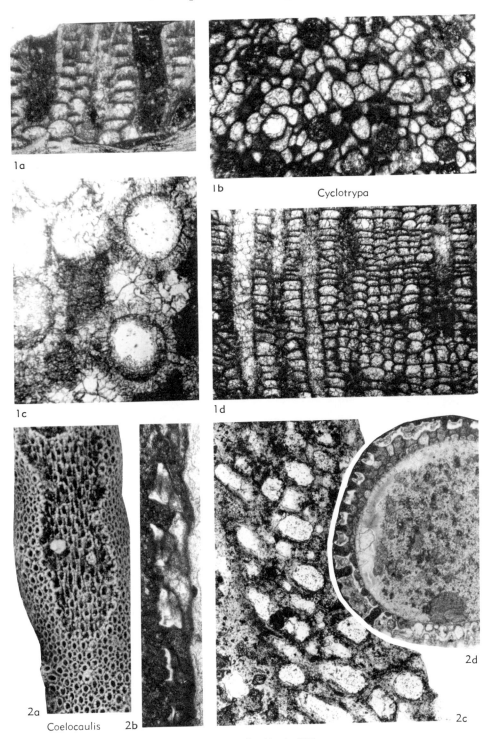

1a

1b   Cyclotrypa

1c

1d

2a

Coelocaulis   2b

2c

2d

Fig. 177.   Fistuliporidae (p. 385).

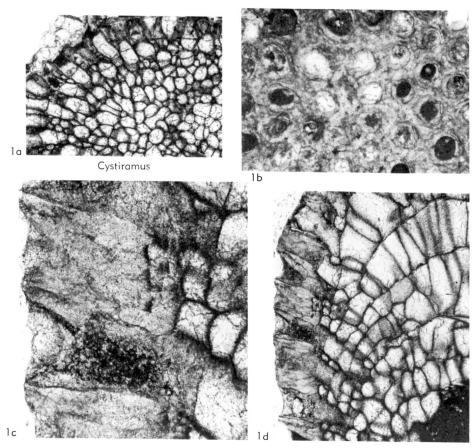

FIG. 178.   Fistuliporidae (p. 385).

198/12, ×20; *b,* dark boundaries in subcircular autozooecia with lunaria of same radius of curvature on right side, stereom (with ?tubuli) isolates autozooecia; tang. sec., paralectotype, PIN 918/316a, ×30; *c,* thick granular-prismatic walls in autozooecia isolated by boxlike vesicles and stereom in exozone; long. sec., paralectotype, PIN 918/317B, ×100; *d,* hemispherical cycles of thin-walled autozooecia in endozone; vesicular tissue and thick-walled autozooecia in exozone; long. sec., paralectotype, PIN 918/317B, ×30.

Diamesopora HALL, 1852, p. 158 [**D. dichotoma;* M; "Niagaran Ls.," M. Sil. (Niag.), Lockport, N.Y., USA] [=*Coeloclema* ULRICH, 1883, p. 258, *non* ULRICH, NICKLES, & BASSLER, 1900, p. 24, 211; *nec* NICKLES & BASSLER, ELIAS, 1954, p. 53]. Zoarium ramose, hollow. Monticules absent. Basal layer granular-prismatic with indistinct tubuli. Autozooecia contiguous at basal layer, hemispherical in cross section, recumbent; keel and sinuses well developed in endozone; autozooecia elongate diamond shaped, rhombically packed in deep tangential view; diameter nar-

rowing in exozone with distal spur in outer exozone; cavity subelliptical, isolated by stereom and in diagonally intersecting rows. Walls transversely laminated; boundary zone indistinct; tubuli outside boundary. Lunaria in exozone, small, most with small hyaline center; laminated distal lining continuous with lining of wall. Vesicular tissue in exozone; low, broad blisters, polygonal in cross section; laminated stereom with tubuli or small acanthostyles; stereom at lower level than autozooecial walls at surface. Diaphragms and subterminal diaphragms few. [*Trematopora osculum* HALL, 1876, is the type species of *Coeloclema* ULRICH, 1883, by subsequent designation (UTGAARD, 1968b, p. 1454). It does not differ significantly from *Diamesopora dichotoma* (HALL).] *M.Sil.,* N.Am.——FIG. 179,*1a–c. D. osculum* (HALL), Waldron Sh., M. Sil., Ind., USA, lectotype, AMNH 1916; *a,* basal layer (left) with thin, dark, lower zone and granular-prismatic upper layer with tubuli, vesicles containing laminated stereom and tubuli isolating autozooecia with subterminal diaphragms;

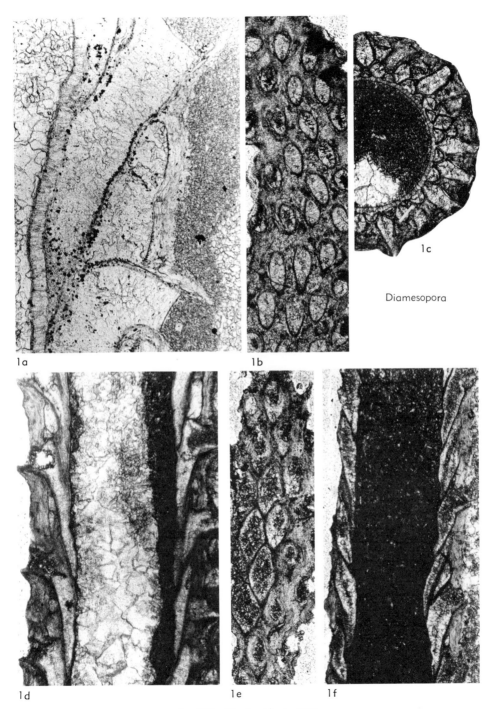

Diamesopora

FIG. 179.   Fistuliporidae (p. 388).

long. sec., ×100; *b,* laminated autozooecial walls, lunaria (lower side of autozooecia), vesicles in inner exozone (below) and stereom in exozone; tang. sec., ×30; *c,* hollow zoarium, granular-

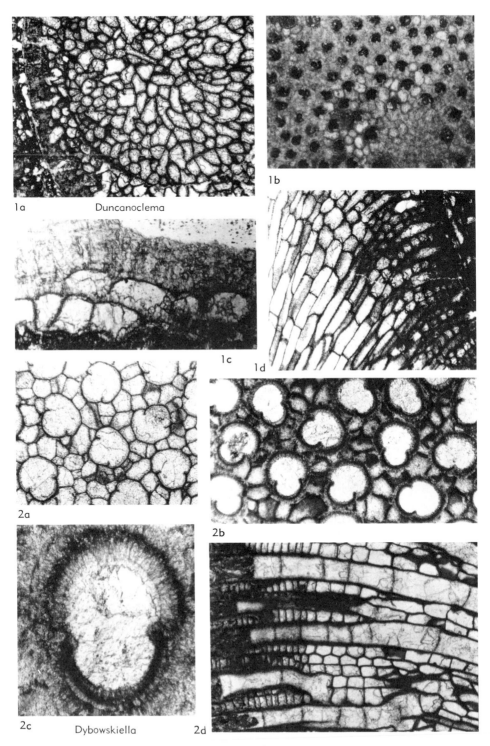

1a   Duncanoclema

1b

1c

1d

2a

2b

2c   Dybowskiella   2d

Fig. 180.   Fistuliporidae (p. 391).

prismatic basal layer, hemispherical autozooecia at basal layer, keel and sinuses in outer endozone, vesicles and laminated stereom in exozone; transv. sec., ×30.——Fig. 179, *1d–f. *D. dichotoma; d,* hollow zoarium, narrowing of autozooecia in exozone, vesicular tissue as low blisters in inner exozone; long. sec., topotype, USNM 159806, ×30; *e,* rhombic arrangement of autozooecia and smaller diameter in exozone than in endozone; tang. sec., holotype, AMNH 1760, ×30; *f,* hollow ramose zoarium, vesicular tissue becoming laminated stereom in exozone; long. sec., holotype, ×30.

Duncanoclema Bassler, 1952, p. 381 [*Fistuliporella marylandica* Ulrich & Bassler, 1913, p. 266; OD; Keyser Ls., U. Sil. (Cayug.), "L. Dev., Helderbergian," Cash Valley, Md., USA]. Zoarium ramose, with encrusting overgrowths. Monticules flush, central cluster of small vesicles; ring of larger zooecia with lunaria radially arranged. Autozooecia isolated at basal layer or in endozone of ramose zoaria by irregular, long, tubelike vesicles with curved roof. Autozooecia subcircular in cross section in endozone; wall thin, granular or granular-prismatic; thicker walled in exozone; diaphragms few. Lunaria in exozone; solid or discontinuous hyaline; some with cores; radius of curvature short and ends inflecting autozooecia. Vesicles becoming lower and more blisterlike in outer endozone and inner exozone; granular-prismatic; roofs thickening to stereom in outer exozone; small acanthostyles or tubuli in walls, roofs, and stereom; widely isolating autozooecia. *U.Sil., N.Am.*——Fig. 180, *1a–d. *D. marylandica* (Ulrich & Bassler); *a,* small subrounded autozooecia isolated by large irregular vesicles in endozone, encrusting overgrowth with hemispherical autozooecia at basal layer, stereom in outer exozone; transv. sec., paralectotype, USNM 159750, ×20; *b,* monticular center of smaller vesicles (lower right) and ring of larger zooecia with lunaria radially arranged, rhombic arrangement of isolated autozooecia; tang. sec., lectotype, USNM 159749, ×20; *c,* granular-prismatic vesicles and stereom with small acanthostyles or tubuli in outer exozone; long. sec., paralectotype, USNM 159751, ×100; *d,* long, tubelike vesicles in endozone and shorter, blisterlike vesicles with thicker roofs in exozone; long. sec., lectotype, ×20.

Dybowskiella Waagen & Wentzel, 1886, p. 916, *nom. subst. pro Dybowskia* Waagen & Pichl, 1885, p. 771, *non* Dall, 1877 [*Dybowskiella grandis*; OD; mid.-up. *Productus* Ls., Perm., Salt Ra., Pak.] [=*Triphyllotrypa* Moore & Dudley, 1944, p. 291]. Zoarium ramose, hollow ramose, encrusting, hemispherical, or massive. Monticules elevated or flush, central cluster of small polygonal vesicles surrounded by larger zooecia with lunaria partly to completely radial in arrangement. Autozooecia full width, hemi-

spherical in cross section at basal layer or where budded on vesicular tissue. Basal layer with dark, granular, primary layer and thick, light-colored, granular-prismatic layer. Autozooecia subcircular in cross section in endozone and exozone; isolated by many small polygonal vesicles; wall with granular boundary zone and light-colored granular-prismatic cortex. Diaphragms straight, curved, oblique, or incomplete. Lunaria in endozone and exozone, ends inflecting autozooecial cavity; dark granular zone not continuous with dark boundary zone in wall; thick distal and proximal granular-prismatic zones. Vesicular tissue in endozone and exozone; isolating autozooecia; vesicles subrectangular with straight superimposed or zigzag walls and flat to slightly curved roofs at same level in adjacent vesicles; small and polygonal in cross section. Roofs thickening into stereom at zoarial surface in some species. Small acanthostyles or tubuli in some vesicle roofs or stereom (a few appear to be central "pores"). Vesicles granular-prismatic. [*Triphyllotrypa speciosa* Moore & Dudley, 1944, p. 291, is the type species of *Triphyllotrypa* Moore & Dudley, 1944, by original designation. Like other species assigned to *Triphyllotrypa,* it differs from *Dybowskiella grandis* Waagen & Wentzel, mainly in growth habit: encrusting, hemispherical, or massive in *T. speciosa* and ramose or hollow ramose in *D. grandis.* In addition, species assigned to *Triphyllotrypa* may have a thick prolongation on the proximal side of lunaria in monticular zooids, and little or no development of stereom. These differences are judged to be too minor for generic separation, and *Triphyllotrypa* is considered to be a synonym of *Dybowskiella.*] *Perm.*, Asia, N.Am., Australia.——Fig. 180, *2a–d. *D. grandis,* "?Carb.," USNM 61314; *a,* thin-walled autozooecia with inflecting ends of lunaria, isolated by small polygonal vesicles in inner exozone; tang. sec., ×30; *b,* thick-walled autozooecia with lunaria, dark acanthostyles in some vesicle roofs (above center); tang. sec., ×30; *c,* dark granular central zone in lunarium (below) not continuous into dark granular autozooecial boundary, thick, light-colored, granular-prismatic layers in autozooecial wall and lunarium; tang. sec., ×100; *d,* diaphragms in autozooecia in endozone and exozone (left) (above), autozooecia budded on vesicular tissue (below) in endozone and exozone, polygonal to boxlike vesicles isolate autozooecia, vesicles decrease in height in outer exozone; long. sec., ×20.

Eofistulotrypa Morozova, 1959b, p. 9 [*E. manifesta*; OD; U. Dev. (*Frasn.*), Kuznetsk basin, USSR]. Zoarium ramose. Endozone with thin-walled polygonal autozooecia. Autozooecia with granular wall, sparse diaphragms in exozone; subrounded in cross section. Lunarium in exozone, large, thin, generally indistinct. Vesicular

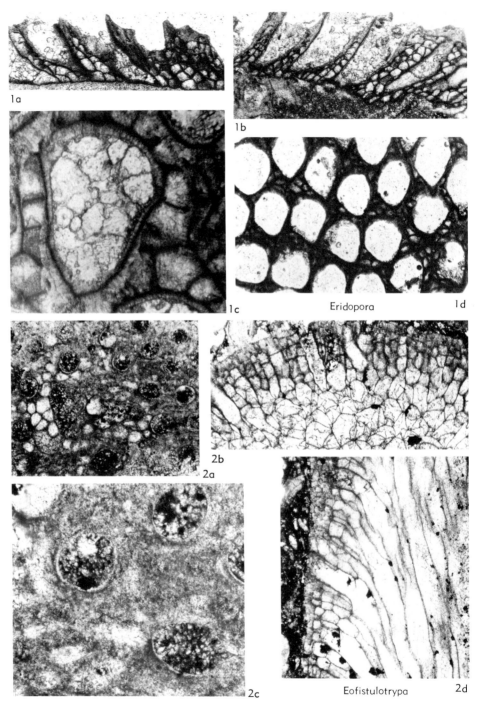

FIG. 181.   Fistuliporidae (p. 391–393).

tissue in exozone; small, subrounded in cross section; isolating autozooecia; boxlike, decreasing in height outward in exozone; thin stereom at zoarial surface. *U. Dev. (Frasn.),* USSR.——FIG.

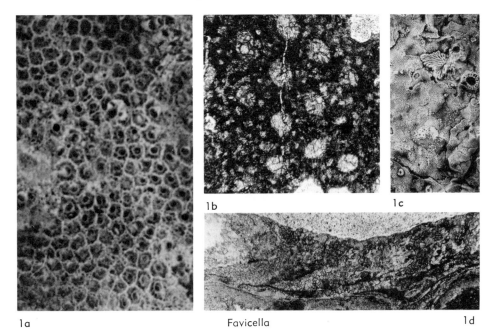

1a Favicella 1d

FIG. 182. Fistuliporidae (p. 393).

181,*2a–d*. *E. manifesta*, paratype, PIN 918/175; *a*, autozooecia with large lunaria (lower left) isolated by subrounded vesicles and stereom (top); tang. sec., ×30; *b*, polygonal autozooecia in endozone (below) and boxlike to polygonal vesicles isolating autozooecia in exozone (above); transv. sec., ×30; *c*, large lunaria (left) in autozooecia isolated by small vesicles and stereom; tang. sec., ×100; *d*, thin-walled autozooecia with no vesicular tissue in endozone (right), sparse diaphragms, angular vesicles becoming shorter in exozone and thin stereom at zoarial surface (left); long. sec., ×30.

Eridopora ULRICH, 1882, p. 137 [**E. macrostoma*; OD; Glen Dean F., U. Miss. (*Chester.*), Sloans Valley, Ky., USA] [=*Erydopora* NIKIFOROVA, 1927, p. 256, incorrect subsequent spelling]. Zoarium encrusting. Monticules small, flush; cluster of vesicles central. Basal layer granular-prismatic. Autozooecia in cross section hemispherical at basal layer, isolated by vesicular tissue; wall with dark, granular, boundary zone and granular-prismatic cortex; many distal and lateral walls formed of simple, superimposed vesicle walls; in exozone, autozooecia either oblique to zooarial surface and opening pyriform, or subperpendicular to surface and opening more circular; narrowly isolated by small vesicles. Lunarium in endozone and exozone, large; dark central zone continuing into dark zone in autozooecial wall; proximal granular-prismatic layer thick, may have irregular nodes. Vesicles in endozone and exozone; small, low blisters; stereom thin at surface, inner dark granular layer and outer granular-prismatic layer with tubuli. [*Eridopora* (*Discotrypella*) *stellata* ELIAS, 1957, p. 393, is the type species, by original designation, of the subgenus. The type of *E.* (*D.*) *stellata*, from the Eskeridge Shale (L. Perm.) at Roca, Nebraska, cannot be located and the status of this subgenus is doubtful.] *U.Miss.*, N.Am., USSR.——FIG. 181,*1a–d*. **E. macrostoma*; *a*, vesicular tissue isolating autozooecia from basal layer to zoarial surface; long. sec., lectotype, USNM 159738, ×30; *b*, recumbent to oblique autozooecia lacking diaphragms and isolated by small, blisterlike vesicles; long. sec., paralectotype, USNM 159737, ×30; *c*, simple autozooecial walls (above) and compound lunarial deposit (below) with thick, proximal, light-colored layer, simple to (a few) compound vesicle walls; tang. sec., lectotype, ×100; *d*, large autozooecia with large lunaria, in intersecting rows and narrowly isolated by small vesicles and stereom; tang. sec., paralectotype, USNM 159736, ×30.

Favicella HALL & SIMPSON, 1887, p. xviii [**Thallostigma inclusa* HALL, 1881, p. 188; OD; "Hamilton beds," M. Dev., York, N.Y., USA]. Zoarium encrusting. Monticules elevated; cluster of vesicles in center and encircling ring of slightly larger zooecia; lunaria obscure, radially arranged. Peristome elevated, ridges of vesicle walls in polygon shape surrounding each zooecium. Autozooecia with long recumbent portion at

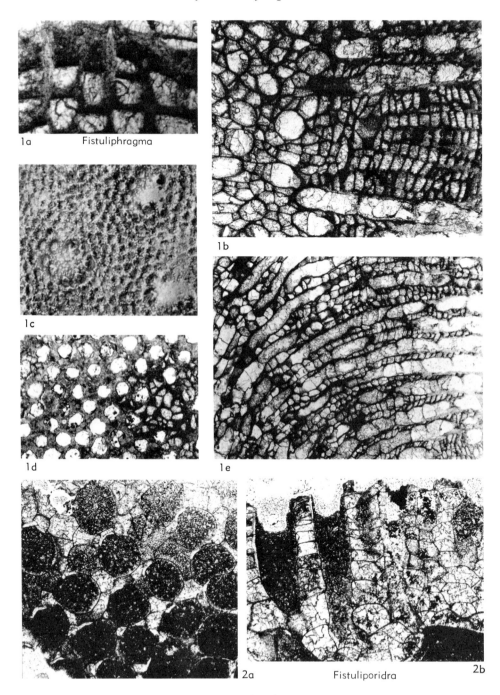

1a   Fistuliphragma

1b

1c

1d   1e

2a   Fistuliporidra   2b

Fɪɢ. 183. Fistuliporidae (p. 395).

basal layer; contiguous; hemispherical cross section with keel and sinuses. Wall indistinctly granular-prismatic. Diaphragms not seen. Lunaria obscure. Vesicles small, blisterlike, widely isolating autozooecia. [The holotype is silicified and many characters are obscure, but the polygonal ridges of vesicle walls that surround each zooecium are distinctive.] *M.Dev.,*

N.Am.——Fig. 182,*1a–d.* *F. inclusa* (Hall), holotype, NYSM 655; *a,* monticule (left, center), obscure lunaria and peristomal ridges and prominent polygonal ridge of vesicle walls surround each zooecium; ×10; *b,* subcircular autozooecia with obscure lunaria and granular-prismatic walls widely isolated by small vesicles; tang. sec., ×30; *c,* elevated monticules, ×1; *d,* recumbent autozooecia in endozone and vesicles in exozone; long. sec., ×30.

Fistuliphragma Bassler, 1934, p. 407 [*Fistulipora spinulifera* Rominger, 1866, p. 121; OD; Traverse Gr., M. Dev., Thunder Bay, Mich., USA]. Zoarium ramose, some encrusting overgrowths. Monticules elevated, cluster of vesicular tissue central, lunaria radially arranged. Autozooecia subrounded to subangular in cross section in endozone, isolated by vesicular tissue; subrounded in exozone. Diaphragms common. Hemiphragms few in endozone, closely spaced in exozone, alternating on proximal and distal sides; spinelike and curved to platelike. Walls indistinctly laminated. Lunaria small in endozone, large in exozone; light-colored dense deposit, some ends inflecting autozooecial cavity. Vesicles laminated, high blisters in endozone, subquadrate to low blisterlike to boxlike in exozone; acanthostyles in vesicle walls; walls superimposed; in some with thin stereom at zoarial surface. *M.Dev., N. Am.*——Fig. 183,*1a–e.* *F. spinulifera* (Rominger); *a,* large, light-colored acanthostyles in vesicle walls in monticule, walls superimposed, roofs flat, indistinctly laminated; long. sec., paralectotype, USNM 159740, ×100; *b,* ovate autozooecia isolated by vesicles in endozone (left), boxlike vesicles with superimposed walls and acanthostyles in monticule (right, center); transv. sec., paralectotype, USNM 159740, ×30; *c,* elevated monticules with stereom at center, lunaria radially arranged, spinelike acanthostyles; lectotype, USNM 159742, ×10; *d,* monticule (right), autozooecia narrowly isolated by stereom; tang. sec., lectotype, ×20; *e,* vesicular tissue decreasing in height from endozone to exozone (right), spinelike hemiphragms and diaphragms in autozooecia, light-colored lunarial deposits on proximal side; long. sec., lectotype, ×20.

Fistuliporella Simpson, 1897, p. 560 [*Lichenalia constricta* Hall, 1883b, p. 183; OD; Hamilton Gr., M. Dev., Leroy, N.Y., USA]. Zoarium encrusting. Monticules elevated, cluster of vesicles or stereom central; ring of larger zooecia with lunaria radially arranged. Autozooecia isolated by vesicular tissue at granular-prismatic basal layer. Autozooecia subcircular in cross section; walls thin, granular or granular-prismatic; distal and lateral parts commonly made of superimposed vesicle walls; diaphragms straight to curved; mural spines in some species. Lunarium in endozone and exozone, of dense hyaline cal-

cite. Vesicles high blisters in endozone, becoming low blisters in exozone; walls and roofs thin, granular or granular-prismatic, small acanthostyles in vesicle walls, zones of thicker vesicle roofs local; vesicles small, subangular to subrounded in cross section. *Sil.-Dev.,* N.Am., Eu. ——Fig. 184,*1a–f.* *F. constricta* (Hall), holotype, NYSM 736; *a,* solid, elevated monticular center with lunaria radially arranged; ×10; *b,* subcircular autozooecia isolated by small subangular vesicles; tang. sec., ×30; *c,* hyaline lunarial deposit (lower right side of autozooecia) and granular vesicular tissue; tang. sec., ×50; *d,* granular-prismatic basal layer, autozooecia isolated by vesicular tissue; long. sec., ×30; *e,* autozooecia bounded by superimposed vesicle walls, hyaline lunarial deposit (left, center), vesicles with thin granular walls and roofs, small acanthostyle in vesicle walls (center); transv. sec., ×50; *f,* autozooecia isolated at basal layer by vesicular tissue, vesicles decrease in height outward in exozone; transv. sec., ×30.

Fistuliporidra Simpson, 1897, p. 606 [*Lichenalia tessellata* Hall & Simpson, 1887, p. 207; OD; Hamilton Gr., M. Dev., Genesee Valley, N.Y., USA]. Zoarium encrusting, monticules elevated; cluster of vesicular tissue central, lunaria radially arranged. Autozooecia partially isolated at basal layer by vesicular tissue. Autozooecia subcircular in cross section in exozone, narrowly isolated; walls thin, indistinctly laminated or granular. Superimposed vesicle walls making up lateral and distal zooecial walls. Diaphragms not seen. Lunaria in endozone and exozone, small, of dense hyaline calcite, short and barlike with little curvature. Vesicular tissue blisterlike or boxlike; vesicles decreasing in height outward in exozone; walls and roofs thin, granular or indistinctly laminated. *M.Dev., N.Am.*——Fig. 183,*2a,b.* *F. tessellata* (Hall & Simpson), holotype, NYSM 5060/1; *a,* subcircular autozooecia with thin walls narrowly isolated by moderately large, thin-walled vesicles, light-colored lunarial deposit (upper left side of autozooecia); tang. sec., ×30; *b,* thin granular-prismatic basal layer, autozooecia isolated by vesicular tissue; oblique long. sec., ×30.

Fistuliramus Astrova, 1960b, p. 362 [*F. sinensis*; OD; U. Sil., Arctic Urals, USSR]. Zoarium ramose. Monticules with ring of larger zooecia; lunaria in part radially arranged. Autozooecia thin-walled in endozone, isolated by vesicular tissue; subcircular in cross section in exozone, narrowly isolated by vesicles; walls laminated; thick laminated lining in exozone, some mural tubuli in lining. Diaphragms closely spaced in outer endozone and exozone, thin, flat to concave. Lunaria in outer endozone and exozone, of light-colored and dense calcite. Vesicular tissue long blisters in endozone, becoming more subquadrate at zooecial-bend region, decreasing

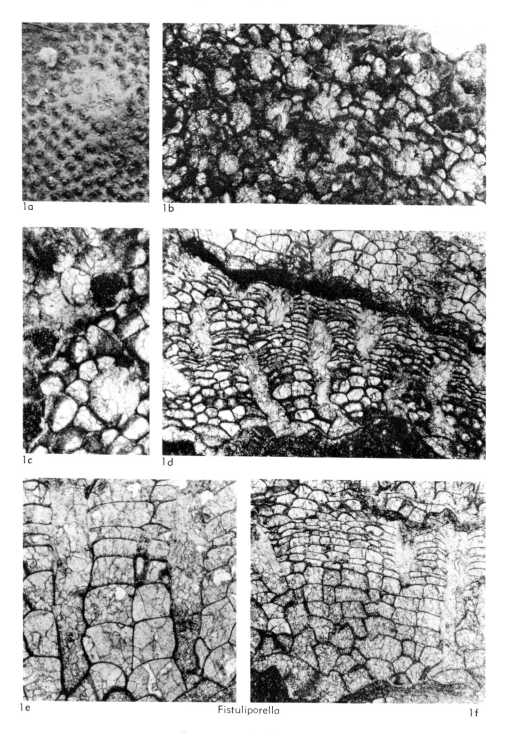

FIG. 184.　Fistuliporidae (p. 395).

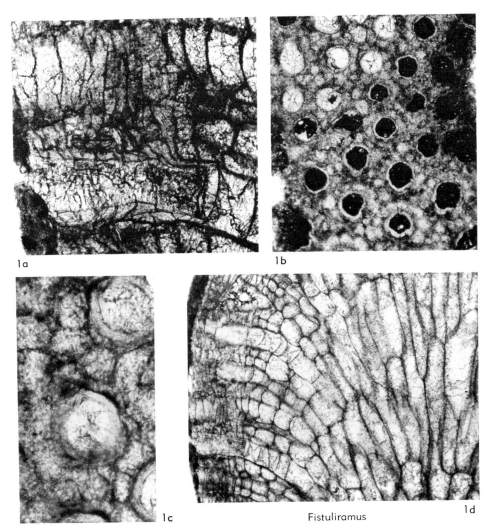

la

lb

lc

Fistuliramus

ld

FIG. 185. Fistuliporidae (p. 395).

in height outward in exozone; walls and roofs laminated, containing small acanthostyles or tubuli. *U.Sil.-L.Dev.,* USSR.——FIG. 185,*1a–d.* **F. sinensis,* paratype, PIN 124-7/32; *a,* autozooecium with closely spaced diaphragms and laminated lining with tubules in living chamber (below, left), low, blisterlike vesicles with zone of thicker roofs (above, center); long. sec., ×100; *b,* autozooecia with light-colored lining and hyaline lunaria isolated by small subcircular vesicles; tang. sec., ×30; *c,* laminated autozooecial lining, light-colored lunaria, and small vesicles; tang. sec., ×100; *d,* thin-walled autozooecia and long, blisterlike vesicles in endozone (right), closely spaced diaphragms in inner exozone and decrease in vesicle height in exozone; long. sec., ×30.

Fistulocladia BASSLER, 1929, p. 49 [**F. typicalis*;

OD; Perm., Noil Boewan, Timor]. Zoarium slender ramose; perpendicular branches form by encrusting main branch. Monticules flush, inconspicuous; lunaria not radially arranged. Central endozone a cylinder of narrow, round tubelike vesicles with flat roofs and cyclic zones of stereom. Autozooecia circular in cross section and narrowly isolated by stereom at origin on central cylinder; walls thick, granular-prismatic; basal diaphragms lacking, terminal diaphragms common; ovate in cross section in endozone and widely isolated by vesicles or stereom. Lunaria in endozone and exozone; dark distal layer continuing into dark zone in wall, proximal granular-prismatic layer thick, with irregular proximal projections in some. Vesicular tissue in small blisters; lower layer dark, granular, gran-

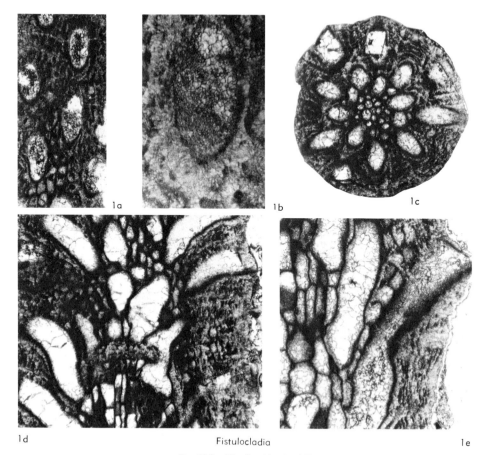

1a   1b   1c

1d   Fistulocladia   1e

FIG. 186.   Fistuliporidae (p. 397).

ular-prismatic layer light colored and with small tubuli; vesicle height very low in exozone, producing nearly solid stereom. *Perm.,* Timor.——FIG. 186, *1a–e.* *F. typicalis; a,* ovate autozooecia with lunaria (below) isolated by small vesicles and stereom; tang. sec., topotype, USNM 159745, ×27; *b,* granular-prismatic autozooecial wall, lunarium with thick granular-prismatic layer (below) and granular-prismatic stereom; tang. sec., topotype, USNM 159748, ×90; *c,* axial cylinder (center), autozooecia subcircular, narrowly isolated by vesicular tissue and with large lunaria in endozone, stereom in exozone; transv. sec., topotype, USNM 159744, ×27; *d,* axial cylinder of tubelike vesicles (below) and zone of stereom in axis, autozooecia budded from axial cylinder, low vesicle height in nearly solid stereom in exozone, thick terminal diaphragm (top, right); long. sec., topotype, USNM 159743, ×27; *e,* axial vesicles (top left) and autozooecia budded from axial cylinder (right and bottom), blisterlike vesicles and stereom, terminal diaphragm in autozooecium; long. sec.,

topotype, USNM 159746, ×45.

Fistulotrypa BASSLER, 1929, p. 48 [**F. ramosa;* OD; Perm., Basleo, Timor]. Zoarium ramose; monticules flush, center of stereom; peristomes elevated. Autozooecia in endozone contiguous; walls thin, granular; diaphragms few. Autozooecia in exozone isolated by vesicles; oval in cross section; diaphragms closely spaced. Wall with dark, granular, boundary zone and thick, granular-prismatic cortex. Lunaria in exozone; inconspicuous; dark, granular zone commonly not continuous into dark, zooecial boundary; distal granular-prismatic zone thick. Vesicles in outer endozone and exozone; low, broad blisters, large and subangular in tangential view, decreasing in height outward; stereom at surface; granular-prismatic structure with tubuli in vesicle roofs and stereom. *Perm.,* Timor.——FIG. 187, *2a–c.* **F. ramosa,* type, USNM 159822; *a,* dark granular boundary, granular-prismatic cortex and lunarium (lower side) in autozooecium; tang. sec., ×100; *b,* dark granular zooecial boundaries and subcircular vesicles; tang. sec., ×30; *c,* large,

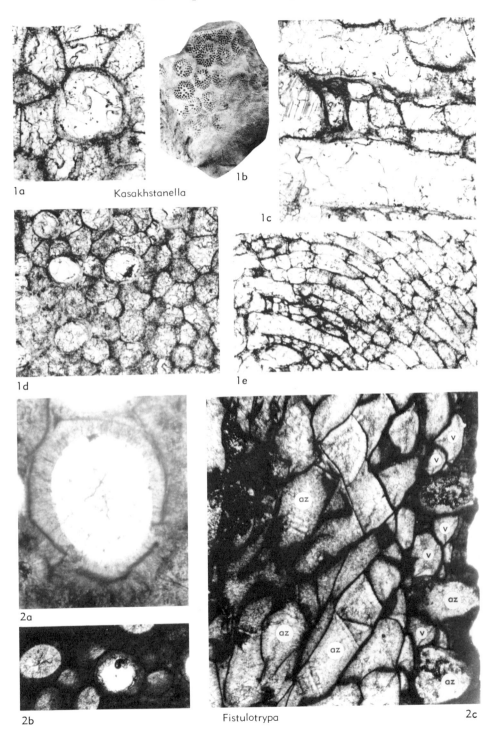

1a
Kasakhstanella
1b
1c
1d
1e
2a
2b
Fistulotrypa
2c

FIG. 187.   Fistuliporidae (p. 398–400).

thin-walled autozooecia (az) in endozone (left), thicker-walled autozooecia (az) and vesicular tissue (v) in exozone (right); oblique long. sec., ×30.

Kasakhstanella NEKHOROSHEV, 1956a, p. 42 [*K. ramosa*; OD; Dev., D2², C. Kazakh., USSR]. Zoarium ramose, encrusting overgrowths. Monticules elevated; cluster of vesicular tissue in center and ring of larger zooecia with lunaria radially arranged. Large subhexagonal ridges in vesicular tissue surrounding each monticule. Autozooecia in endozone isolated by vesicular tissue; walls thin, laminated; diaphragms uniformly spaced. Autozooecia in exozone subcircular in cross section and isolated by vesicles; walls thin, laminated. Lunaria in endozone and exozone; thin, hyaline. Vesicular tissue in endozone and exozone; blisters high, decreasing in height in exozone; local zones of thicker vesicle roofs; laminated. *M.Dev.-U.Dev.,* USSR.——FIG. 187,*1a–e*. *K. ramosa*, paratype, USNM 158360; *a,* indistinctly laminated vesicle walls and autozooecium, hyaline lunarium (lower left); tang. sec., ×100; *b,* monticules surrounded by ridges forming a large subhexagon; ×2; *c,* laminated autozooecial walls, hyaline lunarial deposit, laminated vesicle walls and roofs; long. sec., ×100; *d,* monticular center (right), autozooecia with hyaline lunaria (near monticule), isolated by large vesicles; tang. sec., ×50; *e,* high, narrow vesicles in endozone (right), lower vesicles in exozone (left); long. sec., ×30.

Lichenotrypa ULRICH, 1886b, p. 23 [*L. cavernosa*; M; "U. Helderberg?," M. Dev., Falls of the Ohio, Jeffersonville, Ind., USA]. Zoarium encrusting; monticules flush, central cluster of vesicular tissue ringed by larger zooecia with lunaria radially arranged; peristome elevated; ridges in vesicular tissue in irregular curved to polygonal shapes surrounding most autozooecia. Autozooecia hemispherical in cross section, isolated by vesicles at granular-prismatic basal layer. Subcircular autozooecia with lunate to spinelike hyaline lunaria isolated by small subangular vesicles in exozone. Autozooecial boundary dark granular zone; cortex granular-prismatic. Diaphragms few. Vesicles in endozone and exozone; boxlike to low blisters, decreasing in height in exozone; thin stereom at surface; granular-prismatic. Large hyaline acanthostyles in vesicle walls or autozooecial walls; some sublunate and in position of lunarium. *M.Dev.,* N.Am.——FIG. 188,*1a,c,d*. *L. cavernosa*; *a,* monticular center (right), lunaria radially arranged, peristomes and irregular ridges on vesicular tissue; topotype, USNM 159812, ×10; *c,* subcircular autozooecia and large acanthostyles; tang. sec., topotype, USNM 159813, ×30; *d,* low blisterlike vesicles and large acanthostyles; long. sec., topotype, USNM 159813, ×30.——FIG. 188,*1b,e,f*. *L.* sp. cf. *L. cavernosa*, Jef-

fersonville Ls., M.Dev., Ind., USA, SIUC 3007; *b,* large acanthostylelike lunaria; tang. sec., ×100; *e,* granular-prismatic vesicular tissue, tubuli in vesicle roofs; long. sec., ×100; *f,* autozooecia with lunaria and acanthostyles replacing lunaria isolated by small subangular vesicles; tang. sec., ×30.

Metelipora TRIZNA, 1950, p. 99 [*M. monstrata*; OD; L. Perm., Ural Mts., USSR]. Zoarium encrusting, discoidal. Monticules with central cluster of small vesicles surrounded by slightly larger zooecia with lunaria on side nearest monticular center. Autozooecia large, subcircular in cross section; lunaria with shorter radius of curvature. Vesicles in endozone and exozone; walls thick, roofs distantly spaced; superimposed, producing tubelike vesicles; small and subcircular in cross section. One to three vesicles isolating autozooecia in exozone. [Specimens of *Metelipora* were not available for study. V. P. NEKHOROSHEV informed me that existing thin sections of *Metelipora monstrata* TRIZNA, 1950 are thick and microstructure of the skeleton is unknown.] *?U.Carb., L.Perm.,* USSR.——FIG. 188,*2a,b*. *M. monstrata,* holotype, VNIGRI 2/135; *a,* autozooecia, indistinct wall structure, small tubelike vesicles; long. sec., ×40; *b,* rhombic arrangement of subcircular autozooecia isolated by vesicles, monticule in lower left, indistinct lunaria; tang. sec., ×40 (photographs courtesy of L. Nekhorosheva).

Odontotrypa HALL 1886, pl. 30 [*Lichenalia alveata* HALL, 1883b, p. 152; SD HALL & SIMPSON, 1887, p. xvii; M. Dev., Falls of the Ohio, Jeffersonville, Ind., USA]. Zoarium encrusting; monticules low, central cluster of solid vesicular tissue surrounded by ring of slightly larger zooecia with lunaria radially arranged. Autozooecia hemispherical in cross section at basal layer, isolated by vesicular tissue. Autozooecia subcircular in cross section in exozone. Lunarium in outer endozone and exozone. Vesicular tissue low blisters in endozone and exozone. [Holotype and available topotypes of *Odontotrypa alveata* (HALL) are silicified. This poorly known genus resembles *Buskopora* ULRICH, 1886 in many respects but lacks well-developed peristomes and deep inflection of the ends of the lunaria. With study of better material, it may prove to be a synonym of *Buskopora*.] *M.Dev.,* USA.——FIG. 189,*1a–d*. *O. alveata* (HALL); *a,* thin colony encrusting fenestrate bryozoan; holotype, WM 13991, ×1; *b,* monticules with lunaria radially arranged, autozooecia narrowly isolated; holotype, ×10; *c,* hemispherical autozooecia on basal layer, blisterlike vesicles; transv. sec., topotype, USNM 67689, ×30; *d,* subcircular autozooecia with lunaria narrowly isolated by vesicles; tang. sec., topotype, USNM 67689, ×30.

?Pholidopora GRUBBS, 1939, p. 552 [*P. concen-*

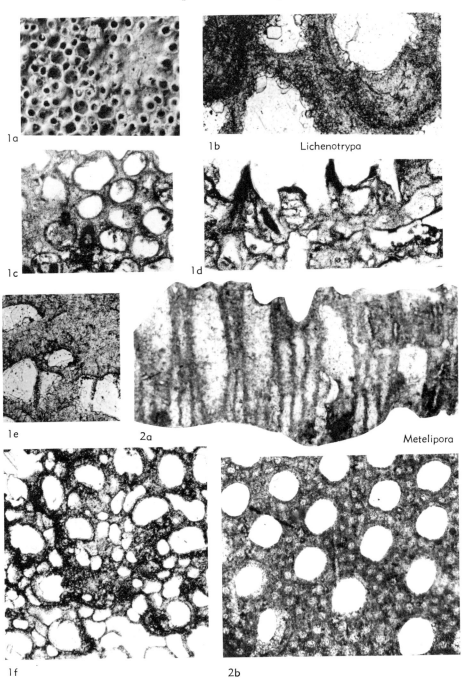

FIG. 188. Fistuliporidae (p. 400).

trica; OD; "Niagaran ls. nodules," M. Sil., Chicago, Ill., USA]. Zoarium encrusting; monticules depressed, cluster of vesicular tissue central, lunaria radially arranged. Autozooecia elongate with low peristomes and small, highly elevated lunaria; diaphragms closely spaced, concave. Vesicles isolating autozooecia. [The type specimens of *Pholidopora concentrica* are silicified and

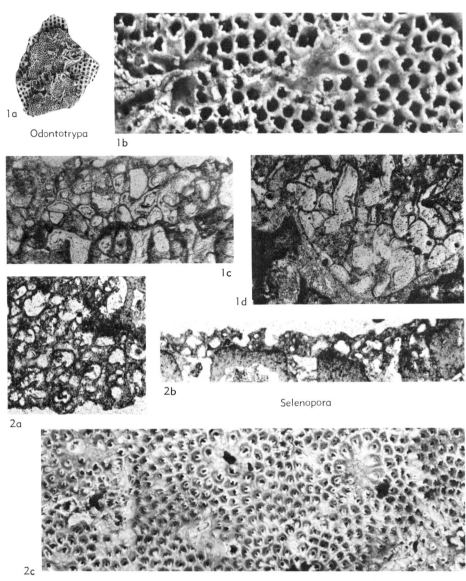

1a

Odontotrypa

1b

1c

1d

2b

Selenopora

2a

2c

FIG. 189.   Fistuliporidae (p. 400–404).

small. Virtually nothing is known of the internal anatomy of this genus, which is questionably placed in the Fistuliporidae.] *M.Sil.,* USA.——FIG. 190,*2*. **P. concentrica,* holotype, WM 46033; monticular center with lunaria radially arranged, elongate autozooecia, with highly elevated lunaria, isolated by vesicular tissue; ×10.

Pileotrypa HALL, 1886, pl. 30 [**Lichenalia denticulata* HALL, 1883a, pl. 24; SD HALL & SIMPSON, 1887, p. xvi; M. Dev., Falls of the Ohio, Jeffersonville, Ind., USA]. Zoarium encrusting; monticules elevated; lunaria highly elevated, radially

arranged. Autozooecia isolated at basal layer by vesicular tissue, elongate; hemiphragms in outer endozone and exozone; lunarium large with markedly shorter radius of curvature. Vesicular tissue large, irregular blisterlike; no stereom. [Cotypes and available topotypes of *Pileotrypa denticulata* (HALL) are silicified and nothing is known of the microstructure.] *M.Dev.,* N.Am.——FIG. 190,*1a–d.* **P. denticulata* (HALL); *a,* autozooecia widely isolated by vesicular tissue at basal layer; transv. sec., topotype, USNM 55071, ×30; *b,* monticule (lower right) with

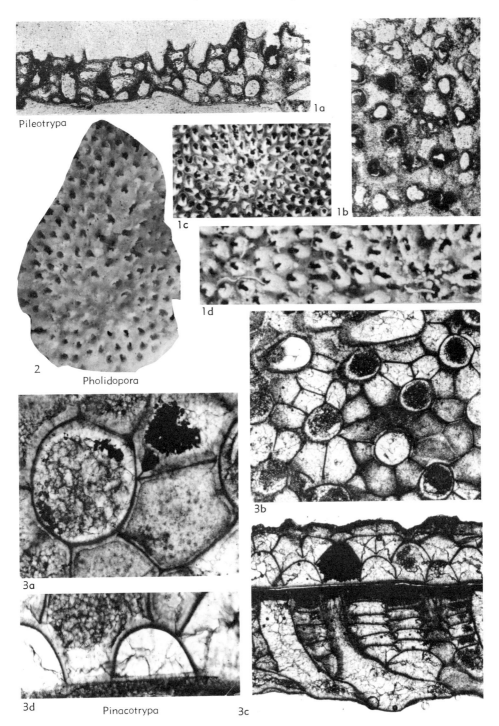

Fig. 190. Fistuliporidae (p. 400–404).

lunaria radially arranged, elongate autozooecia with large lunaria isolated by large subangular vesicles; tang. sec., topotype, USNM 55071, ×30; *c,* monticule with lunaria radially arranged; cotype, WM 13993, ×10; *d,* highly elevated, hood-shaped lunaria; cotype, WM 13993, ×20.

Pinacotrypa ULRICH in MILLER, 1889, p. 315 [**Fistulipora elegans* ROMINGER, 1866, p. 122; OD; Hamilton Gr., M. Dev., Hamburg, N.Y., USA]. [=*Fistulicella* SIMPSON, 1897, p. 606; *Fistuliporina* SIMPSON, 1897, p. 555]. Zoarium encrusting. Basal layer thin, wrinkled; lower primary layer granular, upper layer granular-prismatic. Monticules flush, center with large vesicles. Autozooecia with peristome, circular, isolated by six to eight large polygonal vesicles in petal-like array around zooecia. Autozooecia with long recumbent portion, keel and sinus, erect in exozone; diaphragms sparse, thin. Autozooecial walls having granular boundary zone and granular-prismatic cortex; cortex in some with light-colored, pustulelike areas that form minute nodes on peristome. Vesicular tissue partly isolates zooecia at basal layer; vesicles becoming shorter and more regular outward in exozone, boxlike with superimposed walls, some walls compound, protruding as ridges above vesicle roofs at zoarial surface. Vesicle roofs flat, simple, at same level in adjacent vesicles. Small tubuli (?acanthostyles) in vesicle roofs (some appear porelike in tangential section). Lunaria poorly developed, slightly thicker than remainder of wall, or absent. [*Thallostigma plana* HALL, 1881, p. 187, the type species of *Fistulicella* SIMPSON, by original designation, does not differ significantly from *Pinacotrypa elegans* (ROMINGER). Therefore, *Fistulicella* SIMPSON is here considered to be a junior subjective synonym of *Pinacotrypa* ULRICH. *P. plana* has a weakly developed lunarium, or no lunarium, and minute pustules in the granular-prismatic cortex of autozooecial walls that can be expressed as protuberances on the peristome. *Thallostigma serrulata* HALL, 1883b, p. 185, the type species of *Fistuliporina* SIMPSON by original designation, does not differ significantly from *Pinacotrypa plana* or *P. elegans.* Thus, *Fistuliporina* SIMPSON, 1897, is here considered to be a junior subjective synonym of *Pinacotrypa* ULRICH. *Pinacotrypa serrulata* (HALL) has several zooecia with abundant, pustulelike, light-colored areas and a poorly developed lunarium or no discernable lunarium. In addition, *P. serrulata* may have operculumlike covers at the zooecial orifice.] *M.Dev.,* N.Am.——FIG. 190,*3a–d.* **P. elegans* (ROMINGER); *a,* dark granular zooecial boundary in subcircular autozooecia, granular-prismatic layer, compound construction of some vesicle walls, and vesicle roofs with minute tubuli; tang. sec., paralectotype, UMMP 6667-1, ×100; *b,* petal-like array of vesicles around subcircular autozooecia, poorly

developed lunarium (thicker portion of autozooecial wall) on proximal (left) side of autozooecia; tang. sec., lectotype, UMMP 6667-3, ×30; *c,* basal layer and contiguous autozooecia in endozone, autozooecia isolated by low, boxlike vesicles in exozone (of lower encrusting sheet), hemispherical autozooecial cross section at basal layer and keel and sinus in outer endozone (in upper encrustation); long. sec. (below) and transv. sec. (above), lectotype, ×30; *d,* basal layer with dark, granular, primary layer and lighter colored, secondary, granular-prismatic layer; hemispherical and keel-and-sinus cross-sectional shapes in endozone, note discontinuity in granular layers in autozooecial walls (upper left); transv. sec., lectotype, ×100.

Selenopora HALL, 1886, pl. 25 [**Lichenalia circincta* HALL, 1883b, p. 153; M; M. Dev., Falls of the Ohio, Jeffersonville, Ind., USA]. Zoarium encrusting; monticules flush, lunaria radially arranged. Lunaria highly elevated; subcircular to polygonal pattern of ridges on vesicular tissue surround each autozooecium. Autozooecia isolated by small vesicles in endozone and exozone. Small lunaria indent autozooecial cavity. [Cotypes and available topotypes of *Selenopora circincta* (HALL) are silicified and little is known about internal anatomy of this genus. Presence of subcircular ridges in the vesicular tissue that surrounds each autozooecium distinguishes *Selenopora* from *Buskopora* ULRICH. The small lunarium with inflecting ends separates it from *Favicella* HALL & SIMPSON.] *M.Dev.,* USA.——FIG. 189,*2a–c.* **S. circincta* (HALL); *a,* elongate autozooecia, with indenting lunaria, isolated by vesicles; tang. sec., topotype, USNM 2935, ×30; *b,* autozooecia isolated by small vesicles at basal layer; transv. sec., topotype, USNM 2935, ×30; *c,* monticules with lunaria radially arranged, subcircular to polygonal ridges on vesicles surround autozooecia; cotype, WM 13975, ×10.

Strotopora ULRICH in MILLER, 1889, p. 326 [**S. foveolata* ULRICH, 1890, p. 487; SD ULRICH, 1890, p. 383; Keokuk Gr., L. Miss., Bentonsport, Iowa, Warsaw, Ill., USA]. Zoarium ramose to subfrondose, endozone commonly crushed; anastomosing branches; encrusting overgrowths. Monticules elevated, lunaria radially arranged. Autozooecia subcircular in cross section, isolated by vesicular tissue; wall laminated, boundary dark to obscure, tubuli in cortex perpendicular to boundary. Diaphragms few. Hemiphragms in exozone; thick and spinelike to platelike. Lunaria in outer endozone and exozone; hyaline with irregular proximal projections. Gonozooecia in exozone; funnel- to blister-shaped expansions of autozooecia; subrounded to polygonal in cross section; roofs with abundant tubuli. Vesicular tissue in endozone and exozone; vesicles blister-like and polygonal to subrounded in cross sec-

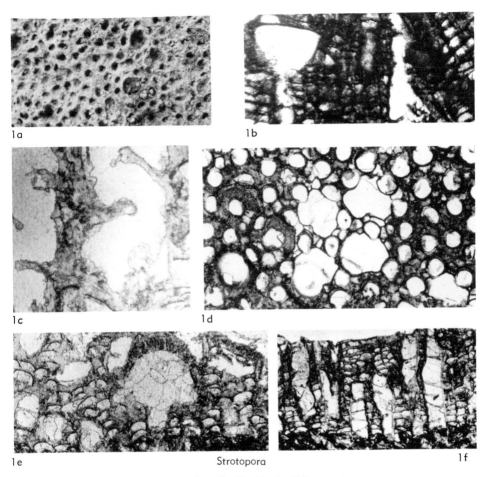

1a
1b
1c
1d
1e
Strotopora
1f

Fig. 191. Fistuliporidae (p. 404).

tion; decreasing in height in exozone; zones of thick vesicle roofs or stereom. Acanthostyles in laminated vesicle walls and roofs; some appearing porelike in tangential section. [*Strotopora* differs from the closely related genus *Cliotrypa* ULRICH & BASSLER by having gonozooecia that are more polygonal in cross section and vesicular tissue throughout the endozone. Crushed endozones in some zoaria may indicate that they were originally hollow ramose.] *L.Miss.*, N.Am.——
Fig. 191,*1a–f.* **S. foveolata*; *a,* monticule (lower right), autozooecia isolated by vesicular tissue, and large, open (broken or eroded) gonozooecia; topotype, USNM 55060, ×10; *b,* funnel-shaped gonozooecium and vesicular tissue with acanthostyles; long. sec., topotype, USNM 159815, ×30; *c,* hemiphragms in autozooecia; transv. sec., "topotype" (Warsaw, Ill.), USNM 159795, ×100; *d,* gonozooecia, one with shelf-like partition (upper left), hyaline lunaria in autozooecia, stereom with acanthostyles; tang.

sec., topotype, USNM 159814, ×20; *e,* blister-like gonozooecium having thick roof with acanthostyles, blisterlike vesicular tissue; transv. sec., "topotype," USNM 159795, ×30; *f,* autozooecia with hemiphragms and diaphragms, blister-like vesicles with zones of thicker vesicle roofs; long. sec., topotype, USNM 159814, ×20.

Family RHINOPORIDAE
Miller, 1889

[Rhinoporidae MILLER, 1889, p. 290]

Zoaria thin; encrusting or bifoliate fronds. Monticules lacking. Autozooecia elongate in cross section, with hyaline lunaria; isolated at surface by vesicular tissue. Tunnels covered by rounded roof on vesicular tissue; standing as elevated, anastomosing ridges on zoarial surface. Walls laminated or granular-

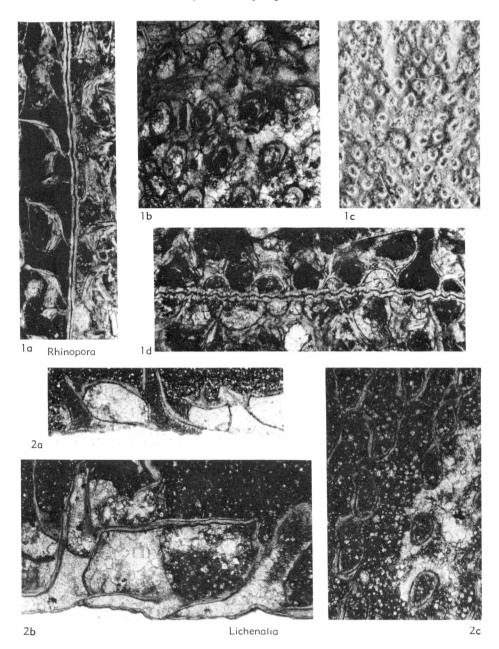

FIG. 192. Rhinoporidae (p. 406–407).

prismatic. Autozooecia narrowing from endozone to exozone, with semirecumbent distal spur in endozone. Vesicular tissue in outer endozone and exozone; large blisters; generally only one vesicle high. *L.Sil.-M.Sil.*

Characters of particular importance are: few, large vesicles; hyaline lunarium; tunnels; and lack of monticules and acanthostyles.

Rhinopora HALL in SILLIMAN, SILLIMAN, & DANA, 1851, p. 399 [*R. verrucosa* HALL, 1852, p. 48; SD; Clinton Gr., Sil., Hill Mill, N.Y., USA].

Zoarium bifoliate fronds. Lunaria highly elevated. Mesotheca thick, crenulated; median layer dark and poorly laminated to granular-prismatic layers light colored. Autozooecia short, almost completely contiguous at mesotheca; hemispherical in cross section; keel and sinuses in outer endozone. Diaphragms lacking. Wall indistinctly laminated. Large hyaline lunarium in exozone. Large blisterlike vesicles isolating autozooecia in exozone; laminated with tubuli in vesicle roofs. *L.Sil.,* E.N.Am.——FIG. 192,*1a–d.* *R. verrucosa; a,* trilayered mesotheca, distal prong on autozooecia in endozone, hyaline lunaria and laminated walls and large blisterlike vesicles; long. sec. of specimen from Ohio, USNM 79326, ×30; *b,* elongate autozooecia with laminated walls and hyaline lunaria isolated by vesicles, branched tunnel in exozone; tang. sec., USNM 79326, ×30; *c,* elevated lunaria and elevated anastomosing covered tunnels; lectotype, AMNH 1492/2-1, ×10; *d,* thick, crenulated mesotheca, hemispherical autozooecia in endozone; transv. sec., USNM 79326, ×30.

Lichenalia HALL in SILLIMAN, SILLIMAN, & DANA, 1851, p. 401 [*L. concentrica* HALL, 1852, p. 171; SM; Rochester Sh., M. Sil., Lockport, N.Y., USA]. Zoarium encrusting. Lunaria highly elevated. Autozooecia with long recumbent portion on laminated basal layer; walls thin, laminated. Diaphragms few. Lunaria hyaline; variable in size and shape. Large blister- to boxlike vesicles in outer endozone and exozone; partly isolating autozooecia; thick, laminated roofs with indistinct tubuli. *M.Sil.,* E.N.Am., Eu.——FIG. 192,*2a–c.* *L. concentrica; a,* autozooecia elevated above large blisterlike vesicles; long. sec., topotype, USNM 159817, ×20; *b,* thin laminated basal layer (below), recumbent autozooecia with distal spur in endozone, thin diaphragm, large laminated vesicles; long. sec., topotype, USNM 159816, ×30; *c,* elongate autozooecia isolated by vesicular tissue and tunnel (filled with clear calcite); tang. sec., topotype, USNM 159816, ×30.

Family BOTRYLLOPORIDAE
Miller, 1889

[Botrylloporidae MILLER, 1889, p. 290]

Zoaria encrusting circular discs with central monticule or encrusting sheets of coalesced discs. Primary and, in some forms, secondary and tertiary elevated fascicles of two rows of small autozooecia radiating from depressed central region of monticule. Vesicular tissue in interfascicle areas, monticular center, and coalesced margins of monticules. Walls laminated. No lunaria or acanthostyles. Proximal hemiseptum at zooecial

bend; diaphragms few, thin. *M.Dev.*

Characters of particular importance are: radiating fascicles of autozooecia; fascicles of two rows of autozooecia; narrow, long autozooecia; large vesicles; proximal hemiseptum; and lack of acanthostyles.

Botryllopora NICHOLSON, 1874a, p. 133 [*B. socialis* NICHOLSON, 1874c, p. 160; SM; Hamilton F., M. Dev., Arkona, Ont., Can.]. Zoarium encrusting, single disc or multiple, coalesced discs each with depressed central monticule. Approximately 10 primary, raised fascicles of autozooecia radiating from central depressed area; secondary and tertiary fascicles in larger zoaria. In each fascicle, an elevated dense, light-colored, median rib separates two rows of zoecia. Autozooecia small, ovate in cross section, with few thin diaphragms; proximal hemiseptum at zooecial bend. Walls transversely laminated; minute porelike structures perpendicular to laminae. Vesicular tissue from basal layer to surface of colony in monticular centers, interrays, and coalesced margins of multimonticular colonies. Vesicles short, wide, blisterlike, and polygonal in cross section to boxlike and subcircular. Monticular vesicles larger than interray vesicles; vesicles at coalesced margins very large. Vesicle roofs thicker and nearly at same level near zoarial surface. Vesicles longitudinally laminated; laminae lapping distally on autozooecial walls. *M.Dev.,* E.N. Am.——FIG. 193,*1a–e.* *B. socialis;* all but specimen in *e* from "Widder Beds"; *a,* fascicle with two rows of autozooecia and median rib (above), laminated wall microstructure and vesicular tissue (below); tang. sec., USNM 66192, ×100; *b,* small, coalescing discoidal colonies; USNM 159797, ×1; *c,* monticular center and radiating autozooecia; tang. sec., USNM 159796, ×20; *d,* fascicles of autozooecia and large blisterlike vesicles between discoidal monticules; oblique transv. sec., USNM 159798, ×20; *e,* monticular center with vesicles flanked by fascicles of autozooecia; long. sec., topotype, USNM 96862, ×30.

Family ACTINOTRYPIDAE
Simpson, 1897

[Actinotrypidae SIMPSON, 1897, p. 479]

Zoaria bifoliate or encrusting. Autozooecia isolated by vesicular tissue; granular-prismatic structure. Lunaria lacking. Acanthostyles in autozooecial walls inflecting autozooecial cavity, producing petaloid appearance. *L.Miss., Perm.*

Characters of particular importance are autozooecia isolated by vesicles, large acan-

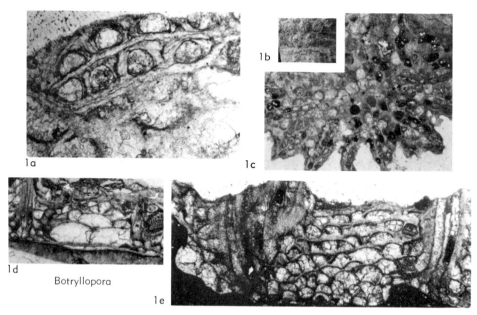

Fɪɢ. 193. Botrylloporidae (p. 407).

thostyles (canaliculi) indenting autozooecia, small acanthostyles in vesicle roofs, and no lunaria.

Actinotrypa Uʟʀɪᴄʜ in Mɪʟʟᴇʀ, 1889, p. 291 [*Fistulipora peculiaris Rᴏᴍɪɴɢᴇʀ, 1866, p. 123; OD; Keokuk Gr., L. Miss. (Osag.), La Grange, Mo., USA]. Zoarium bifoliate or encrusting. Monticules raised, with central cluster of vesicular tissue and encircling ring of slightly larger zooecia. Basal layer granular-prismatic. Mesotheca thin to thick; central granular layer dark, flanking granular-prismatic layers lighter colored. Autozooecia with long recumbent portion in endozone, erect in exozone, basically circular in cross section but idented by 5 to 11 acanthostylelike septa (canaliculi) extending through exozone; canaliculi centers granular to hyaline, with indistinct laminated sheath; diaphragms few and thin; funnel-cystiphragms off center. Autozooecia isolated by blisterlike vesicular tissue with granular inner and granular-prismatic outer layer. Vesicle walls and roofs one thick curved plate; zones of thicker roofs local; roofs with small acanthostyles expressed as bumps on surface. *L.Miss.(Osag.),* C.N.Am.——Fɪɢ. 194,*1a–e.* *A. peculiaris* (Rᴏᴍɪɴɢᴇʀ); *a,* autozooecia with peristomes and canaliculi isolated by vesicular tissue with projecting acanthostyles; paralectotype, UMMP 6409-7, ×9; *b,* dark vesicle walls, vesicle roofs with obscure acanthostyles, and indented autozooecia; tang. sec., lectotype, UMMP 6409-3, ×30; *c,* crushed

endozone with obscure mesotheca, recumbent autozooecium (right) in endozone, blisterlike vesicular tissue with thick roofs near surface, obscure acanthostyles in vesicle roofs; long. sec., lectotype, ×30; *d,* granular- to hyaline-centered acanthostylelike canaliculi in autozooecial wall (left) and small acanthostyles in vesicle roof (right); tang. sec., topotype, USNM 97238, ×100; *e,* well-developed indenting acanthostylelike canaliculi, some with hyaline centers, dark, curved vesicle walls and dense vesicle roofs; tang. sec., specimen from Keokuk F., L. Miss. (Osag.), Iowa, USA, USNM 159761, ×30.

Actinotrypella Gᴏʀʏᴜɴᴏᴠᴀ, 1972, p. 149 [*A. mira*; OD; Sebisurkhskaya suite, L. Perm., Darvaz Ra., USSR]. Zoarium bifoliate, lenticular in cross section, parallel sided, ribbonlike, branched in some. Monticules lacking. Mesotheca thin, trilayered. Autozooecia isolated by dense stereom in exozone, in ranges and rows that intersect diagonally. Perforated operculum in some zooecia. Autozooecia recumbent at mesotheca, short proximal hemiseptum at zooecial bend, erect in exozone. Seven to 10 acanthostyles (canaliculi) in each autozooecium in exozone inflecting living chamber as would septa; having hyaline core and laminated sheath. Vesicular tissue small, blisterlike vesicles in endozone; nearly solid stereom in exozone. Small acanthostyles in thick vesicle roofs and stereom. *L.Perm.,* USSR. ——Fɪɢ. 195,*1a–d.* *A. mira,* holotype, PIN 2351/418; *a,* hemispherical autozooecia on mesotheca, vesicles in endozone, stereom in exozone; transv. sec., ×10; *b,* trilayered mesotheca,

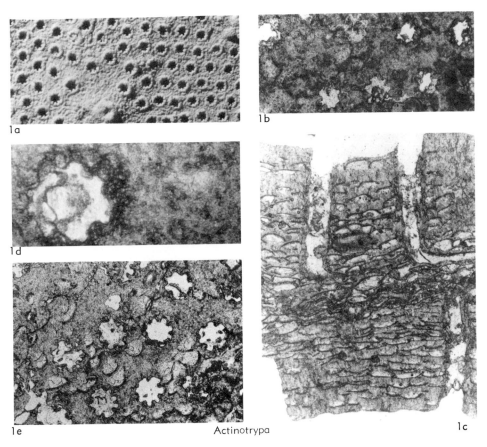

1a   1b   1d   1e   Actinotrypa   1c

FIG. 194. Actinotrypidae (p. 408).

autozooecium with short recumbent portion and hemiseptum at zooecial bend (upper left), vesicles in endozone; long. sec., ×30; *c,* autozooecia indented by canaliculi and widely isolated by stereom with tubules; tang. sec., ×40; *d,* autozooecia partially isolated by vesicles in endozone, isolated by stereom in exozone; long. sec., ×10 (photographs in *a,c,d* courtesy of R. V. Goryunova, in *b,* courtesy of Alan Horowitz).

Epiactinotrypa KISELEVA, 1973, p. 68 [*E. flosculosa*; OD; Chandalazy suite, U. Perm., Partizanskiy distr., Maritime Terr., USSR]. Zoarium encrusting. Autozooecial apertures with peristome, isolated in exozone by vesicular tissue, 12 to 16 acanthostylelike canaliculi in exozone inflecting living chamber. Monticules lacking. Autozooecia recumbent in endozone, proximal hemiseptum at zooecial bend, erect with sparse, thin diaphragms in exozone. Small, angular vesicles isolating autozooecia. Vesicle walls straight, thin to thick, commonly superimposed; roofs flat to curved, thin; boxlike vesicles wider than high. Local zones of thicker roofed vesicles. *U.Perm.,*

USSR.——FIG. 195,2*a,b*. **E. flosculosa,* holotype, PGU 187/48, drawings made from KISELEVA, 1973, pl. 6, fig. 2; *a,* tubular autozooecia isolated by boxlike to blisterlike vesicles, autozooecium budded on vesicular tissue (left); long. sec., ×25; *b,* subcircular autozooecia indented by acanthostyles (canaliculi), isolated by small subangular vesicles; tang. sec., ×50.

Family HEXAGONELLIDAE Crockford, 1947

[*nom. transl.* BASSLER, 1953, p. G87, *ex* Hexagonellinae CROCKFORD, 1947, p. 7]

Zoaria if bifoliate, then frondose or narrow and regularly to irregularly branched or cribrate; if trifoliate, then regularly or irregularly branched; if multifoliate, then with bifoliate branches radiating from multifoliate center of colony. Monticules subcircular, elongate, or absent. Some genera with noncelluliferous,

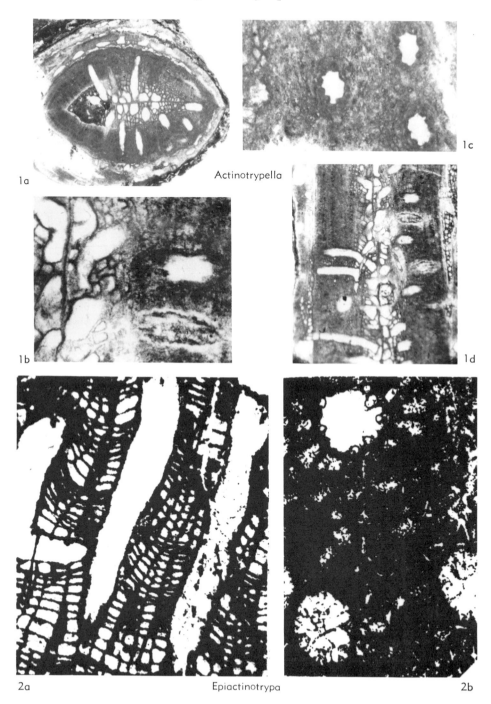

FIG. 195. Actinotrypidae (p. 408–409).

solid branch margins or with a hexagonal pattern of ridges in vesicular tissue surrounding each monticule. Symmetrical arrangement of lunaria on branches in many genera. Mesotheca thin, trilayered; median tubules rare. Few, short, longitudinal ridges on mesotheca in a few genera. Autozooecia recumbent on mesotheca, in endozone hemispherical in cross section; erect in exozone and circular in cross section. Autozooecia clavate to teardrop shaped on mesotheca in a few genera; partly contiguous to isolated by vesicular tissue at mesotheca. Distal hemiseptum at zooecial bend in one genus; diaphragms absent or few. Walls granular-prismatic or laminated. Lunaria well developed to inconspicuous; granular-prismatic or hyaline or laminated. Vesicular tissue low blisters; zones of stereom or solid stereom in exozone; many genera with small acanthostyles in vesicle roofs or stereom. *L.Dev.-U.Perm.*

Characters of particular importance are: bifoliate, trifoliate, or multifoliate colonies; vesicular tissue in endozone; zones of stereom or solid stereom in exozone; lunaria; and small acanthostyles in vesicle roofs and stereom.

Hexagonella WAAGEN & WENTZEL, 1886, p. 911 [*H. ramosa*; SD NICKLES & BASSLER, 1900, p. 291; *Productus* Ls., Perm., Salt Ra., Pak.]. Zoarium bifoliate, compressed to subcylindrical. Monticules flush, central cluster of vesicular tissue and radiating rows of autozooecia with lunaria radially arranged. Each monticule surrounded by elevated hexagonal pattern of ridges in vesicular tissue. Mesotheca thin, straight, granular in middle endozone; thicker, trilayered, with granular central layer and granular-prismatic outer layers in lateral endozone. Autozooecia recumbent, widely isolated by vesicular tissue from mesotheca to zoarial surface. Wall granular-prismatic; diaphragms sparse, planar. Lunaria from outer endozone to zoarial surface; proximal layer thick, granular-prismatic; dark central zone not continuous into zooecial boundary in some autozooecia. Vesicles large, irregular in endozone, boxlike in inner exozone, blisterlike and low in outer exozone. Zones of thicker vesicle roofs and stereom at surface; granular-prismatic; tubuli in vesicle roofs and stereom. Vesicles small, subangular in cross section. *Perm.*, Asia, Australia. ——FIG. 196,*2a–d*. *H. ramosa*; Kalabagh Mbr., Wargal Ls., Salt Ra.; *a*, solid monticular centers, hexagonal ridge surrounds each monticule; USNM 159831, ×1.7; *b*, subcircular autozooecia with lunaria, isolated by small subangular vesicles; tang. sec., USNM 159829, ×30; *c*, thin mesotheca in central endozone, thick mesotheca in lateral endozone (right), autozooecia isolated at mesotheca by vesicular tissue, vesicle height decreases outward in exo-

zone, sparse diaphragms in autozooecia; transv. sec., USNM 159833, ×30; *d*, autozooecium (upper), cycles of decreasing vesicle height upward in exozone, thin zones of stereom; long. sec., USNM 159832, ×30.

Ceramella HALL & SIMPSON, 1887, p. xix [*C. scidacea*; OD; Hamilton Gr., M. Dev., Cayuga Lake, Darien Center, N.Y., USA] [=*Caramella* MOROZOVA, 1960, p. 87, incorrect subsequent spelling]. Zoarium bifoliate fronds. Monticules elongate, depressed, with center of vesicular tissue; zooecia larger around margin. Mesotheca thin; central zone dark and outer layers laminated. Autozooecia hemispherical in cross section, partially isolated, some with keel and sinus; recumbent portion long; slightly oblique at zoarial surface; diaphragms few; walls with dark, thin boundary and laminated cortex. Lunaria in exozone; radius of curvature short; laminated, with thick, light-colored distal layer. Vesicular tissue in endozone and exozone; vesicles blisterlike, of moderate size, thin stereom at surface; isolating autozooecia in exozone; laminated; tubuli in vesicle roofs. *M.Dev.*, N.Am. ——FIG. 196,*1a–e*. *C. scidacea*; *a*, bifoliate frondose zoarium; lectotype, NYSM 623, ×1; *b*, elongate, depressed monticule; lectotype, ×10; *c*, pyriform autozooecia (lunaria to bottom) isolated by moderately large vesicles; tang. sec., paralectotype, NYSM 622, ×30; *d*, autozooecia partly isolated at mesotheca, blisterlike vesicles in endozone and exozone, monticule (upper left); transv. sec., lectotype, ×30; *e*, undulatory mesotheca, long recumbent portion of autozooecia, vesicles in exozone isolate autozooecia; long. sec., lectotype, ×30.

Coscinium KEYSERLING, 1846, p. 191 [*C. cyclops*; SD ULRICH, 1884, p. 38; L. Perm., Timan, USSR]. Zoarium bifoliate, cribrate. Autozooecia with peristome and lunarium, isolated at surface by stereom. Mesotheca trilayered; central zone granular and outer layers granular-prismatic; few, short longitudinal ridges on mesotheca. Autozooecia recumbent, in rows at mesotheca; hemispherical in cross section, some with keel and sinuses; teardrop shaped and mostly contiguous at contact with mesotheca. Diaphragms sparse. Autozooecial walls with granular boundary and granular-prismatic cortex. Lunaria in outer endozone and exozone; granular-prismatic. Vesicular tissue forming small, low blisters; stereom through most of exozone; some discontinuous compound range walls with dark granular boundary; numerous tubuli in vesicle roofs and stereom. *Perm.*, USSR.——FIG. 197,*1a–f*. *C. cyclops*, paratype, USNM 171739; *a*, dark, granular, autozooecial boundary continuous into granular layer in lunarium (below); tang. sec., ×100; *b*, elongate to circular autozooecia isolated by vesicular tissue; tang. sec., ×30; *c*, mesotheca, stereom with tubuli, terminal diaphragm in

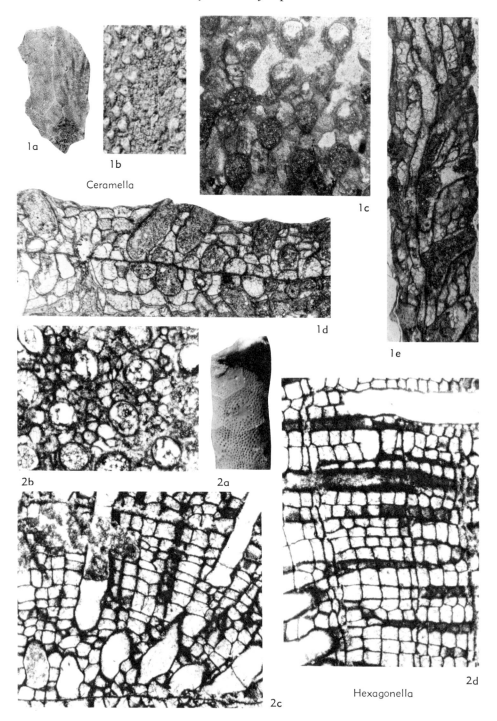

1a

1b

Ceramella

1c

1d

1e

2b

2a

2c

2d

Hexagonella

Fig. 196. Hexagonellidae (p. 411).

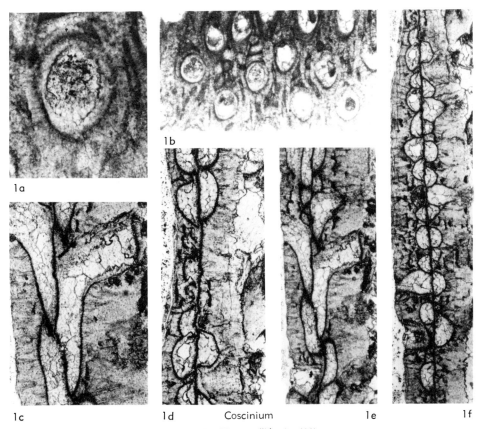

FIG. 197. Hexagonellidae (p. 411).

autozooecium; long. sec., ×50; *d,* thin, granular-prismatic mesotheca, autozooecia mostly contiguous at mesotheca, dark, granular boundary in some range walls in stereom and tubuli in stereom; transv. sec., ×50; *e,* undulatory mesotheca, recumbent autozooecia in endozone, small vesicles in inner exozone, stereom in exozone; long. sec., ×30; *f,* autozooecia contiguous to isolated at mesotheca, stereom in exozone; transv. sec., ×30.

Coscinotrypa HALL, 1886, pl. 29 [**Clathropora carinata* HALL, 1883a, pl. 26; M; M. Dev., Falls of the Ohio, Jeffersonville, Ind., USA]. Zoarium bifoliate, cribrate; fenestrules bordered by rim of solid stereom; lunaria radially arranged around fenestrules; branches symmetrical with lunaria on side of autozooecia away from branch center. Mesotheca trilayered; median layer dark, lateral layers laminated and having tubuli. Mesotheca regularly undulatory in transverse view. Autozooecia contiguous at mesotheca; hemispherical in cross section, with keel and sinuses; size decreasing from endozone to exozone. Walls laminated, with tubuli. Lunaria in outer endozone and exozone; radius of curvature short, pro-

ducing trilobed autozooecial cross section; hyaline or distinctly laminated. Vesicular tissue forming large blisters in outer endozone, low blisters in inner exozone; stereom in most of exozone; laminated, with numerous small acanthostyles. *M.Dev.,* N.Am.——FIG. 198,*1a–g.* **C. carinata* (HALL); *a,* frond with small fenestrules and broken branch; cotype, WM 13986:no fragment number, ×1; *b,* frond with large fenestrules; cotype, WM 13986:f.33, ×1; *c,* branch with lunaria on side of autozooecium away from branch centerline; cotype, WM 13986:f.32, ×10; *d,* undulatory mesotheca, branch at high angle to frond, stereom in exozone; transv. sec., cotype, WM 13986:f.35, ×30; *e,* thick, undulatory mesotheca, hemispherical autozooecia with keel and sinuses in outer endozone; transv. sec., Jeffersonville F., Ind., SIUC 3010, ×30; *f,* laminated walls, laminated stereom with acanthostyles, lunaria on side of autozooecia away from branch midline (horizontal); tang. sec., Jeffersonville F., SIUC 3009, ×30; *g,* thick mesotheca, autozooecia decrease in diameter from endozone to exozone; long. sec., cotype, WM 13986:f.35, ×30.

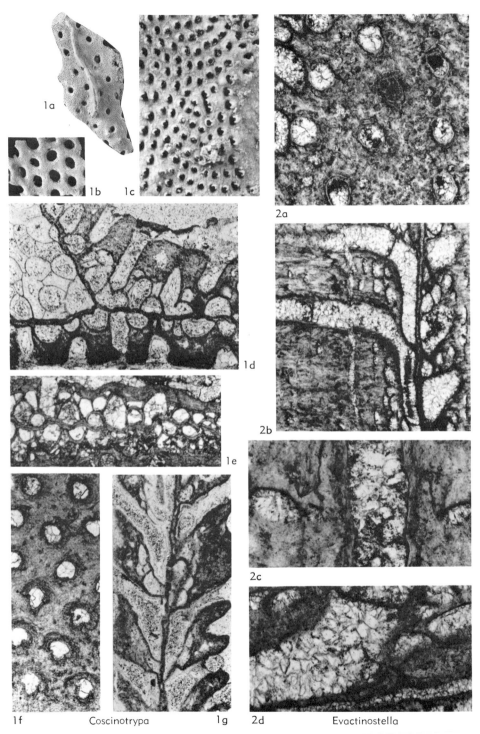

1a

1b   1c

1d

1e

2a

2b

2c

1f   Coscinotrypa   1g   2d   Evactinostella

FIG. 198.   Hexagonellidae (p. 413–417).

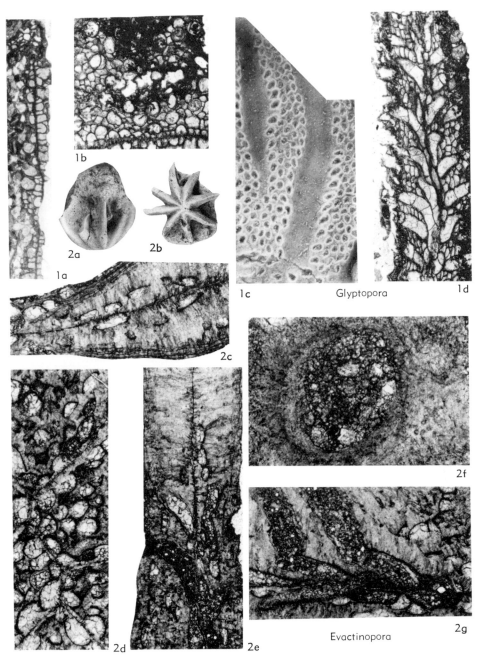

FIG. 199.   Hexagonellidae (p. 415–417).

Evactinopora MEEK & WORTHEN, 1865, p. 165 [*E. radiata*; OD; Miss., Mo., USA]. Zoarium multifoliate; 4 to 8 vertical bifoliate branches radiating from center; base and some branch tips of solid stereom. Monticules lacking. Autozooecial cavity subcircular, widely isolated by stereom. Mesotheca thin, laminated in central endozone; thick, trilayered in branches, with central layer dark and laminated layers light-colored with tubuli near margins resembling granular-prismatic microstructure. Autozooecia isolated at mesotheca by vesicular tissue; recumbent portion

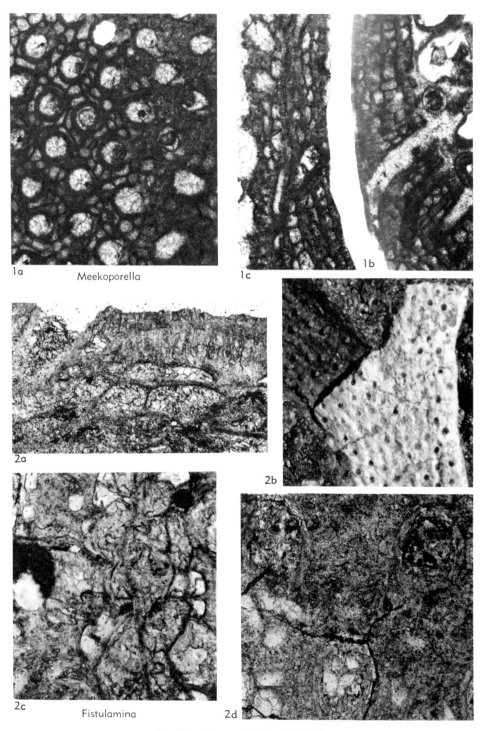

1a   Meekoporella   1c   1b

2a   2b

2c   Fistulamina   2d

FIG. 200.   Hexagonellidae (p. 417–418).

long; short hemiseptum at zooecial bend in some autozooecia. Wall with dark boundary zone and laminated cortex having tubuli. Lunarium in exozone; hyaline or laminated. Vesicular tissue blisterlike in central endozone; stereom throughout most of branch endozones and exozone; laminated, with numerous small acanthostyles. *Miss.,* N.Am.——Fɪɢ. 199,*2a,b.* *E. radiata,* holotype, ISGS(ISM) 10784; *a,* noncelluliferous base and branch margins, ×1; *b,* bifoliate branches radiating from center of colony, ×1. ——Fɪɢ. 199,*2c–g.* *E. sexradiata* Mᴇᴇᴋ & Wᴏʀ-ᴛʜᴇɴ, St. Joe Mbr., Burlington Ls., L. Miss., Mo., USA; *c,* thick vesicle roofs and stereom with acanthostyles near branch margin; transv. sec., USNM 159819, ×30; *d,* branching mesothecae and large, blisterlike vesicles in central endozone; transv. sec., USNM 159820, ×30; *e,* autozooecia, vesicles (below), and stereom (above, near branch margin) adjacent to mesotheca; long. sec., USNM 159820, ×30; *f,* laminated lunarium (below), laminated wall with tubuli, acanthostyles in stereom; tang. sec., USNM 159818, ×100; *g,* autozooecia with short hemiseptum at zooecial bend, thick walls and stereom in exozone; long. sec., USNM 159819, ×30.

Evactinostella Cʀᴏᴄᴋғᴏʀᴅ, 1957, p. 27 [*Evactinopora crucialis* Hᴜᴅʟᴇsᴛᴏɴ, 1883, p. 593; OD; L. Perm., Fossil Ra., Australia]. Zoarium multifoliate, 4 or 5 bifoliate branches radiating from center; branches bifurcating in plane of mesotheca. Monticules low, substellate central cluster of small vesicles or stereom surrounded by larger zooecia with lunaria radially arranged. Low, broad ridges on vesicular tissue surrounding each autozooecium. Mesotheca thick, central layer dark and outer layers laminated. Autozooecia with long recumbent portion, isolated by vesicular tissue at mesotheca; diaphragms few; walls thick, laminated, with indistinct boundary. Lunaria large, laminated, light colored; thicker than remainder of autozooecial wall, with proximal tubuli. Vesicular tissue with vesicles blisterlike, small in cross section; cyclic zones of decreasing vesicle height and stereom in exozone; inner layer dark and outer layer thick, laminated, with tubuli and large, indistinct acanthostyles. *L.Perm.(Artinsk.),* Australia.——Fɪɢ. 198, *2a–d.* *E. crucialis* (Hᴜᴅʟᴇsᴛᴏɴ), Callytharria F. (low. Artinsk.), W. Australia, USNM 159823; *a,* autozooecia widely isolated by small vesicles and stereom; tang. sec., ×30; *b,* mesotheca, recumbent autozooecia, sparse diaphragms, vesicles, and stereom with acanthostyles; long. sec., ×30; *c,* laminated autozooecial walls, dark lower layer and laminated outer layer in stereom; long. sec., ×100; *d,* indistinct laminations in autozooecial walls and vesicular tissue at zooecial bend; long. sec., ×100.

Fistulamina Cʀᴏᴄᴋғᴏʀᴅ, 1947, p. 10, 28 [*F. inornata;* OD; Miss. (low Burindi), Glen William,

Australia]. Zoarium ribbonlike, bifoliate, branching in plane of mesotheca; monticules lacking; margins thin, solid. Autozooecia with fairly long recumbent portion, isolated by vesicular tissue; diaphragms lacking; walls indistinctly laminated. Lunaria laminated, indistinct; on proximal side in endozone and inner exozone, toward lateral side near branch margin and at zoarial surface. Vesicular tissue low blisters; stereom through most of exozone; laminated with tubuli. *Miss.(Visean),* Australia, USSR.——Fɪɢ. 200,*2a–d.* *F. inornata;* *a,* autozooecial walls (left), laminated vesicles and stereom with tubuli (right); long. sec., topotype, USNM 147232, ×100; *b,* bifurcating ribbonlike branches, autozooecia in rows isolated by vesicular tissue, lunaria on side of autozooecia near branch margin (left); holotype, SU 6431, ×10 (photograph by Robin E. Wass); *c,* mesotheca (vertical), vesicles in endozone, laminated stereom in exozone; transv. sec., topotype, USNM 147232, ×100; *d,* ovate autozooecia, laminated walls, indistinct lunaria (below), small vesicles and laminated stereom isolate autozooecia; tang. sec., topotype, USNM 147232, ×100.

Glyptopora Uʟʀɪᴄʜ, 1884, p. 39 [*Coscinium plumosum* Pʀᴏᴜᴛ, 1860, p. 572; OD; St. Louis Gr., U. Miss. (Meramec.), Warsaw, Ill., Barretts Station, Mo., USA] [=*Glyptotrypa* Mɪʟʟᴇʀ, 1889, p. 307, incorrect subsequent spelling]. Zoarium bifoliate fronds, branches nearly at right angles; "monticules" low, long and narrow, solid stereom; autozooecia elongated oblique to monticular margin; lunaria on end nearest monticule. Mesotheca thin to thick, central zone dark with median tubuli, poorly laminated layers lighter colored; narrow longitudinal ridges on mesotheca paralleling budding direction. Autozooecia semirecumbent in endozone, slightly oblique to zoarial surface; diaphragms sparse, thin; walls indistinctly laminated. Lunaria hyaline or laminated. Vesicular tissue isolating autozooecia in endozone and exozone; blisters low; stereom in outer exozone; laminated with numerous tubuli or small acanthostyles. *U.Miss.,* N.Am., USSR, Australia.——Fɪɢ. 199, *1a–d.* *G. plumosa* (Pʀᴏᴜᴛ), Warsaw F., Ill., USA; *a,* thin bifoliate frond, mesotheca with longitudinal ridges, autozooecia and vesicles at mesotheca; transv. sec., USNM 159824, ×20; *b,* mesotheca with median tubuli and longitudinal ridges (below), autozooecia isolated by vesicular tissue (above); deep tang. sec., USNM 159825, ×20; *c,* depressed, elongate monticules of solid stereom, rows of elongate autozooecia with lunaria on end nearest monticule; USNM 159827, ×10; *d,* autozooecia with sparse diaphragms, blisterlike vesicles, stereom at surface; long. sec., USNM 159825, ×20.

Meekopora Uʟʀɪᴄʜ in Mɪʟʟᴇʀ, 1889, p. 312 [*Fistulipora? clausa* Uʟʀɪᴄʜ, 1884, p. 47; M; U.

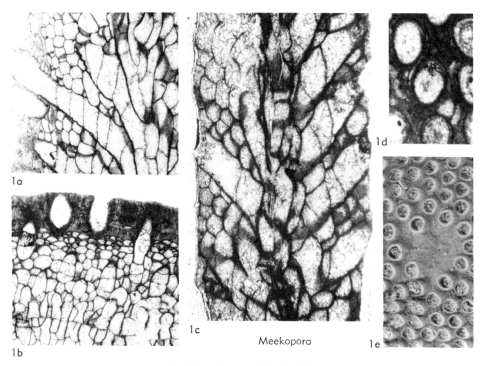

1a

1b

1c

1d

1e

Meekopora

FIG. 201.   Hexagonellidae (p. 417).

Miss., Sloans Valley, Ky., USA]. Zoarium narrow, bifoliate, branching. Monticules depressed; central cluster of vesicular tissue or stereom and ring of slightly larger zooecia. Autozooecia with peristomes and lunaria. Mesotheca thin, granular-prismatic. Autozooecia at mesotheca narrow, boxlike to hemispherical in cross section, partially to completely isolated by vesicular tissue. Walls granular-prismatic, tubuli in cortex; locally, lateral and distal walls replaced by superimposed vesicle walls. Diaphragms thin, closely spaced in endozone and inner exozone; planar to cystoidal; some off-centered funnel cystiphragms. Lunaria in endozone and exozone; dark granular boundary continuous into boundary of autozooecial wall; granular-prismatic layer slightly thicker proximally, may have tubuli. Vesicular tissue decreasing in height to low blisters in exozone; local zones of thicker vesicle roofs and thin stereom; granular-prismatic, with tubuli in roofs and stereom. *Miss.,* N.Am., Asia. ——FIG. 201,*1a–e.* *M. clausa* (ULRICH); *a,* closely spaced diaphragms in endozone and inner exozone, vesicles decreasing in height outward, stereom at surface; long. sec., lectotype, USNM 159834, ×20; *b,* thin mesotheca, boxlike to hemispherical autozooecia, large vesicles in endozone, stereom at surface; transv. sec., lectotype, ×20; *c,* crushed endozone, autozooecia isolated by blisterlike vesicles in exozone, distal

"wall" of autozooecium (right center) of superimposed vesicle walls; long. sec., paralectotype, USNM 159835, ×30; *d,* elongate autozooecia narrowly isolated by small vesicles; tang. sec., paralectotype, USNM 159836, ×30; *e,* solid monticular center, autozooecia with peristomes and lunaria; paralectotype, USNM 159837, ×10.

Meekoporella MOORE & DUDLEY, 1944, p. 304 [**M. dehiscens*; OD; Stanton Ls., U. Penn. (Missour.), Fredonia, Kans., USA]. Zoarium bifoliate sheets joining at 120° and diverging distally. Monticules with central cluster of vesicles and ring of larger zooecia; lunaria, in part, radially arranged. Mesotheca thick to thin; central layer granular, outer layers granular-prismatic. Autozooecia isolated at mesotheca by vesicles, curving gently into exozone. Wall granular-prismatic; diaphragms few. Lunaria large, slightly indenting, thick, granular-prismatic. Vesicular tissue with vesicles box- to blisterlike; granular-prismatic; local zones of thicker vesicle roofs or stereom. Vesicles small in cross section, locally forming ring around autozooecium. Acanthostyles at some junctions of vesicle walls and in vesicle roofs. *U.Penn.-L.Perm.,* N.Am.——FIG. 200,*1a–c.* **M. dehiscens*; *a,* autozooecia with lunaria, isolated by small vesicles, vesicles locally forming ring around autozooecium (upper left); tang. sec., paratype, KUMIP 32393, ×30; *b,*

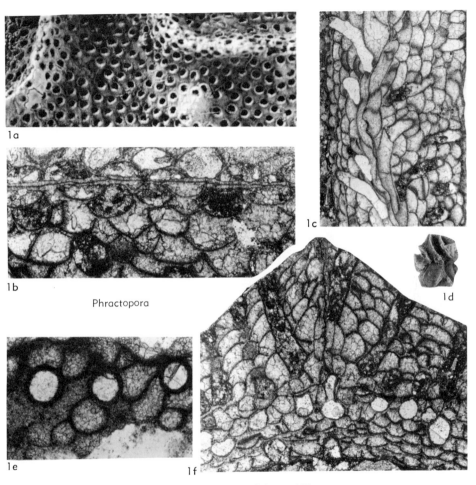

la

lb

Phractopora

lc

ld

le

lf

FIG. 202. Hexagonellidae (p. 419).

autozooecium, parted mesotheca (left), zones of thick vesicle roofs with acanthostyles (left); long. sec., paratype, KUMIP 32393, ×30; *c,* thick mesotheca (vertical), granular-prismatic microstructure, boxlike to blisterlike vesicles; long. sec., paratype, KUMIP 32396, ×30.

Phractopora HALL, 1883b, p. 154 [**Lichenalia (Phractopora) cristata*; OD; M. Dev., Falls of the Ohio, Jeffersonville, Ind., USA]. Zoarium bifoliate, bifurcating branches nearly at right angles to main branch. Monticules low; central cluster of large vesicles or stereom and ring of slightly larger zooecia. Autozooecia with peristomes and lunaria. Mesotheca with granular central zone and granular-prismatic outer layers; thin. Autozooecia isolated at mesotheca by vesicular tissue; walls granular-prismatic; diaphragms sparse; locally, distal side made of superimposed vesicle walls. Lunaria in outer endozone and exozone; radius of curvature slightly smaller than that of

lateral-distal autozooecial wall; inconspicuous; granular-prismatic. Vesicular tissue in endozone and exozone; large, low blisters decreasing in height outward in exozone; granular-prismatic; thin stereom at surface. *M. Dev.,* N.Am.——FIG. 202, *1a–f. *P. cristata; a,* branches expressed as ridges, monticular center, peristomes and lunaria; syntype, WM 14002, ×10; *b,* granular-prismatic mesotheca, blisterlike vesicles isolate autozooecia in endozone of branch; transv. sec., topotype, USNM 159838, ×50; *c,* undulatory mesotheca, autozooecia with sparse diaphragms, isolated by vesicular tissue; long. sec., topotype, USNM 159838, ×20; *d,* bifurcating branches diverge from main branch at nearly right angles; syntype, WM 14002, ×1; *e,* indistinct lunaria (lower side) in subcircular autozooecia, large vesicles; tang. sec., topotype, USNM 159838, ×50; *f,* undulatory mesotheca, granular-prismatic vesicles (below), low branch

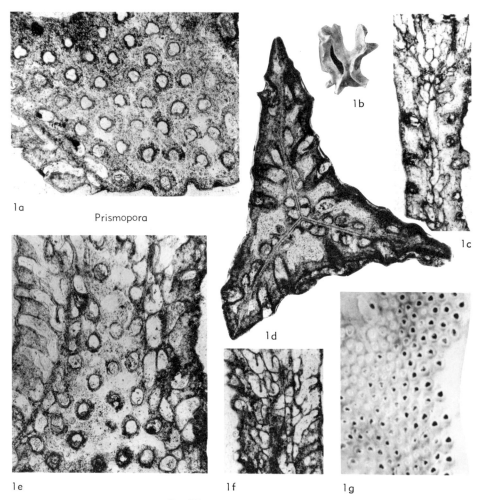

Fig. 203. Hexagonellidae (p. 420).

(above); transv. sec., topotype, USNM 159838, ×30.

Prismopora HALL, 1883b, p. 158 [*P. triquetra*; SD HALL & SIMPSON, 1887, p. xxi; M. Dev., Falls of the Ohio, Jeffersonville, Ind., USA]. Zoarium trifoliate, irregularly branching. Branches narrow, parallel-sided; faces concave; margins solid, noncelluliferous. Monticules lacking. Autozooecia with peristomes and lunaria; lunaria on proximal side of autozooecia in row in center of branch; rotated to progressively more lateral position in rows of autozooecia toward branch margin. Mesotheca with central layer granular, outer layers granular-prismatic. Autozooecia partially isolated at mesotheca by vesicular tissue; clavate at contact with mesotheca, expanding distally, with recurved distal hemiseptum near zooecial bend; subcircular in cross section in exozone. Wall granular-prismatic; tubuli in outer granular-prismatic layer. Lunaria in exozone; radius of curvature short; ends may inflect slightly; microstructure granular-prismatic. Vesicular tissue forming small, low blisters in endozone and inner exozone; granular-prismatic; stereom in most of exozone, with acanthostyles. *Dev.-Perm.*, N.Am., Australia.——Fig. 203, *1a–g. *P. triquetra*; a,* autozooecia isolated by stereom, lunaria on proximal side of autozooecia near branch center, on lateral sides of autozooecia near branch margins; tang. sec., topotype, USNM 159839, ×30; *b,* bifurcating trifoliate branches; cotype, WM 13985:f.10, ×1; *c,* undulatory mesotheca, autozooecia and vesicles in endozone, stereom in exozone; long. sec., topotype, USNM 159839, ×20; *d,* granular central layer in mesothecae, trifoliate zoarium, stereom in exozone; transv. sec., topotype, USNM 159841, ×30; *e,* two mesothecae flank tangential view of concave

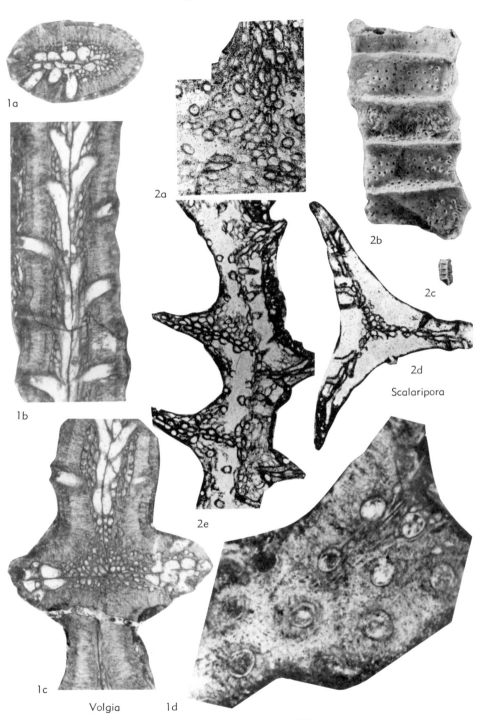

1a

1b

1c

Volgia   1d

2a

2b

2c

2d

Scalaripora

2e

FIG. 204.   Hexagonellidae (p. 422).

branch face; distal hemisepta in longitudinal view
of autozooecia (left); tang. and long. sec., topotype, USNM 159839, ×30; *f,* recumbent autozooecia and vesicles in endozone, distal hemi-

septa (left); long. sec., topotype, USNM 159840, ×20; *g*, peristomes and lunaria, solid branch margin (right); cotype, WM 13985, ×10.

Scalaripora HALL, 1883b, p. 159 [**S. scalariformis*; OT, also SD HALL & SIMPSON, 1887, p. xxi; M. Dev., Falls of the Ohio, Jeffersonville, Ind., USA]. Zoarium trifoliate, branches parallel-sided with concave faces; short branches transverse to main branch. Monticules lacking. Autozooecia with peristomes; widely isolated; lunaria on side of autozooecium nearest center of branch. Mesotheca thin. Autozooecia partly isolated at mesotheca by vesicular tissue; recumbent portion long; diaphragms lacking. Lunaria obscure. Vesicular tissue forming low blisters; stereom at surface thin. [The types and available topotypes are silicified and little is known of the microstructure of this genus.] *M.Dev.,* N.Am.——FIG. 204,*2a–e.* **S. scalariformis*; *a,* mesotheca in transverse branch, vesicular tissue, subcircular autozooecia; tang. sec., topotype, USNM 55275, ×24; *b,* transverse, short branches, autozooecia with peristomes and lunaria on side nearest branch center; cotype WM 13990, ×8; *c,* small, triangular branch; cotype, WM 13990, ×1; *d,* triangular branch with concave faces, thin mesotheca, autozooecia; transv. sec., topotype, USNM 55275, ×16; *e,* main branch, short, side branches with mesotheca and blisterlike vesicles; long. sec., topotype, USNM 55275, ×16.

Volgia STUCKENBERG, 1905, p. 31 [**Coscinium arborescens* STUCKENBERG, 1895, p. 173; OD; U. Carb., Samarskaya Luka, USSR] [=*Ramiporina* SHULGA-NESTERENKO, 1933, p. 40, obj.]. Zoarium bifoliate, branches slightly compressed; secondary branches and some short tertiary branches, with mesotheca at right angles to main branch. Monticules lacking. Mesotheca thin, granular-prismatic. Autozooecia partly isolated at mesotheca, curving gently into exozone. Short, discontinuous compound range walls in endozone. Walls with dark granular boundary and granular-prismatic cortex; tubuli in outer granular-prismatic layer. Diaphragms lacking. Lunaria obscure, granular-prismatic; granular middle layer continuous into autozooecial boundary. Vesicular tissue forming small, low blisters in endozone; granular-prismatic stereom throughout exozone, with small acanthostyles. *Penn.,* USSR. ——FIG. 204,*1a–d.* **V. arborescens* (STUCKENBERG), River Don, USSR, PIN 436/119; *a,* hemispherical autozooecia at mesotheca, stereom in exozone; transv. sec., ×16; *b,* thin mesotheca, autozooecia partially isolated by low, small, blisterlike vesicles in endozone, stereom with acanthostyles in exozone; long. sec., ×16; *c,* secondary branch at right angles to primary branch, mesothecae and small vesicles at branch junction, autozooecia at both ends of primary branch shown in transverse view, vesicles in endozone and stereom in exozone in secondary

branch shown in longitudinal view; transv. sec., ×16; *d,* granular-prismatic autozooecial walls, obscure lunaria, autozooecia; widely isolated by stereom with acanthostyles; small vesicles and discontinuous compound range wall in endozone (right); tang. sec., ×32 (photographs courtesy of G. G. Astrova).

Family CYSTODICTYONIDAE Ulrich, 1884

[Cystodictyonidae ULRICH, 1884, p. 34] [=Arcanoporidae VINE, 1884, p. 203 (part); Acrogeniidae SIMPSON, 1897, p. 480; Thamnotrypidae SIMPSON, 1897, p. 480; Sulcoreteporidae BASSLER, 1935, p. 21]

Zoaria variable, many bifoliate, compressed, with straplike dichotomous or trichotomous branches in plane of mesotheca. One genus jointed at dichotomous branchings of straplike branches in plane of mesotheca. Some bifoliate frondose or trifoliate with branches nearly at right angles to main branch. One bifoliate with anastomosing branches producing large fenestrules. Monticules absent except in *Dichotrypa.* Bipartite or tripartite branch symmetry in most genera. Mesotheca planar, undulatory, or folded into sharp, zigzag folds; no median tubules. Low vertical plates extending from mesotheca into zooecial cavities in a few genera. Compound range walls generally well developed in endozone; well developed and thick, with radiating arrays (libria) of branched dark zones (valvae) and tubules in exozone in most genera; commonly one, three, or more range walls protruding as conspicuous ridges at zoarial surface. Autozooecia recumbent on mesotheca, oblique or direct at zoarial surface; short, diaphragms lacking; proximolateral hemiseptum in a few genera. Autozooecia generally teardrop shaped at contact with mesotheca; right- and left-handed forms; walls laminated. Laminated lunarium in some genera. Vesicles sparse in endozone, small, generally adjacent to proximal tips of autozooecia. Laminated stereom with acanthostyles or tubules in most of exozone in most genera. *M.Dev.-L.Perm.*

Characters of particular importance are: bifoliate or trifoliate colonies; compound range walls in endozone; thick, compound range walls with branched dark zones in exozone; teardrop or club-shaped autozooecial outline at contact with mesotheca; vesicles in endozone; laminated stereom in exozone; and small acanthostyles in stereom.

Cystodictya ULRICH, 1882, p. 152 [**C. ocellata*; OD; New Providence F., L. Miss., Somerset, Ky., USA]. Zoarium bifoliate, straplike, branching in plane of mesotheca. Autozooecia with peristomes and lunaria on side nearest branch margin. Ridges between ranges of autozooecia lacking. Mesotheca thin to moderately thick; indistinctly laminated to granular-prismatic; with low ridges running parallel to ranges of autozooecia. Autozooecia teardrop-shaped at contact with mesotheca; right- and left-handed in form; quadrate in cross section; partly isolated by boxlike vesicles; recumbent portion short; blunt proximolateral hemiseptum at zooecial bend, indenting zooecial cavity and producing slight hook-shaped appearance of autozooecia in deep tangential section. Diaphragms lacking. Walls laminated; boundary serrated; tubuli in cortex. Lunarium in exozone; light colored, laminated, some with core and proximal rib. Compound range walls thin in endozone with dark boundary continuous into dark central layer of mesotheca; thick in exozone with many flexures and irregular tubuli. Vesicles small, boxlike in endozone; low blisters in inner exozone; stereom in exozone; laminated, with tubuli and flexures. *M.Dev.-U.Miss.*, N.Am.——FIG. 205, *1a–h*. **C. ocellata*; *a,* laminated autozooecial walls, light-colored laminated lunaria (right), each with a core and a proximal projection, tubuli in laminated stereom and range walls; tang. sec., holotype, USNM 43650, ×100; *b,* range walls with elongate tubules, lunaria on lateral sides of autozooecia; tang. sec., holotype, ×30; *c,* mesotheca (below), blunt hemiseptum at zooecial bend, laminated walls, larger vesicles in endozone, small, blisterlike vesicles in inner exozone, laminated stereom with flexures and tubuli; long. sec., holotype, ×100; *d,* short, recumbent autozooecia and vesicles in endozone, erect autozooecia isolated by stereom in exozone; long. sec., holotype, ×30; *e,* short recumbent portion of autozooecia in endozone, blunt hemiseptum at zooecial bend, no diaphragms, vesicles in endozone, smaller vesicles in inner exozone, stereom in exozone; long. sec., specimen from U. Miss. (Merimec.), "Lower Division," St. Louis Gp.," Mo., USA, USNM 159843, ×30; *f,* low ridges on mesotheca (upper left), thin range walls and boxlike vesicles in endozone (upper left), autozooecia slightly hooked at zooecial bend where proximolateral hemiseptum is present; tang. sec., USNM 159843, ×30; *g,* planar mesotheca with short ridges, boxlike autozooecia and vesicles in endozone, flexures and tubuli in range

walls; transv. sec., USNM 159843, ×30; *h,* mesotheca with possibly recrystallized granular-prismatic outer layers, range walls continuous with dark central layer of mesotheca, boxlike vesicles in endozone, stereom with flexures and tubuli in range walls (left, right) in exozone; transv. sec., USNM 159843, ×100.

Acrogenia HALL, 1883b, p. 193 [**A. prolifera*; M; "Hamilton Beds," M. Dev., Vincent, N.Y., USA]. Zoarium bifoliate, narrow, compressed; jointed at dichotomous branches, in plane of mesotheca; base of colony with cylindrical branches, noncelluliferous, longitudinally striated, jointed. Autozooecia in ranges; arrangement reticulate to slightly rhombic; branches bilaterally symmetrical; midrib more prominent than other range walls. Border narrow, noncelluliferous. Lunaria elevated, on proximal side of autozooecia; slightly rotated to side of autozooecium away from branch center. Mesotheca thick, trilayered; median zone dark and outer layers laminated. Autozooecia subtriangular, with rounded distal portion, in contact with mesotheca; forms right- and left-handed. Autozooecia triangular in cross section in endozone, near proximal tip; hemispherical to subquadrate toward distal margin. Compound range walls with dark boundary zone continuous with boundary zone in mesotheca; dark zones branched in range walls in exozone. Autozooecia erect in exozone; walls laminated, with tubuli; diaphragms lacking. Lunaria in outer endozone and exozone; hyaline with laminated lining. Vesicles irregular in endozone; between range walls adjacent to proximal tip of autozooecia partially isolating autozooecia; low blisters at zooecial bend; stereom with acanthostyles in exozone. *M.Dev.*, N.Am.——FIG. 206, *1a–f.* **A. prolifera,* holotype, NYSM 594; *a,* stereom in exozone (lower left), vesicles in endozone (upper right), slightly elevated range walls and lunaria near zoarial surface (below); tang. sec., ×30; *b,* subcylindrical, longitudinally striated branches with joints near base of colony; ×2; *c,* thick, trilayered mesotheca, triangular to hemispherical autozooecia in endozone, range wall boundary continuous to dark, central zone of mesotheca, stereom with small acanthostyles in exozone; transv. sec., ×100; *d,* laminated mesotheca, vesicles at zooecial bend, laminated stereom with acanthostyles in exozone; transv. sec., ×100; *e,* planar mesotheca, hemispherical to subquadrate autozooecia in endozone, small vesicles at zooecial bend, stereom in exozone; transv. sec., ×30; *f,* slightly undulatory mesotheca, long recumbent portion of autozooecia in endozone, vesicles in inner exozone, stereom; long. sec., ×30.

Dichotrypa ULRICH in MILLER, 1889, p. 300 [**Fistulipora flabellum* ROMINGER, 1866, p. 122; M; "Warsaw Ls.," U. Miss. (Meramec.), Spergen Hill, Ind., USA]. Zoarium bifoliate fronds; in

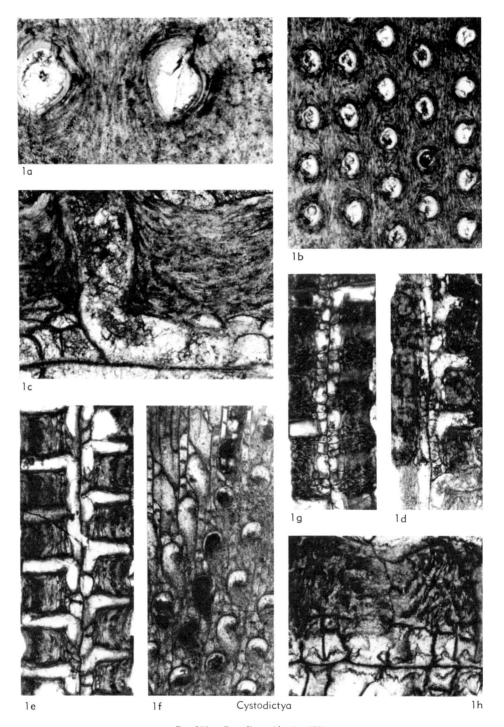

FIG. 205. Cystodictyonidae (p. 423).

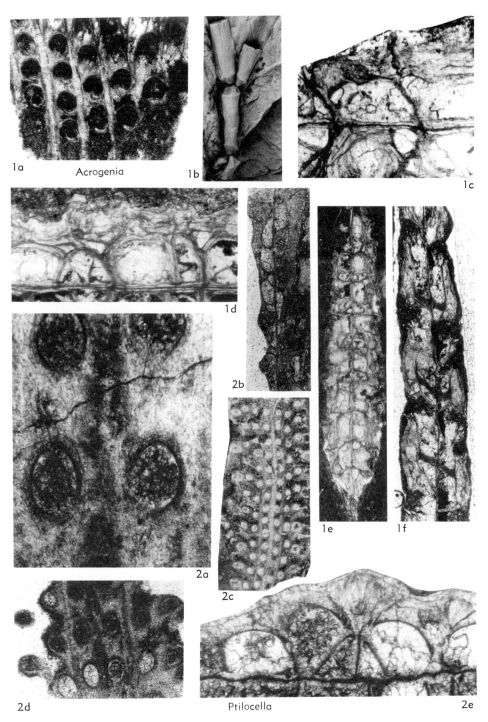

la
Acrogenia
1b
1c
1d
2b
2c
1e
1f
2a
2d
Ptilocella
2e

Fig. 206. Cystodictyonidae (p. 423–427).

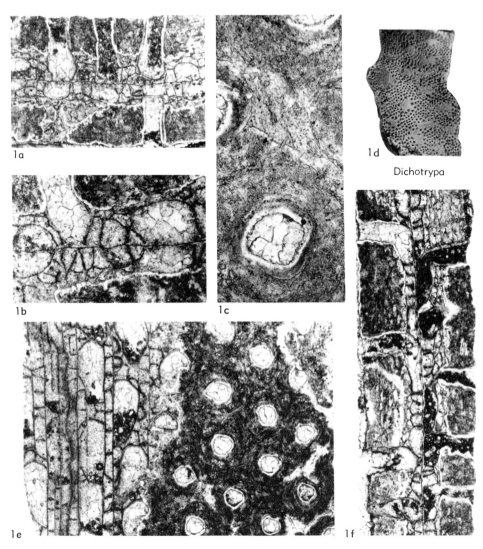

FIG. 207.　Cystodictyonidae (p. 423).

some with irregular branches. Monticules elevated or flush; central cluster of small, angular vesicles or stereom surrounded by ring of slightly larger zooecia with lunaria radially arranged. Peristomes low and lunaria elevated. Mesotheca thin, trilayered, granular-prismatic; with several low, longitudinal ridges per autozooecium. Autozooecia in ranges on mesotheca, rhombically arranged over several ranges, teardrop shaped in outline at junction with mesotheca, right- and left-handed forms. Autozooecia in endozone triangular to hemispherical to quadrate in cross section; partly isolated by vesicles; recumbent portion moderately long, lacking diaphragms; walls granular-prismatic in inner endozone; proximolateral hemiseptum at zooecial bend blunt. Autozooecia erect in exozone; walls laminated with lining; basal diaphragms lacking; terminal diaphragms in some. One or two thin simple range walls separating ranges of autozooecia and subquadrate vesicles in endozone; granular-prismatic; range walls losing identity and continuity in exozone in most species; indistinct range walls and ranges of vesicles forming sinuous trace between ranges of autozooecia in some. Lunaria in exozone, hyaline, with thick distal lining continuous with zooecial lining. Vesicles in endozone high, boxlike to blisterlike, quadrate in cross section, in ranges and adjacent to proximal end of autozooecia, granular-prismatic in inner exozone, low, blisterlike, angular in cross section, laminated. Laminated stereom in exozone, with small acanthostyles. [*Dichotrypa* lacks well-defined compound range walls in the exozone and has monticules. In these characters it resembles members of the Hexagonellidae. In all other characters, it more closely resembles members of the Cystodictyonidae; hence, it is retained in this family with some reservation.] *?M.Dev.,Miss.*, N.Am., USSR. ——FIG 207,*1a–f.* **D. flabellum* (ROMINGER); *a,* mesotheca with short vertical plates, subquadrate autozooecia and vesicles in endozone; transv. sec., topotype, USNM 159857, ×30; *b,* thin, granular-prismatic mesotheca, autozooecia partially isolated by vesicles; transv. sec., topotype, USNM 159857, ×50; *c,* autozooecium with lunarium, stereom with acanthostyles; tang. sec., topotype, USNM 159857, ×100; *d,* zoarium with large monticules; syntype, UMMP 6505-14, ×2 (negative courtesy of A. Horowitz); *e,* teardrop-shaped autozooecia, vesicles and range walls in endozone (left), subcircular autozooecia in exozone (right); tang. sec., topotype, USNM 159857, ×30; *f,* moderately long recumbent portion of autozooecia in endozone, autozooecia partially isolated by vesicles in endozone, smaller vesicles and stereom in exozone; long. sec., topotype, USNM 159857, ×30.

Filiramoporina FRY & CUFFEY, 1976, p. 4 [**F. kretaphilia*; OD; Wreford Ls., L. Perm. (Wolfcamp.), Kans., USA]. Zoarium bifoliate; dichotomous branching in plane of mesotheca; branches slender, slightly compressed. Ridges low, present only locally between ranges of autozooecia, with sparse nodes. Mesotheca relatively thin. Autozooecia in endozone irregularly hemispherical in cross section, recumbent portion moderately long, microstructure granular-prismatic, zooecial bend sharp. Terminal diaphragms sparse, others absent. In exozone, autozooecia ovate in cross section, lunaria apparently absent. Compound range walls thin in endozone; thick, with radiating dark, granular zones in exozone. Vesicles in outer endozone irregular, small, blisterlike. Stereom with small ?tubules in exozone. *L.Perm.*, N.Am.——FIG. 208,*1a–f.* **F. kretaphilia*; *a,* mesotheca, compound range walls in endozone (left and right), vesicles in outer endozone; transv. sec., paratype, PSU PT06Aa-p-7012, ×188; *b,* recumbent autozooecia on mesotheca in endozone, sharp zooecial bend, vesicles in outer endozone, stereom in exozone; long. sec., paratype, PSU PT06Aa-p-7012, ×20; *c,* mesotheca, small vesicles in outer endozone, stereom in exozone; long. sec., holotype, PSU CH10Ab-p-7043, ×38; *d,* compound range walls, ranges of ovate autozooecia isolated by stereom; tang. sec., paratype, PSU CH10Ab-p-7105, ×33; *e,* mesotheca, subhemispherical cross section of autozooecia in endozone, com-

pound range walls with branching dark zones (valvae); transv. sec., paratype, PSU CH10Ab-p-7001b, ×56; *f,* mesotheca, small vesicles in outer endozone, stereom in exozone; transv. sec., paratype, PSU PT06Aa-p-7019, ×56.

Lophoclema MOROZOVA, 1955, p. 567 [**L. semichatovae*; OD; U. Carb., River Don, USSR]. Zoarium bifoliate; branches narrow, lenticular, with marginal noncelluliferous keels, anastomosing in plane of mesotheca; fenestrules large, subpolygonal; median range wall forming prominent ridge. Mesotheca thick, planar; dark (?granular) central layer and (?granular-prismatic) light-colored outer layers. Autozooecia subtriangular in contact with mesotheca; proximal end pointed, distal end rounded to subangular; hemispherical to subquadrate in cross section; diaphragms sparse in exozone. Autozooecia ovate in cross section in exozone; lunaria obscure. Compound range walls thin in endozone with dark boundary zone and light-colored (?granular-prismatic) lateral zones; branching dark zones and tubuli in exozone. Vesicles small, near proximal end of autozooecia in endozone; partially isolating autozooecia; small blisters in inner exozone; stereom (?laminated) with acanthostyles through most of exozone. *U.Carb.*, USSR.——FIG. 208,*2a–c.* **L. semichatovae,* holotype, PIN 436/57; *a,* planar, trilayered mesotheca, hemispherical to subquadrate autozooecia, stereom and range walls with branching dark zones in exozone; transv. sec., ×14; *b,* narrow branches, large fenestrules, prominent medial range wall, ovate autozooecia; deep tang. sec., ×14; *c,* compound range walls, subtriangular autozooecia in endozone, small subangular vesicles, ovate autozooecia and stereom with acanthostyles in exozone; tang. sec., ×38 (photographs courtesy of G. G. Astrova).

Ptilocella SIMPSON, 1897, p. 605 [**Ptilodictya parallela* HALL & SIMPSON, 1887, p. 270; OD; Hamilton Gr., M. Dev., Fall Brook, N.Y., USA] [=*Stictoporidra* SIMPSON, 1897, p. 532, NICKLES & BASSLER, 1900, p. 425; *Stictoporina* SIMPSON, 1897, p. 532 (*non* HALL & SIMPSON, 1887, p. xx), multiple original spelling]. Zoarium straplike, bifoliate; dichotomously branched in plane of mesotheca. Range walls protruding as one prominent median rib or as median rib with one rib on either side. Autozooecia with peristomes, no lunaria; reticulate arrangement in central two or three ranges and rhombic arrangement on lateral margins; margins thin, wide to narrow, noncelluliferous. Mesotheca slightly undulatory; central layer dark and outer layers light colored, laminated to granular-prismatic. Autozooecia subtriangular near branch center to hemispherical in cross section in endozone; circular in exozone; teardrop shaped at junction with mesotheca. Range walls compound; boundary zone continuing into dark zone of mesotheca in some;

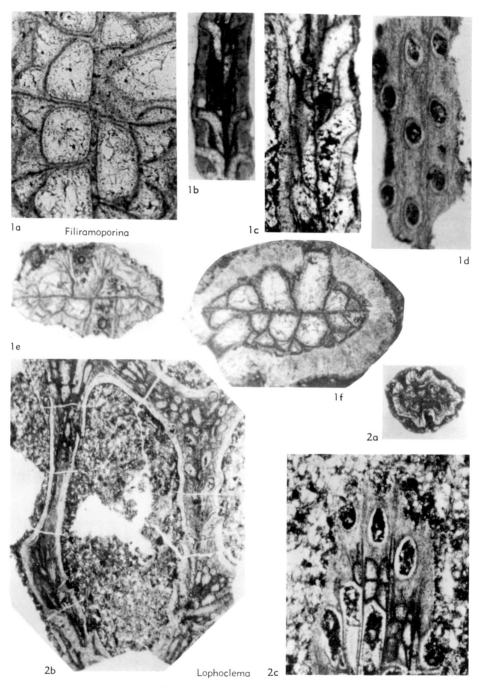

Fig. 208. Cystodictyonidae (p. 427).

indistinct branched dark zones and tubuli in range walls in exozone. Vesicles on mesotheca near proximal tips of autozooecia; small, blister-like; stereom laminated, with tubuli in exozone. [*P. subcarinata* (HALL & SIMPSON), the type species of *Stictoporidra* SIMPSON, 1897 by original designation, is consistent with the emended definition of *Ptilocella*.] *M.Dev.*, N.Am.——FIG. 206,*2a-e*. **P. parallela* (HALL & SIMPSON), holotype, NYSM 889; *a,* thick range walls with tubuli, laminated autozooecial walls, stereom; tang. sec., ×100; *b,* autozooecia and vesicles in endozone, stereom in exozone; oblique long. sec., ×30; *c,* midrib and two other prominent ribs at range walls, rhombic arrangement of lateral autozooecia; ×10; *d,* three prominent range walls at branch center, rhombically arranged and slightly oblique autozooecia near branch margins; tang. sec., ×30; *e,* trilayered mesotheca, triangular autozooecia near branch center, few vesicles in endozone, laminated stereom with tubules in exozone, indistinct branched dark zones and tubules in range walls in exozone; transv. sec., ×100.

Semiopora HALL, 1883b, p. 193 [**S. bistigmata*; M; Hamilton Gr., M. Dev., W. Williams, Ont., Can.]. Zoarium straplike, bifoliate; branches dichotomous or trichotomous in plane of mesotheca. Range walls prominent, distal and proximal ridges elevated; lunaria on side of autozooecia away from middle of branch. Mesotheca thin, undulatory, trilayered, laminated. Autozooecia subtriangular or subquadrate in cross section in endozone, aligned across mesotheca; teardrop shaped outline at contact with mesotheca, right- or left-handed in form. Walls laminated, with minute tubules perpendicular to thin, dark, zooecial boundary. Proximolateral, blunt hemiseptum at sharp zooecial bend. Lunaria laminated, ends projecting slightly. Compound range walls with dark boundary continuous into dark middle zone in mesotheca; dark zones branching in inner exozone, obscure in outer exozone; small tubules in range walls. Sparse, high vesicles near proximal tips of autozooecia in endozone; stereom laminated, with tubules in exozone. *M.Dev.*, N.Am.——FIG. 209,*2a-c*. **S. bistigmata,* holotype, NYSM 958; *a,* laminated walls and lunaria with ends projecting, prominent range walls with dark boundaries and tubules; tang. sec. ×100; *b,* thin mesotheca, subtriangular to subquadrate autozooecia and sparse vesicles in endozone, laminated stereom with tubules in exozone, range walls with branched dark zones in inner exozone; transv. sec., ×100; *c,* undulatory mesotheca, proximolateral hemiseptum at zooecial bend (upper left), sparse vesicles in endozone, laminated stereom in exozone; long. sec., ×100.

Stictocella SIMPSON, 1897, p. 532 [**Stictopora sinuosa* HALL, 1883b, p. 190; OD; Hamilton Gr., M. Dev., Union Springs, N.Y., USA]. Zoarium straplike, bifoliate. Range walls elevated, sinuous; peristomes present. Mesotheca thin; median layer dark, laminated or granular-prismatic layers light colored. Autozooecia hemispherical in cross section in endozone, some with

sinuses, ovate in cross section in exozone. Walls with dark boundary and perpendicular tubules in laminated cortex. Range walls with dark boundary branching in exozone. No discernable lunaria but proximal portion of autozooecial wall thicker and with more tubules. Vesicles in outer endozone, boxlike. Stereom laminated, with tubules in exozone. *M.Dev.*, N.Am.——FIG. 209,*1a-c*. **S. sinuosa* (HALL), holotype, NYSM 1005; *a,* thin mesotheca, low vesicles in outer endozone, stereom in exozone; long. sec., ×100; *b,* laminated autozooecial walls with tubules, range walls and stereom with tubules; tang. sec., ×100; *c,* hemispherical autozooecia, some with keels, in endozone, low vesicles in outer endozone, stereom with tubules in exozone, range walls with branched dark zones; transv. sec., ×100.

Sulcoretepora D'ORBIGNY, 1849, p. 501 [**Flustra? parallela* PHILLIPS, 1836, p. 200; OD; L. Carb., Yorkshire, Eng.] [=*Arcanopora* VINE, 1884, p. 204, obj.; *Acanthopora* VINE, MOROZOVA, 1960, p. 86, incorrect subsequent spelling; *Mstaina* SHULGA-NESTERENKO, 1955, p. 175]. Zoarium narrow bifoliate ribbons, dichotomously branched in plane of mesotheca; elongate, rounded autozooecia in ranges, rhombically arranged on lateral sides of branch; lunaria elevated on proximolateral side of autozooecia; range walls elevated; monticules absent; branch margins narrow, noncelluliferous. Mesotheca with dark central layer and laminated outer layers; sharply folded in center, undulatory near branch margins. Autozooecia full width and rectangular to parallelogram-shaped in deep tangential section; contiguous; alternating across mesotheca. Compound range walls with dark median zone continuous into boundary zone in mesotheca; lateral zones laminated; branched dark zones and tubules in thickened range walls in exozone. Autozooecia subquadrate to subhemispherical in cross section at mesotheca; angular teardrop shaped in deep tangential section in mid exozone and partially isolated between range walls by small, blisterlike vesicles. Walls laminated with minute tubules. Lunaria laminated, indistinct. Vesicles adjacent to mesotheca only at noncelluliferous branch margins. Stereom laminated with tubules in exozone. [*Mstaina laminicurvis* SHULGA-NESTERENKO, 1955, p. 176, from the Lower Carboniferous (Visean) at the River Msta, Vitsa, USSR, is the type species of *Mstaina* SHULGA-NESTERENKO, 1955, by original designation. Because it is similar to species of *Sulcoretepora*, *Mstaina* is considered to be a junior subjective synonym of *Sulcoretepora*.] *L.Carb.*, Eu., USSR.——FIG. 210,*1a-f*. **S. parallela* (PHILLIPS), Carboniferous Ls., Scot.; *a,* undulatory mesotheca, sharp zooecial bend; thin exozone; long. sec., HM D-113-2, ×30; *b,* ovate autozooecia between range walls in exozone;

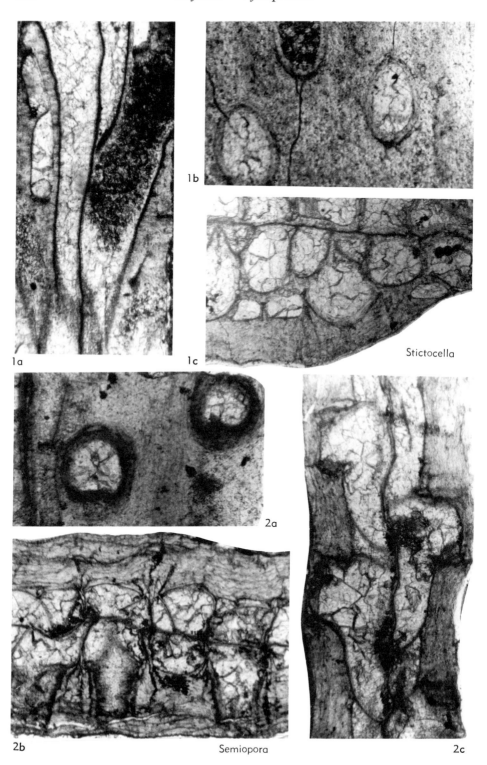

FIG. 209. Cystodictyonidae (p. 429).

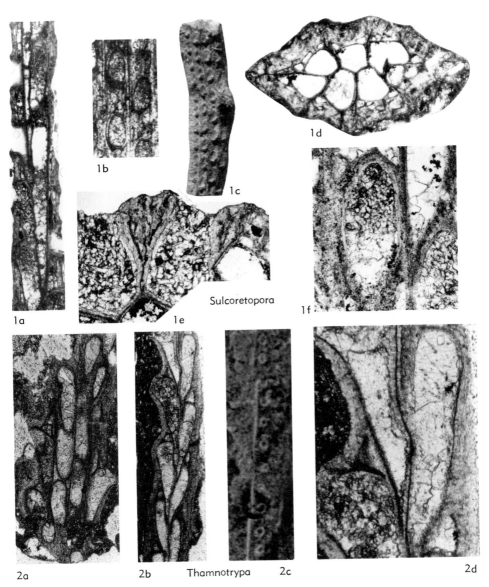

FIG. 210. Cystodictyonidae (p. 429–432).

tang. sec., HM D-113-4, ×30; *c,* straplike zoarium with elevated range walls and lunaria; HM
D-113, ×3; *d,* autozooecia alternating across
folded mesotheca, vesicles in outer endozone and
inner exozone, stereom in exozone; transv. sec.,
HM D-113-3, ×50; *e,* mesotheca (lower left),
range wall with branched dark zones and tubules;
transv. sec., HM D-113-2, ×100.

Taeniopora NICHOLSON, 1874a, p. 133 [**T. exigua*
NICHOLSON, 1874b, p. 122; SD HALL & SIMPSON,
1887, p. xii; Hamilton Gr., M. Dev., Bartlett's
Mills, Arkona, Ont., Can.] [=*Pteropora* HALL,

1883b; p. 192, *non* EICHWALD, 1860, p. 395].
Zoarium straplike, bifoliate; diamond shaped in
cross section; branching trichotomous. Median
rib on each branch prominent, elevated,
rounded, noncelluliferous; branch margin narrow, noncelluliferous. Peristomes elevated.
Mesotheca moderately thick, trilayered; central
zone dark and laminated or granular-prismatic
outer layers light colored. Midrib a prominent,
thick, compound range wall, branching dark
zones and tubules in exozone. Other range walls
loosing identity in exozone. Autozooecia hemi-

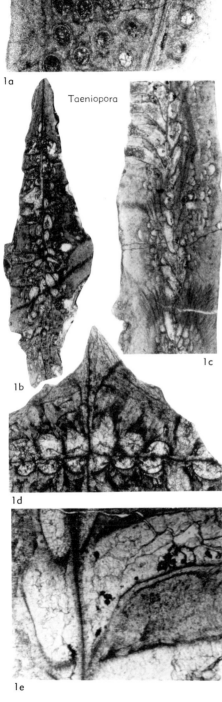

Taeniopora

1a

1b

1c

1d

1e

Fig. 211.   Cystodictyonidae (p. 431).

spherical in cross section in endozone, teardrop shaped at contact with mesotheca, right- or left-handed in form. Autozooecial walls with dark boundary and laminated cortex. Lunaria in outer endozone and exozone; thick, laminated; rotated toward side of autozooecia nearest branch margin away from center of branch. Vesicles on mesotheca near proximal tips of autozooecia; small, low, blisterlike vesicles in outer endozone and inner exozone; laminated stereom with minute acanthostyles and tubules through most of exozone. [Types of *Pteropora duogeneris* HALL, 1883b, p. 192, the type species of *Pteropora* HALL by original designation, are listed as hypotypes of *Taeniopora exigua* NICHOLSON, 1874b. Two specimens from Unadilla Forks, which possibly were the specimens originally described by HALL as *P. duogeneris,* are impressions. In external characters, they agree well with *Taeniopora exigua,* and *Pteropora* HALL, 1883b, is considered to be a junior synonym of *Taeniopora.*] *M.Dev.,* N.Am.——FIG. 211,*1a–e.* **T. exigua*; *a,* prominent midrib (right), circular autozooecia, small vesicles, no well-defined range walls lateral to midrib; tang. sec., topotype, USNM 159844, ×28; *b,* planar mesotheca (vertical), thick midrib, stereom in exozone; oblique transv. sec., topotype, USNM 159844, ×19; *c,* autozooecia (upper left), stereom with acanthostyles and vesicles in exozone; long. sec., topotype, USNM 159845, ×19; *d,* mesotheca (horizontal), midrib (vertical) with several dark zones, hemispherical autozooecia in endozone, laminated stereom in exozone; transv. sec., topotype, USNM 159847, ×47; *e,* trilayered mesotheca with granular-prismatic outer layers, vesicles in endozone (left), laminated autozooecial walls and stereom (right); long. sec., topotype, USNM 159846, ×94.

Thamnotrypa HALL & SIMPSON, 1887, p. xxi, *nom. subst. pro Thamnopora* HALL, 1883b, p. 158, *non* STEININGER, 1831, p. 10 [**Thamnopora divaricata* HALL, 1883a, pl. 26; OD; Onondaga Ls., M. Dev., Buffalo, N.Y., USA]. Zoarium narrow, trifoliate, subtriangular in cross section; branches nearly at right angles to main portion. Median rib rounded, conspicuous; other range walls not expressed as ridges. Autozooecia in rows, with peristomes; lunaria not observed. Mesotheca thin, undulatory along branch axis; trilayered, central zone dark and lateral zones light colored. Autozooecia subhemispherical in cross section in endozone; ovate teardrop shaped in outline at contact with mesotheca, right- or left-handed in form. Walls with thin dark boundary and laminated cortex with tubules perpendicular to boundary. Compound range walls with branching dark zones in exozone. Vesicles high blisters, subtriangular in cross section in endozone. Stereom laminated, with tubules in exozone. *M.Dev.,* N.Am.——FIG. 210,*2a–d.* **T. divar-*

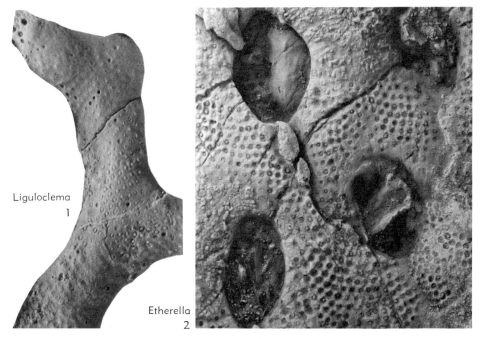

Liguloclema
1

Etherella
2

Fig. 212.   Etherellidae (p. 433).

icata (Hall), holotype, NYSM 1039; *a,* right- and left-handed autozooecia aligned between compound range walls; deep tang. sec., ×30; *b,* undulatory mesotheca, autozooecia, sparse vesicles, stereom in exozone; long. sec., ×30; *c,* prominent midrib, autozooecia with peristomes; ×10; *d,* trilayered mesotheca, sparse vesicles in endozone, laminated stereom with tubules in exozone; long. sec., ×100.

Family ETHERELLIDAE
Crockford, 1957

[Etherillidae Crockford, 1957, p. 30]

Zoaria bifoliate, compressed; form cribrate or branching straplike. Monticules lacking. Margins on branches narrow, noncelluliferous. Autozooecia recumbent on mesotheca, recurved with hook-shaped appearance at zooecial bend; erect in exozone. Lunaria present or possibly absent. Small vesicles in endozone, stereom in exozone. [Genera in the Etherellidae are poorly known owing to lack of thin-sectioned specimens. The recurved, hook-shaped appearance of the autozooecia at the zooecial bend is presumed to be the diagnostic feature of

the family, but further study may reveal a close relationship to the Cystodictyonidae.] *Perm.*

Etherella Crockford, 1957, p. 32 [**E. porosa*; OD; Noonkanbah F., Perm., Fitzroy basin, W. Australia]. Zoarium bifoliate, cribrate. Mesotheca trilayered. Autozooecia partially isolated at mesotheca by vesicular tissue; recumbent portion relatively long; hooked around oblique vertical plate at zooecial bend. Lunaria not apparent. Stereom narrowly isolating autozooecia in exozone. [Internal features and microstructure of this genus are poorly known.] *Perm.*, Australia. ——Fig. 212,*2.* **E. porosa,* holotype, CPC 1102 A; cribrate zoarium, rhombic arrangement of autozooecia narrowly isolated by stereom, hook-shaped autozooecia in weathered portion (above), ?lunaria in unweathered portion (lower right); ×5 (photograph courtesy of R. E. Wass).

Liguloclema Crockford, 1957, p. 35 [**L. typicalis*; OD; Noonkanbah F., Perm., Fitzroy basin, Australia]. Zoarium bifoliate, compressed, straplike, irregularly branching. Autozooecia with lunaria. Stereom widely isolating autozooecia in exozone. [Internal anatomy of this genus is poorly known.] *Perm.*, Australia.——Fig. 212,*1.* **L. typicalis,* holotype, CPC 1106 A; zoarium, indistinct lunaria in less weathered portion (below); ×5 (photograph courtesy of R. E. Wass).

Family GONIOCLADIIDAE
Waagen & Pichl, 1885

[*nom. transl.* NIKIFOROVA, 1938, p. 195, *ex* Goniocladiinae WAAGEN & PICHL, 1885, p. 775]

Zoaria bifoliate, mesotheca vertical in cylindrical to laterally compressed branches. Primary branches with paired secondary branches at right angles, or distolaterally directed; or with alternating distolaterally directed secondary branches, some with tertiary branches; secondary or tertiary branches fused in some genera to produce large fenestrules. Obverse side with one to six rows of autozooecia on flanks dipping moderately to steeply away from median carina. Reverse side noncelluliferous, with or without keel. Monticules absent. Autozooecia with or without sparse diaphragms; walls generally well laminated, with laminated lunarium in most genera. Vesicles small, generally sparse in inner endozone, more abundant in outer endozone; replaced by generally laminated stereom, with tubules or acanthostyles, in exozone or outer exozone. *Miss.-Perm.*

Characters of particular importance are the vertical mesotheca, the celluliferous obverse side, and the noncelluliferous reverse side.

Goniocladia ETHERIDGE, 1876, p. 522, *nom. subst. pro Carinella* ETHERIDGE, 1873, p. 433; *non* JOHNSTON, 1833, p. 232 [*Carinella cellulifera* ETHERIDGE, 1873, p. 433; OD; "Low. Ls. Gr.," L. Carb., Braidwood, Eng.]. Zoarium bifoliate; narrow, curved dichotomous branches in some anastomosing to form large fenestrules; vertical mesotheca protruding as ridge on rounded, noncelluliferous reverse side and as sharp keel on peaked obverse side. Autozooecia in two to three rows on either side of median carina; subcircular apertures opening upward and indented by ends of lunaria on side of autozooecia away from branch center. Mesotheca thin; median layer dark, outer layers laminated. Autozooecia hemispherical in cross section at mesotheca, partially isolated by small vesicles; curving upward and outward; diaphragms sparse. Walls with thin, dark, serrated boundary and laminated cortex having minute tubules perpendicular to boundary. Lunaria laminated, with numerous tubules in proximal side. Vesicles forming small blisters, laminated; stereom laminated, with numerous tubules and small, indistinct acanthostyles in most of exozone and noncelluliferous reverse side. *Miss.-Perm.*, Eu., Asia, Australia.——FIG. 213,1a–h. *G. cellulifera* (ETHERIDGE); *a,*

branching zoarium, elevated lunaria; topotype, BMNH D32637, ×9; *b,* vertical mesotheca, vesicles and stereom in exozone, stereom on reverse side (below); transv. sec., topotype, BMNH D32637-3, ×50; *c,* laminated walls and lunaria with tubules on outer side (right), stereom with numerous acanthostyles; tang. sec., topotype, BMNH D32637-2, ×100; *d,* solid reverse side (left), autozooecia and vesicles in endozone; oblique long. sec., topotype, BMNH D32637-7, ×50; *e,* mesotheca, autozooecia curve distolaterally, vesicles in outer endozone, stereom in exozone; tang. sec., topotype, BMNH D32637-5, ×30; *f,* mesotheca (left), autozooecia isolated by stereom and vesicles, lunaria on outer (right) side; tang. sec., topotype, BMNH D32637-4, ×50; *g,* subcircular autozooecia; tang. sec., topotype, BMNH D32637-7, ×30; *h,* wall microstructure, vesicles and stereom; long. sec., topotype, BMNH D32637-7, ×100.

Aetomacladia BRETNALL, 1926, p. 21 [*A. ambrosioides;* M; Carb., Fossil Hill, Australia]. Zoarium of slender branches, each with vertical mesotheca; paired secondary branches at right angles to primary branch, perpendicular to mesotheca. Obverse with prominent keel at mesotheca, steep flanks; autozooecia in rows with lunarium on side away from branch center. Reverse side rounded with weak striations and median rib where mesotheca protrudes. Mesotheca trilayered; median tubules (?acanthostyles) clear, vertical. Autozooecia hemispherical in cross section at mesotheca; curving outward and upward toward flanks of obverse side. Walls laminated with minute tubules; diaphragms sparse in endozone. Lunaria laminated, thick, with tubules on proximal side. Vesicles low blisters in endozone and inner exozone; stereom laminated, with acanthostyles and tubules through most of exozone. *Carb.-Perm.*, Australia.——FIG. 214,2a–e. *A. ambrosioides,* Callytharra F., Perm., W. Australia; *a,* primary branch with secondary branches nearly at right angles, autozooecia with lunaria, median rib on each branch; USNM 159850, ×10; *b,* vertical mesotheca running from obverse (top) to reverse (bottom), hemispherical autozooecia in endozone; transv. sec., USNM 159849, ×30; *c,* autozooecia and vesicles in endozone, stereom with acanthostyles and tubules in exozone; oblique long. sec., USNM 159851, ×30; *d,* autozooecia with diaphragms in endozone, blisterlike vesicles in outer endozone, stereom in exozone; long. sec., USNM 159848, ×50; *e,* median rib formed by mesotheca with median tubules (below), rows of autozooecia, with lunaria on side away from median rib, widely isolated by stereom; tang. sec., USNM 159848, ×30.

Goniocladiella NEKHOROSHEV, 1953, p. 166 [*G. kasakhstanica;* M; L. Carb., Kazakh., USSR]. Zoarium bifoliate, vertical mesotheca, noncel-

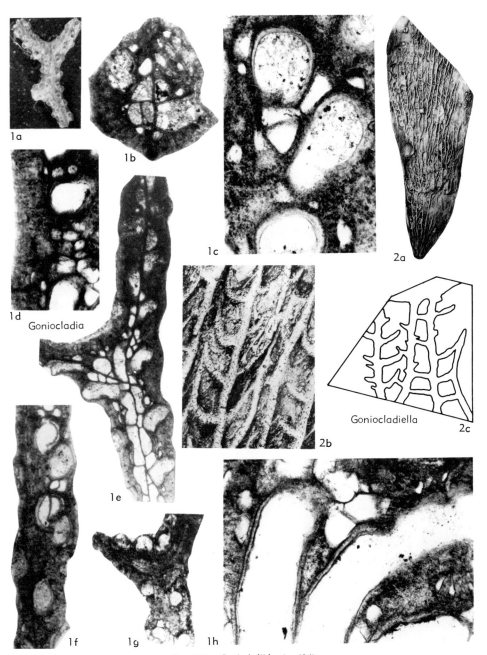

1a

1b

1c

2a

1d

Goniocladia

1e

2b

Goniocladiella

2c

1f

1g

1h

FIG. 213. Goniocladiidae (p. 434).

luliferous reverse and few rows of autozooecia on rounded obverse side. Main branches subparallel, undulatory; side branches diverging distolaterally, fusing to form frond with polygonal fenestrules in some. [Types of the type species of *Goniocladiella* are impressions and the internal anatomy is not known. *G. parallela* NEKHORO-

SHEV, 1956, has been erroneously cited as the type species, but it too is known only from surface features. Thin sections of species of this genus were not available for study.] *L.Carb.*, USSR. ——FIG. 213,*2a,b*. *G. parallela* NEKHOROSHEV, L. Carb., C^2_1, Altai, USSR, type; *a,* fenestrate zoarium with polygonal fenestrules;

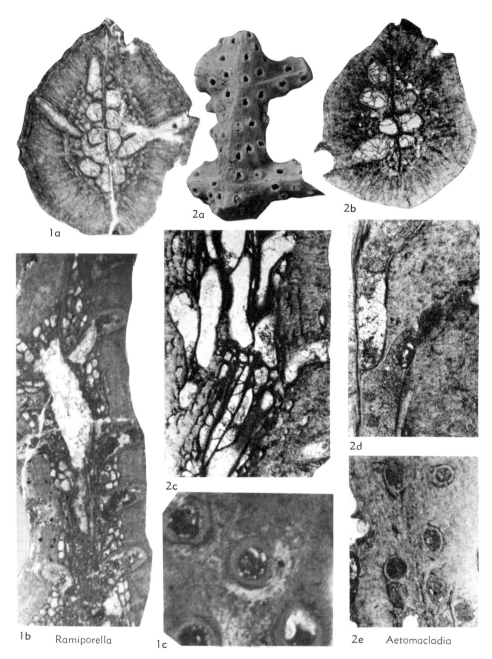

1a

2a

2b

1b   Ramiporella

1c

2c

2d

2e   Aetomacladia

FIG. 214.   Goniocladiidae (p. 434–438).

×0.5; *b,* undulatory branches, irregular fenestrules, few rows of autozooecia; ×3.0 (photographs courtesy of V. P. Nekhoroshev).——FIG. 213,*c.* **G. kasakhstanica,* holotype; fenestrate colony; ×1.5 (after Nekhoroshev, 1953, pl. 24, fig. 4a).

Ramipora TOULA, 1875, p. 230 [**R. hochstetteri;* OD; Perm.-Carb. (?L. Perm.), Spits.]. Zoarium bifoliate; branches rounded, with vertical mesotheca forming keel on noncelluliferous reverse side and on celluliferous obverse side. Main branches with secondary and tertiary branches

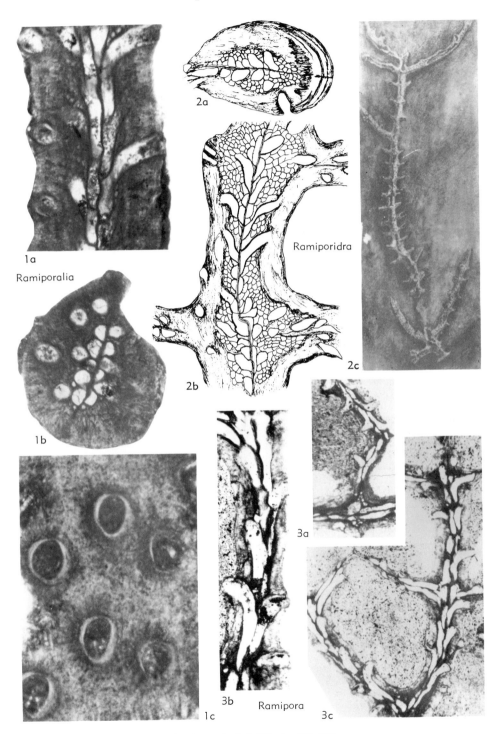

Fig. 215. Goniocladiidae (p. 436–438).

diverging distolaterally, commonly in pairs, and joining to form frond with large, polygonal fenestrules. Autozooecia partially isolated at mesotheca, hemispherical in cross section; recumbent portion long, erect in exozone; diaphragms sparse. Few rows of subcircular autozooecia with lunaria, isolated by stereom. Vesicles in endozone and exozone, blisterlike; stereom laminated, with tubules through most of exozone. *Perm.*, USSR, Spits.——Fig. 215,*3a–c.* *R. hochstetteri,* U. Perm., Starostinskaja Suite, Spits.; *a,* fusion of secondary branches, laminated stereom isolates autozooecia in exozone; deep tang. sec., PIN 2237/254, ×10; *b,* long recumbent portion of autozooecia, sparse vesicles at zooecial bend, stereom in exozone; deep tang. sec., PIN 2237/255, ×15; *c,* branched zoarium with large polygonal fenestrules, vertical mesotheca with one range of autozooecia on each side; deep tang. sec., PIN 2237/255, ×10 (photographs courtesy of I. P. Morozova).

Ramiporalia SHULGA-NESTERENKO, 1933, p. 42 [*R. dichotoma;* OD; L. Perm., N. Urals, USSR]. Zoarium bifoliate; branches dichotomous and subcylindrical, each with vertical mesotheca protruding as low carina on celluliferous obverse side, reverse side noncelluliferous. Autozooecia hemispherical in cross section at mesotheca, partially isolated by tiny vesicles; recumbent portion long, erect in exozone; diaphragms not seen; walls laminated, with laminated lunarium on side of zooecia away from branch center. Vesicles extremely small blisters in endozone and inner exozone. Laminated stereom with tubules in exozone isolating intersecting rows of autozooecia with subcircular apertures. *L.Carb.-U.Perm.,* USSR, Australia.——Fig. 215,*1a–c.* *R. dichotoma,* holotype, PIN 2985/0636; *a,* undulatory mesotheca, long recumbent portion of autozooecia, small vesicles in endozone and inner exozone, stereom in exozone; long. sec., ×20; *b,* mesotheca, hemispherical autozooecia and small vesicles in endozone, stereom with numerous tubules (reverse side below); transv. sec., ×20; *c,* subcircular autozooecia, with laminated walls and lunaria (left), isolated by laminated stereom with tubules; tang. sec., ×40 (photographs courtesy of G. G. Astrova).

Ramiporella SHULGA-NESTERENKO, 1933, p. 39 [*R. asimmetrica;* OD; L. Perm., N. Urals, USSR]. Zoarium bifoliate; branches subcylindrical; primary branch straight or undulatory, with alternating right and left secondary branches and some tertiary branches. Mesotheca vertical, thin, protruding as low carina on subrounded obverse side; reverse side more rounded. Autozooecia in endozone hemispherical in cross section, partially isolated by vesicles; erect in exozone, subcircular in cross section. Walls laminated, with laminated lunarium on side of auto-

zooecia away from branch center. Vesicles in endozone and exozone; small, low blisters; stereom laminated, with numerous tubules, in outer exozone. *?L.Carb.,U.Carb.-U.Perm.,* USSR, Australia.——Fig. 214,*1a–c.* *R. asimmetrica,* holotype, PIN 2985/115; *a,* hemispherical autozooecia partially isolated by vesicles in endozone, laminated stereom in exozone; transv. sec., ×20; *b,* autozooecia isolated by moderately small vesicles in endozone and inner exozone, laminated stereom in outer exozone; long. sec., ×20; *c,* thick, laminated walls and lunaria in autozooecia, isolated by laminated stereom with tubules; tang. sec., ×40 (photographs courtesy of G. G. Astrova).

Ramiporidra NIKIFOROVA, 1938, p. 197 [*Ramipora uralica* STUCKENBERG, 1895, p. 169; OD; L. Perm., N. Urals, USSR]. Zoarium bifoliate; branches narrow, with paired second and third order branches; branch ends fused in some, forming fenestrate zoarium. Mesotheca vertical, protruding as carina on obverse side of branch; reverse side of branch rounded. Autozooecia in 5 or 6 ranges on each side of carina; hemispherical in cross section and partially isolated by vesicles in endozone; diaphragms absent; erect in endozone, subcircular in cross section with indistinct lunaria. Vesicles low blisters in endozone, decreasing in height into exozone. Stereom possibly laminated, with tubules, in outer exozone, widely isolating autozooecia. *?L.Carb., U.Carb.-U.Perm.,* USSR.——Fig. 215,*2a–c.* *R. uralica* (STUCKENBERG); *a,* vertical mesotheca, vesicles in endozone, stereom in exozone; transv. sec., TsGM No. 982/305, ×11; *b,* autozooecia isolated by vesicular tissue at mesotheca, laminated stereom in outer exozone; deep tang. sec., TsGM No. 982/305, ×11; *c,* primary, secondary, and tertiary branches, median carina, few ranges of autozooecia on obverse side; holotype, Museum, State University of Kazan, ×1 (photograph courtesy of V. P. Nekhoroshev).

GENERIC NAMES OF INDETERMINATE OR UNRECOGNIZABLE STATUS ASSIGNED TO CYSTOPORATA

Anellina GREGORIO, 1930, p. 33 [*Eschara (Anellina) parvula;* M]. *Perm.,* Italy.

Archaeotrypa FRITZ, 1947, p. 435 [*A. prima;* OD]. *U.Cam.,* Alberta, Can.

Cycloidotrypa CHAPMAN, 1920, p. 366 [*C. australis;* OD]. *L.Carb.,* Moorowarra, Australia.

Diptheropora DEKONINCK, 1873, p. 13 [*D. regularis;* OD]. *Carb.,* Bleiberg, Ger.

Tuberculopora RINGUEBERG, 1886, p. 21 [*T. inflata;* OD]. *M.Sil.,* Lockport, N.Y., USA.

GENERIC NAMES ERRONEOUSLY ASSIGNED TO CYSTOPORATA

Bolopora Lewis, 1926, p. 420. Chemogenic or bacteriogenic dubiofossil (Hofmann, 1975).

Cambroporella Korde, 1950, p. 371. Alga.

Coenites Eichwald, 1829, p. 179. Tabulate coral. Specimens of this genus lack vesicular tissue and the "lunaria" are proximal projections of oblique corallites developed during late ontogeny.

Dianulites Eichwald, 1829, p. 180. Trepostomate.

Glossotrypa Hall, 1886, pl. xxxi. Trepostomate.

Revalotrypa Bassler, 1952, p. 382. Trepostomate.

Solenopora Dybowski, 1878, p. 124. Alga.

Spatiopora Ulrich, 1882, p. 155. Trepostomate.

NOMINA NUDA

Diaphragmopora McFarlan, 1926, p. 223.

Didymopora Ulrich, 1882, p. 156.

Pakridictya Männil, 1959, p. 38.

THE ORDER CRYPTOSTOMATA

By Daniel B. Blake

[University of Illinois, Urbana]

The Cryptostomata was first proposed as a bryozoan suborder by Vine (1884). Only five genera were assigned to the suborder. Of these, two have since been considered bifoliate cryptostomates (*Ptilodictya* and *Stictoporella*) and two, rhabdomesine cryptostomates (*Glauconome,* now *Glauconomella,* and *Rhabdomeson*). One is now assigned to the Cystoporata (*Arcanopora,* now *Sulcoretepora*). The fenestellines were not included in the original concept. According to Vine, cryptostomates are distinguished by the tubular to subtubular longitudinal outline of zooecia, their angular cross section, and the vestibule concealing the orifice.

In ensuing years, Ulrich (1890, 1893) placed additional families including the fenestellids in the group, which he raised to ordinal rank. He provided a diagnosis and a list and descriptions of component families. Ordinal diagnoses since Ulrich have stressed a limited number of characters, the most important being zooecial shape; presence of a vestibule; a well-developed, generally abruptly arising exozone; and, in many taxa, the presence of hemisepta.

Ulrich was much concerned with relationships among cryptostomates. Many morphological similarities mentioned here were first noted by him (1890, 1893), and he expressed such relationships as those between phylloporinids and fenestellines in phylogenetic terms. Ulrich (1890) stressed the importance of what now would be called a polythetic classification (". . .in the aggregate of characters is found the true test of relationship," p. 329); and in various points in his text, he suggested phylogenetic relationships among taxa (p. 357) and the problems evolution imposes upon the recognition of taxa (p. 328).

McNair (1937) informally recognized three zoarial types in the cryptostomates, a unilaminate group including the phyllopo-

---

FIG. 216. Morphology of the Cryptostomata.——*1. Rhabdomeson* sp., Rhabdomesina; Cathedral Mt. F., M. Perm., Texas; cryptostomate holdfast, rapid development of erect growth habit; USNM 222618, ×5.6.——*2. Arthrostyloecia nitida* Bassler, Rhabdomesina; Edinburg F., M. Ord., Va.; reverse, ridge-bearing surface to left, obverse surface to right; apertures enclosed by prominent peristomes; syntype, USNM 240802, ×9.2.——*3. Ulrichostylus* sp., Rhabdomesina; Bromide F., M. Ord., Okla.; a primitive cryptostomate showing cylindrical growth habit, apertural alignment, longitudinal ridges, and extrazooecial skeleton developed as a basal articulation process; USNM 214196, ×9.2.——*4. Orthopora regularis* (Hall), Rhabdomesina; M. Dev., N.Y.; ramose dichotomous growth habit; cylindrical branches, aligned apertures, and stylet development; syntype, AMNH 35758B, ×3.6.——*5. Saffordotaxis incrassata* (Ulrich), Rhabdomesina; New Providence F., L. Miss., Ky.; spiral arrangement and elliptical outlines of apertures, stylet development, UI X-5380, ×5.6.——*6. Rhabdomeson* sp., Rhabdomesina; Word F., Perm., Texas; conical growth habit, extreme monticule development, apertural arrangement; USNM 222620, ×2.3.——*7. Rhinidictya grandis* Ulrich, Ptilodictyina; Platteville Gr., M. Ord., Ill.; ramose growth habit, flattened branch outline; linear and spiral arrangement of apertures; syntype, USNM 43606, ×9.2.——*8. Hemitrypa proutana* Ulrich, Fenestellina; Warsaw F., mid. Miss., Ill.; reticulate growth habit and a fragment of netlike extrazooidal skeletal superstructure (arrow); ISGS 2818-1, ×5.6.——*9. Phylloporina clathrata* (Miller & Dyer), Fenestellina; Mt. Hope F., U. Ord., Ky.; fenestrate growth and apertural arrangement in a primitive fenestellid; USNM 214213, ×9.2.——*10. Phylloporina variolata* Ulrich, Fenestellina; Eden Gr., U. Ord., Ohio; reverse surface in a phylloporinid; syntype, USNM 214214, ×9.2.——*11. Archimedes* sp., Fenestellina; Pennington F., U. Miss., Ky.; extrazooidal skeleton in form of an axial column and basal struts (arrow); UI X-5381, ×1.6.——*12. Lyropora divergens* Ulrich, Fenestellina; U. Miss., Ill.; fenestrate growth habit and extrazooecial basal keel; syntype, ISGS 2783, ×3.6. ——*13. Acanthocladia fruticosa* Ulrich, Fenestellina; Penn., Ill.; pinnate growth habit, development of surficial ridges, small peristomes enclosing apertures; ?syntype, ISGS 4471-1, ×3.6.

FIG. 216. *(For explanation, see facing page.)*

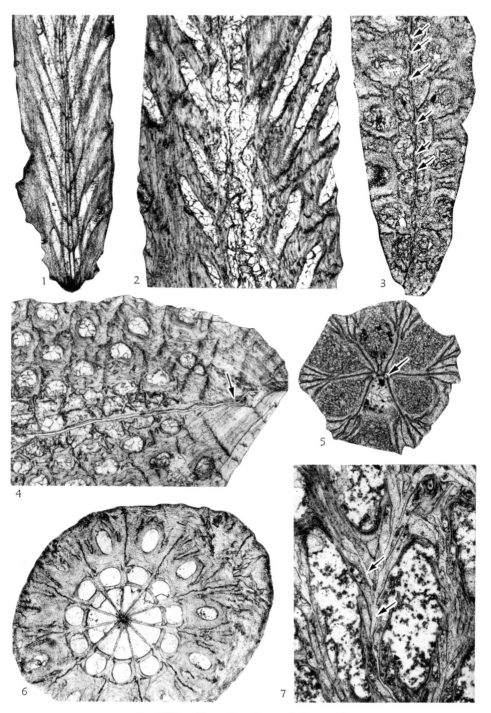

F_{IG}. 217.   *(For explanation, see facing page.)*

rinids and fenestellids, a cylindrical group including the arthrostylids and rhabdomesids, and a bifoliate group including the rhinidictyids, ptilodictyids, and sulcoreteporids. McNair (1937, p. 154) did not consider the three divisions to be taxonomically significant: "There is little evidence to indicate that these three divisions based on zoarial forms should be considered taxonomic units or that members within such divisions are related phylogenetically; it seems more probable that similar zoaria were evolved by a number of stocks."

Astrova and Morozova (1956) formally recognized the three groups of McNair (1937) as suborders, interpreting each as a natural phylogenetic branch. Subsequently, Morozova (1966) established the Timanodictyoidea for certain Permian bryozoans. Termier and Termier (1971) interpreted the three original suborders as phylogenetically distinct, and they thought that the order Cryptostomata represents an evolutionary grade.

The Cryptostomata was subdivided first through the recognition of the order Fenestrata by Elias and Condra (1957) and then by the recognition of the order Rhabdomesonata by Shishova (1968). These changes left only the Ptilodictyoidea and the Timanodictyoidea in the Cryptostomata. The Fenestrata of Elias and Condra was based largely upon the presence of a "colonial plexus," expressed as a clear granular calcite layer. Also important in the concept of the Fenestrata were inferred homologies between the colonial plexus and the cyclostomate common bud. Shishova (1968), in considering the fenestellids, further emphasized zooecial shape, budding pattern, microstructure, the presence of peristomes, and in some cases, lunaria and ovicells. The Rhabdomesonata of Shishova was based primarily upon zoarial form, zooecial shape, and budding pattern. Shishova did not consider the three traditional cryptostomate suborders to be closely related.

The new ordinal concepts of Fenestrata and Rhabdomesonata have not been universally accepted. For example, following the classification of Astrova and Morozova (1966), Tavener-Smith and Williams (1972) considered the three major groups to be suborders. Cuffey (1973), in a numerical taxonomic study, retained the unified Cryptostomata. Within the order, he concluded that the rhabdomesines and ptilodictyines were closer to each other than either were to the fenestellines. He recognized two suborders with the rhabdomesines and ptilodictyines as infraorders in one and the fenestellines alone in the other.

Blake (1975, 1980) argued for retention of the three traditional branches as suborders within an order Cryptostomata because of morphological similarities among the Ordo-

---

Fig. 217. Comparison of Arthrostylidae and Stictoporellidae.——*1. Ulrichostylis* sp., Rhabdomesina; Bromide F., M. Ord., Okla.; long, slightly curved zooecia in an early arthrostylid; long. sec., USNM 214205, ×28.——*2. Stictoporellina gracilis* (Eichwald), Ptilodictyina; Ranicips Ls., L. Ord., Öland, Swed.; long, slightly curved zooecia in an early ptilodictyid; long. sec., USNM 214207, ×28.——*3. Stictopora* sp., Ptilodictyina; McLish F., M. Ord., Okla.; median wall bearing mural rods (arrows), granular zones, and fine laminae; transv. sec., USNM 222627, ×28.——*4. Stictoporellina gracilis* (Eichwald), Ptilodictyina; same data as *2*; median wall, granular zones, extrazooecial skeleton at lateral margin of zoarium, small interval of laminated median wall (arrow), and extensive wall constructed of fine laminae; prominent clear line is a crack following median wall; transv. sec., USNM 214207, ×28.——*5. Nematopora lineata* Billings, Rhabdomesina; Sil., Anticosti Is., Can.; median rods (arrow); granular zones along midline of endozonal wall, radiating in the exozone; fine laminae and trend toward development of a median wall; transv. sec., USNM 43384, ×92.——*6. Ulrichostylus* sp., Rhabdomesina; same data as *1*; sharply defined central axis; granular zones along zooecial boundaries in endozone, then radiating in exozone; fine laminae; transv. sec., USNM 214211, ×28.——*7.* Unidentified genus and species, Phylloporinidae, Fenestellina; Bromide F., M. Ord., Okla.; long attenuated zooecia, thickened and non-laminated layer (arrows); long. sec., USNM 214217, ×28.

FIG. 218.   *(For explanation, see facing page.)*

vician members of the Arthrostylidae (Rhabdomesoidea), Phylloporinidae (Fenestelloidea) and some genera of the Rhinidictyidae and Stictoporellidae (Ptilodictyoidea).

THE ORDER CRYPTOSTOMATA

The Cryptostomata are an order of almost entirely erect stenolaemate bryozoans with generally limited bases of attachment (Fig. 216,1). Zoaria may be unbranched, bushlike, pinnately branched, or reticulated (Fig. 216,2–13). Stem cross sections are approximately cylindrical or flattened. Zooecia generally are arranged in regular longitudinal or spiral rows, and the apertures are elliptical, subcircular, or rectangular in outline (Fig. 216,2,3,5–9). Zooecial apertures may be present on all surfaces, or a barren reverse surface may be developed (Fig. 216,2,10). Surfaces are commonly ornamented with striae or ridges, peristomes (Fig. 216,2–4), and stylets or similar structures (Fig. 216,5). Extrazooecial skeletal material may be extensive, developed as rootlike attachment structures, as thickened deposits between zooecia or along the reverse surfaces of zoaria, in the form of a netlike superstructure, or in other patterns (Fig. 216,2,3,8,11,12). Multizooecial skeletal deposits are present within the reverse wall of some taxa.

Budding took place from linear or planar loci (Fig. 217,3–6). Zooecia generally are short (Fig. 218,2), but can be long and attenuated, especially in earlier representatives (Fig. 217,1,2). In most genera, the endozonal walls are thin and the exozonal walls relatively thick (Fig. 218,1). The transition between the two zones usually is abrupt and marked by a distinct zooecial bend, a change in orientation in which the zooecial axes turn from approximately parallel to the branch axis to essentially perpendicular to the branch surface (Fig. 218,1,6). The body cavity within the exozone, the so-called vestibule, commonly is constricted. One or more hemisepta are common on the proximal wall at the zooecial bend, and one or more may be present on the distal wall, in the outer endozone (Fig. 218,6). Other intrazooecial structures include mural spines and diaphragms (Fig. 219,2), the latter most commonly in genera with elongate zooecia.

Cryptostomate walls are constructed primarily of microscopically laminated deposits, but microscopically nonlaminated material usually is present, and may be extensively developed in the Fenestellina (Fig. 217,7; 221,1). Nonlaminated material forms the interior of walls, usually along zooecial boundaries. It may be discontinuous, appearing as intermittent granules. Nonlaminated material forms the axes of various types of stylets (Fig. 219,4–6,8,9). Walls in erect portions of zoaria apparently were entirely compound in nature, and diverse polymorphs and nonpolymorphic openings were developed (Fig. 216,6; 218,3–5; 219,1,3).

FIG. 218. Comparison of Arthrostylidae and later Rhabdomesina.——*1. Cuneatopora lindstroemi* ULRICH, Rhabdomesina; Sil., Gotl.; arthrostylid morphology; note zooecial shape and well-defined central axis; ?paratype, USNM 214193, ×28.——*2. Nematopora fragilis* ULRICH, Rhabdomesina; Sil., Ill.; short, robust zooecia and median rod representative of some arthrostylids; long. sec., ?syntype, USNM 214222, ×91.——*3. Streblotrypa* cf. *S. marmionensis* ETHERIDGE, Rhabdomesina; Perm. (Callytharra), W. Australia; multiple metapores (arrows) associated with each zooecium; long. sec., USNM 112466, ×91.——*4. Acanthoclema scutulatum* HALL, Rhabdomesina; Ludlowville Sh., M. Dev., N.Y.; stylet and single metapore (arrows) associated with each zooecium; long. sec., USNM 214200, ×91.——*5. Helopora* sp., Rhabdomesina; Jupiter River F., Sil., Anticosti Is., Can.; single metapore (arrows) with each zooecium; long. sec., USNM 214199, ×55.——*6. Orthopora* sp., Rhabdomesina; Keyser Ls., Sil.-Dev., W. Va.; zooecial form in early member of the suborder; well-developed zooecium in the plane of section (arrow); in distal and proximal parts of section, central axis is passing out of plane of view; well-developed hemisepta and acanthostyles also evident; long. sec., USNM 222625, ×68.——*7. Ulrichostylus* sp., Rhabdomesina; Bromide F., M. Ord.; irregular nonlaminated skeletal layer (arrow) near colony axis and in the midline of the wall, enclosed by thicker laminated layer; transv. sec., USNM 222626, ×182.

Space-enclosing vesicles occur in bifoliates (Fig. 219,2). Structurely diverse stylets are developed (Fig. 219,4–9). Cryptostomates only rarely show development of overgrowths, and interzooecial pores are lacking.

A traditional argument for grouping the three cryptostomate suborders is reflected in the ordinal name, "hidden mouth." Various earlier workers believed the terminal membrane to be beneath the skeletal surface of the branch at the base of the vestibule, which is that portion of the zooecium within the exozone. The only evidence for a terminal-membrane position at the base of the vestibule appears to be the complex shape of the zooecium. Although recessed terminal diaphragms have been recognized in modern tubuliporates, these diaphragms are near the aperture in most taxa. There is no strong evidence that cryptostomate terminal membranes were recessed.

The order Cryptostomata contains the suborders Rhabdomesina, Fenestellina, Ptilodictyina, and Timanodictyina. Some subordinal grouping, such as that suggested by CUFFEY (1973), seems desirable but is deferred until completion of *Treatise* revision of stenolaemate taxa. Affinities of the Cryptostomata within the phylum have been a matter of some controversy, and in different classifications the order has been included within both the stenolaemates and the gymnolaemates. BOARDMAN (this revision) discusses the subject in his historical review of stenolaemate studies.

The Timanodictyina is not further considered here because of its exclusively upper Paleozoic range.

COMPARISON OF EARLY CRYPTOSTOMATE FAMILIES

Subordinal status for each of the three long-ranging groups of cryptostomates is based largely on the distinctive nature of later families; however, the morphology of early cryptostomates strongly suggests a common ancestry for the three. As stenolaemate history is presently understood, these affinities should be reflected in classification by inclusion of all branches in a single order. The early families are compared in Table 5.

The ancestral cryptostomate probably separated from earlier bryozoan stocks through the evolution of either a one- or two-dimensional budding locus (Table 5, no. 4; see BLAKE, 1980). The three major cryptostomates lineages subsequently were founded on the development of individual budding-loci patterns and growth habits (Table 5, no. 1). Phylogenetic relationships among the three have not been studied, and their approximately simultaneous appearance in the stratigraphic record does not aid in the determination of sequence. The Arthrostylidae, however, is morphologically intermediate and therefore the other two stocks are compared to it. Because of their diversity, the arthrostylids may be of disparate ancestry. As presently understood, this seems unlikely because the family is unified by a number of common features, most important of which

---

FIG. 219. Comparison of Rhabdomesina and Ptilodictyina.——*1. Streblotrypa* cf. *S. marmionensis* ETHERIDGE, Rhabdomesina; Perm. (Callytharra), W. Australia; metapore appearance (arrow); tang. view, USNM 112446, ×91.——*2.* Unidentified genus and species, Rhinidictyidae, Ptilodictyina; McLish F., M. Ord., Okla.; slender, elongate zooecia, diaphragms, and vesicles; long. sec., USNM 222621, ×28. ——*3. Helopora fragilis* HALL, Rhabdomesina; Clinton Gr., Sil., Hamilton, Ont.; tabulated metapore (arrow); USNM 222622, ×68.——*4. Nicklespora elegantula* (ULRICH), Rhabdomesina; Keokuk Gr., Miss., Ky.; small stylet; syntype, USNM 168365, ×360.——*5. Stictoporellina gracilis* (EICHWALD), Ptilodictyina; Ranicips Ls., L. Ord., Öland, Swed.; small stylets; USNM 214207, ×91.——*6. Ulrichostylus* sp., Rhabdomesina; Bromide F., M. Ord., Okla.; stylets in an early arthrostylid; USNM 222623, ×180. ——*7. Rhombopora* cf. *R. lepidodendroides* MEEK, Rhabdomesina; Catacora Marl, L. Perm., Yauichambi, Bol.; stylet; UI X-5382, ×180.——*8. Nikiforovella* sp., Rhabdomesina; Ludlowville coral bed, Dev., Ont.; small stylets; USNM 222619, ×912.——*9. Acanthoclema scutulatum* HALL, Rhabdomesina; Ludlowville Sh., M. Dev., N.Y.; stylets (arrow); USNM 222624, ×360.

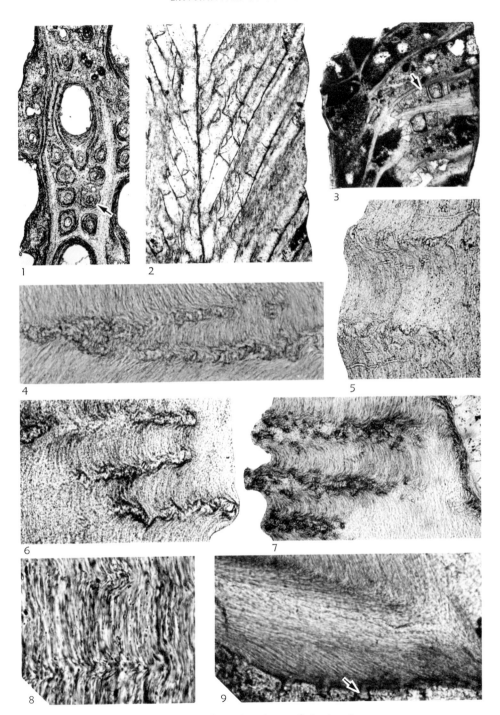

FIG. 219. *(For explanation, see facing page.)*

TABLE 5. *Comparison of Early Cryptostomate Families.*

Character	Arthrostylidae (Rhabdomesina)	Phylloporinidae (Fenestellina)	Rhinidictyidae Stictoporellidae (Ptilodictyina)
1. Zoarial habit	Erect; most species branching but not anastomosing; stems cylindrical, some species articulated throughout	Erect; anastomosing; stems cylindrical, not articulated	Erect; most species branching, anastomosing in some; stems flattened, some species with basal articulation
2. Stem diameter	Early representatives may be under 0.5 mm; to 1.0 mm	Commonly 0.5 to 0.75 mm	Early representatives may be 0.75 to 1.0 mm
3. Zooecial distribution	Around stem axes in most species; one surface lacking apertures in some species; these with 2 to 5 apertural rows	One surface lacking apertures; with 2 to about 8 apertural rows	Two dimensional
4. Budding locus	One dimensional; at least locally in some species, the axis widens to form a two-dimensional budding surface	One dimensional or narrow two dimensional	Two dimensional
5. Zooecial shape	Elongate in early species; shorter in most later species; zooecia angular, with distinctive pattern of wall thickening (Fig. 220)	Elongate in most species; somewhat shortened in a few; zooecia angular, with distinctive pattern of wall thickening (Fig. 220)	Elongate in certain early species, somewhat shortened in others; zooecia angular with distinctive pattern of wall thickening (Fig. 220)
6. Nature of apertures	Elliptical, peristomes present in some	Elliptical to circular; peristomes present in some	Elliptical, subcircular, rectangular; peristomes present in some
7. Intrazooecial structures	Diaphragms and hemisepta present in some	Diaphragms and hemisepta present in some	Diaphragms and hemisepta present in some
8. Polymorphs or similar structures	Lacking in most species, metapores in a few later species	Lacking in some species, mesoporelike structures in others	Vesicular skeletal material and mesoporelike structures in a few species
9. Secondary skeletal structures	Diverse stylets; longitudinal surficial ridges present in some	Simple stylets; longitudinal surficial ridges present in some	Usually simple stylets; longitudinal surficial ridges present in some
10. Skeletal microstructure	Laminated wall dominant, nonlaminated wall commonly thick but discontinuous; laminae fine, bearing radiating, planar, irregular zones in some species; median rods in a few species	Laminated wall dominant, nonlaminated wall in some well developed, continuous; laminae fine, bearing radiating, planar, irregular zones in some species; median rods lacking	Laminated wall dominant; nonlaminated wall commonly thick but discontinuous; laminae fine, bearing radiating, planar irregular zones in some species; median rods in some species

are zoarial form and the nature of budding loci.

Among arthrostylids and bifoliates, two of the most similar taxa are the oldest known representatives of each, *Ulrichostylus spiniformis* (ULRICH) (Arthrostylidae) and *Stictoporella gracilis sensu* BASSLER, 1911 (Stictoporellidae). Similarities include zooecial

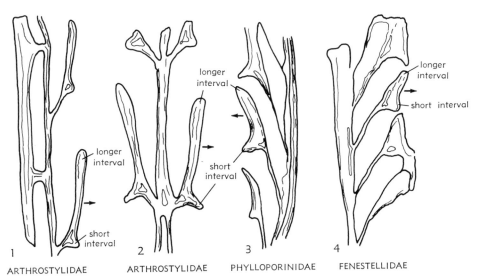

Fig. 220. Cryptostomate exozone development, based on specimens in Figure 221,6–9. *1,2,* Arthrostylidae; *3,* Phylloporinidae; *4,* Fenestellidae. Zooecial shapes are similar among the specimens, and endozonal walls are thin. Endozonal walls branch at the base of the exozone; the branch is marked by thickening of nonlaminated skeletal material. A short interval, thickened on both sides of the zooecial boundary, is directed proximally and forms the distal wall of a zooecium. A longer interval is directed distally, approximately paralleling the zooecial and zoarial axes. This wall is thickened primarily on the outer skeletal surfaces (arrows).

shape (Fig. 217,*1,2*), skeletal structures (Fig. 217,*4,6;* 218,*7*), and the presence of simple stylets (Fig. 219,*5,6*). Major differences are in zoarial size and budding pattern (Fig. 217,*4,6*).

Other shared characters among some early members of the two stocks (Table 5) include an overlapping range of stem diameters, presence of zoarial articulation, nature of apertural development, similar zooecial shapes (Fig. 220), presence of median rods (Fig. 217,*3,5*), and, in some arthrostylids, an indication of a two-dimensional budding locus (Fig. 217,*5*).

Each lineage developed its own distinctive characters, including vesicular skeletal structures and anastomosing growth habits in the bifoliates and extensive articulation in the arthrostylids.

The Phylloporinidae is poorly understood. Ross (1963, p. 592) pointed out, "All the genera. . .in the Family Phylloporinidae . . .require critical study to determine their systematic positions." In order to avoid

inconsistency with future taxonomic arrangement, no attempt has been made here to update names used in illustrations.

Characters that phylloporinids and arthrostylids have in common (Table 5) include the presence of a reverse surface (Fig. 216,*2,10;* 221,*1,2*), similar numbers of apertural rows, and the presence of longitudinal ridges and peristomes. Members of both families have well-developed nonlaminated skeletal deposits along the midlines of lateral and reverse walls and forming axes of ridges on zoarial surfaces. The layer, however, is weak or absent from the front of zooecia (Fig. 217,*7;* 221,*1*). Thin, dark, irregular skeletal zones occur in both (Fig. 217,*6;* 221,*2*), and similar stylets may occur (Fig. 221,*3,4*). Distinctive, irregular polymorphs are present in certain phylloporinids; however, small polymorphlike structures are present in a few arthrostylids (Fig. 219,*3*). The Phylloporinidae is most readily distinguished on the basis of growth habit.

COMMENTS ON OTHER BRYOZOANS

Some authors (e.g., ULRICH, 1890; ROSS, 1964b; SHULGA-NESTERENKO & others, 1972) have described similarities between the Phylloporinidae and the Fenestellidae, placed them together in classifications, and noted the possibility of a close phylogenetic relationship; however, other authors (for example BASSLER, 1953) considered the two not to be closely related. DUNAEVA and MOROZOVA (1975) removed the Phylloporinidae from their Fenestelloidea, emphasizing different zooecial shapes, the presence of "specific heterozooids" in the phylloporinids, and the absence of ovicells from that family. Thus, a brief review of the similarities between the two families is necessary because the phylloporinids occupy a critical position in present interpretations.

Characters common to the phylloporinids and the fenestellids include zoarial habit, skeletal-layer development, wall thicknesses, and stylet development. Zooecial similarities between the phylloporinids and fenestellids are important. Some early fenestellids possess elongate zooecia, shorter than but suggestive of shapes seen in the phylloporinids (*Fenestella granulosa* WHITFIELD, Fig. 220; 221,5,9; compare with Fig. 221,8). Zooecia in a few phylloporinids appear quite short and resemble those of more typical fenestellids, but this pattern is uncommon in the family. In both families, the zooecia are basally attenuated, endozonal walls are thin, and the exozone is relatively narrow and sharply differentiated from the endozone. Exozonal thickening of the wall is almost entirely on the branch-surface side of the zooecial boundary line. Reduction in zooecial length appears to have evolved primarily after the origin of the Fenestellidae.

Fenestella granulosa is a fenestellid with many similarities to phylloporinids; however, this species is correctly assigned to the Fenestellidae because: (1) its zooecia are relatively short compared to those of typical phylloporinids (Fig. 221,8); (2) it is of regular growth habit with subparallel branches of constant diameter linked at frequent intervals by crossbars lacking apertures; (3) the nonlaminated deposits along the branch midline are relatively regular in development, as in other fenestellids and unlike typical phylloporinids (Fig. 217,7); and (4) the spiny keel separating the rows of zooecia is similar to that in typical members of *Fenestella*.

Thus, members of the Phylloporinidae and Fenestellidae are very similar, yet readily distinguished. A hypothesis of close phylogenetic relationships is accepted here, and linking the Phylloporinidae and Arthrostylidae links the Fenestellina to the Arthrostylidae.

I have argued that the earliest known cryptostomate families share many characters, thus implying a close common ancestry. In

---

FIG. 221. Comparison of Arthrostylidae and Fenestellina.——*1. Arthrostylus* cf. *A. obliquus* ULRICH, Rhabdomesina; Sevier Sh., M. Ord., Tenn.; well-developed nonlaminated skeleton (arrow) in an arthrostylid, reverse surface directed down; transv. sec., USNM 222628, ×92.——*2. Phylloporina aspera* (HALL), Fenestellina; Chazy Ls., M. Ord., Can.; granular bands, finely laminated skeletal wall, and nonlaminated skeletal wall (arrow); transv. sec., USNM 43438, ×56.——*3.* Unidentified genus and species, Phylloporinidae, Fenestellina; Bromide F., M. Ord., Okla.; reverse surface illustrating stylet development and spacing; tang. sec., USNM 214217, ×56.——*4. Nematopora* sp., Rhabdomesina; Dev., Ohio; basal attachment, fusion of branches, apertural shape, stylet development; external view, USNM 214215, ×13. ——*5. Fenestella granulosa* WHITFIELD, Fenestellina; Waynesville F., U. Ord., Ohio; parallel branches, barren crossbars, regular development of median nonlaminated wall (arrows), and elongate zooecia in the Fenestellidae; deep tang. sec., USNM 214221, ×56.——*6. Arthrostylus tenuis* (JAMES), Rhabdomesina; Eden Gr., U. Ord., Ky.; zooecial shape, wall development; long. sec. oriented perpendicular to reverse surface (left), USNM 222629, ×184.——*7. A. tenuis* (JAMES), Rhabdomesina; same data as 6; zooecial shape and wall development; long. sec. oriented parallel to reverse surface, USNM 222630, ×184.—— *8. Phylloporina dawsoni* ULRICH, Fenestellina; Trenton Gr., M. Ord., Quebec, Can.; zooecial shape, wall development; long. sec., USNM 222631, ×184.——*9. Fenestella granulosa* WHITFIELD, Fenestellina; Whitewater Sh., U. Ord., Ind.; zooecial shape, wall development; long. sec., USNM 222632, ×92.

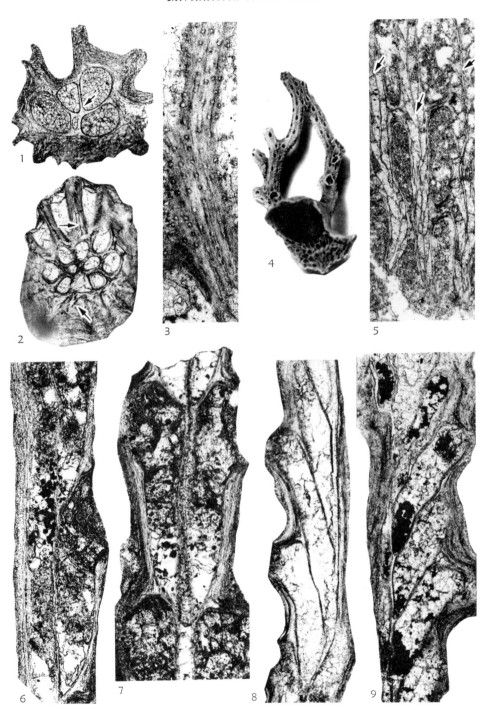

Fig. 221.   *(For explanation, see facing page.)*

contrast, members of other stenolaemate orders appear quite distinctive, thus suggesting isolation from the Cryptostomata.

A number of Lower Ordovician trepostomate genera have been reported from the Baltic region (MÄNNIL, 1959). Zoaria of these tend to be massive and zooecial arrangements, irregular. Budding usually took place across the growing surface rather than from a restricted locus. Exozonal walls tend to be thinner and diaphragms more abundant than in cryptostomates. Somewhat younger trepostomates, from the lower Middle Ordovician Simpson Group of Oklahoma, are generally similar to the Lower Ordovician trepostomates, although exozones are more varied.

ASTROVA (1965) and ROSS (1966a) described members of the order Cystoporata from the Lower Ordovician of the Soviet Union and North America. Zoaria in these species are large and the zooecia elongate. Exozonal walls are thin, and generally well-developed vesicular material is present.

The Tubuliporata, the fourth order of Paleozoic stenolaemates, consists of a relatively small number of inadequately known genera and cannot be readily compared to the cryptostomates.

Order CRYPTOSTOMATA
Vine, 1884

[Cryptostomata VINE, 1884, p. 184, suborder]

Zoaria almost always erect; unbranched, bushlike, pinnately branched, or reticulated; jointing rare; stem cross section cylindrical or flattened; rarely developing overgrowths. Apertures commonly on all stem surfaces, or some surfaces barren; apertures generally in regular longitudinal or spiral rows, apertural outlines elliptical, subcircular, or rectangular. Striae, ridges, peristomes, stylets, polymorphs, small nonpolymorphic depressions commonly well developed on surface. Budding loci linear or planar. Zooecia generally short, rarely elongate; usually with zooecial bend at endozonal-exozonal boundary; hemisepta, diaphragms, mural spines present in some taxa, interzooecial pores lacking. All erect walls compound, exozonal walls generally much thicker than endozonal walls; walls primarily laminated, nonlaminated material generally present (extensive in Fenestellina); in some taxa, extrazooecial skeletal material extensive and multizooecial skeletal material present. Vesicles rare. *Ord.-Perm.*

INTRODUCTION TO THE SUBORDER PTILODICTYINA

By Olgerts L. Karklins

[U.S. Geological Survey, Washington, D.C.]

The Ptilodictyina are a suborder of the Cryptostomata characterized chiefly by an erect, bifoliate growth habit. The earliest known Ptilodictyina are from upper Lower Ordovician strata of the Estonian region. They apparently diversified and dispersed globally during the Middle and Late Ordovician and then gradually diminished in diversity from the Silurian to the Carboniferous periods. Except for a few studies in North America and the Soviet Union, little is known about the distribution of ptilodictyines in the Silurian and Devonian systems. Even less is known about their distribution in the Carboniferous, and they apparently became extinct during the Late Carboniferous or Early Permian. Approximately 260 species have been described, mostly from the Ordovician and Silurian systems.

The Ptilodictyina were exclusively marine, and generally are found in carbonates deposited in shallow basins of continental shelves and inland seas. They commonly are an appreciable component of lower and middle Paleozoic bryozoan assemblages. In places, they are the dominant group of an assemblage, as in the Middle Ordovician of the Siberian region (NEKHOROSHEV, 1961; ASTROVA, 1965).

In carbonate shelf deposits of Middle Ordovician age, ptilodictyines are present in a wide variety of lithologies. They are abundant in such argillaceous and calcareous shales as the Decorah Shale of Minnesota (ULRICH, 1893; KARKLINS, 1969; WEBERS, 1972, p. 479). They form a large and significant part of bryozoan assemblages in irregularly alternating and intertonguing limestone and shale deposits in the Ontario basin of New York and Ontario (Ross, 1970, 1972; WALKER, 1972), and in reef tracts of northern New York and Vermont (Ross, 1963b) and Tennessee (WALKER & FERRIGNO, 1973; ALBERSTADT, WALKER, &

ZURAWSKI, 1974). From these examples, it is apparent that ptilodictyines thrived in a wide variety of relatively shallow environments; however, distinct ptilodictyine associations within larger bryozoan assemblages have not been recognized in different carbonate lithologies.

Most ptilodictyines had an erect growth habit that is variously expressed, and they attached to the substrate by relatively small encrusting bases. An encrusting growth habit, which is common among other stenolaemates, is rare among ptilodictyines. Ross (1963b) indicated that some ptilodictyines occasionally evolved encrusting growth habits in reef tracts. WALKER and FERRIGNO (1973) and ALBERSTADT, WALKER, and ZURAWSKI (1974) noted that encrusting ptilodictyines are indeed rare, being found only in reef cores; throughout a reef tract, other ptilodictyines retained the erect growth habit. These are the only examples in the group demonstrating a relatively clear relationship between growth habit and environment. The taxonomic significance of the changing growth habit is not known and needs study.

Global and stratigraphic distribution of the Ptilodictyina is summarized in Figure 222. At the family level, Ptilodictyina have most commonly been reported from Ordovician rocks in North America and the Soviet Union. Less commonly, they have been reported from western Europe, Asia other than the Soviet Union, Africa, and Australia. Apparently they have not been reported from South America and Antarctica.

Subject to taxonomic assignment of the earliest known species, *Stictoporellina gracilis* (EICHWALD), the family Stictoporellidae appears to have originated during the late Early Ordovician in the Estonian region. It became dispersed globally during the Middle Ordovician and ranged into the Silurian. The

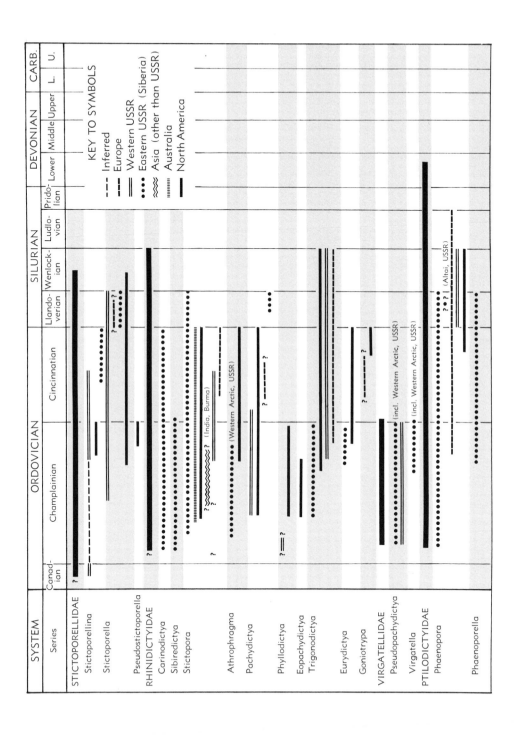

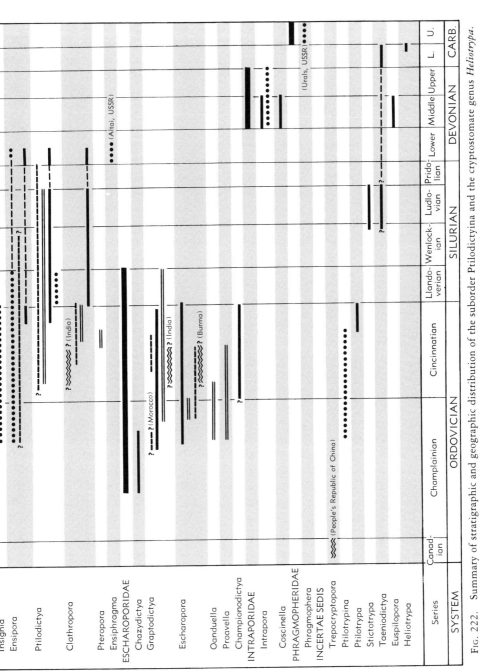

Fig. 222. Summary of stratigraphic and geographic distribution of the suborder Ptilodictyina and the cryptostomate genus *Heliotrypa*.

Stictoporellidae apparently became extinct in Europe during the Early Silurian and in North America during the Middle Silurian.

The earliest occurrence of the Rhinidictyidae is uncertain stratigraphically and geographically. The group apparently dispersed globally during the early Middle Ordovician. In North America, Europe, and Soviet regions of Asia members ranged into the Early Silurian; in Australia they ranged through the Middle and Late Ordovician. The rhinidictyids became extinct in the Siberian region during the Early Silurian. In North America and Europe they apparently became extinct during the Middle Silurian.

The Ptilodictyidae reportedly originated in the Siberian region at least by the early Middle Ordovician and then dispersed globally (NEKHOROSHEV & MODZALEVSKAYA, 1966, p. 101). The Ptilodictyidae in North America appear first in deposits of early Late Ordovician age. They ranged through the Middle Silurian in North America and Europe and became extinct during the Early Devonian in North America and the Soviet Union.

The Escharoporidae, new family, appears to have originated in North America during the early Middle Ordovician and then spread globally. In North America they ranged through the Late Ordovician. In the Estonian region, the family ranged through the Middle Ordovician into the Silurian and apparently became extinct during the Middle Silurian.

The family Virgatellidae is known only from the Middle Ordovician of the Western Arctic and Siberian regions of the Soviet Union.

The family Intraporidae occurs in the Lower Devonian of North America and Siberia. In North America the intraporids ranged only through the Middle Devonian, but in Siberia apparently became extinct during the Late Devonian.

The family Phragmopheridae is known only from late Carboniferous strata of the Uralian region of the Soviet Union.

Occurrences of certain genera of the families Ptilodictyidae, Escharoporidae, and Rhinidictyidae in Ordovician and Silurian strata of the Indian subcontinent (Fig. 222) need to be verified.

The common occurrence of many ptilodictyines in globally separated regions and the presence of comparable ptilodictyine taxa in lithologically different deposits significantly enhances their biostratigraphic value in Ordovician and Silurian rocks. Most ptilodictyines possess a uniform morphology throughout the erect portions of their colonies. Therefore, small fragments of a broken and scattered colony can be easily identified to genus and species, even in randomly oriented thin sections. In addition, generally delicate and fragile ptilodictyine skeletons could not have withstood significant transportation, and scattered parts are likely found near growth sites. Recent paleoecological investigations (ROSS, 1970, p. 361; 1972; WALKER, 1972, p. 2509, 2511) of fragile colonies in erect-growth positions indicate quiet, clear, and relatively shallow environments, thus adding to their value in paleoecological interpretations.

Taxonomic position.—The suborder Ptilodictyina was erected by ASTROVA and MOROZOVA (1956), who revised the Cryptostomata by regrouping families into the new suborders Fenestelloidea, Rhabdomesoidea, and Ptilodictyoidea. The Ptilodictyoidea, the smallest of these suborders, included the Stictoporellidae NICKLES and BASSLER, 1900, Ptilodictyidae ZITTEL, 1880, Rhinidictyidae ULRICH, 1893, Hexagonellidae CROCKFORD, 1947 (removed from the Cyclostomata of BASSLER, 1953), Goniocladiidae NIKIFOROVA, 1938 (removed from the Cyclostomata of BASSLER, 1953), and Rhiniporidae MILLER, 1889.

In 1966, MOROZOVA erected another suborder of Cryptostomata, the Timanodictyoidea, on the basis of the contrasting morphology of *Timanodictya* NIKIFOROVA, 1938, and *Timanotrypa* NIKIFOROVA, 1938, which she removed from the family Rhinidictyidae. The Timanodictyina (=Timanodictyoidea) comprise a special group of

Permian Bryozoa (SHISHOVA, 1968, p. 129) consisting of genera with erect bifoliate and radial modes of zoarial construction, or of single layered encrusting zoaria.

HISTORY OF CLASSIFICATION

In early works, the ptilodictyines were usually grouped, perhaps because of their small size and delicate form, with similarly delicate modern forms now classed as cheilostomate Bryozoa, or less commonly as gorgonacean Coelenterata. GOLDFUSS (1829, p. 104, pl. 37, fig. 2) published what is probably the earliest illustration of a ptilodictyine, which he named *Flustra lanceolata*. Soon after the formal separation of Bryozoa from the Coelenterata (EHRENBERG, 1831), the zoological affinities of the ptilodictyines began to be questioned. MILNE-EDWARDS (in LAMARCK, 1836, v. 2, p. 229) retained *F. lanceolata* in the Coelenterata, but considered the assignment to be doubtful. LONSDALE (in MURCHISON, 1839, p. 676, pl. 15, fig. 11) noted similarities between *F. lanceolata* and certain other fossils in MURCHISON's collections, and he established the genus *Ptilodictya* to accommodate his and the GOLDFUSS specimens. *F. lanceolata* became the type species of *Ptilodictya,* which LONSDALE considered to be a unique form of Paleozoic coelenterate. Because the figured specimens of GOLDFUSS have not been located (ROSS, 1960a, p. 440), the concept of *Ptilodictya* is based on LONSDALE's specimens.

EICHWALD (1842, p. 39, and subsequent publications) reported ptilodictyines from Ordovician strata in Estonia and northwestern Russia, assigning them to the escharid- and flustrid-like cheilostomates or to the gorgonacean coelenterates. In North America HALL erected several ptilodictyine genera, including the well-known *Stictopora* and *Escharopora*. HALL (1847, p. xxii; 1852, p. 354) noted the unusual external morphology of these genera and considered them to be coelenterates or bryozoans. Later, in an endeavor to classify known Bryozoa, D'ORBIGNY (1849, p. 499) recognized *Stictopora* and *Ptilodictya* as fossil bryozoans of undoubted Paleozoic age.

McCoy (1851–1855, p. 45), in his work on British Paleozoic fossils, revised the diagnosis of *Ptilodictya* and assigned it to the family Escharidae (not recognized now) of the class Polyzoa. McCoy noted the subtubular construction of the autozooecia and described the wall structure as being uniformly continuous from the mesotheca to the zoarial surface.

Since the investigations of D'ORBIGNY and McCoy, assignment of *Ptilodictya* and *Stictopora* to the Bryozoa has not been challenged. These genera have been regarded as typical fossil bryozoans, and numerous, superficially similar specimens were assigned to them until the introduction of thin-sectioning methods. NICKLES and BASSLER (1900) listed approximately 100 species of *Ptilodictya* and 80 species of *Stictopora* that were included in these genera at one time or another. Most have since been reassigned.

During the latter part of the nineteenth century, paleontologists studying Bryozoa were involved in a controversy concerning zoological affinities of the fossil Trepostomata, then included with the tabulate corals. NICHOLSON, a leading worker on fossil Bryozoa, at the time was actively involved in the study of the Trepostomata and noted (1874b, p. 123) that *Clathropora,* a ptilodictyine, was a bryozoan, and (1875b, p. 34) that *Ptilodictya* might be a transitional form between tabulate corals and bryozoans. In a subsequent publication, NICHOLSON (1879, p. 11) considered *Ptilodictya* and related forms to be undoubted fossil bryozoans, and later published (1881, fig. 15) what is probably the first illustration of a thin section from a ptilodictyine.

ZITTEL (1880, p. 603) established the family Ptilodictyidae with *Ptilodictya* as the type genus. According to ZITTEL, the family comprised genera in which autozooecia grew back to back. It included known ptilodictyines, some bifoliate fistuliporines, an arthrostylid, and some poorly known and generally unrelated genera. ZITTEL assigned the Ptilodictyidae to the suborder Cyclostomata.

ULRICH, in a series of outstanding papers on fossil Bryozoa, argued (see 1882, p. 121) for the classification of the Monticuliporidae, consisting mostly of the trepostomates with some fistuliporid and ceramoporid genera, as Bryozoa. He discussed the comparative morphologies of a recent tubuliporate, *Heteropora,* and selected trepostomatous bryozoans, and compared them with *Ptilodictya* and *Stictopora.* Using thin-sectioning methods introduced by NICHOLSON, ULRICH (1882, p. 151–152) recognized morphologic differences between *Ptilodictya* and *Stictopora* by making a new family Stictoporidae (now Rhinidictyidae) that, in addition to *Stictopora,* included his genera *Pachydictya* and *Phyllodictya.*

ULRICH (1882) also erected the suborder Trepostomata to accommodate some Bryozoa of controversial phylum affinities, and thereby differentiated them from the suborder Cyclostomata (now order Tubuliporata). The original concept of the suborder Trepostomata included the Ptilodictyidae and Stictoporidae as first and second groups. The Monticuliporidae, Fistuliporidae, and Ceramoporidae completed the suborder.

During the same period in England, VINE published several papers on the British Paleozoic Bryozoa, which included reviews of the existing bryozoan classification. VINE (1884, p. 184, 196) accepted ULRICH's suborder Trepostomata, but he proposed a new suborder, the Cryptostomata, with *Ptilodictya* as the type genus. VINE's (1884, fig. 4) published illustrations show *Ptilodictya* with what appear to be exilazooecia, suggesting that his material may also have included specimens of *Phaenopora* (ULRICH, 1890, p. 344). Exilazooecia are generally lacking in *Ptilodictya* (mesopores of ROSS, 1960c, p. 1064). VINE (1884, p. 203) also

proposed a new family, Arcanoporidae (not recognized now) for *Ptilodictya* and two other genera, *Arcanopora* (now *Sulcoretepora*) and *Glauconome* (now *Glauconomella*). The family Rhabdomesidae (Rhabdomesontidae of VINE, 1884), a group of small branching forms, completed the Cryptostomata.

ULRICH (1890) emended the diagnoses of the Ptilodictyidae ZITTEL and Rhabdomesidae VINE, and enlarged the Cryptostomata by adding the families Acanthocladiidae, Fenestellidae, Phylloporinidae, and Cystodictyonidae.

In 1893, ULRICH published a monograph on Paleozoic Bryozoa from Minnesota. In this work he erected the family Rhinidictyidae, comprising most genera that were previously assigned to the Stictoporidae. This monograph is probably the most important single work on skeletal morphology of bifoliate cryptostomates and still has a bearing on taxonomy of the families.

NICKLES and BASSLER (1900) presented a synopsis of the classification of known Bryozoa, and added the family Stictoporellidae to the Cryptostomata. The Stictoporellidae together with the Ptilodictyidae and Rhinidictyidae constitute the basic group of the Ptilodictyina, as it is used now. These three ptilodictyine families contain the most abundant and widely distributed genera of the oldest known Cryptostomata (ASTROVA & MOROZOVA, 1956; ASTROVA, 1965).

After the extensive work by ULRICH, the concept of the Cryptostomata was not reexamined until 1937, when McNAIR noted critically that the Cryptostomata comprised phylogenetically unrelated groups. He indicated that the order included three general zoarial types. The first is unilaminate with autozooecia opening only on one side, and is found in the Fenestellidae, Phylloporinidae, and Acanthocladiidae. The second type, found in the Arthrostylidae and Rhabdomesidae, has autozooecia opening on all sides. The third type is bifoliate with autozooecia opening on two sides and growing back to back from the mesotheca. This third type, found in the Ptilodictyidae, Stictoporellidae, Rhinidictyidae, Sulcoreteporidae, and Actinotrypidae, was considered to be most characteristic of the Cryptostomata.

In the first edition of this *Treatise,* BASSLER (1953) presented a synopsis of the bifoliate cryptostomate genera. The order Cryptostomata and the contents of the bifoliate families remained essentially unchanged from the arrangement of NICKLES and BASSLER (1900), except for transfer of the Phylloporinidae from the Fenestellidae to the Trepostomata.

In 1956, ASTROVA and MOROZOVA proposed a major revision of the Cryptostomata. They cited the work of MCNAIR (1937) and concluded that the three types of Cryptostomata recognized by him differed not only in zoarial growth but also in microstructure of skeletal walls and internal morphology of autozooecia. ASTROVA and MOROZOVA considered these to be naturally segregated groups that diverged phylogenetically during Late Cambrian or Early Ordovician time. They redefined the order Cryptostomata and subdivided it on morphological and possible phylogenetic grounds into three suborders: the Fenestelloidea (including the Phylloporinidae), Ptilodictyoidea, and Rhabdomesoidea.

ASTROVA (1960a) reviewed the general history of bryozoan studies, including investigations in the Soviet Union, and gave a synopsis of the families and genera of the Ptilodictyoidea. She illustrated genera occurring in the Soviet Union, and others were mentioned and assigned to appropriate families.

ROSS (then PHILLIPS, 1960) restudied type material and additional specimens from several ptilodictyine genera, including *Stictopora* HALL, 1847, *Escharopora* HALL, 1847, and *Pachydictya* ULRICH, 1882. On the basis of zooecial construction in the type species of these genera, she established three informal taxonomic categories: the escharoporids, stictoporids, and pachydictyids. These categories were differentiated on the configuration of laminae in zooecial walls, mode of growth of zooecia from the mesotheca (budding surface), presence of exilazooecia (mesopores of

ROSS), and features in zooecial chambers. In subsequent publications, ROSS (1960a,b,c, 1961a,b; 1963b, 1964a,b) considered other ptilodictyine genera, recognized some new genera, and included them in her classification.

On the basis of zoarial morphologies, ROSS (1964b) interpreted the informal escharoporid, stictoporid, and pachydictyid groups to be phylogenetic lineages within the early ptilodictyines. Genera forming the three lineages overlapped families in the classification of BASSLER (1953), ASTROVA and MOROZOVA (1956), and ASTROVA (1960a). ROSS (1964b) briefly reviewed some biological features of Bryozoa and considered the Hexagonellidae CROCKFORD, 1947, and Goniocladiidae NIKIFOROVA, 1938 (placed by ASTROVA & MOROZOVA, 1956, in the Ptilodictyoidea) as families of the then cyclostomate suborder Ceramoporina BASSLER, 1913.

In a comprehensive work, NEKHOROSHEV (1961) described the Middle Ordovician to Lower Silurian bryozoans of the Siberian platform. The Cryptostomata are represented by the fenestelline Phylloporinidae and Fenestellidae, the rhabdomesine Arthrostylidae, and the ptilodictyine Ptilodictyidae, Stictoporellidae, and Rhinidictyidae. NEKHOROSHEV recognized about 70 cryptostomate species, of which 61 were assigned to ptilodictyine genera. Of the ptilodictyines, *Phaenopora* and *Phaenoporella* are the most common genera; both diversify at the base of the upper Middle Ordovician (Mangazeian Stage), with many species ranging through the Upper Ordovician and some into the Lower Silurian. NEKHOROSHEV noted that the Middle and Upper Ordovician of the Siberian platform is not only remarkable in having a large and diversified ptilodictyid component, but also in having a relatively small trepostomatous component (ASTROVA described the Trepostomata in 1965). NEKHOROSHEV concluded that the Middle and Upper Ordovician fauna of the Siberian platform is distinctly different from that of North America and the Baltic region, and must have inhabited a separate marine basin, which he designated as the "*Phaenopora* province."

Astrova (1965), in a major work on the Paleozoic Bryozoa of Siberia, discussed at length the ptilodictyine families Ptilodictyidae, Stictoporellidae, and Rhinidictyidae because they are the most abundant representatives of the earliest cryptostomates and constitute a large part of the Ordovician and Silurian fauna in Siberia. She described the astogeny and ontogeny of zoaria and the structure of skeletal matter, and suggested possible functions for zoarial and zooecial structures. She also outlined the phylogeny of the Ptilodictyidae, recognizing two closely related evolutionary lineages that she used as the basis for the new subfamilies Ptilodictyinae and Phaenoporinae. The Ptilodictyinae, originating in the upper Lower Ordovician of the Soviet Union, is the older of the two subfamilies. Morphologically, Astrova's Ptilodictyinae and Phaenoporinae appear to form a part of the escharoporid lineage of Ross (1964b). Astrova also recognized the family Virgatellidae, which she considered to be related to the Rhinidictyidae.

Goryunova (1969) erected the family Phragmopheridae, found in Carboniferous strata, and assigned it to the Ptilodictyina. The Phragmopheridae is distinctly different from older families of the Ptilodictyina and its relationship to them is uncertain.

The suborder Ptilodictyina as described herein includes seven families: Ptilodictyidae Zittel, 1880, Escharoporidae new, Intraporidae Simpson, 1897, Phragmopheridae Goryunova, 1969, Rhinidictyidae Ulrich, 1893, Stictoporellidae Nickles and Bassler, 1900, and Virgatellidae Astrova, 1965. The Intraporidae of Simpson is restored.

In addition, the Ptilodictyina also includes six genera that are presently unassigned to families: *Euspilopora* Ulrich in Miller, 1889, *Ptilotrypa* Ulrich in Miller, 1889, *Ptilotrypina* Astrova, 1965, *Stictotrypa*

Ulrich, 1890 (removed from Rhinoporidae Miller, 1889), *Taeniodictya* Ulrich, 1888, and *Trepocryptopora* Yang, 1957. These genera are ptilodictyine in general mode of budding and zoarial morphology but differ from representatives of established families in the shape and arrangement of zooecia, configuration of wall laminae, and other skeletal structures. The insufficient material on which these genera are based does not warrant assignment to suprageneric categories. *Heliotrypa* Ulrich, 1883, is also discussed herein, but is not considered to be a ptilodictyine and therefore is removed from the Stictoporellidae.

Acknowledgments.—I am grateful to R. S. Boardman, John Utgaard, W. C. Banta, D. B. Blake, A. H. Cheetham, T. G. Gautier, J. P. Ross, and Curt Teichert for helpful discussions or technical reviews. L. J. Vigil, R. H. McKinney, H. E. Mochizuki, R. C. Widger, and Janine Higgins gave technical assistance with preparation of fossils, photography, or illustrations. R. L. Batten (American Museum of Natural History), E. S. Richardson, Jr. (Field Museum of Natural History), L. S. Kent (Illinois State Geological Survey), B. M. Bell (New York State Museum and Science Service), C. F. Kilfoyle (formerly New York State Museum and Science Service), D. B. Macurda, Jr. (formerly Museum of Paleontology, University of Michigan), M. D. Dash and Solene Whybrow (Peabody Museum of Natural History, Yale University), Vickie Kohler (Museum of Comparative Zoology, Harvard University), C. L. Forbes (Sedgwick Museum, Cambridge University), and R. F. Wise (British Museum, Natural History) searched or arranged for loan of numerous specimens. I thank V. P. Nekhoroshev (Geological Institute of the USSR, Leningrad) for providing primary specimens, and the late G. G. Astrova (Paleontological Institute, Academy of Sciences of the USSR, Moscow) for loan of specimens.

MORPHOLOGY

BASAL ZOARIAL ATTACHMENTS

Basal attachments of zoaria in the Ptilodictyina are of two basically different kinds. In one, fully developed zoaria are skeletally continuous with their bases (Fig. 223,*2–4*) and are rigidly fixed in their growth position. This type of basal attachment prevails among the ptilodictyines and, in general, is characteristic of the families Intraporidae, Rhinidictyidae, and Stictoporellidae. In the other kind of basal attachment, fully developed zoaria include skeletally detached bases (Fig. 224,*2*; 225,*2*) that are thought to have been separated from distal parts of the zoaria by flexible articulating joints (ULRICH, 1882, p. 151; BASSLER, 1927, p. 164, pl. 11, fig. 9; PHILLIPS, 1960, p. 18; ROSS, 1960a, p. 441; NEKHOROSHEV, 1961, p. 69; ASTROVA, 1965, p. 95). Flexible joints are known in recent species of both gymnolaemates and stenolaemates. They also occur in the family Ptilodictyidae, *Escharopora* of the Escharoporidae, and probably in other fossil bryozoans.

The basal attachment in both rigid and jointed colonies consists of two parts, an encrusting expansion on the substrate (Fig. 223,*2–4*; 224,*2,5*) and a vertical extension (Fig. 223,*2–4*; 225,*2,5*) between the fully developed zoarium and its base, here called the **connecting segment**. General shape of the encrusting base appears to be determined in part by the nature of the substrate, but areal extent of the base may be related to height and thickness of a fully developed zoarium (Fig. 223,*2,3*). Connecting segments rise vertically from the approximate centers of the bases and expand distally into fully grown zoaria of bifoliate growth habit.

Basal attachments of both types are fragile. The rigid type is only rarely found unbroken. Jointed attachments are not known to be connected, even if bases and connecting segments are found together in the same deposits. The flexible material between bases and connecting segments in the unjoined type is thought to have consisted of organic tissues (ASTROVA, 1965, p. 95; others) that were not preserved.

In available ptilodictyines, only a few specimens possess reasonably well-preserved basal zoarial attachments that are suitable for thin sectioning or peeling. Several serial peels and a few thin sections were prepared from bases of species of *Stictopora* (Fig. 223, *1,5,6*) and *Trigonodictya* (Fig. 223,*3,4*; 225,*4*; 226,*1*), both Rhinidictyidae, and from an encrusting base that may belong to an escharoporid (Fig. 224,*1–3,5*) and from a connecting segment of a ptilodictyid (Fig. 225,*1–3,5*).

Microstructure of zoarial attachments is the same as in the main erect part of a zoarium. The laminae are similar in size and configuration, but granular zones or zooecial boundaries are relatively indistinct or not visible. The three-dimensional arrangement and shape of basal zooecia were determined from a series of peels taken at intervals between 0.1 and 0.5 mm upward from the encrusting base and tangentially to it. Because the encrusting bases slope away from their centers, zooecia are ontogenetically younger toward the edge of the encrustation. Because of small size, longitudinal sections paralleling the mesotheca could not be obtained from the specimen of *Stictopora,* but a few, slightly oblique sections are available from *Trigonodictya.*

By analogy to modern bryozoans (RYLAND, 1970), encrusting bases in the ptilodictyines probably contain ancestrulae. TAVENER-SMITH (1975) has speculated on possible growth in the basal attachment of a generalized ptilodictyine.

Basal attachments of rigid zoaria.— Encrusting bases in most species are irregularly explanate, rarely subcircular to indistinctly elongate (Fig. 223,*2–4*). A group of zooecia constituting the center of a base, as determined from serial peels, are regarded as the first generation of basal zooecia because of their position and alignment in the base (Fig. 223,*1d*; 226,*1b*). The first generation of zooecia arise concurrently with the formation of a basal layer and align in a preferred direction (Fig. 223, *1c,d*). Orientation of the first generation establishes the mesothecal plane for a zoarium and preferred elongation for the encrusting base.

Growth of the mesotheca in the base appears to have been initiated between a pair of aligned zooecia in the first generation, and not from the basal layer. Once growth of the mesotheca was initiated, subsequent zooecia arose both from the mesotheca as it extended laterally and vertically (Fig. 223, *1b–d*) and from the basal layer (Fig. 223, *1c,d*). The basal layer is laminar, very thin, and generally of uniform thickness through the central part of the base.

In *Trigonodictya fenestelliformis* (NICHOLSON), the basal layer consists of a thin laminar part at the base and a vesicular part above (Fig. 225, *4*; 226, *1c*). The first generation of zooecia in this species arises from the vesicular structure and aligns in an indistinct spiral pattern (Fig. 226, *1b*). Subsequent generations of zooecia reoriented in a preferred direction while initiating the mesotheca (Fig. 226, *1a*) and also reoriented while arising

from the laterally expanding basal layer (Fig. 226, *1d*). The vesicular structure fills most of the space between basal zooecia and extends into the connecting segment (Fig. 225, *4*; 226, *1a*).

Connecting segments are generally variable in length, subtubular (Fig. 223, *2*) to flattened, and without distinct zoarial margins. They formed with the vertical extension of the mesotheca. Zooecia in the connecting segment arise from the mesotheca in patterns characteristic for species (Fig. 223, *1a,b*), but are commonly closed by thickened walls or extrazooecial stereom in the exozone.

In general, zooecia in basal attachments of rigid zoaria are similarly shaped but smaller in size than autozooecia in the main erect part of the same zoarium. Endozones are poorly defined and of variable width in the base (Fig. 223, *1c,d*; 225, *4*), but are distinct and of uniform width in the connecting segments (Fig. 223, *1a,b*; 226, *1a*).

Basal attachments of jointed zoaria.—Encrusting bases of jointed colonies are generally irregularly subcircular (ULRICH, 1879, p. 29; 1882, p. 151; BASSLER, 1927, pl. 11, fig. 9; NEKHOROSHEV, 1961, p. 69) with

---

FIG. 223.   Basal attachments of rigid zoaria (*1–4*) and autozooecial budding pattern at mesotheca in erect part of zoarium (*5,6*).——*1a–d*. *Stictopora* sp., Lexington Ls., M. Ord., Burgin, Ky.; serial peels of basal attachment; *a*, distal part of connecting segment; mesotheca with median rods and regularly arranged zooecia proximally of fully expanded zoarium; transv. peel; *b*, narrowest part of connecting segment, approximately in middle between encrusting base and expanded zoarium; median rods in mesotheca generally visible, zooecia indistinctly aligned (toward left) and irregularly shaped; transv. peel; *c*, distal portion of first generation of zooecia (upper right) from which mesotheca extends and subsequent zooecia align in preferred direction (toward left), median rods not visible; peel parallel to basal layer of zoarial base; *d*, first generation of zooecia (ancestrula inferred) from which mesotheca (upper right) extends laterally and vertically; subsequent zooecia align in preferred direction toward left; extensive extrazooecial deposits containing numerous mural styles form margin of base; peel in plane of base of first generation of zooecia; all USNM 242610, ×30.——*2*. *Intrapora puteolata* HALL, Onondaga Ls., M. Dev., Ohio; basal attachment, consisting of encrusting base and connecting segment, expands into branched zoarium; external view, USNM 242611, ×5.——*3,4*. *Trigonodictya pumila* (ULRICH), Decorah Sh., M. Ord., Cannon Falls, Minn.; *3*, subcircular base and subtubular connecting segment of zoarium; external view, USNM 163096, ×5; *4*, irregular encrusting base with preferred orientation toward right, connecting segment expands into branched zoarium; external view, USNM 163095, ×5.——*5*. *Stictopora mutabilis* ULRICH, Decorah Sh., M. Ord., Rochester, Minn.; transverse walls arise from mesotheca, both sides of longitudinal wall (center), and one side (facing zoarial margins) of subsequently added longitudinal walls; curved transverse walls incline toward zoarial margins (to left and right from center); tang. sec. of endozone of fully developed branch, USNM 242612, ×50.——*6*. *Stictopora neglecta* (ULRICH), Lexington Ls., M. Ord., Burgin, Ky.; area of branching, longitudinal walls arise from mesotheca and junction of transverse walls of enlarged zooecium (lower right), marginal zooecia and extrazooecial stereom in margin (upper right); tang. sec., USNM 242613, ×50.

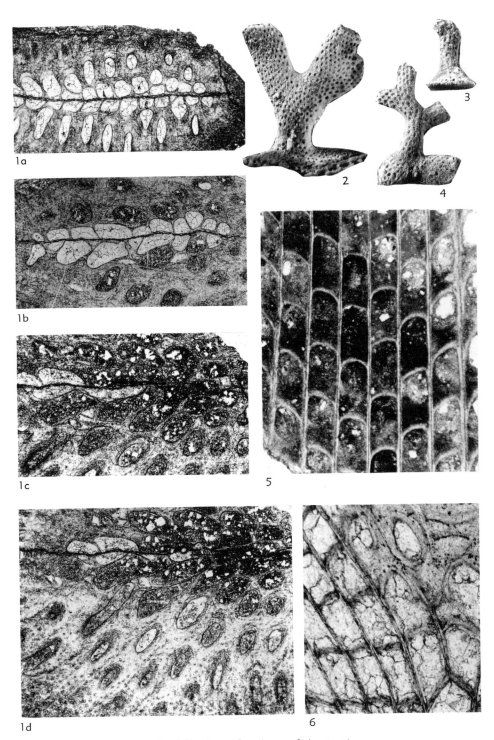

FIG. 223. *(For explanation, see facing page.)*

locally lobate lateral margins. The central part of the base is slightly raised and contains a circular depression (Fig. 224,*2,5*) surrounded by a rim. The first generations of zooecia apparently constitute this depression as they arise from the basal layer. Subsequent zooecia arise in a relatively regular radial pattern from the basal layer as it extends radially (Fig. 224,*5*). Zooecia constitute fine **striae** on the surfaces of these bases (Fig. 224,*1,2*) and are locally overlapping in the margins (Fig. 224,*3a,5*). In general, zooecia forming these bases are indistinctly delineated within the same range (Fig. 224,*1*) and irregularly shaped in cross section (Fig. 224,*3a,5*).

The mesotheca, or similar structure, is absent in the encrusting bases of jointed zoaria. The basal layer consists of indistinctly laminar calcite and is uneven in thickness (Fig. 224,*3*).

Connecting segments of the main part of a zoarium taper to proximal tips (PHILLIPS, 1960, p. 17; ROSS, 1960a, p. 440; 1960c, p. 1064; ASTROVA, 1965, p. 95), which fit into approximately matching circular depressions of their bases. The proximal part of connecting segments is generally subtubular. Connecting segments apparently originated at or within the depression of a corresponding base. A mesotheca is present in the tip (ROSS, 1960a, p. 441) (Fig. 225,*1,3,5*) and extends throughout the connecting segment, but does not link skeletally with the depression of an encrusting base. The mesotheca originates with the inception of the connecting segment.

For each species, zooecia in the connecting segment are in a regular pattern at the mesotheca, but are smaller than those in the fully developed zoarium of the same specimen (PHILLIPS, 1960, p. 18; ROSS, 1960a, p. 441). They are commonly closed by thickened walls or stereom (Fig. 225,*3,5*), as in rigid attachments. In some species, an annular ridge (ROSS, 1960a, p. 441) surrounds the connecting segment at some distance above its tip. This ridge consists of extrazooecial stereom that closes the zooecia (Fig. 225,*5*); however, regular arrangement of zooecia is generally maintained across the ridge (Fig. 225,*3,5*).

MAIN ERECT PARTS OF ZOARIA

The main erect part of a zoarium is bifoliate, consisting of two layers of autozooecia facing in opposite directions. Each layer forms one side of a zoarium and the proximal walls of the autozooecia constitute the mesotheca (Fig. 227), which extends throughout the erect part of the zoarium.

Mesotheca.—The mesotheca, or multi-

FIG. 224.   Basal attachments in jointed zoaria (*1–3,5*) and autozooecial budding pattern in erect part of zoarium (*4,6*) of *Escharopora.*——*1. Escharopora?* sp., U. Ord. (Maysvill.), Cincinnati, Ohio; encrusting zoarial base; longitudinal walls, serrated zooecial boundaries along median of longitudinal walls, elongated and relatively narrow zooecial chambers with few cross partitions, basal walls (portions of skeletal calcite in center); tang. sec., USNM 242600, ×100.——*2a,b. E. acuminata* (JAMES)?, U. Ord. (Eden.), Covington, Ky.; *a,* encrusting base with striae radiating from rim of hollow depression, ×15; *b,* approximate extent of base encrusting a monticuliporine trepostomate, ×5; both external views, USNM 242601 from USNM 56077.——*3a,b. E. pavonia* (D'ORBIGNY)?, U. Ord. (Maysvill.), Cincinnati, Ohio; *a,* basal zooecia in cross section at edge of an encrusting base, massive calcitic layer forms basal layer; peel at right angle to edge of base, ×50; *b,* undifferentiated bottom layer of base; indistinct contact between zoarial base and underlying monticuliporine trepostomate; sec. at right angle to edge of base, ×100; both USNM 242602.——*4. E. subrecta* (ULRICH), Decorah Sh., M. Ord., Minn.; slightly curved continuous longitudinal walls and transverse walls at mesotheca in main erect part of zoarium; sec. just above mesotheca, USNM 163176, ×50.——*5,6. E. falsiformis* (NICHOLSON)?; *5,* U. Ord. (Cincinnat.), Cincinnati, Ohio; zoarial base; zooecia radiate from hollow depression in center, becoming less regularly arranged toward margin; tang. sec., USNM 242603, ×30; *6,* U. Ord. (Maysvill.), Covington, Ky.; regularly curved longitudinal walls, autozooecial chambers widen and narrow alternately in adjacent ranges at junctions between longitudinal and transverse walls (walls at mesotheca, upper right), and along endozone-exozone boundary (left); deep tang. sec., USNM 242604, ×30.

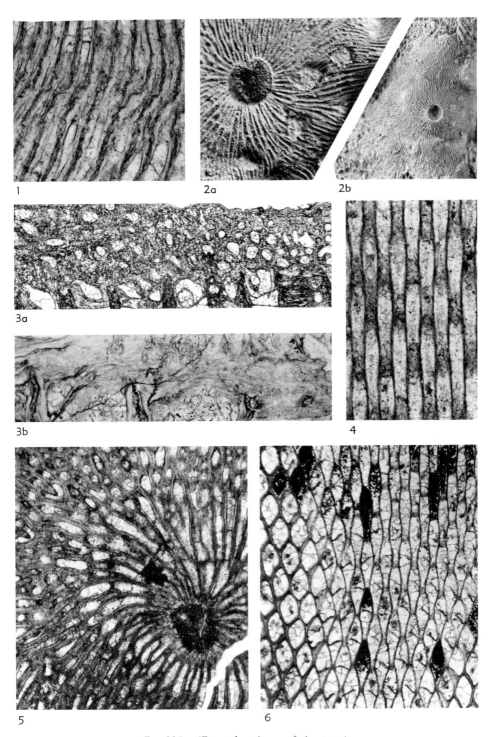

FIG. 224. *(For explanation, see facing page.)*

zooecial median wall, forms the budding surface from which autozooecia arise in characteristic patterns. Except in the zoarial margin, the mesotheca consists of laminated layers that are separated by a **median granular zone** (Fig. 227, 228; see also 240, *1h*).

In laminated layers of the mesotheca, laminae parallel the granular zone as far as the margins, where they adjoin in a broadly or narrowly serrated zone (see Fig. 240, *1f*). Both the granular zone and the adjacent laminated layers appear to have been secreted concurrently while the mesotheca extended vertically and laterally. Part of the mesotheca forms the basal walls of the autozooecia, but laminae of the mesotheca do not appear to be continuous with those of the autozooecial wall.

The mesotheca in most ptilodictyines developed in one growth plane, but it may bifurcate (ULRICH, 1893, p. 160; PHILLIPS, 1960, p. 16; ROSS, 1964a, p. 23) (Fig. 229, 3). The mesotheca is generally straight (see Fig. 240, *1e*) to slightly undulating (see Fig. 244, *1d*); in a few taxa it is zigzag in transverse section. Thickness of mesotheca in zoarial midregions averages between 0.01 and 0.03 mm.

The median granular zone is sheetlike and may be locally discontinuous. At zoarial margins it coalesces with skeletal laminae (Fig. 230, 2). The zone consists of irregularly shaped crystalline particles that are approximately three microns in diameter (TAVENER-SMITH & WILLIAMS, 1972, p. 149).

The granular zone in the mesotheca may contain rodlike structures called **median rods** (Fig. 227, 231), which are segregated from the crystalline particles. These rods (median tubuli of ULRICH, 1893, p. 98; BASSLER, 1953, p. G12; median tubuli or acanthopores of PHILLIPS, 1960, p. 3; median tubuli of ROSS, 1964a, p. 24; 1964b, p. 934; KARKLINS, 1969, p. 17; capillaries of ASTROVA, 1965, p. 101; KOPAYEVICH, 1968, p. 128; zoarial canals of KOPAYEVICH, 1973, p. 59; acanthopores, lenticles of TAVENER-SMITH & WILLIAMS, 1972, p. 149; acanthopores of authors) occur in most genera of the family Rhinidictyidae.

Median rods consist of calcite cores enclosed by dark-colored, laminated sheaths (see Fig. 251, *1d,h*). Calcite in the cores is finely crystalline and the particles are densely packed (TAVENER-SMITH & WILLIAMS, 1972, pl. 28, fig. 184). As observed in the light microscope, the particles are equidimensional and smaller than those in the granular zone. In some specimens, the rods merge locally into a continuous, thin layer (TAVENER-SMITH, 1975, p. 3) that replaces or merges with particles of the granular zone.

Laminated sheaths enclose individual cores. The laminae (see Fig. 251, *1h*; 252, *1a*; 255, *2a*) are generally thinner than those of the mesotheca. The sheath laminae merge indistinctly with those of the mesotheca, but appear to be discontinuous with particles forming the granular zone.

The median rods are elliptical to subrounded in cross section and generally less than 0.01 mm in diameter. They appear to originate within the upper part of encrusting zoarial bases. As the zoarium grows distally

FIG. 225. Basal attachments of jointed (*1–3,5*) and rigid (*4*) zoaria.——*1–3,5. Clathropora frondosa* HALL, Niagara Gr., M. Sil., N.Y.; *1,* mesotheca and regularly arranged zooecia in tapered connecting segment slightly distal of proximal tip; transv. sec., USNM 242605, ×30; *2,* tapered connecting segment (proximal tip broken or abraded), expands into fully developed zoarium; external view, USNM 242606, ×5; *3,* intrazooecial deposits constitute annular ridge around connecting segment a short distance above its proximal tip; zooecia closed by thickened walls or stereom deposits in exozone; transv. sec., USNM 242607, ×30; *5,* extrazooecial stereom in annular ridge (upper part), regularly arranged zooecia throughout the connecting segment, mesotheca in the proximal tip (lower part); long. peel, USNM 242608, ×30.——*4. Trigonodictya fenestelliformis* (NICHOLSON), U. Ord. (Richmond.), Wilmington, Ill.; margin of encrusting zoarial base; basal layer laminar adnate to substrate (a brachiopod, bottom) and vesicular above it; a few zooecia arise from vesicular portion, vesicles in indistinct zones; peel at right angle to edge of encrusting base, USNM 242609, ×30.

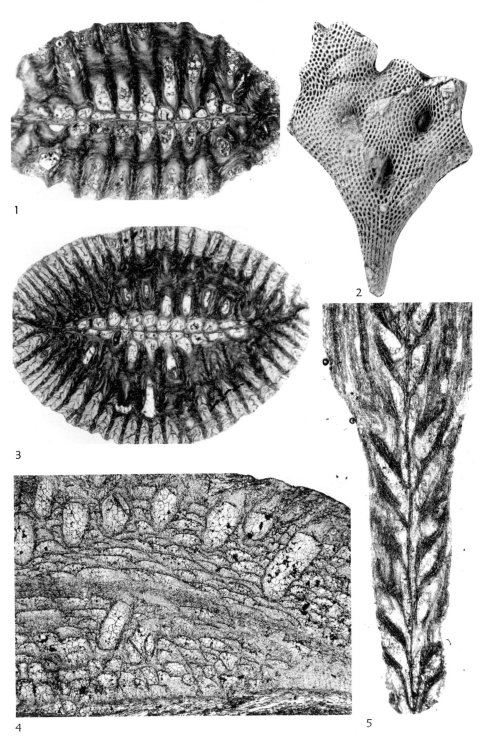

FIG. 225. *(For explanation, see facing page.)*

while expanding laterally, the rods are continuously added along the margins and between previously formed rods. They arise within the granular layer and do not bifurcate. As observed in transverse (see Fig. 252,*1c*) and deep tangential (see Fig. 251,*1h*; 257,*1e*) sections, the rods are evenly spaced and are continuous for appreciable distances. They curve broadly (Fig. 231) toward zoarial margins, where many terminate. In areas of zoarial branching, rods may curve into a new branch or terminate within the granular zone as new rods arise in the new branch. In general outline, the median rod complex in a zoarium is fan shaped and confined to the median granular zone of the mesotheca. I have not observed structural breaks in laminae of a mesotheca, which could indicate extension of the median rods into zooecial walls in the endozone, as reported by PHILLIPS (1960, p. 3, 5). Thus, the median rods do not appear to be associated directly with individual autozooecia, but appear to be extrazooecial.

Autozooecia.—In the Ptilodictyina, autozooecia are comparable in structure, size, shape, and distribution to those in the trepostomates (BOARDMAN, 1971), cystoporates (UTGAARD, 1973), rhabdomesines (BLAKE, 1973c), and cyclostomates (BORG, 1926b, 1933; BROOD, 1972). They are also similar to those in the fenestellines (TAVENER-SMITH, 1969a, 1975) but are larger.

In most ptilodictyines, autozooecia arise from the mesotheca in linear ranges of uniform width (Fig. 223, 231, 232). Junctions between mesothecal and autozooecial wall laminae are usually irregular discontinuities (see Fig. 240,*1h*; 257,*1e*).

Autozooecia are usually delineated laterally by **longitudinal walls** and distally by transverse walls (Fig. 228, 231, 232). Longitudinal walls are continuous until they bifurcate or a new wall arises from within a widened range (Fig. 228). Longitudinal walls delineate the zooecial ranges at precise lateral intervals and either preceded the transverse walls (Fig. 231) or arose concurrently with them (Fig. 228, 232). Transverse walls (Fig. 228, 231, 232) separate successive autozooecia within a range at regular intervals and alternate in position with those of adjacent ranges (rhombic budding of BOARDMAN & UTGARRD, 1966, p. 1083). The shape of a resulting autozooecium in cross section is that of a subrectangle or subparallelogram of constant size (Fig. 228, 231). Width of the autozooecium averages between one-third and half its length at the mesotheca, with longer dimension in the zoarial growth direction. With few exceptions, this ratio is maintained in the endozone.

In a few genera, longitudinal and transverse walls are only partially contiguous, and the longitudinal walls are not linearly continuous. Extrazooecial skeletal deposits between autozooecia are laminar or vesicular

---

FIG. 226.   Basal attachment of a rigid zoarium (*1*) and autozooecial budding pattern in erect zooecia (*2,3*). ——*1a–d. Trigonodictya fenestelliformis* (NICHOLSON), U. Ord. (Richmond.), Wilmington, Ill.; serial peels of basal attachment; *a,* proximal part of connecting segment with fully developed mesotheca containing abundant median rods and regularly aligned zooecia, narrow endozone, vesicular extrazooecial stereom arises locally from the mesotheca; transv. peel, ×50; *b,* encrusting zoarial base; first generation of zooecia in spiral pattern (lower center), zooecia generally separated by vesicles of basal layer, mesotheca not present; peel from base of first generation of zooecia just above bottom of basal layer, ×50; *c,* vesicular portion of basal layer from which first generation of zooecia arises, subsequent zooecia in margin of encrusting base (upper left); peel in plane of vesicular portion of basal layer parallel to substrate (at right angle to margin of encrusting base, Fig. 225,*4*), ×50; *d,* regularly arranged basal zooecia in margin of encrusting base; peel in plane parallel to substrate, ×30, all USNM 242609.——*2. Athrophragma foliata* (ULRICH), Decorah Sh., M. Ord., St. Paul, Minn., paralectotype; subtubular autozooecia surrounded by vesicular and laminar extrazooecial stereom in inner exozone, distinct autozooecial boundaries; tang. sec., USNM 163112, ×30.——*3. Athrophragma grandis* (ULRICH), U. Ord. (Richmond.), Wilmington, Ill.; subtubular autozooecia in relatively straight ranges surrounded by vesicular extrazooecial stereom just above mesotheca, distinct autozooecial boundaries, median rods (near top); tang. sec., USNM 242614, ×30.

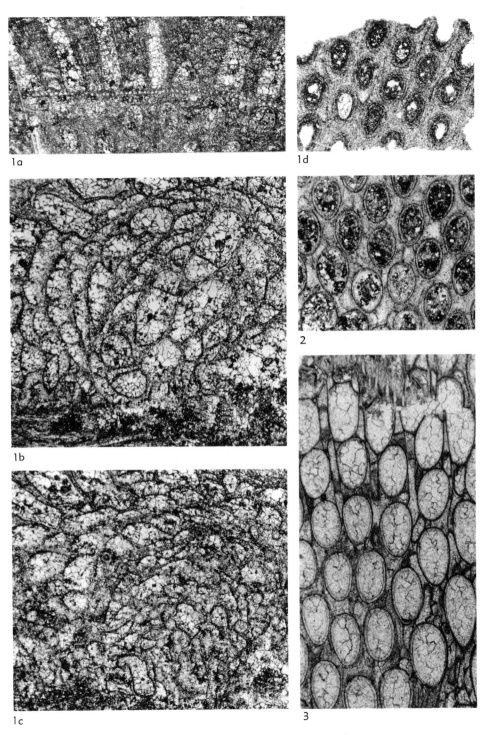

FIG. 226. *(For explanation, see facing page.)*

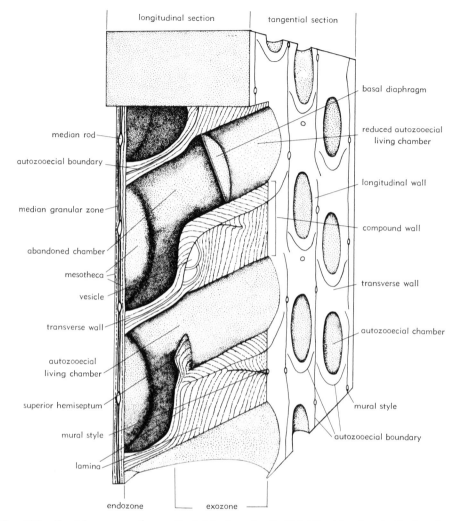

FIG. 227. Zoarial structure of generalized *Stictopora,* family Rhinidictyidae. Longitudinal section at right angle to mesotheca and parallel to length of zooecia from mesotheca to zoarial surface in midregion of erect zoarium. Tangential section parallel to and slightly under surface of zoarium and parallel to plane of mesotheca. Only one side of bifoliate zoarium is shown.

(Fig. 226,*3;* 233; 234; see also 252,*1a*). Generally, the autozooecia are irregularly subelliptical to subcircular in cross section at the mesotheca (Fig. 234).

Ontogenetically, autozooecia formed two distinct growth stages, an early endozonal stage followed by an exozonal stage. In the endozone, autozooecial walls are thin and may curve (Fig. 223,*5;* 224,*4;* 232) regularly as they extend from the mesotheca. Width of the endozone is generally constant

and may be characteristic of a group of species, as in *Stictopora*. With beginning of the exozonal growth stage, zooecial walls thicken considerably as autozooecia diverge from the general growth direction of the mesothecal plane. Morphological changes at the endozone-exozone boundary include adjustments in space between autozooecial chambers and thickened walls, modifications in configuration of wall laminae (Fig. 227), inception of other skeletal structures within

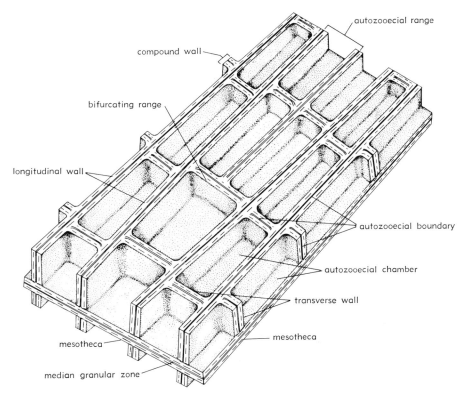

FIG. 228. Arrangement and shape of autozooecia in the endozone of generalized *Ptilodictya*, family Ptilodictyidae. Budding pattern at mesotheca in midpart of erect zoarium is characteristic of the ptilodictyines. Longitudinal walls are structurally continuous; transverse walls may be straight, as shown, or convex distally. One bifurcating range is shown.

laminae (Fig. 235,*3*), and the appearance of polymorphs (see Fig. 242,*2a*). Features typical of the zooecial endozone overlap with those of the exozone so that the endozonal-exozonal boundary describes a narrow, indistinct zone that is irregular in thickness (see Fig. 244,*1d*). At the zoarial margin, the boundary becomes indistinguishable (see Fig. 240,*1f*).

Linearity of zooecial ranges and skeletal continuity of longitudinal walls (see Fig. 240,*1d*) are maintained in the exozone of many taxa but may be obscured by extrazooecial stereom or modified by monticular areas (see Fig. 254,*1a,d*). In other taxa (see Fig. 245,*2b*), zooecial walls that are longitudinal in the endozone lack this alignment in the exozone, and autozooecia in some gen-

era are in a distinct rhombic pattern (Fig. 224,*6*). Autozooecial chambers, regardless of zooecial shape, are usually subtubular with a subelliptical (see Fig. 253,*2a*) to subcircular (see Fig. 254,*1d*) cross section in the exozone.

Ptilodictyine autozooecia consist of structurally different granular and laminar calcareous materials. Granular zones (Fig. 227) constitute only a small part of the total skeletal mass, but they form the basic framework in ptilodictyine zoaria. The significance of the granular component and its bearing on possible phylogenetic relationships between ptilodictyines and other cryptostomate suborders has been only recently recognized (TAVENER-SMITH, 1975; BLAKE, this revision).

In most ptilodictyines, autozooecial

boundaries are granular in the endozone (Fig. 223, 231, 232), and are identical in microstructure to the granular zone in the mesotheca. Laminae in endozonal walls generally parallel the granular autozooecial boundaries, although this relationship is not always clear (see Fig. 259,2a). In most taxa, the interface between granular zones and adjacent laminae is abrupt.

With the beginning of the exozonal growth stage, zooecial boundaries may be obscured by merging zooecial wall laminae (see Fig. 240,1b), become broadly serrated zones (see Fig. 245,1f), or become narrowly serrated zones (see Fig. 259,2a) consisting of irregular discontinuities formed by intertonguing of all laminae along the approximate median between autozooecia (Fig. 229,4,5).

In the exozone, laminae in compound walls form different patterns in different families. In general, the laminae are either in an inverted U-shape (Fig. 229,5; 230,1,2,4) or V-shape (Fig. 227; 229,4; 236) toward the zoarial surface. Walls having a broadly U-shaped configuration possess autozooecial boundaries that are broadly serrated as a result of intertonguing with laminae in adjacent zooecia (see Fig. 240,1b). These boundaries vary from being well defined (see Fig. 245,1f) to indeterminate (see Fig. 242,2a), and are not visibly connected with the boundaries in the endozone. Walls having narrowly (see Fig. 255,1d) to broadly V-shaped laminae (see Fig. 259,2a) possess autozooecial boundaries that are on the average narrowly serrated. These boundaries are well defined in most genera and are continuous with boundaries in the endozone. In a few taxa, dark zones other than autozooecial boundaries may arise in autozooecial walls in the exozone and extend through the walls at approximately right angles to the zoarial surface.

In some ptilodictyines, autozooecial walls are divided into a relatively thick outer part and a much thinner inner part (Fig. 237,1,3). Configuration of laminae in the outer part of a wall is characteristic of a taxon. In the inner part of a wall, adjacent to the zooecial chamber, laminae may be parallel to the long axis of a zooecial chamber (Fig. 229,5) or oblique to the chamber axis (see Fig. 255,1b). The thin part of a wall lines zooecial chambers. Generally, both sets of wall laminae intertongue along a dark discontinuity, suggesting that they were secreted concurrently.

BOARDMAN (1971, fig. 1) described similarly divided zooecial walls in certain Trepostomata as consisting of a cortex (outermost unit of zooecial wall) and lining (wall unit between cortex and zooecial chamber).

The lining of zooecial chambers in ptilodictyines is generally indistinct. In autozooecia, it may be present or absent. When present, it is commonly discontinuous within a chamber.

Configuration of autozooecial wall laminae, together with relative thickness of lam-

FIG. 229. Miscellaneous morphology of ptilodictyines.——*1. Phaenopora twenhofeli* BASSLER, Becscie F., L. Sil., Anticosti Is., Can., holotype; probable remnants of soft parts occurring as elongated brown bodies around superior and inferior hemisepta (center and upper, right of mesotheca); long. sec., USNM 143032, ×50.——*2,3. Trigonodictya conciliatrix* (ULRICH), Decorah Sh., M. Ord., Cannon Falls, Minn., paralectotypes; *2*, extrazooecial stereom in autozooecial range partitions with median dark zones containing distinct mural styles in exozone; tang. sec., USNM 242615, ×50; *3*, mesotheca bifurcating at right angle (lower center), distinct median rods in granular zone, extrazooecial vesicles at base of exozone; transv. sec., USNM 242616, ×100.——*4. Pseudostictoporella typicalis* ROSS, float, M. Ord., Martinsburg, N.Y., paratype; compound autozooecial wall with V-shaped laminae, narrowly serrated and nongranular autozooecial boundary in exozone, indistinct dark layers between laminae; transv. sec., YPM 25455, ×400. ——*5. Phaenoporella transenna mesofenestralia* (SCHOENMANN), M. Ord. (Mangaze.), Podkamennaya Tunguska River, Sib., USSR; U-shaped laminae and broadly serrated autozooecial boundaries, exilazooecium (center) with lining of laminae parallel to chamber, indistinct dark layers between laminae; transv. sec., USNM 171741, ×400.

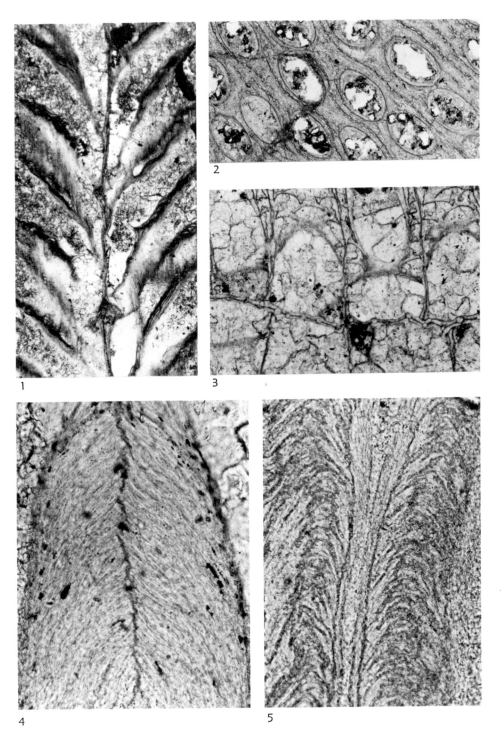

FIG. 229. *(For explanation, see facing page.)*

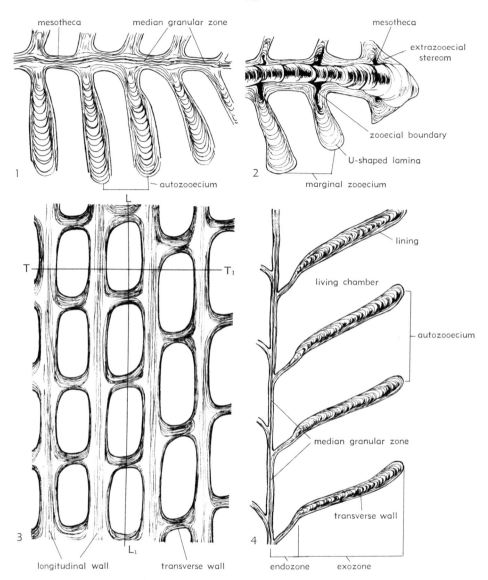

Fig. 230.   Generalized *Ptilodictya*, family Ptilodictyidae.——*1*. Zoarial midregion, transv. sec. (along T–T$_1$ in *3*).——*2*. Zoarial margin and curved laminae in mesotheca, transv. sec.——*3*. Exozone in zoarial midregion, tang. sec.——*4*. Zoarial midregion, long. sec. (along L–L$_1$ in *3*).

inae, characterize various ptilodictyine families.

Autozooecia within a single erect zoarium are identically constructed and show only minor differences in shape, size, and ontogeny. In areas of zoarial branching, autozooecia are generally modified in shape from those in segments between branches. These auto-

zooecia are commonly larger or smaller than regular zooecia and are subcircular in cross section (see Fig. 247,*1c*; 251,*1e*). If the branching results from widening of zooecial ranges at the mesotheca, the autozooecia are commonly larger (Fig. 223,*6*). If the branching results from bifurcating longitudinal walls at the mesotheca, the autozooecia are

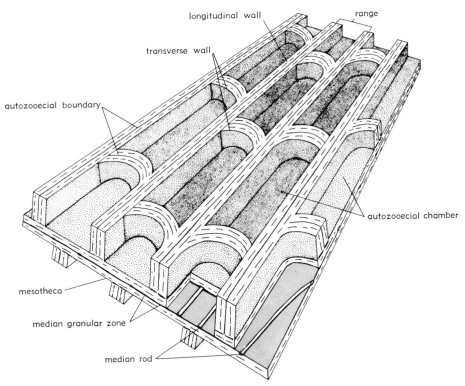

longitudinal wall

range

transverse wall

autozooecial boundary

autozooecial chamber

mesotheca

median granular zone

median rod

FIG. 231. Arrangement and shape of autozooecia in the endozone of a generalized *Stictopora,* family Rhinidictyidae. Structural relationship between autozooecial walls in the endozone suggests that longitudinal walls probably precede transverse walls during formation of a zooecium. This budding pattern appears to characterize some species of *Stictopora.*

commonly smaller and more elliptical in cross section than those in segments between branches (see Fig. 242,*2c*).

These are just a few examples of modified autozooecia that occur along with regularly formed autozooecia. All of these slightly modified zooecia are considered to be autozooecia because they are comparable to regular autozooecia in chamber size, structure, and origin.

Chambers of autozooecia.—Extant tubuliporate bryozoans possess subtubular chambers that house functional soft parts. In most Ptilodictyina, autozooecial chambers differ in size and shape from endozone to exozone. Endozonal segments are generally subrectangular to subrhomboid in cross section parallel to the mesotheca and are only rarely subelliptical (Fig. 227, 228, 231, 232, 234). Exo-

zonal segments are subtubular throughout and subelliptical to subcircular in cross section.

Size, shape, and postulated growth of autozooecia and living chambers in the Ptilodictyina are comparable to those in autozooecia of the Tubuliporata and assumed autozooecia in Trepostomata (BOARDMAN, 1971, p. 18), Cystoporata (UTGAARD, 1973, p. 327), and Rhabdomesidae (BLAKE, 1973c, p. 363).

Basal diaphragms are common in the Ptilodictyina, but occur irregularly. They may be present in some autozooecia of a zoarium or in most autozooecia in some groups of taxa. Basal diaphragms consist of a few laminae that are structurally continuous with autozooecial wall laminae. Diaphragms may be straight or curved (Fig. 227; see also

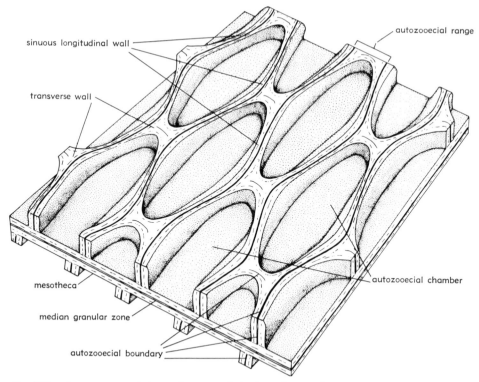

FIG. 232.   Arrangement and shape of autozooecia in the endozone of a generalized *Escharopora*, family Escharoporidae. Autozooecia expand and narrow alternately in adjacent ranges. Development in the endozone characterizes the family and approximates that of the Intraporidae and Stictoporellidae.

252, *1b*). At junctions with autozooecial walls they curve sharply outward toward zoarial surfaces (Fig. 233, 236).

In some species, basal diaphragms are regularly spaced (see Fig. 252, *1b*). In others they are scattered. In a few taxa they are present only at the base of the exozone (see Fig. 246, *1b*), and in others they may occur anywhere in a chamber. In general, variation among taxa in spacing of basal diaphragms suggests that some species differ in length of their living chambers.

In numerous species, diaphragms are absent and there is no evidence for their presence at any time during the growth of a zoarium (Fig. 230, 237–239). Therefore, it is assumed that in these taxa the mesotheca formed the basal proximal wall (floor) of the living chamber throughout the growth of a zoarium (Fig. 229, *1*).

If abandoned chambers (see Fig. 252, *1b*) can be considered as part of the evidence for degeneration-regeneration cycles in colonies, these cycles are irregular in the Ptilodictyina. However, the general lack or rare occurrence of abandoned chambers may not be indicative of the absence of cycles. Perhaps the basal diaphragms were not preserved, or degeneration-regeneration cycles occurred without secretion of diaphragms.

Autozooecial chambers in the Ptilodictyina contain several kinds of skeletal features that project from the walls into the chambers without forming a complete cross structure. Similar structures in the trepostomatous bryozoans have been termed lateral structures (BOARDMAN, 1971, p. 18). In the ptilodictyines lateral structures include hemisepta (Fig. 227), mural spines of varying shapes (see Fig. 240, *1g*; 245, *2c*), and such

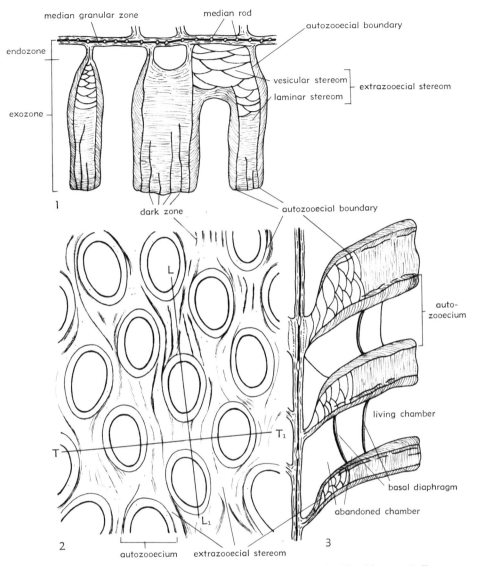

FIG. 233. Generalized zoarial midregion in *Athrophragma*, family Rhinidictyidae.——*1*. Transverse section (along T–T$_1$ in *2*).——*2*. Tangential section of exozone.——*3*. Longitudinal section (along L–L$_1$ in *2*).

uncommon structures as cysts (Ross, 1960a, p. 441) and cystiphragms (see Fig. 241,*2c*). Lateral structures modify the sizes and shapes of autozooecial living chambers.

Hemisepta are the shelflike, straight or curved projections (Fig. 237,*3*) that extend from the wall partway into autozooecial chambers. The projections are formed as

extensions of wall laminae and are of varying shapes. They generally arise from walls at the base of the exozone and project into the endozonal part of a chamber. In Ptilodictyina, a hemiseptum that projects from a proximal wall into the endozonal part of a chamber is called a superior hemiseptum (Fig. 227); one extending from the mesotheca or a distal wall

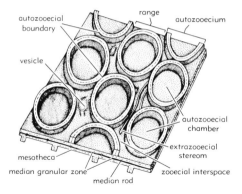

Fig. 234. Arrangement and shape of autozooecia in the endozone of generalized *Pachydictya* and *Athrophragma,* family Rhinidictyidae. Autozooecia are subtubular, partly separated by extrazooecial skeleton or space, and laterally without structurally continuous longitudinal walls.

is called an inferior hemiseptum (Fig. 237,3). Both kinds of hemisepta may occur in the same chamber, arranged alternately, or they may occur singly. Inferior hemisepta are commonly straight (see Fig. 242,2a). Superior hemisepta vary in shape, and may be straight or curved (see Fig. 251,1f). Most superior hemisepta are slightly curved, hooklike in cross section, or knoblike (see Fig. 244,1b). Although hemisepta conventionally have been regarded as one of the most characteristic features of the ptilodictyines, their distribution is uneven within a zoarium or a species, and numerous taxa lack them. In general, hemisepta are more common in taxa that lack basal diaphragms.

Mural spines are thin extensions of zooecial wall laminae that project into autozooecial chambers. They generally curve proximally (see Fig. 240,1g), but may extend at a right angle to the chamber axis. Spines have not been observed in endozonal parts of chambers. In exozones, spines generally occur irregularly; however, in some species (Ross, 1960c, p. 1066) they arise from walls at regular alternating intervals and are closely spaced. In specimens with closely spaced spines, the volume of each living chamber is reduced and soft parts probably curved around tips of the spines.

Cystiphragms are uncommon and of minor significance in ptilodictyines. They are present in one or two Ordovician genera (see Fig. 262,1c), one Devonian genus (see Fig. 241,2c), and one Carboniferous genus (*Phragmophera*).

Cysts are laminated, hollow, irregularly shaped spheres that project inward from zooecial walls. They are generally rare in the ptilodictyines. Ross (1960a, p. 441) described cysts in species of *Ptilodictya* and suggested a possible association with reproductive functions.

Polymorphic zooecia.—Ptilodictyine zoaria contain several kinds of polymorphic zooecia that differ from the autozooecia in ontogeny, size, or shape. They include zooecia near margins of zoaria and fenestrules, exilazooecia (exilapores of Duneava & Morozova, 1967, p. 87; term modified by Utgaard, 1973, p. 339), mesozooecia (Ross, 1964b, p. 940), and zooecia near or in monticules.

Marginal zooecia are those near free margins of branching (see Fig. 251,1e) or explanate zoaria, or along margins of fenestrules in cribrate zoaria (see Fig. 244,1c). These zooecia are considered to be polymorphic because they arise from the mesotheca without forming distinct endozonal chamber portions. Thus, they commonly are shorter and have only an exozonal stage of development (Fig. 230). Marginal zooecia are subtubular and generally oriented oblique to the vertical growth direction of a branch. Because of the different orientation, cross-sectional shape of the zooecia is commonly elongated distally and narrowed laterally. In some taxa, however, marginal zooecia are more rounded than autozooecia (see Fig. 255,1a). Such features as basal diaphragms are also present in marginal zooecia, if present in autozooecia of a taxon. In general, marginal zooecia vary more in shape than do regular autozooecia in a specimen.

In microstructure, marginal zooecia are the same as autozooecia. As marginal zooecia arise from the mesotheca, they form thicker walls than those in a regular endozone, and the walls thicken gradually while forming the

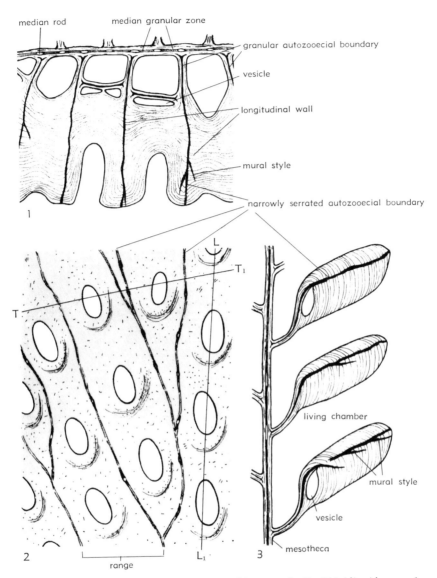

FIG. 235. Generalized zoarial midregion in *Stictopora,* family Rhinidictyidae.——*1.* Transverse section (along T–T$_1$ in *2*).——*2.* Tangential section of exozone showing bifurcating ranges.——*3.* Longitudinal section (along L–L$_1$ in *2*).

margins (see Fig. 245,*1a*). Wall laminae are in structural continuity with laminae of extrazooecial deposits in the margins (see Fig. 249,*1a*).

In general appearance, marginal zooecia are similar to autozooecia forming areas of bifurcation and to zooecia forming the basal attachments of a zoarium. Soft parts in the marginal zooecia probably performed all normal functions but at different rates, which resulted in a modified mode of development and appearance of the marginal zooecia.

Exilazooecia (DUNAEVA & MOROZOVA, 1967, p. 87) are polymorphs that are appreciably smaller than regular autozooecia in the same zoarium (Fig. 229,*5;* 237; 239,*2*).

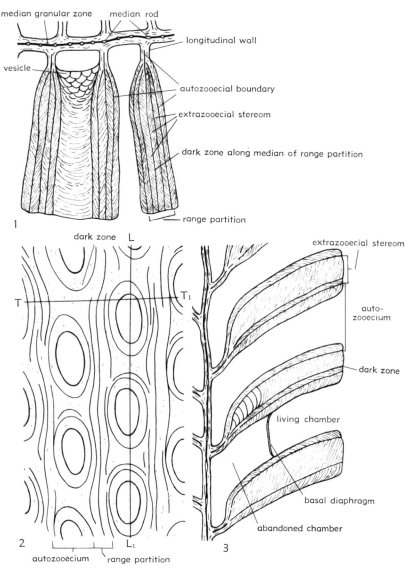

FIG. 236.   Generalized zoarial midregion in *Trigonodictya*, family Rhinidictyidae. Extra-zooecial stereom between range partitions and autozooecial walls consists of laminae that are reversed in direction from those in autozooecial walls.——*1*. Transverse section (along T–T$_1$ in *2*).——*2*. Tangential section of exozone.——*3*. Longitudinal section (along L–L$_1$ in *2*).

They are subtubular skeletal sacs that are irregularly polygonal to subrounded in cross section (see Fig. 259, *1f*). Diaphragms generally are lacking or few. Exilazooecia developed in exozones, but in certain taxa they budded in endozones (see Fig. 264, *1b*). Exi-lazooecia commonly constitute monticules or occur throughout a zoarium. They are common in some ptilodictyine families, but are absent in the Rhinidictyidae and Virgatelli-dae.

Exilazooecia in the ptilodictyines have

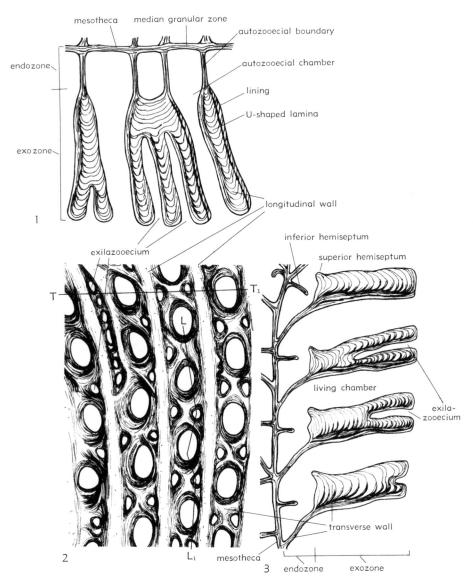

FIG. 237. Generalized zoarial midregion of *Phaenopora,* family Ptilodictyidae.——*1.* Transverse section (along T–T$_1$ in *2*).——*2.* Tangential section of exozone.——*3.* Longitudinal section (along L–L$_1$ in *2*).

been commonly described as mesopores (ULRICH, 1890, 1893; BASSLER, 1953; PHILLIPS, 1960; ROSS, 1960c), as mesozooecia (ROSS, 1964b), and as pseudomesopores (ASTROVA, 1965; KOPAYEVICH, 1972). KOPAYEVICH (1972) in a study of polymorphism in the family Ptilodictyidae recognized at least two kinds of pseudomesopores, which

she distinguished as primary and substituting. Both kinds appear to be skeletally identical but differ in their locus of origin. Primary pseudomesopores originated at the base of the exozone and continued to grow concurrently with the autozooecia. Substituting pseudomesopores originated from living chambers of autozooecia in inner exozones

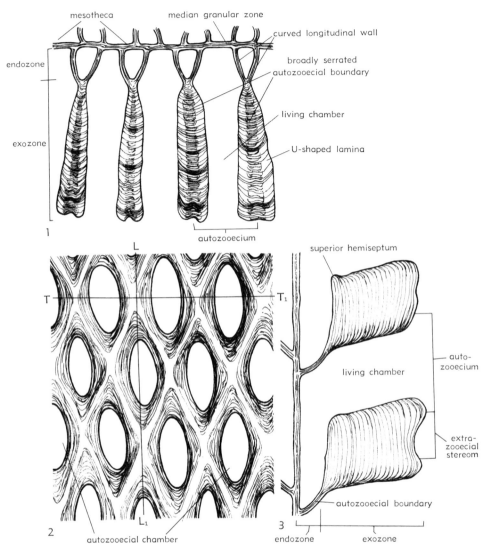

FIG. 238. Generalized zoarial midregion in *Escharopora,* family Escharoporidae.——*1.* Transverse section (along T–T$_1$ in *2*).——*2.* Tangential section of exozone.——*3.* Longitudinal section (along L–L$_1$ in *2*).

and replaced or aborted them in outer exozones or at zoarial surfaces. KOPAYEVICH also reported that groups of substituting pseudomesopores resulted in slightly raised areas at zoarial surfaces, and postulated that these could have been associated with brooding functions in certain ptilodictyids.

Only one kind of exilazooecia is recognized here, and it appears to be the equivalent of

KOPAYEVICH's primary pseudomesopore. The second kind could not be distinguished with certainty in available material.

Mesozooecia are comparable to exilazooecia in size, locus of origin, and pattern of distribution in a zoarium, but possess numerous diaphragms. ROSS (1964b, p. 940) proposed the term mesozooecium for the small zooecia, with or without diaphragms, that are

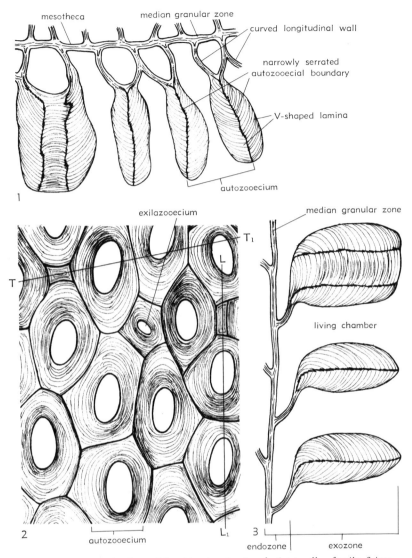

FIG. 239. Generalized zoarial midregion in *Pseudostictoporella,* family Sticto-porellidae.——*1.* Transverse section (along T–T$_1$ in *2*).——*2.* Tangential section of exozone.——*3.* Longitudinal section (along L–L$_1$ in *2*).

generally known as mesopores in the Trepostomata, Cystoporata, and some Cryptostomata. Herein, the term is used only for small zooecia with numerous diaphragms.

In the Ptilodictyina, mesozooecia are known in only three genera. Two of these, *Ensiphragma* (see Fig. 241,*2c*), a ptilodictyid, and *Intrapora* (see Fig. 248,*1d*), an intraporid, both Devonian in age, are from

two morphologically unrelated families. The third, *Ptilotrypina,* is without present family assignment but is of Middle Ordovician age.

In numerous ptilodictyine species the regular zooecial arrangement is commonly modified by clusters of polymorphs that are irregular in size and shape. These clusters may be flat, slightly depressed, or projected above the zoarial surface (see Fig. 254, *1a*;

262, *1b*), and are termed monticules. Their diameter is generally small, ranging from less than 1 mm to rarely more than 2 mm. Commonly, monticules are irregularly subcircular or elongate (see Fig. 255, *1a*), and are irregularly conical or ridgelike when projected above the zoarial surface. Their zooecia differ in shape and size from intermonticular autozooecia, exilazooecia, or mesozooecia, and they may consist of extrazooecial stereom (see Fig. 254, *1d*). Monticules with different constituents may occur separately or in various combinations in zoaria, and are generally differentiated only in exozones. Similar modified zoarial segments occur in numerous other Paleozoic stenolaemate bryozoans and analogous structures appear to be present in recent tubuliporates (BANTA, MCKINNEY, & ZIMMER, 1974).

Monticular zooecia are larger than regular autozooecia but have larger or smaller chambers. Most occur around outer edges of monticules and generally are gradational in size with adjacent intermonticular autozooecia (see Fig. 252, *1a*). They rarely occur in the middle of a monticule. Monticular zooecia may also be less regular in shape than intermonticular autozooecia.

In monticules with mostly exilazooecia (see Fig. 242, *2c*) in the exozone, autozooecia arose from the mesotheca, but were replaced at the base of the exozone by exilazooecia. In some genera of Ptilodictyidae, exilazooecia reportedly arose from autozooecia in the midpart of the exozone (substituting pseudomesopores of KOPAYEVICH, 1972).

Monticules consisting of extrazooecial stereom in the exozone began apparently as zooecia at the mesotheca and then were replaced or filled with the extrazooecial skeletal deposits (see Fig. 257, *1f*) in exozones.

Monticules are common zoarial features in the Ptilodictyina, but their presence is irregularly variable in most taxa. In general, monticules are more abundant in larger or more robust zoaria in the same taxon. Monticules are generally present in explanate zoaria, but are absent in small zoaria with delicate branches. The pattern of distribution of monticules within zoaria appears to be determined by the areal extent of the zoaria. In zoaria with an explanate growth habit, monticules may occur at alternate, regular intervals, forming a rhombic pattern (see Fig. 254, *1a*). In zoaria with a branching growth habit in which branches are relatively wide, monticules are arranged along the median of branches at relatively regular intervals. In some taxa with explanate growth habits, distances between adjacent monticules may vary from 2 to 4 mm, measured from center to center of a monticule. Single monticules commonly occur in branching areas or at random in zoarial branches.

Extrazooecial skeletal deposits.—Extrazooecial skeleton is common in the Ptilodictyina (Fig. 233, 234, 236), its extent being variable in different families. Skeletal deposits comprise zoarial segments that are generally laminated, but may include vesicular structures (vesicular tissue of ULRICH, 1890, p. 298), such structures as median rods in the mesotheca, and thin dark zones in some exozones. Except for median rods in the mesotheca and some vesicular structure in the endozone, skeletal deposits are generally formed during the exozonal growth stage.

Extrazooecial vesicles in ptilodictyines occur in overlapping series (Fig. 233). Most vesicles originated near bases of exozones. In the few genera where autozooecia are only partially contiguous at the budding surface, vesicles arise at the mesotheca (see Fig. 252, *1c*; 255, *1d*) because longitudinal walls are lacking in the endozone. Vesicles terminate in the exozone by merging with stereom. Vesicles are common in the Rhinidictyidae and are absent or rare in other ptilodictyine taxa.

Dark zones in extrazooecial stereom generally arise in exozones (Fig. 233), and only rarely in endozones (Fig. 234). In exozones the dark zones delineate extrazooecial stereom between and around zooecia (see Fig. 257).

In *Trigonodictya* of the Rhinidictyidae, extrazooecial stereom forms distinct longitudinal partitions between autozooecial ranges (Fig. 236; see also 257, *1b*). These straight to slightly sinuous structures are termed **range partitions** (KARKLINS, 1969, p. 26). Extrazooecial stereom between aligned ranges of autozooecia consists of laminae that are reversed in direction in adjacent dark zones (Fig. 236). Such reversal of laminae is uncommon among other Paleozoic bryozoans.

Zooecial wall laminae intertongue with those of the extrazooecial skeleton. Resulting boundaries are variable in shape, or the laminae may be continuous, without visible structural discontinuities (Fig. 238). In a few ptilodictyines, extrazooecial stereom encrusts the zoarial surface (Fig. 225,5). In these forms, and in those with relatively distinct autozooecial boundaries, extent of the extrazooecial deposit can generally be delineated. In zoaria where zooecial boundaries are not visible, extent of the stereom can be approximately inferred from the symmetry of the autozooecia and the generally uniform thickness of their walls. Thus, wider than usual interzooecial spaces in some regions of a zoarium, accompanied by reversal in direction of laminae, suggest the presence of extrazooecial skeleton, although no precise physical boundaries exist (Fig. 238).

Extrazooecial stereom probably constitutes a considerable part of the skeletal mass in zoarial basal attachments (Fig. 225,4). In all zoaria, stereom constitutes the main part of lateral zoarial and fenestrule margins. There it merges with the mesotheca (Fig. 230,2) and forms edges of the margins. Intermittent discontinuities between laminae of the stereom and mesotheca outline approximately the extent of stereom in lateral margins of some taxa (Fig. 230,2). In other taxa, stereom is confluent with the mesotheca, but extent of the mesotheca in a margin is generally indicated by the median granular zone.

In general, extent of extrazooecial stereom in exozones in zoarial midregions can be related to the mode of autozooecial budding at the mesotheca, and to subsequent zooecial adjustments in space along endozone-exozone boundaries. Because development of zoaria is slightly different in different families, distribution of extrazooecial stereom characterizes families to a degree.

Other skeletal structures.—Pustules (mural lacunae of BOARDMAN, 1960, p. 22; pustules or granules of ROSS, 1964b, p. 939; small capillaries of ASTROVA, 1965, p. 103; granules of authors), as understood here, are the very small, irregular dark spots in skeletal laminae, approximately 0.01 mm or less in diameter (ROSS, 1963b, p. 588). These structures lack cores and are not clearly separable in wall laminae. As seen in the light microscope, pustules are crinkled parts (ROSS, 1964b, p. 940) of a few laminae (see Fig. 246, *1a*), as if resulting from a local discontinuity within laminae. These discontinuities may reflect minute changes in the laminae during secretion. Pustules may also be remnants of impurities that were entrapped in skeletons during the process of secretion; however, they are more prevalent and more regularly arranged in some taxonomic groups than in others. They commonly occur within zooecial boundaries or align along the boundaries (see Fig. 246, *1c*). Pustules are also scattered at random, mostly in the outer exozone (see Fig. 259, *1e*). In tangential sections, pustules rarely may resemble mural styles; however, they differ from mural styles in being equidimensional and in lacking distinguishable cores.

Mural styles are elongate, straight to slightly curved, small, rodlike, somewhat irregular structures in the zooecial skeleton. In the Ptilodictyina, mural styles are found in most genera of the Rhinidictyidae and Virgatellidae as well as some genera without family assignment.

With few exceptions (see Fig. 263, *1a*), mural styles arise in the exozone of a zoarium. They may arise from zooecial boundaries (see Fig. 255, *2c*) or appear within walls at random (see Fig. 253, *2a*). They occur singly or are aligned in short, dark zones (see Fig. 251, *1c*). Mural styles originating in outer exozones commonly terminate at zoarial surfaces as small, low protuberances in well-preserved specimens. Mural styles originating in inner exozones generally terminate within the skeleton (see Fig. 255,2*a*). They do not connect with zooecial chambers.

Mural styles consist of indistinct cores surrounded by sheaths of tightly curved, very thin laminae. Because of small size of the mural styles, it is not always possible to establish the presence of a core when using a light microscope. Cores, when distinguishable, appear to be finely laminated, but configuration of the laminae cannot be clearly observed (see Fig. 251,*1a–c*). Cores may also be absent or consist of discontinuous segments separated by a few skeletal laminae or crossing sheath laminae.

Sheath laminae merge with regular skeletal laminae, abut cores, or extend irregularly across them. In tangential sections (see Fig. 251,*1b,c*) sheaths appear as dark rims surrounding the cores.

Mural styles are generally of similar size in a zoarium and vary little within a group of taxa. In most genera, the diameter of a style is less than 0.01 mm, but may range to 0.02 mm. ASTROVA (1965, p. 102) described mural styles in which the diameter ranged between 0.008 and 0.005 mm, with a few less than 0.005 mm. The smallest mural styles probably lack cores (see Fig. 261,*1e*).

Biological significance of median rods, mural styles, and pustules.—Differences in structure and distribution of median rods, mural styles, and pustules suggest different biological functions. Median rods and mural styles have commonly been compared with or regarded as kinds of acanthostyles (acanthopores of ROSS, then PHILLIPS, 1960; others). BLAKE (1973a,b; this revision) reviewed the various functions that have been postulated, and agreed with CUMINGS and GALLOWAY (1915) and TAVENER-SMITH (1969b) that acanthostyles could perform a protective function for soft parts enveloping the zoaria. Whatever the function, it could not have been a major one in the Ptilodictyina, because acanthostyles are rare in representatives of the suborder. Median rods and mural styles,

however, differ from the acanthostyles in abundance, pattern of distribution, and general structure, suggesting differences that are presently unknown.

ASTROVA (1965, p. 102) and KOPAYEVICH (1973, p. 59) suggested that median rods together with mural styles or pustules formed parts of a capillary or canal system in colonies. ASTROVA compared these capillaries with similar structures in the fenestellids (SHULGA-NESTERENKO, 1931, p. 77; 1949) and considered the capillaries in the rhinidictyids as being more primitive than those in the fenestellids. Structure of these capillaries in the fenestellids and other cryptostomates were considered to be granular by several investigators (in ASTROVA, 1965, p. 102), including V. P. NEKHOROSHEV, A. N. NIKIFOROVA, and M. I. SHULGA-NESTERENKO. ASTROVA postulated that the inferred capillaries might be analogous to pseudopores in the Tubuliporata and may have functioned as part of a communication system among zooids and between zooids and the external environment.

KOPAYEVICH (1968, p. 127; 1973, p. 60) described mural styles as capillaries that formed parts of a zoarial canal system. In the Rhinidictyidae, median rods in the mesotheca and dark zones in the exozone complete the suggested canal system. In the family Ptilodictyidae, mural pustules in the exozone, a median granular zone in the mesotheca, and granular zones in endozones (autozooecial boundaries herein) constitute a similar system of zoarial canals. Both authors suggested that this canal complex may have connected individual autozooecia within the zoarium and the zoarium to the external environment, and may have functioned for the passage of biologic, probably gaseous, substances.

Detailed structure of median rods, mural styles, and pustules indicates that they are uniformly solid and may not have been tube-like. Structure of median rods suggests (TAVENER-SMITH & WILLIAMS, 1972, p. 149) that they are original deposits that were formed concurrently with secretion of growing edges of mesothecae and granular zones.

Mural styles in most genera occur only in exozones where they arise from autozooecial boundaries, or from within dark zones or wall laminae. The regular intergrowth of mural styles with skeletal laminae indicates that mural styles also developed concurrently with laminated walls and were solid from their point of origin to their terminal end. That they were not tubelike is also suggested by the general absence of internal, clay-sized, terrigenous particles.

Mural styles and median rods are not structurally connected.

The only taxon in which mural styles could have terminated in autozooecial chambers appears to be the Virgatellidae (see Fig. 260,2a). Although mural styles in this family are distinctly different in growth form, size, and possibly structure, additional specimens are needed to verify a possible connection at the chamber-wall interface. In some taxa (see Fig. 252,1a), mural styles are not present in autozooecial walls but only in extrazooecial stereom in exozones. Thus, they do not connect autozooecia.

Pustules are structurally discontinuous within exozonal laminae (see Fig. 246,1a) and do not extend to zooecial walls in endozones or to granular zones in the mesotheca.

It seems that no median rods, mural styles, or pustules could have been parts of an interconnected zoarial canal system for passage of biological substances. Furthermore, it is unlikely that these structures, consisting of solid calcitic material, could have transmitted biologic substances, even by diffusion (Boardman & Cheetham, 1973, p. 130).

Median rods, mural styles and, to a degree, pustules are interpreted to be discrete zoarial structures. Median rods are extrazooecial structures confined to the mesotheca, and they do not extend into zooecial walls. Mural styles and pustules are zooecial and extrazooecial structures that, except in *Taeniodictya* (see Fig. 263,1a), are confined to the exozone. Their biological significance remains conjectural.

In the ptilodictyines, median rods and mural styles characterize the Rhinidictyidae and some unassigned genera. Relatively large mural styles characterize the Virgatellidae. Pustules occur mostly in the Ptilodictyidae, Escharoporidae, Stictoporellidae, and some unassigned genera.

TAXONOMIC CHARACTERS OF FAMILIES

The suborder Ptilodictyina, as described herein, includes the families Ptilodictyidae, Escharoporidae (new), Intraporidae (restored), Phragmopheridae, Rhinidictyidae, Stictoporellidae, Virgatellidae, and the unassigned genera *Euspilopora, Ptilotrypa, Ptilotrypina, Stictotrypa, Taeniodictya*, and *Trepocryptopora*. *Heliotrypa* is not considered to be a ptilodictyine but is retained tentatively in the Cryptostomata.

Until a few decades ago (Bassler, 1953), certain growth habits were commonly used as the main criteria for differentiating families, although skeletal structures were reasonably well known in several genera. Nicholson (1881, fig. 15) showed the laminar nature of a ptilodictyine skeleton.

Ulrich (1890, p. 308–331; 347–349; 1893, p. 124–187) described and illustrated the main kinds of skeletal structures in several ptilodictyines, mostly Rhinidictyidae, which partly form the morphological basis for families described herein.

Ross reviewed the ptilodictyine genera in a series of publications beginning in 1960. Skeletal morphology in *Ptilodictya* (Ross, 1960a, p. 441, 444, text-fig. 1; 1960c, p. 1062–1072, text-fig. 2) and *Phaenopora* (Ross, 1961a, p. 332; 1962, text-fig. 4) is characteristic of the family Ptilodictyidae. Astrova (1965) considered these genera to be representative of two distinct phylogenetic lineages within the family.

The revised concept (herein) for the family

Stictoporellidae is largely based on skeletal development in *Stictoporella* (PHILLIPS, 1960, p. 23; ROSS, 1960c, p. 1072–1074; 1964a, p. 19), *Pseudostictoporella* (ROSS, 1970, p. 378), and *Stictoporellina* (KARKLINS, 1970).

The concept for the Escharoporidae, new family, is based on skeletal development in *Championodictya* (ROSS, 1964a, p. 18), *Chazydictya* (ROSS, 1963b, p. 587), *Escharopora* (PHILLIPS, 1960, p. 17–19, text-fig. 1; ROSS, 1964a, p. 17), and *Graptodictya* (PHILLIPS, 1960, p. 19–23; ROSS, 1960b, p. 859).

In the Rhinidictyidae, reinterpreted or new material has been described for *Stictopora* (PHILLIPS, 1960, p. 6–8, text-fig. 2; ROSS, 1961a, p. 336; 1961b, p. 76–83; 1964a, p. 24–28); *Eopachydictya* (ROSS, 1963b, p. 591), *Eurydictya* (PHILLIPS, 1960, p. 12), and *Pachydictya* (PHILLIPS, 1960, p. 13–17; ROSS, 1961a, p. 337–342; 1964a, p. 21–24).

Revised concepts of families are mostly based on erect parts of zoaria. The following characters have been used extensively: (1) structure of mesothecae and median granular zones; (2) modes of budding of autozooecia from mesothecae, and autozooecial shapes in endozones; (3) changes in autozooecial shape from endozone through exozone, including shape of living chambers; (4) changes in microstructure of zooecial walls, including autozooecial boundaries from endozone through exozone; (5) nature of any such additional structures in autozooecial walls as pustules or mural styles; (6) kinds of any polymorphic zooecia in exozones; and (7) relative amounts and structures of any extrazooecial skeletal deposits in exozones.

Zoarial growth habits and basal attachments are, to a degree, characteristic of families. However, care should be exercised in the taxonomic use of these features.

Zoarial development and structures in those genera not included in families are ptilodictyine in general appearance, but differ in the modified growth modes of autozooecia at the mesotheca, in the exozone, and in modified skeletal structures.

Until recently, descriptions and interpretations of skeletal structures have been based on observations using the light microscope. Introduction of the scanning electron microscope in the study of the stenolaemate bryozoans (BOARDMAN & TOWE, 1966; BOARDMAN in BOARDMAN & CHEETHAM, 1969; TAVENER-SMITH, 1969a,b, 1975; ARMSTRONG, 1970; TAVENER-SMITH & WILLIAMS, 1972; BROOD, 1972; BLAKE, 1973a,b) has just begun. In general, data obtained on ptilodictyines by using the scanning electron microscope (TAVENER-SMITH & WILLIAMS, 1972; TAVENER-SMITH, 1975) appear to confirm interpretations made by using the light microscope. Further use of the scanning electron microscope will undoubtedly provide additional information of value in refining taxonomy.

SYSTEMATIC DESCRIPTIONS FOR THE SUBORDER PTILODICTYINA

By Olgert L. Karklins

[U.S. Geological Survey, Washington, D.C.]

Suborder PTILODICTYINA
Astrova & Morozova, 1956

[*nom. correct.* herein, *pro* Ptilodictyoidea Astrova & Morozova, 1956, p. 663, suborder]

Zoaria are erect and characterized by bifoliate growth habit. Autozooecia usually are in linear ranges and offset in adjacent ranges, forming rhombic pattern. Basal attachments of zoaria are either skeletally continuous or, rarely, flexibly jointed. Zoaria expand from basal attachments, generally bifurcating in mesothecal plane. Mesothecae are usually planar, having a median granular zone and laminated layers; the median granular zone may contain median rods. Mesothecae are partitioned to form basal autozooecial walls.

Autozooecia consist of compound walls and generally include two distinct growth zones, endozones and exozones. In endozones, most autozooecial walls have distally elongated, subrectangular to subrhomboid shapes in cross section at junction with the mesothecae, but may be subelliptical to subcircular in cross section. Autozooecial boundaries in endozones are thin, rarely discontinuous, granular zones. In exozones, autozooecia generally form angles between 40 and 90 degrees with the mesothecae and are subelliptical, subcircular, subrectangular, or hexagonal in cross section. Autozooecia are contiguous or may be separated by polymorphic zooecia or extrazooecial skeleton.

Autozooecial wall laminae are either broadly U-shaped and form broadly serrated autozooecial boundaries or broadly to narrowly V-shaped and form narrowly serrated autozooecial boundaries. Autozooecial living chambers extend either from mesothecae or from skeletal diaphragms forming basal walls (mostly in exozones) to autozooecial apertures. In endozones, autozooecial living chambers are generally subrectangular to subelliptical in cross section; in exozones, living chambers are subtubular and may contain abandoned chambers proximal to living chambers. Autozooecial chambers may contain various lateral structures; inferior and superior hemisepta, rarely mural spines, are characteristic of some taxa. Such lateral structures are lacking in many taxa.

Polymorphism is expressed by modified zooecia in zoarial margins, zoarial basal attachments, and monticular zooecia. In exozones, exilazooecia are common, small polymorphs having few or no diaphragms; mesozooecia having numerous diaphragms are rare. Monticules consisting of polymorphic zooecia and extrazooecial stereom in various combinations are common.

Extrazooecial skeletal deposits of laminated stereom or laminated stereom and vesicles form connective skeleton between zooecia, margins of zoaria, and basal zoarial attachments. Vesicular extrazooecial skeleton is generally present in inner exozones. Pustules and mural styles are common in zooecial walls and extrazooecial stereom. Acanthostyles are rare. *Ord.-Carb.*

Family PTILODICTYIDAE
Zittel, 1880

[*nom. correct.* Bassler, 1953, p. G136, *pro* Ptilodictyonidae Zittel, 1880, p. 603] [=Clathroporidae Simpson, 1897, p. 543; Ptilodictyinae Zittel (*nom. transl.* Astrova, 1965, p. 251); Phaenoporinae Astrova, 1965, p. 254]

Zoaria unbranched and commonly lanceolate, or explanate and cribrate, or branched; commonly tapering proximally. Mesothecae straight to sinuous, rarely zigzag in transverse section. Median granular zones extend discontinuously through most of mesothecae, terminate near thickened mesothecal margins. In endozones, autozooecia in straight to curving ranges, aligned on opposite sides of mesothecae, subrectangular in cross section parallel to mesotheca, contiguous, with generally straight transverse walls, continuous longitudinal walls. Boundaries become broadly serrated, or laminae from adjacent autozooecia merge so that boundaries are not visible at base of exozone. In exozones, autozooecia form angles with mesotheca ranging from 50° to 80°; in straight to curving ranges; subrectangular, elliptical, or subhexagonal in cross section; generally contiguous laterally, and contiguous or separated transversely within ranges by exilazooecia. Longitudinal walls continuous, extending into ridges at zoarial surfaces. Wall laminae broadly curved and U-shaped. Boundaries broadly serrated or not visible. Pustules rare to common and scattered throughout exozonal walls. Living chambers subrectangular in cross section in endozones; elliptical, subelliptical, or subcircular in cross section in exozones. Basal diaphragms rare. Chamber lining and superior and inferior hemisepta common. Mural spines and cysts generally rare and scattered in zoaria. Cystiphragms present in one genus. Polymorphs marginal, monticular, and basal; exilazooecia abundant to rare or lacking; mesozooecia in one genus. Monticules rare to common in most genera, lacking in some; distributed irregularly; consisting in varying combinations of exilazooecia, larger or smaller zooecia, and extrazooecial stereom. Extrazooecial deposits laminated and irregularly delineated; sparse in zoarial midregions and distally. *M.Ord.-L.Dev.*

Ptilodictya LONSDALE in MURCHISON, 1839, p. 676 [*Flustra lanceolata* GOLDFUSS, 1829, p. 104; OD; glacial drift, "encrinite limestone," ?U. Sil., Groningen, Ger.] [=*Heterodictya* NICHOLSON, 1875, p. 33, L. Dev., Ont., Can.]. Zoarium lanceolate with tapering proximal segment. Mesothecae straight, rarely zigzag locally. In endozones, autozooecia in straight ranges, subrectangular to subhexagonal in cross section. In exozones, autozooecia in straight ranges, arranged in rhombic to reticulate pattern in adjacent ranges; contiguous; commonly subrectangular in cross section, few irregularly polygonal in lateral regions. Autozooecial boundaries generally not visible; pustules rare. Living chambers elliptical to subrectangular in cross section; lining common in endozones, discontinuous or lacking in exozones. Superior hemisepta few,

blunt, short, thick; inferior hemisepta few, thin, short. Both hemisepta scattered in a zoarium. Spines curved proximally; cysts rare at mesotheca. Exilazooecia few, generally lacking. Monticules irregularly distributed, flat to raised, indistinct; consisting of slightly larger, possible autozooecia. [Two syntypes of *P. lanceolata* are at the Geologisch-Paläontologisches Institut, Bonn, Germany, and are poorly preserved (ROSS, 1960a, p. 440). The specimen figured by GOLDFUSS is lost and its original locality is unknown. ROSS (1960a, p. 440) redescribed and subjectively defined *P. lanceolata* on the basis of material from the Wenlock Limestone (Silurian) of Dudley, England; from calcareous clay, lower Ludlovian Series (Silurian) at Mulde, near Klinteham, Gotland, Sweden; and from the upper Llandoverian Series at Roneham, Gotland. According to ROSS (1960a, p. 444), LONSDALE described *Ptilodictya* and its type species, *P. lanceolata*, on the basis of material from the Wenlock Limestone, Malvern Hills, England.] *U.Ord.-L.Dev.*, USSR, Swed., Eng., N.Am., India.——FIG. 240, *1a-h*. **P. lanceolata* (GOLDFUSS), Wenlock Ls., U. Sil., Dudley, Eng.; *a*, mesotheca, straight longitudinal walls, slightly flexed transverse walls; transv. sec., USNM 137913, ×30; *b*, autozooecia in distinct linear ranges, reticulate in lateral regions, smaller living chambers in mid zoarium; external view, USNM 137913, ×4; *c*, indistinctly subhexagonal autozooecia in endozone, indistinct brown bodies near mesotheca; deep tang. sec., USNM 137913, ×30; *d*, elliptical to subrectangular living chambers between structurally continuous longitudinal walls, autozooecial boundaries not visible; tang. sec., USNM 137913, ×30; *e*, mesotheca, shape of living chambers parallel to growth direction; oblique long. sec., USNM 137913, ×30; *f*, broadly curved laminae of mesotheca in zoarial margin, reduced endozone in zoarial margin; transv. sec., USNM 137912, ×30; *g*, median granular zone along middle of mesotheca, U-shaped laminae in transverse walls, discontinuous lining, recurved mural spine in living chamber; long. sec., USNM 137911, ×50; *h*, granular zone in mesotheca, broadly curved laminae in longitudinal walls, zooecial boundaries indistinct, zooecial lining in endozone and exozone; transv. sec., USNM 137911, ×100.

Clathropora HALL in SILLIMAN, SILLIMAN, & DANA, 1851, p. 400 [*C. frondosa* HALL, 1852, p. 159; SD ULRICH, 1890, p. 392; Rochester Sh., M. Sil., Lockport, N.Y., USA]. Zoarium branched or unbranched and cribrate with tapering, connecting segments. Fenestrules ovate to subcircular, varying in size in cribrate zoaria, generally aligned in growth direction; marginal zooecia in indistinct ranges. Mesothecae slightly sinuous in longitudinal section. In endozones, autozooecia

1a

1f

1b

1g

1c

1d

1h

Ptilodictya

1e

Fig. 240. Ptilodictyidae (p. 490).

in straight or curving ranges. In exozones, autozooecia subhexagonal in cross section, contiguous, in straight ranges in midregions between fenestrules, in curving ranges near fenestrules and margins. Autozooecial boundaries broadly serrated, rarely with pustules. Living chambers broadly elliptical in cross section, generally with distinct lining in cribrate zoaria. Superior hemisepta common, short, blunt or irregularly shaped, regularly arranged. Inferior hemisepta lacking. Exilazooecia rare to lacking in midregions, scattered at bifurcations; singly or in scattered groups near fenestrules and zoarial margins. Extrazooecial stereom rarely fills fenestrules in proximal regions, commonly forming annular ridge around distal parts of connecting segments. Monticules not observed. *U.Ord.-L.Dev.,* Eu. (Est., France), USA.——Fig. 241,*1a–d.* *C. frondosa,* lectotype, AMNH 1734/2; *a,* arrangement of autozooecial ranges, shape of fenestrules; external view, ×5; *b,* alignment of autozooecia across mesotheca; transv. sec., ×30; *c,* subhexagonal autozooecia in endo-exozone, sinuous longitudinal walls, exilazooecia near fenestrule (lower right); tang. sec., ×30; *d,* sinuous mesotheca, thick lining on distal sides of zooecial walls; long. sec., ×50.

Ensiphragma Astrova in Astrova & Yaroshinskaya, 1968, p. 61 [*E. mirabilis*; OD; Kireyev stratum, L. Dev., Solov'ikha River basin, Altai Mts., USSR]. Zoarium unbranched. Mesotheca straight. Autozooecia in straight ranges throughout ontogeny. In exozones, autozooecia subrectangular in cross section, contiguous laterally, partly separated within ranges by mesozooecia. Autozooecial boundaries not visible; pustules indistinct, scattered in exozonal walls. Living chambers elliptical in cross section, variable in length; lining thin, locally discontinuous. Superior hemisepta common, blunt, thick, straight; inferior hemisepta common, thin, straight, projecting from mesothecae or distal walls. Cystiphragms regularly arranged, generally open with irregularly curved proximal tips. Basal diaphragms thin, slightly curved, relatively uniform in spacing. Mesozooecia common, subcircular in cross section, regularly arranged in pairs between successive autozooecia in midregions; abundant along zoarial margins. Mesozooecial diaphragms closely spaced; chamber linings thin, discontinuous. Monticules absent. *L.Dev.,* USSR (Altai Mts.).——Fig. 241,*2a–d.* *E. mirabilis*; *a,* alignment of autozooecia across mesotheca; transv. sec., holotype, PIN 2218/508, ×40; *b,* continuous laminae of longitudinal walls (left), basal diaphragms, abandoned chambers, open cystiphragms, mesozooecia between autozooecia; long. sec., ×40; *c,* autozooecial lining along walls, open cystiphragms in autozooecial chambers, diaphragms in mesozooecia, hemisepta (chamber in endozone, mid-

dle right); long. sec., paratype, PIN 2218/514, ×100; *d,* straight autozooecial ranges, mesozooecia between successive autozooecia; tang. sec., paratype, PIN 2218/510, ×30 (photographs courtesy G. G. Astrova).

Ensipora Astrova, 1965, p. 263 [*Escharopora tenuis* Hall, 1874, p. 99; OD; low. Helderberg Gr., L. Dev., Clarksville, N.Y., USA]. Zoarium unbranched and lanceolate. Mesothecae generally straight. Autozooecia in straight ranges throughout ontogeny; subrectangular in cross section of exozone, generally contiguous, with relatively thin walls. Living chambers elliptical in cross section, relatively large. Superior hemisepta thin, long, straight, regularly arranged. Inferior hemisepta shorter, extending from distal walls. Exilazooecia and monticules probably absent. [The concept of *Ensipora,* to which numerous species have been assigned (Astrova, 1965; Astrova in Astrova & Yaroshinskaya, 1968), is unclear because the type material of *E. tenuis* is poorly preserved. Hall did not designate a holotype and primary types cannot be related to subsequently figured specimens (Hall, 1883a, pl. 13, fig. 14, pl. 17, fig. 7–13; Hall, 1887, pl. 13, fig. 14, pl. 17, fig. 7–12, pl. 23A, fig. 15). The budding pattern, cross-sectional shape of autozooecia, straight longitudinal walls in zoaria, and hemisepta in exozones resemble ptilodictyids; however, microstructure of laminae and presence of exilazooecia cannot be verified in primary material.] *M.Ord.-L.Dev.,* USSR, ?Baltic region, ?Eng., N. Am.——Fig. 242,*1a–e.* *E. tenuis* (Hall); *a,* shape, alignment of autozooecia across mesotheca; transv. sec., lectotype, AMNH 2309/2310, ×50; *b,* arrangement of autozooecial ranges along middle and margins of zoarium; external view, lectotype, ×5; *c,* shape of living chambers; tang. sec., lectotype, ×50; *d,* hemisepta; long. sec., lectotype, ×50; *e,* zoarium with distal, tapered connecting segment and partly closed (?encrusted) basal zooecia in narrow ranges; external view, paralectotype, NYSM 893, from New Scotland Ls., N.Y., ×5.

Insignia Astrova, 1965, p. 271 [*Phaenopora insignis* Nekhoroshev, 1961, p. 89; OD; Nishnyaya Chunka River, U. Ord., Sib., USSR]. Zoarium branched or unbranched and subcylindrical to irregularly explanate, relatively large and variable in thickness. Unbranched zoaria subcylindrical with conical, proximal tips. Branched zoaria with approximately parallel branches and tapering proximal segments. Zoarial midregions slightly raised, subcylindrical in transverse section, tapering to flattened lateral regions. Mesothecae slightly sinuous in longitudinal section. In endozone, autozooecia in straight and variably curving ranges. In exozones, autozooecia ontogenetically subrectangular to subelliptical in cross section; in straight ranges for varying distances in zoarial midre-

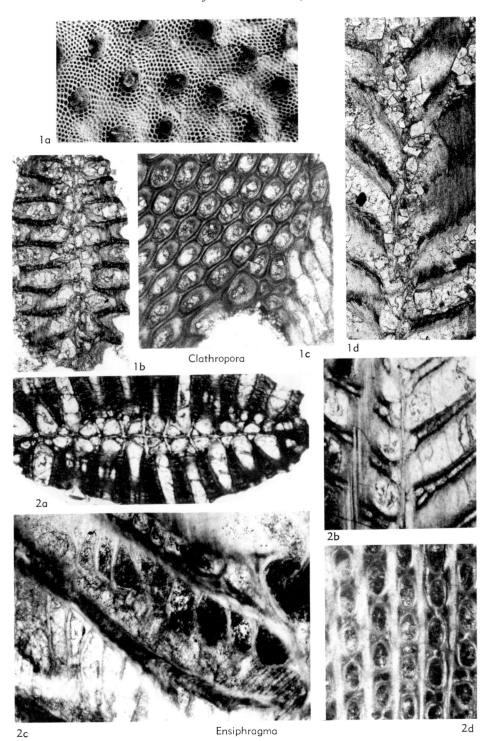

Clathropora

1a
1b
1c
1d

2a
2b

2c   Ensiphragma   2d

FIG. 241. Ptilodictyidae (p. 490–492).

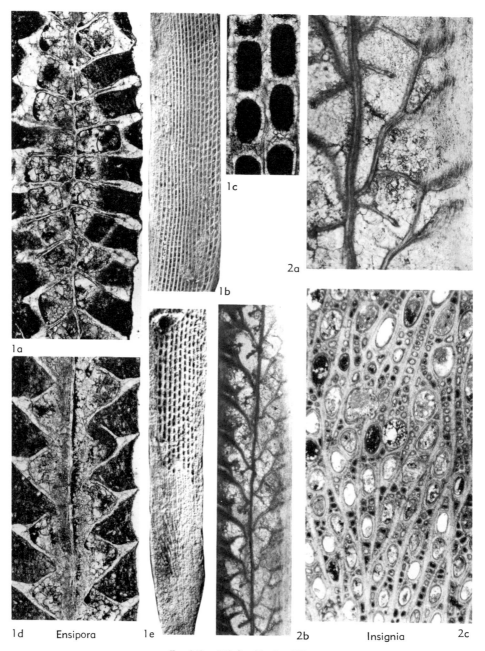

1a
1d　　Ensipora　　1e

1c

1b

2a

2b　　　　Insignia　　　2c

FIG. 242.　Ptilodictyidae (p. 492).

gions; in irregularly curving, converging, or bifurcating ranges in greater part of zoarium. Autozooecia contiguous or separated laterally and within ranges by exilazooecia, commonly replaced by groups of exilazooecia at irregular intervals. Autozooecial boundaries generally not visible; pustules scattered in exozonal walls. Liv-

ing chambers broadly elliptical to subcircular in cross section, lining thick to lacking. Superior and inferior hemisepta common, long, straight, relatively thick, and regularly arranged. Inferior hemisepta projecting from mesothecae or distal walls; superior hemiseptum locally a basal diaphragm in some species. Exilazooecia abundant,

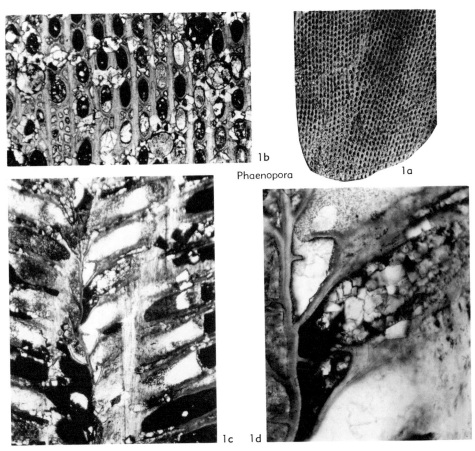

FIG. 243.   Ptilodictyidae (p. 495).

irregularly subcircular or varying in cross section; slightly variable in size, commonly with narrower chambers in outer exozones; generally arranged in irregular groups, rarely in pairs between successive autozooecia, or in one or two relatively straight rows of variable length in areas of replaced autozooecia. Monticules common, consisting of several exilazooecia and scattered zooecia in varying combinations; pustules common in walls. [*Insignia* is closely related to *Phaenopora* but differs from it in having a modified autozooecial budding pattern, in abundance and distribution of exilazooecia, in having massive zoaria, and in having unbranched zoaria of variable growth habits that probably result from the irregular autozooecial budding pattern. According to ASTROVA (1965, p. 271), variations in growth habits in *Insignia* do not seem to have been controlled by changes in depositional environments.] *M.Ord.-U.Ord.,* USSR (Sib.).—— FIG. 242,*2a–c.* **I. insignis* (NEKHOROSHEV), Podkamennaya Tunguska River, Sib.; *a,* mesotheca with discontinuous median granular zone, aban-doned chambers with superior hemisepta as basal diaphragms and inferior hemisepta in endozone, exilazooecial chambers at base of exozone; long. sec., PIN 1242/81, ×100; *b,* living chambers with alternating hemisepta, curved transverse walls in endozone; long. sec., PIN 1242/81, ×30; *c,* irregularly aligned ranges, monticule with larger zooecium, (upper left); tang. sec., PIN 1242/87, ×30.

Phaenopora HALL, in SILLIMAN, SILLIMAN, & DANA, 1851, p. 399 [**P. explanata* HALL, 1852, p. 46; SD ULRICH, 1890, p. 392; ?Cataract F., L. Sil., Flamborough Head, Ont., Can.]. Zoarium branched or unbranched and explanate. Mesothecae sinuous in longitudinal section. Autozooecia in straight ranges throughout ontogeny. In exozones, autozooecia subrectangular in cross section, generally contiguous laterally, partially separated within ranges by exilazooecia; may be replaced by exilazooecia. Autozooecial boundaries generally not visible; pustules indistinct, scattered in zoaria. Living chambers elliptical in cross section, generally without lining. Superior and inferior hemisepta common, regularly arranged. Superior hemisepta curved proximally, relatively long and thick; inferior hemisepta extending from mesotheca, relatively thick, variable in length. Exilazooecia common, irregularly triangular to subcircular or elongate longitudinally in cross section; regularly arranged in pairs or may be in short rows between successive autozooecia; singly or in short rows in areas of bifurcation and margins, and in areas of replaced autozooecia. Monticules common in some zoaria, raised, consisting of exilazooecia and sparse zooecia. [Internal structure in type specimens of *P. explanata* is poorly preserved (Ross, 1960c, p. 1072; 1961a, p. 332); however, it is reasonable to assume that the internal structure in *P. explanata* was closely similar to that in *P. constellata* HALL and to that in the other ptilodictyid genera.] *M.Ord.-U.Sil.,* USSR, N.Am., Eng., Swed.——FIG. 243,*1a.* **P. explanata,* lectotype, AMNH 1490; alignment of autozooecial ranges; external view, ×5.——FIG. 243,*1b–d.* *P. constellata* HALL, ?Cataract F., Ont.; *b,* straight autozooecial ranges, structurally continuous longitudinal walls, pairs of exilazooecia between successive autozooecia, and exilazooecia in groups between longitudinal walls; tang. sec., USNM 242618, ×30; *c,d,* shape of living chambers, exilazooecia and sinuous mesotheca, superior and inferior hemisepta projecting into living chambers in endozone; long. sec., USNM 242617, ×30, 100.

Phaenoporella NEKHOROSHEV, 1956a, p. 48 [**Phaenopora transenna* SCHOENMANN, 1927, p. 788; OD; M. Ord. (Mangaze.), Podkamennaya Tunguska River, Sib., USSR]. Zoarium cribrate, commonly fan shaped. Fenestrules ovate to subcircular, variable in size, irregularly arranged or in indistinct rhombic pattern, rarely delineated transversely by relatively straight cross segments of zooecia; in proximal regions, may be closed by extrazooecial stereom, exilazooecia, or both. Autozooecial ranges generally curve around fenestrules. Low expansions at right angles to zoarial surface may result in irregular, three-dimensional, cribrate growth. Mesotheca irregularly sinuous in longitudinal section. Autozooecia in straight to curving ranges throughout ontogeny. In exozones, autozooecia subelliptical to irregularly subelliptical in cross section, generally contiguous laterally across pronounced longitudinal walls, partially separated within ranges, and probably replaced locally by exilazooecia. Autozooecia aligned irregularly in cross segments between fenestrules. Autozooecial boundaries broadly serrated or not visible; pustules common, irregularly arranged in exozonal walls. Living chambers elliptical to irregularly subcircular in cross section; lining common, variable in thickness. Superior and inferior hemisepta common, somewhat irregularly arranged.

Superior hemisepta generally short, blunt; inferior hemisepta long, straight, and may curve proximally from mesothecae or distal walls. Exilazooecia common to abundant, irregularly triangular to elongate subcircular or variable in cross section, commonly with narrower chambers in outer exozones, rarely with lining; generally in pairs, rarely in groups of three or more between successive autozooecia; in groups in scattered areas of replaced autozooecia, and in groups or curving rows in fenestrule margins. Monticules absent. *M.Ord.-L.Sil.,* USSR (Tuva).——FIG. 244,*1a–d.* *P. transenna mesofenestralia,* paratype, USNM 171741; *a,* autozooecia aligned across mesotheca, endozone narrows toward margins, broadly curved laminae in walls in exozone; transv. sec., ×30; *b,* mesotheca with median granular zone, distinct lining along distal wall, blunt superior hemiseptum; long. sec., ×100; *c,* exilazooecia in ranges in margin surrounding fenestrule, curving autozooecial ranges in midregion; tang. sec., ×30; *d,* sinuous mesotheca, hemisepta, exilazooecia with wide chambers at base of exozone; long. sec., ×30.

Pteropora EICHWALD, 1860, p. 395 [**P. pennula*; OD; Pirgu and Porkuni horizons at Haapsalu and Seli-Metskula respectively, U. Ord., Est., USSR]. Zoarium unbranched; consisting of straight midsegments and lateral ribs diverging obliquely from midsegments at regular intervals. Mesothecae straight in midsegments, probably merging with extrazooecial stereom between lateral ribs. In exozones, autozooecia in straight ranges in midsegment, subrectangular in cross section, contiguous laterally, partially separated within ranges by exilazooecia. In lateral ribs, autozooecia irregularly rhombic to subcircular in cross section, arranged in rhombic pattern, generally contiguous without continuous longitudinal walls. Autozooecial boundaries not visible; pustules probably absent. Living chambers elliptical in cross section in midsegment, irregularly polygonal to subcircular in cross section in lateral ribs. Chamber lateral structures probably absent. Exilazooecia common to abundant, subcircular in cross section, variable in size; arranged singly, in pairs, or in short rows between successive autozooecia in midregions; scattered to lacking in lateral ribs. Exilazooecial and extrazooecial stereom common to abundant between ribs. [*Pteropora* is characterized by a ribbed growth habit, which is unusual among ptilodictyines. It is included in the Ptilodictyidae because of the linear arrangement and shape of autozooecia in the zoarial midsegment. In the lateral ribs, autozooecia are in a rhombic pattern instead of linear ranges and are indistinctly polygonal in cross section. This autozooecial arrangement and shape is somewhat similar to that near the margins in generalized ptilodictyids; however, structural relationship between the mesotheca and autozooecia, or mar-

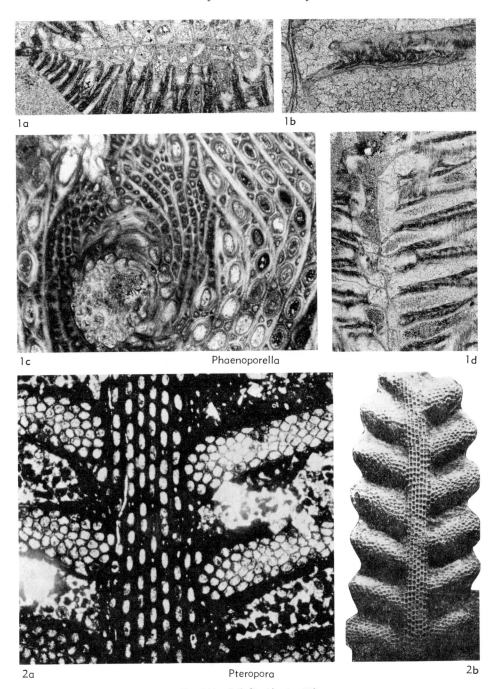

la   1b

1c   Phaenoporella   1d

2a   Pteropora   2b

FIG. 244.   Ptilodictyidae (p. 496).

ginal zooecia, in areas between the midsegment and the ribs is not determinable from available illustrations. This diagnosis is summarized from MÄNNIL (1958, p. 344), because type material was not available.] *U.Ord.*, USSR (Est.).——FIG. 244,*2a,b.* *P. pennula*; *a*, autozooecia in linear ranges between structurally continuous longitudinal walls in midsegment, exilazooecia in midsegment and ribs; tang. sec., ×20; *b,* ribbed zoarium, arrangement of autozooecia in midsegment and ribs, extrazooecial stereom between diverging ribs, external view, ×5 (Männil, 1958).

Family ESCHAROPORIDAE
Karklins, new

Zoaria branched or unbranched and lanceolate, explanate, or cribrate. Basal attachments continuous with erect parts of zoaria in some genera; with tapering proximal connecting segments, which probably articulated with encrusting zoarial bases, in other genera. Mesothecae straight to sinuous, rarely zigzag in transverse section. Mesothecae in zoarial margins thickened, consisting of broadly curved laminae in transverse section, forming serrated zones along middle of zoarial margins beyond median granular zones. Median granular zones discontinuous through most of mesothecae, terminating near thickened zoarial margins. In endozones, autozooecia in straight ranges, aligned or alternating on opposite sides of mesothecae, contiguous, with sinuous continuous longitudinal walls, expanded and narrowed alternately in adjacent ranges, rectangular to subrhomboidal in cross section at mesothecae, generally subelliptical in later endozones. Autozooecial boundaries broadly serrated in later endozones. In exozones, autozooecia forming angles with mesothecae ranging from 45° to 90°, subpolygonal and elliptical to subcircular in cross section, contiguous or separated by extrazooecial stereom. Autozooecia in rhombic arrangement such that lateral walls restricted to individual autozooecia and longitudinal walls and ranges not formed. Autozooecial wall laminae broadly curved and U-shaped; zooecial boundaries broadly serrated. Pustules common to abundant throughout exozonal walls and stereom. Living chambers subrectangular to subelliptical in cross section of endozones, subelliptical to subcircular in cross section of exozones. Basal diaphragms rare to common in some genera, absent in others. Chamber lining absent to common. Superior hemisepta common in most genera; inferior hemisepta and mural spines absent to common in some genera, lacking in others. Exilazooecia few to absent. Monticules consisting of polymorphs and extrazooecial stereom in varying combinations. Extrazooecial stereom between autozooecia laminated, abundant to absent, irregularly delineated. Stereom laminae may be slightly sinuous, locally crinkled, generally parallel to zoarial surface, commonly forming striae at zoarial surfaces. *M.Ord.-L.Sil.*

Distinguishing features of the Escharoporidae are the mode of arrangement and cross-sectional shape of autozooecia parallel to mesothecae in endozones and exozones, skeletal microstructure in exozones, and distribution and relative sparsity of exilazooecia in the erect parts of zoaria. Arrangement of autozooecia in the Escharoporidae is similar to that in the Intraporidae and Stictoporellidae; however, in those families, autozooecia in endozones are generally subrectangular in cross section, with less sinuous longitudinal walls, and autozooecia in exozones are polygonal to subcircular in cross section. Compound autozooecial walls in exozones of the Escharoporidae have U-shaped laminae like those in the Ptilodictyidae and Intraporidae, but differ in having well-defined boundaries between autozooecia (Fig. 245,*1f*), numerous pustules throughout exozones (Fig. 246,*1c*), and crinkled laminae in parts of the exozonal skeleton (Fig. 246,*1a*; 247,*1a*). Ptilodictyids also differ in having autozooecia in distinct linear ranges throughout zoarial midregions. Stictoporellids differ from escharoporids in having autozooecia with broadly V-shaped laminae and narrowly serrated autozooecial boundaries in exozones. The Escharoporidae, Ptilodictyidae, and Stictoporellidae all have exilazooecia; however, in the Escharoporidae they are sparse or may be absent. Where present, the exilazooecia are mostly along zoarial margins and in proximal zoarial parts, but generally are uncommon in zoarial midregions.

Escharopora HALL, 1847, p. 72 [**E. recta*; OD; Trenton Gr., M. Ord., Trenton, Middleville, N.Y., USA]. Zoarium generally unbranched and lanceolate, rarely branched; connecting segments with tapered proximal tips, probably articulating with encrusting bases. Mesothecae straight to sinuous; autozooecial ranges aligned or alternating across mesotheca. In exozones, autozooecia form angles between 50° and 80° with mesothecae, subelliptical in cross section. Autozooecial wall and stereom laminae form sinuous striae at zoarial surface. Pustules common along autozooecial boundaries, striae scattered in exozonal walls and stereom. Living chambers subelliptical in cross section. Superior hemisepta common, generally blunt and short, rarely thin and long, curving proximally, usually scattered in zoaria, but may be regularly arranged. Mural spines absent to common, irregularly shaped, scattered in zoaria; may be regularly arranged. Exilazooecia few, subelliptical to subcircular in cross section, commonly closed at zoarial surfaces by thickened walls, sparse in zoarial margins and in proximal zoarial parts, generally absent in zoarial midregions. Monticules absent to common, flat to slightly raised, irregularly shaped or may form annular ridges at regular intervals across zoaria. *M.Ord.-U.Ord.*, N.Am., USSR(Eu.), Greenland, Burma.——Fig. 245,*1a–f. *E. recta*; *a*, laminae on opposite sides of mesotheca intertongue along broadly serrated zone in zoarial margin, closed and open zooecia without endozone in margin; transv. sec., lectotype, AMNH 668/1, ×30; *b*, autozooecia with thickened walls, elongated and closed exilazooecia, extrazooecial stereom in proximal part of zoarium; tang. sec., lectotype, ×50; *c*, autozooecia in distinct rhombic pattern; external view, lectotype, ×5; *d*, subelliptical autozooecia in inner exozone; tang. sec., lectotype, ×30; *e*, slightly sinuous mesotheca with median granular zone, blunt hemisepta (fragment crushed); long. sec., lectotype, ×30; *f*, microstructure of serrated autozooecial boundaries in exozone; transv. sec., paralectotype, NYSM 654, ×100.

Championodictya Ross, 1964a, p. 18 [*C. pleasantensis*; OD; up. "Denmark?" F., low. "Cobourg?" F., ?U. Ord., Pleasant Lake, N.Y., USA]. Zoarium branched or unbranched and explanate. Mesothecae straight, locally crenulated in longitudinal section. Autozooecial ranges aligned across mesotheca. Endozones relatively wide. In exozones, autozooecia generally at right angles with mesothecae, locally sloping proximally, irregularly subpolygonal in cross section. Pustules abundant in exozonal walls, locally aligned in series at right angles to zoarial surface. Living chambers subelliptical to subcircular in cross section. Basal diaphragms common, scattered in zoaria; relatively thick, irregularly curved, incomplete locally. Lining common locally, variable in thickness. Superior hemisepta common, regularly arranged, relatively thick, irregularly shaped; locally with thin, proximally

curved terminal edges. Spines common, blunt, relatively thick, scattered in zoaria. Exilazooecia absent to few, subelliptical to polygonal in cross section, scattered in outer exozone, commonly closed by thickened walls. Monticules absent to rare, indistinct. *?M.Ord., U.Ord.*, N.Am.——Fig. 245,*2a–d. *C. pleasantensis*, holotype, YPM 25462; *a*, autozooecia aligned across mesotheca, relatively wide endozone; transv. sec., ×30; *b*, subrhomboidal autozooecia in endozone, subelliptical to subcircular autozooecia in exozone, mesotheca below; deep to shallow tang. sec., ×30; *c*, irregular and blunt spines, lining in some chambers, pustules in autozooecial boundary (left); tang. sec., ×100; *d*, crenulated mesotheca, shape of autozooecial chambers, superior hemisepta, thick basal diaphragms, autozooecial walls slope proximally in exozone; long. sec., ×30.

Chazydictya Ross, 1963b, p. 587 [*C. chazyensis*; OD; Chazy Ls., M. Ord., Isle La Motte, Vt., USA]. Zoarium branched or unbranched and explanate. Mesothecae generally straight; autozooecial ranges partly aligned across mesotheca. In exozones, autozooecia form angles between 55° and 65° with mesothecae. Autozooecia subelliptical in cross section. Pustules abundant along autozooecial boundaries and in extrazooecial stereom, scattered in autozooecial walls. Living chambers elliptical to subcircular in cross section. Basal diaphragms common, thin, slightly curved; regularly arranged in outer endozones and base of exozones. Exilazooecia and monticules absent. Locally, extrazooecial stereom laminae irregularly crinkled. *M.Ord.*, USA.——Fig. 246,*1a–c. *C. chazyensis*; *a*, serrated autozooecial boundary, crinkled stereom laminae with pustules in exozone; long. sec., paratype, YPM 22098, ×100; *b*, abandoned chambers near base of exozone, thin basal diaphragm in outer endozone and base of exozone; long. sec., paratype, YPM 22069, ×30; *c*, subcircular living chambers, abundant pustules in extrazooecial stereom; tang. sec., holotype, YPM 22067, ×30.

Graptodictya Ulrich, 1882, p. 165 [*Ptilodictya perelegans* Ulrich, 1878, p. 94; OD; Waynesville Sh., U. Ord., Clarksville, Ohio, USA] [=*Arthropora* Ulrich, 1882, p. 167]. Zoarium branched, anastomosing irregularly in some species. Mesothecae slightly sinuous in longitudinal section, may zigzag in transverse section. Autozooecial ranges generally alternating across mesothecae. In exozones, autozooecia form angles between 80° and 90° with mesothecae, subelliptical in cross section. Pustules abundant along autozooecial boundaries and throughout exozonal walls and extrazooecial stereom. Living chambers subelliptical to subcircular in cross section. Superior hemisepta common, generally short and blunt, rarely thin and long, curving proximally; usually scattered in zoaria, but may

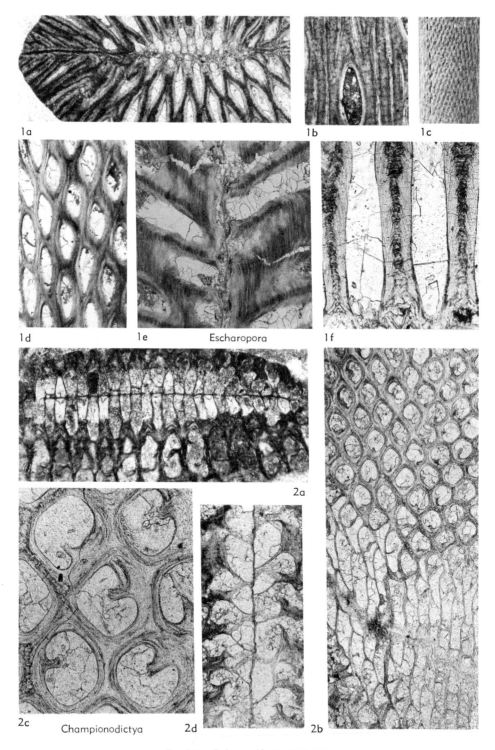

1a

1b

1c

1d   1e   Escharopora   1f

2a

2c   Championodictya   2d   2b

FIG. 245.   Escharoporidae (p. 498–499).

be regularly arranged. Exilazooecia absent to rare, generally subelliptical in cross section, commonly closed by thickened walls. Monticules absent to rare, generally flat. Extrazooecial stereom laminae commonly crinkled, forming abundant and longitudinally sinuous striae between autozooecia, and along zoarial margins and proximal zoarial parts. *M.Ord.-L.Sil.*, USSR(Est.), Morocco, France, ?Austria, India.——FIG. 247,1a–f. *G. perelegans* (ULRICH); *a,* microstructure of autozooecial wall and extrazooecial stereom in exozone; transv. sec., holotype, USNM 137607, ×200; *b,* curved autozooecial walls in endozone; transv. sec., holotype, ×30; *c,* branching pattern, striae along zoarial margins; external view, holotype, ×5; *d,* indistinct blunt hemisepta, shape of living chambers; long. sec., holotype, ×50; *e,* autozooecia in rhombic pattern, sinuous striae between autozooecia and in zoarial margin; tang. sec., holotype, ×30; *f,* sinuous and continuous longitudinal autozooecial walls in endozone, extrazooecial stereom with striae in margin; tang. sec., USNM 242619, ×30.

Oanduella MÄNNIL, 1958, p. 340 [*O. bassleri; OD; Oandu horizon, D3, Oandu bed, M. Ord., Oandu River, Est., USSR]. Zoarium cribrate; fenestrules ovate to subcircular, variable in size, surrounded by exilazooecia or extrazooecial stereom. Mesothecae slightly sinuous in longitudinal section. In exozones, autozooecia form angles between 50° and 70° with mesothecae, subelliptical to irregularly polygonal in cross section. Pustules common in exozonal walls and stereom. Living chambers subelliptical to subcircular in cross section. Inferior hemisepta common, long, thin, extending from mesothecae or distal autozooecial walls, regularly arranged or scattered in zoaria. Basal diaphragms and superior hemisepta absent. Exilazooecia common, subelliptical to polygonal in cross section, scattered in zoarial midregions, regularly arranged in zoarial margins or absent. Monticules probably absent. [Rhombic arrangement and skeletal microstructure of autozooecia in the exozone indicate a zoarial development similar to that in *Escharopora* HALL (MÄNNIL, 1958, p. 341) and other genera herein assigned to the Escharoporidae.] *M.Ord.*, USSR(Est.).——FIG. 247,2a–c. *O. bassleri; a,* arrangement and shape of autozooecia, fenestrules surrounded by zone of exilazooecia; external view, holotype, ×5; *b,* arrangement of autozooecia; tang. sec., ×25; *c,* sinuous mesotheca, inferior hemisepta; long. sec., ×25 (Männil, 1958).

?Proavella MÄNNIL, 1958, p. 345 [*Gorgonia proava* EICHWALD, 1842, p. 44; OD; ?Vasalemma, M. Ord., Est., USSR]. [MÄNNIL (1958, p. 345) erected *Proavella* and designated *Gorgonia proava* as its type species, but did not figure

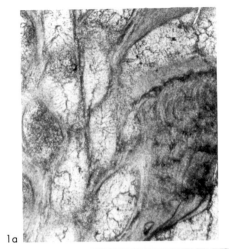

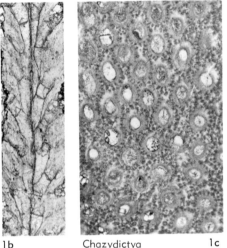

1a

1b   Chazydictya   1c

FIG. 246.   Escharoporidae (p. 499).

a type. According to MÄNNIL, *Proavella* is similar in internal structure to *Graptodictya* ULRICH, but differs from it in having a cribrate growth habit. Ross (1964a, p. 13) questioned the validity of *Proavella* because type material is inadequately documented, and she noted similarities between *Proavella* and *Stictoporellina* NEKHOROSHEV in growth habits and arrangement of exilazooecia. The concept of *Proavella* and its taxonomic assignment will remain questionable until the type material becomes available for description and further comparison. Herein *Proavella* is tentatively assigned to the Escharoporidae because of its similarity in internal zoarial structure to *Graptodictya*, as noted by MÄNNIL (1958)].

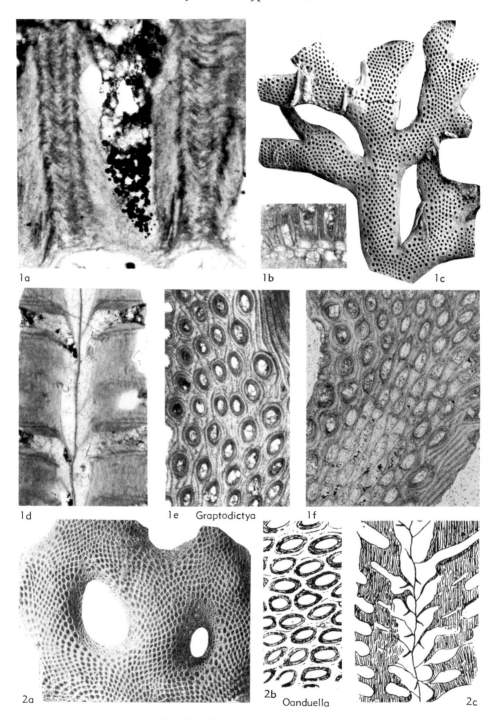

1a 1b 1c

1d 1e Graptodictya 1f

2a 2b Oanduella 2c

Fig. 247. Escharoporidae (p. 499–501).

Family INTRAPORIDAE
Simpson, 1897

[Intraporidae SIMPSON, 1897, p. 543]

Zoaria branched or unbranched and cribrate or explanate. Zoarial attachments generally continuous skeletally. Mesotheca straight with median granular zone extending to edge of zoarial margins. In endozones, autozooecia arranged in straight ranges, generally alternating on opposite sides of mesothecae, subrectangular to rhomboid in cross section, relatively elongate parallel to mesotheca, contiguous with continuous longitudinal walls. In outer endozones and bases of exozones, autozooecia slightly expand and narrow alternately in adjacent ranges. In exozones, autozooecia form angles between 75° and 90° with mesothecae, arranged in rhombic pattern without continuous longitudinal walls, irregularly polygonal to subcircular in cross section, contiguous, partly separated by mesozooecia, or completely separated by pitted extrazooecial stereom. Autozooecial wall laminae curved and broadly U-shaped, may form striae at zoarial surface. Autozooecial boundaries not visible; pustules generally absent. Acanthostyles few. Living chambers subrectangular to subrhomboidal in cross section in endozones, subelliptical to subcircular in cross section in exozones. Superior hemisepta scattered in zoaria; inferior hemisepta lacking; chamber lining generally lacking. Basal diaphragms and abandoned chambers few. Mesozooecia, monticular and basal polymorphs in some genera. Monticules common to absent. *M.Dev.-U.Dev.*

The family Intraporidae SIMPSON, 1897, differs in skeletal microstructure and presence of mesozooecia from the Stictoporellidae NICKLES and BASSLER, 1900, and is removed from synonymy (BASSLER, 1953, p. G137) with that family. The Intraporidae is similar to the Stictoporellidae and Escharoporidae in the rhombic arrangement of autozooecia in the exozone. *M.Dev.-U.Dev.*

Intrapora HALL, 1883b, p. 157 [*I. puteolata*; M; Jeffersonville Ls., M. Dev., Falls of Ohio River, Ky.-Ind., USA]. Zoarium branched or unbranched and explanate. In exozones, auto-

zooecia usually form angle between 75° and 80° with mesothecae, contiguous or separated partially by mesozooecia. Acanthostyles rare to common, consisting of straight cores of cryptocrystalline particles and thin laminar sheaths. Sheath laminae abut cores at low angle. Acanthostyles irregularly arranged, originating at base of exozone, terminating in outer exozones or as low protuberances at zoarial surfaces. Living chambers broadly subelliptical in cross section. Superior hemisepta indistinct, short, blunt. Mesozooecia abundant, polygonal to subcircular in cross section, variable in size. Mesozooecial diaphragms closely spaced, commonly thickening distally, rarely filling mesozooecial chambers. Monticules rare to common, generally raised; consisting of irregularly shaped, somewhat larger zooecia and some mesozooecia; common in species with explanate zoaria. *M.Dev.-U.Dev.*, N.Am., USSR.——FIG. 248,*1a–f.* *I. puteolata*; *a,* median granular zone in mesotheca, autozooecial boundaries in endozone, broadly curved laminae in exozone; transv. sec., USNM 242620 from Alpena Ls., Mich., ×50; *b,* shape and arrangement of autozooecia in endo-exozone, acanthostyles and mesozooecia in exozone; tang. sec., USNM 242620, ×30; *c,* arrangement of autozooecia and mesozooecia, mesozooecia in zoarial margins; external view, syntype, FMNH 13987 from Jeffersonville Ls., Ky.-Ind., ×5; *d,* shape of living chambers, indistinct superior hemiseptum, mesozooecia with diaphragms; long. sec., USNM 242621 from Alpena Ls., Mich., ×30; *e,* broadly curved laminae of autozooecia and mesozooecia, core and sheath of an acanthostyle, median granular zone in mesotheca; long. sec., USNM 242621, ×100; *f,* zooecia and mesozooecia in monticule; tang. sec., USNM 242622 from Alpena Ls., Mich., ×50.

Coscinella HALL, 1887, p. xix [*C. elegantula*; OD; Hamilton Gr., M. Dev., Widder, Ont., Can.]. Zoarium cribrate; fenestrules subelliptical to subcircular or irregularly shaped, generally smaller in middle and proximal regions than in distal and lateral regions, rarely closed partly by extrazooecial stereom. Zoarial and fenestrule margins of pitted extrazooecial stereom. Monticules absent. In exozones, autozooecia usually at right angles to mesothecae, elliptical to subcircular in cross section, generally surrounded by pitted extrazooecial stereom. Living chambers broadly elliptical to subcircular in cross section, may be closed locally by diaphragm at zoarial surface. Superior hemisepta short, blunt, indistinct. [Zoarial surfaces contain numerous pits (Fig. 249,*1e*) formed by concave laminae that are skeletally continuous with those of autozooecia, but they are extrazooecial. Shape and structure seem to indicate concurrent growth of these laminae in the autozooecia and the exozone. Specimens also possess scattered cavities in

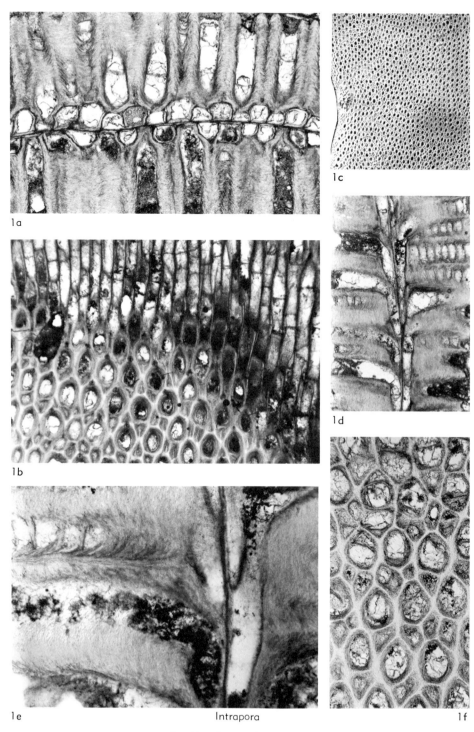

1a

1b

1c

1d

1e Intrapora 1f

Fig. 248. Intraporidae (p. 503).

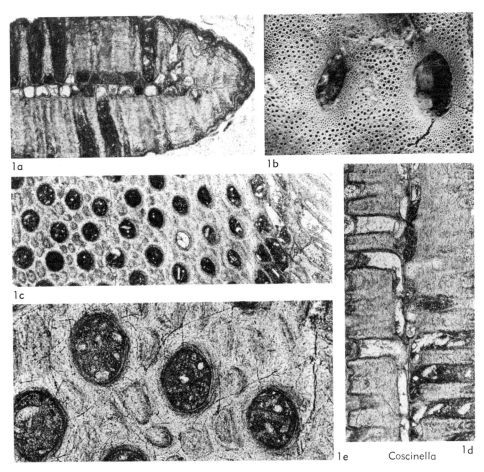

la   1b

1c

1e   Coscinella   1d

FIG. 249.   Intraporidae (p. 503).

walls at the base of the exozone, which may have been unfilled portions of extrazooecial skeleton.] *M.Dev.,* Can.——FIG. 249,*1a–e.* **C. elegantula,* holotype, NYSM 641,6220 / 1; *a,* broadly curved laminae of autozooecia in exozone, extrazooecial stereom of zoarial margin; transv. sec., ×30; *b,* wide zone of pitted extrazooecial stereom surrounding fenestrules; external view, ×5; *c,* shape of autozooecia in endo-exozone, pitted extrazooecial stereom in exozone; tang. sec., ×30; *d,* shape of living chambers, indistinct hemisepta, median granular zone in mesotheca; long. sec., ×30; *e,* pitted extrazooecial stereom surrounding autozooecia, subcircular living chambers; tang. sec., ×100.

Family PHRAGMOPHERIDAE
Goryunova, 1969

[Phragmopheridae GORYUNOVA, 1969, p. 129]

Zoaria branched. In endozones, autozooe-

cia in ranges alternating across mesotheca, contiguous, with continuous longitudinal walls. Walls slightly flexed at base of exozone in transverse section. In exozones, autozooecia arranged in rhombic pattern, without continuous longitudinal walls, generally polygonal to subcircular in cross section. Autozooecial wall laminae curved; autozooecial boundaries narrowly serrated. Mural styles common. Living chambers broadly elliptical to subcircular in cross section, variable in length. Cystiphragms and basal diaphragms common. Polymorphism expressed by mesozooecia. Extrazooecial stereom laminated and irregularly delineated. Acanthostyles present. *U.Carb.*

Capillaries of GORYUNOVA (1969, p. 129,

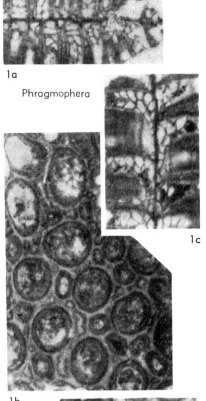

1a

Phragmophera

1c

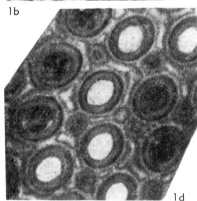

1b

1d

FIG. 250. Phragmopheridae (p. 506).

130) are interpreted here as mural styles, secondary zoarial deposits as extrazooecial stereom, and tubercles as acanthostyles.

Phragmophera GORYUNOVA, 1969, p. 129 [*P. eximia; OD; U. Carb., C. Urals, USSR]. In exozones, autozooecia usually at right angles to mesothecae, irregularly subcircular in cross section, only locally contiguous, generally separated by mesozooecia. Mural styles aligned along autozooecial boundaries. Autozooecial walls extend into peristomes on zoarial surface. Living chambers generally broadly elliptical in cross section. Cystiphragms closed; in late endozones and exozones of autozooecia, in regular series. Basal diaphragms thin, slightly curved, regularly spaced in endozones and exozones. Mesozooecia abundant, irregularly polygonal to subcircular in cross section, variable in size, with few diaphragms, regularly arranged throughout zoaria, locally filled by stereom. Mesozooecial walls extending into peristomes at zoarial surface. Acanthostyles variable in size. *U.Carb.*, USSR(C. Urals).——FIG. 250,*1a–d*. *P. eximia,* holotype, PIN 389/ 654; *a,* mesotheca, autozooecial boundaries, and cystiphragms in chambers in exozone; transv. sec., ×20; *b,* arrangement of autozooecia, mesozooecia, and acanthostyles in exozone; tang. sec., ×40; *c,* abandoned chambers, basal diaphragms in endozone, cystiphragms along distal walls in exozone; long. sec., ×20; *d,* subpolygonal autozooecia, polygonal mesozooecia, indistinct mural styles along autozooecial boundaries, acanthostyles; tang. sec., ×40 (Goryunova, 1969).

Family RHINIDICTYIDAE
Ulrich, 1893

[*nom. correct.* BASSLER, 1953, p. G140, *pro* Rhinidictyonidae ULRICH, 1893, p. 124]

Zoaria branched or unbranched and explanate, rarely cribrate. Basal attachments generally continuous skeletally with erect parts of zoaria. Mesothecae straight, sinuous, or bifurcated. Median granular zones extending throughout mesothecae. Median rods usually present, closely spaced, generally straight; consisting of cryptocrystalline cores and thin, laminated sheaths; subelliptical to circular in cross section; extending throughout median granular zone, diverging gradually into zoarial margins. In endozones, autozooecia in ranges alternating on opposite sides of mesothecae; commonly rectangular to subrhomboid in cross section parallel to mesothecae, contiguous, with continuous longitudinal walls and straight to slightly curved transverse walls; may be partially contiguous, without continuous longitudinal walls and separated by extrazooecial skeleton. Autozooecial boundaries extending into exozone, becoming narrowly serrated. In exozones, autozooecia usually form angle between 50° and 80° with mesothecae; commonly aligned in straight and distinct ranges; subrectangular to subelliptical in cross section, with straight to curved, generally indistinctly delineated longitudinal walls; may be aligned in indistinct ranges, subelliptical to subcircular in cross section, partially contiguous or separated by extrazooecial deposits, and without continuous longitudinal walls. Autozooecial wall laminae in most genera slightly curved and V-shaped. Autozooecial boundaries narrowly serrated. Autozooecial walls commonly vesicular in inner exozone in some genera. Mural styles rare to common, consisting of tightly curved segments of wall laminae; rarely with small, indistinct and discontinuous cores; usually variable in size, may be relatively large. Mural styles may be present in autozooecial boundaries or diverge from them; may be single or aligned in dark zones in zooecial walls and extrazooecial stereom; generally oriented perpendicular to zoarial surface, terminating in walls or at zoarial surface. Dark zones in walls and extrazooecial stereom rare to common, generally aligned longitudinally. Living chambers usually variable in length, elliptical to subcircular in cross section. Superior hemisepta rare to common, generally scattered in zoaria; may be regularly arranged. Inferior hemisepta rare and scattered. Chamber lining usually rare to lacking, but may be common. Intrazooecial cysts rare. Basal diaphragms absent to common, generally scattered in zoaria, may be regularly arranged. Polymorphism expressed by marginal, basal, and monticular zooecia. Monticules absent to common, consisting of extrazooecial skeletal deposits and few zooecia. Exilazooecia and mesozooecia absent. Extrazooecial skeletal deposits rare to common, consisting of laminar and vesicular portions in inner exozone or endozone in some genera. Distribution of extrazooecial skeletal deposits variable. *L.Ord.-M.Sil.*

Stictopora HALL, 1847, p. 73 [*S. fenestrata*; SD ULRICH, 1886a, p. 67; Chazy Gr., M. Ord. (Chazy.), N.Y., USA] [=*Sulcopora* D'ORBIGNY, 1849, p. 499, obj.; *Rhinidictya* ULRICH, 1882, p. 152; *Dicranopora* ULRICH, 1882, p. 166; *Hem-*

idictya CORYELL, 1921, p. 303]. Zoarium branched or unbranched and explanate, rarely cribrate. Mesothecae generally straight, may be locally sinuous in longitudinal section. Median rods subelliptical in cross section. In endozones, autozooecia subrectangular to subrhomboidal in cross section, contiguous, with straight continuous longitudinal walls. In exozones, autozooecia in straight ranges, generally contiguous, with straight to slightly sinuous longitudinal walls, subrectangular in cross section, walls locally may be vesicular in inner exozone. Mural styles common, mostly in autozooecial boundaries or scattered in walls. Living chambers generally elliptical in cross section. Superior hemisepta rare to common, regular, thin, curved proximally, variable in length. Inferior hemisepta in few species; short, thin, generally projecting from mesotheca, scattered in zoaria. Basal diaphragms thin, slightly curved, variable in spacing, absent in some. Monticules common, generally scattered in zoaria. Extrazooecial stereom laminated, may be sparse in zoarial midregion. [The status of *Stictopora* HALL, 1847 and *Rhinidictya* ULRICH, 1882 is controversial. ROSS (then PHILLIPS, 1960; see also ROSS, 1961a, 1966b) reviewed the nomenclature of *Stictopora* and considered *Rhinidictya* to be a synonym. More recently, KOPAYEVICH (1973) has argued for retention of *Rhinidictya* as an independent genus. Because of poor preservation of type specimens, skeletal differences noted by KOPAYEVICH in *S. fenestrata* (type species of *Stictopora*) and *R. nicholsoni* (type species of *Rhinidictya*) cannot be verified, and *Rhinidictya* is retained herein as a synonym of *Stictopora*.] *L.Ord.-L.Sil.*, USSR, N.Am., Australia, India, Burma, G.Brit.——FIG. 251,1a-f. *S. nicholsoni* (ULRICH), Tyrone Ls., High Bridge Gr., M. Ord., Ky.; *a*, indistinct laminae in autozooecial walls, superior hemisepta, mural styles in outer exozone; long. sec., paralectotype, USNM 137615, ×30; *b*, elliptical living chambers, sinuous longitudinal walls, mural styles in boundaries and walls; tang. sec., paralectotype, USNM 137615, ×50; *c*, mural styles in boundary between longitudinal walls; tang. sec., paralectotype, USNM 137615, ×200; *d*, median rods in mesotheca, autozooecial boundaries in longitudinal walls; transv. sec., paralectotype, USNM 137615, ×30; *e*, branching zoarium, autozooecia in linear ranges; external view, lectotype, USNM 137622, ×3; *f*, autozooecial boundaries in transverse walls, shape of living chambers, superior hemisepta; long. sec., USNM 242623, ×50.——FIG. 251,1g-k. *S. fenestrata*; *g*, sinuous mesotheca; long. sec., lectotype, NYSM 915, ×30; *h*, autozooecia in broadly curved ranges, shape of autozooecia, median rods in mesotheca (left); tang. sec., lectotype, ×30; *i*, mural styles in boundaries between longitudinal walls, elliptical living chambers; tang. sec., lec-

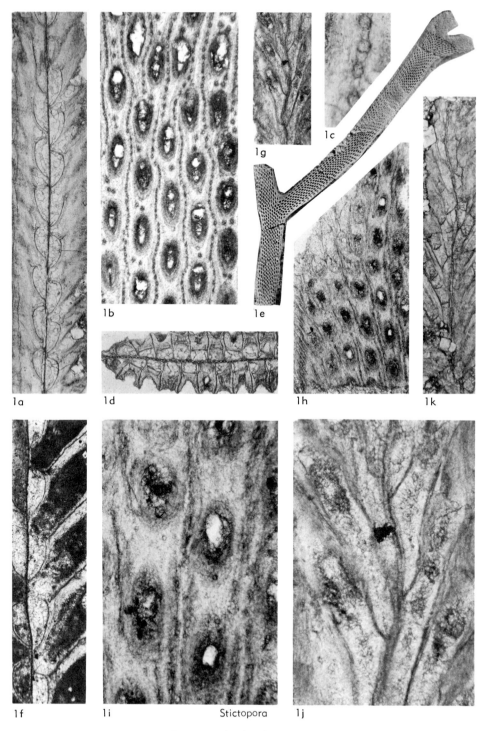

Fig. 251.   Rhinidictyidae (p. 507).

totype, ×100; *j,* granular zone in mesotheca, indistinct autozooecial boundaries in endozone; long. sec., lectotype, ×100; *k,* basal diaphragms in endozone and exozone, segment of median rods in median granular zone of mesotheca; long. sec., YPM 22158, ×30.

Athrophragma KARKLINS, 1969, p. 61 [*Pachydictya foliata* ULRICH, 1886a, p. 73; OD; Spechts Ferry Sh. Mbr., Decorah Sh., M. Ord., St. Paul, Minn., USA]. Zoarium explanate, slightly lobate and undulating. Mesotheca straight to slightly sinuous. Median rods subcircular in cross section, diameter greater than width of median granular zones. In endozone, autozooecia in indistinct ranges, subelliptical to subcircular in cross section, locally contiguous, generally separated by extrazooecial vesicles, and without continuous longitudinal walls. In exozones, autozooecia in indistinct ranges, subelliptical to subcircular in cross section, without continuous longitudinal walls, generally separated by extrazooecial vesicles and stereom. Autozooecial walls relatively thin. Mural styles indistinct or lacking; locally in autozooecial boundaries and in dark, longitudinally aligned, discontinuous zones in extrazooecial stereom. Mural styles generally absent in autozooecial walls. Living chambers broadly elliptical to subcircular in cross section; lateral chamber structures absent. Basal diaphragms straight to slightly curved, regularly spaced. Monticules common, flat and raised, arranged in rhombic pattern; generally vesicular in inner exozones, having stereom in outer exozones. Extrazooecial skeleton common, consisting of stereom and vesicles. Vesicular structures common in endozones and inner exozones; stereom present throughout exozones. *M.Ord.-U.Ord.,* N.Am., USSR(W. Arctic).——FIG. 252,*1a-c.* *A. foliata* (ULRICH), lectotype, USNM 163111; *a,* broadly elliptical autozooecia and their chambers, stereom between autozooecia, monticule with larger zooecium and dark zones in stereom; tang. sec., ×30; *b,* basal diaphragms at regular intervals, vesicles in inner exozone, thin autozooecial walls; long. sec., ×30; *c,* extrazooecial vesicles between autozooecia in endozone and inner exozone, numerous thin, dark zones in exozonal laminar stereom; transv. sec., ×30.

Carinodictya ASTROVA, 1965, p. 287 [*Rhinidictya carinata* ASTROVA, 1955, p. 157; OD; M. Ord. (Mangaze.), Podkamennaya Tunguska River, Sib., USSR]. Zoarium branched. Mesotheca generally straight; median rods indistinctly delineated. In exozones, autozooecia in straight ranges, subelliptical in cross section, contiguous, with regularly sinuous and continuous longitudinal walls. Autozooecial walls generally vesicular in inner exozones. Mural styles common in autozooecial boundaries, rare in wall laminae. Living chambers elliptical in cross section. Supe-

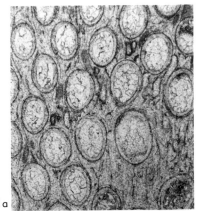

1a

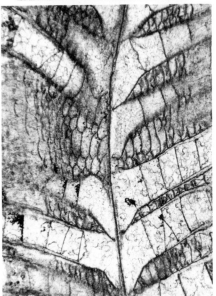

1b

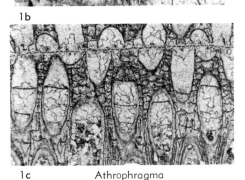

1c   Athrophragma

FIG. 252. Rhinidictyidae (p. 509).

rior hemisepta rare, short, blunt, and scattered. Inferior hemisepta lacking. Basal diaphragms rare, scattered in zoaria, or absent. Monticules rare or absent. Extrazooecial stereom laminated, sparse in zoarial midregions. *M.Ord-U.Ord.,* USSR (Sib.).——Fig. 253,*1a–c.* *C. carinata* (ASTROVA), PIN 1242.150; *a,* autozooecial ranges, sinuous longitudinal walls, elliptical living chambers; tang. sec., ×30; *b,* mesotheca with median granular zone and segments of median rods; long. sec., ×50; *c,* V-shaped laminae of autozooecial walls (mesotheca obscured); transv. sec., ×30.

Eopachydictya Ross, 1963b, p. 591 [*E. gregaria*; OD; Chazy Ls., M. Ord., Isle La Motte, Vt., USA]. Zoarium branched. Mesotheca straight. Median rods subcircular in cross section. In endozones, autozooecia in straight ranges, generally subelliptical in cross section, contiguous, with slightly curved and continuous longitudinal walls. In exozones, autozooecia in indistinct ranges, subelliptical in cross section, partially contiguous or locally separated by extrazooecial stereom. Walls may be vesicular in inner exozones. Mural styles common along autozooecial wall boundaries and in extrazooecial stereom. Living chambers elliptical in cross section; lateral chamber structures absent. Basal diaphragms generally straight, rare to common. Monticules rare, generally flat, locally vesicular in inner exozones, with mural styles in stereom in outer exozones. Monticules scattered in zoaria. Extrazooecial skeleton of stereom and vesicles. Vesicles locally present in inner exozones, stereom irregularly arranged throughout exozones. *M.Ord.,* USA.——Fig. 253,*2a–d.* *E. gregaria*; *a,* indistinct autozooecial boundaries, mural styles in extrazooecial stereom between autozooecia and in monticule; tang. sec., holotype, YPM 22076, ×100; *b,* general shape of living chambers; long. sec., paratype, YPM 22079, ×30; *c,* indistinct vesicles in inner exozone, shape of chambers in endozone; long. sec., paratype, YPM 22079, ×100; *d,* mesotheca with median granular zone and indistinct median rods, autozooecial boundaries and extrazooecial stereom in exozone; transv. sec., paratype, YPM 22080, ×100.

Eurydictya ULRICH in MILLER, 1889, p. 301 [*E. montifera* ULRICH, 1890, p. 521; OD; U. Ord. (Richmond.), Wilmington, Ill., USA]. Zoarium explanate. Mesotheca straight, median rods elliptical in cross section. In endozones, autozooecia rectangular in cross section, contiguous, with straight continuous longitudinal walls and generally straight transverse walls. In exozones, autozooecia usually forming an angle of about 80° with mesothecae. Autozooecia generally in indistinct ranges, contiguous to partly contiguous, locally separated by extrazooecial stereom. Longitudinal walls slightly curved, continuous or merging with extrazooecial stereom. Mural styles

common in autozooecial boundaries and locally in walls. Vesicles absent in walls. Living chambers broadly elliptical to subcircular in cross section; lining thin, generally discontinuous. Superior hemisepta common, short or long, blunt, curving proximally. Basal diaphragms rare and scattered in zoaria. Monticules common, may be arranged in rhombic pattern. Monticular zooecia commonly filled by stereom, which is laminated and contains scattered mural styles. *M.Ord.-U.Ord.,* USA, USSR(Sib.).——Fig. 254,*1a–d.* *E. montifera*; *a,* irregularly conical monticules, alignment of autozooecia; external view, holotype, ISGS 2668, ×5; *b,* autozooecial boundaries (granular zones) in endozone, autozooecial boundaries and mural styles in exozone, median rods in granular zone in mesotheca; transv. sec., USNM 137614, ×50; *c,* shape of living chambers, hemisepta at base of exozone; long. sec., holotype, ×30; *d,* autozooecia in indistinct ranges, shape of living chambers, monticule with extrazooecial stereom, filled zooecia, and mural styles; tang. sec., USNM 137614, ×30.

Goniotrypa ULRICH, 1889, p. 40 [*G. bilateralis*; OD; Stony Mountain F., ?U. Ord., Manitoba, Can.]. Zoarium small, probably unbranched; consisting of 2 to 4 autozooecial ranges and longitudinal ridge along middle of branch. Mesothecae straight; median rods apparently lacking. In endozones, autozooecia in straight ranges, contiguous, subrhomboid in cross section, with straight and continuous longitudinal walls. Endozones relatively wide. In exozones, autozooecia in straight ranges, subcircular in cross section, contiguous; ranges probably separated laterally by longitudinal ridge along center of branch. Longitudinal walls straight and continuous. Mural styles and vesicular structure apparently absent. Exozones relatively narrow. Living chambers with relatively long endozonal and short exozonal portions subcircular in cross section in exozones. Superior hemisepta short, blunt, regularly arranged. Inferior hemisepta and other lateral structures absent. Basal diaphragms and monticules absent. Extrazooecial stereom laminated in longitudinal ridge along middle of zoarium. [*Goniotrypa* is based on poorly preserved material and its assignment to the Rhinidictyidae is tentative. The budding pattern and shape of autozooecia in the endozone is similar to that in *Stictopora*; however, *Goniotrypa* differs from *Stictopora* and other rhinidictyids in having narrow, probably unbranched zoaria, and in having a relatively thin exozone with a median ridge along the middle of the branch.] *?U.Ord.,* Can., ?N.Ire.——Fig. 254,*2a–d.* *G. bilateralis, a,* mesotheca, wide endozone, narrow exozone; long. sec., syntype, USNM 242625, ×100; *b,* continuous longitudinal wall; tang. sec., syntype, USNM 242626, ×100; *c,* autozooecial range, straight longitudinal walls; tang. sec., USNM

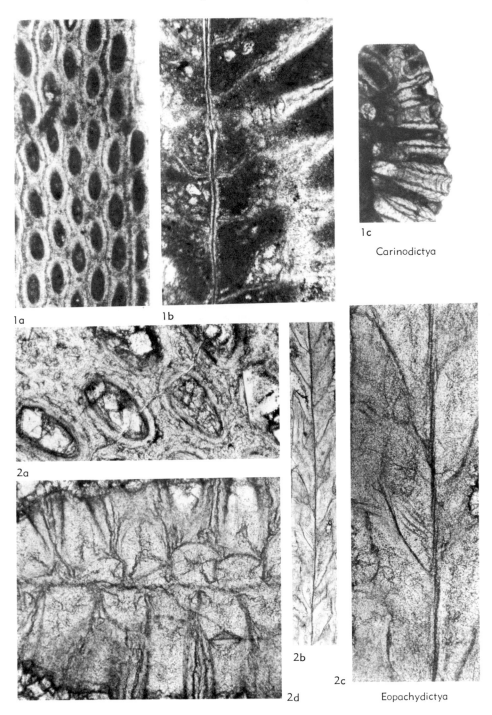

FIG. 253. Rhinidictyidae (p. 509–510).

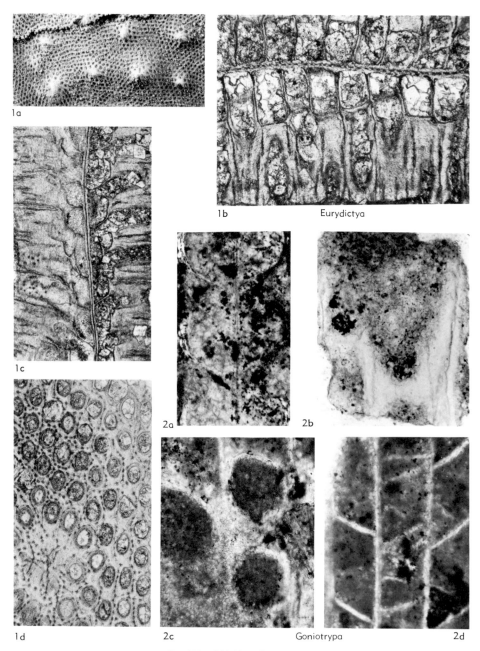

la

lb   Eurydictya

lc

2a   2b

ld   2c   Goniotrypa   2d

FIG. 254.   Rhinidictyidae (p. 510).

242627, ×100; *d,* rhomboid autozooecia in endozone; tang. sec., syntype, USNM 242628, ×100.

Pachydictya ULRICH, 1882, p. 152 [**P. robusta;* OD; "Trenton Gr.," M. Ord., Knoxville, Tenn., USA]. Zoarium branched; branches commonly with wide margins. Mesothecae generally straight. Median rods subcircular in cross section, commonly with diameter greater than width of median granular zone. In endozones, autozooecia in indistinct ranges, subelliptical to subcircular in cross section, partly contiguous, partly separated by extrazooecial stereom, and lacking continuous longitudinal walls. In exozones, autozooecia in relatively distinct ranges, broadly elliptical in cross section, partly contiguous or separated by extrazooecial stereom and lacking continuous longitudinal walls. Autozooecial walls locally vesicular in inner exozones. Mural styles common, mostly in autozooecial boundaries, also scattered in walls and laminar stereom. Living chambers subelliptical to subcircular in cross section. Chamber lining common, relatively thick. Other lateral chamber structures absent. Basal diaphragms straight to curved, generally common. Monticules common, flat or raised, locally with scattered zooecia. Monticules commonly vesicular in inner exozones, with mural styles singly or in indistinct rows in outer stereom. Extrazooecial skeleton common; vesicles localized in endozones and inner exozones, stereom scattered throughout exozones. *M.Ord.-L.Sil.,* USSR, Austria.——Fig. 255,1a-d. **P. robusta;* a, branched zoarium with wide zoarial margins, flat monticules, aligned autozooecia; external view, lectotype, USNM 137608, ×5; b, shape of living chambers, vesicles in inner exozone, chamber lining; long. sec., lectotype, ×30; c, mural styles in autozooecial boundaries, walls and in stereom of outer exozone, distinct lining along chambers, monticule with scattered mural styles; tang. sec., paralectotype, USNM 137609, ×30; d, autozooecial boundaries in exozone, vesicles in endozone; transv. sec., paralectotype, USNM 137625, ×30.

Phyllodictya ULRICH, 1882, p. 153 [**P. frondosa;* OD; High Bridge Gr., M. Ord., High Bridge, Ky., USA]. Zoarium explanate, irregularly lobate locally. Mesotheca slightly sinuous in longitudinal section. Median rods elliptical to subcircular in cross section. In endozones, autozooecia in straight to curving ranges, subrectangular to subrhomboid in cross section, contiguous, with straight to slightly sinuous and continuous longitudinal walls. In exozones, autozooecia commonly form angles between 45° and 50° with mesothecae; in straight to curved and indistinct ranges, contiguous or partly contiguous, separated partly by extrazooecial stereom locally. Longitudinal walls regularly sinuous, generally continuous or locally merging with extrazooecial stereom. Transverse walls slightly raised proximal to autozooecial chambers in some species. Autozooecial walls commonly vesicular in inner exozones. Mural styles common in autozooecial boundaries, walls, and extrazooecial stereom. Living chambers elliptical in cross section, without lateral structures. Basal diaphragms straight to slightly curved, common; scattered, or may be regularly arranged. Monticules rare to common, generally scattered in zoaria. Monticular zooecia commonly filled with stereom. Extrazooecial skeleton consisting of stereom and vesicular portions; vesicular structures local in inner exo-

zones; laminated stereom localized throughout exozones, with mural styles arranged singly or in discontinuous rows. *?L.Ord.-M.Ord., ?USSR*(Est.), USA.——Fig. 255,2a-c. **P. frondosa; a,* median granular zone with median rods, autozooecial boundaries and mural styles in exozone; transv. sec., lectotype, USNM 242630, ×50; b, subelliptical autozooecia, elliptical living chambers, mural styles in boundaries, monticule with open and filled zooecia; tang. sec., lectotype, ×30; c, median rods in granular zone of mesotheca, mural styles in autozooecial walls, shape of chambers; long. sec., paralectotype, USNM 242634, ×50.

Sibiredictya NEKHOROSHEV, 1960, p. 277 [**S. usitata;* OD; M. Ord. (Mangaze.), Rybokupchaya River, Sib., USSR]. Zoarium cribrate; fenestrules irregularly shaped, variable in size, surrounded by extrazooecial stereom with mural styles. Mesothecae sinuous in longitudinal section. Median rods poorly delineated or lacking. In endozones, autozooecia in straight to slightly curving ranges, irregularly subrectangular to subrhomboidal in cross section, with continuous longitudinal walls. In exozones, autozooecia in straight ranges in midregions between fenestrules, in curving ranges around fenestrules, subelliptical to subrectangular in cross section, contiguous to partly contiguous, separated locally by extrazooecial skeleton. Longitudinal walls continuous in midregions, merging locally with extrazooecial stereom in lateral regions. Mural styles rare in zoarial boundaries and walls, common in extrazooecial stereom in zoarial margins. Living chambers subelliptical to subcircular in cross section, without lateral structure. Basal diaphragms not observed. Monticules absent. Extrazooecial skeleton generally of laminated stereom, vesicular locally in inner exozone; vesicle walls relatively thick. Extrazooecial skeleton may encrust proximal parts of zoaria. *M.Ord., USSR*(Sib.).——Fig. 256,1a-e. **S. usitata,* Amutkan Cr., Sib., paratype, USNM 171740; a, median granular zone in mesotheca, autozooecial boundaries, indistinct extrazooecial vesicles in inner exozone; transv. sec., ×30; b, irregular shape of fenestrules, arrangement of autozooecia; external view, ×5; c, longitudinal autozooecial walls, subelliptical living chambers; tang. sec., ×50; d, subrectangular autozooecia in endozone, mural styles in extrazooecial stereom adjacent to fenestrule; tang. sec., ×50; e, sinuous mesotheca, extrazooecial vesicles in inner exozone; long. sec., ×50.

Trigonodictya ULRICH, 1893, p. 160 [**Pachydictya conciliatrix* ULRICH, 1886a, p. 76; OD; Decorah Sh., M. Ord., Cannon Falls, Minn., USA] [= *Astreptodictya* KARKLINS, 1969, p. 49]. Zoarium irregularly branched or unbranched and explanate; ridgelike expansions lateral to general growth planes of zoaria in some. Mesothecae

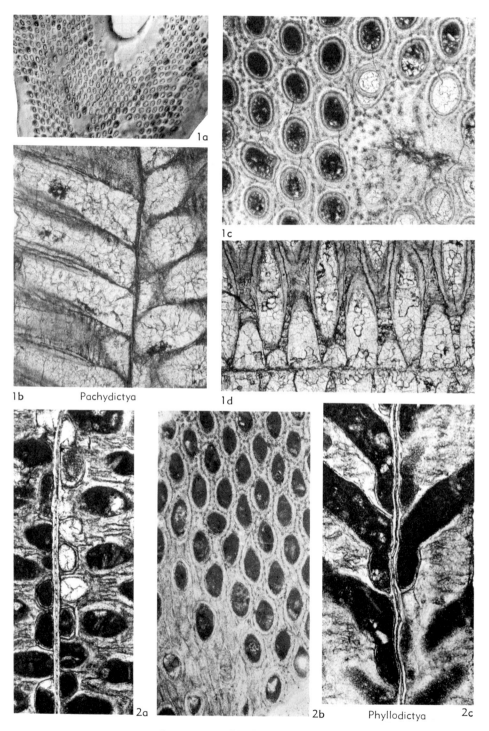

1a

1c

1b   Pachydictya

1d

2a

2b   Phyllodictya   2c

FIG. 255. Rhinidictyidae (p. 512–513).

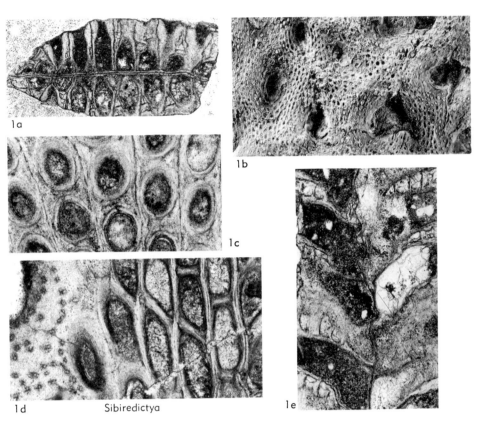

1a

1b

1c

1d   Sibiredictya   1e

Fig. 256. Rhinidictyidae (p. 513).

straight to sinuous in longitudinal section, locally zigzag in transverse section; bifurcating where branches or ridgelike expansions form lateral growth planes. Median rods circular in cross section, commonly with diameter greater than width of median granular zones. In endozones, autozooecia in straight ranges, subrectangular to subrhomboidal in cross section, generally contiguous laterally, locally separated by extrazooecial vesicles within ranges, with straight and generally continuous longitudinal walls. In exozones, autozooecia in straight ranges without continuous longitudinal walls, separated by extrazooecial skeletal deposits, elliptical in cross section, walls generally without vesicular structure. Mural styles indistinct; common in autozooecial boundaries and in dark zones in extrazooecial stereom, generally absent in autozooecial walls. Living chambers elliptical in cross section, without lateral chamber structures. Basal diaphragms straight to slightly curved, generally scattered in zoaria, but may be regularly arranged. Monticules rare to common, flat or raised, may be irregularly ridgelike, generally scattered in zoaria. Monticules commonly vesicular in inner exozones, laminar in outer exozones; laminar part

commonly with mural styles aligned in dark zones, locally discontinuous. Extrazooecial skeletal deposits common, consisting of laminar and vesicular portions. Vesicular structures common in inner exozones, locally in endozones, and between longitudinally aligned autozooecia. Extrazooecial stereom aligned in straight to slightly curving ridgelike range partitions that are delineated laterally by continuous dark zones and autozooecial boundaries. Extrazooecial stereom between range partitions and autozooecia consisting of laminae inclined proximally relative to those in autozooecial walls and range partitions; laminar stereom commonly with dark zones, longitudinally aligned, locally with indistinct mural styles. *M.Ord.-M.Sil.,* N.Am., G.Brit., USSR, Swed. ——Fig. 257,*1a,b. T. acuta* (HALL), Trenton Gr., M. Ord., N.Y., holotype, AMNH 666/1; *a,* autozooecial boundaries, extrazooecial stereom with vesicles at base of exozone, dark zones within laminar stereom in exozone; transv. sec., ×100; *b,* elliptical autozooecia and chambers, microstructurally continuous dark zones along middle of extrazooecial stereom of range partitions; tang. sec., ×100.—— Fig. 257,*1c-e.* **T. conciliatrix* (ULRICH); *c,* sub-

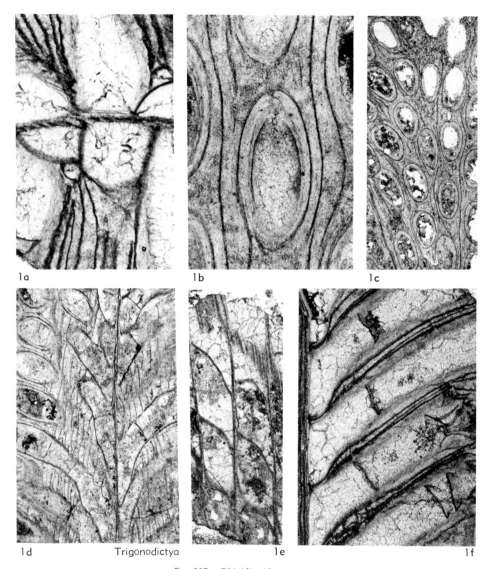

1a   1b   1c

1d   Trigonodictya   1e   1f

FIG. 257.   Rhinidictyidae (p. 513).

elliptical autozooecia, range partitions of extra-zooecial stereom, and mural styles in dark zones in stereom; tang. sec., paralectotype, USNM 242652, ×30; *d,* abandoned chambers, extra-zooecial vesicles in endozone and inner exozone, granular zone with median rods in mesotheca; oblique long. sec., lectotype, USNM 242650, ×30; *e,* median rods in mesotheca (near top), subrhomboidal autozooecia with structurally continuous longitudinal autozooecial walls in endozone; deep tang. sec., paralectotype, USNM 242653, ×50.——FIG. 257,*1f. T. fenestelliformis* (NICHOLSON), U. Ord. (Richmond.), Ill.; basal diaphragms, probable remnants of brown

body in chamber closed by monticule (middle right), extrazooecial stereom with dark zone between autozooecia; long. sec., USNM 242624, ×50.

Family STICTOPORELLIDAE
Nickles & Bassler, 1900

[Stictoporellidae NICKLES & BASSLER, 1900, p. 46]

Zoaria branched or unbranched and cribrate or explanate. Zoarial attachments continuous skeletally with erect parts of zoaria. Mesothecae straight to slightly sinuous with

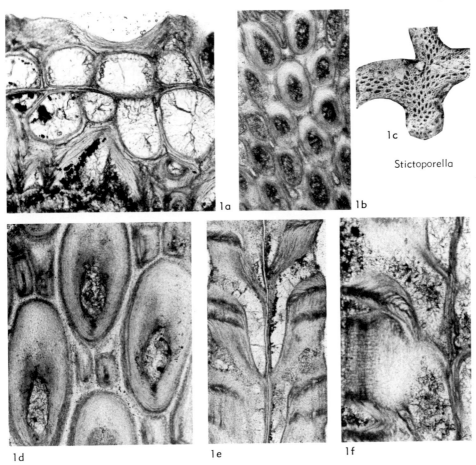

FIG. 258. Stictoporellidae (p. 517).

median granular zone generally extending to zoarial margins. In endozones, autozooecia in straight ranges, alternating on opposite sides of mesothecae; contiguous with continuous longitudinal walls, subrectangular to subrhomboidal in cross section, slightly expanded and narrowed alternately in adjacent ranges in outer endozones and base of exozones. In exozones, autozooecia form angles with mesothecae ranging between 50° and 80°, contiguous or separated by exilazooecia, generally polygonal to subcircular in cross section, not in linear ranges, with lateral walls restricted to individual autozooecia. Autozooecial wall laminae broadly V-shaped. Autozooecial boundaries narrowly serrated. Pustules common in exozonal walls and extrazooecial stereom. Living chambers subrectangular to subrhomboidal in cross section in endozones, elliptical to subelliptical in cross section in exozones. Basal diaphragms and lateral chamber structures absent. Exilazooecia common. Monticules absent to common, consisting of exilazooecia and zooecia of variable sizes. Extrazooecial stereom laminated, sparse in midregions of zoaria. *L.Ord.-M.Sil.*

The Stictoporellidae resemble Escharoporidae and Intraporidae in rhombic arrangement of autozooecia in the exozone, but differ in microstructure, cross-sectional shape of autozooecia, and distribution of exilazooecia in exozones.

Stictoporella ULRICH, 1882, p. 152 [*S. interstincta*; OD; ''Economy'' Mbr., ''Eden'' F., U. Ord., West Covington, Ky., USA; =*Ptilodictya flexuosa* JAMES, 1878, p. 4] [=*Lemmatopora* POČTA, 1894, p. 102]. Zoarium branched. In exozones, autozooecia subpolygonal in cross section, generally contiguous, or locally separated by exilazooecia or extrazooecial stereom. Pustules common; scattered along autozooecial boundaries, in exozonal walls, and extrazooecial stereom. Living chambers elliptical in cross section. Exilazooecia subelliptical to irregularly polygonal in cross section; regularly arranged in pairs or singly between successive autozooecia, or in groups along zoarial margins. Monticules rare to absent, flat or slightly raised, irregularly arranged in zoaria; consisting of exilazooecia and few zooecia of variable size. [The type specimens of *Lemmatopora* POČTA are poorly preserved (PRANTL, 1935a) and were unavailable for study. Thus, I follow BASSLER (1953, p. G138) in considering *Lemmatopora* to be a synonym of *Stictoporella*.] *M.Ord.-M.Sil.*, N.Am., USSR, ?Czech. ——FIG. 258,*1a–f*. **S. interstincta*; *a,* mesotheca with discontinuous median granular zone, microstructure of autozooecial walls; transv. sec., paralectotype, USNM 137613, ×100; *b,* polygonal autozooecia, elliptical living chambers in outer exozone; tang. sec., paralectotype, USNM 137613, ×30; *c,* branching pattern; external view, lectotype, USNM 137612, ×5; *d,* shape of autozooecia and exilazooecia in exozone; tang. sec., USNM 242635, ×100; *e,* shape of living chamber, microstructure of autozooecial walls; long. sec., USNM 242635, ×100; *f,* serrated autozooecial boundaries in exozone; long. sec., paralectotype, USNM 137613, ×100.

Pseudostictoporella Ross, 1970, p. 376 [**P. typicalis*; OD; Selby Mbr., Rockland F., M. Ord., Napanee, Ont., Can.]. Zoarium branched or unbranched and explanate. In exozones, autozooecia irregularly hexagonal in cross section, contiguous or partly separated by exilazooecia. Pustules common along autozooecial boundaries, scattered in exozonal walls. Living chambers subelliptical in cross section. Exilazooecia polygonal to irregularly subcircular in cross section, scattered in zoaria; arranged in groups, singly or in short rows. Exilazooecia commonly closed locally by thickened walls. Monticules common, generally flat; consisting mostly of exilazooecia, few zooecia, and some extrazooecial stereom. Monticules generally scattered in zoaria; may be regularly arranged in some species with explanate zoaria. *M.Ord.*, N.Am.——FIG. 259,*2a–c*. **P. typicalis*; *a,* narrowly serrated autozooecial boundaries in exozone, median granular zone of mesotheca in zoarial margin; transv. sec., paratype, YPM 25455, ×100; *b,* polygonal autozooecia, open and closed exilazooecia in exozone; tang. sec., holotype, YPM 2545, ×100; *c,* sinuous mesotheca, shape of living chambers; long. sec., holotype, ×30.

Stictoporellina NEKHOROSHEV, 1956a, p. 48 [**Stic-*

toporella? *cribrosa* ULRICH, 1886a, p. 69; OD; Decorah Sh., M. Ord., Minneapolis, Minn., USA]. Zoarium cribrate; fenestrules subelliptical to subcircular, generally elongate distally. Fenestrule margins with numerous exilazooecia. In exozones, autozooecia irregularly subpolygonal to subcircular in cross section, contiguous or locally separated by exilazooecia. Autozooecial boundaries locally not visible in some species. Pustules common along autozooecial boundaries, scattered in exozonal walls. Living chambers elliptical to subelliptical in cross section. Exilazooecia irregularly polygonal to subcircular in cross section, variable in size, locally closed by thickened walls, arranged singly or in scattered groups in zoaria. Monticules common to absent, generally raised and irregularly shaped, consisting mostly of exilazooecia and extrazooecial stereom. *Ord.*, USSR(?Est.), USA.——FIG. 259,*1a–f*. **S. cribrosa* (ULRICH); *a,* mesotheca with median granular zone, V-shaped laminae in exozone; transv. sec., paralectotype, USNM 162023, ×50; *b,* irregularly sinuous mesotheca, shape of living chambers; long. sec., lectotype, USNM 162015, ×30; *c,* arrangement of autozooecia and exilazooecia, shape of fenestrules; exterior view, lectotype, ×5; *d,* autozooecia alternating across mesotheca; transv. sec., lectotype, ×30; *e,* subpolygonal autozooecia, pustules in autozooecial boundaries, open and filled exilazooecia; tang. sec., lectotype, ×100; *f,* autozooecia in indistinct rhombic pattern, distribution of exilazooecia in exozone; tang. sec., lectotype, ×30.

Family VIRGATELLIDAE Astrova, 1965

[Virgatellidae ASTROVA, 1965, p. 290]

Zoaria branched or unbranched and explanate. Mesothecae straight to sinuous. Median granular zone discontinuous, without median rods. In exozones, autozooecia arranged in indistinct rhombic pattern, without continuous longitudinal walls or in linear ranges with continuous longitudinal walls partly contiguous or separated by extrazooecial stereom, subelliptical to subcircular in cross section, with indistinct wall laminae. Autozooecial boundaries narrowly serrated. Mural styles abundant, relatively large, consisting of distinct cores and thin sheaths. Living chambers subelliptical to subcircular in cross section, variable in length. Superior and inferior hemisepta common. Basal diaphragms common. Exilazooecia and mesozooecia absent. Monticules common, con-

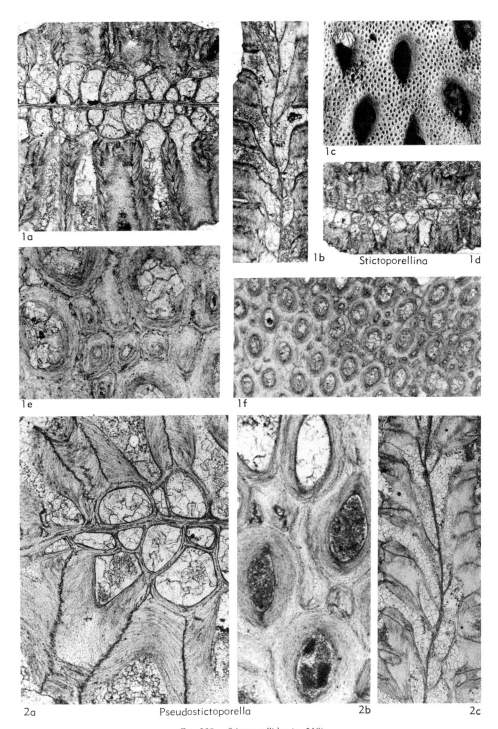

Fig. 259. Stictoporellidae (p. 518).

sisting of extrazooecial stereom. Extrazooecial skeleton common; consisting of vesicular portion in inner exozones, laminar stereom in parts of outer exozones. *M.Ord.*

According to Astrova (1965, p. 290), Virgatellidae are more similar to the Rhinidictyidae than to other ptilodictyines in skeletal microstructure, shape of autozooecia in exozones, and in presence of abandoned chambers. Of the Rhinidictyidae, *Athrophragma* (Fig. 252,*1a*) and *Pachydictya* (Fig. 255,*1c*) possess autozooecia that are similar in arrangement and shape to those in *Virgatella* (Fig. 260,*2b*). *Stictopora* has autozooecia similar to those in *Pseudopachydictya*; however, *Pseudopachydictya* differs in arrangement and shape of autozooecia (Fig. 260,*1a*) in exozones from that in *Virgatella* (Fig. 260,*2b*). Thus, morphological relationship between the Virgatellidae and Rhinidictyidae can be inferred only to a degree because the arrangement and shape of autozooecia at mesothecae and in endozones of the Virgatellidae is not determinable in available material. Virgatellids differ from most other ptilodictyines in having autozooecia commonly surrounded by extrazooecial stereom, in rhombic and linear arrangements of autozooecia in exozones, and in kind and distribution of mural styles (Astrova, 1965, p. 290).

Virgatella Astrova, 1955, p. 158 [**V. bifoliata*; OD; M. Ord. (Mangaze.), Podkamennaya Tunguska River, Sib., USSR]. Zoarium branched or unbranched and irregularly explanate. Mesotheca slightly sinuous in longitudinal section. In exozones, autozooecia subelliptical to subcircular in cross section, aligned in indistinct rhombic pattern, generally separated by extrazooecial stereom. Autozooecial boundaries discontinuous locally. Mural styles common throughout laminar part of autozooecial walls and stereom; arranged singly and in irregularly curving series and clusters, bifurcating frequently. Living chambers subelliptical in cross section. Superior hemisepta thin, long, curving proximally; inferior hemisepta short, blunt, extending from mesothecae; both hemisepta indistinct, regularly arranged. Basal diaphragms scattered near base of exozones, relatively thick, slightly curved, may be incomplete in outer exozone. Monticules flat, irregularly arranged. Extrazooecial stereom arranged regularly in exozones throughout zoaria. *M.Ord.*, USSR(N. Zemlya,Sib.).——Fig. 260,*2a,b*. **V. bifoliata*, PIN 1242/30; *a*, microstructure of bifurcating mural styles in exozone, mesotheca with indistinct median granular zone; long. sec., ×100; *b*, rhombic pattern of subelliptical to subcircular autozooecia in inner exozone; tang. sec., ×30.

Pseudopachydictya Astrova, 1965, p. 293 [**Pachydictya multicapillaris* Astrova, 1955, p. 155; OD; M. Ord. (Mangaze.), Podkamennaya Tunguska River, Sib., USSR]. Zoarium branched. In exozones, autozooecia in straight ranges or laterally in ranges oblique to zoarial midregion. Autozooecia subrectangular to subelliptical in cross section, generally contiguous in midregions, partly separated by extrazooecial stereom in lateral regions. Longitudinal walls slightly sinuous and continuous, locally merging with extrazooecial stereom, relatively thick; vesicular structure absent. Mural styles common, relatively large with distinct cores and thin sheaths, present in autozooecial boundaries and throughout walls. Mural styles gradually curve and some bifurcate. Living chambers elliptical in cross section, relatively small, locally narrowed or closed by thickened walls in outer exozones. Superior hemisepta short, blunt, regularly arranged. Inferior hemisepta thin, long, generally arising from mesothecae, scattered in zoaria. Basal diaphragms generally straight, scattered to common in zoaria. Monticules common, irregularly spaced, generally flat and irregularly shaped. Extrazooecial stereom laminated, common in lateral regions, sparse in midregions; containing mural styles. *M.Ord.*, USSR(W. Arctic, Sib.).——Fig. 260,*1a,b*. **P. multicapillaris* (Astrova), Vaygach, W. Arctic, PIN 1393/269; *a*, autozooecia in linear ranges in midregion (right), ranges in lateral region (left) oriented obliquely to midregion; tang. sec., ×30; *b*, indistinct wall laminae, diverging mural styles; long. sec., ×50.

Family Uncertain

Euspilopora Ulrich in Miller, 1889, *Ptilotrypa* Ulrich in Miller, 1889, *Ptilotrypina* Astrova, 1965, *Stictotrypa* Ulrich, 1890, *Taeniodictya* Ulrich, 1888, and *Trepocryptopora* Yang, 1957, are not assigned to revised families. These genera, however, are retained in the Ptilodictyina because they possess features of the suborder but differ from the type genera of the families and among themselves. Because most include only one species, and each species is represented by only a few specimens in varying states of preservation, these genera are not

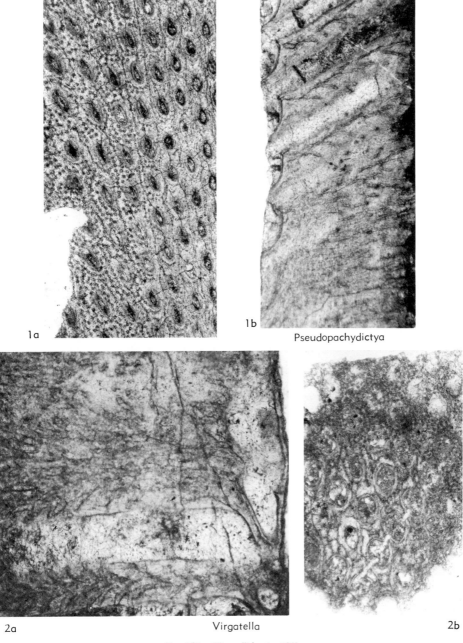

1a

1b

Pseudopachydictya

2a

Virgatella

2b

FIG. 260. Virgatellidae (p. 520).

sufficiently well known to establish new families.

Euspilopora ULRICH in MILLER, 1889, p. 301 [*E. serrata* ULRICH, 1890, p. 526; OD; Cedar Valley Ls., M. Dev., Buffalo, Iowa, USA]. Zoarium branched or unbranched and explanate, margins broadly crenulated. Mesothecae straight or slightly sinuous locally in longitudinal section. Median granular zone extends throughout mesothecae, may contain scattered median rods. In endozones, autozooecia arranged in straight to slightly curving ranges, aligned or alternating on opposite sides of mesothecae, irregularly subelliptical in cross section parallel to mesothecae, mostly contiguous, with continuous longitudinal walls, only locally separated by extrazooecial stereom. In exozones, autozooecia usually forming angles between 80° and 90° with mesothecae; ranges straight to slightly curved; broadly elliptical to subcircular in cross section, contiguous laterally, with continuous longitudinal walls extending into ridges at zoarial surface; separated locally by transverse extrazooecial vesicles. Autozooecial wall laminae indistinct and broadly curved. Autozooecial boundaries narrowly serrated in inner exozones, generally not visible in outer exozones. Mural styles abundant in walls and laminated stereom, closely spaced and very small, consisting of closely curved segments of skeletal laminae or locally containing minute and discontinuous cores. Mural styles generally in indistinct series usually at right angles to zoarial surface. Acanthostyles common, small, generally with straight cores; present in outer exozones, scattered in zoaria; rarely arranged at regular intervals along longitudinal walls. Living chambers broadly elliptical in cross section, without lateral structures. Basal diaphragms absent. Polymorphism expressed by modified marginal, basal, and monticular zooecia. Monticules common, generally depressed; consisting mostly of extrazooecial skeleton and few zooecia; extrazooecial skeleton locally vesicular in inner exozone. Monticules in branched zoaria regularly arranged near zoarial margins, locally extending into margins. Monticules in explanate zoaria furrowlike, elongated parallel to growth direction of zoaria; at relatively regular intervals throughout zoaria. Extrazooecial skeleton common, irregularly delineated, consisting of vesicular and laminar portions. Vesicles with relatively thick walls, variable in size, commonly elongated longitudinally; mostly in inner exozone, rarely in endozone. Laminar stereom with numerous mural styles; acanthostyles rare in outer exozones. [*Euspilopora* is similar to rhinidictyids in having autozooecia in relatively straight ranges and median rods in the mesotheca; however, it differs from rhinidictyids and most other ptilodictyines in cross-sectional shape of autozooecia (Fig. 261,1*b,d*), irregular presence or lack of median rods in some species, general appearance and abundance of mural styles, presence of acanthostyles (Fig. 261,1*d*), and abundance and appearance of monticules. In skeletal microstructure as well as presence and kind of mural styles, *Euspilopora* is similar to *Taeniodictya* ULRICH.] M.Dev., USA.——FIG. 261,1*a–e*. **E. serrata*; *a,* autozooecial boundaries in endozone and inner exozone, extrazooecial vesicles with relatively thick walls in exozone; long. sec., lectotype, USNM 242639, ×100; *b,* acanthostyles aligned in longitudinal walls; tang. sec., paralectotype, USNM 242638, ×30; *c,* mesotheca with median granular zone; transv. sec., paralectotype, USNM 242637, ×50; *d,* autozooecia in curving ranges, monticule in margin, numerous mural styles throughout exozone; tang. sec., paralectotype, USNM 242636, ×50; *e,* living chambers with mesotheca as basal wall, numerous mural styles in autozooecial walls; including same long. sec. as *a,* ×50.

Ptilotrypa ULRICH in MILLER, 1889, p. 320 [**P. obliquata* ULRICH, 1890, p. 531; OD; U. Ord. (Richmond.), Wilmington, Ill., USA]. Zoarium generally large and branched. Mesothecae irregularly crenulated locally in longitudinal section, gradually thickening toward lateral zoarial margins; thickened margins with curved laminae forming narrowly serrated zone beyond central median granular zone. Median rods absent. In endozones, autozooecia generally in straight ranges, alternating on opposite sides of mesothecae, subrectangular to subelliptical in cross section parallel to mesotheca, contiguous, with slightly sinuous and continuous longitudinal walls. In exozones, autozooecia usually forming angles between 40° and 50° with mesothecae, arranged in indistinct rhombic pattern, subelliptical in section, partly contiguous, separated laterally by local extrazooecial stereom, without continuous longitudinal walls. Autozooecial wall and stereom laminae narrowly to broadly curved and U-shaped, locally forming striae at zoarial surface. Autozooecial boundaries narrowly to broadly serrated, walls locally crenulated in inner exozones. Pustules rare or lacking. Mural styles absent. Living chambers elliptical in cross section, variable in length. Chamber lining thin and discontinuous. Mural spines rare, short, and blunt. Cysts rare, spherical to irregularly shaped. Cystiphragms common, variable in size, locally in discontinuous series along distal walls. Polymorphism expressed by modified marginal, basal, and rare monticular zooecia. Monticules rare to common, scattered, generally flat; consisting mostly of extrazooecial stereom, commonly with striae at zoarial surface. [*Ptilotrypa* is somewhat similar to escharoporids in structure of the mesotheca (Fig. 262,1*a*) as well as arrangement and cross-sectional shape of autozooecia (Fig. 262,1*d*); however, it differs from escharoporids and ptilodictyoids in autozooecial angle with the mesotheca, gradual thickening of autozooecia at the base of the exozone (Fig. 262,1*c*), longitudinal shape of autozooecia (Fig. 262,1*c*), presence and shape of scattered cystiphragms, and general appearance of skeletal laminae.] U.Ord., USA.——FIG. 262,1*a–d*. **P. obliquata*, lectotype, USNM 242640; *a,* median granular zone merging with mesothecal laminae in zoarial margin, autozooecial boundaries in endozone; transv. sec., ×30; *b,* arrangement of

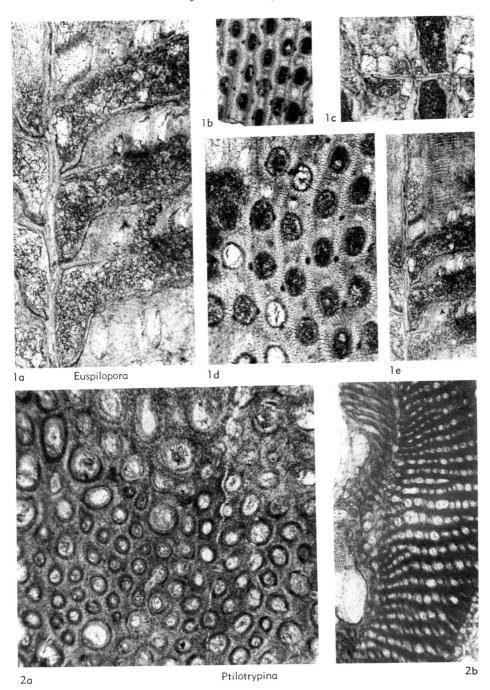

1b

1c

1a   Euspilopora   1d   1e

2a   Ptilotrypina   2b

FIG. 261. Family Uncertain (p. 521–522).

autozooecia, scattered monticules; external view, ×5; *c,* mesotheca locally crenulated, living and few abandoned chambers, irregularly shaped cystiphragms at base of exozone; long. sec., ×30;

d, autozooecia in indistinct rhombic arrangement, extrazooecial stereom in lateral walls between some autozooecia; tang. sec., ×30.

Ptilotrypina ASTROVA, 1965, p. 249 [**P. semibi-*

foliata; OD; M. Ord. (Mangaze.), Podkamennaya Tunguska River, Sib., USSR]. Zoarium generally bifoliate and irregularly explanate with local unifoliate and encrusting segments. Mesothecae broadly sinuous in longitudinal section. Endozones narrow and indistinctly delineated. In exozones, autozooecia relatively few, slightly curving, usually forming angles between 60° and 80° with mesothecae, subelliptical to subcircular in cross section, arranged singly or in groups, irregularly aligned and distributed, partly contiguous, mostly separated by mesozooecia or extrazooecial stereom. Zooecial walls irregularly nodular and variable in thickness, consisting of broadly curved and irregularly U-shaped laminae. Autozooecial boundaries broadly serrated. Pustules abundant, present throughout exozone. Living chambers irregularly subelliptical to subcircular in cross section, variable in length, apparently without lateral structures. Basal diaphragms common, relatively thick, straight to slightly curved. Mesozooecia abundant, constituting most of zoarium, subelliptical to subcircular in cross section, variable in size, commonly with scattered diaphragms, arranged locally in indistinct ranges, apparently arising locally from mesothecae. Extrazooecial stereom common, irregularly delineated; forming irregularly shaped, monticulelike flat areas and low protuberances at zoarial surface. [*Ptilotrypina* is characterized by zoaria with combined bifoliate and unifoliate growth habit, different mode of development of zooecia, and a nodular zooecial wall structure that suggests a different configuration of skeletal laminae (Astrova, 1965, p. 249). Zoaria consist of a few, scattered zooecia with large skeletal apertures and numerous zooecia with small skeletal apertures. Those with large apertures (regular zooecia of Astrova), which I consider to be autozooecia, occur singly or in irregular groups. Those with small apertures (pseudomesoporelike of Astrova), which I consider to be mesozooecia, constitute the major part of a zoarium, and some may arise from the mesotheca. Available material is not adequate for determining microstructure of the mesotheca and zooecial walls, and the structural relationship between autozooecia and mesozooecia in the endozone and base of the exozone.] *M.Ord.-U.Ord.*, USSR(Sib.).——Fig. 261,*2a,b.* *P. semibifoliata,* U. Ord. (Dolbor.), Sib., paratype, PIN 1242/230; *a,* irregular distribution of autozooecia and mesozooecia; tang. sec., ×50; *b,* bifoliate (upper) and local unifoliate (lower) segments of zoarium; oblique long. sec., ×30.

Stictotrypa Ulrich, 1890, p. 393 [*Stictopora similis* Hall, 1876, p. 122; OD; Niagara Gr., M. Sil., Waldron, Ind., USA]. Zoarium branched. Mesothecae slightly sinuous in longitudinal section. Median granular zone discontinuous locally, without median rods. In endozones, autozooecia in straight ranges, alternating on opposite sides of mesothecae, subrectangular to subrhomboid in cross section, contiguous, with slightly sinuous and continuous longitudinal walls. Endozone relatively wide. In exozones, autozooecia usually at right angles with mesothecae, in indistinct ranges, locally sloping proximally, broadly elliptical in cross section, partly contiguous, separated by extrazooecial stereom, without continuous longitudinal walls. Autozooecial walls consisting of broadly curved laminae, locally forming low peristomes around autozooecial apertures. Autozooecial boundaries narrowly serrated. Acanthostyles rare to common, small, generally with straight cores; mostly in outer exozones in extrazooecial stereom near autozooecial boundaries, scattered in zoaria. Living chambers subelliptical in cross section, without lateral structures. Basal diaphragms absent. Polymorphism expressed by modified marginal and basal zooecia. Extrazooecial stereom occurring regularly throughout exozones. Stereom laminae usually broadly curved, concave to zoarial surface. Stereom may contain single cavities at base of exozones. [*Stictotrypa* is similar to genera of the Escharoporidae, Intraporidae, and Stictoporellidae in having longitudinal autozooecial walls that are continuous in the endozone, but become restricted to autozooecia in the exozone. It differs from those genera, however, in having a relatively wide endozone (Fig. 262,*2b*), in microstructure of autozooecial walls and autozooecial boundaries, in shape of autozooecia, in lack of exilazooecia or mesozooecia, and somewhat in distribution of extrazooecial skeleton in the exozone.] *M.Sil.,* USA. ——Fig. 262,*2a–d.* *S. similis* (Hall), lectotype, AMNH 1926-1; *a,* autozooecia alternate across slightly curved mesotheca; transv. sec., ×30; *b,* indistinct autozooecial boundaries, median granular zone in mesotheca; long. sec., ×100; *c,* microstructure of intermittent median granular zone in mesotheca, indistinct serrated autozooecial boundaries in exozone; transv. sec., ×100; *d,* autozooecia in indistinct ranges at base of exozone; tang. sec., ×30.

Taeniodictya Ulrich, 1888, p. 80 [*T. ramulosa* Ulrich, 1890, p. 528; OD; Keokuk Ls., L. Miss., Nauvoo, Ill., USA]. Zoarium branched, rarely unbranched and explanate, with skeletally continuous basal attachments. Mesothecae relatively thick, slightly sinuous, crenulated locally in longitudinal section, containing small and abundant mural styles at right angles to median granular zone, merging with extrazooecial stereom in zoarial margins. Median granular zone discontinuous through mesothecae. In endozones, autozooecia in ranges, alternating or aligned on opposite side of mesothecae, irregularly subrhomboid in cross section, contiguous with slightly sinuous, continuous, relatively

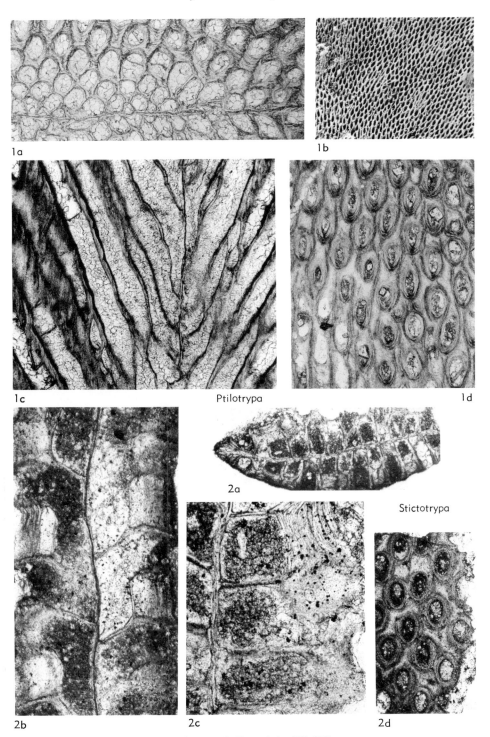

1a

1b

1c

Ptilotrypa

1d

2a

Stictotrypa

2b

2c

2d

Fig. 262. Family Uncertain (p. 522–524).

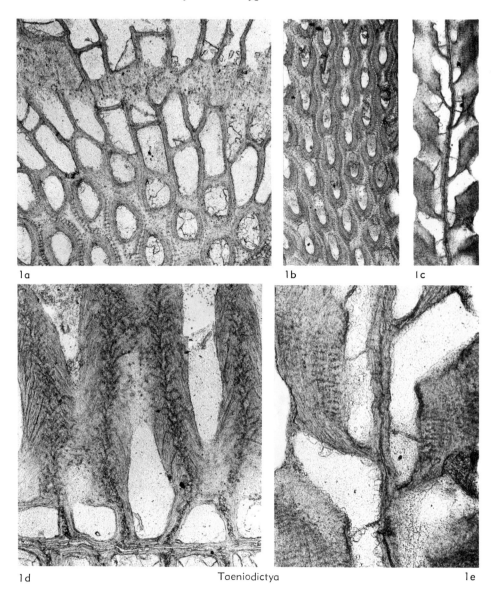

1a   1b   1c

1d   Taeniodictya   1e

FIG. 263.   Family Uncertain (p. 524).

thick, longitudinal walls. In exozones, autozooe-cia usually forming angles between 45° and 60° with mesothecae, in slightly curving to straight ranges, elliptical in cross section, generally con-tiguous, separated partly by extrazooecial ster-eom within ranges. Autozooecial wall and ster-eom laminae curved and broadly U-shaped. Autozooecial boundaries broadly serrated, com-monly absent. Mural styles abundant, small, closely spaced, consisting of tightly curved seg-ments of wall laminae or minute and discontin-uous cores or granules, generally curved,

arranged in diverging pattern along middle of longitudinal walls, terminating in walls at angles to chambers or zoarial surface. Living chambers narrowly elliptical in cross section, variable in length. Superior hemisepta common, short, blunt, generally scattered in zoaria. Basal dia-phragms rare, thin, irregularly curved, spaced irregularly through zoarium. Polymorphism expressed by marginal and basal zooecia. Extra-zooecial stereom common, indistinctly delin-eated, generally between successive autozooecia in zoarial midregions. Stereom laminae commonly forming striae at zoarial surface in margins, containing abundant mural styles. [*Taeniodictya* is similar to rhinidictyids in arrangement and shape of autozooecia in the endozone and to a lesser degree in the exozone, but differs from rhinidictyids and other ptilodictyines in skeletal structure. In microstructure, *Taeniodictya* is similar to *Euspilopora,* but differs in shape of autozooecia and presence of abandoned chambers (Fig. 263,*1c*), in having mural styles in the mesotheca (Fig. 263,*1a*), in configuration of mural styles in the exozone (Fig. 263,*1d*), and lack of acanthostyles and monticules.] *?M.Sil., Miss.,* USA.——FIG. 263,*1a–e.* **T. ramulosa; a,* subrhomboid autozooecia in endozone, mural styles in mesotheca and exozone; tang. sec., paralectotype, USNM 242644, ×50; *b,* contiguous and regularly curved longitudinal walls in exozone; tang. sec., paralectotype, USNM 242642, ×30; *c,* abandoned chambers, thin and irregularly curved basal diaphragms; long. sec., USNM 242642, ×30; *d,* mural styles in diverging pattern along middle of longitudinal walls in exozone; transv. sec., paralectotype, USNM 242645, ×100; *e,* slightly crenulated mesotheca with mural styles perpendicular to the median granular zone, basal diaphragms; long. sec., USNM 242642, ×100.

Trepocryptopora YANG, 1957, p. 7 (English summary) [**T. dichotomata*; OD; up. L. Ord., S. Shaanxi, China]. Zoarium branched or unbranched and explanate. Mesothecae straight to slightly sinuous, may be crenulated locally in transverse section. In endozones, autozooecia arranged in rhombic pattern, elliptical in cross section, contiguous, without continuous longitudinal walls. Endozones narrow, indistinctly delineated; autozooecial walls thickening only slightly at base of exozones. In exozones, autozooecia usually at right angles to mesothecae, arranged in rhombic pattern, elliptical in cross section, only partly contiguous locally, generally separated by exilazooecia. Autozooecial walls relatively thin, wall and stereom laminae indistinct. Autozooecial boundaries apparently narrowly serrated. Living chambers broadly elliptical in cross section, relatively short, with thin and indistinct lining locally. Basal diaphragms thin, straight to curved, commonly cystoidal, at relatively regular intervals throughout zoarium. Exilazooecia common, very small, indistinctly delineated, generally subelliptical to subcircular or irregularly shaped in cross section, may have scattered diaphragms, present throughout exozones. [*Trepocryptopora* appears to be one of the earliest ptilodictyines. It possesses a distinct mesotheca but differs in other ptilodictyine characters, as noted by YANG (1957). The endozone is indistinctly delineated because of different autozooecial growth. Autozooecia are subtubular throughout a zoarium, arising from the meso-

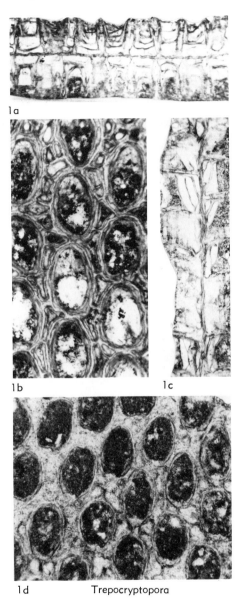

1a

1b   1c

1d   Trepocryptopora

FIG. 264. Family Uncertain (p. 527).

theca at right angles and changing only slightly in shape and thickness at the base of the exozone. Autozooecial living chambers are very short because abandoned chambers are abundant. Basal walls of abandoned chambers are commonly cystoidal diaphragms, which are uncommon in ptilodictyines. *Trepocryptopora* also differs in kind of polymorphism by having very small zooecia (mesopores of YANG) with a few scattered diaphragms. I consider these polymorphs to be exilazooecia. Microstructure of the

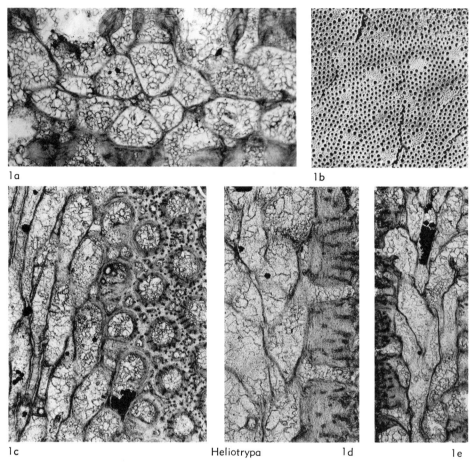

1a
1b
1c
Heliotrypa   1d   1e

FIG. 265.   Suborder Uncertain (p. 528).

mesotheca and configuration of laminae in the exozone are not determinable in available specimens. If the Early Ordovician age of *Trepocryptopora* can be verified, it is a ptilodictyine that differs considerably from those of younger ages.] *up.L.Ord.*, China(S. Shaanxi).——FIG. 264,*1a–c. *T. dichotomata*, holotype, Northwest University Catalogue 8950; *a,* crenulated mesotheca, curved basal diaphragms, indistinct chamber lining; transv. sec., ×20; *b,* partly contiguous autozooecia, irregularly shaped exilazooecia; tang. sec., ×40; *c,* relatively thick, sloping basal diaphragms near mesotheca and in exozone, autozooecia arising from mesotheca without forming distinct endozone; long. sec., ×20 (photographs courtesy of King-Chih Yang). ——FIG. 264,*1d. T. flabelata* YANG, up. L. Ord., S. Shaanxi, holotype, Northwest University Catalogue 8953; subelliptical autozooecia with narrowly serrated boundaries, indistinct exilazooecia; tang. sec., ×40 (photograph courtesy of King-Chih Yang).

Suborder Uncertain

Heliotrypa ULRICH, 1883, p. 277 [**H. bifolia*; M; U. Miss. (Chester.), Ky., USA]. Zoarium irregularly explanate and undulating, may self-encrust locally. Autozooecia budded in relatively straight ranges from medial zones, ranges partly aligned across medial budding zones. Autozooecial basal walls contiguous proximally, irregularly sinuous vertically, alternating in adjacent ranges, continuous planar mesothecae not formed. Granular median zones having discontinuous median rods of variable size, locally coalescing, extending into autozooecial walls of endozones in some. In endozones, autozooecia contiguous with sinuous longitudinal walls, irregularly subrectangular to subelliptical in cross section, elongated parallel to medial budding zones, alternately expanded and narrowed in adjacent ranges, becoming subelliptical and abruptly thickened at base of exozones. In exozones, autozooecia irregularly contiguous and locally separated by exilazooecia or extrazooecial stereom, irregularly subcircular in cross section with lateral walls restricted to individual autozooecia, and longitudinal linear ranges not formed. Autozooecial boundaries broadly serrated. Autozooecial walls slightly variable in thickness, consisting of broadly U-shaped laminae. Mural styles abundant throughout exozones, straight or irregularly diverging, locally indenting autozooecial chambers; consisting of broadly curved segments of skeletal laminae and discontinuous small cores; may bifurcate locally. Living chambers irregularly subrectangular to subelliptical in cross section in endozones, subcircular in cross section in exozones. Basal diaphragms rare to absent, scattered in autozooecia of outer exozones, locally present under selfencrusted zoarial parts. Superior hemisepta common, irregularly shaped with blunt terminal edges. Inferior hemisepta absent. Intrazooecial cysts common, generally scattered in endozones, circular in cross section or irregular in shape. Exilazooecia abundant throughout exozones, variable in size, subcircular or irregularly shaped in cross section. Monticules common, generally flat, irregularly shaped, arranged in rhombic pattern; consisting of numerous exilazooecia with abundant mural styles. Extrazooecial stereom laminated, irregularly delineated, sparse throughout outer exozones. [*Heliotrypa* is similar to ptilodictyines in having flattened and bifoliate zoaria, but differs in mode of budding of autozooecia and in other zoarial features. Autozooecia bud from a medial zone (Fig. 265,*1a,c,d*) without delineating a planar mesotheca because basal autozooecial walls, although contiguous proximally, are limited in structural continuity. As in some ptilodictyines (Rhinidictyidae), median rods are formed in basal walls of autozooecia, but are variable in width, are in an irregular pattern, and may extend into autozooecial walls in endozones. Microstructurally, however, median rods in *Heliotrypa* are similar to those in the Rhinidictyidae. *Heliotrypa* differs from other ptilodictyines in shape of autozooecia and their living chambers, appearance of skeletal laminae, microstructure and abundance of mural styles (Fig. 265,*1c*), and in abundance and kind of exilazooecia. *Heliotrypa* resembles rhabdomesines in mode of autozooecial budding, but differs in having planar medial budding zones instead of axial zones. It also differs from rhadomesines in distribution of autozooecia and exilazooecia in exozones, and in skeletal microstructure. Because of its different growth habit, *Heliotrypa* is not assigned to a suborder, but is tentatively retained in the Cryptostomata until its taxonomic affinities are better established.] *U.Miss.,* USA.——— Fig. 265,*1a–e.* **H. bifolia*; *a,* medial budding zone of autozooecia, irregularly shaped median rods, shape of autozooecia in endozone; transv.

sec., lectotype, USNM 242646, ×100; *b,* arrangement of autozooecia and exilazooecia, distribution of monticules with exilazooecia; external view, paralectotype, USNM 242647, ×5; *c,* shape of autozooecia in endozone and exozone, medial budding zone with rods and intrazooecial cyst (lower left); tang. sec., paralectotype, USNM 242648, ×50; *d,* microstructure of autozooecial walls and mural styles; long. sec., paralectotype, USNM 242648, ×100; *e,* shape of living chambers, superior hemisepta in some chambers, sinuous basal autozooecial walls; long. sec., paralectotype, USNM 242647, ×50.

Invalid and Unconfirmed Generic Names Applied to the Ptilodictyina

The following names are considered to be invalid or unconfirmable, and are either available or unavailable.

Crateriopora ULRICH, 1879, p. 29, *nom. dub.* [**C. lineata*]. Name applied to encrusting parts of zoaria that ULRICH (1882, p. 151) subsequently recognized as encrusting bases of bifoliate cryptostomates with proximally tapering parts of zoaria. *U.Ord.,* USA.

Disteichia SHARPE, 1853, p. 146, *nom. oblit.* [**D. reticulata*; M]. According to NILS SPJELDNAES (pers. commun., March 17, 1971), it is not a ptilodictyine, but probably a phylloporinid. Skeletal microstructure is obliterated and species is not recognizable. *Ord.,* Port.

Fimbriapora ASTROVA, 1965, p. 254, *nom. dub.* [**Ptilodictya fimbriata* JAMES, 1878, p. 8]. Skeletal microstructure of type material is almost obliterated and species is not recognizable. *M.Sil.,* USA.

Graptopora ULRICH, 1882, p. 148, *nom. nud.* (*non* SALTER, 1858, p. 63; *non* LANG, 1916, p. 405). Diagnosis not given, species not named. *Ord.,* USA.

Hemipachydictya KOPAYEVICH, 1968, p. 128, *nom. dub.* [**Stictopora crassa* HALL, 1852, p. 45]. Skeletal microstructure of type material is recrystalized and species is not recognizable (ROSS, 1961a, p. 337). *M.Sil.,* USA.

Nicholsonia WAAGEN & WENTZEL, 1886, p. 874. Specimens not located. *Ord.,* India.

Sladina REED, 1907, p. 208, *nom. dub.* [**S. cateniformis*]. Skeletal microstructure is recrystallized and species is not recognizable. Probably not a ptilodictyine (for contrasting view see SPJELDNAES, 1957, p. 367). *U.Ord.,* G.Brit.

Stictoporina HALL & SIMPSON, 1887, p. xx [**Trematopora claviformis* HALL, 1883b, p. 181]. Repository of species not known; concept of genus not verifiable. *M.Dev.,* USA.

INTRODUCTION TO THE SUBORDER RHABDOMESINA

By Daniel B. Blake

[University of Illinois, Urbana]

The cryptostomate suborder Rhabdomesina includes many of the slender dendroid bryozoans found in Paleozoic marine sediments. Although specimens are known from much of the world, research has been concentrated in North America, the Soviet Union, Australia, and parts of Asia. The suborder ranges from the lower Middle Ordovician to the Upper Permian.

A large number of taxonomic characters are available, but recognition of the limits of the suborder has proven difficult because no unique suite of characters is recognized. Convergence with other groups, in particular the trepostomates, apparently has been common.

Six families are recognized here. The Arthrostylidae is dominant in Ordovician and Silurian rocks, and then declines. A few genera are known from younger Paleozoic rocks of the Soviet Union. Arthrostylids have been described mainly from North America, Europe, and western and central portions of the Soviet Union. They are poorly known because their small size renders them inconspicuous in the field and difficult to study in the laboratory. The Bactroporidae includes a single Devonian genus known only from North America. The remaining families, Rhabdomesidae, Rhomboporidae, Hyphasmoporidae, and Nikiforovellidae, are geographically widespread and primarily found in middle and upper Paleozoic rocks.

Acknowledgments.—I am grateful to the following individuals for making available specimens under their care, or for providing illustrations or information on specimens and publications: Roger Batten, J. L. H. Bemelmans, B. M. Bell, C. Blotwijk, R. S. Boardman, Krister Brood, Robert Conrad, P. L. Cook, Donald Dean, J. M. Edwards, Julia Golden, R. V. Goryunova, W. G. E. Graham, A. H. Kamb, O. L. Karklins, A. Kellermeyer, Lois Kent, Marie Kiepura, W. J. Kilgour, J. S. Lawless, D. B. Macurda, R. V. Melville, I. P. Morozova, V. P. Nekhoroshev, M. H. Nitecki, Eugene Richardson, Paul Siegfried, F. M. Swain, and R. F. Wise. The figures were prepared by D. R. Phillips. R. S. Boardman, R. J. Cuffey, and J. R. P. Ross read portions of the manuscript and made useful suggestions.

GROWTH PATTERNS

Rhabdomesines formed encrusting holdfasts, generally attached to such firm substrate as shell material. The period of encrusting growth was probably brief, for known holdfasts are small relative to overall zoarial size (Fig. 266,10a).

Most members of the suborder developed relatively slender (0.5–3.0 mm) cylindrical branches of fairly constant diameter between bifurcations. In many genera, especially earlier ones, zooecial apertures are arranged at the surface in a rhombic pattern that is mostly uninterrupted by polymorphs, monticules, or apparent microenvironmental influences (Fig. 266, 9,10). However, branches of a number of upper Paleozoic genera (e.g., *Rhombopora,* see Fig. 286,3b) are wider (up to about 5 mm) and have lost the constancy of diameter between bifurcations and regularity of apertural arrangement.

Although large silicified specimens are rare, available material shows that thickening of the exozonal wall and partial or complete closing of apertures was possible during colony life.

The growing tip may be attenuated in zoaria with steeply ascending zooecia (Fig. 266,6), but generally it is blunt. Intracolony

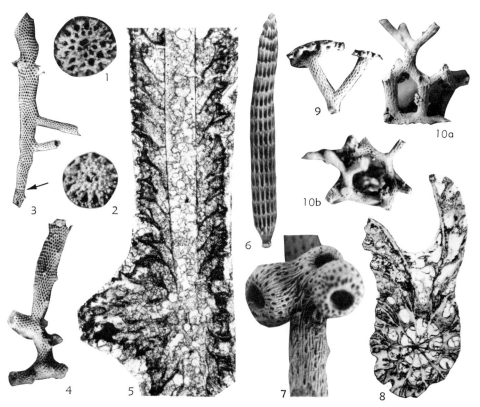

FIG. 266. Growth habits and articulation structures in Rhabdomesina.——*1–4. Rhabdomeson* sp., Perm., Texas; colonies that survived breakage and resumed growth elsewhere; *1,* broken branch base in early stage of healing with stylets developed but zooecia still open; USNM 240829, ×15.0; *2,* later stage of healing with zooecia mostly closed; USNM 240830, ×15.0; *3,* conical branch that resumed growth in former proximal direction at level of arrow after breaking from parent (compare with *5*); USNM 222647, ×2.5; *4,* conical branch arising from cylindrical parent branch; USNM 240831, ×2.5.——*5. Rhabdomeson* sp., Hamilton Gr., M. Dev., N.Y.; hollow axial cylinder; zooecial orientation is reversed at level of branch; specimen was apparently broken but survived to resume growth; long. sec., USNM 249311, ×30.——*6–8. Ulrichostylus* aff. *U. spiniformis* (ULRICH), Bromide F., M. Ord., Okla.; *6,* slender branch with growing tip and weakly developed exozone; proximal tip is spherical surface of joint; USNM 249332, ×10; *7,* parent stem and three cup-shaped proximal joints; USNM 249328, ×10.0; *8,* transverse section of branch and longitudinal section of cup-shaped proximal joint; hollow of cup continuous with one zooecium from main stem, lined with weakly laminated skeletal layer; USNM 249336, ×30.0.——*9. ?Acanthoclema* sp., ?Jeffersonville Ls., M. Dev., Falls of the Ohio near Louisville, Ky.; exterior view of colony with fenestellid colony partially encrusted and used as brace; USNM 178558, ×8.0.——*10a,b. ?Orthopora* sp., ?Jeffersonville Ls., M. Dev., Falls of the Ohio near Louisville, Ky.; interconnected growth habit at colony base provides strength without significant thickening of exozonal walls; *a,* lateral and *b,* top views, USNM 178559, ×8.0.

overgrowths are uncommon. The ramifying growth pattern was maintained throughout life, except where branches encountered foreign objects, which they partially encrusted to brace the colony (Fig. 266,*9*).

Study of sectioned zoarial bases suggests that the zone of astogenetic repetition was established after only a few founding zooids were developed. Zooecia were budded about a varied but generally well-defined, one- or two-dimensional median axis. In some early arthrostylids, the axis is sharply defined (see Fig. 281,*1b–d*), but in many later rhabdomesines it is an irregular, slender budding zone (see Fig. 286, *3c*). In many cross sections, an apparent median budding plate is developed at least locally (see Fig. 274, *1d*; 276, *1c*), but none is structurally differentiated, as in the Ptilodictyina. The median plate in the Rhabdomesina appears to be largely a product of local alignment of zooecial walls during growth.

In other taxa, the axial region may contain a small undifferentiated bundle of zooecia (see Fig. 269), a more or less enlarged and differentiated axial bundle (see Fig. 283, *1a*), or a hollow axial cylinder (Fig. 266, *5*). In these taxa, budding of nonaxial zooecia is from the outer surface of the axial structure.

Zooecia are typically added in a spiral pattern about the axis, and the spiral is reflected in apertural arrangement at the branch surface. Such addition is probably largely a response to geometric growth restraints, as outlined by GOULD and KATZ (1975) for receptaculitids. Locally, on branches showing typical spiral budding, a number of zooecial tiers may be added in an annular pattern.

A few authors have described spiral growth with terms implying that each zooecial tube is helically wrapped about the axis; if dissected from a zoarium, the shape of such a zooecium would broadly resemble that of an openly spiraled gastropod shell. Longitudinal thin sections show spiraling is limited, if it occurs at all. In these sections, individual zooecia commonly can be traced from the median axis to the zoarial surface. That is, the entire length of the tube lies within a single, flat plane of section and does not curve more than a few degrees because it does not pass out of the plane of section. If extensive coiling occurred, sections oriented along the axis should show zooecia in oblique cross section, which has not been observed. Cross sections of zooecia appear in longitudinal sections only near the centers of branches; this results from sections not passing through branch axes, as well as from the irregular nature of many zoaria.

The rhabdomesine zoarium is divided into a thin-walled endozone and a thick-walled exozone. Generally, the boundary is sharply defined. CHEETHAM (1971) argued that peripheral-wall thickening provides colony support in cheilostomates, and such thickening would have been useful in Paleozoic bryozoans as well. Although the exozone is well developed in most rhabdomesines, specimens showing walls peripherally thickened to an unusual degree are rare. Such enlarged stems developed near the base of the zoaria seemingly would have been useful for support of a large colony. Extensively interconnected stems may be present in basal attachment areas (Fig. 266, *10*); perhaps colony support usually was attained in this manner.

Growth generally appears to have been rather continuous with few indications of periodicity. Periodicity in bryozoans is recognized by skeletal banding in the exozone, seen in a few rhabdomesines, or by signs of temporary growth termination and exozone development across a branch axis, as in many trepostomates, or by annular growth banding, as in certain cheilostomates (e.g., *Myriapora*). Lack of such indications suggests that rhabdomesine colonies usually developed in a continuous growth period.

SKELETON

Skeletal wall materials and wall growth sequences in rhabdomesines are little studied but appear generally similar to those in trepostomates and other cryptostomates. TAVENER-SMITH and WILLIAMS (1972) published observations made with the scanning electron microscope. Other authors (e.g., BROOD, 1970; BLAKE, 1973a) have discussed development of specific skeletal features.

Most of the rhabdomesine wall is constructed of laminated calcite (Fig. 267, *3*). The laminae are made of lenticular platelets arranged in clearly defined layers of approximately constant thickness. The layers are usually oriented approximately parallel to the surface of zooecial chambers or parallel to the branch surface, except where locally deflected about such structures as stylets.

Ontogenetic changes in wall thickness and profile, primarily between endozone and exozone, are effected by local addition rather than thickening of laminae. Patterns of wall thickening are generally consistent within single zoaria and within genera, implying genetic control of growth, and thereby providing a useful taxonomic character. Lamellar profile is defined as the outline of a lamellar plane between adjacent zooecial chambers. In rhabdomesines, this outline is basically V-shaped in the early exozone and generally becomes increasingly rounded or flattened as the exozone thickens. Two important factors, wall thickness and stylet development, alter the profile. As walls thicken and spaces expand between chambers, the wall profile becomes flatter. Sheath laminae around stylets are deflected toward the zoarial surface and commonly cause an inflated profile, especially in relatively thin-walled taxa with a narrow stylet field approximately centered between chambers. In diagnoses that follow, emphasis is placed on the outline as seen in transverse view because the distance between laterally adjacent chambers is commonly less than that between longitudinally successive chambers, and the profile is easier to evaluate. Both views are considered in diagnoses of genera in which longitudinal spacing is significantly different from lateral spacing, as in many arthrostylids.

Another skeletal material, nonlaminated calcite, is present but limited in rhabdomesine zoaria (Fig. 267,*2,3*). Under the scanning electron microscope, nonlaminated material differs from laminated wall platelets only in crystal size and shape, the nonlaminated crystals being relatively enlarged and irregular. A well-defined and continuous nonlaminated layer is developed along zooecial boundaries in a few arthrostylids; however, the layer is discontinuous and weakly differentiated from the enclosing laminated skeleton in other families. Small, equidimensional, nonlaminated granules are common

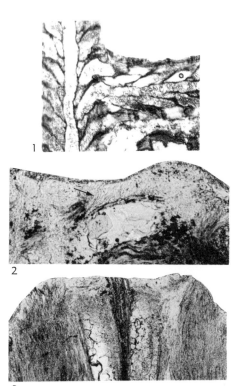

FIG. 267. Growth, wall materials in Rhabdomesina.——*1. Rhabdomeson noinskyi* SHISHOVA, U. Perm. (Kazan.), Nemda River basin, USSR; hollow part of branch continuous with autozooecium, suggesting relatively simple derivation of the axial cylinder; long. sec., ×15.——*2. Cuneatopora bellula* (BILLINGS), Jupiter F., L. Sil., Anticosti Is., Can.; articulation surface of distal branch tip (compare with *3*); wall material generally nonlaminated but some weak growth lines are present (arrow); USNM 249327, ×154.——*3. Arthroclema angulare* ULRICH, Fort Atkinson Ls., U. Ord., Ill.; walls mostly constructed of laminated calcite, distal tip with articulation surface of nonlaminated skeletal material; long. sec., USNM 249326, ×77.

along zooecial boundaries of many rhabdomesines.

Nonlaminated skeletal material also is present in the axial region of stylets and, in some arthrostylids and early hyphasmoporids, as rods along the branch axis (see Fig. 276,*2c*).

Under the light microscope, thin irregular bands termed dark zones (KARKLINS, this revision) are seen in the exozone of some forms. They are longitudinally oriented planar structures, typically clustered. Dark zones arise at or near zooecial boundaries at the base of the exozone, then radiate in the exozone. Distinct skeletal layers are usually lacking although granular and discontinuous nonlaminated intervals are present in some genera. The exozonal wall in *Ulrichostylus* shows changes in lamellar orientation at positions of dark zones, but no disruption of the laminated wall when studied with the scanning electron microscope.

STRUCTURAL CATEGORIES

BOARDMAN (this revision) recognizes three basic structural elements in stenolaemates: zooids, multizooidal structures, and extrazooidal structures.

Zooids are minimally defined as body walls enclosing coelomic space (BOARDMAN & CHEETHAM, 1973), the definition covering ancestrulae, feeding zooids, and polymorphs. Polymorphs are relatively uncommon in the Rhabdomesina; most zooids probably were typical feeding autozooecia. BOARDMAN's (this revision) description of vertical zooidal walls is applicable to the Rhabdomesina. Zooecial boundaries in the suborder, especially in the endozone, may be marked by dark bands, granular or nonlaminated skeletal zones, or laminae apparently extending uninterrupted between zooecial chambers.

Multizooidal structures are grown by the colonies, and eventually become parts of zooecia. These parts include walls from which zooids bud as well as budding zones (BOARDMAN, this revision). Encrusting basal walls, either at the level of the ancestrula or beneath overgrowths, are examples.

Extrazooidal structures grown by colonies are never included within zooecial boundaries (BOARDMAN, this revision). Recognition of extrazooidal material is contingent upon recognition of zooecial boundaries, but a clearly defined boundary is generally lacking in rhabdomesines. The boundary is considered to lie close to the zooecial chamber, and most of the exozonal wall is considered to be extrazooidal.

AUTOZOOECIA

Polymorphism is limited in the Rhabdomesina. Axial and monticular polymorphs are developed in some taxa; all other larger tubes are considered to be autozooecia. Autozooecia bud from a linear or cylindrical locus and are consistently oriented within a genus. Pores through walls linking zooecia are unknown.

Autozooecial shapes are varied in the suborder (Fig. 268) but relatively constant within genera and species, and therefore taxonomically useful. Most zooecia are angular or sigmoidal. In the majority of taxa, the initial axis of the zooecium is oriented approximately normal to the axis of the stem (Fig. 268,*2,3,5*). After a relatively short interval, the axis is deflected in the distal direction (Fig. 268,*1,4*). Most of the endozonal length of the zooecium is in this second interval. In a relatively few species, after deflection, the zooecial axis parallels the budding locus rather than diverging from it, and the zooecium is recumbent in the endozone (Fig. 268,*5*). In some taxa, the zooecial axis is straight in the endozone and the base of the zooecium is attenuated (Fig. 268,*4*) or flattened (Fig. 268,*6,7*) in longitudinal section. At the base of the exozone, the zooecial axis is generally deflected more (Fig. 268,*2*) or less (Fig. 268,*3,4*) abruptly toward the branch surface, a change reflected in both the proximal and distal zooecial outline (Fig. 268,*6*) or in only the proximal outline (Fig. 268,*7*). In a few species, the axis does not change orientation at the exozonal boundary (Fig. 268,*1*).

Position of the zooecial boundary in the endozone of rhabdomesines lies near the midline of the wall and is generally a distinct, irregular, commonly granular zone. The position in the exozone typically is obscure, especially in later genera, but is readily seen

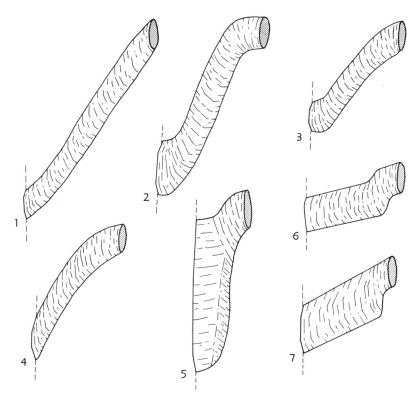

Fɪɢ. 268. Diagrams of zooecial shape in some rhabdomesines. Vertical dashed lines represent position of budding axes, stippled ellipses are apertures, zooecial shape changes from polygonal to rounded at the endozonal-exozonal boundary.——*1.* Sublinear, weakly inflated zooecial base in *Ulrichostylus.*——*2.* Weakly inflated zooecial base and abrupt zooecial bend in *Streblotrypella.*——*3.* Inflated zooecial base and rounded zooecial bend in *Rhabdomeson.*——*4.* Attenuated zooecial base and rounded zooecial bend in *Rhabdomeson.*——*5.* Inflated zooecial base, recumbent endozone, and abrupt zooecial bend in *Nematopora.*——*6.* Linear endozone in *Cuneatopora,* with abrupt deflection at zooecial bend.——*7.* Linear endozone in *Acanthoclema,* with distal wall not deflected at zooecial bend.

near to the autozooecial opening in such genera as *Nematopora* (see Fig. 276,*2*) and *Arthrostylus* (see Fig. 271).

In the exozone of later rhabdomesines, the position of the zooecial boundary appears to correspond to that in *Nematopora* and *Arthrostylus*. Indications of the zooecial boundary are seen in walls of endozones and the proximal part of exozones of such genera as *Orthopora, Tropidopora* (see Fig. 280,*1e*), *Rhabdomeson* (see Fig. 282,*1d*), *Arthroclema* (see Fig. 272,*2e*), *Osburnostylus* (see Fig. 277,*1d*), *Hyphasmopora* (see Fig. 293,*1c,d*), and *Streblotrypa* (see Fig. 293,*2d*). The apparent position of the boundary close to the zooecial chamber is also seen in tangential

views of such genera as *Ascopora* (see Fig. 283,*1c*), *Rhombopora* (see Fig. 286,*3e*), and *Arthroclema* (see Fig. 272,*2g*).

The wall beyond the zooecial boundaries is considered to be extrazooidal. Although apparently limited in a few genera (e.g., *Helopora, Cuneatopora*), extrazooidal wall generally is extensive, forming most of the thick exozonal wall.

Position of the zooecial boundary in rhabdomesines corresponds with that in such ptilodictyines as *Athrophragma* (KARKLINS, 1969, p. 25), close to the zooecial chamber. In *Athrophragma,* at the base of the exozone, the wall between zooecial boundaries is filled by a cystose and hence extrazooidal material.

This cystose interval is followed by the development of laminated wall. In rhabdomesines, the extrazooidal wall is laminated throughout.

Different interpretations can be made of zooecial-boundary position. Where obscure, BOARDMAN (this revision; BOARDMAN & CHEETHAM, 1973) suggested placing the boundary at the first break in lamellar curvature from the zooecial chamber. In rhabdomesines this position typically is dictated by the location of stylets. Following this hypothesis, many zooecial boundaries would be polygonal in outline and extrazooidal wall either absent or localized. BOARDMAN's suggestion is not followed here because of the lack of independent skeletal evidence in the exozone of later rhabdomesines and because the typical irregular placement of stylets would provide an apparently effective tissue support system but an irregular zooecial boundary.

Polypide size must have been quite varied if size was correlated with chamber diameter. DUDLEY (1970) and RYLAND (1970) both suggested that food resources are not the same in modern bryozoan species of different tentacular crown sizes. It seems likely that different rhabdomesine species also exploited different food resources.

AXIAL ZOOECIA

In some genera, **axial zooecia** are elongate polymorphs that may form a distinct **axial bundle** (see Fig. 283, *1a,b*). Four morphologic changes may contribute to the development of an axial bundle: (1) axial zooecia become more slender than neighboring autozooecia; (2) axial zooecia become thinner walled than neighboring autozooecia; (3) axial zooecia infrequently diverge from the axial region; and (4) outer surfaces of marginal axial zooecia become curved to produce the cylindrical axial surface.

That axial zooecia are derived phylogenetically from autozooecia is indicated by the existence of many intermediate morphologies. Axes may be well defined and linear;

zooecia may parallel a poorly defined axis for varying distances; two or three zooecia may be present along the axis, regularly turning toward the surface to be replaced by newly budded individuals; or there may be a well-defined central bundle (e.g., *Ogbinopora*).

As the axial bundle develops, dimorphism of zooecial length appears because shorter, more typical zooecia continuously bud from the outer surface of the bundle. Intrazooecial structures are rare in axial zooecia, although diaphragms may be present. Axial zooecia provide a means of thickening the stem and increasing the area of budding locus, thus increasing the maximum possible number of autozooecia around the branch without requiring major changes in autozooecial shape or orientation. Axial zooecia become more clearly differentiated and the axial bundle better defined during the history of the Rhabdomesina.

Whether or not axial zooecia contained polypides, and if so, their possible functions, are unknown. However, those axial zooecia reaching the lateral surfaces of branches show typical autozooecial morphology.

AXIAL CYLINDERS

The axial cylinder is a hollow tubular polymorph in the axial region of one genus, *Rhabdomeson*. The cylinder is usually wider than neighboring autozooecia and has typical stenolaemate compound walls; diaphragms may be developed. Structural discontinuity has not been recognized between walls of cylinders and of autozooecia. Rather, lamellar planes can be traced from the cylinder into the endozonal walls of autozooecia, proving that the cylinder walls were part of the colony. The presence of diaphragms and compound walls demonstrates that the cylinder was not produced by simple encrustation of a foreign substrate, but was a part of the colony body cavity. True encrusting rhabdomesines, however, have been described (NEWTON, 1971).

As seen in a few ideally oriented specimens, the axial cylinder of a daughter branch was produced as a longitudinal extension of an autozooecium in the parent stem (Fig. 267, *1*). GORYUNOVA and MOROZOVA (1979) have described the growth relationship.

In some specimens, the axial structure is conical rather than cylindrical. These conical specimens previously were assigned to *Coeloconus,* but that genus was synonymized with *Rhabdomeson* after the discovery of conical stems extending as branches from typical cylindrical *Rhabdomeson* zoaria (BLAKE, 1976).

Rhabdomeson branches that were broken and survived to resume growth, with new zooecia directed in the former proximal direction, have been recognized from Devonian (Fig. 266, *5*), Mississippian, and Permian rocks (BLAKE, 1976). It seems likely that cylinder development was linked to fragmentation as a mode of colony increase.

SKELETAL DIAPHRAGMS

Skeletal diaphragms in rhabdomesines are generally thin and calcified on their outer surfaces. They most commonly developed well below zoarial surfaces and in taxa with elongate zooecia. It has not been established if differences in diaphragm frequency were ecologically controlled, for example by recurring unfavorable conditions that induced frequent polypide degeneration.

Thickened terminal diaphragms, commonly clustered, have been observed in some species. Because rhabdomesine zooecia generally reached a maximum length in a given species, the presence of such diaphragms seems to indicate a termination of polypide activity. SILÉN and HARMELIN (1974) have described analogous inactive areas in the central portions of circular encrusting colonies of modern tubuliporates.

HEMISEPTA

Hemisepta in rhabdomesines are centered either on the proximal or distal wall, but not on lateral walls (Fig. 269). They occur singly, in offset pairs, or in multiple proximal wall series. A single hemiseptum on the proximal wall at the zooecial bend is common. In cross section, hemisepta range from low and rounded to thin and extended. Typically, hemisepta on proximal walls are thicker than those on distal walls. Hemisepta were secreted from both sides and apparently originated through a simple fold in epidermal tissues.

Functionally, hemisepta seem related to polypide position because multiple hemisepta occur in taxa with elongate zooecia, suggesting that addition of new hemisepta was associated with the degeneration-regeneration cycle. In *Orthopora,* the zooecial chamber is commonly curved about the hemisepta (see Fig. 285, *2e*), suggesting a polypide position either lateral to or behind these structures. In such positions, the hemisepta may have served to protect the polypide, to provide muscle and ligament attachment points, or to guide the polypide during protrusion. Hemisepta probably were not zooecial floors because: (1) the partitions are incomplete and would not serve to isolate the new polypide; (2) in the case of paired hemisepta, the distal-wall member of the pair would lie behind the zooecial floor and would seem to be without function; and (3) in such taxa as *Rhabdomeson,* the space distal to the hemiseptum appears to be insufficient for a functioning polypide.

STYLETS

The term stylet is applied here to any rodlike skeletal structure oriented approximately perpendicular to the zoarial surface and parallel to the zooecium. Stylets formed more or less prominent spines, or low, hemispherical knobs on zoarial surfaces. Structurally, they possess an axial component, the **core**, and a concentric bundle of **sheath laminae** that enclose the core. The core may be constructed of nonlaminated material or of laminae oriented subparallel to the zoarial surface but arched toward the surface, or a combination of both. The sheath laminae are simply zoarial laminae deflected toward the zoarial sur-

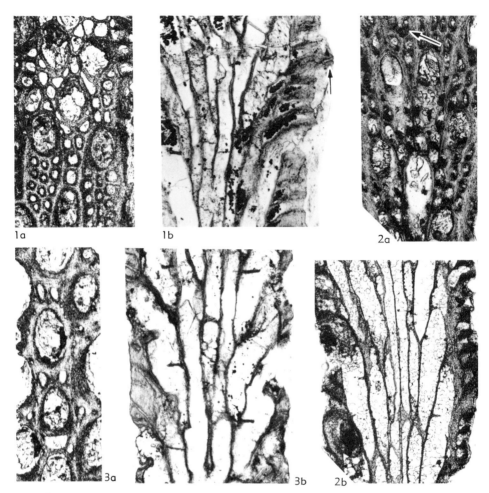

Fig. 269. Population variation in Hyphasmoporidae.——*1–3. Streblotrypa nicklisi* Vine, U. Miss. (Chester.), Ill.; USNM 249315–249317, all ×75; *1a,* metapores in fields, stylets present, tang. peel; *1b,* typical interior, stylets (arrow) present, long. sec.; *2a,* metapores in fields and one questionable stylet (arrow), tang. sec.; *2b,* typical interior, stylets absent, long. sec.; *3a,* stylets present, metapores not in distinct field, tang. peel; *3b,* typical zooecia but shorter than those in *1* and *2,* long. sec.

face about the core. A growth discontinuity is lacking between sheath and zoarial laminae, but a discontinuity is present between the core and the sheath laminae, except for small diaphragmlike lamellae that extend across cores in some forms.

Most, and probably all, stylets were solid during colony life, containing no soft tissues. In rhabdomesines, most stylets are restricted to exozones. Although some stylets arise very close to zooecia, axial structures appear to be isolated from the zooecial chambers by at least a few laminae.

Paurostyles (Fig. 219,*4;* 270,*1*) are the simplest type of stylet. The paurostyle core is an irregular cylinder of nonlaminated material, usually crossed by rare laminae and commonly offset along its length. The sheath lamellar bundle is narrow, and typical lamellae are weakly deflected in the distal direction. Paurostyles are approximately 0.02 to 0.04 mm in diameter. Many of the micracanthopores of earlier workers are pauro-styles; however, morphologically diverse structures were included under the older term.

Heterostyles (Fig. 219,5; 270,2) differ from paurostyles in having a core of distinct lenses of nonlaminated material separated by bands of sheath laminae that arch across the axis. The sheath is narrow and deflection of laminae is weak to strong. Heterostyles are slender (approximately 0.02 to 0.04 mm) and of nearly constant diameter, but irregular outline.

Stylets approximating traditional acantho-pore morphology are termed acanthostyles (Fig. 219,9; 270,1). In these structures, the axial core is a continuous, clearly defined cyl-inder of nonlaminated material, usually uninterrupted, but in places crossed by thin lamellar bands. The sheath laminae are well developed, forming a broad bundle usually strongly deflected away from the zoarial axis (e.g., *Acanthoclema,* Fig. 218,4). Therefore acanthostyles form prominent structures on zoarial surfaces. In a few taxa, the enclosing laminae are only weakly deflected (e.g., *Pamirella,* see Fig. 288,1c) and apparently form only low surficial structures. Most acan-thostyles range in diameter from approxi-mately 0.02 to 0.12 mm, some tapering along their length.

Morphology is transitional between pau-rostyles and acanthostyles, but the axial core is more clearly defined in acanthostyles and the sheath lamellar bundle is wider and more strongly deflected. Paurostyles also are gen-erally smaller.

Aktinotostyles (Fig. 219,7; 270,3) are a type of stylet with a core formed by a broad band of distally arched laminae. Laterally, these laminae are deflected into cones point-ing away from the aktinotostyle axis. Irreg-ular scattered granules of clear material are common along the axis, and in some a dis-tinct cylinder resembles that of acanthostyles. Typically, the sheath lamellar bundle is rel-atively narrow. Surrounding laminae may be either strongly or weakly deflected toward the zoarial surface. The deflected interval is

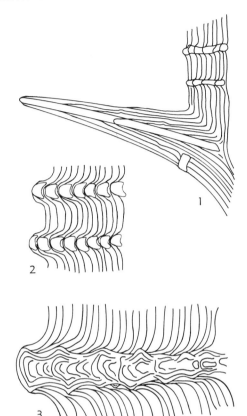

Fig. 270. Diagrams of stylets.——*1.* Two pau-rostyles (above) showing irregular cores and weakly deflected sheath laminae; acanthostyle (below) showing cylindrical core, conically deflected sheath laminae, and one lamellar surface crossing the core; a mural spine is present in the sheath laminae below the acanthostyle.——*2.* Two heterostyles showing discontinuous core of lenticular nonlaminated material, and sheath laminae.——*3.* Aktinoto-style, showing laminated, spinose core containing several nonlaminated fragments near base of the structure; the sheath lamellar bundle is narrow and individual lamellae are strongly deflected. All approximately same scale.

small, and does not prominently affect the surface. Branching aktinotostyles have been observed in few specimens. Aktinotostyle diameter ranges from approximately 0.02 to 0.13 mm, commonly with distal increase.

In a few zoaria, stylets have been observed to change longitudinally from aktinotostyles to acanthostyles. The nonlaminated cylinder of the acanthostyle extends from the lamellar core of the aktinotostyle, and the sheath laminae abruptly change, becoming much more strongly deflected in the outward direction. Although the structural change is abrupt, there is no indication of fracture or other disruption of growth.

The stellatopores of ROMANCHUK (1966) superficially resemble aktinotostyles; however, they were described as hollow structures containing diaphragms, and skeletal walls around the stellatopore contain a ring of capillaries. This description does not fit aktinotostyles and I consider the two structures to be distinct.

Mural spines are styletlike structures that generally grow into zooecial chambers. The core of a mural spine is nonlaminated and clearly differentiated from enclosing laminae, which are weakly deflected in the direction of growth. Unlike true stylets, mural spines are very short, arising within the exozonal wall during late ontogeny. BOARDMAN and CHEETHAM (1969) suggested that mural spines may have functioned as polypide attachment structures.

Usage of the term "capillary" has been reviewed by KARKLINS (this revision). The term has been employed primarily by Soviet authors for a variety of smaller skeletal structures, most of which I here refer to as paurostyles, mural spines, deflections in the sheath laminae of acanthostyles, or deflections in the core laminae of aktinotostyles.

Various Soviet workers have suggested that capillaries may represent a form of open communication system. I have seen no evidence either under the light microscope or with the scanning electron microscope for open passages within the exozonal wall beyond the zooecia or metapores. The small size of most capillarylike structures in rhabdomesines, combined with their irregular outline and common lack of well-defined continuous cores, seem greatly to limit their function as communication structures. In contrast, the open links among neighboring autozooecia and interzooecial areas across the hypostegal coelom would appear to provide effective communication.

Interpretation of stylet morphology, much of it derived from trepostomate species, has differed among various authors. Because stylets are very similar between orders, arguments based on members of one group may be pertinent to others. Important is whether the core of stylets, especially acanthostyles, was solid or open during colony life. If solid, stylets could have functioned only for strengthening support. If hollow, some communication function may have been performed, or a zooidal polymorph may have been present. ASTROVA (1971) found diaphragms and sedimentary particles in larger trepostomate acanthostyles and interpreted these as having been originally open. The distinct, well-defined core may also indicate an open axis. BLAKE (1973b) illustrated scanning electron micrographs showing core material sharply defined from enclosing laminae. BLAKE and TOWE (1971) dissolved the zoarial laminae from a specimen of *Idioclema,* showing that the core material could be removed intact from the skeleton, and therefore was cohesive and not interrupted by laminae or foreign material. ARMSTRONG (1970) believed that calcite in the stylet core of two species of the trepostomate *Stenopora* was secreted by specialized zooidal epithelium, and BROOD (1970) preferred an originally solid core in his interpretation of two species of *Orthopora.* BLAKE (1973c) argued in favor of an originally solid stylet, largely because of the relationships between the sheath laminae and the core. The laminae abut and were intergrown in irregular patterns with the core, without indication of a lining in an open cavity. Laminae rarely enclose organic materials and are directed distally as they cross the core. These layers follow orientation of the sheath laminae and were apparently deposited on a firm (core) substrate.

It has been argued that certain acanthostyles opened into autozooecial chambers. Although sheath laminae are commonly thin near the base of stylets, in most examples the bases are clearly enclosed by laminae. A few are equivocal because of the plane of section. There is no clear evidence that the core of rhabdomesine acanthostyles opened into autozooecia.

Stylet function.—Interpretations of stylet function must be consistent with a number of observations of rhabdomesines. (1) Stylets are present in a majority, but not all rhabdomesine species or all members of some populations. (2) More than one stylet type may be present in a species. (3) Although stylets may be no more numerous than autozooecia, they are usually abundant, and more or less fill the exozonal wall between apertures. (4) Although in many specimens enlarged stylets are immediately proximal to the zooecia, relatively few of these are associated with only a single zooecium. (5) Evidence of other associated structures (e.g., a spine) is lacking. (6) Although usually arising at the base of the exozone, many stylets developed as the exozone enlarged and some originated during endozonal development. (7) All stylets were apparently solid structures during colony life, with no opening into zooecial chambers. Lack of internal coelomic space and physical isolation seem to preclude stylet association with such vital functions as respiration and reproduction.

It has long been argued that prominent acanthostyles would provide protection (e.g., CUMINGS & GALLOWAY, 1915). The inclined orientation of the acanthostyles, and their grooved, hoodlike appearance in a few species (e.g., *Helopora inexpectata* McNAIR) suggest that acanthostyles may also have functioned as a guide during lophophore protrusion. Recognizing that the bryozoan colonies were covered with living tissue during life, TAVENER-SMITH (1975) suggested that low stylets may have provided support for surficial tissue, a function seemingly particularly appropriate for the spinose aktinotostyles.

Because paurostyles seem too small to provide effective protection and acanthostyles more prominent than necessary for simple support of soft tissue, stylets may be an example of bryozoan structures that origi-

nated for one function (tissue support) and became adapted to a new one (protection).

SKELETAL RIDGES

Skeletal ridges are elongate skeletal folds developed on rhabdomesine branches (see Fig. 266, 7; also see 272, 1). Such ridges are common in the suborder, and are developed in several patterns. In some genera, especially earlier ones, straight to somewhat sinuous longitudinal ridges separate rows of zooecia. Short ridges may separate successive apertures or border the proximal margins of apertures. The ridges then flare distally and join to form longitudinal ridges (see Fig. 275, 1). Peristomes, ridges surrounding apertures, are common in older genera. In later genera, ridges are less clearly defined or absent, although the crest of the wall separating neighboring zooecia has at times been referred to as a ridge.

Skeletal ridges and derivation of stylets.—Skeletal ridges are best developed in early Rhabdomesina, in which stylets are weakly developed. Ridges probably supported soft tissue and provided strength for slender-stemmed colonies. Later in the history of the suborder, stylets apparently assumed tissue support functions, whereas thicker branches provided structural support.

The various stylet types appear to be of common origin because they intergrade morphologically and are constructed of similar materials. When they first appear in the fossil record, stylets are found along skeletal ridges, and ridge morphology provides a suitable stylet precursor. Nonlaminated calcite, typical of stylet cores, occurs in thin, locally discontinuous bands along ridge midzones. Laminated layers on either side of wall midzones are directed toward the zoarial surface and abut the midzone, suggesting the enclosing laminae of a stylet. Stylets appear to have originated by development of continuous linear cores and of sheath laminae enclosing the cores. Weakly differentiated, styletlike structures are present along the wall midzone in *Moyerella,* which provides a suitable morphological precursor to the relatively small paurostyles found in many arthrostylids.

METAPORES

Several types of small open cavities have been recognized in Paleozoic bryozoans (Fig. 269). The term "exilapore" of Dunaeva and Morozova (1967) was applied to short, hollow tubes lacking diaphragms, which arise in the exozone of some trepostomates. Utgaard (1973) and Boardman and McKinney (1976) changed the term to "exilazooecia." At least some exilazooecia were budded as a part of the normal autozooecial pattern.

The term **metapore** was applied by Shishova (1965) to tubular cavities arising in the basal exozone in genera of the Hyphasmoporidae (herein the Nikiforovellidae and Hyphasmoporidae). These cavities were described as either filling spaces between apertures or, more rarely, surrounding zooecial apertures. In *Acanthoclema*, only a simple metapore is developed for each autozooecium, and in such genera as *Streblotrypella* and *Nikiforovella*, metapores are scattered in the exozone. Where closely spaced, metapores are angular in transverse outline, but they are circular where widely spaced. In some specimens, metapores appear in the late endozone (Morozova, 1970, pl. 28). Although initially used only for hyphasmoporid genera, the term "metapore" is here also applied to similar structures (i.e., slender, tubular cavities arising near the base of the exozone) in *Trematella* (Rhabdomesidae) and *Rhombopora* and its allies (Rhomboporidae).

Autozooecial budding in the Rhabdomesina generally takes place at or near the axis of the branch, whereas metapores arise at the base of the exozone. Metapores are not a part of the autozooecial budding pattern of the suborder and, as open tubular structures, they are not homologous with autozooecia. Therefore, there is no reason to hypothesize the presence of polypides, a conclusion supported by the large number and small volume of metapores in some genera.

The most reasonable function for metapores seems to be as space-filling structures separating zooecia. Spacing of polypides would provide neighboring individuals with enough room for effective feeding, and skeletal material would be conserved without seriously weakening the zoarium.

The metapore-bearing *Trematella* and *Rhombopora* are not assigned to the Hyphasmoporidae or the Nikiforovellidae because the sum of their characters suggest affinities with other families.

Some metapores (e.g., in *Rhombopora* and most species of *Nikiforovella*) are superficially indistinguishable from trepostomate exilazooecia that arose in the exozone. The term "metapore" is retained in the rhabdomesines because there is no evidence that these structures arose as polymorphs, as they apparently did in trepostomates.

Mesozooecia are typical of older trepostomates. They typically arose in the outer endozone or exozone, and they contain closely spaced diaphragms. Similar structures are present in a few arthrostylids (e.g., *Helopora*); however, the term metapore is retained in the arthrostylids because no clear information is available as to their possible common origin.

OVICELLS

Brood (1970) interpreted large swellings on the stems of some specimens of *Orthopora* (=*Saffordotaxis* of Brood) *ludlowensis* (Brood) to be gonozooids. These bulbous structures are internally open and are covered by stylet-bearing, laminated, skeletal wall. They are linked to the interior of the zoarium by a zooecium. Possible gonozooids are present in other rhabdomesines, but these structures are subject to varied interpretations (e.g., pathological reactions or overgrowths on foreign objects). In most rhabdomesine species, it has not been possible to recognize brood chambers. Goryunova (1975) suggested that this implies direct development, without a brooding phase, in these bryozoans.

MONTICULES

Monticules, broadly defined as areas of modified zooecia, are relatively uncommon in the Rhabdomesina. An example in *Nicklesopora* is a specimen (Fig. 284,*1a*) with a much enlarged polymorph surrounded by a group of enlarged, angular, thin-walled polymorphs. In *Nemataxis* (see Fig. 284,*2f*), monticulelike annular bands are developed in which the zooecia are closed by terminal diaphragms. In *Rhombopora simplex* (ULRICH), described in *Bactropora* by ULRICH (1890), raised, semiannular bands were developed by elongation of exozonal intervals of zooecia, combined with minor deflections in zooecial orientations but no other apparent changes in zooecial shape.

Attenuated and, in some specimens, branchlike monticules consisting of elongate but otherwise typical autozooecia (Fig. 216, 6) have been described in conical branches of *Rhabdomeson* (BLAKE, 1976). The pointed tips of these monticules are apparently formed of fused stylets, and the monticules are regularly spaced in alternating rows along convex surfaces of the curved branches.

BANTA, MCKINNEY, and ZIMMER (1974) hypothesized a chimneylike exhaust function for monticules in some Ordovician bryozoans, a function appropriate in the rhabdomesines. Other functions, or combinations of functions, are possible. For example, the prominent monticules of *Rhabdomeson* may have been protective or may have served as initiation points for branches (BLAKE, 1976). The much enlarged polymorph in *Nicklesopora* may have been reproductive or protective, as in some polymorphs in modern bryozoans, and as suggested by various authors for fossil polymorphs. ANSTEY, PACHUT, and PREZBINDOWSKI (1976) discussed monticules as budding centers and polar points maintaining morphogenetic gradients within sub-colonies.

The morphologically diverse and taxonomically isolated occurrences of monticulelike structures imply phylogenetically independent origins.

ARTICULATED ZOARIA

Articulated zoaria, in which discrete segments were linked by noncalcified material, are typical of the Arthrostylidae. Most colonies were disarticulated prior to burial, but original relationships can be seen in a few specimens (see Fig. 272,*2c*). Jointing developed both along unbranched stems and at dichotomies.

In most arthrostylids, the articulation surface on the proximal end of each zoarial segment is convex, and in most specimens there is a series of low concentric or radial ridges on the apex. The surface on the distal end of the segment is similar but usually flattened or slightly concave. Therefore, the surfaces usually were subparallel and mobility of the joint would have been limited. Articulation surfaces were solid and communication through the skeleton between segments was impossible. Many ends of segments are enlarged, so that branch diameter at the joint was greater than that along the stem. Skeletal material forming the joint surfaces is weakly laminated or nonlaminated in spite of its occurrence in the exozone, where lamination is typical. Although generally nonlaminated, faint growth lines approximately perpendicular to the articulation surface have been detected in a few specimens. These growth lines are subparallel to the zoarial surface and parallel to the overall exozonal growth surface (Fig. 267,*2,3*). This implies that the uncalcified joint material of arthrostylids was secreted in layers parallel to the growing zoarial surface, in the same orientation as that of more typical calcified walls.

Two joint patterns developed in a single Ordovician species of *Ulrichostylus*. Typical, low, conical joint surfaces are associated with a ball-and-socket pattern (Fig. 266, 7). The sockets are on the sides of branches, but have not been observed on ends of branches. A single zooecium from the parent stalk opened to form the hollow interior of the socket joint (Fig. 266,8); hence, the opening is a polymorph of a zooecium. The interior of the cup is lined with weakly laminated skeletal material, and radial ridges are developed on the wall surface. The remainder of the structure appears to be built of autozooecia. The ball joint forms the distal portion of the complete joint structure, and therefore occurs on the proximal end of zoarial segments. The ball consists of a steeply conical surface terminated by a commonly perforated spherical tip (Fig. 266,6). The surface bears linear ridges. The ball-and-socket joint probably was more flexible than the typical joint.

Articulated joints comparable to the arthrostylid patterns are absent from most members of later families, but may be present in Devonian *Bactropora* (see Fig. 290,*1b*) and Mississippian *Rhombopora*.

Borg (1926a) described jointing in the modern stenolaemate *Crisia*. In this genus, organic matrix of the joint is continuous with that of the calcified wall, and joint mobility is limited. Probably, rhabdomesines were similarly constructed.

FRAGMENTATION AND ZOARIAL DIMORPHISM

Some rhabdomesines were capable of increase by means of colony fragmentation. Blake (1976) described *Rhabdomeson* branches of Permian age that had apparently broken free of parent colonies and survived to resume growth elsewhere (Fig. 266,*2,4*). This is shown by healed breakage scars, some extending across individual zooecia, at the bases of many branches (Fig. 266,*1,3*). These specimens show different stages of repair, ranging from fresh breaks to near closure of the zooecial tubes and formation of an exozonal surface. The appearance of a regrown base is very different from that of a surface grown entirely against a foreign substrate. In some specimens, after breakage, a new branch developed in alignment with the axis of the old, but with the growth direction reversed (Fig. 266,2). The hollow axial cylinder, typical of *Rhabdomeson,* appears to be characteristic of increase by fragmentation in rhabdomesines because a Devonian *Rhabdomeson* (Fig. 266,5) also was broken and budding direction reversed.

Colony fragments that were subject to breakage are typically conical with the central axial cavity expanding distally as the remainder of the zoarium, the endozone and exozone, remain constant in dimensions. Such zoarial parts (Fig. 266,*3,4*) were previously assigned to a distinct genus, *Coeloconus*.

Fragmentation may have provided an effective mode of increase in higher energy environments. For example, reproduction could have been accomplished without larval loss to predation in densely populated communities dominated by suspension feeders.

The expanded, conical shape of the branch may have helped trigger fragmentation under higher energy conditions. Commonly, bases of the cones are slightly constricted immediately distal to branching points. This constriction may have provided a mechanism for local weakening of the branch in order to induce fragmentation.

Modes to increase fragmentation have been reported in modern bryozoans. Boardman and Cheetham (1973, p. 173) pointed out that in some cheilostomates, especially such free-living genera as *Cupuladria*, "fragmentation may be so common as to provide an important means of colony reproduction." Some modern fragmentation appears to be environmentally controlled. In the cheilostomate bryozoan *Discoporella umbellata* (Defrance), Marcus and Marcus (1962, p. 301) reported colonies produced by fragmentation only in depths from 3 to 4 meters, and ancestrular colonies only from depths over 70 meters. These authors suggested, "Perhaps the settlement of larvae is difficult in irregularly agitated shallow waters. However by budding the species succeeds to populate this biotope." Following fragmentation in *Discoporella umbellata,* budding begins to occur around the proximal margins of the preexisting colony, with the polarity of the new zooecia reversed from that of the old. This reversed direction of growth is not near the prefragmentation budding zone but in a

previously inactive area, and the new zooecia are somewhat irregular in development. Both of these patterns are present in fragmented *Rhabdomeson.*

BIOLOGY AND DISTRIBUTION OF RHABDOMESINA

Position of epidermal tissue layers.— Growth models of BORG (1926b, 1933) have been profoundly important in reconstructing various Paleozoic bryozoans (see BOARDMAN, KARKLINS, UTGAARD, this revision), including the Rhabdomesina (BROOD, 1970; BLAKE, 1973a). Critical to these interpretations is the observation that skeletal laminae in most Paleozoic bryozoans are so oriented as to necessitate secretion over skeletal surfaces, rather than from the interior of the skeleton, as in brachiopods. In these colonies, the skeleton was apparently covered by two tissue layers that were in turn separated by coelomic space. The tissue layers not only overlay the outer skeletal surface, but lined the zooecia at least to the depth of any basal diaphragm. Because skeletal walls were secreted from both sides, they have been termed compound walls, and comparable growth patterns were described by BORG (1926b) in modern tubuliporates. A second wall type, in which skeletal walls were secreted exclusively from the interior, much as in brachiopods, apparently is present only in basal walls in rhabdomesines. Such walls most commonly form basal attachment surfaces, but may also have developed where rhabdomesines encrusted foreign objects above the base of the colony. This second growth pattern, yielding so-called simple walls, is common in post-Paleozoic tubuliporates.

Apparent epidermal tissues have been discovered in the type suite of *Rhombopora simplex* (ULRICH). These brown traces form a continuous layer over the zoarial surface, lying close to the tips of the stylets in most areas, but somewhat above the tips in others. The organic layer is quite similar to that illustrated by BOARDMAN and CHEETHAM (1973, fig. 36B) in a ramose trepostomate. The

rhabdomesine tissues are not preserved beneath an overgrowth or other protective structure, thereby suggesting crystallization of the enclosing calcite must have taken place very early.

Rhabdomesine polypides.—Brown, apparently organic structures that seemingly represent preserved remnants of polypides (as well as other soft tissues) have been described in diverse Paleozoic bryozoans (CUMINGS & GALLOWAY, 1915; later authors). Such well-preserved polypidelike remains are not known from rhabdomesines, although probable organic matter of unrecognizable shape is present in some zooecia of many zoaria. Fossils with recognizable structures in the zooecia are massive cystoporate and trepostomate bryozoans. In these groups, organic material was protected by overgrowths and diaphragms. Thus, lack of preserved remains in rhabdomesines may have resulted from the small, relatively exposed nature of the zoaria rather than from any basic original difference in polypide structure.

Reconstruction of rhabdomesine polypides must be based on comparisons with Paleozoic and modern materials, combined with information provided by the zooecial outline. Reconstructions have been presented by NEWTON (1971) and TAVENER-SMITH (1974).

Brown, spherical, possibly organic bodies occur in some zooecia. Distribution of these bodies and associated diaphragms suggest that rhabdomesines were subject to degeneration-regeneration cycles, and that polypide position advanced with ontogeny in at least those taxa in which brown bodies occur. Brown bodies have not been observed in metapores.

Astogeny.—Available information on

early astogenetic stages in the Rhabdomesina is limited. TAVENER-SMITH (1974) interpreted some small, conical structures as young rhabdomesine zoaria, although his reasons for this taxonomic assignment are not clear. TAVENER-SMITH's fossils are 1.3 to 3.0 mm in length and 0.7 to 0.9 mm in width. Surfaces of the conical zoarial bases are covered by a wrinkled material that proved to be microgranular when studied with the scanning electron microscope. Zooecia are present within the microgranular wall, where they originated near the pointed tip and grew approximately parallel to the axis of the cone. New individuals were added medially as growth proceeded. The conical, proximal, attachment area was very small. TAVENER-SMITH (1974) believed that these colonies gained added support by encrusting foreign objects as well as by enlargement of the base. In a few available thin sections of basal attachments, I observed no such conical structures as those described by TAVENER-SMITH.

Colony integration.—Following BOARDMAN and CHEETHAM (1973), colony integration in terms of interrelationships among autozooecia was quite high in the Rhabdomesina, but integration was low in terms of astogeny and polymorphism. Moreover, there was no significant change in level of integration during the known history of the group.

Ecology.—Rhabdomesines generally are associated with diverse marine faunas typically dominated by suspension-feeding epifaunal organisms (e.g., crinoids, brachiopods, bryozoans). Generally, the most abundant, largest, and best-preserved fragments occur in slabs of little-disturbed skeletal debris in fine-grained clastic rocks. Because of these associations, rhabdomesines seem to have preferred generally open, marine waters of normal salinity and perhaps of low turbulence. Influx of fine clastic material was common, but not enough sediment was introduced to preclude development of a diverse suspension-feeding fauna. The rhabdomesines are primarily present within patches of abundant shelly epifauna, probably in large part because of a need for firm attachment surfaces. Some rhabdomesines appear to have lived in nearshore, possibly open, lagoonal environments (NEWTON, 1971). BROOD (1975a) described rhabdomesines of Gotland from inferred shallow-water (20 to 50 m), soft-bottom sediments. Algae were believed to have been common.

Biogeography and biostratigraphy.—Lack of comprehensive studies have limited the use of rhabdomesines in biogeographic and biostratigraphic work, but many of the better known genera have been recognized from widely separated areas of the world. Long generic ranges appear to limit biostratigraphic usefulness; however, rhabdomesines have been extensively used in the Soviet Union.

TAXONOMIC CONCEPTS WITHIN THE RHABDOMESINA

Formal taxonomic recognition of the suborder Rhabdomesina (*nom. correct.* herein) has been relatively recent (ASTROVA & MOROZOVA, 1956), although affinities among component families were recognized earlier (e.g., McNAIR, 1937). Most family-level taxa, including five of the six employed here, were proposed during the late nineteenth century. For many years, however, only the names Arthrostylidae and Rhabdomesidae were generally used. The content of the Arthrostylidae here remains basically unchanged from the usage of BASSLER (1953) and other earlier workers. SHISHOVA (1965) redefined the Hyphasmoporidae of VINE (1886), assigning to it those rhabdomesine genera with abundant metapores. GORYUNOVA (1975) recognized the Nikiforovellidae for those hyphasmoporid genera lacking axial zooecia and possessing abundant stylets.

Even with removal of the Hyphasmoporidae and Nikiforovellidae, the Rhabdomesidae remained a morphologically diverse family. The Rhomboporidae of SIMPSON (1897) is recognized here for genera sharing strong similarities of axial development, zooecial shape and arrangement, and stylet development. The Bactroporidae of SIMPSON (1897) is employed for a single genus. The Rhabdomesidae is characterized by a trend toward development of axial zooecia, zooecial shape, and stylet development.

Generic recognition in the suborder has been hampered by a lack of adequate illustration, and as a result, characters have remained obscure. In general, genera have been based on a small number of often rather narrowly defined characters, leaving little room for either population variation or evolutionary convergence.

Species concepts have been largely based on statistics, including average size of apertural openings or aperture spacing along a stem, number of stylets or polymorphs per square millimeter, and so on. Relatively little effort has been made to establish the taxonomic value of such information by careful comparison within and between populations, both at inter- and intraspecific levels. Intraspecific variation can be significant (Fig. 266,3,4; 269). Some characters used in species definition are questionable. For example, because most rhabdomesines are slender and cylindrical, only a relatively few zooecial chambers can be oriented perpendicular to the plane of section; lateral rows will have apparent apertural diameters less than their actual diameter. Even sections slightly offset from perpendicular can have an apparent diameter beyond the accuracy in hundredths of a millimeter cited in many taxonomic descriptions. Stylets, especially aktinotostyles, arise in different parts of the exozone, becoming more abundant near the zoarial surface. Therefore, numbers measured per square millimeter will change with the position of the section in the exozone.

Growth changes can be readily overlooked because available material generally is fragmentary. Relatively complete silicified specimens from the Permian of West Texas (USNM collections) commonly have relatively thin exozonal walls near growing tips, whereas older intervals typically are thicker walled, with apertures commonly either closed or much reduced. Branching in some of the West Texas zoaria is irregular, with one stem passing near to another. In these specimens, facing apertures were nearly closed by skeletal material, presumably because the proximity of the neighboring branch precluded effective lophophore function. Apertures on opposite sides of the branches are of more typical, open outline. Thus, apparent wall development and apertural size may depend not only on position in a zoarium but even on side of the branch.

Morphologic characters stressed here in the recognition of family- and genus-level taxa in the Rhabdomesina include: (1) presence of jointed zoaria (family level), nature of jointing (genus level), and zoarial growth habit (primarily genus level); (2) nature of budding locus and development of axial region (family and genus levels); (3) shape, orientation, and regularity of arrangement of autozooecia (family and genus levels); (4) presence and development of hemisepta (family and genus levels); (5) width of exozone relative to branch radius (of limited value at both family and genus levels); (6) lamellar profile in the exozone (genus level); (7) presence and abundance of metapores (family and genus levels); and (8) development of stylets (family and genus levels). Evolutionary trends are evident within some of these characters; however, taxon boundaries are based on character presence and development rather than on hypotheses of character evolution.

Discussion of evolutionary trends and comments on diagnostic characters follows.

1. Zoarial form and branching patterns provide some useful taxonomic characters, especially in older members of the suborder and within the Arthrostylidae. Various authors have noted that in stenolaemates, zoarial form may vary within genera or species. HARMELIN (1973, 1975) described

significant, microenvironmentally controlled variation in colony form in several modern tubuliporate species. This variation included both encrusting and erect growth habits within single species.

Based on total characters known to me, all zoaria clearly of rhabdomesine affinities are erect, and most are dendroid. Encrusting holdfasts, both basal and in erect portions of zoaria, have been described, but these are apparently small relative to overall colony size. A number of nonramose genera possess characters suggesting both trepostomates and cryptostomates, but more research is needed to determine their affinities.

Articulated colonies may have been primitive in the Arthrostylidae and the growth pattern was largely restricted to this family. Because individual arthrostylid segments were apparently linked only by soft tissues, few colonies are preserved intact and zoarial habit is usually difficult to determine. Development of articulation facets is a guide to zoarial form, and most arthrostylids were jointed at fairly regular intervals. Clear evolutionary trends in jointing are not apparent.

Branches in most earlier genera appear to have approximately constant mature diameters, although enlarged ?basal intervals are known. Also in earlier taxa, zooecial apertures were generally arranged at the surface in regular rhombic or annular patterns. In many later genera, for example in younger rhomboporids, both branch diameter and apertural arrangement were quite irregular, even over short branch intervals. These changes took place within families; for example, diameter is constant and apertural arrangement regular in *Saffordotaxis* (see Fig. 289, *1*) and some older species of *Rhombopora,* whereas later species of *Rhombopora* (see Fig. 286, *3*) and *Megacanthopora* (see Fig. 287) are more irregular in habit. Regularity changed within genera as well; for example, zooecia are quite regular in *Helopora fragilis* Hall (see Fig. 274, *1*) from the Silurian but irregular in *Helopora inexpectata* McNair from the Devonian.

Variations in zoarial habit, noted by Har-

melin (1973, 1975), are correlated with both environmental and microenvironmental changes. In the Rhabdomesina, in general, greater flexibility in zoarial habit is associated with younger genera and species. If zoarial habit in the Rhabdomesina also reflects environmental control and breadth of habitat tolerance, then seemingly later genera possessed generally broader environmental tolerance.

2. The budding locus is linear and well defined in primitive taxa. Evolutionary trends away from a sharply defined axis took place in different, but not all rhabdomesine lineages. The process began with loss of axial regularity and development of weak alignment of basal portions of zooecia within the axial region. The process continued with the development of increasingly well-defined axial bundles of zooecia. These changes are best seen in the Rhabdomesidae and Hyphasmoporidae. In the Rhomboporidae, Nikiforovellidae, and Bactroporidae, axial zooecial development is weak or absent, but the budding locus may be irregular to somewhat planar. Although irregular in some taxa, a more or less well-defined linear, planar, or cylindrical budding locus is one of the unifying conservative features of the Rhabdomesina.

3. Clearly defined evolutionary trends in zooecial shape have not been recognized.

4. Hemisepta developed independently in different lineages, being generally present in the Hyphasmoporidae and Rhabdomesidae but absent in the Arthrostylidae, Rhomboporidae, and Nikiforovellidae.

5. Although exozonal walls thicken with ontogeny, zooecia in many taxa appear to have attained an approximate mature size, as in the cheilostomates. Therefore, mature branch diameter and relative thickness of the exozone is more or less constant within a genus, and exozonal wall thickness provides some indication of affinities. Considerable variation may be seen, however, among species, among zoaria, or among branches within zoaria. Relatively thicker exozonal walls are commonly associated with later Paleozoic genera.

6. Evolutionary trends in lamellar profile have not been recognized.

7. Metapores are present in at least one genus of all recognized families except the Bactroporidae, but they are most prevalent in the Hyphasmoporidae and the Nikiforovellidae. Metapores first appear in *Cuneatopora* (Arthrostylidae) and closely related genera. Once established, metapores were conservative in development and occurrence, for only in *Cuneatopora* have populations been found in which metapores are present in some zooaria, absent in others, although abundance may vary significantly (Fig. 269). In rhabdomesines, metapore presence, especially in large numbers, or in taxa with regular arrangement of zooecia, is therefore considered to be strongly indicative of hyphasmoporid or nikiforovellid affinities.

In the Rhomboporidae and in *Trematella* of the Rhabdomesidae, metapores are developed in relatively small numbers. These genera are of somewhat irregular growth mode and the metapores appear to have functioned largely as space-filling mechanisms.

8. Stylets of different types are believed to have had a common origin along the skeletal ridges as surficial tissue support structures. Simple stylets (paurostyles) are present in primitive arthrostylids, and during the history of the suborder stylets became more strongly differentiated and more clearly defined.

Position of the zooecial boundary has been inferred to be relatively constant within the suborder and generally not of taxonomic value within the group. In some relatively primitive arthrostylid genera, the position of the boundary is clear near to and paralleling the zooecial chamber. In many genera (e.g., *Streblotrypa, Orthopora, Nemataxis*) in later families, a similar position can be detected. Position of the zooecial boundary is interpreted to be constant and to lie close to the zooecial chamber in almost all members of the suborder, which implies that extrazooecial wall is virtually ubiquitous, the amount depending on wall thickness. In *Ulrichostylus* and *Helopora,* zooecial boundaries are atypical for the suborder. In *Ulrichostylus,* the zooecial boundary is clearly defined proximal to the apertures, but the boundaries appear to flare distally, joining longitudinal dark zones. Lamellar orientation in longitudinal section still implies a probable boundary position near the chamber and the development of thick extrazooidal walls. In *Helopora,* the zooecial boundary is near the middle of the wall, and extrazooidal material is limited but does seem to be present near some zooecial junctions.

Genera of uncertain affinities.—A number of genera: *Anisotrypa, Callocladia, Coeloclemis, Dyscritella, Hyalotoechus, Idioclema, Linotaxis, Nikiforopora, Stenocladia,* and *Syringoclemis,* possess characters typical of both trepostomates and cryptostomates. Budding patterns, zooecial shapes and arrangements, zoarial growth habits, and polymorph budding positions tend to resemble trepostomates whereas lamellar profiles and development of stylets and hemisepta resemble cryptostomates. Ordinal assignment must await the future review of the trepostomates and comparative assessment of all Paleozoic genera planned for this *Treatise* revision of Bryozoa.

SYSTEMATIC DESCRIPTIONS FOR THE SUBORDER RHABDOMESINA

By Daniel B. Blake

[University of Illinois, Urbana]

Suborder RHABDOMESINA
Astrova & Morozova, 1956

[*nom. correct.* herein, *pro* Rhabdomesoidea Astrova & Morozova, 1956, p. 664, suborder] [=Rhabdomesonata Shishova, 1968, p. 131, order]

Zoaria erect, generally dendroid, rarely pinnate, some unbranched. Branch jointing common in one family, otherwise rare. Branch diameters 0.1 to 6.0 mm; generally constant between bifurcations, some irregular; in jointed taxa, expanding distally along segments. Most branches subcircular in outline, few polygonal. Apertures generally in rhombic pattern, or in longitudinal rows; rarely confined to one side of branch. Longitudinal and peristomial ridges well developed to absent. Metapores present or absent; where present, ranging from few to densely spaced in exozonal walls between autozooecia. Metapores generally arising at bases of exozones, cross sections rounded where widely spaced, angular where closely spaced, diaphragms may be present. Axial regions containing linear axes, planar walls, axial cylinders, or axial bundles of zooecia. Planar walls, where developed, restricted to endozone. Walls of axial zooecia commonly thinner than neighboring endozonal walls; mural rods parallel to branch length, present in planar walls of some taxa. Zooids budded at or near axial structures or reverse surface. Zooecial bases attenuated, inflated, or flattened in longitudinal profile. Zooecial cross sections generally polygonal in endozone. Zooecia may be recumbent in endozone or diverge from 10° to 70°. Zooecial bends broadly rounded to abrupt. Living chambers generally elliptical in exozones, may be subcircular; living chambers usually oriented 70° to 90° to branch surfaces, but may be as low as 30°. Zooecial lengths 2 to 15 times diameter in late endozone. Longitudinal arrangements of zooecia regular to irregular. Hemisepta present or absent; where present, generally developed near zooecial bend; hemisepta commonly paired. Diaphragms absent to common. Exozonal widths of mature stems ranging from about one-fifth to four-fifths of branch radius. Zooecial boundaries variable; locally not visible, especially in exozone, or marked by irregular, narrow, dark zone; granular or nonlaminated material present in some areas along zooecial boundaries. Dark zones present in exozones of some taxa, similar in structure to zooecial boundaries. Lamellar profiles varying from V-shaped to flattened or concave. Extrazooecial wall material usually well developed in exozones between zooecia. Polymorphs and monticules rare. Stylets usually abundant, more than one type in many taxa; usually approximately paralleling zooecial chambers. Mural spines may be present. *Ord.-Perm.*

KEY TO GENERA OF RHABDOMESINA

Multiple routes are provided for certain genera because of character state variation within taxa and the probability of incomplete information for many fossil suites.

1 Zoaria divided into segments articulated at least terminally, in some taxa also laterally, or zoaria articulated only at base of branch; reverse surface developed in some genera (most genera of the Arthrostylidae) 2
– Zoaria not obviously articulated; reverse surface never developed 16
2(1) Reverse surface developed 3
– Reverse surface not developed 7
3(2) Branching on alternate sides of primary stem, lateral arms developed at regular intervals; articulated only basally, if at all . . . . *Glauconomella* (Fig. 274, *2*)
– Branching varied but not regularly alternating 4
4(3) Articulated rarely if at all; branch cross

section usually polygonal, apertures elliptical, not flaring distally
　　　　　Heminematopora (Fig. 275, *1*)
– Zoarium jointed at regular intervals 5
5(4) Apertures flaring distally, zooecia usually in 4 rows . . *Hemiulrichostylus* (Fig. 275, *2*)
– Apertures elliptical, zooecia usually in 3, sometimes 2 or 4 rows 6
6(5) Peristomes prominent
　　　　　Arthrostyloecia (Fig. 272, *1*)
– Peristomes subdued . . *Arthrostylus* (Fig. 271)
7(2) Articulated only basally, if at all 8
– Articulated at regular intervals 9
8(7) Cross section of zooecia triangular, zooecia arranged in 3-fold annular pattern; zooecia elongate . . . . *Hexites* (Fig. 275, *3*)
– Cross section varied but distinct, triangular, 3-fold pattern lacking; zooecia short
　　　　　Nematopora (Fig. 276, *2*)
9(7) Zooecia arranged in distinct cycles, with prominent peristomes
　　　　　Osburnostylus (Fig. 277)
– Distinct cyclic arrangement of zooecia, prominent peristomes lacking 10
10(9) Individual branch segments flaring strongly in distal direction
　　　　　Sceptropora (Fig. 279)
– Individual zoarial segments flaring weakly, if at all, in distal direction 11
11(10) Zoaria highly branched, with articulated primary, secondary, and tertiary branches (in disarticulated suites, look for different size classes and lateral articulation sockets) . . *Arthroclema* (Fig. 272, *2*)
– Zoaria rarely branched, usually articulated only terminally 12
12(11) Some zoarial segments weakly expanded distally; zooecial apertures arranged in rhombic pattern in which spiral rows appear dominant; zooecia arranged in numerous rows; true acanthostyles, metapores usually present . . . 13
– Zoarial segments generally cylindrical, not expanded; zooecial apertures arranged in either annular or rhombic patterns such that longitudinal rows appear dominant; zooecia arranged in few to numerous rows; metapores absent, paurostyles usually only of stylet type 15
13(12) Diaphragmed metapores, interconnected peristomial ridges present; zoaria articulated basally . . . . *Moyerella* (Fig. 276, *1*)
– Diaphragmed metapores, interconnected peristomial ridges absent 14
14(13) Zooecia generally short, diverging sharply from stem axis; metapores not diaphragmed . . . *Cuneatopora* (Fig. 273, *2*)
– Zooecia generally elongate, gradually divergent; metapores diaphragmed
　　　　　Helopora (Fig. 274, *1*)

15(12) Zoaria robust, some specimens jointed only rarely, apertures flaring; zooecia sublinear, elongate, diverging gradually from central axis . . *Ulrichostylus* (Fig. 281)
– Zoaria slender, closely jointed, apertures elliptical; zooecia short, recumbent in endozone . . . . . . *Nematopora* (Fig. 276, *2*)
16(1) Metapores almost always present 17
– Metapores absent (Arthrostylidae, Bactroporidae, Rhabdomesidae, Rhomboporidae) 27
17(16) One or more metapores present for each autozooecium (Nikiforovellidae, Hyphasmoporidae) 18
– Metapores present, but in numbers smaller than 1 for each autozooecium (Rhomboporidae, Rhabdomesidae) 33
18(17) Central axis linear or axial zooecia weakly developed, but no distinct central bundle of zooecia present; stylets usually present; zooecia generally short, length approximately 5 times diameter; zooecial base inflated (Nikiforovellidae) 19
– Few axial zooecia present or a distinct bundle of axial zooecia (possibly linear axis in one genus); stylets usually lacking; zooecia generally elongate, length approximately 10 or more times diameter, but shorter where a distinct axial bundle is present (Hyphasmoporidae) . . 23
19(18) One metapore for each zooecium; stylets present, median axis well defined
　　　　　Acanthoclema (Fig. 292, *1*)
– Metapores either absent or more than 1 metapore for each zooecium 20
20(19) Zooecia of varied lengths, with some individuals following axial region for varying distances; axis poorly defined . . 21
– Zooecial shapes constant, arrangement regular, metapores always present 22
21(20) Zooecial outline irregular, exozonal walls thin . . . . . . *Pinegopora* (Fig. 291, *2*)
– Zooecial outline rounded, exozone robust
　　　　　Nikiforovella (Fig. 291, *1*)
22(20) Zooecia elongate, exozonal interval of living chamber oriented perpendicular to branch surface, stylets may be lacking, exozone in mature stems relatively narrow . . . . *Streblotrypella* (Fig. 292,*2*)
– Zooecia shorter, may be inclined to surface, stylets apparently always present, exozone in mature stems relatively wide
　　　　　Nikiforovella (Fig. 291, *1*)
23(18) Distinct axial bundle lacking, or axial region formed by about 10 or fewer axial zooecia 24
– Axial region formed by more than about 10 axial zooecia 26
24(23) Axis linear or possibly formed by very few axial zooecia . *Petaloporella* (Fig. 295)

- At least some axial zooecia forming axial
region 25
25(24) Radiating dark zones present in exo-
zone . . . . . . *Hyphasmopora* (Fig. 293, *1*)
- Radiating dark zones absent
Streblotrypa (Streblotrypa) (Fig. 293, *2*)
26(23) Axial bundle large; single hemiseptum
usually present on distal wall in late endozone, proximal wall inflated at zooecial
bend . . . . *Ogbinopora* (Fig. 294, *1*)
- Axial bundle small to moderate in size,
hemisepta absent
Streblotrypa (Streblascopora) (Fig. 294, *2*)
27(16) Branch outlines commonly polygonal;
zooecia commonly radially aligned as
viewed in transverse section; endozonal
interval of zooecia may be recumbent
(Arthrostylidae) 28
- Branch outlines not polygonal; zooecia not
radially aligned as viewed in transverse
section; endozonal interval of zooecia not
recumbent (Bactroporidae, Rhombopor-
idae, Rhabdomesidae) 33
28(27) As viewed in transverse section, zooecia
in well-defined radial rows 29
- Zooecia not in well-defined radial rows . . . 32
29(28) Axial zooecia present
Heloclema (Fig. 273, *1*)
- Axial zooecia absent 30
30(29) Zooecia elongate . . *Hexites* (Fig. 275, *3*)
- Zooecia short 31
31(30) Skeletal cysts present
Pseudonematopora (Fig. 278)
- Skeletal cysts absent . *Nematopora* (Fig. 276, *2*)
32(28) Median axis regular, well defined, linear or planar; zooecia regular in shape and
arrangement . . . *Nematopora* (Fig. 276, *2*)
- Median axis irregular, linear; zooecia somewhat irregular in shape and arrangement
Tropidopora (Fig. 280)
33(27) Axial zooecia or axial cylinder generally
more or less well developed; zooecia generally elongate; zooecial base more or less
attenuated; paurostyles or acanthostyles,
or both, present; aktinotostyles, heterostyles absent; hemisepta usually present
(Rhabdomesidae) 39
- Axial zooecia and axial cylinder absent;
zooecia short, zooecial base more or less
inflated; aktinotostyles usually present,
acanthostyles may be present, or well-
developed acanthostyles present alone;
heterostyles and hemisepta absent
(Rhomboporidae) 34
- Axial zooecia and axial cylinder absent;
zooecia elongate; exozonal intervals of
living chambers inclined to stem surface;
aktinotostyles, paurostyles, acanthostyles
absent; heterostyles present; hemisepta
present or absent (Bactroporidae)
Bactropora (Fig. 290)

34(33) Aktinotostyles absent, well-developed
acanthostyles present . . *Pamirella* (Fig. 288)
- Aktinotostyles present, acanthostyles present or absent 35
35(34) Metapores more or less common
Megacanthopora (Fig. 287)
- Metapores absent or rare 36
36(35) Zooecial arrangement more or less regular; metapores may be present 37
- Zooecial arrangement more or less irregular;
metapores absent 38
37(36) Acanthostyles present, metapores may
be present . . . . . *Rhombopora* (Fig. 286, *3*)
- Acanthostyles and metapores absent
Saffordotaxis (Fig. 289, *1*)
38(36) Exozone generally narrow in mature
stems, endozonal walls thin; zooecia may
be budded from more or less clearly
defined planar axial surface
Klaucena (Fig. 286, *1*)
- Exozone wide in mature stems, endozonal
walls thick . . . . . *Primorella* (Fig. 289, *2*)
39(33) Median axis formed by well-defined
bundle of zooecia . *Ascopora* (Fig. 283, *1*)
- Median axis varied but not formed by well-
defined bundle of zooecia 40
40(39) Median axis formed by well-defined
cylinder of diameter usually greater than
that of zooecia; or axis open, conical
Rhabdomeson (Fig. 282)
- Median axis varied but not enlarged, cylindrical or conical 41
41(40) Branch diameter 2 mm or greater;
zooecia highly elongate, regular in
arrangement; zooecial bend abrupt, monticules may be present 42
- Branch diameter usually 2 mm or less; zooecia of varied lengths but usually not
highly elongate, and when elongate,
zooecia of irregular arrangement; zooecial
bend more or less gradual; monticules
absent 43
42(41) Zooecial rows not separated by well-
defined longitudinal ridges; zooids budded from somewhat irregular axis of one
or more longitudinal zooecia; monticular
areas of enlarged apertures may be pres-
ent . . . . *Nicklesopora* (Fig. 284, *1*)
- Zooecial rows separated by well-defined
longitudinal ridges; zooids budded from
more or less clearly defined longitudinal
axis; zooecia in annular bands and closed
by terminal diaphragms
Nemataxis (Fig. 284, *2*)
43(41) Zooecia short or moderately elongate,
with median axis generally well defined,
linear or somewhat planar; at least one
pair of closely overlapping hemisepta
generally present at zooecial bend
Orthopora (Fig. 285, *2*)
- Zooecia elongate, typically following cen-

tral axis for varying distances before diverging toward exozone; median axis more or less ill defined 44

44(43) One pair of overlapping hemisepta present . . . . *Orthopora* (Fig. 285, *2*)
– Overlapping hemisepta absent 45

45(44) Median axis irregular but quite well defined; zooecial arrangement somewhat irregular . . . . *Trematella* (Fig. 285, *1*)
– Median axis weakly defined, zooecial arrangement very irregular
Mediapora (Fig. 283, *2*)

Family ARTHROSTYLIDAE
Ulrich, 1882

[*nom. correct.* ULRICH, 1888, p. 230, *pro* Arthronemidae ULRICH, 1882, p. 151] [=Arthroclemidae SIMPSON, 1897, p. 546]

Zoaria erect; generally dendroid; some unbranched; rarely planar, branching. Branch jointing usually present. Branch diameters from about 0.1 to 2.5 mm; relatively constant between bifurcations, or expanding distally along segment in jointed taxa. Branch cross sections rounded or polygonal. Apertures in longitudinal rows or rhombic pattern; reverse surfaces may be present. Longitudinal and peristomial ridges usually present; ridge development varied. Metapores may be present, with or without diaphragms. Axial regions formed by well-defined linear axes, except in taxa with reverse surfaces; planar budding surfaces and axial zooecia uncommon in a few taxa. Zooids budded near axial region or from reverse surfaces. Zooecial bases attenuated to inflated in longitudinal profile. Zooecial cross sections in endozone polygonal, usually triangular. Zooecia initially recumbent in many taxa, diverging only at zooecial bend; zooecial divergence in other taxa approximately 15° to 70°. Zooecial bends generally rounded to abrupt, but may be weakly developed. Living chamber usually elliptical in exozone, subcircular in cross section; usually oriented between 70° and 90° to surface, but may be as low as 30°. Zooecial lengths approximately 3 to 12 times diameter. Proximal wall at zooecial bend more angular and inflated than distal wall in some taxa; true hemisepta absent. Diaphragms generally rare to few. Exozonal width varied, commonly about half

branch radius. Zooecial boundaries generally well developed, especially in endozone; usually dark, of granular or nonlaminated material, commonly more or less irregular. Boundaries locally not visible. Radiating dark zones arising at or near zooecial boundaries at base of exozone in some taxa. Lamellar profile in exozone generally V-shaped to rounded in transverse section; V-shaped, rounded, flattened, or concave in longitudinal section. Development of extrazooecial skeleton varied. Paurostyles or acanthostyles usually present, perpendicular to branch surface or parallel to zooecia. *L.Ord.-L.Perm.*

Arthrostylus ULRICH, 1882, *nom. subst.* ULRICH, 1888, p. 230, *pro Arthronema* ULRICH, 1882, p. 151; *non* ESCHSCHOLTZ, 1825 [**Helopora tenuis* JAMES, 1878, p. 3; OD; Economy Mbr., Eden Sh., U. Ord., Cincinnati, Ohio, USA]. Zoarium dendroid, jointed, branching at ends of segments only. Individual segments straight or slightly curved, segment diameters approximately 0.3 mm, diameters usually constant between joints except for terminal enlargement at joint surfaces. Segment cross section polygonal. In segments from single populations, apertures in 2 to 4 longitudinal rows on obverse surfaces, offset in adjacent rows. Prominent longitudinal ridges developed on reverse surfaces, between rows of apertures, and between successive chambers; peristomes present. Lateral zooecia budded from walls of reverse surfaces; medial zooecia, where developed, budded from walls of lateral zooecia. Zooecial bases inflated. Zooecia recumbent in endozone, diverging from reverse surface at zooecial bend; rounded to subpolygonal in cross section. Zooecial bends abrupt, living chambers in exozones about 90° to segment surfaces. Zooecial length from 4 to 6 times diameter; arrangement of zooecia regular. One or two thin diaphragms common near base of zooecia. Exozonal width between ridges approximately one-quarter zooecial diameter. Zooecial boundaries well defined, nonlaminated wall locally well developed between zooidal wall along reverse surface and extrazooidal wall, and along zooecial boundaries in endozone. Nonlaminated material locally forming endozonal wall; zooecial boundaries elsewhere irregular, or locally not visible. Lamellar profiles rounded over longitudinal ridges, between laterally adjacent zooecia; flattened between longitudinally successive zooecia. Paurostyles present, weakly developed. [*Arthrostylus* is distinguished on budding pattern, zooecial shape and orientation, development of nonlaminated wall, and paurostyle development. Although distinctive in growth habit and devel-

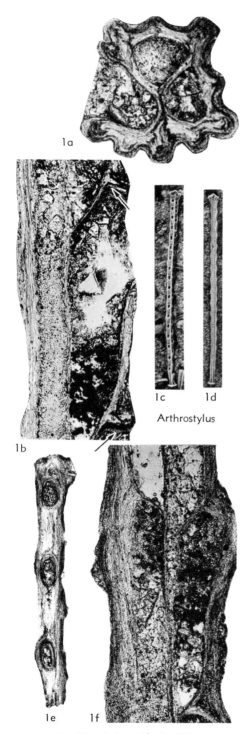

la

lc   ld

Arthrostylus

lb

le   lf

FIG. 271.   Arthrostylidae (p. 553).

opment of nonlaminated walls, it resembles some other arthrostylid genera in jointing pattern and stylet development. SIMPSON (1897) suggested separation of *Arthrostylus* from other arthrostylid genera at the family level. Budding patterns and wall development suggest affinities between *Arthrostylus* and the Phylloporinidae.] *M.Ord.(Blackriv.)-U.Sil.(Wenlock.* or *?Ludlov.),* N.Am., Greenl., Baltic region.——FIG. 271,*1a–f.* **A. tenuis* (JAMES); *a,* obverse surface (top), three zooecial rows, left zooecium at aperture, dark zones; transv. sec., USNM 240789, ×240; *b,* reverse surface (left), one complete zooecium and portions of two others; long. sec., USNM 240790, ×240; *c,* articulation facets, apertural arrangement, longitudinal ridges; external obverse view, USNM 240785, about ×11; *d,* articulation facets, longitudinal ridges; external reverse view, USNM 240786, about ×11; *e,* apertural arrangements; shallow tang. sec., USNM 240787, ×80; *f,* two rows of zooecia; deep tang. sec. parallel to reverse surface and perpendicular to orientation of *b,* USNM 240792, ×160.

Arthroclema BILLINGS, 1865, p. 54 [**A. pulchellum*; M; Trenton Ls., M. Ord., Ottawa, Ont., Can.]. Zoarium branching, with well-defined axial stem and alternate secondary and tertiary branches; jointed longitudinally, laterally. Primary segments up to about 1 mm in diameter. Segment diameters generally constant except for terminal flanges in some specimens; cross sections subcircular. Apertural arrangement predominantly longitudinal, locally weakly rhombic. In most species, sinuous or straight longitudinal ridges separate apertural rows and longitudinally successive apertures. Proximal and lateral margins of aperture commonly bordered by peristome. Metapores absent. Axial region formed by well-defined linear axis. Zooecial bases weakly to moderately inflated longitudinally. In endozone, zooecial cross section subtriangular, rounded; zooecia recumbent, diverging at rounded zooecial bend. Living chambers oriented from 30° to nearly 90° to branch surface; angle increasing with exozonal thickening. Zooecial length varied, commonly 4 to 5 times diameter, up to about 10 times in primary and secondary segments. Longitudinal arrangement of zooecia regular. Diaphragms few in some species. Exozonal width varied, depending in part on segment type. Zooecial boundaries usually narrow, irregular; granular or nonlaminated material locally developed; longitudinal dark zones, similar to zooecial boundaries, developed in exozonal wall between longitudinal rows of apertures and longitudinally successive apertures. Extrazooecial skeleton well developed. Lamellar profile in exozones V-shaped against dark zones, flattened between longitudinally successive apertures. Paurostyles scattered to com-

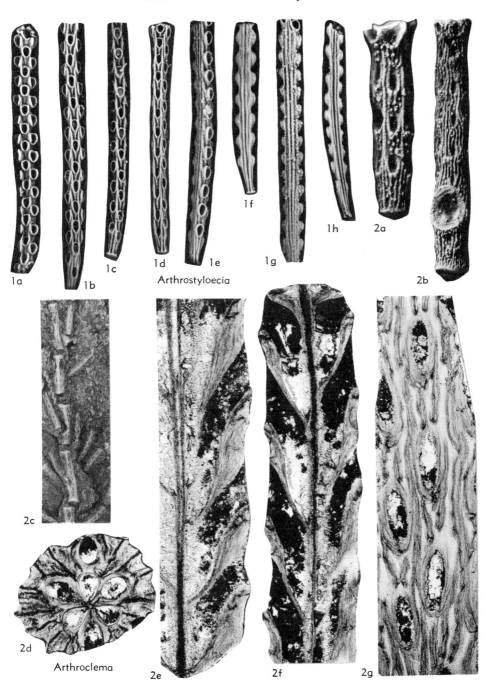

1a 1b 1c 1d 1e 1f 1g 1h 2a 2b

Arthrostyloecia

2c

2d

Arthroclema

2e 2f 2g

FIG. 272. Arthrostylidae (p. 554–557).

mon, usually developed on ridges. [*Arthroclema* is distinguished by zoarial form, zooecial shape, and wall structure. Tertiary segments may resemble branches of *Nematopora* and *Ulrichostylus,* and distinct size classes are necessary for differentiation (ULRICH, 1893). *Arthroclema* resembles

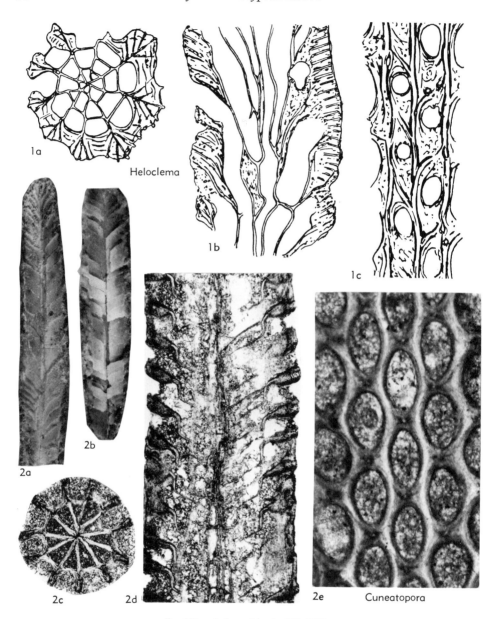

1a

Heloclema

1b

1c

2a

2b

2c   2d

2e   Cuneatopora

FIG. 273.   Arthrostylidae (p. 557–559).

Ulrichostylus in wall structure and presence of lateral articulation joints but differs in zooecial orientation. *Arthroclema* resembles *Nematopora* in zooecial form and orientation but differs in growth habit. BASSLER (1911) reported a few specimens of the Trentonian species *A.* cf. *A. armatum* from the Lower Ordovician (B₂) of Estonia, but that report is questionable.] *?L.Ord., M.Ord.(Blackriv.)-U.Ord.(Richmond.),* N.Am., Baltic region.——FIG. 272,*2a,b. A.* cf.

A. cornutum ULRICH, Decorah Sh., M. Ord., Minn., USA; variation in articulation facets, apertural alignment, longitudinal ridges, stylets; probable secondary segments, USNM 240859 and 240857, both ×17.——FIG. 272,*2c–g.* **A. pulchellum,* USNM 240862; *c,* branching zoarium; exterior view, ×4; *d,* zooecial cross section and dark zones in secondary segment; transv. sec., ×75; *e,* zooecial shapes and boundaries, thick exozone; long. sec., ×75; *f,* zooecial shapes, thin exozone, and distal articulation surface in tertiary segment; long. sec., ×75; *g,* apertural arrangement, dark zones, and stylets in secondary segment; tang. sec., ×75.

Arthrostyloecia BASSLER, 1952, p. 384 [**A. nitida;* OD; Edinburg F., M. Ord., Strasburg Junction, Va., USA]. Zoarium dendroid, jointed, branching at ends of segments only. Individual segments straight or slightly curved. Segment diameters approximately 0.3 mm, usually constant except for terminal enlargement. Apertures alternate in three longitudinal rows on obverse surface. Peristomes prominent; some branching and joining to form longitudinal ridges between apertural rows, between longitudinally successive apertures, and on reverse surface. Zooecia budded at or near reverse surface. [*Arthrostyloecia* is known from silicified material. Externally, it differs from *Arthrostylus* only in peristome development, but it is not synonymized because of lack of information on internal features.] *Ord., E.N.Am., Baltic region.*——FIG. 272,*1a–h. *A. nitida,* all about ×16; *a,* apertural arrangement and peristomes; holotype, USNM 116409; *b–h,* apertural arrangement, longitudinal ridges, reverse surfaces, and peristomes; paratypes, USNM 240795, 240798, 240799, 240802, 240803, 240805, and 240806.

Cuneatopora SIEGFRIED, 1963, p. 138 [**C. erratica;* OD; in clasts of possible M. Ord. age from glacial deposits, Baltic region]. Zoarium erect, not known to branch, jointed longitudinally. Segments usually straight, approximately 0.5 to 2.5 mm in diameter. Segments generally expanding distally; circular in cross section. Apertural arrangement rhombic. Longitudinal ridges and peristomes absent. Metapores present or absent; where developed, arising in exozone, usually at junction of three autozooecia, diaphragms absent. Zooecia inflated or flattened; at bases cross section in endozone polygonal, commonly triangular. Zooecial divergence from axis between 35° and 70°. Zooecial bend abrupt, zooecial axis in exozone commonly parallel to axis of endozone. Living chambers generally inclined between 80° and 90° to segment surface, may be as low as 60°. Zooecial length approximately 3 to 6 times diameter. Longitudinal arrangement of zooecia regular. Diaphragms absent from zooecia. Exozonal width usually about half branch radius. Zooecial boundaries commonly obscure; or narrow, irregular, granular material locally developed. Extrazooecial skeleton limited. Lamellar profile in exozone V-shaped to weakly rounded. Acanthostyles well developed. Prominent conical or cylindrical deflections common in sheath laminae. [*Cuneatopora* is distinguished on zoarial habit, budding pattern, zooecial shape and orientation, presence of metapores and acanthostyles, and reduction of extrazooidal skeleton. Longer zooecia in this genus are similar to the shorter zooecia of *Helopora* in shape, growth habit, stylets, and presence of metapores.] *?M.Ord.,Sil.(Llandov.-Wenlock.),* N.Am., Eu.——FIG. 273,*2a,b. *C. erratica; a,* broken, relatively elongate, proximal zooecia; holotype, Münster B523a, ×10; *b,* zooecial shapes in broken paratype; Münster B523d, ×10.——FIG. 273,*2c. C. bellula* (BILLINGS), Jupiter F., L.Sil., Anticosti Is., Can.; zooecial shapes and exozonal development; transv. sec., USNM 240780, ×70.——FIG. 273,*2d,e. C. lindstroemi* (ULRICH), Sil., Gotl., Swed.; *d,* zooecial outlines; long. sec., ?paratype, USNM 240863, ×35; *e,* apertural arrangement, shapes, stylets; tang. sec., ?paratype, USNM 214193, ×70.

Glauconomella BASSLER, 1952, p. 384 [**Glauconome disticha* GOLDFUSS, 1831, p. 217; OD; Wenlock Ls., equals Dudley of GOLDFUSS, U. Sil., Eng.]. Zoarium pinnate; primary, secondary branches only; jointing unknown. Branch diameters approximately 0.5 mm, constant between bifurcations; branch cross section rounded on reverse side, angular on obverse side. On obverse surface, apertures aligned in 4 longitudinal rows, 2 on each side of median keel; reverse surface bearing fine ridges. Metapores absent. Median zooid rows budded from walls of lateral zooids; lateral zooid rows budded from reverse surface. Zooecial bases inflated. Zooecial cross section in endozone polygonal to rounded. Zooecia recumbent in endozone, diverging only at zooecial bend; apertures large. Zooecia short, flask-shaped, length approximately 4 times diameter, longitudinal arrangement regular. Diaphragms present in some species. Zooecial boundaries well defined, wide, irregular, usually containing granular material, nonlaminated material locally developed, especially between zooecial rows. Planar dark zones, similar to zooecial boundaries, radiating in exozone from zooecial boundaries; exozonal dark zones paralleling branch axis. Extrazooecial skeleton well developed. Lamellar profile in exozone V-shaped against dark zones, rounded to flattened away from dark zones. Stylets absent. [*Glauconomella* is distinguished on branching pattern, presence of a reverse surface, zooecial and apertural shapes, and wall structure. Limits and affinities of the genus are uncertain. Some species of *Glauconomella* and *Penniretepora* are superficially similar and, in general, the post-Silurian species have been assigned to *Penniretepora*. Whether or not these genera are distinct is uncertain, and a Silurian limit to *Glauconomella* is arbitrarily accepted here. *Glauconomella* is very similar to *Nematopora* in zooecial shape and wall structure, but differs in budding and branching patterns.] *U.Ord.-U.Sil.(Wenlock.),* Eu., N.Am.——FIG. 274,*2a–e. *G. disticha* (GOLDFUSS), *a–c* from Eng. and *d,e* from Gotl.; *a,* branching pattern

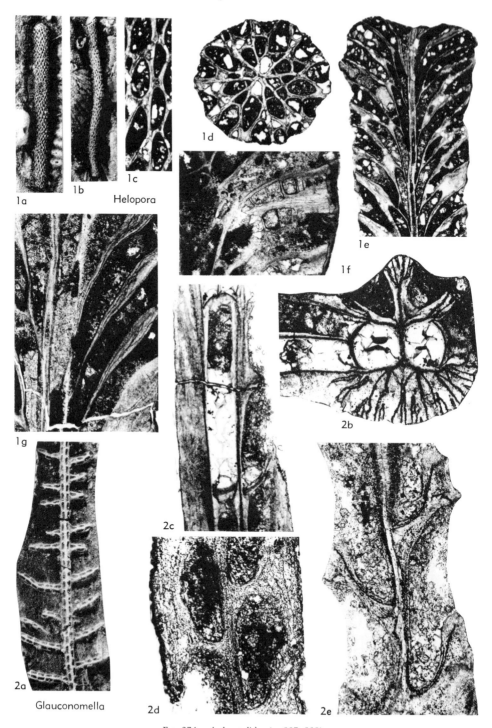

Fig. 274.   Arthrostylidae (p. 557–559).

and apertural arrangement on obverse surface; USNM 240808, ×6; *b,* lateral branch (left), zooecial cross section and arrangement, dark zones; transv. sec., USNM 240810, ×75; *c,* reverse surface (left); deep tang. sec. of left zooecial row, long. sec. of right zooecial row, USNM 240810, ×50; *d,* apertural arrangement and shapes; deep tang. sec., USNM 240811, ×75; *e,* zooecial outlines; long. sec. parallel to reverse surface, USNM 240811, ×75.

Heloclema SHULGA-NESTERENKO, 1955, p. 139 [**H. spiralis*; OD; Steshevskij Level, L. Carb., Oka River, Luzhki Village, Russ. plat., USSR]. Zoarium dendroid, jointing unknown. Branches 0.55 to 0.70 mm in diameter, subcircular in cross section. Apertures in 7 to 9 longitudinal rows, separated by ridges; metapores absent. Axial region formed by few axial zooecia. Zooecial bases attenuated to weakly inflated. Zooecial cross section in endozone polygonal, initially triangular. Zooecial divergence from axial region approximately 15° to 30°; zooecial bend abrupt. Living chambers oriented from 70° to 90° to branch surface. Autozooecial length generally about 10 times diameter. Longitudinal arrangement of autozooecia somewhat irregular. Proximal wall angular at zooecial bend; diaphragms rare. Exozonal width varied, generally about half branch radius. Zooecial boundaries well defined, narrow; nonlaminated material locally developed. Planar, longitudinal, dark zones radiating through exozone from approximate position of zooecial boundaries at base of exozone. Extrazooecial skeleton well developed. Lamellar profile in exozone V-shaped against dark zones. Stylets developed on longitudinal ridges between apertures. [*Heloclema* is distinguished by zooecial shape and orientation, arrangement of stylets, and development of the exozonal wall. It resembles *Nematopora* in development of the exozonal wall and arrangement of stylets.] *L.Carb. (Visean),* USSR.——FIG. 273,*1a–c.* **H. spiralis*; drawings of zooecial shapes and arrangements; *a,* transv. sec.; *b,* long. sec.; *c,* tang. sec., ×55 (Shulga-Nesterenko, 1955).

Helopora HALL in SILLIMAN, SILLIMAN, & DANA, 1851, p. 398 [**H. fragilis*; M; ?Cabot Head Sh., Cataract Gr., equals Clinton Gr. of HALL, L. Sil., Flamborough Township, near ?Hamilton, Ont., Can.]. Zoarium erect, not known to branch, jointed longitudinally. Individual segments generally straight, diameters 0.5 to 2.0 mm. Segments expanding distally along length, or enlarged terminally at joint surfaces; cross sections circular. Apertural arrangement basically rhombic. Longitudinal ridges, peristomes absent. Metapores with diaphragms common in some species, especially at expanded ends of segments. Metapores narrower than zooecia, arising in exozone, paralleling zooecia; diaphragms thickened, irregularly spaced. Axial region

formed by more or less well-defined linear axis. Zooecial bases weakly inflated, or attenuated. Zooecial cross section polygonal in endozone, initially triangular. Zooecial divergence from axis ranging from 25° to 70°; zooecial bend rounded. Living chambers oriented from 60° to 90° to zoarial surface. Zooecial length from 6 to 10 times diameter. Longitudinal arrangement of zooecia regular to irregular. Zooecial diaphragms few. Exozonal width about half segment radius. Zooecial boundaries locally obscure; where developed, usually narrow, irregular, commonly granular; nonlaminated material may be developed. Extrazooecial skeleton limited. Lamellar profile in exozone V-shaped to somewhat rounded. Acanthostyles large, well developed, common on zooecial boundaries, arising near base of exozone; prominent conical deflections common in sheath laminae. [*Helopora* is distinguished on segment shape, apertural arrangement, budding pattern, zooecial shape and orientation, presence of acanthostyles and metapores with diaphragms, and reduction of extrazooecial skeleton.] *L.Sil. (Llandov.)-L.Dev., ?M.Dev.(Eifel.), U.Dev.,* E.N.Am., S.Am., USSR.——FIG. 274,*1a–g.* **H. fragilis*; *a,b,* stem shapes and apertural arrangement; syntypes, AMNH 30718, 30722, both ×6; *c,* apertural arrangement, stylet development; deep tang. sec., USNM 240814, ×50; *d,* zooecial and metapore cross sections; transv. sec., USNM 240814, ×50; *e,* expanded distal end of stem, zooecial shapes, acanthostyles; long. sec., USNM 240818, ×30; *f,* distal end of segment, metapore; long. sec., USNM 222622, ×75; *g,* zooecial shapes and boundaries, lamellar profile; long. sec., USNM 240816, ×75.

Heminematopora BASSLER, 1952, p. 384 [**H. virginiana*; OD; Edinburg F., M. Ord., Strasburg Junction, Va., USA]. Zoarium dendroid, not known to be jointed. Branch diameters approximately 0.25 mm, constant between joints. Apertures alternating in 5 to 7 longitudinal rows on obverse surface; lateral rows discontinuous, some apertures separated by barren intervals continuous with reverse surface. Ridges in varied patterns on obverse surface; peristomial ridges usually complete about apertures; some peristomes branching and joining to form sinuous longitudinal ridges between apertural rows or between longitudinally successive apertures. Converging longitudinal ridges developed on reverse surface. Zooecia budded at or near reverse surface; exozonal interval of living chamber inclined to branch surface. [Known only from silicified material, *Heminematopora* is distinguished on budding pattern, development of ridges, and zooecial orientation.] *Ord.,* E.N.Am., Baltic region.——FIG. 275,*1a–f.* **H. virginiana,* all about ×16; *a,* zooecial arrangement, longitudinal ridges; holotype, USNM 116411; *b–f,* zooe-

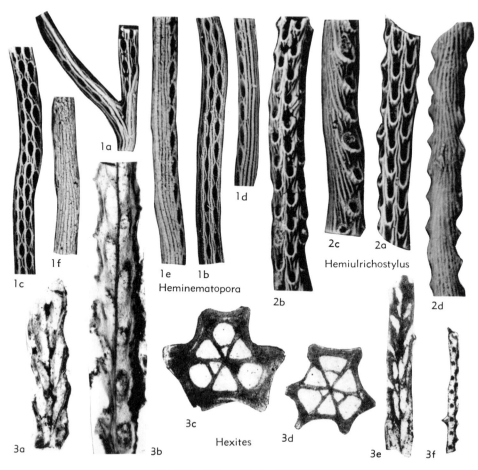

FIG. 275. Arthrostylidae (p. 559–560).

cial arrangement, longitudinal ridges, reverse surfaces; paratypes, USNM 240822–240826.

Hemiulrichostylus BASSLER, 1952, p. 384 [*H. lineatus*; OD; Edinburg F., M. Ord., Strasburg Junction, Va., USA]. Zoarium erect, not known to branch or to be jointed; diameters approximately 0.5 mm, constant. Apertures alternating in 4 longitudinal rows on obverse surface. Peristomes complete in some specimens or apertures proximally and laterally bordered by prominent ridges flaring distally to join longitudinal ridges between zooecial rows; some ridges converging distally on reverse surface. Ridges on sloping surfaces between longitudinally successive apertures. Exozonal interval of living chamber inclined to surface. Plane of aperture inclined to stem axis. [Known only from silicified material, *Hemiulrichostylus* is distinguished by budding pattern, development of ridges, and zooecial arrangement.] *M.Ord.(Blackriv.),* E.N.Am.——FIG. 275,*2a–d*. **H. lineatus,* all about ×20; *a,*

zooecial arrangement, ridges; holotype, USNM 116412; *b–d,* obverse and reverse surfaces, zooecial arrangement, longitudinal ridges; paratypes, USNM 240819–240821.

Hexites SHULGA-NESTERENKO, 1955, p. 137 [*H. triangularis*; OD; Tul'skij level, L. Carb., Chekhurskij Village, Russ. plat., USSR]. Zoarium dendroid, jointing unknown. Branches with diameters 0.18 to 0.38 mm, cross sections polygonal. Apertures with peristomes in six longitudinal rows, separated by prominent ridges; metapores absent. Axial region formed by well-defined linear axis. Zooids budded around axis in groups of three. Zooecial bases attenuated or weakly inflated. Zooecial cross sections triangular in endozone. Zooecia initially recumbent in endozone, then diverging at approximately 25°. Zooecial bend weakly defined, proximal wall profile angular, distal wall profile more or less straight. Zooecial length about 7 times diameter. Longitudinal arrangement of zooecia regular.

Proximal wall swollen at zooecial bend, but true hemisepta not developed; diaphragms absent. Exozonal width approximately half branch radius at longitudinal ridge positions, one-fourth radius over zooecia or less. Extrazooecial skeleton well developed. Lamellar profile in exozone flattened between longitudinally successive zooecia. Small stylets present on longitudinal ridges. [*Hexites* is distinguished on budding pattern and zooecial shape.] *L.Carb.(Visean),* USSR.
——Fig. 275,*3a–f.* **H. triangularis; a,* zooecial outlines, linear axis; long. sec., holotype, PIN 309/66, ×30; *b,* apertural outlines, longitudinal ridges; tang. to deep tang. sec., holotype, ×40; *c,* zooecial cross sections; transv. sec., holotype, ×80; *d,* zooecial cross sections; transv. sec., paratype, PIN 309/80, ×80; *e,* zooecial outlines; long. sec., paratype, PIN 309/32, ×25; *f,* external view; paratype, PIN 309/32, ×10 (Shulga-Nesterenko, 1955).

Moyerella NEKHOROSHEV, 1956b, p. 45 [*M. stellata;* OD; L. Sil. (Llandov.), Mojero, Kurejka rivers, Sib. plat., USSR]. Zoarium erect, not known to branch, jointed longitudinally; segment diameters from less than 0.5 to more than 1.5 mm. Segments expanding distally along length; circular in cross section. Apertural arrangement rhombic. Longitudinal ridges lacking except at tapered segment base; zooecial apertures lacking in base. Peristomes prominent, at least in type species; deflected distally and usually intersecting with peristome of next aperture. Metapores with diaphragms, parallel to zooecia, arising near base of exozone, usually at juncture of three zooecia. Axial region formed by linear axis or planar budding surface. Zooecial bases attenuated. Zooecial cross sections polygonal in endozone, commonly triangular. Zooecial divergence from axial region 30° to 45°. Zooecial bend rounded to abrupt; zooecial axis in exozone commonly subparallel to axis in endozone, offset distally. Living chambers usually oriented from 60° to 70° to zoarial surface. Zooecial length usually 3 to 5 times diameter. Longitudinal arrangement of zooecia generally regular. Diaphragms rare. Exozonal width varied, commonly about half stem radius. Zooecial boundaries typically narrow, locally irregular, granular. Nonlaminated material thin, discontinuous in endozone, somewhat thickened in exozone in peristomial ridges. Lamellar profile in exozone V-shaped. Large acanthostyles developed at junction of peristomes; prominent conical or cylindrical deflections present in sheath laminae. Small paurostyles present along peristomial ridges. [*Moyerella* is distinguished on zooecial shape and orientation, as well as metapore, stylet, and peristomial ridge development. It resembles *Cuneatopora* in zooecial shape and orientation and *Helopora* in budding pattern and metapore development.] *L.Sil.(Llandov.),* USSR.——FIG.

276,*1a–e.* **M. stellata,* paratypes; *a,* apertural arrangement, peristomial ridges; USNM 240832, ×20; *b,* apertural arrangement, peristomial ridges; tang. peel, USNM 240833, ×75; *c,* zooecial outlines, arrangement; transv. sec., USNM 240833, ×75; *d,* peristomial ridges; tang. sec., USNM 240833, ×200; *e,* zooecial shapes, arrangement; long. sec., USNM 240833, ×75.

Nematopora ULRICH, 1888, p. 231 [**Trematopora minuta* HALL, 1876, pl. 11; OD; Waldron Sh., M. Sil., Waldron, Ind., USA]. Zoarium dendroid, usually jointed only at base; unjointed in some species (specimens?); closely jointed in at least one species. Branch or segment diameters 0.1 to 0.7 mm; usually constant between bifurcations or joints. Branch cross section polygonal to subcircular. Apertures in 4 to 10 longitudinal rows. Prominent longitudinal ridges usually present between apertural rows and longitudinally successive apertures. Peristomes commonly present, metapores absent. Axial region usually formed by well-defined linear axis; planar median wall developed locally in some species. Two or three median rods developed in walls of some specimens. Zooecial bases inflated. Zooecial cross sections in endozone triangular. Zooecia recumbent in endozone, diverging from axis at zooecial bend; zooecial bend abrupt. Living chambers oriented 90° to branch surface. Zooecial length from 4 to 6 times diameter. Longitudinal arrangement of zooecia regular. Diaphragms absent. Exozonal width approximately half branch radius at ridges. Zooecial boundaries generally well defined, narrow; locally with granular or nonlaminated material. Planar, longitudinal dark zones, similar to zooecial boundaries, radiate through exozone from approximate position of zooecial boundaries. Lamellar profile in exozone V-shaped against dark zones in transverse section, flattened to slightly concave in longitudinal section. Extrazooecial wall material well developed. Paurostyles common on ridges in many species. [*Nematopora* is distinguished on branch and zooecial shapes, surficial features, and development of zooecial boundaries and exozonal dark zones. Its branches commonly resemble tertiary segments of *Arthroclema.*] *M.Ord.-L.Perm.,* N.Am., Eu., USSR, Asia.——FIG. 276,*2a.* **N. minuta* (HALL); zoarial form; syntype, AMNH 1919, about ×8.——FIG. 276,*2b. N. granosa* ULRICH, ?Decorah Sh., M. Ord., Minn., USA; branch shape, apertural arrangement, stylets, longitudinal ridges; USNM 240834, about ×20.——FIG. 276,*2c–e. N. lineata* (BILLINGS), Ellis Bay F., L. Sil., Anticosti Is., Can., all ×75; *c,* zooecial cross section, dark zones, median rods; transv. sec., USNM 240836; *d,* apertural arrangement, zooecial boundaries, dark zones; tang. sec., USNM 240838; *e,* zooecial shape, zooecial boundaries; long. sec. through apertures, USNM 240840.

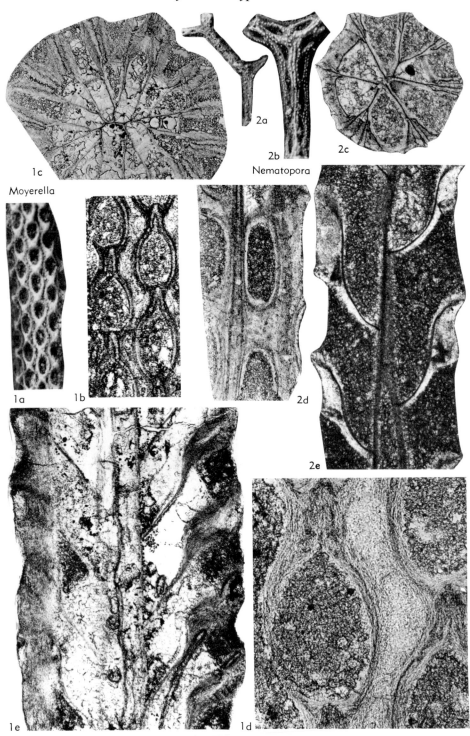

1c

Moyerella

2a

2b

Nematopora

2c

1a 1b 2d

2e

1e 1d

Fig. 276. Arthrostylidae (p. 561).

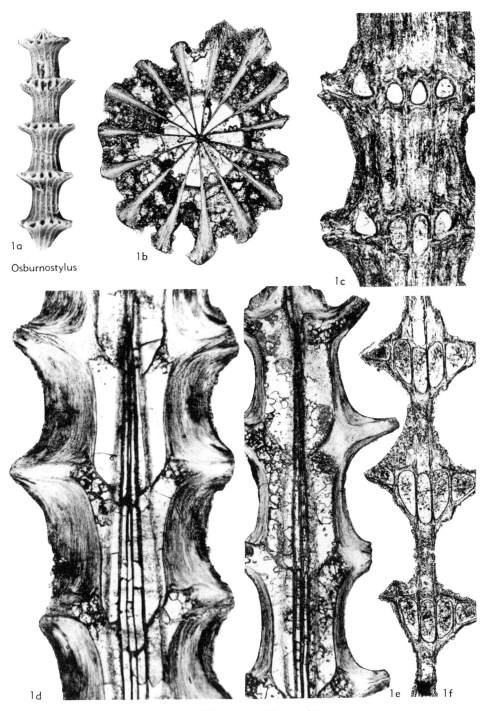

la
Osburnostylus
1b
1c
1d
1e 1f

FIG. 277. Arthrostylidae (p. 563).

Osburnostylus Bassler, 1952, p. 381 [*O. articulatus; OD; Benbolt F., M. Ord., Rye Cove, Va., USA]. Zoarium erect, jointed longitudinally; branching not observed. Segments straight or

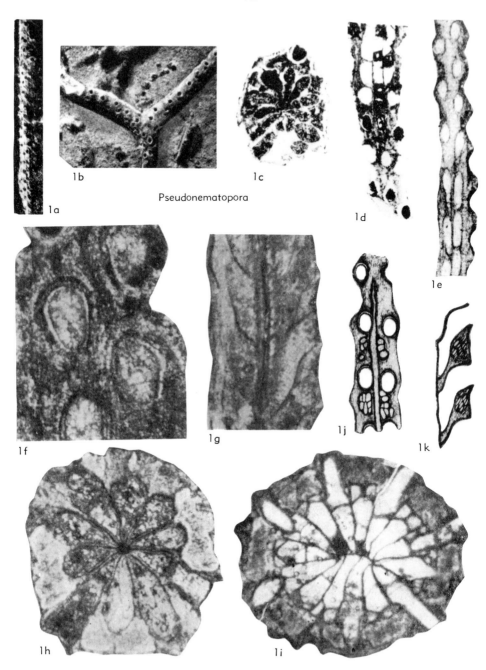

Pseudonematopora

FIG. 278.   Arthrostylidae (p. 565).

curved, diameters 0.5 to 1.0 mm, cross sections approximately circular. Apertures aligned in approximately 15 longitudinal rows; transversely, in prominent annular bands. Prominent longitudinal ridges separating apertural rows.

Metapores absent. Axial region formed by well-defined linear axis. Zooecial bases inflated. Zooecial cross sections triangular in endozone. Zooecia recumbent in endozone, diverging from axis at zooecial bend; zooecial bend rounded. Living chambers oriented from 70° to 90° to segment surface. Zooecial length about 7 to 10 times diameter. Longitudinal arrangement of zooecia regular. Diaphragms present near base of some zooecia. Exozonal width varied. Zooecial boundaries usually well defined, locally obscure, narrow. Extrazooecial skeleton well developed. Lamellar profile in exozone V-shaped in transverse section at level of apertures; flattened to concave in longitudinal section. Acanthostyles well developed lateral to zooecial chambers. [*Osburnostylus* is distinguished by growth habit, budding pattern, zooecial shape and orientation, and presence of acanthostyles. It resembles *Nematopora* in zooecial orientation and shape, and *Helopora* in the large number of zooecia about the axis and the presence of acanthostyles.] *M.Ord.(Blackriv.),* E.N.Am.——Fig. 277,*1a–f.* **O. articulatus*; *a,* annular zooecia; paratype, USNM 240841, about ×15; *b,* zooecial outlines, linear axis; transv. sec., paratype, USNM 240842, ×75; *c,* thick exozone, apertural arrangement; tang. peel, paratype, USNM 240843, ×50; *d,* thick exozone, zooecial outlines and arrangement; long. sec., paratype, USNM 240843, ×50; *e,* thin exozone, zooecial outlines and arrangement; long. sec., paratype, USNM 240844, ×50; *f,* thin exozone; deep tang. peel, paratype, USNM 240845, ×50.

Pseudonematopora BALAKIN, 1974, p. 130 [**Nematopora? turkestanica* NIKIFOROVA, 1948, p. 39; OD; L. Carb. (low. Visean), W. Talas Alatau Ra., S. Kazakh., USSR]. Zoarium dendroid, unjointed. Branch diameters 0.8 to 2.8 mm; usually constant between bifurcations. Branch cross sections subcircular. Apertures in 8 to 16 longitudinal rows. Longitudinal ridges present or absent; peristomes complete or only on proximal sides of apertures, tapering distally. Metapores absent. Axial region formed by well-defined linear axis or planar median wall; median wall present in most branches between 1.3 and 2.6 mm in diameter; median rods absent. Zooecial base weakly inflated. Zooecial cross sections in endozone triangular (where budded along linear axis, or at ends of median wall) to polygonal (where budded along median wall). Zooecia initially recumbent, diverging from axis near gradual zooecial bend. Living chambers oriented from 55° to 90° to branch surface. Zooecial length about 3 to 5 times diameter. Longitudinal arrangement of zooecia regular. Diaphragms absent. Exozonal width approximately one-third of branch radius. Zooecial boundaries dark, well defined, narrow. Planar, longitudinal dark zones similar to zooecial boundaries, probably present in exozone. Lamellar profile in exozone flattened to slightly concave; extrazooecial wall material well developed. Skeletal cysts may be well developed at endozonal-exozonal boundary. Stylets absent. [*Pseudonematopora* is distinguished on

the usual presence of skeletal cysts and a planar median wall. BALAKIN (1974) cited NIKIFOROVA (1948) as the original publication of *N.? turkestanica,* but NIKIFOROVA (1948) cited a 1936 date and the title "Lower Carboniferous Bryozoa from the western extremity of the Talaskian Alatau." No journal was cited, and I have been unable to trace the source.] *L.Carb.(low. Tournais.-low. Visean),* USSR.——FIG. 278,*1a–k.* **P. turkestanica* (NIKIFOROVA); *a,b,* Kassin strata, Karaganda region, and Irsu-Kazanchukur divide, respectively; exterior surfaces and zooecial arrangement; TsGM 5648, both ×4 (Nekhoroshev, 1953); *c–e,* Irsu-Kazanchukur divide; transv., deep tang., and tang. to deep tang. secs.; paratypes, TsGM 6548/808, all ×20; *f–i,* Kshikainda Suite, low. Visean (Balakin, 1974); *f,* tang. sec., MGU 409/513, ×59; *g,* long. sec., MGU 409/1950, ×20; *h,* linear axis, transv. sec., MGU 409/8552, ×30; *i,* planar median wall, transv. sec., MGU 409/1121, ×39; *j,k,* Dzhaltyrsky F., up. Visean; skeletal cysts; tang., long. secs., TsGM 184a, both ×20 (Nekhoroshev, 1956b; illustrated as *Nematopora peregrina*).

Sceptropora ULRICH, 1888, p. 228 [**S. facula*; M; Stony Mountain F., U. Ord., Stony Mt., Manit., Can.]. Zoarium dendroid, jointed longitudinally, bifurcations rare. Segments straight; slender proximally, expanding more or less abruptly in distal direction to form bulbous or discoid end; subcircular in cross section. Apertures aligned in 12 to 20 distal rows, absent proximally. Prominent longitudinal ridges separating apertural rows. Angular metapores with diaphragms may enclose zooecial chambers. Axial region formed by well-defined linear axis. Zooecial bases weakly inflated. Zooecial cross sections polygonal in endozone, commonly triangular. Zooecia recumbent in proximal interval of segment, diverging abruptly if budded distally. Zooecial outline broadly rounded, zooecial bend not distinct. Living-chamber orientation dependent on zooecial position. Zooecial length varied with position, from less than 4 to more than 10 times diameter. Longitudinal arrangement of zooecia regular to somewhat irregular. Zooecial diaphragms absent. Exozone forming most of branch radius near segment base, relatively narrow distally. Zooecial boundaries usually well defined, locally obscure; granular or nonlaminated material in some intervals. Extrazooecial skeleton well developed. Lamellar profile V-shaped in exozone. Paurostyles abundant on ridges. [*Sceptropora* is distinguished on segment shape, zooecial shape and orientation, and metapore and stylet development. It is similar to *Helopora* in budding pattern, zooecial shape and orientation, and presence of metapores, but differs in segment shape.] *U.Ord.(Richmond.)-L.Sil.(Llandov.);?U.Sil. (?Wenlock.),* N.Am.,

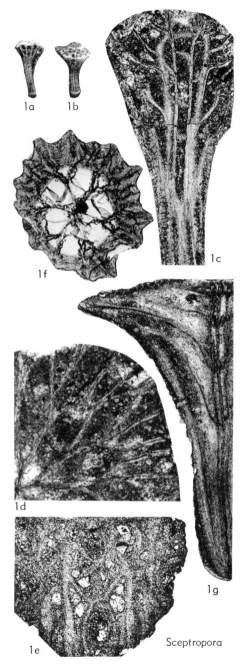

FIG. 279. Arthrostylidae (p. 565).

Baltic Region, USSR.——FIG. 279,1a–f. *S. facula, a–e* syntypes; *a,* segment shape, longitudinal ridges, apertures; USNM 240848, about ×6; *b,* surface of expanded distal end of segment; USNM 240849, about ×6; *c,* zooecial arrange-

ment; deep tang. sec., USNM 240851, ×60; *d,* zooecial orientation, median axis; transv. sec. near distal end of segment, USNM 240850, ×60; *e,* zooecial apertures at distal end of segment; tang. peel, USNM 240846, ×80; *f,* U. Ord., Anticosti Is., Can.; zooecial arrangement; transv. sec. near proximal end of segment, USNM 240847, ×160; *g,* Elkhorn F., U. Ord., Ohio, USA; zooecial outlines and arrangement; long. sec., USNM 240852, ×60.

Tropidopora HALL, 1886, pl. 25 [**T. nana*; M; Onondaga Ls., equals Helderberg Gr. of HALL, M. Dev., Onondaga Valley, Erie Co., N.Y., USA]. Zoarium dendroid, jointing unknown. Branch diameters approximately 0.4 mm, constant between bifurcations; branch cross section subcircular. Apertural arrangement rhombic, somewhat irregular. Weak longitudinal ridges separating apertural rows; peristomes, metapores absent. Axial region formed by irregular linear axis. Zooecial bases attenuated to weakly inflated. Zooecial cross sections triangular, rounded in endozone. Zooecia recumbent in endozone, diverging from axis at abrupt zooecial bend. Living chambers oriented from 80° to 90° to branch surface. Zooecial length from 5 to 6 times diameter. Longitudinal arrangement of zooecia somewhat irregular. Diaphragms absent. Exozonal width about half branch radius. Zooecial boundaries narrow, granular; nonlaminated material locally developed. Extrazooecial skeleton well developed. Lamellar profile flattened in exozone. Paurostyles scattered on or near longitudinal ridges. Mural spines common in exozone. [*Tropidopora* is distinguished on branch size, surficial features, and zooecial shape and orientation. It is known from a single specimen.] *M.Dev.(Erian),* E.N.Am.——FIG. 280, *1a–e.* **T. nana,* holotype, NYSM 1053; *a,* paurostyles; long. sec., ×400; *b,* zooecial outline, lamellar profile; transv. peel, ×100; *c,* apertural arrangement; exterior, ×15; *d,* apertural arrangement, stylets; tang. sec., ×100; *e,* median axis, zooecial outlines and arrangement; long. sec., ×100.

Ulrichostylus BASSLER, 1952, p. 384 [**Helopora divaricatus* ULRICH, 1886a, p. 59; OD; ?Decorah Sh., equals Trenton shales of ULRICH, M. Ord., Minneapolis, Minn., USA]. Zoarium dendroid in some species, possibly unbranched in others. Jointed longitudinally, also laterally in dendroid branches; joint surfaces generally weakly concave-convex; ball-and-socket pattern present in at least one species. Segments straight or curved; diameters 0.5 to 1.0 mm, usually constant between joints; cross sections polygonal to subcircular. Apertures in 6 to 8 longitudinal rows. Prominent longitudinal ridges separating apertural rows; apertures bordered proximally and laterally by strong ridges that flare distally to join longitudinal ridges, forming inverted V-pattern.

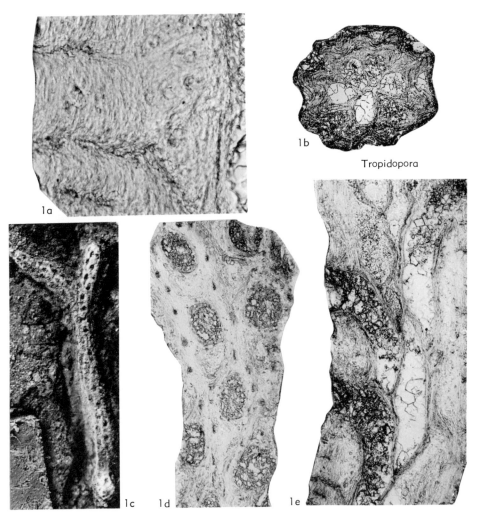

1b

Tropidopora

1a

1c   1d   1e

FIG. 280.   Arthrostylidae (p. 566).

Ridges proximal to apertures sloping gradually into aperture below; 1 or 2 longitudinal ridges may be present on sloping surface. Metapores absent. Axial region formed by well-defined linear axis. Zooecial bases attenuated to weakly inflated. Zooecial cross sections triangular in endozone. Zooecial divergence from axis approximately 20° to 40°; zooecial bend weakly developed, broadly rounded. Living chambers in exozone elliptical in cross section, oriented from 60° to 70° to branch surface. Zooecial length 5 to 12 times diameter. Longitudinal arrangement of zooecia regular. Diaphragms scattered in elongate zooecia. Exozonal width more than half branch radius. Zooecial boundaries generally narrow in endozone, irregular, commonly with granular or nonlaminated material, locally not

visible; well developed in exozone near proximal and lateral margins of chambers, positions obscure distal to chambers. Planar longitudinal dark zones radiating through exozone from near zooecial boundaries at base of exozone; appearance similar to that of zooecial boundaries. Lamellar profile in exozone rounded in transverse section, interrupted by radiating boundary zones; flattened in longitudinal section, sloping toward zoarially proximal zooecium. Exozonal wall material well developed. Paurostyles scattered, weakly developed, concentrated in wall between longitudinally successive zooecial chambers. [*Ulrichostylus* is distinguished on budding pattern, zooecial shape and orientation, and wall structure.] *M.Ord.(Chazy.-Blackriv.), ?U.Ord. (Richmond.),* E.N.Am., Baltic region.——Fɪɢ.

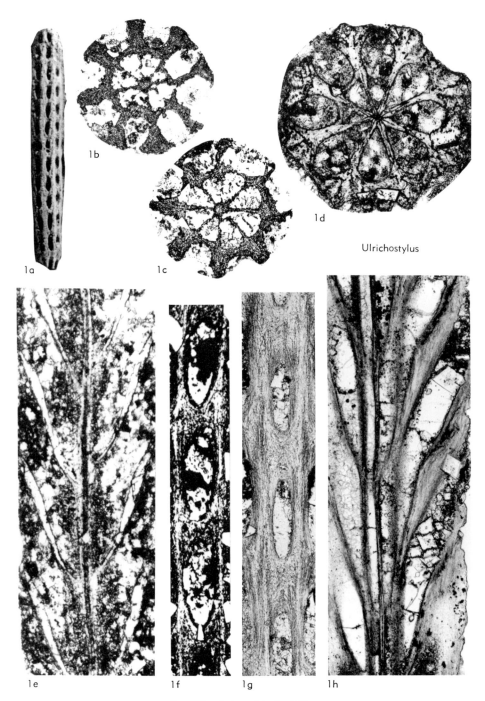

Ulrichostylus

FIG. 281.   Arthrostylidae (p. 566).

281,1a–f. *U. divaricatus* (ULRICH), syntypes (*a*, Univ. Minnesota 5928A; *b–f*, 5928B); *a*, apertural arrangement, longitudinal ridges; ×11; *b,c*, zooecial cross sections and arrangements, linear axes; transv. peels, ×70; *d*, radiating dark zones; transv. sec., ×94; *e*, zooecial shapes; long. sec.,

×85; *f,* apertural outlines, longitudinal ridges; tang. sec., ×85.——FIG. 281, *g,h. U. spiniformis* (ULRICH), Lebanon Ls., M. Ord., Tenn., USA, all USNM 240853, ×70; *g,* apertural outlines, longitudinal ridges; tang. sec.; *h,* zooecial outlines, lamellar profile; long. sec.

Nomen Nudum

Oandupora MÄNNIL, 1959, p. 39. Name not accompanied by characters differentiating taxon, Article 13a, ICZN.

Family RHABDOMESIDAE
Vine, 1884

[*nom. correct.* BASSLER, 1953, p. G130, *pro* Rhabdomesontidae VINE, 1884, p. 205]

Zoaria generally dendroid; some zoaria, branches, or parts of branches conical; jointing unknown. Branch diameters approximately 0.5 to 6.0 mm, constant or varied between bifurcations. Branches subcircular in cross section. Apertural arrangement basically rhombic, somewhat irregular in some taxa or areas of some branches; ridges absent. Metapores rare. Axial region variable, from linear axes or a few weakly defined axial zooecia to large bundles of axial zooecia or enlarged axial cylinder. Diaphragms may be present in axial cylinders. Autozooecial bases attenuated to inflated in longitudinal profile; zooecial cross sections polygonal in endozone. Zooecial divergence from axial bundle 15° to 45°. Zooecial bends rounded to abrupt. Living chambers in exozone generally elliptical in cross section, rarely subcircular. Living chambers usually oriented from 80° to 90° to surface, may be as low as 50°. Autozooecial lengths approximately 4 to 15 times diameter. Hemisepta usually paired, mostly developed near zooecial bend; may be in multiple series or absent. Diaphragms few in most species, scattered in some; terminal diaphragms present in one genus. Exozonal width usually less than half branch radius, may range from about one-third to two-thirds branch radius in mature stems. Zooecial boundary narrow, dark; may contain granular or nonlaminated material; or locally, may not be visible. Lamellar profiles V-shaped in exozone, more or less flattened between stylets. Monticules of enlarged,

thin-walled polymorphs rare. Paurostyles or acanthostyles always present, both in many species; few to common, but usually not densely spaced in exozonal walls. Stylets generally arising in exozone, paralleling zooecial chambers in most species. Mural spines present in some species. *U.Sil.-U.Perm.*

Rhabdomeson YOUNG & YOUNG, 1874, p. 337 [**Millepora gracilis* PHILLIPS, 1841, p. 20; OD; Carb., N. Devon, Eng.} [=*Coeloconus* ULRICH in MILLER, 1889, p. 298]. Zoarium usually dendroid; some zoaria, branches, or parts of branches conical. Cylindrical branch diameters 0.7 to 6.0 mm, constant between bifurcations. Apertural arrangement rhombic. Axial region formed by hollow, regular to somewhat irregular, axial cylinder; diameter of cylinder usually greater than that of autozooecia. Wall thickness of axial cylinder usually comparable to that of other endozonal walls; diaphragms present in axial cylinders of some species. Zooecial bases usually attenuated, inflated in some species; zooecia in endozone initially triangular in cross section, becoming hexagonal. Zooecial divergence 20° to 45° from axial surface in cylindrical stems; ascending along surface of axial cylinder to zooecial bend in conical stems. Zooecial bend generally abrupt, somewhat rounded in some species; living chambers oriented from 80° to 90° to branch surface. Zooecial length in cylindrical stems generally from 4 to 7 times diameter; longitudinal arrangement of zooecia usually regular. Inflated, recurved hemiseptum usually on proximal wall at zooecial bend; may be rare or multiple. Diaphragms rare. In cylindrical branches, exozonal width generally about half branch radius or less, rarely more. Zooecial boundary generally narrow, irregular, granular in some areas; locally not visible. Lamellar profile in exozone narrowly to broadly V-shaped. One or two acanthostyles occurring proximal to zooecial chambers, paurostyles few to common. Mural spines may be present in exozonal living chambers. Stylets arising in exozone, paralleling autozooecia. Monticulelike structures formed of fused stylets are present in at least one species. [*Rhabdomeson* has commonly been recognized on the presence of an axial cylinder. It resembles *Ascopora* in development of autozooecia, exozone, stylets, and hemisepta, but differs primarily in structure of the axial region. The synonymy of *Rhabdomeson* and *Coeloconus* has been discussed by BLAKE (1976). Location of the types of *R. gracilis* is unknown and they may be lost (SHERBORN, 1940).} *M.Dev.(Erian)-U.Perm.(Dzhulf.),* N.Am., USSR, Asia, Australia.——FIG. 282, *1a,d–f. R. rhombicus* (ULRICH), Warsaw Sh., mid. Miss., Ill., USA; *a,* branch shape, apertural

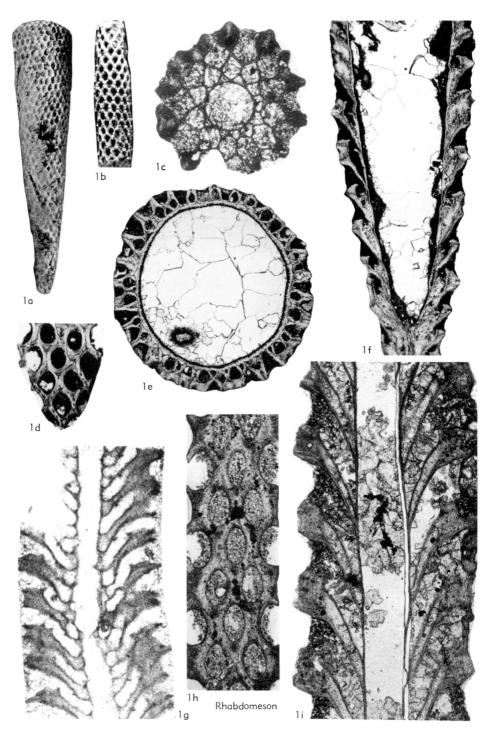

Fig. 282. Rhabdomesidae (p. 569).

arrangement; syntype, USNM 240827, ×6; *d,* apertural and stylet arrangements; tang. sec., syntype, USNM 43335, ×28; *e,* central cone, arrangement of zooecia; transv. sec., syntype, USNM 240828, ×28; *f,* zooecial shapes and arrangement, hemisepta, stylets; long. sec., syntype, USNM 43335, ×28.——Fɪɢ. 282,*1b,c,h,i.* **R. gracilis* (Pʜɪʟʟɪᴘs); *b,* aperture and stylet arrangements; USNM 240771, about ×8; *c,* axial cylinder, zooecial cross sections; transv. sec., USNM 121681, ×47; *h,* living chamber outlines, stylets; tang. sec., USNM 121681, ×47; *i,* zooecial shapes, axial cylinder; long. sec., USNM 121681, ×47.——Fɪɢ. 282,*1g. R. kansasensis* Sᴀʏʀᴇ, Drum Ls., U. Penn., Mo., USA; irregular axial cylinder, inflated zooecial base, wide exozone; long. sec., syntype, KUMIP 125167, ×38.

Ascopora Tʀᴀᴜᴛscʜoʟᴅ, 1876, p. 367 [**Millepora rhombifera* Pʜɪʟʟɪᴘs, 1836, p. 199; M; Carb., Yorkshire, Eng.]. Zoarium dendroid. Branch diameters 1.0 to 5.5 mm, usually constant between bifurcations. Apertural arrangement rhombic. Axial region formed by weakly to well-defined cylindrical bundle of 4 to 30 axial zooecia. Axial zooecia polygonal in cross section, walls commonly thinner than those of autozooecia; diaphragms usually absent, may be rare. Autozooecial bases attenuated to weakly inflated; autozooecia in endozone initially triangular in cross section, becoming hexagonal. Autozooecial divergence from axial bundle mostly between 20° and 45°. Zooecial bend generally abrupt; living chambers commonly oriented about 90° to branch surface. Autozooecial length mostly 5 to 10 times diameter. Longitudinal arrangement of zooecia usually regular. Single, massive, recurved hemiseptum may be present on proximal wall at zooecial bend; single, slender hemiseptum rarely present on distal wall in late endozone; multiple hemisepta rarely present on proximal wall; or hemisepta may be absent. Autozooecial diaphragms generally absent, may be rare. Exozonal width ranging from less than half branch radius in slender species to about two-thirds branch radius in robust species. Zooecial boundary generally narrow, irregular, granular in some areas; locally not visible; lamellar profile V-shaped in exozone. One or two acanthostyles proximal to each zooecial chamber; orientation relative to zoarial surface may be greater than zooecial angle. Paurostyles common to densely spaced; in single or double rows between apertures, or stylet fields may be present. Mural spines may be present in exozonal living-chamber wall. Stylets arising in exozone. [*Ascopora* is distinguished by an axial bundle of zooecia, zooecial shape, stylet development, and lack of metapores. Location of the types of *A. rhombifera* is unknown, and they may be lost (Sʜᴇʀʙoʀɴ, 1940).] *L.Carb. (Tournais.* or *Visean)-L.Perm.(?Artinsk.),* USSR, N.Am., Asia.——Fɪɢ. 283,*1a,b. A. mag-*

niseptata Sʜᴜʟɢᴀ-Nᴇsᴛᴇʀᴇɴᴋo, U. Carb., Russ. plat., USSR, holotype, PIN 136/95; *a,* axial bundle, zooecial outlines; transv. sec., ×20; *b,* zooecial shapes, hemisepta, axial zooecia; long. sec., ×20.——Fɪɢ. 283,*1c. Ascopora* sp., Penn.-Perm., Nev., USA; living chamber outlines, acanthostyles, paurostyles; tang. sec., USNM 240854, ×20.——Fɪɢ. 283,*1d. Ascopora* sp., Earp F., Cisco Gr., U. Penn-Perm., Ariz., USA; zooecial shapes, hemisepta, axial zooecia; long. sec., USNM 240855, ×30.——Fɪɢ. 283,*1e. Ascopora* sp., ?Brazier Ls., Carb. (?Penn.), Idaho, USA; zooecial shapes, stylets, axial zooecia; long. sec., USNM 240856, ×30.

Mediapora Tʀɪᴢɴᴀ, 1958, p. 209 [**M. injaensis;* OD; Taidonskaya and Fominskaya zones, L. Carb. (Tournais.–Visean), Inya and Tykhta rivers, Kuznetsk basin, USSR]. Zoarium dendroid. Branch diameters 1.4 to 2.3 mm. Apertural arrangement rhombic, locally irregular. Metapores present in at least some species. Axial region formed by a few axial zooecia or possibly ill-defined linear axis. Axial zooecia not in distinct bundle; ascending variable distances along axial region before diverging toward surface. Axial zooecia weakly differentiated from autozooecia, except in length and probably diameter; diaphragms lacking. Autozooecial base attenuated to inflated; zooecial cross sections polygonal in endozone, irregular or hexagonal. Autozooecial divergence from axial region 20° to 40°. Zooecial bend usually rounded; living chamber outlines irregular, varied within single zoarium. Outer interval of exozone oriented about 90° to branch surface. Autozooecial length 8 to 15 times diameter; longitudinal arrangement of autozooecia irregular to highly irregular. Hemisepta absent, diaphragms scattered in most species. Exozonal width usually one-third to half branch radius. Lamellar profile probably V-shaped in exozone. Acanthostyles common, varied in size, arranged in linear series. [*Mediapora* is distinguished by the presence of axial zooecia and by zooecial shape and arrangement.] *?U.Sil.(Ludlov.), L.Carb.(Tournais.-Visean),* USSR.——Fɪɢ. 283,*2a-c.* **M. injaensis,* all ×20; *a,* irregular zooecial arrangement; biased long. sec., holotype, VNIGRI 263/913; *b,* zooecial cross sections, development of exozone; transv. sec., holotype; *c,* apertural shapes, arrangement; tang. sec., paratype, VNIGRI 264/913 (Tʀɪᴢɴᴀ, 1958).——Fɪɢ. 283,*2d. M. fragilis* Tʀɪᴢɴᴀ, L. Carb.; axial zooecia; long. sec., paratype, VNIGRI 266/913, ×20 (Tʀɪᴢɴᴀ, 1958).

Nemataxis Hᴀʟʟ, 1886, pl. 25 [**N. fibrosus;* M; lithology suggests Springvale Ss. (W. A. Oʟɪᴠᴇʀ, 1974, pers. commun.), equals up. Helderberg Gr. of Hᴀʟʟ, L. Dev., Ont., Can.]. Zoarium dendroid. Branch diameters approximately 3 to 4 mm, constant between bifurcations. Apertural

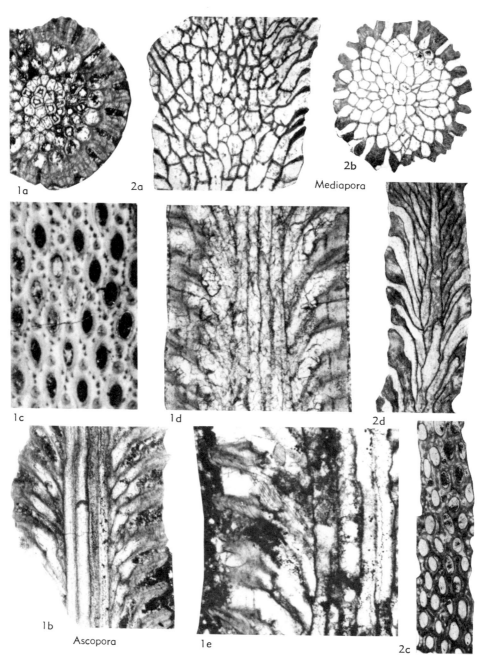

1a　　　　2a　　　　Mediapora

2b

1c　　　　1d　　　　2d

1b

Ascopora　　　1e　　　　2c

FIG. 283. Rhabdomesidae (p. 571).

arrangement rhombic. Axial region formed by linear axis. Zooecial bases inflated; zooecia initially polygonal, irregular in cross section, becoming hexagonal, then subrectangular. Zooecial divergence from axis about 45°. Zooecial

bend abrupt. Living chambers oriented at 80° to 90° to branch surface. Zooecial length about 15 times diameter; longitudinal arrangement of zooecia regular. Single, massive, short, rarely recurved hemiseptum on proximal wall at zooe-

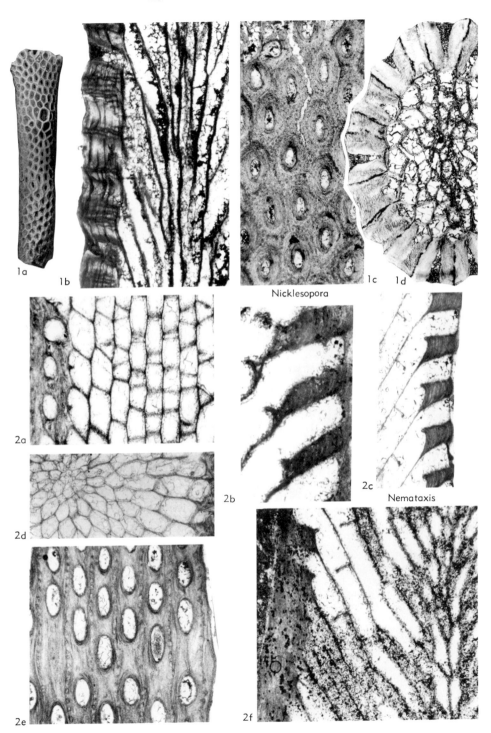

Nicklesopora

Nemataxis

Fig. 284.   Rhabdomesidae (p. 571–574).

cial bend in most zooecia; single, slender, straight hemiseptum with recurved margin on distal wall in late endozone; multiple pairs of hemisepta on either wall in some zooecia. Small chambers may be developed under hemisepta. Diaphragms scattered in endozone; thick terminal diaphragms locally in annular bands around branch. Exozonal width about one-third branch radius. Zooecial boundary narrow, irregular, locally not visible. Lamellar profile flattened in exozone, except where orientation of sheath laminae around closely spaced stylets has chevron profile. Small acanthostyles common, aligned between apertural rows; arising near base of exozone, approximately parallel to zooecia. Sheath laminae flaring, not closely parallel to core axis. [*Nemataxis* is distinguished on size of zoarium, presence of terminal diaphragms in annular bands, shape of zooecia, and development of hemisepta and stylets.] *L.Dev.(Ulster.),* E.N.Am. ——Fig. 284,*2a–f.* **N. fibrosus*; *a,* zooecial outlines, alignment; deep tang. sec., syntype, FMNH UC 23803, ×28; *b,* terminal diaphragms, lamellar profile, hemisepta, chambers beneath hemiseptum; long. sec., syntype, FMNH UC 23803, ×46; *c,* lamellar profile, hemisepta; long. sec., USNM 240766, ×28; *d,* exozone (right), zooecial cross sections, axial region of branch; transv. sec., FMNH UC 23803, ×28; *e,* terminal diaphragms, stylet arrangement, living chamber outlines; tang. sec., USNM 240766, ×28; *f,* zooecial outlines, branch axis; long. sec., syntype, FMNH UC 23803, ×28.

Nicklesopora BASSLER, 1952, p. 384 [**Rhombopora elegantula* ULRICH, 1884, p. 33; OD; New Providence Sh., L. Miss., Kings Mt. at Halls Gap, Lincoln Co., Ky., USA]. Zoarium dendroid. Branch diameters 0.7 to more than 2.5 mm, may vary somewhat between bifurcations. Apertural arrangement basically rhombic; somewhat irregular, especially near monticules. Axial region formed by few axial zooecia, not in distinct bundle. Axial zooecia may be more slender, thinner walled than autozooecia. Axial zooecia ascending five or more zooecial ranks before diverging from axial region, assuming morphology of autozooecia. Zooecial bases attenuated; zooecia initially polygonal, irregular in cross section, becoming hexagonal. Autozooecial divergence from axial region approximately 15° to 30°. Zooecial bend abrupt; living chambers usually oriented about 90° to branch surface, some as little as 75°. Length of zooecia not arising in axial region 10 or more times diameter; longitudinal outline, arrangement of zooecia somewhat irregular. Single, short, commonly massive hemiseptum usually on proximal wall near zooecial bend; slender short hemiseptum may be present on distal wall in late endozone. Diaphragms rare. Exozonal width approximately one-third to half branch radius. Zooecial boundary generally narrow,

irregular, granular; locally not visible. Lamellar profile in exozone broadly V-shaped. Monticules rare, consisting of one much enlarged, thin-walled polymorph surrounded by smaller, enlarged, thin-walled polymorphs. Paurostyles common, most in single well-defined linear series either enclosing zooecial apertures in polygonal pattern or extending longitudinally between rows of zooecia. Paurostyles arising in exozone paralleling zooecial chambers. [*Nicklesopora* is distinguished on zoarial size, zooecial shape, development of the exozone, and presence of monticules and paurostyles.] *U.Dev.-L.Carb.,?U.Perm.,* N.Am., Australia, USSR.—— Fig. 284,*1a–d.* **N. elegantula* (ULRICH); *a,* zooecial arrangement and monticule; syntype, USNM 240768, about ×6; *b,* axial zooecia, zooecial shapes; long. sec., syntype, USNM 43716, ×28; *c,* living chamber outlines, stylet alignment; tang. sec., syntype, USNM 168365, ×28; *d,* zooecial cross sections, lamellar profile; transv. sec., syntype, USNM 168365, ×28.

Orthopora HALL, 1886, pl. 25 [**Trematopora regularis* HALL, 1874, p. 106; SD HALL & SIMPSON, 1887, p. xiv; New Scotland Ls., equals up. Helderberg Gr. of HALL, L. Dev., Clarksville, Albany Co., N.Y., USA]. Zoarium dendroid. Branch diameters 0.5 to 1.0 mm, constant between bifurcations. Apertural arrangement rhombic. Axial region usually formed by more or less well-defined linear axis. Axial bundle absent, but some autozooecia may ascend along axial region for short interval in endozone before diverging toward surface. Zooecial bases attenuated to inflated; zooecia initially polygonal, irregular in cross section, becoming hexagonal. Zooecial divergence usually 25° to 40°. Zooecial bend generally abrupt; living chamber oriented about 90° to branch surface. Zooecial length usually 4 to 6 times diameter, greater in species with zooecia ascending parallel to axis in endozone. Longitudinal arrangement of zooecia usually regular. Straight, moderately massive hemiseptum usually present on proximal wall at zooecial bend; slender, straight hemiseptum usually on distal wall in late endozone; overlap of hemisepta and changes in wall orientation commonly producing U-shaped zooecial outline near zooecial bend; second hemiseptum may be present on proximal wall in late endozone; rarely, hemisepta lacking. Diaphragms rare. Exozonal width one-third to half branch radius. Zooecial boundary narrow, irregular; granular, nonlaminated material may be present in intervals of endozone; locally not visible. Lamellar profile in exozone flattened. Paurostyles and acanthostyles commonly occurring together, with paurostyles more abundant. Stylets scattered to common in more or less well-defined rows; may be confined to longitudinal rows between lines of zooecial chambers; arising in exozone, parallel to zooecia. Sheath laminae

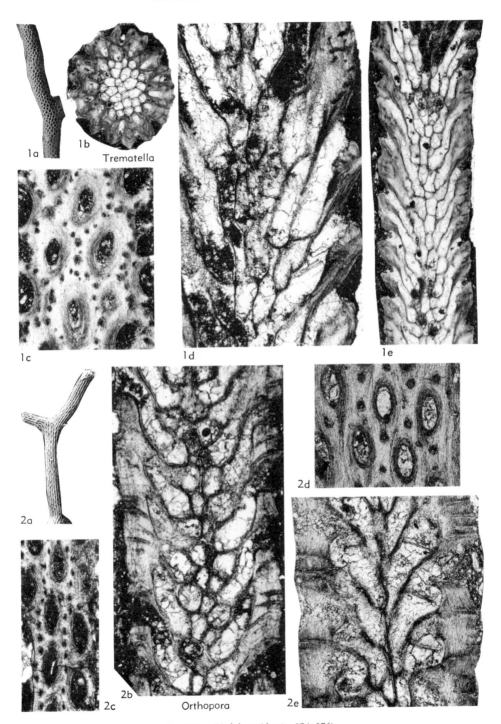

FIG. 285. Rhabdomesidae (p. 574–576).

commonly steeply ascending, oriented subparallel to core, distinct from zoarial laminae. [*Orthopora* is distinguished on apertural arrangement, zooecial shape, and development of hemisepta and stylets.] *U.Sil.(Ludlov.)- M.Dev.(Givet., Erian),* N.Am., Eu., USSR——Fig. 285,*2a–c.* **O. regularis* (HALL); *a,* branching pattern, apertural arrangement; syntype, AMNH 35758A, ×4; *b,* branch axis, zooecial shapes; long. sec., syntype, NYSM 942, ×50; *c,* apertural and stylet arrangements; tang. sec., syntype, NYSM 942, ×50.——Fig. 285,*2d,e. O. tonolowayensis?* BASSLER, Keyser Ls., L. Dev., W. Va., USA; *d,* living chamber outlines, stylet arrangement; tang. sec., USNM 214197, ×75; *e,* zooecial shapes, hemisepta, stylets, lamellar profile; long. sec., USNM 214194, ×75.

Trematella HALL, 1886, pl. 25 [**T. glomerata*; SD DUNCAN, 1949, p. 133; Onondaga Ls., equals up. Helderberg Gr. of HALL, M. Dev., Onondaga Valley, Erie Co., N.Y., USA]. Zoarium dendroid. Branch diameters 1.2 to 2.0 mm, somewhat varied between bifurcations. Apertural arrangement rhombic, locally irregular. Metapores rare. Axial region generally formed by weakly defined linear axis; axial zooecia absent but autozooecia may ascend along axis in endozone for varying intervals before diverging from axis. Zooecial bases usually inflated, rarely attenuated; zooecia initially polygonal, irregular in cross section, becoming hexagonal. Zooecial divergence from axial region approximately 20° to 30°. Zooecial bend rounded; living chambers oriented 50° to 70° to branch surface. Zooecial length about 5 to 12 times diameter; longitudinal arrangement of zooecia usually irregular. Single, small to massive hemiseptum common on proximal wall at zooecial bend. Diaphragms rare in most species, scattered in some. Exozonal width approximately one-third to half branch radius. Zooecial boundary generally narrow, irregular, granular, with nonlaminated material in places; locally not visible. Lamellar profile in exozone narrowly to broadly V-shaped. Paurostyles and acanthostyles present, with paurostyles more abundant. Stylets in more or less well-defined rows, or scattered; arising in exozone, approximately parallel to autozooecia; two types gradational in form. Sheath laminae in some acanthostyles steeply ascending, oriented subparallel to core, distinct from zoarial laminae. [*Trematella* is distinguished on zooecial shape and arrangement, development of stylets, metapores and hemisepta.] *L.Dev.(Ulster.)- M.Dev.(Erian),* E. N.Am.——Fig. 285,*1a–e.* **T. glomerata,* holotype, NYSM 1040; *a,* apertural arrangement; ×5; *b,* zooecial arrangement, exozonal development; transv. sec., ×20; *c,* living chamber outlines, acanthostyles, paurostyles, metapores; tang. sec., ×50; *d,* zooecial outlines and arrangement; long. sec., ×50; *e,* zooecial

outlines and arrangement; long. sec., ×20.

Family RHOMBOPORIDAE
Simpson, 1895

[Rhomboporidae SIMPSON, 1895, p. 549]

Zoaria dendroid, jointing rare. Branch diameters 0.5 to about 4.5 mm, relatively constant or varied between bifurcations; branches subcircular in cross section. Apertural arrangement basically rhombic, somewhat irregular in some taxa or areas of some branches; ridges absent. Metapores may be present. Axial region generally formed by linear axis; median planar surfaces discontinuous in some species; weak trend toward development of axial zooecia in few species. Zooecial bases attenuated to inflated in longitudinal profile. Zooecial cross sections polygonal in endozone, irregular to triangular near budding locus, hexagonal away from locus. Zooecial divergence from axial region 20° to 30°. Zooecial bends rounded to abrupt; living chambers in exozones elliptical to subcircular in cross section, usually oriented about 90° to branch surfaces, but may be as low as 60°. Autozooecial lengths approximately 5 to 10 times diameters. Hemisepta absent; diaphragms rare to common. Exozonal width one-fifth to four-fifths branch radius in mature stems. Zooecial boundaries locally not visible; or narrow, dark; granular material and nonlaminated material in some areas. Lamellar profiles V-shaped to broadly rounded in exozones, more or less flattened between stylets. Aktinostyles or acanthostyles always present, both in many species; stylets common to abundant, mostly arising in exozone; stylets usually parallel to zooecial chambers. Mural spines may be present. *?U.Dev.,L.Miss.- U.Perm.*

Rhombopora MEEK, 1872, p. 141 [**R. lepidodendroides;* OD; ?Willard Sh., Penn., Nebraska City, Otoe Co., Neb., USA]. Zoarium with jointed branches in at least one species; branch diameters 0.7 to 4.5 mm, may vary between bifurcations. Apertural arrangement approximately rhombic, locally irregular. Metapores uncommon in some species; typically fewer than 1 metapore for every 15 zooecia. Axial region usually formed by irregular linear axis; intraspecifically, some zooecia parallel axis for short

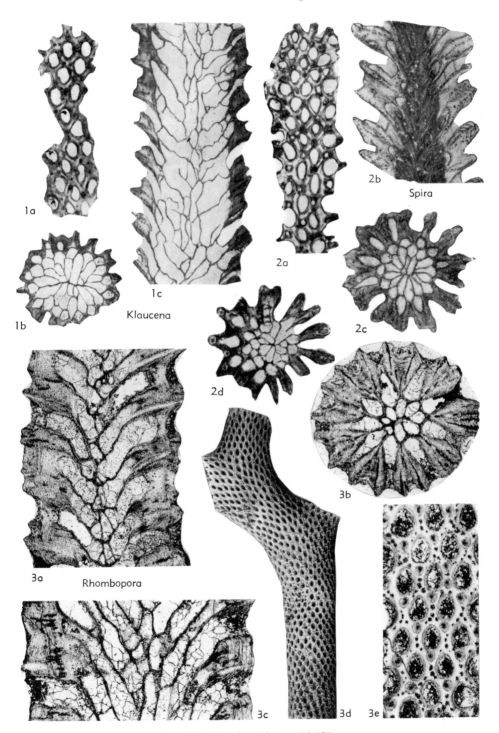

1a

1b

1c
Klaucena

2a

2b
Spira

2c

2d

3a
Rhombopora

3b

3c

3d

3e

FIG. 286.   Rhomboporidae (p. 576–578).

intervals, true axial zooecia not developed. Zooecial bases inflated; zooecia initially polygonal and irregular in cross section, becoming hexagonal. Zooecial divergence from axis approximately 30° to 50°. Zooecial bend generally rounded; living chambers oriented 80° to 90° to branch surface. Zooecial length varied, usually 5 to 9 times diameter; longitudinal arrangement of zooecia regular to somewhat irregular. Diaphragms uncommon. Exozonal width from one-fifth to more than half branch radius. Lamellar profile V-shaped in exozone. In one species, semiannular monticulelike ridge developed by elongation of some autozooecia; other polymorphs absent in annulations. One or two acanthostyles proximal to each zooecial chamber; most stylets parallel to zooecia, a few less steeply inclined to surface than zooecia. Aktinotostyles common to abundant; diameters generally constant through exozone. Mural spines may be present in exozonal wall of zooecia. Some acanthostyles arising in endozone, most stylets arising in exozone. [Many more species have been assigned to *Rhombopora* than can be readily justified by comparison with the type species, bearing in mind ranges of variation in other rhabdomesine genera. *Rhombopora* is distinguished on branch size and shape, zooecial shape and orientation, and presence of acanthostyles, aktinotostyles, and only a few metapores. In many later species of *Rhombopora*, size increases, zooecial arrangement is locally irregular, the median axis becomes less well defined, there is some tendency toward development of axial zooecia, the zone of budding is somewhat broadened, most endozonal walls become relatively thin, and the exozone becomes relatively narrow. These characters are similar to those of some trepostomates. *Rhombopora* and its allies are here considered to be rhabdomesines because of the restricted nature of the budding locus, zooecial shape, and similarity to such unequivocal rhabdomesines as *Orthopora*.] *?U.Dev.,L.Miss.(Osag.)-U.Perm. (Dzhulf.)*, N.Am., Asia, S.Am.——Fig. 286,3a–e. *R. lepidodendroides*; *a*, irregular branch axis, zooecial shapes, broad exozone; long. sec., lectotype, USNM 168360, ×30; *b*, zooecial cross sections, lamellar profile; transv. sec., paralectotype, USNM 168360, ×30; *c*, elongate zooecia near axis, zooecial shapes, narrow exozone, stylet arrangement; long. sec., paralectotype, USNM 168359, ×30; *d*, irregular growth habit, apertural arrangement; paralectotype, USNM 240773, ×10; *e*, living chamber outlines, acanthostyles, paurostyles; tang. sec., paralectotype, USNM 168359, ×30.

Klaucena Trizna, 1958, p. 213 [**K. immortalis*; OD; Taidonskaya zone, L. Carb. (up. Tournais.), Kondoma River, Kuznetsk basin, USSR]. Zoarium dendroid, jointing unknown. Branch diameters 0.7 to 2.0 mm. Apertural arrangement rhombic, locally irregular. Metapores unknown. Axial region formed by well-defined linear axis or by median plate similar in structure to other endozonal walls. Zooecial bases inflated; zooecia initially polygonal, irregular in cross section, becoming hexagonal. Zooecial bend rounded; some living chamber outlines deflected by acanthostyles; chambers oriented 60° to 80° to branch surface, orientation varied within single specimens. Zooecial length approximately 6 times diameter; longitudinal arrangement of zooecia irregular. Hemisepta absent, diaphragms rare to scattered in some species. Exozonal width approximately one-third to half branch radius. Zooecial boundaries, lamellar profile unknown. Large acanthostyles may be scattered in exozone, possible aktinotostyles may be developed proximal and distal to zooecial chambers. [Trizna (1958) based *Klaucena* largely on the presence of a planar, median, budding surface in species of general rhabdomesine character. Two subgenera, *Klaucena* and *Spira*, were recognized. Similar, discontinuous, median plates are not unusual in the Rhabdomesina and, therefore, this feature by itself is a weak generic criterion. Available information does not permit full reassessment of affinities, but zooecial shape and stylet development, especially in *K. (Spira)*, suggest affinities with the Rhomboporidae.] *L.Carb. (Tournais.-Visean)*, USSR.

K. (Klaucena). Species of *Klaucena* with branch diameters 1.3 to 2.0 mm; axial region formed by median plate similar in structure to other endozonal walls. Scattered diaphragms may be present; exozonal width approximately one-third branch radius; large acanthostyles scattered in exozone. [Distinguished on the presence of a median plate, zooecial shape and arrangement, and development of acanthostyles.] *L.Carb.(Tournais.)*, USSR.——Fig. 286,1a–c. **K. (K.) immortalis*, holotype, VNIGRI 271/913, all ×20; *a*, living chamber outlines and arrangement, large stylets; tang. sec.; *b*, planar median plate, zooecial cross sections, transv. sec.; *c*, zooecial outlines and arrangement; long. sec. (Trizna, 1958).

K. (Spira) Trizna, 1958, p. 218 [**K. (S.) altinodata*; OD; Taidonskaya zone, L. Carb. (up. Tournais.), Kondoma River, Kuznetsk basin, USSR]. Species of *Klaucena* with branch diameters 0.7 to 2.0 mm; axial region apparently formed by linear well-defined axis in which local alignment of zooecial walls forms a weak axial plate in some intervals. Diaphragms rare; exozonal width approximately half branch radius; possible aktinotostyles developed proximal and distal to zooecial chambers. [Distinguished by zooecial shape and arrangement, and stylet development]. *L.Carb.(Tournais.-Visean)*, USSR.——Fig. 286,2a–d. **K. (S.) altinodata*, all ×20; *a*,

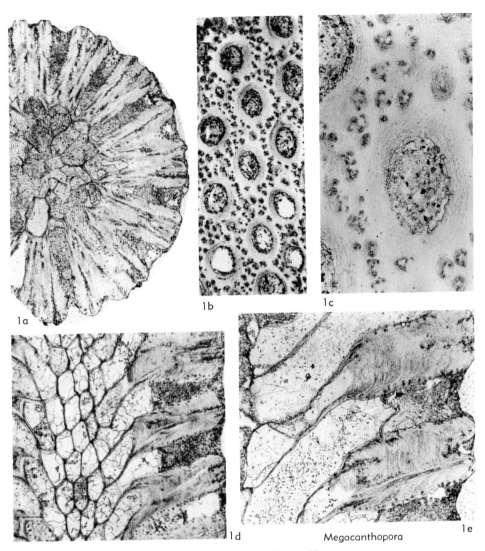

1a

1b

1c

1d

Megacanthopora

1e

Fig. 287. Rhomboporidae (p. 579).

apertural arrangement; tang. sec., paratype, VNIGRI 276/913; *b,* zooecial shapes and arrangement, stylets; long. sec., holotype, VNIGRI 275/913; *c,* arrangement of zooecia, exozonal development; transv. sec., holotype; *d,* zooecial cross sections and arrangement; transv. sec., paratype, VNIGRI 277/913? (Trizna, 1958).

Megacanthopora MOORE, 1929, p. 10 [**M. fallacis*; M; Wayland Sh. Mbr., Graham F., Cisco Gr., U. Penn., Cisco, Texas, USA] [=*Neorhombopora* SHISHOVA, 1964, p. 55]. Zoarium with jointing unknown; branch diameters 0.7 to 4.5 mm, somewhat varied between bifurcations. Apertural arrangement basically rhombic, locally irregular. Metapores common. Axial region formed by irregular linear axis. Zooecial bases inflated; zooecia initially polygonal, irregular in cross section, becoming hexagonal. Zooecial divergence from axis commonly about 45°, rarely less. Zooecial bend generally rounded; living chambers oriented 80° to 90° to branch surface. Zooecial length usually 5 to 8 times diameter; longitudinal arrangement of zooecia varied, irregular. Diaphragms uncommon. Exozonal width about one-third to two-thirds branch radius. Lamellar profile broadly V-shaped in exozone. Acanthostyles uncommon, not localized, angle relative to zoarial surface may be greater than zooecial angle. Aktinotostyles abundant, generally closely spaced; diameter approximately constant through length. Mural spines regularly arranged in exozonal living-chamber wall. Stylets arising in exozone. [DUNAEVA (1973) assigned *Megacanthopora* to the Stenoporidae, apparently on the presence of two types of stylets and narrow tubular exozonal cavities, here termed metapores. More than one stylet type occurs within single specimens elsewhere in the Rhabdomesina (e.g., most genera of Rhabdomesidae), and slender, apparently nonhomologous cavities or depressions are widely distributed in the class Stenolaemata. Therefore, I do not consider these structures in *Megacanthopora* to indicate affinities with the Trepostomata, and I assign the genus to the Rhabdomesina because it possesses the restricted budding pattern and basic angular zooecial shape and orientation typical of the suborder. Distinctive generic features include zooecial shape and arrangement, development of the exozone, and presence and development of acanthostyles, aktinotostyles, and metapores. *Megacanthopora* is similar to *Rhombopora,* differing only in metapore abundance, and possibly relative exozonal width. *Neorhombopora* SHISHOVA (1964) was named for species lacking large acanthostyles; however, its type species, *Rhombopora crassa* (ULRICH), possesses large acanthostyles and differs little from *M. fallacis.* Therefore, *Neorhombopora* is here synonymized with *Megacanthopora.*] *U.Carb.(Namur.-Stephan.),* N.Am., USSR.——FIG. 287,*1a–e.* **M. fallacis;* *a,* lamellar profile, stylets; transv. sec., paratype, KUMIP 58441, ×30; *b,* metapores, stylets; tang. sec., paratype, KUMIP 58441, ×30; *c,* aktinotostyles, acanthostyle, mural spines, metapores; tang. sec., paratype, KUMIP 58441, ×100; *d,* autozooecial shapes; biased long. sec., paratype, KUMIP 58441, ×30; *e,* autozooecial shapes, complexly arranged stylets; long. sec., paratype, KUMIP 58438, ×50.

Pamirella GORYUNOVA, 1975, p. 62 [**P. nitida;* OD; Bezardarinska F., L. Perm. (Artinsk.), Kur-Teka River, Pamir, USSR]. Zoarium with jointing unknown; branch diameters 0.5 to 2.5 mm, generally constant between bifurcations. Apertural arrangement rhombic. Metapores unknown. Axial region usually formed by well-defined linear axis; endozonal zooecia may parallel axis for short intervals, but true axial zooecia not developed. Zooecial bases attenuated or weakly inflated; zooecial cross sections polygonal, irregular in endozone. Zooecial divergence from axis approximately 20° to 40°. Zooecial bend rounded, living chamber outlines may be deflected by stylets, chambers oriented 70° to 90° to branch surface. Zooecial length generally ranging to about 10 times diameter; longitudinal arrangement of zooecia somewhat irregular. Diaphragms may be common. Exozonal width approximately half to two-thirds branch radius.

Lamellar profile V-shaped in exozone. Acanthostyles common to abundant, filling exozone in some species, usually not aligned in well-defined series. Acanthostyles arising in exozone, parallel to zooecia; core typically large, well developed; sheath laminae commonly subparallel to core, sharply defined. [As originally described, *Pamirella* included the type species and *P.* (ex *Rhombipora*) *pulchra* (BASSLER); it is here extended to include *P. orientalis* (BASSLER), *P. nicklesi* (ULRICH), *P. minor* (ULRICH), and *P. asperula* (ULRICH), all previously assigned to *Rhombopora. Pamirella* is distinguished on development of the axial region, zooecial shape and orientation, acanthostyle development, and lack of hemisepta.] *L.Carb.(Osag.)-L.Perm.(Artinsk.),* USSR, Timor, N.Am.—— FIG. 288,*1a–d.* **P. nitida; a,* zooecial cross sections, stylet development, lamellar profile; transv. sec., holotype, PIN 2351/215, ×25; *b,* branch axis, zooecial shapes, stylet development; long. sec., PIN 2351/99, ×25; *c,* branch axis, zooecial shapes, stylet development; long. sec., holotype, ×25; *d,* apertural and stylet arrangements; tang. sec., holotype, ×40.

Primorella ROMANCHUK & KISELEVA, 1968, p. 57 [**P. polita;* OD; Barabash Suite, U. Perm., Bol'shoy Mangugay River, Maritime Terr., USSR]. Zoarium with jointing unknown; branch diameters 1.0 to 1.5 mm, probably varied between bifurcations. Apertural arrangement basically rhombic, somewhat irregular. Metapores unknown. Axial region formed by linear, generally well-defined axis. Zooecial bases weakly inflated; zooecial cross sections polygonal, irregular in endozone. Zooecial divergence approximately 20° to 30°. Zooecial bend rounded. Living chambers generally oriented about 90° to branch surface. Zooecial length 5 to 8 times diameter; longitudinal arrangement of zooecia irregular. Diaphragms scattered. Exozone irregular, wide, ranging to four-fifths branch radius. Lamellar profile broadly V-shaped in exozone. Aktinotostyles abundant, in single or double series in exozone; arising near base of exozone, with diameters relatively constant with growth, parallel to zooecia. [*Primorella* was originally assigned to the Trepostomata (ROMANCHUK & KISELEVA, 1968), apparently because of overall growth habit. GORYUNOVA (1975) reassigned the genus to the Rhabdomesoidea (=Rhabdomesina), noting it differed from *Pamirella* only in stylet development. *Primorella* is here included in the Rhabdomesina because of the nature of the axial region, zooecial shape, and development of stylets. It is distinguished on apertural arrangement, zooecial shape and arrangement, development of exozone, and development of aktinotostyles as the only stylet type.] *U.Carb.(Stephan.)-U.Perm.,* USSR.——FIG. 289,*2a–c.* **P. polita,* PIN 2210/386; *a,* axial

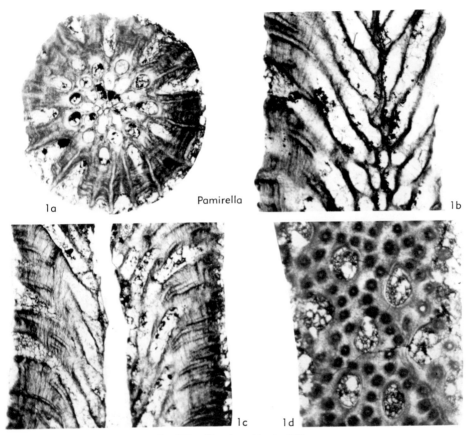

la   Pamirella   1b

lc   ld

FIG. 288.   Rhomboporidae (p. 580).

region, zooecial shapes and arrangement; long. sec., ×24; *b,* zooecial arrangement, lamellar profile, stylets; transv. sec., ×40; *c,* apertural arrangement, stylet development; tang. sec., ×40.

Saffordotaxis BASSLER, 1952, p. 385 [*Rhombopora incrassata* ULRICH, 1888, p. 89; OD; New Providence Sh., L. Miss., Kings Mt. at Halls Gap, Lincoln Co., Ky., USA]. Zoarium with jointing unknown; branch diameters 0.7 to 2.0 mm, constant between bifurcation. Apertural arrangement rhombic. Metapores unknown. Axial region usually formed by well-developed, linear axis; alignment of zooecial walls forming planar median surface in some intervals. Zooecial bases weakly inflated; zooecia initially polygonal, irregular in cross section, becoming hexagonal. Zooecial divergence approximately 20° to 30°. Zooecial bend abrupt; living chambers in exozone oriented about 90° to branch surface. Zooecial length about 8 times diameter; longitudinal arrangement of zooecia regular. Diaphragms rare. Exozonal width one-third to half branch radius. Lamellar profile broadly rounded in exozone. Aktinotostyles common to abundant, in single or multiple rows, arising near base of exozone, rarely with nonlaminated core near stylet base, typically expanding with growth, paralleling zooecia. [*Saffordotaxis* is distinguished on zooecial shape, and arrangement and presence of aktinotostyles as the only stylet type. Its characters are very similar to those in early species of *Rhombopora,* differing largely in stylet development. Intervals of nonlaminated core, as in acanthostyles, are present in very few stylets of *Saffordotaxis,* but otherwise, only typical aktinotostyles are present. In contrast, *Rhombopora* has one or two enlarged acanthostyles proximal to each zooecium.] *L.Miss.(Kinderhook.-Osag.),* E.N.Am.——FIG. 289, *1a–e.* *S. incrassata* (ULRICH); *a,* axial region, zooecial shapes, stylets; long. sec., syntype, USNM 43345, ×40; *b,* axial region, zooecial shapes, aktinotostyles; long. sec., syntype, USNM 240774, ×24; *c,* zooecial cross sections, exozonal development; transv. sec., syntype, USNM 240774, ×24; *d,* living chamber outlines, thick exozonal walls, stylets; tang. sec., syntype, USNM 43345, ×24; *e,* apertural arrangement, stylets; UI X-5272, about ×5.

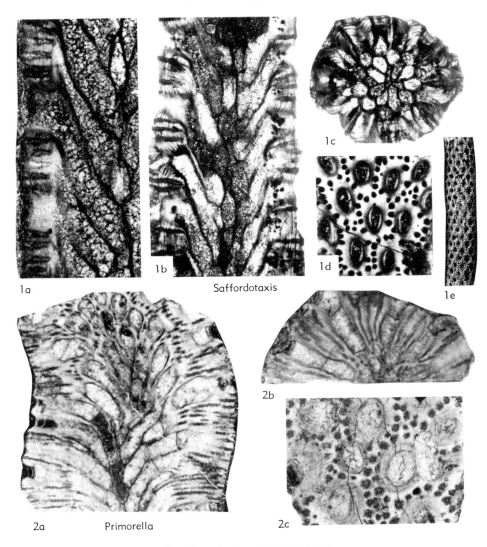

1a   1b   1c   1d   1e

Saffordotaxis

2a   Primorella   2b   2c

Fig. 289.   Rhomboporidae (p. 580–581).

Family BACTROPORIDAE
Simpson, 1897

[Bactroporidae SIMPSON, 1897, p. 553]

Zoaria erect, not known to branch; articulated basally in at least one species, proximal side of joint unknown; tapered basal interval lacking apertures, bearing nodose, discontinuous ridges. Stem diameters 1 to 2 mm, constant along length except for tapered base; cross sections subcircular. Apertural arrangement rhombic, ridges absent. Metapores absent. Axial region formed by narrow median plate; median rods questionably developed in some species. Zooids budded from median plate, divergence about 30°. Zooecial bases inflated in longitudinal sections. Zooecial cross sections polygonal, regular in endozone. Zooecial bends rounded; living chambers elliptical in cross section in exozone, oriented about 50° to 60° to stem surface. Zooecial lengths 7 to 10 times diameter. Single, prominent, straight hemiseptum on proximal wall at zooecial bend in one species; absent in others. Diaphragms absent. Exozonal width about half stem radius.

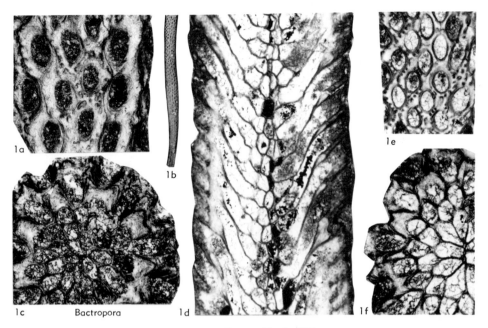

1a 1b 1c Bactropora 1d 1e 1f

FIG. 290.　Bactroporidae (p. 583).

Zooecial boundaries may not be visible; or narrow, irregular, dark with granular material in some areas. Lamellar profiles V-shaped in exozone, flattened between stylets. Heterostyles abundant, filling exozonal wall between zooecia, in linear series or irregular groups, arising in exozone, approximately parallel to zooecial chambers. Mural spines regularly arranged in exozone of one species. *M.Dev.*

Bactropora HALL & SIMPSON, 1887 p. xv [*?*Trematopora granistriata* HALL, 1881, p. 182; OD; ?Ludlowville Sh., Hamilton Gr., M. Dev., Darien Center, Genesee Co., N.Y., USA]. Characters of family. [*Bactropora* is distinguished on zoarial and zooecial shapes, development of exozone, and presence of heterostyles. It is similar to *Nematopora* and *Orthopora* in development of the axial region and zooecial shape, but distinctive in zooecial orientation, stylet development, and presence of basal articulation joints.] *M.Dev.(Erian)*, E.N.Am.——FIG. 290,*1a–c*. *B. granistriata* (HALL), holotype, NYSM 599; *a*, living chamber outlines, stylet development; tang. sec., ×45.0; *b*, external form, apertural arrangement, stylet development; ×13.5; *c*, zooecial cross sections, lamellar profile; transv. sec., ×45.0——FIG. 290,*1d–f*. *B. simplex* (HALL) (named in *Nematopora* by ULRICH, 1886), ?Ludlowville Sh., Hamilton Gr., M. Dev., N.Y.,

USA, holotype, NYSM 817; *d*, axial region, zooecial shapes and arrangement; long. sec., ×27.0; *e*, living chamber outlines, stylet development; tang. sec., ×27.0; *f*, median plate, zooecial cross sections; transv. sec., ×45.0.

Family NIKIFOROVELLIDAE Goryunova, 1975

[Nikiforovellidae GORYUNOVA, 1975, p. 67]

Zoaria dendroid, jointing unknown. Branch diameters approximately 0.5 to 2.0 mm, relatively constant between bifurcations; branches subcircular in cross section. Apertural arrangement rhombic, longitudinal ridges present or absent. Metapores scattered, or closely spaced in exozonal walls between zooecia, or absent; metapores arising at base of exozone, diaphragms absent. Axial region formed by linear axes or planar walls; elongate zooecia may parallel axis, true axial zooecia not developed. Zooecial bases inflated to flattened in longitudinal section. Zooecial cross section usually polygonal in endozone, may be subhexagonal or rounded. Zooecial divergence from axial region 20° to 70°. Zooecial bends mostly rounded, may be abrupt or lacking. Living chambers in exozone usually elliptical in cross section, may be subcircular; outline may be deflected inward by stylets. Living chambers oriented at 50° to 90° to branch surface, orientation varied in some branches. Zooecial length 2 to 10 times diameter. Hemisepta usually absent, weakly developed in some species. Diaphragms generally few; terminal diaphragms rare. Exozonal width one-third to more than half branch radius. Zooecial boundaries locally not visible; or dark, irregular, with granular or nonlaminated wall material in some areas of most taxa. Lamellar profiles V-shaped or rounded in exozone; becoming flattened between widely spaced apertures. Paurostyles and acanthostyles common to abundant; aktinotostyles present in one genus; stylets generally arising at base of exozone, parallel to zooecial chambers. Mural spines may be regularly arranged in exozonal living chambers. *?L.Dev.,M.Dev.-U.Perm.*

Nikiforovella NEKHOROSHEV, 1948a, p. 56 [**N. alternata*; OD; L. Carb.; near Lake Baikal, USSR]. Branch diameters about 1 mm. Longitudinal ridges absent. Metapores relatively few in most species, densely spaced between autozooecia in few species. Axial region usually formed by well-defined linear axis or local planar wall. Axial zooecia absent, but endozone of zooecia may ascend near branch axis for short intervals before diverging; otherwise similar to typical autozooecia. Zooecial bases usually weakly inflated. Zooecial cross sections polygonal, commonly hexagonal in endozone. Zooecial divergence from axis approximately 30° to 45°. Zooecial bend usually rounded. Living-chamber walls may be deflected into chambers by stylets; chambers oriented 70° to 90° to branch surface, orientation varied within zoaria in some species. Zooecial length 4 to 8 times diameter; longitudinal arrangement of zooecia regular to irregular. Hemisepta usually absent; may be weakly developed. Diaphragms rare. Exozonal width mostly greater than half branch radius. Zooecial boundary commonly irregular, locally with granular or nonlaminated wall material, or locally not visible. Lamellar profile in exozone rounded between closely spaced apertures. Paurostyles and acanthostyles common to abundant, scattered; some acanthostyles large, well developed. Mural spines may be regularly arranged in exozonal living-chamber wall. *M.Dev.-L.Perm.*, USSR, S.E.Asia, N.Am.——FIG. 291,*1a–c.* **N. alternata,* holotype, TsGM 201, all ×40; *a,* zooecial shapes and

orientation; long. sec.; *b,* scattered stylets, metapores; tang. sec.; *c,* linear axis, zooecial cross sections; transv. sec. (Nekhoroshev, 1948b).

Acanthoclema HALL, 1886, pl. 25 [**Trematopora alternata* HALL, 1883b, p. 148; OD; Onondaga Ls., equals up. Helderberg Gr. of HALL, M. Dev., Onondaga Valley, N.Y., USA]. Branch diameters 0.7 to 1.5 mm. Longitudinal ridges absent. Single metapore proximal to each zooecium. Axial region formed by generally well-defined axis. Zooecial bases flattened. Zooecial cross section polygonal, in endozone subhexagonal. Zooecial divergence from axis approximately 70°. Zooecial bend not developed, proximal wall of zooecium deflected abruptly at metapore, distal wall not deflected. Living chamber oriented approximately 70° to branch surface. Zooecial length 2 to 3 times diameter; longitudinal arrangement of zooecia regular. Hemisepta absent; shallow or terminal diaphragms may be developed, other diaphragms absent. Exozonal width about half branch radius. Zooecial boundary commonly irregular, locally not visible, or locally with granular or nonlaminated wall material. Lamellar profile in exozone rounded in transverse view, near metapore; flattened in longitudinal view. Paurostyles or acanthostyles few, concentrated between longitudinally successive apertures. Mural spines regularly arranged in exozonal living-chamber wall. [Differentiation of *Acanthoclema, Nikiforovella, Streblotrypa,* and *Streblotrypella* depends largely on the development of metapores, stylets, and median axes; however, these features may vary significantly within populations. Because of inadequate illustration, *Acanthoclema* was largely ignored during development of the concepts of *Nikiforovella* and *Streblotrypella. Acanthoclema* is here restricted to those nikiforovellid species with a single metapore developed proximal to each autozooecium. It is further characterized by regular arrangement of zooecia and metapores, development of a linear axis, zooecial shape, and nature of the stylets. *Acanthoclema* is also similar to the arthrostylid *Cuneatopora* in typical mature branch diameter, zooecial shape and orientation, axial definition, and presence of metapores, but *Cuneatopora* is articulated and its metapores are lateral to the autozooecia.] *M.Dev.(Erian),* E.N.Am.——FIG. 292,*1a–c.* **A. alternatum* (HALL), holotype, NYSM 579; *a,* linear axis, zooecial shapes, stylets; long. sec., ×47; *b,* median axis, zooecial cross sections; transv. sec., ×47; *c,* zooecial shapes, mural spines; long. sec., ×94.——FIG. 292,*1d,e. A. scutulatum* HALL, Hamilton Gr., M. Dev., N.Y., USA; *d,* linear axis, zooecial shapes, metapores; long. sec., USNM 240782, ×47; *e,* apertural arrangement, stylets, metapores; tang. sec., USNM 168344, ×94.

Pinegopora SHISHOVA, 1965, p. 60 [**P. delicata*; OD; U. Perm. (Kazan.), Pinega River, Arkhan-

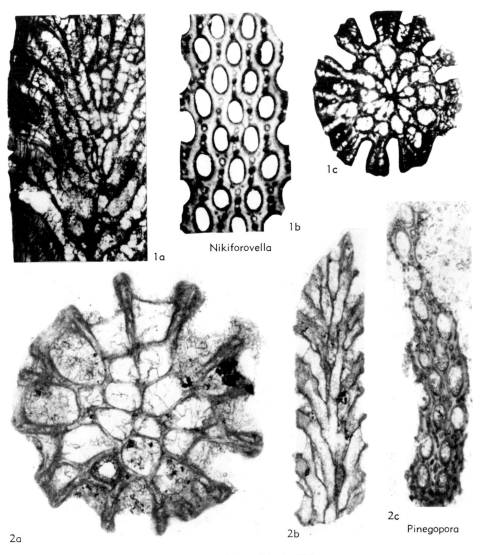

Nikiforovella

1a
1b
1c

2a
2b
2c

Pinegopora

FIG. 291. Nikiforovellidae (p. 584).

gel'sk Prov., Russ. plat., USSR]. Branch diameters 0.6 to 0.7 mm. Longitudinal, tuberculate ridges separating apertural rows. Metapores scattered in exozone. Axial region formed by elongate endozonal intervals of some zooecia, undifferentiated except in length; shorter zooecia budded from surfaces of longer zooecia. Zooecial bases weakly inflated. Zooecial cross sections polygonal in endozone, usually irregular. Zooecial divergence from axis approximately 25°. Zooecial bend rounded. Living chamber orientation varied, usually about 50° to surface. Zooecial length varied, shorter zooecia approximately 7 times diameter. Longitudinal arrangement of zooecia irregular. Hemisepta absent, diaphragms

scattered. Exozonal width approximately one-third branch radius; endozonal-exozonal boundary gradational. Zooecial boundary irregular, granular, locally not visible. Lamellar profile V-shaped in exozone. Acanthostyles scattered. [*Pinegopora* is distinguished on the nature of the axial region, zooecial shape and orientation, and development of the exozone.] *U.Perm.(Kazan.),* USSR.——FIG. 291,*2a–c*. **P. delicata,* holotype, PIN 1692/275; *a,* zooecial cross sections, lamellar profile; transv. sec., ×100; *b,* zooecial outlines, arrangement; long. sec., ×30; *c,* apertural arrangement; tang. sec., ×30.

Streblotrypella NIKIFOROVA, 1948, p. 41 [**Streblotrypa major* ULRICH in MILLER, 1889, p. 326; OD;

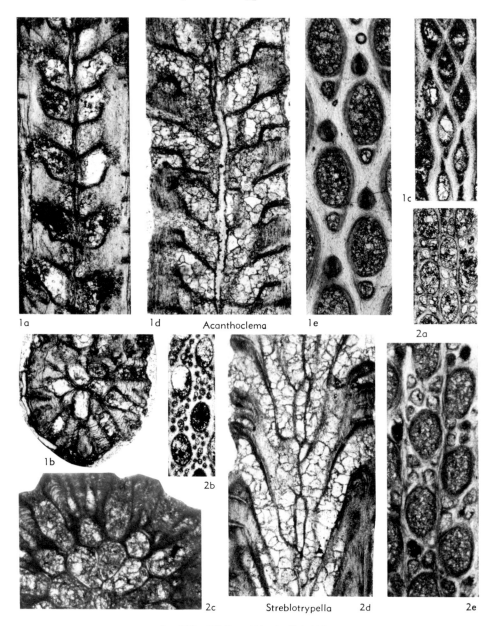

FIG. 292.   Nikiforovellidae (p. 584–585).

New Providence Sh., L. Miss., Kings Mt. at Halls Gap, Lincoln Co., Ky., USA]. Branch diameters 0.7 to 1.5 mm. Longitudinal ridges may separate rows of apertures. Metapores few to densely spaced between autozooecia. Axial region formed by well-defined linear axis. Zooecial base weakly inflated. Zooecial cross sections polygonal in endozone, usually hexagonal. Zooecial divergence from axis approximately 20° to 30°. Zooe-

cial bend abrupt. Living chambers usually oriented 80° to 90° to zoarial surface. Zooecial length 5 to 10 times diameter; longitudinal arrangement of zooecia generally regular. Hemisepta absent, diaphragms rare. Exozonal width approximately one third to half branch radius. Zooecial boundary commonly dark, irregular; locally granular, or with nonlaminated wall material, or not visible. Lamellar profile broadly flattened to V-shaped in exozone. Aktinotostyles or acanthostyles scattered, or stylets absent. [Compared to *Nikiforovella, Streblotrypella* usually has more elongate and steeply ascending zooecia, living chambers in the exozone are more perpendicular to the zoarial surface, and the exozone generally is narrower. *Streblotrypella* sometimes lacks stylets and has a correlated concentration of metapores in a cluster proximal to the zooecia.] *?Dev.,L.Carb.(Osag.)-L.Perm.,?U. Perm.,* N.Am., USSR, S.E.Asia, Japan, Australia.——Fig. 292,*2a–e.* **S. major* (ULRICH); *a,* apertural and metapore arrangements; tang. peel, USNM 240790, ×28; *b,* apertural arrangement, stylets, metapores; tang. sec., USNM 240791, ×28; *c,* zooecial cross sections, lamellar profile; transv. sec., syntype, USNM 44095, ×47; *d,* zooecial shapes, metapore arrangement; long. sec., syntype, USNM 44095, ×47; *e,* apertural and metapore arrangements; tang. sec., USNM 240789, ×47.

Family HYPHASMOPORIDAE
Vine, 1886

[Hyphasmoporidae VINE, 1886, p. 95] [=Streblotrypidae ULRICH, 1890, p. 365]

Zoaria dendroid, jointing unknown. Branch diameters 0.2 to 5.5 mm, relatively constant between bifurcations in most species; branches subcircular in cross section. Apertural arrangement rhombic, longitudinal ridges commonly separating rows of apertures. Metapores generally filling exozonal wall between autozooecia, but may be scattered; usually in longitudinal rows between successive apertures, present or absent beyond distolateral margins of zooecial apertures; arising in late endozone or at base of exozone; diaphragms absent. Axial region formed by weak to well-defined axial zooecia, or well-defined bundles of axial zooecia. Axial zooecia, especially those in bundles, typically with narrower and thinner walls than in endozones of autozooecia. Zooids budded from surfaces of axial zooecia, or near branch axis. Autozooecial bases attenuated to inflated in longitudinal section. Autozooecial cross sections polygonal in endozone, irregular to hexagonal. Zooecial divergence from axial region approximately 10° to 30°. Zooecial bends rounded to abrupt. Living chambers in exozone usually subcircular to elliptical in cross section, may be flattened proximally, usually oriented about 90° to branch surface. Autozooecial length 8 or more times diameter. Single, slender hemiseptum usually on distal wall in late endozone; proximal wall at zooecial bend commonly inflated, but not developed as true hemiseptum; hemisepta may be absent. Diaphragms scattered or absent. Exozonal width from one-third to over half branch radius. Zooecial boundaries narrow, dark, irregular, with granular or nonlaminated material in some areas; locally not visible. Lamellar profile V-shaped to rounded in exozone, zooecial lining may be present. Stylets rare, parallel to autozooecial chambers. *L.Carb.-U.Perm.*

Hyphasmopora ETHERIDGE, 1875, p. 43 [**H. buskii*; M; L. Carb., E. Kilbride, Scot.]. Branch diameters 0.2 to 0.4 mm, usually constant between bifurcations. Weakly developed longitudinal ridges separating apertural rows. Metapores densely spaced between autozooecia, absent beyond distolateral margins of zooecial apertures; arising at base of exozone. Axial region formed by weakly differentiated, narrow, axial zooecia paralleling axis for varying intervals before diverging toward surface, assuming autozooecial morphology. Zooecial bases attenuated. Zooecial cross sections polygonal in endozone, irregular. Zooecial divergence from axial region 10° to 20°. Zooecial bend rounded to abrupt. Living chamber orientation varied, inclined to branch surface. Zooecial length generally more than 10 times diameter; longitudinal arrangement of zooecia irregular. Single, straight, slender hemiseptum on distal wall in late endozone; proximal wall at zooecial bend inflated, true hemisepta lacking. Diaphragms apparently absent. Exozonal width approximately one-third to half branch radius. Zooecial boundary well defined, narrow, irregular, granular, ramifying into multiple planar dark zones at base of longitudinal ridges in exozone; thickened, nonlaminated wall material locally developed, especially at base of exozone. Lamellar profile in exozone V-shaped to subrounded between dark zones, rounded between metapores. Stylets absent. [*Hyphasmopora* is distinguished on zooecial shape and arrangement, and wall structure. It resembles *Streblotrypa* in nature of the zooecia, exozone, and metapores, but is distinctive in the presence of weakly defined axial zooecia and well-defined zooecial boundaries. Location of the primary types of *H. buskii* is unknown.] *L.Carb.,* Scot.——Fig. 293,*1a–f.* **H. buskii; a,* zooecial aperture, metapore arrangement; tang. sec., USNM 240779, ×75; *b,* metapores, zooecial

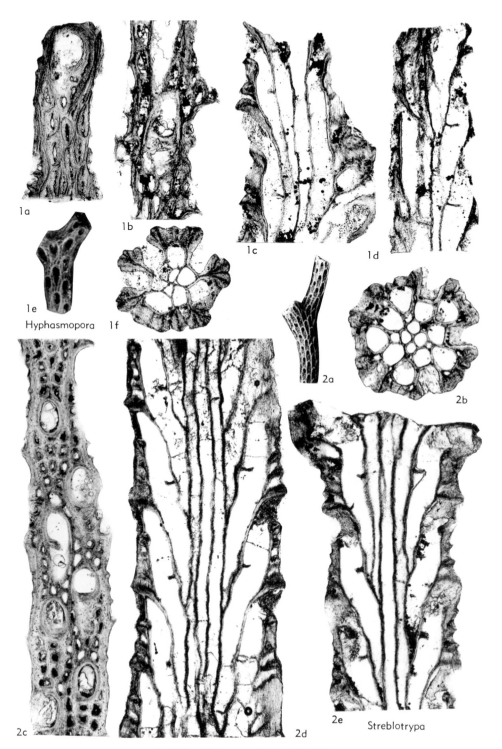

1a

1b

1c

1d

1e

Hyphasmopora   1f

2a

2b

2c

2d

2e   Streblotrypa

Fig. 293.   Hyphasmoporidae (p. 587–590).

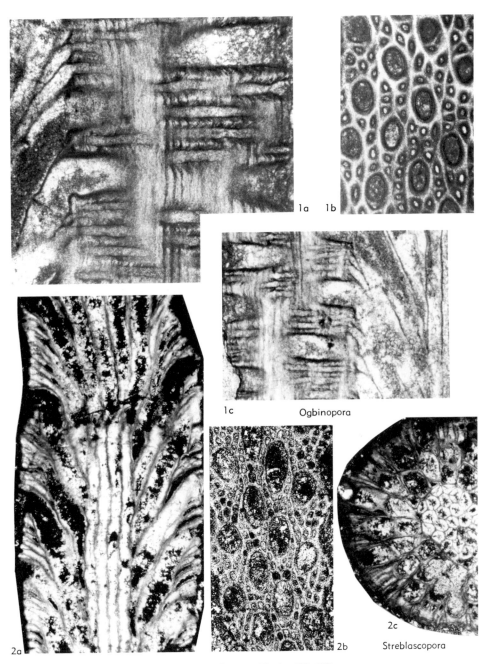

FIG. 294.   Hyphasmoporidae (p. 589–590).

boundaries; tang. sec., USNM 240778, ×75; *c*, zooecial shapes and boundaries; long. sec., USNM 240777, ×75; *d*, zooecial shapes and arrangement, hemisepta; long. sec., USNM 240775, ×75; *e*, zooecial aperture and metapore arrangements; USNM 240780, ×20; *f*, dark zones in exozone; transv. sec., USNM 240779, ×75.

Ogbinopora SHISHOVA, 1965, p. 59 [**O. armeniensis*; OD; Gnishik horizon, U. Perm. (Guadalup.), Ogbin Village, Transcauc., USSR]. Branch diameters 2.5 to 5.5 mm, varied between bifurcations. Irregular longitudinal ridges separating apertural rows. Metapores densely spaced between autozooecia, present beyond distolateral margins of zooecial apertures, arising in late endozone or base of exozone. Axial region formed by large bundle of axial zooecia. Autozooecial base attenuated. Autozooecial cross sections hexagonal in endozone. Zooecial divergence from axial region approximately 20° to 30°. Zooecial bend abrupt. Living chambers oriented approximately 90° to branch surface. Autozooecial length approximately 12 times diameter; longitudinal arrangement of autozooecia regular. Single, straight hemiseptum usually on distal wall in late endozone; proximal wall at zooecial bend usually inflated and may form massive hemiseptum. Diaphragms sparse. Exozonal width one-third to half branch radius. Zooecial boundary generally not visible; locally a discontinuous dark zone. Lamellar profile generally rounded in exozone; zooecial lining in autozooecia thick, sharply defined, absent from metapores. Stylets absent. [*Ogbinopora* is distinguished by zoarial size, presence of a large bundle of axial zooecia, zooecial shape, broad exozone, and development of hemisepta.] *Perm.(Artinsk.-Guadalup.),* USSR, S.E.Asia.——FIG. 294, *1a–c.* **O. armeniensis,* holotype, PIN 1613/126; *a,* lamellar profile, paired hemisepta at zooecial bend; long. sec., ×47; *b,* zooecial aperture and metapore arrangements; tang. sec., ×28; *c,* zooecial outlines, long. sec., ×28.

Streblotrypa VINE, 1885, p. 391 [**S. nicklisi;* M; Carb.; Yorkshire, Eng.] [=*Lanopora* ROMANCHUK, 1975, p. 77]. Branch diameters 0.7 to 2.5 mm, usually constant between bifurcations. Weak to well-developed longitudinal ridges separating apertural rows. Metapores usually densely spaced between autozooecia, rarely scattered in exozone; present or absent beyond distolateral margins of zooecial apertures, arising in exozone or rarely in late endozone. Axial region varied; ranging from few axial zooecia to large, well-defined axial bundles. Individual zooecia rarely diverging from well-defined axial bundles, but commonly diverging and developing morphology typical of autozooecia in species with few axial zooecia. Autozooecial bases attenuated to weakly inflated. Autozooecial cross sections polygonal in endozone, irregular or hexagonal. Zooecial divergence from axial region approximately 20° to 30°. Zooecial bend generally abrupt. Living chamber flattened proximally in exozone, chamber oriented about 90° to branch surface. Autozooecial length usually 8 to 12 times diameter. Longitudinal arrangement of autozooecia usually regular. True hemisepta rare or lacking; single, straight, slender hemiseptum may be present on distal wall in late endozone; proximal wall at zooecial bend inflated. Scattered diaphragms may be present. Exozonal width

usually between one-third and half branch radius, rarely greater. Zooecial boundary usually well defined, irregular, rarely not visible; discontinuous, nonlaminated wall material may be present. Lamellar profile rounded in exozone. Stylets usually absent; paurostyles and weakly developed acanthosyles may be present. [VINE's (1885) one specimen of *S. nicklisi* from England is lost. DUNCAN (1949) recommended replacement of VINE's specimen by a suite of fossils in collections of the U.S. National Museum, but such replacement does not fulfill ICZN requirements for designation of a neotype. Nevertheless, the concept of *S. nicklisi* has been generally based on the North American specimens illustrated here. *Lanopora* ROMANCHUK (1975) differs from *Streblotrypa* only in presence of swellings on longitudinal ridges, a feature I consider to be of no generic significance, and *Lanopora* is herein synonymized with *Streblotrypa.* Some species of *Streblotrypa* and *Streblascopora* BASSLER, 1952, are distinct, but others combine features of both genera; therefore, *Streblascopora* is herein reduced to subgenus rank under *Streblotrypa. Streblotrypa* is similar to the nikiforovellid *Streblotrypella* in zooecial orientation, trend toward stylet loss, and concentration of metapores proximal to zooecial apertures.] *U.Miss.(Meramec.)-U.Perm.,* Eu., Asia, Australia, N.Am., S.Am.

S. (Streblotrypa). Species of *Streblotrypa* lacking distinct bundle of axial zooecia; about 10 or fewer axial zooecia at any level in branch. Hemisepta usually present, metapores usually restricted to rows between zooecial apertures, commonly absent beyond distolateral margins of apertures. [*S. (Streblotrypa)* is distinguished by its axial region, narrow exozone, and usual lack of stylets. It differs from *S. (Streblascopora)* primarily in development of hemisepta and axial zooecia. In *S. (Streblotrypa),* polymorphs are relatively less numerous than in *S. (Streblascopora)* and they are not set off in a distinct axial bundle.] *U.Miss.(Meramec.)-U.Perm.,* Eu., Asia, Australia, N.Am., S.Am.——FIG. 293, *2a–e.* **S. (Streblotrypa) nicklisi* VINE, U. Miss. (Chester.), Ill., Ala., USA; *a,* apertural and metapore arrangements, longitudinal ridges; USNM 240786, ×10; *b,* zooecial cross sections, lamellar profile; transv. sec., USNM 240788, ×75; *c,* apertural and metapore arrangements; tang. sec., USNM 240788, ×75; *d,* axial zooecia, autozooecial outlines, hemisepta, metapores; long. sec., USNM 240784, ×75; *e,* axial zooecia, autozooecial outlines, hemisepta, metapores; long. sec., USNM 240788, ×75.

S. (Streblascopora) BASSLER, 1952, p. 385 [**Streblotrypa fasciculata* BASSLER, 1929, p. 66; OD; Perm., Soefa, Timor, Indon.]. Species of *Streblotrypa* with more or less clearly defined bundle of axial zooecia and more than

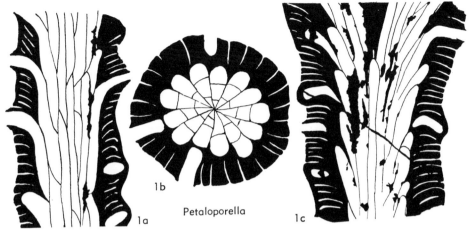

Fɪɢ. 295. Family Uncertain (p. 591).

about 10 axial zooecia at any level in branch. Hemisepta rare or absent, metapores common beyond distolateral margins of zooecial apertures. *L.Carb.-U.Perm.,* USSR, S.E.Asia, Japan, Australia, N.Am.——Fɪɢ. 294,*2a-c.* *S. (Streblascopora) fasciculata* Bᴀssʟᴇʀ, holotype, Delft 12340KA, all ×28; *a,* axial zooecia, autozooecial outlines, metapores; long. sec.; *b,* apertural and metapore arrangements; tang. sec.; *c,* axial bundle, zooecial cross sections, lamellar profile; transv. sec.

Family Uncertain

Petaloporella Pʀᴀɴᴛʟ, 1935b, p. 4 [*P. bohemica*; M; Branik Ls., M. Dev., Branik, Czech.]. Zoarium dendroid, jointing unknown. Branch diameters 1.1 to 1.8 mm, apparently relatively constant between bifurcations. Apertural arrangement rhombic, longitudinal ridges absent, metapores more or less densely spaced in exozone between autozooecia. Axial region formed by linear axis or axial zooecia. Zooids budded around axis or from surface of axial zooecia. Zooecial base weakly inflated. Zooecial cross section in endozone triangular. Zooecial divergence from branch axis 10° to 20°. Zooecial bend rounded. Living chambers oriented about 90° to branch surface. Autozooecial length approximately 10 times diameter. Longitudinal arrangement of zooecia somewhat irregular. Hemisepta, diaphragms, stylets absent. [Type specimens of *Petaloporella bohemica* could not be located in the Narodini Museum (Prague) and may be lost (W. A. Oʟɪᴠᴇʀ, pers. commun. to R. S. Bᴏᴀʀᴅᴍᴀɴ). The concept of *Petaloporella* cannot be refined because some features are unclear in the original illustrations. One drawing shows a well-defined median axis (Fig. 295,*1b*), whereas one appears to show axial zooecia (Fig. 295,*1c*), and another

(Fig. 295,*1a*) is difficult to interpret. If axial zooecia are present, *Petaloporella* may be a synonym of *Streblotrypa*.] *M.Dev.,* Czech.——Fɪɢ. 295,*1a-c.* *P. bohemica*; *a,* zooecial shapes, metapores; drawing, biased long. sec.; *b,* budding; drawing, transv. sec.; *c,* axial zooecia; drawing, long. sec., approx. ×20 (Prantl, 1935b).

Summary of Recent Important Taxonomic Changes in Rhabdomesina

Acanthoclema Hᴀʟʟ, 1886. Herein transferred from the Rhabdomesidae to the Nikiforovellidae; generic concept restricted in scope.

Bactropora Hᴀʟʟ & Sɪᴍᴘsᴏɴ, 1887. Herein returned from the Rhabdomesidae to the Bactroporidae, following Sɪᴍᴘsᴏɴ (1897).

Bactroporidae Sɪᴍᴘsᴏɴ, 1897. Family concept accepted herein; not in general usage since original description.

Coeloconus Uʟʀɪᴄʜ, 1889. Synonymized with *Rhabdomeson*; see Bʟᴀᴋᴇ (1976).

Cuneatopora Sɪᴇɢꜰʀɪᴇᴅ, 1963. Some species have been transferred from *Helopora*; see Kᴏᴘᴀʏᴇᴠɪᴄʜ (1975).

Hyphasmoporidae Vɪɴᴇ, 1886. Restricted in content by reassignment of some genera to the Nikiforovellidae by Gᴏʀʏᴜɴᴏᴠᴀ (1975).

Klaucena Tʀɪᴢɴᴀ, 1958. Herein transferred from the Rhabdomesidae to the Rhomboporidae.

Lanopora Rᴏᴍᴀɴᴄʜᴜᴋ, 1975. Herein synonymized with *Streblotrypa*.

Megacanthopora Mᴏᴏʀᴇ, 1929. Considered by some Soviet authors to belong to the Trepostomata, herein assigned to the Rhomboporidae.

Nematopora Uʟʀɪᴄʜ, 1888. Generally regarded as including only species articulated basally or not at all, herein considered to include *N. harrisi* (JAMES), a highly segmented species formerly assigned to *Helopora*.

Neorhombopora SHISHOVA, 1964. Herein synonymized with *Megacanthopora*.

Osburnostylus BASSLER, 1952. Herein transferred from the Tubuliporata to the Rhabdomesina.

Petaloporella PRANTL, 1935b. Herein transferred from the Tubuliporata to the Rhabdomesina, family uncertain.

Primorella ROMANCHUK & KISELEVA, 1968. Originally assigned to the Trepostomata, herein assigned to the Rhabdomesina, following GORYUNOVA, 1975.

Rhabdomesidae VINE, 1884. Herein restricted in scope by reassignment of some previously included genera to the Bactroporidae, Rhomboporidae, and Arthrostylidae.

Rhabdomesina [=Rhabdomesoidea] ASTROVA & MOROZOVA, 1956. Recognized herein as a suborder; generally given ordinal rank by Soviet authors.

Rhabdomeson YOUNG & YOUNG, 1874. Includes conical branches previously assigned to *Coeloconus*; see BLAKE (1976).

Rhombopora MEEK, 1872. Herein transferred from the Rhabdomesidae to the Rhomboporidae; concept of genus here much restricted.

Rhomboporidae SIMPSON, 1897. Family concept accepted herein; not in general usage since original description.

Saffordotaxis BASSLER, 1952. Herein transferred from the Rhabdomesidae to the Rhomboporidae.

Streblascopora BASSLER, 1952. Herein reduced to subgeneric rank and assigned to *Streblotrypa*.

Streblotrypella NIKIFOROVA, 1948. Herein transferred from the Hyphasmoporidae to the Nikiforovellidae.

Trematella HALL, 1886. Herein transferred from the Trepostomata to the Rhabdomesidae.

Tropidopora HALL, 1886. Herein transferred from the Rhabdomesidae to the Arthrostylidae.

Vetofistula ETHERIDGE, 1917. Transferred to the Coelenterata; see ROSS (1961).

Different authors have included genera in the Rhabdomesina that I consider to have other or uncertain affinities. Some of these genera were reassigned by various authors prior to this work, and include: *Archaeomeson* ASTROVA, 1965; *Clausotrypa* BASSLER, 1929; *Denmeadopora* FLEMING, 1969; *Goldfussitrypa* BASSLER, 1952; *Hayasakapora* SAKAGAMI, 1960; *Hyalotoechus* MCNAIR, 1942; *Idioclema* GIRTY, 1910; *Linotaxis* BASSLER, 1952; *Maychella* MOROZOVA, 1970; *Mongoloclema* SHISHOVA, 1970; *Nemacanthopora* TERMIER & TERMIER, 1971; *Nemataxidra* BASSLER, 1952; *Nematotrypa* BASSLER, 1911; *Ottoseetaxis* BASSLER, 1952; *Pesnastylus* CROCKFORD, 1942; *Rhombocladia* ROGERS, 1900; *Rhomboporella* BASSLER, 1936; *Spirillopora* GURICH, 1896; *Streblocladia* CROCKFORD, 1944; and *Syringoclemis* GIRTY, 1910.

REFERENCES

Compiled by JoAnn Sanner

[Smithsonian Institution, Washington, D.C.]

Alberstadt, L. P., Walker, K. R., & Zurawski, R. P., 1974, *Patch reefs in the Carters Limestone (Middle Ordovician) in Tennessee, and vertical zonation in Ordovician reefs:* Geol. Soc. Am. Bull., v. 85, p. 1171–1182, 10 text-fig.

Allman, G. J., 1856, *A Monograph of the Freshwater Polyzoa, including all the known species, both British and foreign:* vii + 119 p., 17 text-fig., 11 pl., The Ray Society (London).

Annoscia, Enrico, 1968, *Briozoi; introduzione allo studio con particolare riguardo ai Briozoi italiani e mediterranei:* 397 p., 24 pl., Palaeontographia Italica (Pisa).

Anstey, R. L., Pachut, J. F., & Prezbindowski, D. R., 1976, *Morphogenetic gradients in Paleozoic bryozoan colonies:* Paleobiology, v. 2, p. 131–146, 13 text-fig.

Armstrong, John, 1970, *Zoarial microstructures of two Permian species of the bryozoan genus Stenopora:* Palaeontology, v. 13, p. 581–587, pl. 112–117.

Astrova, G. G., 1955, *Mshanki:* in E. A. Ivanova *et al.,* Fauna ordovika i gotlandiya nizhnego techeniya r. podkamennoj Tunguski, ee ekologiya i stratigraficheskoe znachenie, Tr. Akad. Nauk SSSR Paleontol. Inst., v. 56, p. 128–161, text-fig. 11–22, pl. 14–21. [*Bryozoa:* in *Fauna of the Ordovician and Gotlandian of the lower reaches of the Podkamennaya Tunguska River, its ecology and stratigraphic significance.*]

——, 1960a, *Obshchaya chast':* in Osnovy Paleontologii, p. 15–42, text-fig. 1–27. [*General part:* in *Fundamentals of paleontology;* see T. G. Sarycheva, 1960 for complete citation.]

——, 1960b, *Silurijskie Fistuliporidy iz severnykh rajonov RSFSR:* Tr. Akad. Nauk SSSR Kom. Fil. Syktyvkar Inst. Geol., v. 1, p. 352–376, 1 text-fig., 5 pl. [*Silurian fistuliporids from the northern regions of the RSFSR.*]

——, 1964, *O novom otryade paleozojskikh mshanok:* Paleontol. Zh., 1964, no. 2, p. 22–31, 3 text-fig. [*A new order of Paleozoic Bryozoa.*]

——, 1965, *Morfologiya, istoriya razvitiya i sistema ordovikskikh i silurijskikh mshanok:* Tr. Akad. Nauk SSSR Paleontol. Inst., v. 106, 432 p., 52 text-fig., 84 pl. [*Morphology, evolutionary history, and systematics of Ordovician and Silurian bryozoans.*]

——, 1971, *K voprosu o morfologii i funktsional'nom znachenii akantopor u trepostomat (Bryozoa):* Paleontol. Zh., 1971, no. 3, p. 72–79, pl. 7,8. [*Morphology and functional significance of acanthopores in the Trepostomata (Bryozoa).*]

——, 1973, *Polymorphism and its development in the trepostomatous Bryozoa:* in G. P. Larwood (ed.), Living and Fossil Bryozoa, p. 1–10. 2 pl., Academic Press (London).

——, & Morozova, I. P., 1956, *K sistematike mshanok otryada Cryptostomata:* Dokl. Akad. Nauk SSSR, n.ser., v. 110, p. 661–664. [*Systematics of Bryozoa of the order Cryptostomata.*]

——, & Yaroshinskaya, A. M., 1968, *Rannedevonskie i eyfel'skie mshanki Salaira i Gornogo Altaya:* Tr. Tomsk. Gos. Univ., v. 202, p. 47–62, 2 text-fig., 5 pl. [*Early Devonian and Eifelian Bryozoa of Salair and Gorno-Altai.*]

Audouin, J. V., & Milne-Edwards, Henri, 1828, *Résumé des recherches sur les animaux sans vertèbres, faites aux îles Chausey:* Ann. Sci. Nat., v. 15, p. 5–19.

Balakin, G. V., 1974, *Pseudonematopora, novyj rod rannekamennougol'nykh mshanok:* Paleontol. Zh., 1974, no. 4, p. 130–132, 1 text-fig. [*Pseudonematopora, a new Early Carboniferous bryozoan genus;* transl. Paleontol. J., v. 8, p. 557–559.]

Banner, F. T., & Wood, G. V., 1964, *Recrystallization in microfossiliferous limestones:* Geol. J., v. 4, p. 21–34, 1 text-fig. pl. 1–6.

Banta, W. C., 1968, *The body wall of cheilostome Bryozoa, I. The ectocyst of Watersipora nigra (Canu and Bassler):* J. Morphol., v. 125, p. 497–508, 6 text-fig., 1 pl.

——, 1969, *The body wall of cheilostome Bryozoa, II. Interzoidal communication organs:* J. Morphol., v. 129, p. 149–170, 35 text-fig., 1 pl.

——, 1970, *The body wall of cheilostome Bryozoa, III. The frontal wall of Watersipora arcuata Banta, with a revision of the Cryptocystidea:* J. Morphol., v. 131, p. 37–56, 26 text-fig.

——, 1971, *The body wall of cheilostome Bryozoa, IV. The frontal wall of Schizoporella unicornis (Johnston):* J. Morphol., v. 135, p. 165–184, 6 pl.

——, 1972, *The body wall of cheilostome Bryozoa, V. Frontal budding in Schizoporella unicornis floridana:* Mar. Biol., v. 14, p. 63–71, 3 text-fig.

——, 1975, *Origin and early evolution of cheilostome Bryozoa:* in S. Pouyet (ed.), Bryozoa 1974, Doc. Lab. Geol. Fac. Sci. Lyon, hors-série 3, fasc. 2, p. 565–582, 25 text-fig., 2 pl.

——, McKinney, F. K., & Zimmer, R. L., 1974, *Bryozoan monticules; excurrent water outlets?:* Science, v. 185, p. 783–784, 2 text-fig.

Barnes, D. J., 1970, *Coral skeletons; an explanation of their growth and structure:* Science, v. 170, p. 1305–1308, 2 text-fig.

Barrois, Jules, 1877, *Recherches sur l'embryologie des Bryozoaires:* Trav. Inst. Zool. Lille, fasc. 1, 305 p., 16 pl.

——, 1882, *Embryogénie des Bryozoaires; essai*

d'une théorie générale du développement basée sur l'étude de la métamorphose: J. Anat. Physiol. Norm. Pathol. Homme Anim., v. 18, p. 124–157, pl. II.

Bassi, Ferdinando, 1757, *De quibusdam exiguio madreporis Agri Bononiensis:* De Bononiensis Scientiarum et Artium Instituto atque Academia Commentarii, 49 p., pl. 1–4 (Bononia).

Bassler, R. S., 1911, *The early Paleozoic Bryozoa of the Baltic Provinces:* U.S. Natl. Mus. Bull. 77, 382 p., 226 text-fig., 13 pl.

———, 1913, *Bryozoa:* in K. A. von Zittel & C. R. Eastman (eds.), Text-book of Paleontology, 2nd ed., v. 1, p. 314–355, text-fig. 436–525, Macmillan and Co. (London).

———, 1927, *Bryozoa:* in W. H. Twenhofel, Geology of Anticosti Island, Can. Geol. Surv. Mem. 154 [Geol. Ser. 135], p. 143–168, pl. 5–14.

———, 1929, *The Permian Bryozoa of Timor:* Paläontologie von Timor, Lief. 16, no. 28, p. 36–89, pl. 225–247 (Stuttgart).

———, 1934, *Notes on fossil and recent Bryozoa:* J. Wash. Acad. Sci., v. 24, p. 404–408.

———, 1935, *Bryozoa:* in W. Quenstedt (ed.), Fossilium Catalogus, 1: Animalia, pt. 67, 229 p., W. Junk ('s-Gravenhage).

———, 1936, *Nomenclatorial notes on fossil and recent Bryozoa:* J. Wash. Acad. Sci., v. 26, p. 156–162, 12 text-fig.

———, 1952, *Taxonomic notes on genera of fossil and recent Bryozoa:* J. Wash. Acad. Sci., v. 42, p. 381–385, 27 text-fig.

———, 1953, *Bryozoa:* in R. C. Moore (ed.), Treatise on Invertebrate Paleontology, Part G, xiv + 253 p., 175 text-fig., Geological Society of America & University of Kansas Press (New York & Lawrence).

Bathurst, R. G. C., 1975, *Carbonate Sediments and their Diagenesis:* Developments in Sedimentology, v. 12, 2nd ed., 658 p., 359 text-fig., Elsevier (New York).

Beerbower, J. R., 1960, *Search for the Past; An Introduction to Paleontology:* 562 p., Prentice-Hall, Inc. (Englewood Cliffs).

Beneden, P.-J. Van, 1848, *Recherches sur les Bryozoaires fluviatiles de Belgique:* Mem. Acad. R. Bruxelles, v. 21, p. 1–33, 2 pl.

Berner, R. A., 1971, *Principles of Chemical Sedimentology:* 240 p., McGraw-Hill Book Co. (New York).

Billings, Elkanah, 1865, *Palaeozoic Fossils, Vol. 1; containing descriptions and figures of new or little known species of organic remains from the Silurian rocks:* Geol. Surv. Can., 426 p., 401 text-fig.

Blainville, H. M. de, 1820, *Flustre:* in G. Cuvier, Dictionnaire des Sciences Naturelles, v. 17, p. 171–179 (Paris).

———, 1834, *Manuel d'Actinologie et de Zoophytologie:* viii + 694 p., 103 pl. (Paris).

Blake, D. B., 1973a, *Acanthopore ultrastructure in*

the Paleozoic bryozoan family Rhabdomesidae: in G. P. Larwood (ed.), Living and Fossil Bryozoa, p. 221–229, 2 pl., Academic Press (London).

———, 1973b, *Acanthopore morphology and function in the bryozoan family Rhabdomesidae:* J. Paleontol., v. 47, p. 421–435, 1 text-fig., 4 pl.

———, 1973c, *Coloniality and polymorphism in Bryozoa of the families Rhabdomesidae and Hyphasmoporidae (order Rhabdomesonata):* in R. S. Boardman, A. H. Cheetham, & W. A. Oliver, Jr. (eds.), Animal Colonies, p. 361–376, 37 text-fig., Dowden, Hutchinson, & Ross (Stroudsburg).

———, 1975, *The order Cryptostomata resurrected:* in S. Pouyet (ed.), Bryozoa 1974, Doc. Lab. Geol. Fac. Sci. Lyon, hors-série 3, fasc. 1, p. 211–223, 3 pl.

———, 1976, *Functional morphology and taxonomy of branch dimorphism in the Paleozoic bryozoan genus Rhabdomeson:* Lethaia, v. 9, p. 169–178, 11 text-fig.

———, 1980, *Homeomorphy in Paleozoic bryozoans; a search for explanations:* Paleobiology, v. 6, p. 451–465, 5 text-fig.

———, & Towe, K. M., 1971, *Acanthopore ultrastructure in the Paleozoic bryozoan Idioclema insigne Girty:* J. Paleontol., v. 45, p. 913–917, 2 text-fig.

Boardman, R. S., 1960, *Trepostomatous Bryozoa of the Hamilton group of New York State:* U.S. Geol. Surv. Prof. Pap. 340, iv + 87 p., 27 text-fig., 22 pl.

———, 1968, *Colony development and convergent evolution of budding pattern in "rhombotrypid" Bryozoa:* in E. Annoscia (ed.), Proceedings of the First International Conference on Bryozoa, Atti Soc. Ital. Sci. Nat. Mus. Civ. Stor. Nat. Milano, v. 108, p. 179–184.

———, 1971, *Mode of growth and functional morphology of autozooids in some recent and Paleozoic tubular Bryozoa:* Smithson. Contrib. Paleobiol., no. 8, 51 p., 6 text-fig., 11 pl.

———, 1973, *Body walls and attachment organs in some recent cyclostomes and Paleozoic trepostomes:* in G. P. Larwood (ed.), Living and Fossil Bryozoa, p. 231–246, 4 text-fig., 3 pl., Academic Press (London).

———, 1975, *Taxonomic characters for phylogenetic classifications of cyclostome Bryozoa:* in S. Pouyet (ed.), Bryozoa 1974, Doc. Lab. Geol. Fac. Sci. Lyon, hors-série 3, fasc. 2, p 595–606, 4 pl.

———, & Cheetham, A. H., 1969, *Skeletal growth, intracolony variation, and evolution in Bryozoa; a review:* J. Paleontol., v. 43, p. 205–233, 8 text-fig., pl. 27–30.

———, & ———, 1973, *Degrees of colony dominance in stenolaemate and gymnolaemate Bryozoa:* in R. S. Boardman, A. H. Cheetham, & W. A. Oliver, Jr. (eds.), Animal Colonies, p. 121–220, 40 text-fig., Dowden, Hutchinson, & Ross (Stroudsburg).

———, ———, & Cook, P. L., 1970, *Intracolony variation and the genus concept in Bryozoa:* North American Paleontological Convention, Chicago, 1969, Proc., v. 1, pt. C, p. 294–320, 12 text-fig.

———, & McKinney, F. K., 1976, *Skeletal architecture and preserved organs of four-sided zooids in convergent genera of Paleozoic Trepostomata (Bryozoa):* J. Paleontol., v. 50, p. 25–78, 18 text-fig., 16 pl.

———, & Towe, K. M., 1966, *Crystal growth and lamellar development in some recent cyclostome Bryozoa (abstr.):* Geol. Soc. Am. Program 1966 Annu. Meet., p. 20.

———, & Utgaard, John, 1964, *Modifications of study methods for Paleozoic Bryozoa:* J. Paleontol., v. 38, p. 768–770.

———, & ———, 1966, *A revision of the Ordovician bryozoan genera Monticulipora, Peronopora, Heterotrypa, and Dekayia:* J. Paleontol., v. 40, p. 1082–1108, 9 text-fig., pl. 133–142.

Bobin, Geneviève, 1958a, *Structure et genèse des diaphragmes autozoéciaux chez Bowerbankia imbricata (Adams):* Arch. Zool. Exp. Gen., v. 96, p. 53–99, text-fig. 1–9.

———, 1958b, *Histologie des bourgeons autozoéciaux et genèse de leurs diaphragmes chez Vesicularia spinosa (Linné) (Bryozoaire Cténostome):* Bull. Soc. Zool. Fr., v. 83, p. 132–144, 4 text-fig.

———, 1964, *Cytologie de rosettes de Bowerbankia imbricata (Adams) (Bryozoaire Cténostome, vésicularine). Hypothèse sur leur fonctionnement:* Arch. Zool. Exp. Gen., v. 104, p. 1–44, 8 text-fig.

———, 1971, *Histophysiologie du système rosettes— funicule de Bowerbankia imbricata (Adams) (Bryozoaire Cténostome). Les lipides:* Arch Zool. Exp. Gen., v. 112, p. 771–792, 3 text-fig.

———, 1977, *Interzooecial communications and the funicular system:* in R. M. Woollacott & R. L. Zimmer (eds.), Biology of Bryozoans, p. 307–333, 5 text-fig., Academic Press (New York).

———, & Prenant, Marcel, 1957, *Les cellules cystidiennes et les phénomènes d'histolyse et de phagocytose chez Alcyonidium gelatinosum (L.) (Bryozoaire Cténostomes):* Bull. Biol. Fr. Belg., v. 91, p. 203–224, 9 text-fig., 1 pl.

———, & ———, 1968, *Sur le calcaire des parois autozoéciales d'Electra verticillata (Ell. et Sol.), Bryozoaire Chilostome, Anasca. Notions préliminaires:* Arch. Zool. Exp. Gen., v. 109, p. 157–191, 9 text-fig.

———, & ———, 1972, *Sur les cellules cavitaires de quelques Vésicularines (Bryozoaires Cténostomes):* Cah. Biol. Mar., v. 13, p. 479–510, 7 text-fig.

Boekschoten, G. J., 1970, *On Bryozoan borings from the Danian at Fakse, Denmark:* in T. P. Crimes & J. C. Harper (eds.), Trace Fossils, Geol. J., spec. issue no. 3, p. 43–48, 1 text-fig., Seel House Press (Liverpool).

Borg, Folke, 1926a, *Studies on recent cyclostomatous Bryozoa:* Zool. Bidr. Uppsala, bd. 10, p. 181–507, 109 text-fig., 14 pl.

———, 1926b, *On the body-wall in Bryozoa:* Q. J. Microsc. Sci., v. 70, p. 583–598, 6 text-fig.

———, 1933, *A revision of recent Heteroporidae (Bryozoa):* Zool. Bidr. Uppsala, bd. 14, p. 253–394, 29 text-fig., 14 pl.

———, 1941, *On the structure and relationships of Crisina (Bryozoa Stenolaemata):* Ark. Zool., v. 33A, no. 11, p. 1–44, 16 text-fig., 4 pl.

———, 1944, *The stenolaematous Bryozoa:* in S. Bock (ed.), Further Zoological Results of the Swedish Antarctic Expedition 1901–1903. . ., v. 3, no. 5, 276 p., 26 text-fig., 16 pl., P. A. Norstedt & Söner (Stockholm).

———, 1965, *A comparative and phyletic study on fossil and recent Bryozoa of the suborders Cyclostomata and Trepostomata:* Ark. Zool., ser. 2, v. 17, p. 1–91, pl. 1–14 [posthumously edited by Lars Silén and Nils Spjeldnaes].

Braem, Fritz, 1890, *Untersuchungen über die Bryozoen des süssen Wassers:* Bibl. Zool., v. 2, Heft 6, 134 p., 15 pl.

———, 1896, *Die geschlechtliche Entwicklung von Paludicella ehrenbergii:* Zool. Anz., v. 19, p. 54–57, 4 text-fig.

———, 1897, *Die geschlechtliche Entwickelung von Plumatella fungosa:* Zoologica, v. 10, no. 23, p. 1–96, 9 text-fig., 8 pl.

Bretnall, R. W., 1926, *Descriptions of some Western Australian fossil Polyzoa:* West. Aust. Geol. Surv. Bull. 88, ser. 7, no. 13, p. 7–33, 4 pl.

Brien, Paul, 1936, *Contribution à l'étude de la reproduction asexuée des Phylactolémates:* Mus. R. Hist. Nat. Belg. Mem., ser. 2, fasc. 3, p. 570–625, 28 text-fig.

———, 1953, *Étude sur les Phylactolémates. Evolution de la zoécie—Bourgeonnement d'accroisement—Bourgeonnement statoblastique—Embryogénèse—l'Ontogénèse multiple:* Ann. Soc. R. Zool. Belg., v. 84, p. 301–444, 66 text-fig.

———, 1954, *À propos des Bryozoaires phylactolémates:* Bull. Soc. Zool. Fr., v. 79, p. 203–239, 19 text-fig.

———, 1960, *Classe des Bryozoaires:* in P. Grassé (ed.), Traité de Zoologie, v. 5, fasc. 2, p. 1054–1335, text-fig. 876–1223, Masson (Paris).

Bromley, R. G., 1970, *Borings as trace fossils and Entobia cretacea Portlock, as an example:* in T. P. Crimes & J. C. Harper (eds.), Trace Fossils, Geol. J., spec. issue no. 3, p. 49–90, 4 text-fig., 5 pl., Seel House Press (Liverpool).

Bronstein, Georges, 1937, *Étude du système nerveux de quelques Bryozoaires gymnolémides:* Stat. Biol. Roscoff. Trav., v. 15, p. 155–174, 12 text-fig.

Brood, Krister, 1970, *On two species of Saffordotaxis (Bryozoa) from the Silurian of Gotland, Sweden:* Stockholm Contrib. Geol., v. 21, p. 57–68, 4 text-fig., 9 pl.

————, 1972, *Cyclostomatous Bryozoa from the Upper Cretaceous and Danian in Scandinavia:* Stockholm Contrib. Geol., v. 26, 464 p., 148 text-fig., 78 pl.

————, 1973, *Paleozoic Cyclostomata (a preliminary report):* in G. P. Larwood (ed.), Living and Fossil Bryozoa, p. 247–256, 2 text-fig., 2 pl., Academic Press (London).

————, 1975a, *Paleoecology of Silurian Bryozoa from Gotland (Sweden):* in S. Pouyet (ed.), Bryozoa 1974, Doc. Lab. Geol. Fac. Sci. Lyon, hors-série 3, fasc. 2, p. 401–414, 4 text-fig., 4 pl.

————, 1975b, *Cyclostomatous Bryozoa from the Silurian of Gotland:* Stockholm Contrib. Geol., v. 28, p. 45–119, 19 text-fig., 16 pl.

————, 1976, *Wall structure and evolution in cyclostomate Bryozoa:* Lethaia, v. 9, p. 377–389, 11 text-fig.

Brown, C. J. D., 1933, *A limnological study of certain fresh-water Polyzoa with special reference to their statoblasts:* Trans. Am. Microsc. Soc., v. 52, p. 271–316, 3 text-fig., pl. 39–40.

Brown, D. A., 1958, *The relative merits of the class names "Polyzoa" and "Bryozoa":* Bull. Zool. Nomencl., v. 15, p. 540–542.

Bryan, W. H., 1941, *Spherulites and allied structures, I:* Proc. R. Soc. Queensl., v. 52, p. 41–53, 10 text-fig., pl. 3–5.

————, & Hill, Dorothy, 1941, *Spherulitic crystallization as a mechanism of skeletal growth in the hexacorals:* Proc. R. Soc. Queensl., v. 52, p. 78–91, 2 text-fig.

Brydone, R. M., 1929, *Further notes on new or imperfectly known Chalk Polyzoa, pt. 1:* 40 p., 14 pl., Dulau & Co. (London).

Buge, Émile, 1952, *Classe des Bryozoaires:* in J. Piveteau (ed.), Traité de Paléontologie, v. 1, p. 688–749, 142 text-fig., Masson (Paris).

————, 1957, *Les Bryozoaires du Néogène de l'ouest de la France et leur signification stratigraphique et paléobiologique:* Mem. Mus. Natl. Hist. Nat. Paris, Ser. C. t. 6, 435 p., 53 text-fig., 12 pl.

Bushnell, J. H., 1966, *Environmental relations of Michigan Ectoprocta, and dynamics of natural populations of Plumatella repens:* Ecol. Monogr., v. 36, p. 95–123, 14 text-fig.

————, 1974, *Bryozoans (Ectoprocta):* in C. W. Hart, Jr. & S. L. H. Fuller (eds.), Pollution Ecology of Freshwater Invertebrates, p. 157–194, Academic Press (New York).

————, & Rao, K. S., 1974, *Dormant or quiescent stages and structures among the Ectoprocta; physical and chemical factors affecting viability and germination of statoblasts:* Trans. Am. Microsc. Soc., v. 93, p. 524–543, 18 text-fig.

————, & Wood, T. S., 1971, *Honeycomb colonies of Plumatella casmiana Oka (Ectoprocta: Phylactolaemata):* Trans. Am. Microsc. Soc., v. 90, p. 229–231, 1 text-fig.

Busk, George, 1852, *An account of the Polyzoa, and sertularian Zoophytes. . . :* in J. MacGillivray, Narrative of the Voyage of H.M.S. Rattlesnake, . . .during the years 1846–1850, v. 1, p. 343–402, 1 pl., T. & W. Boone (London).

————, 1859, *A monograph of the fossil Polyzoa of the Crag:* xiv + 136 p., 7 text-fig., 22 pl., Palaeontographical Society (London).

————, 1884, *Report on the Polyzoa (Part I.—The Cheilostomata):* in J. Murray, Report on the Scientific Results of the Voyage of H.M.S. Challenger during the years 1873–76. . . , Zoology, v. 10, pt. 30, xxiv + 216 p., 59 text-fig., 36 pl., H.M. Stationery Office (London).

————, 1886, *Report on the Polyzoa (Part II.—The Cyclostomata, Ctenostomata, and Pedicellinea):* in J. Murray, Report on the Scientific Results of the Voyage of H.M.S. Challenger during the years 1873–76. . . , Zoology, v. 17, pt. 50, viii + 47 p., 2 text-fig., 10 pl., H.M. Stationery Office (London).

Calvet, Louis, 1900, *Contributions à l'histoire naturelle de Bryozoaires Ectoproctes marins:* Trav. Inst. Zool. Univ. Montpellier, n.ser., mem. 8, 488 p., 45 text-fig., 13 pl.

Canu, Ferdinand, 1900, *Revision des Bryozoaires du Crétacé figurés par d'Orbigny, II. Cheilostomata:* Bull. Soc. Geol. Fr., ser. 3, t. 28, p. 334–463, 71 text-fig., pl. 4–7.

————, & Bassler, R. S., 1917, *A synopsis of American Early Tertiary cheilostome Bryozoa:* U.S. Natl. Mus. Bull. 96, 87 p., 6 pl.

————, & ————, 1920, *North American Early Tertiary Bryozoa:* U.S. Natl. Mus. Bull. 106, 879 p., 279 text-fig., 162 pl.

Caster, K. E., 1934, *The stratigraphy and paleontology of northwestern Pennsylvania, part 1—Stratigraphy:* Bull. Am. Paleontol., v. 21, no. 71, 185 p., 12 text-fig.

Chapman, Frederick, 1920, *Lower Carboniferous limestone fossils from New South Wales:* Proc. Linn. Soc. N.S.W., v. 45, p. 364–367, pl. 24.

Cheetham, A. H., 1963, *Late Eocene zoogeography of the eastern Gulf Coast region:* Geol. Soc. Am. Mem. 91, xii + 113 p., 34 text-fig., 3 pl.

————, 1968, *Morphology and systematics of the bryozoan genus Metrarabdotos:* Smithson. Misc. Collect., v. 153, no. 1, viii + 122 p., 24 text-fig., 18 pl.

————, 1971, *Functional morphology and biofacies distribution of cheilostome Bryozoa in the Danian Stage (Paleocene) of southern Scandinavia:* Smithson. Contrib. Paleobiol., no. 6, 87 p., 29 text-fig., 7 pl.

————, 1972, *Cheilostome Bryozoa of late Eocene age from Eua, Tonga:* U.S. Geol. Surv. Prof. Pap. 640-E, vii + 26 p., 7 text-fig., 7 pl.

————, 1975a, *Preliminary report on early Eocene cheilostome bryozoans from Site 308–Leg 32, Deep Sea Drilling Project:* in R. L. Larson, R. Moberly et al., Initial Reports of the Deep Sea Drilling Project, v. 32, p. 835–851, 2 text-fig., 4 pl. U.S. Government Printing Office (Washington).

————, 1975b, *Taxonomic significance of autozooid size and shape in some early multiserial cheilostomes from the Gulf Coast of the U.S.A.:* in S. Pouyet (ed.), Bryozoa 1974, Doc. Lab. Geol. Fac. Sci. Lyon, hors-série 3, fasc. 2, p. 547–564, 11 text-fig., 3 pl.

————, & Håkansson, Eckart, 1972, *Preliminary report on Bryozoa (Site 117):* in A. S. Laughton, W. A. Berggren *et al.,* Initial Reports of the Deep Sea Drilling Project, v. 12, p. 432–441, text-fig. 29. pl. 13–28, U.S. Government Printing Office (Washington).

————, & Lorenz, D. M., 1976, *A vector approach to size and shape comparisons among zooids in cheilostome Bryozoa:* Smithson. Contrib. Paleobiol., no. 29, vi + 56 p., 37 text-fig.

————, Rucker, J. B., & Carver, R. E., 1969, *Wall structure and mineralogy of the cheilostome bryozoan Metrarabdotos:* J. Paleontol., v. 43, p. 129–135, 1 text-fig., pl. 26.

Cook, P. L., 1963, *Observations on live lunulitiform zoaria of Polyzoa:* Cah. Biol. Mar., v. 4, p. 407–413, 1 text-fig., 1 pl.

————, 1964, *Polyzoa from West Africa, I. Notes on the Steganoporellidae, Thalamoporellidae and Onychocellidae (Anasca, Coilostega):* Ann. Inst. Oceanogr. (Paris), t. 41, p. 43–78, 13 text-fig., 1 pl.

————, 1968a, *Polyzoa from West Africa, the Malacostega, Part I:* Bull. Br. Mus. Nat. Hist. Zool., v. 16, p. 115–160, 20 text-fig., 3 pl.

————, 1968b, *Bryozoa (Polyzoa) from the coasts of tropical West Africa:* Atl. Rep., no. 10, p. 115–262, 2 text-fig., pl. 8–11.

————, 1968c, *Observations on living Bryozoa:* in E. Annoscia (ed.), Proceedings of the First International Conference on Bryozoa, Atti Soc. Ital. Sci. Nat. Mus. Civ. Stor. Nat. Milano, v. 108, p. 155–160, 3 text-fig.

————, 1973a, *Settlement and early colony development in some Cheilostomata:* in G. P. Larwood (ed.), Living and Fossil Bryozoa, p. 65–71, 4 text-fig., Academic Press (London).

————, 1973b, *Preliminary notes on the ontogeny of the frontal body wall in the Adeonidae and Adeonellidae (Bryozoa, Cheilostomata):* Bull. Br. Mus. Nat. Hist. Zool., v. 25, p. 245–263, pl. 1–3.

————, 1975, *The genus Tropidozoum Harmer:* in S. Pouyet (ed.), Bryozoa 1974, Doc. Lab. Geol. Fac. Sci. Lyon, hors-série 3, fasc. 1, p. 161–168, 3 text-fig., 3 pl.

————, 1977, *Colony-wide water currents in living Bryozoa:* Cah. Biol. Mar., v. 18, p. 31–47, 1 text-fig., 1 pl.

————, & Lagaaij, Robert, 1976, *Some Tertiary and recent conescharelliniform Bryozoa:* Bull. Br. Mus. Nat. Hist. Zool., v. 29, p. 319–376, 7 text-fig., 8 pl.

Cori, C. J., 1941, *Ordnung der Tentaculata; Bryozoa:* in W. Kükenthal & T. Krumbach (eds.), Handbuch der Zoologie, Bd. 3, Heft 2, p. 263–502, text-fig. 272–611, W. de Gruyter & Co. (Berlin).

Corneliussen, E. F., & Perry, T. G., 1973, *Monotrypa, Hallopora, Amplexopora, and Hennigopora (Ectoprocta) from the Brownsport Formation (Niagaran), western Tennessee:* J. Paleontol., v. 47, p. 151–220, 34 text-fig., 10 pl.

Coryell, H. N., 1921, *Bryozoan faunas of the Stones River group of central Tennessee:* Indiana Acad. Sci., 1919, p. 261–340, 3 text-fig., 14 pl.

Crockford, Joan, 1941, *Bryozoa from the Silurian and Devonian of New South Wales:* J. Proc. R. Soc. N.S.W., v. 75, p. 104–114, 1 text-fig., pl. 5.

————, 1944, *Bryozoa from the Permian of Western Australia:* Proc. Linn. Soc. N.S.W., v. 69, p. 139–175, 49 text-fig., pl. 4, 5.

————, 1947, *Bryozoa from the Lower Carboniferous of New South Wales and Queensland:* Proc. Linn. Soc. N.S.W., v. 72, p. 1–48, 51 text-fig., 6 pl.

————, 1957, *Permian Bryozoa from the Fitzroy Basin, Western Australia:* Aust. Bur. Miner. Resour. Geol. Geophys. Bull. 34, 136 p., 13 text-fig., 21 pl.

Crowe, J. H., 1971, *Anhydrobiosis; an unsolved problem:* Am. Nat., v. 105, p. 563–573.

Cuffey, R. J., 1969, *Bryozoa versus Ectoprocta—The necessity for precision:* Syst. Zool., v. 18, p. 250–251.

————, 1973, *An improved classification, based upon numerical-taxonomic analyses, for the higher taxa of entoproct and ectoproct bryozoans:* in G. P. Larwood (ed.), Living and Fossil Bryozoa, p. 549–564, 4 text-fig., Academic Press (London).

Cumings, E. R., 1904, *Development of some Paleozoic Bryozoa:* Am. J. Sci., ser. 4, v. 17, p. 49–78, 83 text-fig.

————, 1905, *Development of Fenestella:* Am. J. Sci., ser. 4, v. 20, p. 169–177, pl. 5–7.

————, 1906, *Descriptions of the Bryozoa of the Salem Limestone of southern Indiana:* Indiana Dept. Geol. Nat. Res. 30th Annu. Rep., p. 1274–1296, pl. 27–40.

————, 1912, *Development and systematic position of the monticuliporoids:* Geol. Soc. Am. Bull., v. 23, p. 357–370, pl. 19–22.

————, & Galloway, J. J., 1913, *The stratigraphy and paleontology of the Tanner's Creek section of the Cincinnati Series of Indiana:* Indiana Dept. Geol. Nat. Res. 37th Annu. Rep., p. 353–478, 20 pl.

————,& ————, 1915, *Studies of the morphology and histology of the Trepostomata or monticuliporoids:* Geol. Soc. Am. Bull., v. 26, p. 349–374, pl. 10–15.

Cutler, J. F., 1973, *Nature of "acanthopores" and related structures in the Ordovician bryozoan Constellaria:* in G. P. Larwood (ed.), Living and Fossil Bryozoa, p. 257–260, Academic Press (London).

Dall, W. H., 1877, *Note on "Die Gasteropoden Fauna Baikalsees"*: Proc. Boston Soc. Nat. Hist., v. 19, p. 43–47.

Dana, J. D., 1846, *Zoophytes:* in United States Exploring Expedition . . .1838–42, under the command of Charles Wilkes, v. 7, 740 p., C. Sherman (Philadelphia).

———, 1849, *Atlas. Zoophytes:* in United States Exploring Expedition. . .1838–42, under the command of Charles Wilkes, 61 pl., Lea & Blanchard (Philadelphia).

Davenport, C. B., 1890, *Cristatella; the origin and development of the individual in the colony:* Bull. Mus. Comp. Zool. Harvard Univ. [Coll.], v. 20, p. 101–151, text-fig. a, b, 11 pl.

———. 1891, *Observations on budding in Paludicella and some other Bryozoa:* Bull. Mus. Comp. Zool. Harvard Univ. [Coll.], v. 22, p. 1–114, 12 pl.

Dawydoff, Constantin, & Grassé, P.-P., 1959, *Classe des phoronidiens:* in P. Grassé (ed.), Traité de Zoologie, v. 5, fasc. 1, p. 1008–1053, text-fig. 835–875, pl. 5, Masson (Paris).

Delage, M. Y., & Herourard, Edgard, 1897, *Traité de zoologie concrète; Les Vermidiens (v. 5):* xii + 372 p., 523 text-fig., 46 pl. (Paris).

DeVries, Hugo, 1890, *Die Pflanzen und Thiere in den dunklen Räumen der Rotterdamer Wasserleitung. Bericht über die biologischen Untersuchungen der Crenothrix-Commission zu Rotterdam. . .1887:* 73 p., illus., (Jena).

Dodd, J. R., 1963, *Paleoecological implications of shell mineralogy in two pelecypod species:* J. Geol., v. 71, p. 1–11, 8 text-fig.

———, 1967, *Magnesium and strontium in calcareous skeletons; a review:* J. Paleontol., v. 41, p. 1313–1329, 4 text-fig.

Dollfus, G. F., 1875, *Observations critiques sur la classification des Polypiers paléozoïques:* C. R. Acad. Sci. Paris, v. 80, p. 681–683.

Dudley, J. W., 1970, *Differential utilization of phytoplankton food resources by marine ectoprocts:* Biol. Bull. Woods Hole Mass., v. 139, p. 420.

Duméril, A. M. C., 1806, *Zoologie analytique, ou méthode naturelle de classification des animaux, rendue plus facile à l'aide de tableaux synoptiques:* xxxii + 344 p. (Paris).

Dunaeva, N. N., 1968, *On the mode of sexual reproduction of some trepostomatous Bryozoa:* in E. Annoscia (ed.), Proceedings of the First International Conference on Bryozoa, Atti Soc. Ital. Sci. Nat. Mus. Civ. Stor. Nat. Milano, v. 108, p. 61–63, 1 text-fig.

———, 1973, *Mshanki roda Megacanthopora iz kamennougol'nykh otlozhenij Bol'shogo Donbassa:* Paleontol. Zh., 1973, no. 4, p. 56–61, pl. 9. [*Bryozoans of the genus Megacanthopora from Carboniferous deposits of the Greater Donbass.*]

———, & Morozova, I. P., 1967. *Osobennosti razvitiya i sistematicheskoe polozhenie nekotorykh*

pozdne-Paleozojskikh Trepostomat: Paleontol. Zh., 1967, no. 4, p. 86–94, 2 text-fig., pl. 5. [*Evolutionary features and systematic position of some late Paleozoic Trepostomata.*]

———, & ———, 1975, *Revision of the suborder Fenestelloidea:* in S. Pouyet (ed.), Bryozoa 1974, Doc. Lab. Geol. Fac. Sci. Lyon, hors-série 3, fasc. 1, p. 225–233.

Duncan, Helen, 1949, *Genotypes of some Paleozoic Bryozoa:* J. Wash. Acad. Sci., v. 39, p. 122–136.

———, 1957, *Bryozoans:* in H. S. Ladd (ed.), Treatise on Marine Ecology and Paleoecology, v. 2, Geol. Soc. Am. Mem. 67, p. 783–799.

Dybowski, Władisław, 1877, *Die Chaetetiden der ostbaltischen Silur-formation:* 134 p., 4 pl., Kaiserlichen Akademie der Wissenschaften (St. Petersburg).

Dzik, Jerzy, 1975, *The origin and early phylogeny of the cheilostomatous Bryozoa:* Acta Paleontol. Pol., v. 20, p. 395–423, 12 text-fig., pl. 15–18.

Ehrenberg, C. G., 1831, *Symbolae Physicae, seu Icones et descriptiones Corporum Naturalium novorum aut minus cognitorum, quae ex itineribus per Libyam, Aegyptum, Nubiam, Dongalam, Syriam, Arabiam et Habessiniam. . .studio annis 1820–25 redierunt. . .Pars Zoologica:* v. 4, Animalia Evertebrata exclusis Insectis, 10 pl. (Berolini).

Eichwald, C. E., 1829, *Zoologia specialis quam expositis Animalibus tum vivis, tum fossilibus potissimum Rossiae in universum, et Poloniae in specie, etc.:* v. 1, vi + 314 p., 5 pl. (Vilna).

———, 1842, *Die Urwelt Russlands, durch Abbildungen erläutert:* Heft 2, 184 p., 4 pl., Kaiserlichen Akademie der Wissenschaften (St. Petersburg).

———, 1860, *Lethaea Rossica, ou Paléontologie de la Russie. . .:* v. 1 (l'ancienne période), p. 355–518, E. Schweizerbart (Stuttgart).

Eitan, Gabriel, 1972, *Types of metamorphosis and early astogeny in Hippopodina feegeensis (Busk) (Bryozoa—Ascophora):* J. Exp. Mar. Biol. Ecol., v. 8, p. 27–30, 1 text-fig.

Elias, M. K., 1954, *Cambroporella and Coeloclema, Lower Cambrian and Ordovician bryozoans:* J. Paleontol., v. 28, p. 52–58, pl. 9–10.

———, 1957, *Late Mississippian fauna from the Redoak Hollow formation of southern Oklahoma, part 1:* J. Paleontol., v. 31, p. 370–427, pl. 39–50.

———, & Condra, G. E., 1957, *Fenestella from the Permian of West Texas:* Geol. Soc. Am. Mem. 70, x + 158 p., 17 text-fig., 23 pl.

Ellis, John, 1754, *Observations on a remarkable coralline. . .:* Philos. Trans. R. Soc. London, v. 48, pt. 1, p. 115–117, pl. 5.

———, 1755a, *A letter to Mr. Peter Collinson concerning a particular species of coralline:* Philos. Trans. R. Soc. London, v. 48, pt. 2, p. 504–507, pl. 17–18.

———, 1755b, *A letter from Mr. John Ellis to Mr. Peter Collinson concerning the animal life of cor-*

allines, that look like minute trees. . .: Philos. Trans. R. Soc. London, v. 48, pt. 2, p. 627–633, pl. 22–23.

——, 1755c, *An essay towards a natural history of the corallines and other marine productions of the like kind, commonly found on the coasts of Great Britain and Ireland:* xvii + 103 p., 38 pl. (London).

Erben, H. K., 1974, *On the structure and growth of the nacreous tablets in gastropods:* Biomineralisation Forschungsber., Bd. 7, p. 14–27, 5 pl.

Eschscholtz, J. F. von, 1825, *Bericht über die zoologische Ausbeute während der Reise von Kronstadt bis St. Peter und Paul:* Isis von Oken, 1825, col. 733–747, pl. 5.

Etheridge, Robert, Jr., 1873, *Description of Carinella, a new genus of Carboniferous Polyzoa:* Geol. Mag., v. 10, p. 433–434, pl. 15.

——, 1875, *Note on a new provisional genus of Carboniferous Polyzoa:* Ann. Mag. Nat. Hist., ser. 4, v. 15, p. 43–45, pl. 4.

——, 1876, *Carboniferous and post-Tertiary Polyzoa:* Geol. Mag., n.ser., dec. 2, v. 3, p. 522–523, 1 text-fig.

——, 1917, *"Vetofistula," a new form of Palaeozoic Polyzoa, allied to "Rhabdomeson" Young and Young, from Reid's Gap, near Townsville:* Geol. Surv. Queensl. Publ. no. 260, p. 17–20, pl. 4.

——, & Foord, A. H., 1884, *Descriptions of Paleozoic corals in the collections of the British Museum (Natural History):* Ann. Mag. Nat. Hist., ser. 5, v. 13, p. 472–476, pl. 17.

Farmer, J. D., Valentine, J. W., & Cowen, Richard, 1973, *Adaptive strategies leading to the ectoproct ground-plan:* Syst. Zool., v. 22, p. 233–239, 1 text-fig.

Farre, Arthur, 1837, *Observations on the minute structure of some of the higher forms of Polypi, with views of a more natural arrangement of the class:* Philos. Trans. R. Soc. London, v. 127, p. 387–426, pl. 20–27.

Fischer, Paul, 1866, *Étude sur les Bryozoaires perforants de la famille des Térébriporides:* Nouv. Arch. Mus. Natl. Hist. Nat. Paris, t. 2, p. 293–313, pl. 11.

Fleming, P. J. G., 1969, *Fossils from the Neerkol Formation of central Queensland:* in K. S. W. Campbell (ed.), Stratigraphy and Palaeontology; Essays in Honour of Dorothy Hill, p. 264–275, pl. 16, 17, Australian National University Press (Canberra).

Foerste, A. F., 1887, *Letters to the editor: recent methods in the study of Bryozoa:* Science, v. 10, no. 248, p. 225–226.

Franzén, Åke, 1970, *Phylogenetic aspects of the morphology of spermatozoa and spermiogenesis:* in B. Baccetti (ed.), Comparative Spermatology, p. 29–46, 8 text-fig., Accademia Nazionale dei Lincei & Academic Press (Rome & New York).

Fritz, M. A., 1947, *Cambrian Bryozoa:* J. Paleontol., v. 21, p. 434–435, pl. 60.

Fry, H. C., & Cuffey, R. J., 1976, *Filiramoporina kretaphilia—A new genus and species of bifoliate tubulobryozoan (Ectoprocta) from the Lower Permian Wreford Megacyclothem of Kansas:* Univ. Kans. Paleontol. Contrib. Pap. 84, 12 p., 3 fig., 2 pl.

Fyfe, W. S., & Bischoff, J. L., 1965, *The calcite-aragonite problem:* in L. C. Pray & R. C. Murray (eds.), Dolomitization and Limestone Diagenesis—A Symposium, Soc. Econ. Paleontol. Mineral. Spec. Publ. 13, p. 3–13.

Gabb, W. M., & Horn, G. H., 1862, *Monograph of the fossil Polyzoa of the secondary and tertiary formations of North America:* J. Acad. Nat. Sci. Philadelphia, v. 5, p. 111–179, pl. 19–21.

Gautier, T. G., 1970, *Interpretive morphology and taxonomy of bryozoan genus Tabulipora:* Univ. Kans. Paleontol. Contrib. Pap. 48. p. 1–21, 9 text-fig., 8 pl.

——, 1972, *Growth, form, and functional morphology of Permian acanthocladiid Bryozoa from the Glass Mountains, West Texas:* Unpubl. Ph.D. dissert., University of Kansas, 187 p., 37 text-fig., 21 pl.

——, 1973, *Growth in bryozoans of the order Fenestrata:* in G. P. Larwood (ed.), Living and Fossil Bryozoa, p. 271–274, 2 text-fig., Academic Press (London).

Geiser, S. W., 1937, *Pectinatella magnifica Leidy, an occasional river-pest in Iowa:* Field Lab., v. 5, p. 65–76, 1 text-fig.

Gervais, Paul, 1837, *Recherches sur les polypes d'eau douce des genres Plumatella, Cristatella, et Paludicella:* Ann. Sci. Nat. Zool. Biol. Anim., ser. 2, v. 7, p. 74–93, 1 pl.

Gerwerzhagen, Adolf, 1913, *Beiträge zur Kenntnis der Bryozoen I. Das Nervensystem von Cristatella mucedo Cuv.:* Z. Wiss. Zool., v. 107, p. 309–345, 3 text-fig., pl. 12–14.

Girty, G. H., 1910, *New genera and species of Carboniferous fossils from the Fayetteville Shale of Arkansas:* Ann. N.Y. Acad. Sci., v. 20, no. 3, pt. 2, p. 189–238.

Goldfuss, G. A., 1829, *Petrefacta Germaniae, Abbildung und Beschreibungen der Petrefacten Deutschlands und der angreuzenden Länder:* Bd. 1, lief. 2, p. 77–164, pl. 26–50, Arnz & Co. (Düsseldorf).

——, 1831, *Petrefacta Germaniae . . .:* Bd. 1, lief. 3, p. 165–240, pl. 51–71, Arnz & Co. (Düsseldorf).

Gordon, D. P., 1968, *Zooidal dimorphism in the polyzoan Hippopodinella adpressa (Busk):* Nature (London), v. 219, p. 633–634, 1 text-fig.

——, 1971a, *Colony formation in the cheilostomatous bryozoan Fenestrulina malusii var. thyreophora:* N.Z. J. Mar. Freshwater Res., v. 5, p. 342–351, 6 text-fig.

——, 1971b, *Zooidal budding in the cheilostomatous bryozoan Fenestrulina malusii var. thy-*

reophora: N.Z. J. Mar. Freshwater Res., v. 5, p. 453–460, 5 text-fig.

——, 1975, *Ultrastructure of communication pore areas in two bryozoans:* in S. Pouyet (ed.), Bryozoa 1974, Doc. Lab. Geol. Fac. Sci. Lyon, hors-série 3, fasc. 1, p. 187–192, 1 text-fig., 2 pl.

Goryunova [Gorjunova], R. V., 1969, *O novom rode mshanok iz verkhnego karbona Srednego Urala:* Paleontol. Zh., 1969, no. 2, p. 129–131, 1 text-fig. [*New Upper Carboniferous bryozoan genus from the Central Urals.*]

——, 1972, *Semejstvo Actinotrypidae, ego sostav i sistematicheskoe polozhenie:* Paleontol. Zh., 1972, no. 2, p. 147–150, 1 text-fig. [*The family Actinotrypidae, its composition and systematic position.*]

——, 1975, *Permskie mshanki Pamira:* Tr. Akad. Nauk SSSR Paleontol. Inst., v. 148, 128 p., 20 text-fig., 29 pl. [*Permian Bryozoa of the Pamirs.*]

——, & Morozova, I. P., 1979, *Pozdnepaleozojskie mshanki Mongolii:* Trudy Sovmestnaya Sovetski-Mongol'skaya Paleontologicheskaya Ekspeditsiya, v. 9, 139 p., 17 text-fig., 27 pl. [*Late Paleozoic Bryozoa of Mongolia: Proceedings of the Joint Soviet-Mongolian Paleontological Expedition.*]

Gould, S. J., & Katz, Michael, 1975, *Disruption of ideal geometry in the growth of receptaculitids; a natural experiment in theoretical morphology:* Paleobiology, v. 1, p. 1–20, 12 text-fig.

Grant, R. E., 1827, *Observations on the structure and nature of Flustrae:* Edinburgh New Philos. J., ser. 2, v. 3, p. 107–118, 337–342.

Greeley, Ronald, 1967, *Natural orientation of lunulitiform bryozoans:* Geol. Soc. Am. Bull., v. 78, p. 1179–1182, 2 text-fig., 2 pl.

——, 1969, *Basally "uncalcified" zoaria of lunulitiform Bryozoa:* J. Paleontol., v. 43, p. 252–256, 2 text-fig., pl. 33–34.

Gregorio, Antonio de, 1930, *Sul Permiano di Sicilia:* Ann. Geol. Paleontol. Palermo, v. 52, p. 1–69, 21 pl.

Gregory, J. W., 1893, *On the British Palaeogene Bryozoa:* Trans. Zool. Soc. London, v. 13, p. 219–279, pl. 29–32.

——, 1896, *Catalogue of the fossil Bryozoa in the Department of Geology of the British Museum (Natural History). The Jurassic Bryozoa:* 239 p., 22 text-fig. 11 pl. (London).

——, 1909, *Catalogue of the fossil Bryozoa (Polyzoa) in the Department of Geology, British Museum (Natural History). The Cretaceous Bryozoa (Polyzoa):* v. 2, 346 p., 75 text-fig., 9 pl. (London).

Grubbs, D. M., 1939, *Fauna of the Niagaran nodules of the Chicago area:* J. Paleontol., v. 13, p. 543–560, 2 text-fig., pl. 61–62.

Gürich, Georg, 1896, *Das Palaeozoicum im polnischen Mittelgebirge:* Zap. Vses. Mineral. O-va., v. 32, 539 p.

Gurley, W. F. E., 1884, *New Carboniferous fossils:* Bulletin no. 2, 12 p., Privately published (Danville, Ill.).

Håkansson, Eckart, 1973, *Mode of growth of the Cupuladriidae (Bryozoa, Cheilostomata):* in G. P. Larwood (ed.), Living and Fossil Bryozoa, p. 287–298, 2 text-fig., 2 pl., Academic Press (London).

Häntzschel, Walter, 1975, *Trace fossils and problematica:* in Curt Teichert (ed.), Treatise on Invertebrate Paleontology, Part W, Supplement 1. xxii + 269 p., 110 text-fig., Geological Society of America & University of Kansas (Boulder & Lawrence).

Hall, James, 1847, *Organic remains of the lower division of the New York system:* Natural History of New York, Pt. 6, Paleontology of New York, v. 1, xxiv + 338 p., 88 pl.

——, 1852, *Organic remains of the lower middle division of the New York system:* Natural History of New York, Pt. 6, Paleontology of New York, v. 2, viii + 362 p., 85 pl.

——, 1874, *Descriptions of Bryozoa and corals of the Lower Helderberg group:* N.Y. State Mus. Nat. Hist., 26th Annu. Rep., p. 93–116.

——, 1876, *The fauna of the Niagara group in central Indiana:* N.Y. State Mus. Nat. Hist., 28th Annu. Rep., 32 pl. and explanations.

——, 1881, *Bryozoans of the Upper Helderberg and Hamilton groups:* Trans. Albany Inst., v. 10 (1883), 36 p. [Issued as abstract in 1881 (from Nickles & Bassler, 1900).]

——, 1883a, *Fossil corals and bryozoans of the Lower Helderberg group, and fossil bryozoans of the Upper Helderberg group:* State Geologist of New York, Rep. for 1882, pl. 1–33.

——, 1883b, *Bryozoans of the Upper Helderberg and Hamilton groups:* Trans. Albany Inst., v. 10, p. 145–197.

——, 1886, *Bryozoa of the Upper Helderberg group; plates and explanations:* State Geologist of New York, 5th Annu. Rep. for 1885, pl. 25–53.

——, & Simpson, G. B., 1887, *Corals and Bryozoa; text and plates containing descriptions and figures of species from the Lower Helderberg, Upper Helderberg, and Hamilton groups:* Natural History of New York, Pt. 6, Paleontology of New York, v. 6, xxvi + 298 p., 66 pl.

Hallam, Anthony, & O'Hara, M. J., 1962, *Aragonitic fossils in the Lower Carboniferous of Scotland:* Nature (London), v. 195, p. 273–274.

Harmelin, J.-G., 1973, *Morphologic variations and ecology of the recent cyclostome bryozoan "Idmonea" atlantica from the Mediterranean:* in G. P. Larwood (ed.), Living and Fossil Bryozoa, p. 95–106, 6 text-fig., Academic Press (London).

——, 1974, *Les Bryozoaires Cyclostomes de Méditerranée, écologie et systématique:* Thèse présentée à l'Université d'Aix-Marseille, 2 vol., 365 p., 53 text-fig., 38 pl.

——, 1975, *Relations entre la forme zoariale et l'habitat chez les Bryozoaires Cyclostomes. Conséquences taxonomiques:* in S. Pouyet (ed.), Bryozoa 1974, Doc. Lab. Geol. Fac. Sci. Lyon, hors-série 3, fasc. 2, p. 369–384, 6 text-fig., 3 pl.

————, 1976, *Le sous-ordre de Tubuliporina (Bryozoaires Cyclostomes) en Méditerranée, écologie et systématique:* Mem. Inst. Oceanogr. Monaco no. 10, 326 p., 50 text-fig., 38 pl.

Harmer, S. F., 1893, *On the occurrence of embryonic fission in cyclostomatous Polyzoa:* Q. J. Microsc. Sci., n.ser., v. 34, p. 199–241, pl. 22–24.

————, 1896, *On the development of Lichenopora verrucaria Fabr.:* Q. J. Microsc. Sci., n.ser., v. 39, p. 71–144, pl. 7–10.

————, 1898, *On the development of Tubulipora, and on some British and northern species of this genus:* Q. J. Microsc. Sci., n.ser., v. 41, p. 73–157, pl. 8–10.

————, 1901, *On the structure and classification of the cheilostomatous Polyzoa:* Proc. Cambridge Philos. Soc., v. 11, p. 11–17.

————, 1902, *On the morphology of the Cheilostomata:* Q. J. Microsc. Sci., n.ser., v. 46, p. 263–350, pl. 15–18.

————, 1915, *The Polyzoa of the Siboga Expedition (Part I.—Entoprocta, Ctenostomata, and Cyclostomata):* Siboga-Expeditie, v. 28a, p. 1–180, pl. 1–12, E. J. Brill (Leiden).

————, 1926, *The Polyzoa of the Siboga Expedition (Part II.—Cheilostomata anasca) with additions to previous reports:* Siboga-Expeditie, v. 28b, p. 181–501, text-fig. 1–23, pl. 13–34, E. J. Brill (Leiden).

————, 1930, *Polyzoa (Presidential address):* Proc. Linn. Soc. London, v. 141, p. 68–118, 11 text-fig., 1 pl.

————, 1931, *Recent work on Polyzoa (Presidential address):* Proc. Linn. Soc. London, v. 143, p. 113–168, 6 text-fig.

————, 1934, *The Polyzoa of the Siboga Expedition (Part III.—Cheilostomata Ascophora. I. Family Reteporidae):* Siboga-Expeditie, v. 28c, p. 503–640, text-fig. 24–48, pl. 35–41, E. J. Brill (Leiden).

————, 1947, *On the relative merits of the names Bryozoa and Polyzoa as the name for the class in the animal kingdom now known by one or the other of these names:* Bull. Zool. Nomencl., v. 1, p. 230–231.

————, 1957, *The Polyzoa of the Siboga Expedition (Part IV.—Cheilostomata Ascophora. II. Ascophora, except Reteporidae) with additions to Part II, Anasca:* Siboga-Expeditie, v. 28d, p. 641–1147, text-fig. 49–118, pl. 42–74, E. J. Brill (Leiden).

Hastings, A. B., 1943, *Polyzoa (Bryozoa) I.—Scrupocellariidae, Epistomiidae, Farciminariidae, Bicellariellidae, Aeteidae, Scrupariidae:* Discovery Rep., v. 22, p. 301–510, 66 text-fig., pl. 5–13.

————, 1957. *Explanatory note:* in S. F. Harmer, The Polyzoa of the Siboga Expedition, Siboga-Expeditie, v. 28d, p. xiii–xv, E. J. Brill (Leiden).

Hatschek, Berthold, 1888, *Lehrbuch der Zoologie, eine morphologische Übersicht des Thierreiches zur Einführung in das Studium dieser Wissenschaft:* v. 1, 144 p., 155 text-fig., Gustav Fischer (Jena).

Hay, W. W., Wise, S. W., Jr., & Stieglitz, R. D., 1970, *Scanning electron microscope study of fine grain size biogenic carbonate particles:* Trans. Gulf Coast Assoc. Geol. Soc., v. 20, p. 287–302, 23 fig.

Healey, N. D., & Utgaard, John, 1979, *Ultrastructure of the skeleton of the cystoporate bryozoans Ceramophylla, Crassaluna and Cystodictya:* in G. P. Larwood & M. B. Abbott (eds.), Advances in Bryozoology, p. 179–194, 2 pl., Academic Press (New York).

Herwig, Ernst, 1913, *Beiträge zur Kenntniss der Knospung bei den Bryozoen:* Arch. Naturgesch., Abt. A, v. 79. p. 1–24, 29 text-fig.

Hiller, Stanisław, 1939, *The so-called colonial nervous system in Bryozoa:* Nature (London), v. 143, p. 1069–1070, 1 text-fig.

Hillmer, Gero, 1968, *Artificial moulds for studying the internal structure of paleontological objects:* in E. Annoscia (ed.), Proceedings of the First International Conference on Bryozoa, Atti Soc. Ital. Sci. Nat. Mus. Civ. Stor. Nat. Milano, v. 108, p. 37–42, pl. 2.

————, 1971, *Bryozoen (Cyclostomata) aus dem Unter-Hauterive von Nordwestdeutschland:* Mitt. Geol. Palaeontol. Inst. Univ. Hamburg, v. 40, 106 p., 30 text-fig., 22 pl.

Hilton, W. A., 1923, *A study of the movements of entoproctan bryozoans:* Trans. Am. Microsc. Soc., v. 42, p. 135–143, 9 text-fig.

Hincks, Thomas, 1887, *Critical notes on the Polyzoa:* Ann. Mag. Nat. Hist., ser. 5, v. 19, p. 150–164.

————, 1890, *Critical notes on the Polyzoa:* Ann. Mag. Nat. Hist., ser. 6, v. 5, p. 83–103.

Hinds, R. W., 1973, *Intrazooecial structures in some tubuliporinid cyclostome Bryozoa:* in G. P. Larwood (ed.), Living and Fossil Bryozoa, p. 299–306, 1 text-fig., 1 pl., Academic Press (London).

————, 1975, *Growth mode and homeomorphism in cyclostome Bryozoa:* J. Paleontol., v. 49, p. 875–910, 18 text-fig., 4 pl.

Hofmann, H. J., 1975, *Bolopora not a bryozoan, but an Ordovician phosphatic, oncolitic accretion:* Geol. Mag., v. 112, p. 523–526, 1 text-fig., 1 pl.

d'Hondt, J.-L., 1975, *Bryozoaires Cténostomes bathyaux et abyssaux de l'Atlantique Nord:* in S. Pouyet (ed.), Bryozoa 1974, Doc. Lab. Geol. Fac. Sci. Lyon, hors-série 3, fasc. 2, p. 311–333, 8 text-fig.

Hubschman, J. H., 1970, *Substrate discrimination in Pectinatella magnifica Leidy (Bryozoa):* J. Exp. Biol, v. 52, p. 603–607, 1 text-fig.

Hudleston, W. H., 1883, *Notes on a collection of fossils and of rock-specimens from West Australia, north of the Gascoyne River:* Q. J. Geol. Soc. London, v. 39, p. 582–595, 2 text-fig., pl. 23.

Hudson, J. D., 1962, *Pseudo-pleochroic calcite in recrystallized shell-limestones:* Geol. Mag., v. 99, p. 492–500, pl. 21.

Huxley, T. H., 1853, *On the morphology of the cephalous Mollusca, as illustrated by the anatomy of certain Heteropoda and Pteropoda, collected during the voyage of H.M.S. Rattlesnake in 1846–50:* Philos. Trans. R. Soc. London, v. 143, p. 29–66, pl. 2–5.

Hyman, L. H., 1958, *The occurrence of chitin in the lophophorate phyla:* Biol. Bull. Woods Hole Mass., v. 114, p. 106–112.

————, 1959, *The Invertebrates; Smaller Coelomate Groups:* v. 5, 783 p., 241 text-fig., McGraw-Hill Book Co. (New York).

————, 1966, *Further notes on the occurrence of chitin in invertebrates:* Biol. Bull. Woods Hole Mass., v. 130, p. 94–95.

Imperato, Ferrante, 1599, *Dell'Historia Naturale . . .libri XXVIII, nella quale ordinatamente di tratta della diversa condition de Miniere, e Pietre. Con alcune historie de Piante ed Animali; sin hora non date in luce:* xxiv + 791 p., illus., C. Vitale (Naples).

James, U. P., 1878, *Descriptions of newly discovered species of fossils from the lower Silurian formation, Cinncinnati group:* The Palaeontologist, no. 1, 8 p.

Jebram, Diethardt, 1973a, *Stolonen-Entwicklung und Systematik bei den Bryozoa Ctenostomata:* Z. Zool. Syst. Evolutionsforsch., v. 11, p. 1–48, 7 text-fig., 3 pl.

————, 1973b, *The importance of different growth directions in the Phylactolaemata and Gymnolaemata for reconstructing the phylogeny of the Bryozoa:* in G. P. Larwood (ed.), Living and Fossil Bryozoa, p. 565–576, 1 text-fig., Academic Press (London).

————, 1975, *Dauerknospen ("Hibernacula") bei den Bryozoa Ctenostomata in mesohalinen und vollmarinen Gewässern:* Mar. Biol., v. 31, p. 129–137, 3 text-fig.

Jeuniaux, Charles, 1963, *Chitine et chitinolyse; un chapitre de la biologie moléculaire:* 181 p., 28 text-fig., Masson (Paris).

————, 1971, *Chitinous structures:* in M. Florkin & E. H. Stotz (eds.), Comprehensive Biochemistry, v. 26, pt. C, p. 595–632, 11 text-fig., Elsevier (New York).

Johnston, George, 1833, *Illustrations in British zoology:* Mag. Nat. Hist., v. 6, p. 232–235, text-fig. 24–25.

————, 1838, *A History of the British Zoophytes:* xii + 341 p., 44 pl., W. H. Lizars (Edinburgh).

————, 1847, *A History of the British Zoophytes:* 2nd ed., v. 1, xvi + 488 p., 87 text-fig.; v. 2, 74 pl., John Van Voorst (London).

Jullien, Jules, 1881, *Note sur une nouvelle division de Bryozoaires cheilostomiens:* Bull. Soc. Zool. Fr., v. 6, p. 271–285, 5 text-fig.

————, 1885, *Monographie des Bryozoaires d'eau douce:* Bull. Soc. Zool. Fr., v. 10, p. 91–207, 250 text-fig.

————, 1888a, *Bryozoaires:* in Mission Scientifique du Cap Horn, 1882–1883, v. 6, Zoologie, pt. 3, 92 p., 15 pl. (Paris).

————, 1888b, *Observations anatomiques sur les Caténicelles:* Mem. Soc. Zool. Fr., v. 1, p. 274–280, pl. 11.

————, 1888c, *Sur la sortie et la rentrée du polypide dans les zoécies chez les Bryozoaires cheilostomiens monodermiés:* Bull. Soc. Zool. Fr., v. 13, p. 67–68.

Karklins, O. L., 1969, *The cryptostome Bryozoa from the Middle Ordovician Decorah Shale, Minnesota:* Minn. Geol. Surv. Spec. Publ. Ser., SP-6, vi + 121 p., 11 text-fig., 18 pl.

————, 1970, *Restudy of type species of the Ordovician bryozoan genus Stictoporellina:* J. Paleontol., v. 44, p. 133–139, pl. 32–34.

Kaufmann, K. W., 1971, *The form and functions of the avicularia of Bugula (phylum Ectoprocta):* Postilla, no. 151, 26 p., 9 text-fig.

Keyserling, A. F. M. von, 1846, *Wissenschaftliche Beobachtungen auf einer Reise in das Petschora-Land im Jahre 1843:* iv + 465 p., 22 pl. (St. Petersburg).

Kiseleva, A. V., 1973, *Nekotorye novye i redkie Cystoporata iz verkhnej Permi Yuzhnogo Primor'ya:* Paleontol. Zh., 1973, no. 3, p. 65–70, pl. 5–6. [*Some new and rare Cystoporata from the upper Permian of southern Primorye.*]

Kobayashi, Iwao, 1969, *Internal microstructure of the shell of bivalve molluscs:* Am. Zool., v. 9, p. 663–672, 3 text-fig., 1 pl.

————, 1971, *Internal shell microstructure of recent bivalvian molluscs:* Niigata Univ. Sci. Rep., ser. E, no. 2, p. 27–50.

de Koninck, L. G., 1873, *Recherches sur les animaux fossiles, Pt. 2.—Monographie des fossiles Carbonifères de Bleiberg en Carinthie:* 116 p., 4 pl. (Bruxelles).

Kopayevich, G. V., 1968, *O novom rode mshanok semejstva Rhinidictyidae iz Silura Estonii:* Byull. Mosk. O-va. Ispyt. Prir. Otd. Geol., v. 43, no. 3, p. 127–129, 1 text-fig. [*A new bryozoan genus of the family Rhinidictyidae from the Silurian of Estonia.*]

————, 1972, *O polimorfizme mshanok semejstva Ptilodictyidae:* Paleontol. Zh., 1972, no. 1, p. 57–63, pl. 9, 10. [*On polymorphism in Bryozoa of the family Ptilodictyidae.*]

————, 1973, *Rody Stictopora i Rhinidictya semejstva Rhinidictyidae (Bryozoa, Cryptostomata):* Paleontol. Zh., 1973, no. 3, p. 59–64, 2 text-fig. [*The genera Stictopora and Rhinidictya of the family Rhinidictyidae (Bryozoa, Cryptostomata).*]

————, 1975, *Silurijskie mshanki Estonii i Podolii*

(*Cryptostomata, Rhabdomesonata*): Tr. Akad. Nauk SSSR Paleontol. Inst., v. 151, 155 p., 13 text-fig., 28 pl. [*Silurian Bryozoa of Estonia and Podolia.*]

Korde, K. B., 1950, *Dasycladaceae iz kembriya Tuvy:* Dokl. Akad. Nauk SSSR, v. 73, no. 2, p. 371–374, 3 text-fig. [*Dasycladaceae from the Cambrian of Tuva.*]

Koschinsky, Carl, 1885, *Ein Beitrag zur Kenntniss der Bryozoenfauna der älteren Tertiärschichten des südlichen Bayerns, I. Abtheilung. Cheilostomata:* Palaeontographica, v. 32, 73 p., 7 pl.

Kraepelin, K. M. F., 1886, *Die Fauna der Hamburger Wasserleitung:* Abh. Naturwiss. Ver. Hamburg, v. 9, no. 1, p. 1–15.

——, 1887, *Die deutschen süsswasser-Bryozoen. Eine Monographie. I. Anatomisch-systematischer Teil:* Abh. Naturwiss. Ver. Hamburg, v. 10, no. 9, 168 p., 7 pl.

——, 1892, *Die deutschen Süsswasser-Bryozoen. Eine Monographie. II. Entwickelungsgeschichtlicher Teil:* Abh. Naturwiss. Ver. Hamburg, v. 12, no. 2, 68 p., 5 pl.

Labracherie, Monique, 1973, *Functional morphology and habitat of Bryozoa in the Eocene of the northern Aquitaine basin, France:* in G. P. Larwood (ed.), Living and Fossil Bryozoa, p. 129–138. 6 text-fig., Academic Press (London).

——, & Sigal, Jacques, 1975, *Les Bryozoaires Cheilostomes des formations Eocène inférieur du Site 246 (croisière 25, Deep Sea Drilling Project):* in S. Pouyet (ed.), Bryozoa 1974, Doc. Lab. Geol. Fac. Sci. Lyon, hors-série 3, fasc. 2, p. 449–466, 2 text-fig.

Lacourt, A. W., 1968, *A monograph of the freshwater Bryozoa—Phylactolaemata:* Zool. Verh. Rijksmus. Nat. Hist. Leiden, no. 93, 159 p., 13 text-fig., 18 pl.

Lagaaij, Robert, & Cook, P. L., 1973, *Some Tertiary to recent Bryozoa:* in A. Hallam (ed.), Atlas of Paleobiogeography, p. 489–498, 4 text-fig., 1 pl., Elsevier (Amsterdam).

——, & Gautier, Y. V., 1965, *Bryozoan assemblages from marine sediments of the Rhône delta, France:* Micropaleontology, v. 11, p. 39–58, 34 text-fig.

Land, L. S., MacKenzie, F. T., & Gould, S. J., 1967, *Pleistocene history of Bermuda:* Geol. Soc. Am. Bull., v. 78, p. 993–1006, 5 text-fig., 4 pl.

Lang, W. D., 1916, *A revision of the "cribrimorph" Cretaceous Polyzoa:* Ann. Mag. Nat. Hist., ser. 8, v. 18, p. 381–410.

Lankester, E. R., 1885, *Polyzoa:* in Encyclopedia Britannica, 9th ed., v. 19, p. 429–441, University of Chicago Press (Chicago).

Larwood, G. P., 1969, *Frontal calcification and its function in some Cretaceous and recent cribrimorph and other cheilostome Bryozoa:* Bull. Br. Mus. Nat. Hist. Zool., v. 18, p. 171–182, 10 text-fig.

——, 1975, *Preliminary report on early (pre-Cenomanian) cheilostome Bryozoa:* in S. Pouyet (ed.), Bryozoa 1974, Doc. Lab. Geol. Fac. Sci. Lyon, hors-série 3, fasc. 2, p. 539–545, 1 text-fig.

Lea, Isaac, 1833, *Contributions to geology:* vi + 227 p., 6 pl., Carey, Lea, & Blanchard (Philadelphia).

Leveaux, Marcelle, 1939, *La formation des gemmules chez les Spongillidae:* Ann. Soc. R. Zool. Belg., v. 70, p. 53–96, 21 text-fig.

Levinsen, G. M. R., 1902, *Studies on Bryozoa:* Vidensk. Medd. Naturhist. Foren. Kjøbenhavn, 1902, p. 1–31.

——, 1909, *Morphological and Systematic Studies on the Cheilostomatous Bryozoa:* viii + 431 p., 6 text-fig., 27 pl., Nationale Forfatteres Forlag (Copenhagen).

——, 1912, *Studies on the Cyclostomata operculata:* Acad. R. Sci. Let. Danemark Mem., ser. 7, sect. sci., v. 10, p. 1–52, 2 text-fig., 7 pl.

Lewis, H. P., 1926, *On Bolopora undosa gen. et sp. nov.; a rock-building bryozoan with phosphatized skeleton, from the basal Arenig rocks of Ffestiniog (North Wales):* Q. J. Geol. Soc. London, v. 82, p. 411–427, 7 text-fig., pl. 27–29.

Lindström, Gustav, 1876, *On the affinities of the Anthozoa Tabulata:* Ann. Mag. Nat. Hist., ser. 4, v. 18, p. 1–17.

Linné, Carl, 1758, *Systema naturae per regna tria naturae. secundum classes, ordines, genera, species, cum characteribus, differentiis synonymis, locis:* 10th ed., v. 1, 823 p., Laurentii Salvii (Holmiae).

Lippmann, Friedrich, 1973, *Sedimentary Carbonate Minerals:* 228 p., 54 text-fig., Springer-Verlag (Berlin).

Lister, J. J., 1834, *Some observations on the structure and functions of tubular and cellular Polypi, and of Ascidiae:* Philos. Trans. R. Soc. London, v. 124, p. 365–388, pl. 8–12.

Lonsdale, William, 1839, *Corals:* in R. I. Murchison, The Silurian System, Part II.—Organic remains, p. 675–694, pl. 25, 26, John Murray (London).

Lowenstam, H. A., 1954, *Factors affecting the aragonite-calcite ratios in carbonate-secreting marine organisms:* J. Geol., v. 62, p. 284–322, 15 text-fig.

——, 1963, *Biologic problems relating to the composition and diagenesis of sediments:* in T. W. Donnelly (ed.), The Earth Sciences—Problems and Progress in Current Research, p. 137–195, 14 text-fig., 4 pl., University of Chicago Press (Chicago).

——, 1964a, *Sr/Ca ratio of skeletal aragonites from the recent marine biota at Palau and from fossil gastropods:* in H. Craig, S. L. Miller, & G. J. Wasserburg (eds.), Isotopic and Cosmic Chemistry, p. 114–132, 2 text-fig., North Holland Publishing Co. (Amsterdam).

——, 1964b, *Coexisting calcites and aragonites from skeletal carbonates of marine organisms and their strontium and magnesium contents:* in Y. Miyake & T. Komaya (eds.), Recent Research in the Fields of Hydrosphere, Atmosphere, and Nuclear Geochemistry, p. 373–404, 5 text-fig., 1 pl., Nagoya University, Water Research Laboratory (Suguwara Festival Volume).

Lutaud, Geneviève, 1959a, *Étude cinématographique du bourgeonnement chez Membranipora membranacea (Linné), Bryozoaire chilostome. I.— Le développement du polypide:* Bull. Soc. Zool. Fr., v. 84, p. 167–173, 1 text-fig.

————, 1959b, *Application de la microcinématographie à l'étude de la croissance et du bourgeonnement chez les Bryozoaires chilostomes:* Research Film, v. 3, p. 193–196, 6 text-fig.

————, 1961, *Contribution à l'étude du bourgeonnement et de la croissance des colonies chez Membranipora membranacea (Linné), Bryozoaire chilostome:* Ann. Soc. R. Zool. Belg., v. 91, p. 157–300, 28 text-fig., 8 pl.

————, 1969, *Le "plexus" pariétal de Hiller et la coloration du système nerveux par le bleu de méthylène chez quelques Bryozoaires chilostomes:* Z. Zellforsch., v. 99, p. 302–314, 4 text-fig.

————, 1973, *L'innervation du lophophore chez le Bryozoaire chilostome Electra pilosa (L.):* Z. Zellforsch., v. 140, p. 217–234, 8 text-fig.

————, 1974, *Le plexus pariétal des Cténostomes chez Bowerbankia gracilis Leidy (Vésicularines):* Cah. Biol. Mar., v. 15, p. 403–408, 1 text-fig., 1 pl.

————, 1977, *The bryozoan nervous sytem:* in R. M. Woollacott & R. L. Zimmer (eds.), Biology of Bryozoans, p. 377–410, 6 text-fig., 5 pl., Academic Press (New York).

————, 1979, *Étude ultrastructurale du "plexus colonial" et recherche de connexions nerveuses interzoïdiales chez le Bryozoaire chilostome Electra pilosa (Linné):* Cah. Biol. Mar., v. 20, p. 315–324, 1 text-fig., pl. 14.

————, 1981, *The innervation of the external wall in the carnosan ctenostome Alcyonidium polyoum (Hassall):* in G. P. Larwood & C. Nielsen (eds.), Recent and Fossil Bryozoa, p. 143–150, 3 text-fig., Olsen & Olsen (Fredensborg).

————, & Painleve, J., 1961, *Le bourgeonnement et le croissance des colonies chez Membranipora membranacea (L.), Bryozoaire Chilostome:* Film, black-and-white 16 mm, Institut de Cinématographie Scientifique, Paris.

MacClintock, Copeland, 1967, *Shell structure of patelloid and bellerophontoid gastropods (Mollusca):* Peabody Mus. Nat. Hist. Yale Univ. Bull., v. 22, x + 140 p., 128 text-fig., 32 pl.

McCoy, Frederick, 1849, *On some new genera and species of Palaeozoic corals and Foraminifera:* Ann. Mag. Nat. Hist., ser. 2, v. 3, p. 119–136, 6 text-fig.

————, 1851–1855, *A systematic description of the British Palaeozoic fossils in the Geological Museum*

of the University of Cambridge: in Adam Sedgwick, A Synopsis of the Classification of the British Palaeozoic Rocks: 611 p., 25 pl., J. W. Parker (London). [p. 1–184, 1851; p. 185–406, 1852; p. 407–611, 1855.]

McFarlan, A. C., 1926, *The bryozoan faunas of the Chester series of Illinois and Kentucky:* University of Chicago, Thesis Abstracts.

McKinney, F. K., 1969, *Organic structures in a Late Mississippian trepostomatous ectoproct (bryozoan):* J. Paleontol., v. 43, p. 285–288, 1 text-fig., pl. 50.

————, 1975, *Autozooecial budding patterns in dendroid stenolaemate bryozoans:* in S. Pouyet (ed.), Bryozoa 1974, Doc. Lab. Geol. Fac. Sci. Lyon, hors-série 3, fasc. 1, p. 65–76, 2 text-fig., 4 pl.

————, 1977a, *Functional interpretation of lyre-shaped Bryozoa:* Paleobiology, v. 3, p. 90–97, 5 text-fig.

————, 1977b, *Autozooecial budding patterns in dendroid Paleozoic bryozoans:* J. Paleontol., v. 51, p. 303–329, 6 text-fig., 9 pl.

————, 1977c, *Paraboloid colony bases in Paleozoic stenolaemate bryozoans:* Lethaia, v. 10, p. 209–217, 3 text-fig.

McNair, A. H., 1937, *Cryptostomatous Bryozoa from the Middle Devonian Traverse group of Michigan:* Mich. Univ. Mus. Paleontol. Contrib., v. 5, no. 9, p. 103–170, 1 text-fig., 14 pl.

————, 1942, *Upper Devonian Bryozoa:* J. Paleontol., v. 16, p. 343–350, pl. 45–49.

Männil, R. M., 1958, *Novye mshanki otryada Cryptostomata iz Ordovika Estonii:* Eesti NSV Tead. Akad. Toim. Fuus.-Mat. Tehnikatead. Seer., v. 7, no. 4, p. 330–347, 6 text-fig., 8 pl. [*New bryozoans of the order Cryptostomata from the Ordovician of Estonia.*]

————, 1959, *Voprosy stratigrafii i mshanki Ordovika Estonii:* 40 p., Akademiya Nauk Estonskoj SSR, Otdelenie Tekhnicheskikh i Fiziko-Matematicheskikh Nauk (Tallin). Dissertatsii na soiskanie uchenoj stepeni kandidata geologo-mineralogicheskikh nauk.) [*Problems of stratigraphy and Bryozoa of the Ordovician of Estonia. Dissertations for the degree of Candidate in Geological-Mineralogical Sciences.*]

Majewski, O. P., 1969, *Recognition of Invertebrate Fossil Fragments in Rocks and Thin Sections:* 101 p., 106 pl., E. J. Brill (Leiden).

Marcus, Ernst, 1926, *Beobachtungen und Versuche an lebenden Süsswasserbryozoen:* Zool. Jahrb. Abt. Syst. Oekol. Geogr. Tiere, Bd. 52, p. 279–350, 34 text-fig., 6 pl.

————, 1934, *Über Lophopus crystallinus (Pall.):* Zool. Jahrb. Abt. Anat. Ontog. Tiere, Bd. 58, p. 501–606, 66 text-fig.

————, 1938a, *Bryozoarios marinhos brasileiros— II:* Univ. Sao Paulo Fac. Filos. Cienc. Let. Bol., no. 4, Zool., no. 2, p. 1–137, pl. 1–29.

———, 1938b, *Bryozoarios perfuradores de conchas:* Arq. Inst. Biol. Sao Paulo, v. 9, p. 273–296, 7 text-fig.

———, 1941, *Sôbre Bryozoa do Brasil:* Univ. Sao Paulo Fac. Filos. Cienc. Let. Bol., no. 12, Zool., no. 5, p. 3–208, 9 text-fig., 18 pl.

———, 1958, *On the evolution of the animal phyla:* Q. Rev. Biol., v. 33, p. 24–58, 1 text-fig.

Marcus, Evelina, 1955, *Polyzoa:* in The Percy Sladen Trust Expedition to Lake Titicaca. . ., Trans. Linn. Soc. London, ser. 3, v. 1, pt. 3, p. 355–357.

———, & Marcus, Ernst, 1962, *On some lunulitiform Bryozoa:* Univ. Sao Paulo Fac. Filos. Cienc. Let. Bol., no 261, Zool., no. 24, p. 281–324, 5 pl.

Matricon, Isabelle, 1973, *Quelques données ultrastructurales sur un myoépithélium; le pharynx d'un Bryozoaire:* Z. Zellforsch., v. 136, p. 569–578, 4 pl.

Maturo, F. J. S., Jr., 1973, *Offspring variation from known maternal stocks of Parasmittina nitida (Verrill):* in G. P. Larwood (ed.), Living and Fossil Bryozoa, p. 577–584, 1 pl., Academic Press (London).

Mawatari, Shizuo, 1965, *Tentaculata:* in T. Uchida (ed.), Systematic Zoology, Dōbutsu keitō bunruigaku, v. 8-I, p. 9–258, 190 text-fig. [In Japanese.]

Mayr, Ernst, 1968, *Bryozoa versus Ectoprocta:* Syst. Zool., v. 17, p. 213–216.

Medd, A. W., 1964, *On the musculature of some Cretaceous membranimorph Polyzoa:* Ann. Mag. Nat. Hist., ser. 13, v. 7, p. 185–187, 3 text-fig.

Meek, F. B., 1872, *Report on the paleontology of eastern Nebraska:* in F. V. Hayden, Final Report of the United States Geological Survey of Nebraska and Portions of the Adjacent Territories. . ., p. 81–239, 11 pl. U.S. Government Printing Office (Washington).

———, & Worthen, A. H., 1865, *Descriptions of new species of Crinoidea, etc., from the Paleozoic rocks of Illinois and some of the adjoining states:* Proc. Acad. Nat. Sci. Philadelphia, 1865, p. 143–166.

Merida, J. E., & Boardman, R. S., 1967, *The use of Paleozoic Bryozoa from well cuttings:* J. Paleontol., v. 41, p. 763–765, pl. 100.

Miller, S. A., 1889, *North American Geology and Paleontology for the Use of Amateurs, Students, and Scientists:* 664 p., 1194 text-fig., Western Methodist Book Concern (Cincinnati).

Milliman, J. D., 1974, *Marine carbonates:* in Recent Sedimentary Carbonates, part 1, 375 p., 94 text-fig., 39 pl., Springer-Verlag (New York).

Milne-Edwards, Henri, 1836, *Histoire des polypes:* in J. B. P. A. de Lamarck, Histoire naturelle des animaux sans vertèbres, 2nd ed., v. 2 683 p., Baillière (Paris).

———, 1843, *Élémens de zoologie, ou Leçons sur l'anatomie, la physiologie, la classification et les moeurs des animaux:* 2nd ed., v. 3 [Animaux sans vertèbres], 360 p., text-fig. 432–853, Fortin, Masson et Cie (Paris).

———, & Haime, Jules, 1850[-1855], *A Monograph of the British Fossil Corals:* lxxxvi + 322 p., 72 pl. [p. i–lxxxv, 1–71, pl. 1–11 (1850), p. 72–146, pl. 12–30 (1851), p. 147–210, pl. 31–46 (1852), p. 211–244, pl. 47–56 (1853), p. 245–299, pl. 57–72 (1855)], Palaeontographical Society (London).

Moore, R. C., 1929, *A bryozoan faunule from the Upper Graham formation, Pennsylvanian, of north central Texas:* J. Paleontol., v. 3, p. 1–27, 3 text-fig., 3 pl.

———, & Dudley, R. M., 1944, *Cheilotrypid bryozoans from Pennsylvanian and Permian rocks of the midcontinent region:* Kansas State Geol. Surv. Bull. 52, pt. 6, p. 229–408, 48 pl.

Morozova, I. P., 1955, *O nakhodke novogo roda mshanok v verkhnem Karbone Donskoy luki:* Dokl. Akad. Nauk SSSR, v. 100, p. 567–569, 1 text-fig. [*A new bryozoan genus from the Upper Carboniferous of the Don bend.*]

———, 1959a, *Novyj rod mshanok semejstva Fistuliporidae iz Devona Kuznetskogo Bassejna:* Paleontol. Zh., 1959, no. 2, p. 79–81, 1 text-fig. [*A new bryozoan genus of the family Fistuliporidae from the Devonian of the Kuznetsk basin.*]

———, 1959b, *Devonskie mshanki otryada Cyclostomata iz Kuznetskoj i Minusinskikh Kotlovin:* Akad. Nauk SSSR Paleontol. Inst., Materialy k "Osnovam Paleontologii," v. 3, p. 7–11, pl. 2. [*Devonian Bryozoa of the order Cyclostomata from Kuznetsk and the Minusinsk basins.*]

———, 1960, *Otryad Cryptostomata; Sem. Hexagonellidae, Sulcoreteporidae, i Goniocladiidae:* in Osnovy Paleontologii, p. 84–88, text-fig. 156–167. [*Order Cryptostomata, families Hexagonellidae, Sulcoreteporidae, and Goniocladiidae; see T. G. SARYCHEVA, 1960 for complete citation.*]

———, 1966, *Novyj podotryad pozdnepaleozojskikh mshanok otryada Cryptostomata:* Paleontol. Zh., 1966, no. 2, p. 33–41, pl. 5, 6. [*A new suborder of late Paleozoic bryozoans of the order Cryptostomata.*]

———, 1970, *Mshanki pozdnej Permi:* Tr. Akad. Nauk SSSR Paleontol. Inst., v. 122, 314 p., 43 text-fig., 64 pl. [*Late Permian bryozoans.*]

Morton, S. G., 1829, *Note; containing a notice of some fossils recently discovered in New Jersey:* J. Acad. Nat. Sci. Philadelphia, v. 6, p. 120–129, pl. 7.

———, 1834, *Synopsis of the Organic Remains of the Cretaceous Group of the United States:* vi + 88 p., 19 pl., Key & Biddle (Philadelphia).

Mukai, Hideo, 1974, *Germination of the statoblasts of a freshwater bryozoan, Pectinatella gelatinosa:* J. Exp. Zool., v. 187, p. 27–40, 3 text-fig.

Mutvei, Harry, 1972, *Ultrastructural relationships*

between the prismatic and nacreous layers in Nautilus (Cephalopoda): Biomineralisation Forschungsber., Bd. 4, p. 80–86, 2 text-fig., 2 pl.

Nekhoroshev, V. P., 1948a, *Devonskie mshanki Altaya:* Akad. Nauk SSSR Paleontol. Inst., Paleontol. SSSR, v. 3, ch. 2, vyp. 1, 172 p., 34 text-fig., 48 pl. [*Devonian bryozoans of the Altai.*]

————, 1948b, *Kamennougol'nye mshanki Severo-Vostochnogo Pribalkhash'ya:* Akad Nauk Kaz. SSR, Alma-Ata, 70 p., 11 pl. [*Carboniferous Bryozoa of Northeast Pribalkhash.*]

————, 1953, *Nizhnekamennougol'nye mshanki Kazakhstana:* Tr. Vses. Neft. Nauchno-Issled. Geol. Razved. Inst. (VNIGRI), 234 p., 25 pl. [*Lower Carboniferous Bryozoa of Kazakhstan.*]

————, 1956a, *Bryozoa:* in Materialy po paleontologii, novye semejstva i rody, Vses. Nauchno-Issled. Geol. Inst. (VSEGEI), Materialy, n.ser., vyp. 12, p. 42–49, 1 pl. [*Bryozoa in Materials for paleontology, new families and genera.*]

————, 1956b, *Nizhnekamennougol'nye mshanki Altaya i Sibiri:* Tr. Vses. Nauchno-Issled. Geol. Inst. (VSEGEI), n.ser., v. 13, 420 p., 72 text-fig., 57 pl. [*Lower Carboniferous Bryozoa of the Altai and Siberia.*]

————, 1960, *Nekotorye vidy Paleozojskikh cryptostomat SSSR:* in B. P. Markovskiy (ed.), Novye vidy drevnikh rastenij i bespozvonochnykh SSSR, Ch. 1, p. 268–283, pl. 67–71, Vses. Nauchno-Issled. Geol. Inst. [VSEGEI], (Moscow). [*Some species of Paleozoic Cryptostomata of the USSR:* in B. P. Markovskiy (ed.), *New species of ancient plants and invertebrates of the USSR.*]

————, 1961, *Ordovikskie i Silurijskie mshanki Sibirskoj Platformy, Otryad Cryptostomata:* Tr. Vses. Nauchno-Issled. Geol. Inst. (VSEGEI), n.ser., v. 41, vyp. 2, 246 p., 37 pl. [*Ordovician and Silurian Bryozoa of the Siberian Platform, Order Cryptostomata.*]

————, & Modzalevskaya, E. A. 1966, *Stratigraficheskie rubezhi na primere evolyutsionnogo razvitiya Paleozojskikh mshanok:* in Paleontologicheskie kriterii ob'ema i ranga stratigraficheskikh podrazdelenij, Tr. Vses. Paleontol. Obshch., 8 sess., p. 93–112. [*Stratigraphic boundaries based on the evolutionary development of Paleozoic Bryozoa:* in Paleontologic criteria of the size and rank of stratigraphic subdivisions.*]

Neviani, Antonio, 1894, *Terza contribuzione alla conoscenza dei Briozoi fossili italiani; di alcuni Briozoi pliocenici del Rio Landa, illustrati da Ferdinando Bassi nel 1757:* Boll. Soc. Geol. Ital. Roma, v. 12 (1893), p. 659–668, 4 text-fig., pl. 1–12.

Newton, G. B., 1971, *Rhabdomesid bryozoans of the Wreford Megacyclothem (Wolfcampian, Permian) of Nebraska, Kansas, and Oklahoma:* Univ. Kans. Paleontol. Contrib. Artic. 56 (Bryozoa 2), 71 p., 19 text-fig., 2 pl.

Nicholson, H. A., 1874a, *Summary of recent* researches on the paleontology of the Province of Ontario, with brief descriptions of some new genera:* Can. J. Sci. Lit. Hist., n.ser., v. 14, p. 125–136.

————, 1874b, *Descriptions of new fossils from the Devonian formations of Canada West:* Geol. Mag., n.ser., dec. 2, v. 1, p. 117–126, text-fig. 15, pl. 6.

————, 1874c, *Descriptions of new fossils from the Devonian formations of Canada West:* Geol. Mag., n.ser., dec. 2, v. 1, p. 159–163, pl. 9.

————, 1875, *Descriptions of new species and of a new genus of Polyzoa from the Paleozoic rocks of North America:* Geol. Mag., n.ser., dec. 2, v. 2, p. 33–38.

————, 1876, *Notes on the Paleozoic corals of the State of Ohio:* Ann. Mag. Nat. Hist., ser. 4, v. 18, p. 85–95, pl. 5.

————, 1879, *On the structure and affinities of the "tabulate corals" of the Paleozoic Period, with critical descriptions of illustrative species:* 342 p., 44 text-fig., 15 pl., William Blackwood & Sons (Edinburgh).

————, 1881, *On the structure and affinities of the genus Monticulipora and its subgenera, with critical descriptions of illustrative species:* 240 p., 50 text-fig., 6 pl., William Blackwood & Sons (Edinburgh).

Nickles, J. M., & Bassler, R. S., 1900, *A synopsis of American fossil Bryozoa including bibliography and synonymy:* U.S. Geol. Surv. Bull. 173, 663 p.

Nielsen, Claus, 1970, *On metamorphosis and ancestrula formation in cyclostomatous bryozoans:* Ophelia, v. 7, p. 217–256, 41 text-fig.

————, 1971, *Entoproct life-cycles and the entoproct/ectoproct relationship:* Ophelia, v. 9, p. 209–341, 82 text-fig.

————, & Pedersen, K. J., 1979, *Cystid structure and protrusion of the polypide in Crisia (Bryozoa, Cyclostomata):* Acta Zool. Stockholm, v. 60, p. 65–88, 24 text-fig.

Nikiforova, A. I., 1927, *Materialy k poznaniyu nizhne-kamennougol'nykh mshanok Donetskogo bassejna:* Izv. Geol. Kom., v. 46, p. 245–268, 2 text-fig., pl. 12–14. [*Materials for the understanding of Lower Carboniferous Bryozoa of the Donets Basin.*]

————, 1938, *Tipy kamennougol'nykh mshanok Evropejskoj chasti SSSR:* Akad. Nauk SSSR Paleontol. Inst., Paleontol. SSSR, v. 4, ch. 5, vyp. 1, 290 p., 80 text-fig., 55 pl. [*Types of Carboniferous Bryozoa of the European part of the USSR.*]

————, 1948, *Nizhnekamennougol'nye mshanki Karatau:* Akad. Nauk Kaz. SSR, Alma Ata, 53 p., 15 pl. [*Lower Carboniferous Bryozoa of Karatau.*]

Nitsche, Hinrich, 1869, *Beiträge zur Kenntniss der Bryozoen:* Z. Wiss. Zool., Bd. 20, heft 1, p. 1–36, pl. 1–3.

————, 1871, *Beiträge zur Kenntniss der Bryozoen:* Z. Wiss. Zool. Bd. 21, Heft 4, p. 37–119, 4 text-fig., pl. 4–6.

Nye, O. B., Jr., 1968, *Aspects of microstructure in post-Paleozoic Cyclostomata:* in E. Annoscia (ed.), Proceedings of the First International Conference on Bryozoa, Atti Soc. Ital. Sci. Nat. Mus. Civ. Stor. Nat. Milano, v. 108, p. 111–114.

———, 1976, *Generic revision and skeletal morphology of some cerioporid cyclostomes (Bryozoa):* Bull. Am. Paleontol., v. 69, no. 291, 222 p., 20 text-fig., 51 pl.

———, Dean, D. A., & Hinds, R. W., 1972, *Improved thin section techniques for fossil and recent organisms:* J. Paleontol., v. 46, p. 271–275, 1 pl.

Oda, Shuzitu, 1959, *Germination of the statoblasts in freshwater Bryozoa:* Sci. Rep. Tokyo Kyoiku Daigaku Sect. B, v. 9, no. 135, p. 90–131, 16 text-fig.

Oka, Asajiro, 1891, *Observations on freshwater Polyzoa (Pectinatella gelatinosa nov. sp.):* J. Coll. Sci. Imp. Univ. Japan, Tokyo, v. 4, p. 89–150, pl. 17–20.

d'Orbigny, A. D., 1839, *Zoophytes:* in Voyage dans l'Amérique méridionale (le Brézil. . .l'Uruguay, la République Argentine, la Patagonie. . . Chili. . .Bolivia. . .Pérou), exécuté pendant. . .1826–33, v. 5, pt. 4, 28 p., 13 pl.

———, 1849, *Description de quelques genres nouveaux de Mollusques Bryozoaires:* Rev. Mag. Zool., ser. 2, v. 1, p. 499–504.

———, 1851-1854, *Bryozoaires:* in Paléontologie française. Description des animaux invertébrés. . .continuée. . .sous la direction d'un comité spécial, Terrains crétacés, v. 5, p. 1–188 (1851); p. 185–472 (1852); p. 473–984 (1853); p. 985–1192, pl. 600–800 (1854), Masson (Paris).

Ostroumov [Ostroumoff], A. A., 1886a, *Opyt izsledovanya mshanok Sevastopol'skoj bukhty v systematicheskom i morfologicheskom otnosheniyakh:* Tr. Imp. Univ. Kazan Obshchest. Estestvo., v. 16, p. 1–124, pl. 1–5. [*Studies on the Bryozoa of Sevastopol Bay with reference to their systematics and morphology.*]

———, 1886b, *Contribution à l'étude zoologique et morphologique des Bryozoaires du Golfe de Sébastopol (Pt. 1.—Données systématiques. Pt. 2.—Données anatomiques. Pt. 3.—Données sur l'histoire du développement.):* Arch. Slaves Biol., v. 1, p. 557–569; v. 2, p. 8–25, 184–190, 329–355, 5 pl.

———, 1903, *Sur le développement du cryptocyst et de la chambre de compensation:* Zool. Anz., Bd. 27, p. 96–97.

Perry, T. G., & Hattin, D. E., 1958, *Astogenetic study of fistuliporoid bryozoans:* J. Paleontol., v. 32, p. 1039–1050, 1 text-fig., pl. 129–131.

Phillips, John, 1836, *Illustrations of the Geology of Yorkshire; or, a description of the strata and organic remains of the Yorkshire coast, etc. Part II.—The Mountain Limestone district:* xx + 253 p., 25 pl.

(London).

———, 1841, *Figures and Descriptions of the Palaeozoic Fossils of Cornwall, Devon, and West Somerset:* 231 p., 60 pl., Longmans, Brown, Green, & Longmans (London).

Phillips, J. R. P. [J. R. P. Ross], 1960, *Restudy of types of seven Ordovician bifoliate Bryozoa:* Palaeontology, v. 3, p. 1–25, 2 text-fig., pl. 1–10.

Pinter Morris, P. A., 1975, *A comparative study of decalcification of Mollusca shells by various bryozoans:* in S. Pouyet (ed.), Bryozoa 1974. Doc. Lab. Geol. Fac. Sci. Lyon, hors-série 3, fasc. 1, p. 109–113, 2 pl.

Pitt, L. J., 1976, *A new cheilostome bryozoan from the British Aptian:* Proc. Geol. Assoc., v. 87, p. 65–67, 1 pl.

Počta, Philippe, 1894, *Bryozoaires, hydrozoaires et partie des anthozoaires:* in Joachim Barrande, Système Silurien du Centre de la Bohême, v. 8, x + 230 p., 21 pl. (Prague).

Pohowsky, R. A., 1973, *A Jurassic cheilostome from England:* in G. P. Larwood (ed.), Living and Fossil Bryozoa, p. 447–461, 3 text-fig., 1 pl., Academic Press (London).

———, 1974, *Notes on the study and nomenclature of boring Bryozoa:* J. Paleontol., v. 48, p. 556–564, 1 pl.

———, 1975, *Boring Bryozoa:* in S. Pouyet (ed.), Bryozoa 1974, Doc. Lab. Geol. Fac. Sci. Lyon, hors-série 3, fasc. 1, p. 255–256.

Poluzzi, Angelo, & Sartori, Renzo, 1973, *Carbonate mineralogy of some Bryozoa from Talbot Shoal (Strait of Sicily, Mediterranean):* G. Geol., ser. 2a, v. 39, p. 11–15.

———, & ———, 1975, *Report on the carbonate mineralogy of Bryozoa:* in S. Pouyet (ed.), Bryozoa 1974, Doc. Lab. Geol. Fac. Sci. Lyon, hors-série 3, fasc. 1, p. 193–210, 5 text-fig.

Pouyet, Simone, 1971, *Schizoporella violacea (Canu et Bassler, 1930) (Bryozoa Cheilostomata); variations et croissance zoariale:* Geobios, v. 4, fasc. 3, p. 185–197, 3 text-fig., pl. 13–14.

Powell, N. A., 1970, *Schizoporella unicornis—An alien bryozoan introduced into the Strait of Georgia:* J. Fish. Res. Board Can., v. 27, p. 1847–1853, 5 text-fig.

Prantl, Ferdinand, 1935a, *Über Gattung Lemmatopora Počta:* Zentralbl. Mineral Geol. Palaeontol., Jahrg. 1935, Abt. B, no. 2, p. 40–45, 3 text-fig.

———, 1935b, *O nových mechovkách z vápencu bránických gα:* Rozpr. Ceska Akad. Ved Rada Mat. Prir. Ved, v. 45, no. 5, 12 p., 5 text-fig., 1 pl. [*New cyclostomatous Bryozoa from the Branik limestone gα.*]

Prenant, Marcel, & Bobin, Geneviève, 1966, *Bryozoaires, deuxième partie, Chilostomes Anasca:* Faune de France, v. 68, p. 1–647, 210 text-fig.

Prigogine, Ilya, Nicolis, Gregoire, & Babloyantz,

Agnes, 1972, *Thermodynamics of evolution, I:* Phys. Today, 1972 (November), v. 25, no. 11 p. 23–28., 2 text-fig.

Prout, H. A., 1860, *Fourth series of descriptions of Bryozoa from the Paleozoic rocks of the western states and territories:* Trans. St. Louis Acad. Sci., v. 1, p. 571–581.

Purdy, E. G., 1968, *Carbonate diagenesis; an environmental survey:* Geol. Rom., v. 7, p. 183–228, 10 text-fig., 6 pl.

Quenstedt, F. A., 1878–1881, *Petrefactenfunde Deutschlands, etc.; Band VI.—Korallen (Röhren- und Sternkorallen):* lief. 6, p. 1–144 (1878); lief. 7–9, p. 145–624 (1879); lief. 10–11, p. 625–912 (1880); lief. 12, p. 913–1094 (1881), Atlas containing 42 pl. (1881), Fues's Verlag, R. Reisland (Leipzig).

Rafinesque, C. S., 1831, *Enumeration and Account of some Remarkable Natural Objects of the Cabinet of Prof. Rafinesque, in Philadelphia; being animals, shells, plants, and fossils, collected by him in North America between 1816 and 1831:* 8 p. (Philadelphia).

Raup, D. M., 1966, *Geometric analysis of shell coiling; general problems:* J. Paleontol., v. 40, p. 1178–1190, 10 text-fig.

Reed, F. R. C., 1907, *New fossils from Haverford-west. VII:* Geol. Mag., ser. 2, dec. 5, v. 4, p. 208–211, pl. 6.

Repiachoff, W. M., 1880, *Zur Kenntnis der Bower-bankia-Larven:* Zool. Anz., v. 3, p. 260.

Reuss, A. E., 1851, *Ein Beitrag zur Paläontologie der Tertiärschichten Oberschlesiens:* Z. Dtsch. Geol. Ges. Berlin, v. 3, p. 149–184, pl. 8–9.

———, 1864, *Die Fossilen Foraminiferen, Anthozoen und Bryozoen von Oberburg in Steiermark. Ein Beitrag zur Fauna der Oberen Nummuliten-schichten:* Akad. Wiss. Wien Math. Naturwiss. Kl. Denkschr., v. 23, p. 1–38, 10 pl.

Richards, R. P., 1974, *A Devonian Immergentia (Ectoprocta Ctenostomata) from Ohio:* J. Paleontol., v. 48, p. 940–946, 2 text-fig., 1 pl.

Ringueberg, E. N. S., 1886, *New genera and species of fossils from the Niagara shales:* Bull. Buffalo Soc. Nat. Sci., v. 5, p. 5–22, 2 pl.

Rogers, A. F., 1900, *New bryozoans from the coal measures of Kansas and Missouri:* Kans. Univ. Q., v. 9, ser. A. p. 1–12, pl. 1–4.

Rogick, M. D., 1935, *Studies on freshwater Bryozoa. II. The Bryozoa of Lake Erie:* Trans. Am. Microsc. Soc., v. 54, p. 245–263, pl. 40–42.

———, 1937, *Studies on freshwater Bryozoa. VI. The finer anatomy of Lophopodella carteri var. typica:* Trans. Am. Microsc. Soc., v. 56, p. 367–396, 7 pl.

———, 1938, *Studies on freshwater Bryozoa. VII. On the viability of dried statoblasts of Lophopodella carteri var. typica:* Trans. Am. Microsc. Soc., v. 57, p. 178–199, 3 pl.

———, & Brown, C. J. D., 1942, *Studies of fresh-water Bryozoa. XII. A collection from various sources:* Ann. New York Acad. Sci., v. 43, p. 123–143, 4 pl.

Romanchuk, T. V., 1966, *Novye permskie mshanki Khabarovskogo kraya:* Paleontol. Zh., 1966, no. 2, p. 42–48, pl. 7, 8. [*New Permian bryozoans from Khabarovsk Krai.*]

———, 1975, *Pervye nakhodki kamennougol'nykh mshanok v Tuguro-Chumikanskom rajone Khabarovskogo kraya:* Paleontol. Zh., 1975, no. 2, p. 69–78, pl. 5, 6. [*First records of Carboniferous Bryozoa in the Tuguro-Chumikansk region of the Khabarovsk Krai.*]

———, & Kiseleva, A. V., 1968, *Novye pozdne-permskie mshanki Dal'nego Vostoka:* Paleontol. Zh., 1968, no. 4, p. 55–60, 2 text-fig. [*New Late Permian bryozoans of the Far East.*]

Rominger, Carl, 1866, *Observations on Chaetetes and some related genera, in regard to their systematic position; with an appended description of some new species:* Proc. Acad. Nat. Sci. Philadelphia, [v. 18], p. 113–123.

Rose, Gustav, 1859, *Über die heteromorphen Zustände der kohlensauren Kalkerde. II. Vorkommen des Aragonits und Kalkspaths in der organischen Natur:* K. Akad. Wiss. Berlin Abhandl., 1858, p. 63–111, 3 pl.

Ross, J. R. P. [J. R. P. Phillips], 1960a, *Type species of Ptilodictya—Ptilodictya lanceolata (Goldfuss):* J. Paleontol., v. 34, p. 440–446, 1 text-fig., pl. 61, 62.

———, 1960b, *Re-evaluation of the type species of Arthropora Ulrich:* J. Paleontol., v. 34, p. 859–861, pl. 108.

———, 1960c, *Larger cryptostome Bryozoa of the Ordovician and Silurian, Anticosti Island, Canada—Part 1:* J. Paleontol., v. 34, p. 1057–1076, 2 text-fig., pl. 125–128.

———, 1961a, *Larger cryptostome Bryozoa of the Ordovician and Silurian, Anticosti Island, Canada—Part 2:* J. Paleontol., v. 35, p. 331–344, pl. 41–45.

———, 1961b, *Ordovician, Silurian, and Devonian Bryozoa of Australia:* Aust. Bur. Miner. Resour. Geol. Geophys. Bull. 50, 1972 p., 13 text-fig., 28 pl.

———, 1963a, *Constellaria from the Chazyan (Ordovician), Isle la Motte, Vermont:* J. Paleontol., v. 37, p. 51–56, 2 text-fig., pl. 5, 6.

———, 1963b, *Ordovician cryptostome Bryozoa, standard Chazyan Series, New York and Vermont:* Geol. Soc. Am. Bull., v. 74, p. 577–607, 7 text-fig., 10 pl.

———, 1964a, *Champlainian cryptostome Bryozoa from New York State:* J. Paleontol., v. 38, p. 1–32, 11 text-fig., pl. 1–8.

———, 1964b, *Morphology and phylogeny of early Ectoprocta (Bryozoa):* Geol. Soc. Am. Bull., v. 75, p. 927–948, 10 text-fig.

———, 1966a, *Early Ordovician ectoproct from*

Oklahoma: Okla. Geol. Notes, v. 26, p. 218–224, 3 pl.

———, 1966b, *Stictopora Hall 1847 (Ectoprocta, Cryptostomata), a valid name:* J. Paleontol., v. 40, p. 1400–1401.

———, 1970, *Distribution, paleoecology and correlation of Champlainian Ectoprocta (Bryozoa), New York State, part III:* J. Paleontol., v. 44, p. 346–382, 8 text-fig., pl. 67–74.

———, 1972, *Paleoecology of Middle Ordovician ectoproct assemblages:* Rep. Int. Geol. Congr. 24th Sess., sec. 7, Paleontol., p. 96–102, 2 text-fig. (Montreal).

———, 1976, *Body wall ultrastructure of living cyclostome ectoprocts:* J. Paleontol., v. 50, p. 350–353, 2 text-fig.

Rucker, J. B., 1967, *Carbonate mineralogy of recent cheilostome Bryozoa (abstr.):* Geol. Soc. Am. Abstr. Program, 1967, p. 191–192.

———, & Carver, R. E., 1969, *A survey of the carbonate mineralogy of cheilostome Bryozoa:* J. Paleontol., v. 43, p. 791–799, 5 text-fig.

Ryland, J. S., 1970, *Bryozoans:* 175 p., 21 text-fig., Hutchinson University Library (London).

———, 1976, *Physiology and ecology of marine bryozoans:* Adv. Mar. Biol., v. 14, p. 285–443, 44 text-fig.

Sakagami, Sumio, 1960, *Hayasakapora, a new Permian bryozoan genus from Iwaizaki, Miyagi Prefecture, Japan:* Trans. Proc. Paleontol. Soc. Jpn. New Ser., no. 39, p. 321–323, pl. 37.

Salter, J. W., 1858, *On Graptopora, a new genus of Polyzoa, allied to the graptolites:* Proc. Am. Assoc. Adv. Sci., v. 11, pt. 2, p. 63–66.

Sandberg, P. A., 1971, *Scanning electron microscopy of cheilostome bryozoan skeletons; techniques and preliminary observations:* Micropaleontology. v. 17, p. 129–151, 1 text-fig., pl. 1–4.

———, 1973, *Degree of individuality in cheilostome Bryozoa; skeletal criteria:* in R. S. Boardman, A. H. Cheetham, and W. A. Oliver, Jr. (eds.), Animal Colonies, p. 305–315, 17 text-fig., Dowden, Hutchinson, & Ross (Stroudsburg).

———, 1975a, *Bryozoan diagenesis; bearing on the nature of the original skeleton of rugose corals:* J. Paleontol., v. 49, p. 587–606, 4 pl.

———, 1975b, *New interpretations of Great Salt Lake ooids and of ancient non-skeletal carbonate mineralogy:* Sedimentology, v. 22, p. 497–537, 24 text-fig.

———, 1976, *Ultrastructural clues to skeletal development in cheilostome Bryozoa:* Stockholm Contrib. Geol., v. 30, p. 1–14, 2 pl.

Sandberg, P. A. et al., 1969, *Relationship between ultrastructure and mineralogy in some cheilostome bryozoans:* Geol. Soc. Am. Abstr. 1969, part 4 (southeast sect.), p. 72–73.

———, Schneidermann, Nahum, & Wunder, S. J., 1973, *Aragonitic ultrastructural relics in cal-cite-replaced Pleistocene skeletons:* Nat. London Phys. Sci., v. 245, no. 148, p. 133–134, 1 text-fig.

Sanders, J. E., & Friedman, G. M., 1967, *Origin and occurrence of limestones:* in G. V. Chilingar, H. J. Bissell, & R. W. Fairbridge (eds.), Carbonate Rocks: Origin, Occurrence and Classification, Developments in Sedimentology, v. 9A, p. 169–265, 10 text-fig., Elsevier (Amsterdam).

Sarycheva, T. G. (ed.), 1960, *Osnovy paleontologii, spravochnik dlya paleontologov i geologov SSSR; Mshanki, brakhiopody; prilozhenie: Foronidy:* 343 p., illus., Izdatelstvo Akademii Nauk SSSR (Moscow). *Fundamentals of Paleontology, a reference book for paleontologists and geologists of the USSR: Bryozoa, Brachiopoda; Appendix: Phoronidae;* English transl., 1972, Fundamentals of Paleontology, v. 7, Bryozoa: Indian Natl. Sci. Doc. Cent., New Delhi, 233 p.]

Schneider, Dietrich, 1957, *Orientiertes Wachstum von Calcit-Kristallen in der Cuticula mariner Bryozoen:* Verh. Dtsch. Zool. Ges., v. 21, p. 250–255, 7 text-fig.

———, 1958, *Calcitwachstum und Phototropismus bei Bugula (Bryozoa):* Inst. Wiss. Film, Göttingen, film B 762, with text, 14 p., 5 text-fig.

———, 1963, *Normal and phototropic growth reactions in the marine bryozoan Bugula avicularia:* in E. C. Dougherty et al. (eds.), The Lower Metazoa, p. 357–371, 13 text-fig., University of California Press (Berkeley).

Schneidermann, N., 1970, *Genesis of some Cretaceous carbonates in Israel:* Isr. J. Earth Sci., v. 19, p. 97–115, 5 text-fig., 13 pl.

Schoenmann [Sheynmann], Yu. M., 1927, *Mshanki verkhnego Silura r. Srednej Tunguzki:* Izv. Geol. Kom., t. 45, p. 783–794, pl. 25. [*Upper Silurian Bryozoa from the Middle Tunguzka River.*]

Schopf, T. J. M., 1967, *Names of phyla: Ectoprocta and Entoprocta, and Bryozoa:* Syst. Zool., v. 16, p. 276–278.

———, 1968, *Ectoprocta, Entoprocta, and Bryozoa:* Syst. Zool., v. 17, p. 470–472.

———, 1969a, *Paleoecology of ectoprocts (bryozoans):* J. Paleontol., v. 43, p. 234–244, 5 text-fig.

———, 1969b, *Geographic and depth distribution of the phylum Ectoprocta from 200 to 6,000 meters:* Proc. Am. Philos. Soc., v. 113, p. 464–474, 4 text-fig.

———, 1973, *Ergonomics of polymorphism; its relation to the colony as the unit of natural selection in species of the phylum Ectoprocta:* in R. S. Boardman, A. H. Cheetham, & W. A. Oliver, Jr. (eds.), Animal Colonies, p. 247–294, 3 text-fig., Dowden, Hutchinson, & Ross (Stroudsburg).

———, & Bassett, E. L., 1973, *F. A. Smitt, marine Bryozoa, and the introduction of Darwin into Sweden:* Trans. Am. Philos. Soc., n.ser., v. 63, pt. 7, p. 3–30, 1 text-fig.

————, & Manheim, F. T., 1967, *Chemical composition of Ectoprocta (Bryozoa):* J. Paleontol., v. 41, p. 1197–1225, 6 text-fig.

Sharpe, Daniel, 1853, *Description of the new species of Zoophyta and Mollusca from the Lower Silurian formations of the Serra de Bussaco:* Q. J. Geol. Soc. London, v. 9, p. 146–158, pl. 7–9.

Sherborn, C. D., 1940, *Where is the ———— Collection? An account of the various natural history collections which have come under the notice of the compiler between 1880 and 1939:* 149 p., University Press (Cambridge).

Shishova, N. A., 1964, *Novye pozdnepermskie rahdomezonidy Sovetskogo Soyuza:* Paleontol. Zh., 1964, no. 3, p. 52–57. [*New Late Permian Rhabdomesonidae of the Soviet Union.*]

————, 1965, *O sistematicheskom polozhenii i ob'eme semejstva Hyphasmoporidae:* Paleontol. Zh., 1965, no. 2, p. 55–62, 1 text-fig., pl. 6. [*The systematic position and size of the family Hyphasmoporidae.*]

————, 1968, *New order of Paleozoic bryozoans:* Paleontol. J., v. 2, no. 1, p. 117–121. [Transl. of *Novyj otryad Paleozojskikh mshanok:* Paleontol. Zh., 1968, no. 1, p. 129–132.]

————, 1970, *Nektorye novye silurijskie i devonskie mshanki Mongolii:* in Novye vidy paleozojskikh mshanok i korallov, p. 28–31, illus., Akademiya Nauk SSSR (Moscow). [*Some Silurian and Devonian bryozoans from Mongolia:* in *New species of Paleozoic Bryozoa and corals.*]

Shulga-Nesterenko, M. I., 1931, *Novyj rod Lyrocladia iz nizhnepermskikh mshanok Pechorskogo kraya:* Vses. Paleontol. Obshch. Ezhegodnik, t. 9, p. 47–94. [*A new Lower Permian bryozoan genus, Lyrocladia from Pechora Krai.*]

————, 1933, *Mshankovaya fauna uglenosnykh i poduglenosnykh otlozhenij Pechorskogo kraya:* Tr. Vses. Geol. Razved. Obedin., vyp. 259, 64 p., 39 text-fig., 9 pl. [*Bryozoan fauna from the coal-bearing and subjacent deposits of Pechora Krai.*]

————, 1949, *Funktsionalnoe, filogeneticheskoe i stratigraficheskoe zhachenie mikrostruktury skeletnykh tkanej mshanok:* Tr. Akad. Nauk SSSR Paleontol. Inst., t. 23, 67 p., 26 text-fig., 12 pl. [*Functional, phylogenetic, and stratigraphic significance of the microstructure of bryozoan skeletal material.*]

————, 1955, *Kamennougol'nye mshanki Russkoj platformy:* Tr. Akad. Nauk SSSR Paleontol. Inst., v. 57, 207 p., 62 text-fig., 32 pl. [*Carboniferous bryozoans of the Russian platform.*]

————, Nekhoroshev, V. P., Morozova, I. P., Astrova, G. G., & Shishova, N. A., 1960, *Otryad Cryptostomata:* in Osnovy Paleontologii, p. 72–93, text-fig. 112–184. [*Order Cryptostomata;* see T. G. Sarycheva, 1960 for complete citation.]

Siegfried, Paul, 1963, *Bryozoen in Steinkernerhaltung aus ordovizischem Geschiebe:* Paleontol. Z., v. 37, p. 135–146. 4 text-fig., pl. 4–7.

Silén, Lars, 1938, *Zur Kenntnis des Polymorphismus der Bryozoen. Die Avicularien der Cheilostomata Anasca:* Zool. Bidr. Uppsala, Bd. 17, p. 149–366, 80 text-fig., 18 pl.

————, 1942, *Origin and development of the cheilo-ctenostomatous stem of Bryozoa:* Zool. Bidr. Uppsala, v. 22, p. 1–59, 64 text-fig.

————, 1944a, *The anatomy of Labiostomella gisleni Silén (Bryozoa Protocheilostomata) with special regard to the embryo chambers of the different groups of Bryozoa and to the origin and development of the bryozoan zoarium:* K. Sven. Vetenskapsakad. Handl., ser. 3, v. 21, no. 6, 111 p., 85 text-fig., 5 pl.

————, 1944b, *On the formation of the interzooidal communications of the Bryozoa:* Zool. Bidr. Uppsala, v. 22, p. 433–488, 59 text-fig., 1 pl.

————, 1947, *On the anatomy and biology of Penetrantiidae and Immergentiidae (Bryozoa):* Ark. Zool., v. 40A, no. 4, 48 p., 70 text-fig.

————, 1966, *On the fertilization problem in the gymnolaematous Bryozoa:* Ophelia, v. 3, p. 113–140, 15 text-fig.

————, 1972, *Fertilization in the Bryozoa:* Ophelia, v. 10, p. 27–34.

————, & Harmelin, J. G., 1974, *Observations on living Diastoporidae (Bryozoa Cyclostomata), with special regard to polymorphism:* Acta Zool. Stockholm, v. 55, p. 81–96, 22 text-fig.

Silliman, Benjamin, Silliman, Benjamin, Jr., & Dana, J. D., 1851, *New genera of fossil corals from the report of James Hall, on the paleontology of New York:* Am. J. Sci. Arts, ser. 2, v. 11, p. 398–401.

Simma-Krieg, Brigitte, 1969, *On the variation and special reproduction habits of Aetea sica (Couch):* Cah. Biol. Mar., v. 10, p. 129–137, 5 text-fig., 2 pl.

Simpson, G. B., 1897 (1895), *A Handbook of the Genera of the North American Paleozoic Bryozoa; with an introduction upon the structure of living species:* State Geol. N.Y., 14th Annu. Rep., p. 403–669, pl. A–E, 1–25.

Smitt, F. A., 1865, *Kritisk förteckning öfver Skandinaviens Hafs-Bryozoer, I.:* Ofvers. K. Vetenskapsakad. Forh. Stockholm, v. 22, p. 115–142, pl. 16.

————, 1866, *Kritisk förteckning öfver Skandinaviens Hafs-Bryozoer, II.:* Ofvers. K. Vetenskapsakad. Forh. Stockholm, v. 23 (suppl.), p. 395–533, p. 3–13.

————, 1867, *Kritisk förteckning öfver Skandinaviens Hafs-Bryozoer, III.:* Ofvers. K. Vetenskapsakad. Forh. Stockholm, v. 24, p. 279–429, pl. 16–20.

————, 1868, *Bryozoa marina in regionibus arcticis et borealibus viventia recensuit:* Ofvers. K. Vetenskapsakad. Forh. Stockholm, v. 24, p. 443–487.

————, 1872, *Remarks on Dr. Nitsche's researches on Bryozoa:* Q. J. Microsc. Sci. n.ser., v. 12, p. 246–248.

Sneath, P. H. A., & Sokal, R. R., 1973, *Numerical Taxonomy, the Principles and Practice of Numerical Classification:* 573 p., W. H. Freeman (San Francisco).

Söderqvist, S. T., 1968, *Observations on extracellular body wall structures in Crisia eburnea L.:* in E. Annoscia (ed.), Proceedings of the First International Conference on Bryozoa, Atti Soc. Ital. Sci. Nat. Mus. Civ. Stor. Nat. Milano, v. 108, p. 115–118, pl. 4.

Sorauf, J. E., 1971, *Microstructure in the exoskeleton of some Rugosa (Coelenterata):* J. Paleontol., v. 45, p. 23–32, 3 text-fig., pl. 5–11.

————, 1974, *Growth lines on tabulae of Favosites (Silurian, Iowa):* J. Paleontol., v. 48, p. 553–555, 1 pl.

Sorby, H. C., 1863, *On the cause of the difference in the state of preservation of different kinds of fossil shells:* Br. Assoc. Adv. Sci. Rep. 32 (1862), Notices & Abstr., p. 95–96.

————, 1879, *On the structure and origin of limestones. The anniversary address of the president:* Proc. Geol. Soc. London, v. 35, p. 39–95.

Soule, J. D., 1954, *Post-larval development in relation to the classification of the Bryozoa Ctenostomata:* Bull. South. Calif. Acad. Sci., v. 53, p. 13–34, pl. 5–7.

————, & Soule, D. F., 1968, *Perspectives on the Bryozoa: Ectoprocta question:* Syst. Zool., v. 17, p. 468–470.

————, & ————, 1969, *Systematics and biogeography of burrowing bryozoans:* Am. Zool., v. 9, p. 791–802, 6 text-fig.

————, & ————, 1975, *Spathipora, its anatomy and phylogenetic affinities:* in S. Pouyet (ed.). Bryozoa 1974, Doc. Lab. Geol. Fac. Sci. Lyon, hors-série 3, fasc. 1, p. 247–253, 2 text-fig., 4 pl.

Spjeldnaes, Nils, 1957, *A redescription of some type specimens of British Ordovician Bryozoa:* Geol. Mag., v. 94, p. 364–376, 1 text-fig., pl. 12, 13.

Stach, L. W., 1936, *Correlation of zoarial form with habitat:* J. Geol., v. 44, p. 60–65, 1 text-fig.

Steen, E. B., 1971, *Dictionary of Biology:* Barnes & Noble (New York).

Stehli, F. G., 1956, *Shell mineralogy in Paleozoic invertebrates:* Science, v. 123, p. 1031–1032.

Steininger, Johann, 1831, *Bemerkungen über die Versteinerungen, welche in dem Übergange-Kalkgebirge der Eifel gefunden werden:* Gymnasium zu Trier, Jahresberichte, p. 1–44.

Stieglitz, R. D., 1972, *Scanning electron microscopy of the fine fraction of recent carbonate sediments from Bimini, Bahamas:* J. Sediment. Petrol., v. 42, p. 211–226.

Stratton, J. F., 1975, *Ovicells in Fenestella from the Speed Member, North Vernon Limestone (Eifelian, Middle Devonian) in southern Indiana, U.S.A.:* in S. Pouyet (ed.), Bryozoa 1974, Doc. Lab. Geol. Fac. Sci. Lyon, hors-série 3, fasc. 1, p. 169–177, 3 text-fig.

Stuckenberg, A. A., 1895, *Korally i mshanki kamennougol'nykh otlozhenij Urala i Timana:* Tr. Geol. Kom., t. 10 (1890–1895), no. 3, p. 1–244, 24 pl. [*Corals and Bryozoa from Carboniferous deposits of the Urals and Timan;* German summary p. 179–244.]

————, 1905, *Fauna verkhne-kamennougol'noj tolshchi Samarskoj luki:* Tr. Geol. Kom., n.ser., vyp. 23, p. 1–144, 13 pl. [*Fauna of the Upper Carboniferous strata of the Samara bend;* German summary p. 111–144.]

Tavener-Smith, Ronald, 1966, *Ovicells in fenestrate cryptostomes of Visean age:* J. Paleontol., v. 40, p. 190–198, 2 text-fig., pl. 25.

————, 1968, *Skeletal structure and growth in the Fenestellidae (Bryozoa) (preliminary report):* in E. Annoscia (ed.), Proceedings of the First International Conference on Bryozoa, Atti Soc. Ital. Sci. Nat. Mus. Civ. Stor. Nat. Milano, v. 108, p. 85–92.

————, 1969a, *Skeletal structure and growth in the Fenestellidae (Bryozoa):* Palaeontology, v. 12, p. 281–309, 9 text-fig., pl. 52–56.

————, 1969b, *Wall structure and acanthopores in the bryozoan Leioclema asperum:* Lethaia, v. 2, p. 89–97, 7 text-fig.

————, 1971, *Polypora stenostoma: A Carboniferous bryozoan with cheilostomatous features:* Palaeontology, v. 14, p. 178–187, 5 text-fig., pl. 25.

————, 1973, *Some aspects of skeletal organization in Bryozoa:* in G. P. Larwood (ed.), Living and Fossil Bryozoa, p. 349–359, 5 text-fig., 1 pl., Academic Press (London).

————, 1974, *Early growth stages in rhabdomesoid bryozoans from the Lower Carboniferous of Hook Head, Ireland:* Palaeontology, v. 17, p. 149–164, 4 text-fig., pl. 16, 17.

————, 1975, *The phylogenetic affinities of fenestelloid bryozoans:* Palaeontology, v. 18, p. 1–17, 5 text-fig., 2 pl.

————, & Williams, Alwyn, 1970, *Structure of the compensation sac in two ascophoran bryozoans:* Proc. R. Soc. London Ser. B, v. 175, p. 235–254, 35 text-fig., pl. 42–48.

————, & ————, 1972, *The secretion and structure of the skeleton of living and fossil Bryozoa:* Philos. Trans. R. Soc. London Ser. B, v. 264, p. 97–159, pl. 6–30.

Taylor, J. D., 1973, *The structural evolution of the bivalve shell:* Palaeontology, v. 16, p. 519–534, 5 text-fig., pl. 60.

————, Kennedy, W. J., & Hall, A., 1969, *The shell structure and mineralogy of the Bivalvia. Introduction, Nuculacea-Trigonacea:* Bull. Br. Mus. Nat. Hist. Zool., suppl. 3, 125 p., 77 text-fig., 29 pl.

Taylor, P. D., 1975, *Monticules in a Jurassic cyclostomatous bryozoan:* Geol. Mag., v. 112, p. 601–

606, 2 text-fig., 1 pl.

Termier, Henri, & Termier, Geneviève, 1971, *Bryozoaires du Paléozoïque supérieur de l'Afghanistan:* Doc. Lab. Geol. Fac. Sci. Lyon, no. 47, 52 p., 6 text-fig., 32 pl.

Thomas, H. D., & Larwood, G. P., 1956, *Some "uniserial" membraniporine polyzoan genera and a new American Albian species:* Geol. Mag., v. 93, p. 369–376, 3 text-fig.

——, & ——, 1960, *The Cretaceous species of Pyripora d'Orbigny and Rhammatopora Lang:* Palaeontology, v. 3, p. 370–386, 4 text-fig., pl. 60–62.

Thompson, J. V., 1830, *On Polyzoa, a new animal discovered as an inhabitant of some Zoophites, with a description of the newly instituted genera of Pedicellaria and Vesicularia, and their species:* in Zoological researches, and illustrations; or natural history of nondescript or imperfectly known animals, in a series of memoires, Mem. 5, p. 89–102, 3 pl., King & Ridings (Cork).

Thorpe, J. P., 1975, *Behaviour and colonial activity in Membranipora membranacea (L.):* in S. Pouyet (ed.), Bryozoa 1974, Doc. Lab. Geol. Fac. Sci. Lyon, hors-série 3, fasc. 1, p. 115–121, 3 text-fig.

——, Shelton, G. A. B., & Laverack, M. S., 1975a, *Colonial nervous control of lophophore retraction in cheilostome Bryozoa:* Science, v. 189, p. 60–61, 2 text-fig.

——, ——, & ——, 1975b, *Electrophysiology and co-ordinated behavioural responses in the colonial bryozoan Membranipora membranacea (L.):* J. Exp. Biol., v. 62, p. 389–404, 14 text-fig.

Tillier, Simon, 1975, *Recherches sur la structure et revision systématique des hétéroporides (Bryozoa, Cyclostomata) des faluns de Touraine:* Trav. Univ. Paris Fac. Sci. Orsay Lab. Paleontol., 100 p., 31 text-fig., 8 pl.

Toriumi, Makoto, 1956, *Taxonomical study on freshwater Bryozoa, XVII.—General consideration: interspecific relation of described species and phylogenic consideration:* Tohoku Univ. Sci. Rep., ser. 4, v. 22, p. 57–88. 18 text-fig.

Toula, Franz, 1875, *Permo-Carbon-Fossilien von der Westküste von Spitzbergen:* Neues Jahrb. Mineral. Geol. Palaeontol., Jahrg. 1875, p. 225–264, 3 text-fig., pl. 5–10.

Towe, K. M., 1972, *Invertebrate shell structure and the organic matrix concept:* Biomineralisation Forschungsber., Bd. 4, p. 1–14, 2 text-fig., 7 pl.

——, & Cifelli, Richard, 1967, *Wall ultrastructure in the calcareous Foraminifera—Crystallographic aspects and a model for calcification:* J. Paleontol., v. 41, p. 742–762, pl. 87–99.

——, & Hemleben, Christoph, 1976, *Diagenesis of magnesian calcite; evidence from miliolacean Foraminifera:* Geology, v. 4, p. 337–339, 1 text-fig.

Trautschold, H. A., 1876, *Die Kalkbrüche von Mjatschkowa. Eine Monographie des oberen Bergkalks:* Moskov. O-vo. Ispyt. Prir. Nov. Mem., v. 13, p. 325–374, pl. 32–38.

Trembley, Abraham, 1744, *Mémoires pour servir à l'histoire d'un genre de Polypes d'eau douce, à bras en forme de cornes:* xv + 324 p., 13 pl., J. & H. Verbeck (Leiden).

Trizna, V. B., 1950, *K. kharakteristike rifovykh i sloistykh fatsij tsentral'noj chasti Ufimskogo plato:* Tr. Vses. Nauchno-Issled. Geol. Inst. (VNIGRI), n.ser., vyp. 50, p. 47–144, pl. 1–16. [*Characterization of reef and bedded facies of the central region of the Ufimsk plateau.*]

——, 1958, *Rannekamennougol'nye mshanki Kuznetskoj Kotloviny:* Tr. Vses. Nauchno-Issled. Geol. Inst. (VNIGRI), vyp. 122, 436 p., 26 text-fig., 64 pl. [*Early Carboniferous Bryozoa of the Kuznetsk Basin.*]

Turner, R. F., 1975, *A new Upper Cretaceous cribrimorph from North America with calcareous opercula:* in S. Pouyet (ed.), Bryozoa 1974, Doc. Lab. Geol. Fac. Sci. Lyon, hors-série 3, fasc. 2, p. 273–279, 1 pl.

Ulrich, E. O., 1878, *Descriptions of some new species of fossils from the Cincinnati group:* J. Cincinnati Soc. Nat. Hist., v. 1, p. 92–100.

——, 1879, *Descriptions of new genera and species of fossils from the Lower Silurian about Cincinnati:* J. Cincinnati Soc. Nat. Hist., v. 2, p. 8–30, pl. 7.

——, 1882, *American Paleozoic Bryozoa:* J. Cincinnati Soc. Nat. Hist., v. 5, p. 121–175, pl. 6–8; p. 232–257, pl. 10–11.

——, 1883, *American Paleozoic Bryozoa:* J. Cincinnati Soc. Nat. Hist., v. 6, p. 245–279, pl. 12–14.

——, 1884, *American Paleozoic Bryozoa:* J. Cincinnati Soc. Nat. Hist., v. 7, p. 24–51, pl. 1–3.

——, 1886a, *Report on the Lower Silurian Bryozoa with preliminary descriptions of some of new species:* Minn. Geol. Nat. Hist. Surv. Annu. Rep. 14, p. 57–103.

——, 1886b, *Description of new Silurian and Devonian fossils:* in Contributions to American Paleontology, v. 1, p. 3–35, pl. 1–3, E. O. Ulrich (Cincinnati).

——, 1888a, *On Sceptropora, a new genus of Bryozoa, with remarks on Helopora Hall, and other genera of that type:* Am. Geol., v. 1, p. 228–234, 2 text-fig.

——, 1888b, *A list of the Bryozoa of the Waverly Group in Ohio, with descriptions of new species:* Denison Univ. Sci. Lab. Bull., v. 4, p. 62–96, pl. 13, 14.

——, 1889, *Contributions to the micro-palaeontology of the Cambro-Silurian rocks of Canada, Part II:* Geol. Nat. Hist. Surv. Can., p. 22–57, pl. 8, 9.

——, 1890, *Palaeozoic Bryozoa:* Ill. Geol. Surv.,

v. 8, p. 283–688, pl. 29–78.

——, 1893 (1895), *On Lower Silurian Bryozoa of Minnesota:* Minn. Geol. Nat. Hist. Surv. Final Rep., v. 3, pt. 1, p. 96–332, text-fig. 8–20, 28 pl. [Author's separate issued 1893.]

——, 1896, *Bryozoa:* in K. A. von Zittel & C. R. Eastman (transl. & ed.), Text-book of Paleontology, v. 1, p. 257–291, text-fig. 411–488, Macmillan & Co. (New York).

——, & Bassler, R. S., 1913, *Systematic paleontology of the Lower Devonian deposits of Maryland; Bryozoa:* Md. Geol. Surv., Lower Devonian volume, text, p. 259–290.

Utgaard, John, 1968a, *A revision of North American genera of ceramoporoid bryozoans (Ectoprocta). Part I, Anolotichiidae:* J. Paleontol., v. 42, p. 1033–1041, pl. 129–132.

——, 1968b, *A revision of North American genera of ceramoporoid bryozoans (Ectoprocta). Part II, Crepipora, Ceramoporella, Acanthoceramoporella, and Ceramophylla:* J. Paleontol., v. 42, p. 1444–1455, pl. 181–184.

——, 1969, *A revision of North American genera of ceramoporoid bryozoans (Ectoprocta). Part III, the ceramoporoid genera Ceramopora, Papillalunaria, Favositella, and Haplotrypa:* J. Paleontol., v. 43, p. 289–297, pl. 51–54.

——, 1973, *Mode of colony growth, autozooids, and polymorphism in the bryozoan order Cystoporata:* in R. S. Boardman, A. H. Cheetham, & W. A. Oliver, Jr. (eds.), Animal Colonies, p. 317–360. 74 text-fig., Dowden, Hutchinson, & Ross (Stroudsburg).

Vángel, Eugen, 1894, *Daten zur Bryozoen-fauna Ungarns:* Zool. Anz., v. 17, p. 153–155.

Viganò, Antonio, 1968, *Note su Plumatella casmiana Oka (Bryozoa):* Riv. Idrobiol., v. 7, p. 421–468, 4 text-fig., 22 pl.

Vigelius, W. J., 1884, *Morphologische Untersuchungen über Flustra membranaceo-truncata Smitt:* Biol. Zentralbl., v. 3, p. 705–721.

——, 1886, *Zur Ontogenie der marinen Bryozoen:* Mittheil. Zool. Sta. Neapel, v. 6, p. 499–541. pl. 26–27.

——, 1888, *Zur Ontogenie der marinen Bryozoen:* Mittheil. Zool. Sta. Neapel, v. 8, p. 374–376, pl. 19.

Vinassa de Regny, P. E., 1921, *Sulla classificazione dei Treptostomidi:* Atti Soc. Ital. Sci. Nat. Mus. Civ. Stor. Nat. Milano, v. 59, p. 212–231.

Vine, G. R., 1884, *Fourth report of the committee, consisting of Dr. H. C. Sorby and Mr. G. R. Vine, appointed for the purpose of reporting on fossil Bryozoa:* Br. Assoc. Adv. Sci., 53rd Meeting (Southport, 1883), p. 161–209, 3 text-fig.

——, 1885, *Further notes on new species, and other Yorkshire Carboniferous Polyzoa described by Prof. John Phillips:* Proc. Yorks. Geol. Soc., n.ser., v. 8, p. 377–393, pl. 20, 21.

——, 1886, *Notes on the Yoredale Polyzoa of North Lancashire:* Proc. Yorks. Geol. Soc., n.ser., v. 9, p. 70–98, pl. 10.

Viskova, L. A., 1972, *Pozdnemelovye mshanki Cyclostomata Povolzh'ya i Kryma:* Tr. Akad. Nauk SSSR Paleontol. Inst., v. 132, 90 p., 40 text-fig., 18 pl. [Late Cretaceous cyclostomatous bryozoans of the Volga region and the Crimea.]

——, 1973, *On the morphology and systematics of some Late Cretaceous cyclostomatous Bryozoa:* in G. P. Larwood (ed.), Living and Fossil Bryozoa, p. 497–502, 2 pl., Academic Press (London).

Voigt, Ehrhard, 1959, *La signification stratigraphique des Bryozoaires dans le Crétacé supérieur:* Congrès Soc. Savantes, v. 84, p. 701–707.

——, 1966, *Die Erhaltung vergänglicher Organismen durch Abformung infolge Inkrustation durch sessile Tiere:* Neues Jahrb. Geol. Palaeontol. Abhandl., Bd. 125, p. 401–422, 6 text-fig., pl. 33–37.

——, 1968, *Eine fossile Art von Arachnidium (Bryozoa, Ctenostomata) in der Unteren Kreide Norddeutschlands:* Neues Jarb. Geol. Palaeontol. Abhandl., Bd. 132, p. 87–96, 4 text-fig., pl. 7.

——, 1972a, *Amathia immurata n. sp., ein durch Biomuration erhaltenes ctenostomes Bryozoon aus der Maastrichter Tuffkreide:* Palaeontol. Z., v. 46, p. 87–92, 1 text-fig. pl. 17.

——, 1972b, *Les méthodes d'utilisation stratigraphique des Bryozoaires du Crétacé supérieur:* Mem. Bur. Geol. Min. Fr., no. 77, v. 1, p. 45–53.

——, 1973, *Vinelloidea Canu, 1913 (angeblich jurassische Bryozoa Ctenostomata) = Nubeculinella Cushman, 1930 (Foraminifera):* Palaeontol. Abhandl. Abt. A, Bd. 4, Heft 4, p. 665–670, 1 pl.

——, 1974, *Über Opercula bei der fossilen Bryozoengattung Inversaria v. Hagenow 1851 (Cheilostomata, Ob. Kreide):* Palaeontol. Z., v. 48, p. 214–229, 2 text-fig., pl. 32–34.

——, 1975, *Heteromorphy in Cretaceous Bryozoa:* in S. Pouyet (ed.), Bryozoa 1974, Doc. Lab. Geol. Fac. Sci. Lyon, hors-série 3, fasc. 1, p. 77–95, 10 pl.

——, & Soule, J. D., 1973, *Cretaceous burrowing bryozoans:* J. Paleontol.. v. 47, p. 21–33, 1 text-fig., pl. 1–4.

Vovelle, Jean, 1972, *Sclérotinisation et minéralisation des structures squelettiques chez les Mollusques:* Haliotis, v. 2, p. 133–165, 7 text-fig.

Waagen, W. H., & Pichl, J., 1885, *Salt Range fossils; Productus-limestone fossils:* Mem. Geol. Surv. India (Palaeontol. Indica), ser. 13, v. 1, pt. 5, p. 771–834, pl. 87–96.

——, & Wentzel, Joseph, 1886, *Salt Range fossils; Productus-limestone fossils:* Mem. Geol. Surv. India (Palaeontol. Indica), ser. 13, v. 1, pt. 6, p. 835–924, pl. 97–116.

Wada, Koji, 1972, *Nucleation and growth of aragonite crystals in the nacre of some bivalve molluscs:*

Biomineralisation Forschungsber., Bd. 6, p. 141–159, 34 text-fig.

Walker, K. R., 1972, *Community ecology of the Middle Ordovician Black River Group of New York State:* Geol. Soc. Am. Bull., v. 83, p. 2499–2524, 16 text-fig.

———, & Ferrigno, K. F., 1973, *Major middle Ordovician reef tract in east Tennessee:* Am. J. Sci., v. 273-A (Cooper Memorial Volume), p. 294–325, 6 text-fig., 3 pl.

Walter, Bernard, & Powell, H. P., 1973, *Exceptional preservation in cyclostome Bryozoa from the Middle Lias of Northamptonshire:* Palaeontology, v. 16, p. 219–220, pl. 20.

Warner, D. J., & Cuffey, R. J., 1973, *Fistuliporacean bryozoans of the Wreford Megacyclothem (Lower Permian) of Kansas:* Univ. Kans. Paleontol. Contrib. Pap. 65, 24 p., 5 text-fig., 3 pl.

Wass, R. E., & Yoo, J. J., 1975, *Bryozoa from Site 282 west of Tasmania:* in J. P. Kennett et al., Initial Reports of the Deep Sea Drilling Project, v. 29, p. 809–831, 9 pl., U.S. Government Printing Office (Washington).

Waters, A. W., 1889, *Supplementary report on the Polyzoa:* in J. Murray, Report on the Scientific Results of the Voyage of H.M.S. Challenger During the Years 1873–76. . ., Zoology v. 31, pt. 79, 41 p., 3 pl., H.M. Stationery Office (London).

———, 1913, *The marine fauna of British East Africa and Zanzibar from collections made by Cyril Crossland in the years 1901–1902. Bryozoa Cheilostomata:* Proc. Zool. Soc. London, 1913, p. 458–537, text-fig. 79–82. pl. 64–73.

Webers, G. F., 1972, *Paleoecology of the Cambrian and Ordovician strata of Minnesota:* in P. K. Sims & G. B. Morey (eds.), Geology of Minnesota: A Centennial Volume, p. 474–484, text-fig. 22–31, Minnesota Geological Survey (St. Paul).

Wells, J. W., 1969, *The formation of dissepiments in zoanthinarian corals:* in K. S. W. Campbell (ed.), Stratigraphy and Palaeontology: Essays in Honour of Dorothy Hill, p. 17–26, text-fig. 5, pl. 1, 2, Australian National University Press (Canberra).

Westbroek, Peter, 1967, *Morphological observations with systematic implications on some Palaeozoic Rhynchonellida from Europe, with special emphasis on the Uncilulidae:* Leidse Geol. Meded., deel 41, p. 1–82, 81 text-fig., 14 pl.

Whipple, G. C., 1910, *The Microscopy of Drinking Water:* 2nd ed., xiii + 323 p., 19 pl., John Willey & Sons (New York).

Wiebach, Fritz, 1963, *Studien über Plumatella casmiana Oka (Bryozoa):* Vie Milieu, v. 14, p. 579–596, 4 text-fig.

———, 1975, *Specific structures of sessoblasts (Bryozoa, Phylactolaemata):* in S. Pouyet (ed.), Bryozoa 1974, Doc. Lab. Geol. Fac. Sci. Lyon, hors-série 3, fasc. 1, p. 149–154, 3 pl.

Williams, Alwyn, 1968, *A history of skeletal secretion among articulate brachiopods:* Lethaia, v. 1, p. 268–287, 13 text-fig.

———, 1971a, *Ultrastructure of the growing tip and the process of calcification in some Bryozoa (abstr.):* International Bryozoology Association, 2nd International Conference, University of Durham, England, Abstr. no. 2.

———, 1971b, *Scanning electron microscopy of the calcareous skeleton of fossil and living Brachiopoda:* in V. H. Heywood (ed.), Scanning Electron Microscopy, Systematic and Evolutionary Applications, p. 37–66, 3 text-fig., 5 pl., Academic Press (London).

Winston, J. E., 1976, *Experimental culture of the estuarine ectoproct Conopeum tenuissimum from Chesapeake Bay:* Biol. Bull. Woods Hole Mass., v. 150, p. 318–335, 4 text-fig.

———, 1978, *Polypide morphology and feeding behavior in marine ectoprocts:* Bull. Mar. Sci., v. 28, p. 1–31, 15 text-fig.

Wise, S. W., Jr., 1969, *Study of molluscan shell ultrastructures:* in Scanning electron microscopy/1969, Proceedings of the 2nd Annual Scanning Electron Microscope Symposium Chicago, p. 205–216, 23 text-fig.

———, 1970, *Microarchitecture and deposition of gastropod nacre:* Science, v. 167, p. 1486–1488, 1 text-fig.

———, & de Villiers, Johan, 1971, *Scanning electron microscopy of molluscan shell ultrastructures—Screw dislocations in pelecypod nacre:* Trans. Am. Microsc. Soc., v. 90, p. 376–380, 9 text-fig.

Wood, T. S., 1973, *Colony development in species of Plumatella and Fredericella (Ectoprocta: Phylactolaemata):* in R. S. Boardman, A. H. Cheetham, & W. A. Oliver, Jr. (eds.), Animal Colonies, p. 395–432, 23 text-fig., Dowden, Hutchinson, & Ross (Stroudsburg).

Woollacott, R. M., & Zimmer, R. L., 1972a, *Origin and structure of the brood chamber in Bugula neritina (Bryozoa):* Mar. Biol., v. 16, p. 165–170. 4 text-fig.

———, ———, 1972b, *A simplified placenta-like brooding system in Bugula neritina (Bryozoa):* in C. J. Arceneaux (ed.), 30th Annual Proceedings of the Electron Microscopy Society of America, p. 130–131, 2 text-fig. (Los Angeles)

———, ———, 1975, *A simplified placenta-like system for the transport of extraembryonic nutrients during embryogenesis of Bugula neritina (Bryozoa):* J. Morphol., v. 147, p. 355–378, 2 text-fig., 7 pl.

Yang, King-Chih, 1957, *Some Bryozoa from the upper part of the Lower Ordovician of Liangshan, southern Shensi (including a new genus):* Acta Palaeontol. Sinica, v. 5, p. 1–10, 2 text-fig., 2 pl.

Yaroshinskaya, A. M., 1960, *Tip Bryozoa. Mshanki:* in L. L. Khalfin (ed.), Biostratigrafiya paleozoya

Sayano-Altaiskoj gornoj oblasti, v. 1, Nizhnij paleozoj: Tr. Sib. Nauchno-Issled. Inst. Geol. Geofiz. Miner. Syrya, v. 19, p. 393–400, pl. 14–16. [*Phylum Bryozoa:* in L. L. Khalfin (ed.), Biostratigraphy of the Paleozoic of the Sayan-Altai mountain region, v. 1, Lower Paleozoic.]

Yonge, C. M., 1953, *The monomyarian condition in the Lamellibranchia:* Trans. R. Soc. Edinburgh, v. 62, p. 443–478, 13 text-fig.

Young, John, & Young, John, 1874, *On a new genus of Carboniferous Polyzoa:* Ann. Mag. Nat. Hist., ser. 4, v. 13, p. 335–339, pl. 16.

Zimmer, R. L., 1973, *Morphological and developmental affinities of the lophophorates:* in G. P. Larwood (ed.), Living and Fossil Bryozoa, p. 593–599, Academic Press (London).

Zittel, K. A., 1880, *Handbuch der Palaeontologie, Abt. 1. Palaeozoologie:* Bd. 1, 765 p., 558 text-fig., R. Oldenbourg (München).

INDEX

Italicized names in the following index are considered to be invalid; those printed in roman type are accepted as valid. The names of all taxa above the rank of superfamily are distinguished by the use of full capitals, and authors' names are set in large and small capitals. Page references having chief importance are in boldface type.